# Glassware and Glassworking in Thessaloniki

## 1st Century BC – 6th Century AD

Anastassios Ch. Antonaras

Archaeopress Roman Archaeology 27

ARCHAEOPRESS PUBLISHING LTD
Summertown Pavilion
18-24 Middle Way
Summertown
Oxford OX2 7LG
www.archaeopress.com

ISBN 978 1 78491 679 4
ISBN 978 1 78491 680 0 (e-Pdf)



Cover images: Glass vessels, Thessaloniki, 1st-5th c.

Dedicated to
Mania and Christos

# Contents

# List of Figures

# Acknowledgments

Special thanks are due to the former directors of the Museum of Byzantine Culture, †Eutychia Kourkoutidou-Nikolaidou and Anastasia Tourta for their support at the beginning of my research.

I warmly thank all my colleagues from the archaeological service of Thessaloniki for their help and willingness in letting me publish material from their excavations, and/or for their general support in the long course of my research: H. Anagnostopoulou-Chatzipolychroni, Ch. Bakirtzis, Ch. Gatzolis, D. Grammenos, D. Ignatiadou, N. Kapizioni, I. Kanonidis, †N. Karydas, A. Keramaris, Ch. Koukouli-Chrysanthaki, K. Kousoulakou, E. Lamprothanasi, D. Makropoulou, Ch. Mallia, E. Marki, I. Motsianos, N. Moutsopoulos, M. Besios, D. Nalpantis, M. Pappa, Th. Pazaras, †E. Pelekanidou, M. Polychronaki, S. Serdaris, L. Trakosopoulou, E. Tsigaridas, M. Tsimpidou, Th. Zontanos.

This present book began as research for my Ph.D. and I would also like to thank Professors M. Andronikou, †A. Paliouras, A. Stavropoulou and specially Stella Drougou for her considerable interest and time spent helping me improve the present work, and to my colleague and friend Dr Despina Ignatiadou for their help at that stage.

In addition I would like to name several other colleagues who helped me with discussions on glass-related issues and also in acquiring bibliographic material which was at the time totally inaccessible in Thessaloniki:

Dr S. Auth, Newark Museum.
† Dr M. Bajalović-Hadži-Pešić, director emerita of the Museum of the City of Belgrade.
Professor emeritus, S. Ćurčić, Princeton University for his support.
Professor E. Erten, Mersin Universitesi.
Dr A. -B. Follmann-Schulz, curator emerita, Rheinischen Landesmuseum Bonn.
Dr S. Fünfschilling, Römerstadt Augusta Raurica/Römermuseum Augst.
UCLA Professor and good friend, S. Gerstel, for her hospitality and all the help she gave me with the libraries of Washington DC and UCLA.
Dr Y. Gorin Rosen, Israel Antiquities Authorities.
Mrs M. Križanac, at the time Curator in Muzej Primenjenih Umetnosti Belgrade.
Dr Z. Kotitsa, Marburg University, for her pertinent comments on the grouping my material and on terminology issues.
Dr C. Lightfoot, Metropolitan Museum, NY.
Dr F. Naumann-Steckner, at the time in Römisch-Germanisches Museum Köln, for information concerning the glass bowl from Thessaloniki acquired by the museum.
Dr M.-D. Nenna, Maison de l'Orient, Lyon.
Professor Y. Olcay, Anadolu Universitesi, Eskişehir.
Dr M. Padgett, Curator of Ancient Art, Princeton University Art Museum.
Dr G. Platz, curator emerita, Antikenmuseum Berlin Staatliche Museen Preussischer Kulturbesitz.
Dr D. Papanikola-Bakirtzi, director emerita of the Archaeological Institute of Macedonia and Thrace, with whom, during the years we were both working in the Museum of Byzantine Culture, I shared my thoughts and reflected my concerns on issues relating to the typology and use of glass vessels.
Dr E. Marianne Stern, for her hospitality and for rendering accessible her library, our fruitful discussions and visits to the glass collections in Toledo, Ohio, and generally on the East Coast of America. In addition, I thank her for her significant remarks on the text of the book, especially concerning technology and typology issues.
Dr A.-M. Talbot, director emerita at the Dumbarton Oaks Research Library and Collection, for inviting me to the symposium on Byzantine Glass, the hospitality at DOC, and facilitating the use of the library.
† Dr V. Tatton-Brown, British Museum, for information on the glass finds from the Thessaloniki region held at the British Museum.
Dr P. Triantafyllidis, Archaeological Service, Rhodes.
Mr S. Trifunović, Narodni Muzej Novi Sad.
Dr Ch. Tsochos, Erfurt University.
Dr E. Zečević, Curator, Narodni Muzej Belgrade.

I would like to thank the committee of the Program in Hellenic Studies, Princeton University for awarding me in 2003 the Stanley J. Seeger Visiting Fellowship in Hellenic Studies. During that stay at Princeton I had the chance to improve considerably the present book with new ideas originating from interactions with the community of the University and with additional data from the University's wonderful libraries.

For the current, updated edition of the book in English I want to express my gratitude to the Association for the History of Glass for the grant that covered part of the publication's expenses and mostly Professor Ian Freestone for his support and help in contacting Archaeopress in Oxford who have done a great job in the setting up and printing of this book which I hope that the reader will enjoy using.

# Introduction

Glass was one of the first substances invented by man, perhaps as a very fortunate accident while experimenting with glazed pottery or faience. This probably took place at some time in the 3rd millennium BC, probably in Mesopotamia, and the first objects – monochrome or polychrome translucent or opaque items – were produced in Mesopotamia and Egypt. The first products appear to have been simple beads, and, later, decorative elements and inlays were for a very long period the only glass creations, with vessels as such (polychrome/opaque) only being known from the 15th century BC onwards. The use of glass vessels was extremely restricted and reserved only for elites. In the Mycenaean world of the second half of the second millennium BC, many beads and pendants of dark-blue glass were used and a few small unguentaria appeared employing the mosaic technique. Colourless transparent glass was made for the first time in 8th-century BC Assyria and was used for the production of vessels.

In Classical Greece, from the 6th century BC onwards, core-formed multicoloured vessels, mainly unguentaria, miniature replicas of tableware, were produced; these were apparently expensive items meant only for gods, rulers, and the highest ranks of society. In the 4th century BC in Greece and the Near East, colourless transparent glass was reintroduced and small objects and fancy tableware vessels, seemingly used at official or ceremonial banquets, were made of it. In Hellenistic times technical and artistic advances occurred in glassworking, but glass vessels remained exquisite items, for instance tableware (almost exclusively drinking and pouring vessels, bowls and amphorae), that were still restricted to privileged users. It was only in late Hellenistic times that drinking vessels, simpler in form and decoration, began appearing in greater quantities and now, for the first time, intended for wealthier users.

The invention of the blowing technique, in the 1st century BC, made glass objects accessible for the first time to wider society, introducing glass objects into almost every middle-class household. All classes of tableware, drinking vessels (bowls, beakers), vessels for presenting or offering food (dishes, plates, trays) or pouring liquids/drinks (jugs, decanters, flasks/bottles) were abundantly produced in glass. Unguentaria appear in a great variety of shapes and sizes, containing perfumes, cosmetics, medical and religious substances. In this period the storage, preservation and trade of various products, even in large quantities, was conducted using bulky and totally utilitarian vessels. Furthermore, from the 3rd and 4th centuries onwards, developments in transparent glass led to the wide use of the material for lamps. Some forms of tableware evolved to that end, or other special shapes were devised for society's growing demands for lighting, which, from Late Antiquity onwards, were mainly met by the use of glass receptacles. Throughout the Late Antique, Byzantine and Medieval periods (and even later, depending on social, economic and commercial fluctuations), glass items retained these uses, serving people in all aspects of their mundane or special occasions, before finally accompanying them to their graves (and thus protected from the breakage and consequent recycling that widely occurred), offering a better picture of the very wide range of shapes, sizes and colours that these products appeared in.

This present work deals with the production and decoration of glass in the Roman and Early Byzantine periods; the social and economic aspects of glassworkers; the typology of glass vessels found in Thessaloniki and its environs; and the wide variety of human needs met by such objects. Furthermore, this research sheds light on the find-spots and excavations conducted in this region during the 20th century, which, in the main, have been summarily presented or completely unpublished hitherto.

In contrast to other regions of the Roman Empire, in Greece the history of glass during Roman and Early Christian times is a subject that has only been cursorily investigated.[1] Specifically, glassworking in Thessaloniki remained – before the Greek publication of this book in 2009 – almost entirely unknown and was presented only sporadically

[1] The most important of the studies dedicated exclusively or mainly to glass finds are noted infra, referring to the following sites or collections: Corinth: G. Davidson 1952 – the first systematic study of this kind, where glass finds from the excavations of the American School of Classical Studies in Corinth were presented; Nea Anchialos: Weinberg 1962 – including vessels and probably also a glass-workshop; Samothrace: Dusenbery 1967 and Dusenbery 1998; Galaxidi: Zafeiropoulou 1975; the collections of the Benaki Museum: Clairmont 1977; De Tommaso 1991 – concerning glass unguentaria from Greek territory; Knossos: Price 1992; part of the collection of the National Archaeological Museum: Weinberg, McClellan 1992; Amorgos: Triantafyllidis 1998. Congresses dedicated to the history of glass in Greek territories were organized in Margarites/Mylopotamos in 1997, and in Rhodes in 2001. In the proceedings of these congresses (Themelis (ed.) 2002 and Kordas (ed.) 2002) several studies regarding Roman and Early Christian glassworking were published. A three-day international seminar was organized in Thessaloniki in 2001 and the published proceedings (Antonaras, Kordas (ed.) 2002) include studies presenting for the first time in Greek the history of glassworking from the Mycenaean to Ottoman periods.

in the form of illustrations accompanying excavation reports and entries in exhibition catalogues.[2] Only one article reported specifically on the glass finds from an excavation.[3]

The present study aims to fill this existing gap in the research. It includes nearly all the glass vessels, both Roman and Early Christian, brought to light by excavations conducted in Thessaloniki and its environs from the early 20th century until the year 2002. In addition to the 800 vessels presented in detail in the Catalogue, the author has studied thousands of additional vessels and fragments that were ultimately excluded from publication, either due to their fragmentary state – which made their identification problematic and insecure – or because they are intended for publication by their excavators. However, all the excluded vessels belong to types presented in depth in the chapter on typology through other examples. Thus, the entirety of glass vessel types excavated in the greater Thessaloniki region are illustrated and published here. These types fall within the more general framework of glassworking during the period, and provide a secure basis for the study of the region's material culture and the daily lives of its inhabitants. In addition, the study of this material has enabled us to draw meaningful conclusions about the commercial and economic relations and contacts of Macedonia, essentially the main gateway for commercial goods making their way into the Balkan hinterlands during the Roman and Early Christian periods.

The first chapter here deals with production. It examines the places where raw materials were collected, the location and operation of glassmaking workshops (known only in the Middle East), and those workshops used for secondary production (i.e. the forming of glass objects) and which operated in various cities throughout the Empire. The raw materials and two main types of workshop are presented in such a way as to make the artisanal/glassworking finds from Thessaloniki, which form part of the following chapter, more easily understandable.

All the techniques employed in producing, forming and decorating the glass vessels that comprise the Catalogue are examined. A detailed presentation of these techniques was deemed necessary for several reasons. First, most of the techniques used in ancient glassworking are not known with any precision. This is the result of a break in the related technical tradition, the absence of clear references in ancient historical sources, and the absence of relevant excavation finds. Second, the theories and views presented until recently by archaeologists and art historians, and which have often and uncritically been repeated, were frequently based on arbitrary conclusions arrived at by 19th-century scholars through analogies to metalworking and pottery techniques, or based on observations of the methods of contemporary glassworkers. However the technical means and infrastructure of contemporary glassworkers differ radically from those of their Hellenistic and Roman peers. The second half of the 20th century inaugurated a new period of study in which researchers began to deal steadily and substantively with technical issues, and to attempt to demonstrate empirically the techniques that were employed based on interpretations of the technical details found on the vessels themselves, as well as on the fragments of glass remains from workshops. Some earlier theories were proved credible; others were excluded as technically unfeasible and replaced by new views concerning techniques that were experimentally demonstrated, while for some as-yet unresolved issues the conversation continues, together with efforts to identify the most likely hypotheses.

The following chapter presents and analyzes the excavation data and portable finds that document the presence of glass-workshops in Thessaloniki. We examine the issue of the location of glass-workshops in the Roman world generally, and in Thessaloniki in particular, on the basis of legal texts and excavation data. The various workshop groups with common characteristics are presented, and the corresponding finds from Thessaloniki are included in their respective groups. This chapter concludes with a detailed presentation of the evidence and most important workshop finds. At the end, the vessel types that probably comprised local products are presented.

Next, the historical, commercial, economic and social contexts of production are established. We discuss the social position and economic situation of glassworkers, the place of glassworking in artisanal and artistic production generally, and links are drawn to other crafts and products. The evidence for commercial relations and trade in glass products is examined, as well as that for the development of local production and the means by which glass products were channelled to local markets.

[2] *Thessaloniki Istoria kai Techni 1986*, passim; *Thessaloniki apo ta proïstorika mechri ta christianika chronia*, 1986, passim; Papanikola-Bakirtzi (ed.) 2002, passim. In 2010, a year after publication of the Greek version of this book, the author provided catalogue entries (*Glass Cosmos 2010*) for 134 glass vessels that form part of the assemblage catalogued in this present work. Furthermore, all publications related to Thessalonian and Macedonian glass finds in general have been included in this updated edition, first published in Greek in 2009. Glass-related publications referring to finds from other regions of the Roman Empire, published after 2009, have been added only occasionally, whenever there was a need to shed light on chronological issues that were perhaps unclear in the Greek edition.

[3] Mavromichali 2001.

Vessels are classified based on shape and size as main criteria, these being mainly determined by their use, and which for the most part allow identification. Each type is grouped within existing classifications of vessels and includes in its descriptive title the classification type according to Clasina Isings (1957), which is the most commonly accepted format and makes the identification of Macedonian finds easier. Each type is then described and the technique by which it was formed is explained. International parallels are identified across the entire geographic compass of the Roman Empire and neighbouring regions. Finally, a valid excavation dating for the specific examples from Thessaloniki is cited.

The decoration of vessels is examined in depth in a chapter where each vessel is discussed, if its depicted motif(s) permits or requires it, or each motif category if the motif is simpler and repeated on many vessels. In addition, production and decoration workshops are indicated, together with the provenance of the various decorative motifs.

Next, guided by the specific functional needs of the vessel, we analyze and explain the reasons that allowed, or imposed the choice of glass for its production, e.g. for transport or tableware use.

There follows a chronological overview of the finds. The dating of objects (where this is possible) is done on the basis of their excavation evidence. First, if the find is from a closed tomb or sealed layer (61 vessels, i.e. 8.09% of the total) then its excavation dating is accepted; however, the evidence used for dating was scrutinized and rechecked. In each case the dating was controlled for accuracy on the basis of international typological parallels, which were checked in turn for whether they were dated on the basis of stylistic or excavation evidence. When the original publication mentions it, the evidence for the dating of each piece is cited. If the find came from an excavation or layer only generally dated (116 vessels, i.e. 15.38%), or if the dating does not appear securely established by excavation, or there are no excavation data at all (578, i.e. 76.65%), then the main dating criterion was its parallels. This chapter also cites statistical evidence for vases, which is examined by century, on the basis of vessel use and place of production, in a condensation and decoding of the evidence from the chapter dealing with typology. We avoided any attempt to distinguish material on the basis of whether it was found in a tomb or a habitation site, since, apart from those vessels lacking all excavation evidence (59, i.e. 7.82%), nearly all the material came from cemeteries (556, i.e. 73.7%). The relatively small number of finds coming from public spaces (agora, temples-churches, and baths) (111, i.e. 14.7%) and private residences (28, i.e. 3.71%) came from just a few excavations and may easily be dramatically altered through additional evidence from more recent excavations in the city. Finally, there is an attempt to interpret the evidence that results from the study of the archaeological material in historical terms and incorporate it into the socio-economic processes of the region's political history.

The Catalogue of the vessels under study follows. Each entry consists of seven sections divided by double strokes. Namely, the sections comprise the following information: Cat. no (Registration Number) // Find spot and date // Condition-Preservation // Colour and quality of glass // Dimensions in cm. // Publications // Date. As to Catalogue Numbers, it should be noted that for the fourteen items which were not included in the Greek edition and were only added to the present English edition, instead of changing the numbering of the entire catalogue, they were inserted at the appropriate places with the addition of a letter after their number, namely nos 133a-b, 134a-b, 135-b, 142a-b, 164a-c, 175a, b, c were added.

The Catalogue and a brief history of excavations and find-spots of the objects dealt with in this work are provided in the following chapter. The Catalogue cites in summary form the excavation evidence coming from publications which were fully catalogued and in many cases augmented by evidence discovered in the course of research in the relevant archives of the Directorates of Antiquities. Each excavation history is followed by a chart with the vases found, identified by their sequential number in the general Catalogue, and classification type; in addition, their exact excavation data and date of discovery are also noted. Within each excavation the objects are presented by Catalogue Number, and when more than one campaign of excavation was carried out at the same plot they are presented chronologically. A map of the region is provided, on which the precise locations of excavations have been labelled and annotation provided.

A glossary of terms referring to glassworking technology that are used in the book follows.

A concordance is provided between the Museum or Directorate Inventory Numbers and the ascending numbers for the general Catalogue of this volume in ascending alphanumeric series on the basis of Museum or Directorate Inventory Number. This is followed by a typological-chronological chart containing a drawing of each type, the period during which it was in use, and the Catalogue numbers of the vessels belonging to it.

Finally, we offer drawing and photographic documentation of the material, with illustrations of all the objects in the Catalogue.

A full Index of the 602 geographical locations found in the volume is also provided. The ancient name for each site is noted in parentheses in parallel with its contemporary name to make it easier for readers to locate. Beside each site the name of the country to which it belongs today is cited. For relatively unknown cities and sites, especially in Greece, their prefecture or nearest major municipality is noted. In general the names of ancient Greek sites and monuments are transliterated in the conventional way. Modern Greek sites and street names were transliterated according to the Hellenic Organization for Standardization system of 1982, as adopted by UN 1987.

# 1. Production – Forming – Decoration

## Raw materials

The raw materials required for the production of glass are silicon, soda (sodium carbonate) and calcium.

The main ingredient of glass is silica, which works as a network former. Silica is derived from sand. Since the very high melting point (1720 °C) of silica could not be achieved in ancient glassmaking workshops, flux or modifying agents were used; these agents, however, reduced the water resistance of glass. Soda, a natural sodium sesquicarbonate served as the main modifying agent. The source of soda was natron, a mineral salt also used as a detergent, a mummifying agent, and a fabric dye, as well as for medical purposes.[4] The largest source of natron was in Wadi Natrun,[5] found in the desert northwest of Cairo. Comparable sources of natron, however, are also attested in Macedonia, and specifically at Chalastra (modern Pikrolimni)[6] and at Philippi.[7]

In order to counteract soda's negative effect on the water resistance of glass, stabilizers such as calcium oxide or, less frequently, manganese, aluminum or other similar oxides were added to the batch. Limestone was used as a source of calcium; however, its addition to the batch was not always necessary since there was often enough calcium in the other two raw materials, in the form of crushed shells contained in the sand.[8]

Glass was coloured through the addition of metallic oxides (iron, manganese, cobalt, copper, lead, antimony).[9] The final colour of the glass produced depended on the colouring agent used, the way it was mixed into the batch and the conditions of firing. Because of the small quantities of various metals, especially iron, that exist in sand, the glass took on various hues. It became blue if the firing conditions were oxidizing, and yellowish if they were reducing. If the conditions of firing were mixed, glass took on various shades of green, ranging from blue-green to yellow-green. The specific final hue was produced almost randomly as it was impossible to fully control all of the parameters listed above.[10] In general, colour is not a feature of glass that can be used reliably to identify a particular workshop or place of production. Similarly, in order for glass to become decolourized and lose its natural, greenish tint, antimony was used in reducing conditions or at high firing temperatures. For this procedure to be successful great precision was needed in all steps of the process in order for the required conditions to be maintained; otherwise it could lead to complete failure, i.e. to the production of opaque glass. For this reason it is thought that this difficult work could only be done in specific workshops to high technical standards. This would also explain why this type of glass was used exclusively for the making of luxury objects.[11]

## Production

Glassmaking, that is, the production of glass out of raw materials, and glassworking, the forming of objects out of existing glass, constituted two distinct processes of ancient and Medieval glassworking that took place in different regions and at different times.[12] For example, already by the mid-second millennium BC the glass produced in the East, that is, on the Syro-Palestinian coast and in Egypt, was transported in ingot form and sold in distant western regions, where it could be used much later, depending on the needs and intentions of the glassworker who bought it.

Figure 1. Fragment of a raw glass chunk, probably 4th-7th c. Antonaras 2009, fig. 6, cat. No. 468; Princeton University Art Museum, y, 1928-32. Photo Bruce M. White.

---

4 Ignatiadou *et al.* 2003, 64-65, for an overview of its uses according to Pliny's texts.

5 Nenna, Picon, Vichi 2000, 99.

6 Ignatiadou *et al.* 2003, 64-68.

7 Pliny, *Nat. Hist.* 36.106-122; For a detailed discussion of natron in Macedonia and specifically in the Pikrolimni region, see Ignatiadou 2002, 241-248.

8 Pliny, *Nat. Hist.* 36.192; Ignatiadou 2013, 43.

9 Pliny, *Nat. Hist.* 36.62 and 193; Seneca, *Epistulae morales* 15.90.31 and 105; Trowbridge 1930, 100-101.

10 Brill 1999, 1, 15-17.

11 Ignatiadou 2013, 43-5, with further bibliography.

12 Glassmaking and glassworking had been considered distinct processes since the 14th century BC, as evidence of glass ingots from the Ulu Burun shipwreck demonstrates. See Bass 1986, 281-282; Nicholson, Jackson, Trott 1997, 143-153. For other glass ingots of the 2nd millennium BC from several western Asian sites, see Barag 1985a, 107-113.

The terminology preserved in ancient Greek sources reflects the prevalence of this distinction at that time, that is to say that the individual responsible for the primary production of glass (glassmaking) was not the same as the one responsible for the secondary production of objects out of glass.[13]

The ancient Greek terms *ὑαλουργός/ὑελουργός*[14] (glassworker), and correspondingly, *ὑαλουργεῖον/ὑελουργεῖον*[15] (glassworking workshop) refer to the maker of glass objects and to his workshop respectively. In Latin the corresponding terms are *vitrearius* or *vitriarius*[16] for the glassworker and *vitrarium*[17] for his workshop. The terms *ὑελοψός* and *κάμινος ὑελοψική*[18] refer to the artisans who produced the glass, the glassmakers, and to their furnace respectively. The Latin term *(re)coquere*, considered to be the equivalent of *ἕψειν*, semantically covered the concept of glassmaking,[19] while the specific word *vitrifactor*, which corresponds to *ὑελοψός*, did not appear until the 7th century.[20]

## Glassmaking – primary glass production

Excavation evidence relating to the production of glass is only rarely available for earlier periods since the discovery of glass in Mesopotamia, and it is mainly better documented from the Roman period onwards.[21] It appears that during this period only certain workshops undertook the primary production of glass, that is, the production of glass through the fusion of raw materials. With two possible exceptions,[22] all such workshops have so far been discovered only in the Eastern Mediterranean. In Israel workshops have been found in Beth She'arim,[23] Bet Eli`ezer[24] and Apollonia);[25] in Egypt in the Wadi Natrun region[26] (Beni Salama, Wadi Natrun and Zakik) and on the shore of Lake Mareotis (Mariout)[27] (at Taposiris Magna and Philoxénité). Physicochemical analyses have shown, however, that the glass produced in Egyptian workshops was not exported. It was mainly Syro-Palestinian workshops that supplied the Imperial market with raw glass.[28]

The excavation at Bet Eli`ezer sheds particular light on the operation of workshops.[29] Glass production did not take place all year long. It appears that preparations for the construction and building of furnaces began in the spring and ended in the summer, when the wood used as fuel was also collected and dried. Glassmaking itself began in the autumn, when the winds were conducive to attaining the required temperature in the furnace. The glassmaking process was completed in ten to fifteen days of continuous heating of the furnace; the temperature had to reach and be maintained at around 1100° C. Each furnace was used only once and then destroyed; its bricks, however, were recycled and used in the construction of new furnaces.

[13] Stern 1989, 121.

[14] Strabo, *Geog*, 16.2.25.9; Oribasius, *Medical Compilations*, XIII.α'.9.2; Evagrius, *Ecclesiastical History*, 185.25; Trowbridge 1930, 50, note 12 ff.

[15] Dioscorides, *De Materia Medica*, 5.161.1.2; Paulus Aegineta, *Medical Compendium*, 7.3.1.304; Trowbridge 1930, 50-51.

[16] Seneca, *Epistulae morales* 15.90.31; Codex Theodosianus 13.4.2 *De excusationibus artificum;* Codex Justinianus 10.66.1 *De excusationibus artificum;* Trowbridge 1930, 66, note 2 ff.

[17] Corpus Glossariorum Latinarum, s.v. 'ὑαλουργεῖον'; Trowbridge 1930, 65.

[18] 'Υελέψης = Hesychius of Alexandria *Lexicon*, s.v. ὑελέψης; Olympiodorus, *Commentary on Aristotle's* 'Meterology', 2.228; 'Υελεψός = Didymus the Blind, *Commentary on Psalms 40-44*, 329.13-17; Michael Glycas, *Chronicle of Events*, 4.506.7; Leontius Episcopus Neapolitanus, *Vita S. Simeonis Sali, confessoris, P.G.* 93, col. 1736D; *Acta Sanctorum Iuli.* 1.168.16; 164.52; 'Υαλοψός = John Moschus, *Spiritual Meadow, P.G.* 87, col. 2932°; 'Υελοψός = Georgius Cedrenus, *Synopsis of History*, 1.687.1; Nicephorus Blemmydes, *On Making Gold*, 2.452.13; 'Υάλωψ = John of Alexandria, *Commentary on Hippocrates' 'On the Nature of the child'*, 2.222.5. Κάμινος ὑελοψική = Zosimos of Panopolis, *Le premier livre*, 2.246.18; Michael Glycas, *Chronicle of Events*, 4.506.7; Leontius Episcopus Neapolitanus, *Vita S. Simeonis Sali, confessoris,* P.G. 93, col. 1736D; Acta Sanctorum Iuli. 1.168.16; 164.52; Trowbridge 1930, 50.

[19] Stern 1989, 121.

[20] Bede, *Homilies* 17, in *Natale Sancti Benedicti Episcopi* (PL 94, col. 228A); Trowbridge 1930, 78.

[21] On Mesopotamian finds, see Oppenheim *et al.* 1970, 69-71; Schlick-Nolte, Lierke 2002, 19; Vandiver 1983, 239-247. On Pharaonic finds, see Nicholson, Jackson, Trott 1997, 143-153, pl. XVII; Nicholson 2007, esp. 125, 158; Rehren, Pusch 1997, 127-141; Schlick-Nolte, Lierke 2002, 17-22, figs. 3-4. The oldest excavated glass-workshop, safely dated in the 4th century BC, was active in the city of Rhodes, where vessels and beads had been produced; see Triantafyllidis 2000, 36-39, 193-95.

[22] Chemical analyses of glass in York have demonstrated the incomplete fusion of raw materials; see Jackson, Cool, Wager 1998, passim. Hambach Forest, west of Cologne, has also been considered as a primary glass production centre; see Wedepohl, Gaitzsch, Follmann-Schulz 2001, passim.

[23] Brill 1967, 88-95.

[24] Gorin-Rosen 1995, 42-43.

[25] Gorin-Rosen 2000, 54-55.

[26] Nenna, Picon, Vichi 2000, 99-102.

[27] Nenna, Picon, Vichi 2000, 102-104; Nenna 2008a, 61-66.

[28] Foy, Nenna 2001, 37.

[29] Gorin-Rosen 2000, 52-54.

Figure 2. Remains of a primary glass workshop, Wadi Natrun, Egypt, 1st - 2nd c. A.D.
Antonaras 2009, fig. 3; Courtesy of Mission archéologique du Wadi Natrun.

Figure 3a. Remains of a primary glass workshop, Bet Eli'Ezer, Israel, 6th - 7th c. A.D.
Gorin-Rosen 2000, cover page. Photo Jochai Rosen.

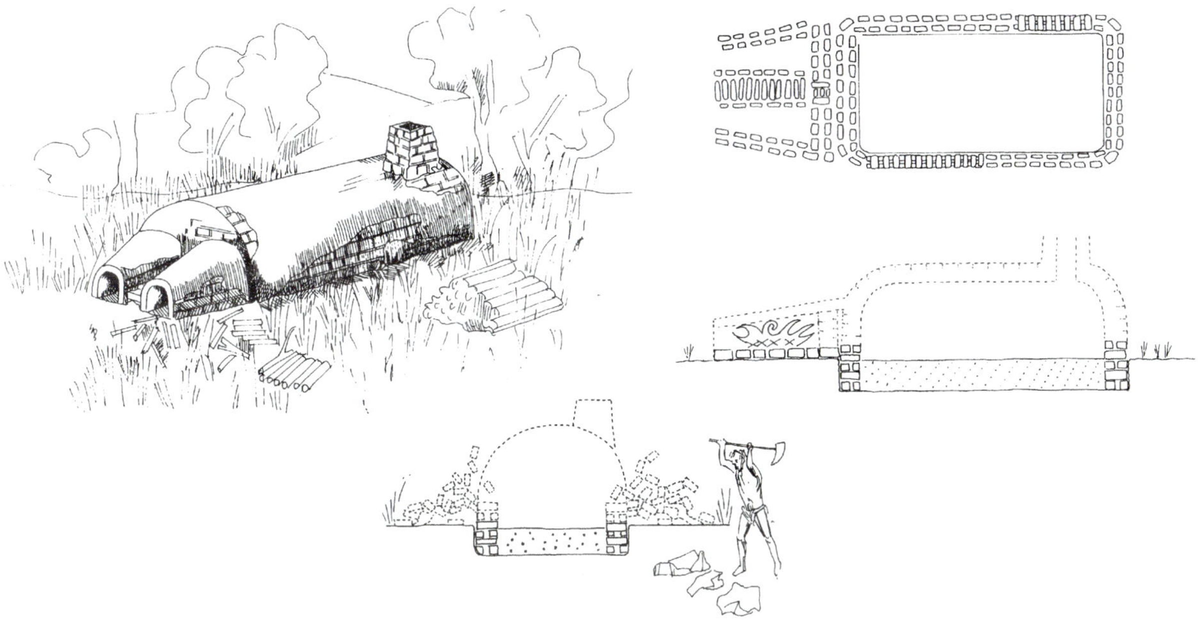

Figure 3b. Reconstruction of a primary glass workshop, Bet Eli'Ezer, israel, 6th - 7th c. A.D.
Gorin-Rosen 2000, p. 53, figs 3, 4. Drawing Michael Miles.

The glass was produced in rectangular brick furnaces. These consisted of a *praefurnium*, which was covered by a semi-conical vault, and a main melting chamber that was covered by a semi-cylindrical or other lower vault.

The firing chamber was divided into two *praefurnia*. The heated air flowed from these compartments over to the melting chamber or reservoir. This movement was facilitated by the constant current of air created as a result of the way the furnace was oriented in relation to local winds. The heated air escaped from a rectangular chimney placed at the far end of the reservoir.[30] The constant drafts of wind and the consequent flow of heat over the entire surface of the melting chamber were absolutely indispensable for the successful production of glass in furnaces of this type, in which the source of heat was at a considerable distance from the end of the furnace and heating occurred by reverberation.

The melting chamber, in which sand and natron gradually fused to form glass, was large and rectangular, measuring approximately 2x4m. It was dug into the ground, at a deeper level than the *praefurnia*; its above-ground component

[30] Gorin-Rosen 2000, 52-54.

would usually be largely destroyed in order to extract the manufactured glass. The content of the reservoir formed a single ingot of glass that was then broken into smaller pieces. Excavation evidence shows that these fragments were grouped together in units of 25kg, which were then transported by boat over long distances, supplying remote markets all around the Empire.[31]

## Glassworking – secondary glass production

As archaeological evidence has shown, secondary glass workshops dedicated to the forming of objects were small units, in principle located at the periphery of cities to minimize the risk of fire and disturbance to local residents. They were, nonetheless, often housed in abandoned public spaces and buildings in the centres of cities as well; it appears, moreover, that workshops also operated in the vicinity of military camps to provide for the army's needs. It has also been observed that glassworkers were often active in close proximity to other artisans who might also require furnaces for their work, such as metalworkers and potters; sometimes they even shared the same space.

Figure 4. Secondary glass workshop, Thessaloniki, Vasileos Irakleiou No. 45, late 6th c. Photo Sofia Akrivopoulou. Copyright Hellenic Ministry of Culture and Tourism

As archaeological evidence has shown, secondary glass-workshops reserved for forming objects were very simple structures. Ethnographic comparisons to contemporary traditional workshops,[32] as well as surviving ancient depictions, also confirm this. On the disks of clay lamps of the mid or late 1st century found in Benkovac (Asseria) in Croatia,[33] in Voghenza, near Ferrara in Italy,[34] and in Spodnje Skofije, near Koper in Slovenia,[35] the following representation has been preserved: a glassworker, named Trellus, seated in front of the arched opening of a vaulted glassworking furnace, is blowing a vessel with the aid of a blowpipe, while a second glassworker, named Athenio<s>,

[31] Gorin-Rosen 2000, 52-54; Fontain, Foy 2007, 241-243.

[32] Nenna 2003, 52-59.

[33] Abramić 1959, 149-151.

[34] Baldoni 1987, 22-29.

[35] Lazar 2004a, 28, fig. 15, cat. no. 25; Lazar 2005, 17-18; Lazar 2006a, 227-234; Lazar 2006b, 337-339.

seated behind, in front of another opening (possibly designating the annealing chamber), is inspecting an object, raising it high, perhaps against the light, in order to spot possible cracks. Scholars have also identified the shape of a glassworking furnace, considerably more summarily rendered, in a clay Egyptian figurine of Eros, dating from the 1st or 2nd century.[36]

The furnace, a small circular or horseshoe-shaped structure with an external diameter of 90-120cm, was the only element required for the operation of the workshop. It appears that it evolved into this shape from the simple, open furnace, sometime in the 1st century AD, in response to the special requirements of glassblowing, i.e. achieving/obtaining a balanced distribution of heat around the blowpipe, which was necessary for the successful and symmetrical inflating of the vessel produced. It was constructed of bricks, rocks and mortar. Its walls were approximately 30cm thick. An entirely different type of furnace was used mainly in the 1st and 2nd centuries, and it survived in places into the 4th century. This type was rectangular, and instead of having a crucible, a large section of its interior was formed into a tank (c. 100x70cm) for melting larger quantities of glass.[37] The compartment housing the fire was often dug partially into the ground; its opening was sometimes oriented toward the entrance of the workshop to take better advantage of the wind currents necessary for achieving high temperatures more quickly and easily. A similar layout has been observed in traditional 20th-century workshops in Afghanistan and Egypt.[38] Ancient literary sources mention that the wood used for fuel should be light and dry (*inter alia* tamarisk and papyrus are specifically mentioned).[39] It is proposed that wood was used as fuel, even in cases when other types of fuel, such as coal, were also available.[40] However what is known from other types of traditional kilns in southern Italy, Sicily and Greece leads to the assumption that every accessible combustible material was used, especially in areas with little or no wood.[41]

Figure 5. Drawing of clay lamp with the representation of a secondary glass workshop showing glass blowers Athenios and Trellus at work, Asseria, Croatia (reg. no. 1094-30), 1st c. AD. Grose 1989, p. 33 fig. 10

Figure 6. Clay lamp with the representation of a secondary glass workshop showing glass blowers at work, Pokrainski Muzej Ptuj, Slovenia, 1st c. AD. Lazar 2005, p. 27, fig. 15

At the desired height a small tank was embedded in the kiln. In other cases along the internal perimeter of the furnace a thick rim-like shelf was formed to hold the clay crucible, in which the batch of glass would melt. It is also possible that in some cases the crucible was instead placed on top of a clay stand. Open-shaped household vessels with a flat base were used as crucibles. However, in order to withstand repeated exposure to very high temperatures they had to be coated with a thick layer of clay. Ceramic vessels that appear to have been originally manufactured for precisely this purpose have also been excavated. There is, finally, some evidence that in rare cases stone vessels were also used as crucibles. Even though surviving specimens vary considerably in size, it seems that crucible rim diameter and height were not greater than 23-30cm and 15cm respectively.[42]

---

36 Price 1988a, 317-319.

37 For an overview, see Nenna 2008b, 135-136. Relevant finds were also excavated at Corinth and published by the author.

38 Reut 1973, 98-99; Heneim 1974, 2, 5-11, figs. 1-8.

39 Trowbridge 1930, 104, notes 18-20, with references to ancient sources.

40 Gaitzsch 1999, 135.

41 Hampe, Winter 1965, 196; Jackson, Nicholson, Gneisinger 1998, 11-23, as noted by Stern 2004, 95, note 22.

42 Price, Cool 1989, 24-25; Sternini 1995b, 73-82, where there is also an annotated bibliography; Van Geesbergen 1999, 112-113; Cool, Jackson, Monaghan 1999, 150-154.

Figure 7. Kiln of a secondary glass workshop, site Louloudies in Pieria, Greece, 6th-7th c. Aggelkou 2002, p. 66. Copyright Hellenic Ministry of Culture and Tourism.

It is thought that the section of the furnace that housed the molten glass was covered by a relatively low vault that trapped the hot air, and, because of its shape, reflected it back to the crucible or tank, thus maximizing the efficiency of the relatively expensive fuel. The worker could access this compartment through the 'glory hole', a small vaulted opening in the furnace wall that was located at the height of the tank's rim, or at the height either of the bottom or rim of the crucible, depending on the way the latter was held up. Through this opening the glassworker could fill the crucible and check on the molten glass; most importantly, however, the opening provided easy and convenient access for his blowpipe, which he used to initially gather the mass of molten glass, as well as to heat the vessel during the forming process. There was a smooth, usually marble, surface attached to the furnace, either in front of or next to the opening, which was used for marvering, that is, for rolling the vessel to achieve its final form, smoothen the surface or attach decorative elements.

Figure 8. Representation of a glassblower in action in front of a kiln. Rabano Mauro, De Universo XVII, X, fol. 89. Montecassino, library of the monastery, 1023. Rabano Mauro, De Universo XVII, X, fol. 89. Montecassino, library of the monastery, 1023.

The glass used in the forming of vessels came in part from ras ingots or chunkslumps acquired through trade, and in part from recycled fragments gathered in the vicinity of the workshop.[43] The practice of using recycled fragments in vessel production is not seen in the pre-Roman era: it appears that it first spread during Flavian times (AD 69-96), in the same period as the technique of free-blowing.[44] Because of the very high temperatures developed the furnaces had a short life span; they were quickly worn down and constantly required small repairs. It is possible, moreover, that they were demolished and rebuilt every year, looking like overlapping rings when excavated. Glass vessels, immediately after being formed, had to be gradually cooled to release the internal stresses created during glassworking; this ensured that the vessels would not fracture as a result of the difference in contraction rates between surface and interior. This process, known as annealing, required 18-20 hours; usually the vessels were placed in a small space, adjacent to the furnace so that the latter's radiating heat could be used.[45] On the basis of observations of 20th-century traditional, primitive workshops in the Eastern Mediterranean and in Africa, it has been hypothesized that small products, such as beads and small unguentaria, were annealed by being placed in hot ash kept in a small container, or in a hole in the ground near the kiln.[46]

The small dimensions of glassworking workshops were primarily determined by economic factors; a single glassworker could usually adequately cover local demand. Moreover, the fact that a separate opening into the firing chamber of the furnace would be required for each worker, thus causing a reduction in the high temperatures that were so difficult and costly to achieve, also contributed to the tendency to build small furnaces and workshops, with only one glassworker active at a time. However, a few workshops are known in which many furnaces operated at the same time, such as a 5th-century workshop in Gradishte, near Gabrovo in Bulgaria, where six furnaces, arranged as The volume of surviving residual materials from glassworking is small, mainly because such materials were constantly being recycled. Tear-shaped test droplets constitute the typical main finds. Every day, before beginning blowing, the glassworker, even in modern times, checks the colouring and viscosity of the molten glass by extracting a small quantity and allowing it to become deformed by gravity, through the pull of its own weight, into a tear-shaped thread. The thread ends up on the floor where it usually remains permanently. In addition to these oblong pieces, many other glass fragments are also often preserved, some bearing marks of jacks and other imprints of the floor, or of another coarse surface with which they came into contact while still hot and malleable. Moreover, semicircular, or more usually ring-shaped fragments, tips of moil cracked off from the blowpipe, are also

Figure 9. Testing droplets, and other glassworking refuse, early imperial period, Ancient Corinth, Greece, East of Theater area. Photo by the author. Copyright Hellenic Ministry of Culture and Tourism, American School of Classical Studies at Athens.

---

43 Pliny, *Nat. Hist.* 36.199; Martial, *Epigr.* 1.41.1-5; Statius, *Silvae.* 1.6.74; Trowbridge 1930, 106-107.

44 Stern 1999, 451.

45 Stern 1999, 454-455.

46 Nenna 2007.

typically found. On rare occasions when the rim of the vessel being formed was cracked off and then polished after annealing, a ring-cone-shaped moil of glass is preserved and provides additional information, such as, for example, rim diameters.[47]

Open-shaped ceramic vessels that were used as crucibles, and which are coated, mostly internally, with a layer of glass, constitute another typical finding. Furthermore, remains of ingots or raw glass lumps, as well as piles of glass fragments intended for recycling, are also typical evidence of glassworking workshops. These fragments came in part from vessels that broke during use and were sold for recycling, and in part from unsuccessfully formed and discarded objects and moils of glass cracked off from products during their separation from the blowpipe in the final stage of forming.

The tools required for glassworking are very few. The only one that is exclusively associated with a glass-workshop is the iron blowpipe, the long, hollow rod used for blowing glass. The rest of the metal tools (jacks, possibly scissors and a solid cane) were used for a variety of workshop tasks. On their own, therefore, they do not constitute sufficient evidence of glassworking activity.[48]

Preserved ceramic vessels with traces of glass can only indicate the existence of a glass- workshop. In contrast, bricks and ceramic vessels with traces of glaze could equally possibly point to metalworking activity. Moulds, made of either stone or clay, are quite rarely found. Square, round and polygonal moulds that formed the bottoms of transport vessels have survived almost exclusively.[49]

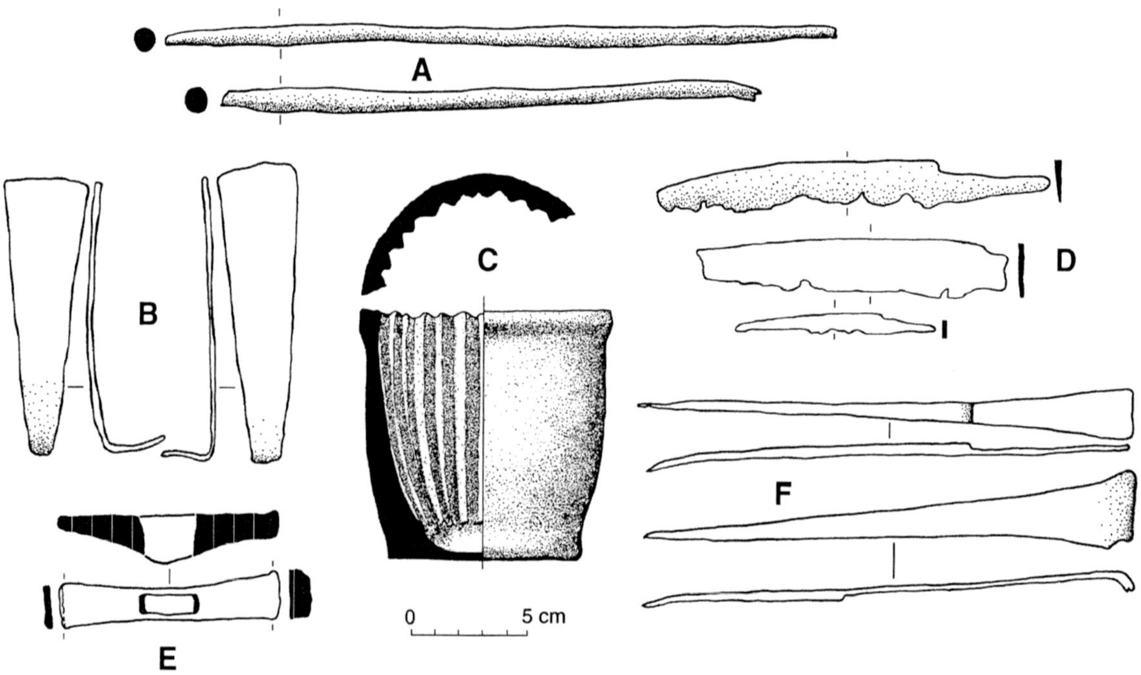


Figure 10. Glass working tools, Komarowo, Ukraine, 3rd-4th c. A.D.
Stern 1995, p. 24, fig. 8. Original drawings by Nina Sorokina: redrawn by Timothy A. Motz

**Glass-forming techniques associated with the vessels under study**

All glass-forming techniques take advantage of the fact that glass becomes liquid if adequately heated. Subsequently, with the help of gravity – and frequently also of rotation, which helps to maintain an object's symmetry – the molten glass is formed into the desired shape, using appropriate (and very basic) tools. During the Hellenistic era there was a push to explore the possibility of using a variety of forming techniques. Even though blowing became

[47] See Foy, Nenna 2001, 40-44. Price 1998, 336.

[48] Bezborodov, Abdurazakov 1964, 64-65; Lang, Price 1975, passim; Auth 1975, 167; Sternini 1995b, 83-95; Foy, Nenna 2001, 76-77, nos. 59-60; Stern 2002, 159-166.

[49] Stern 1995, 46-48; Foy, Nenna 2001, 51; Amrein, Nenna 2006, 491-502; Follmann-Schulz 2010, 235-240, with very illustrative ancient examples and modern replicas.

the dominant technique after the 1st century AD, older techniques continued in use, albeit infrequently, in the forming of objects of a particular aesthetic, e.g. cameo glass, *diatreta* (cage cups), and gold-glass vessels.[50]

### *Casting*

Casting is the technique of forming a glass vessel by the use of a mould filled with *cullet*. It is based on the lost-wax casting technique, widely employed in metalworking.[51] Cullet was used because it was not possible with ancient pyrotechnologies to achieve temperatures high enough for the glass to melt and be poured into the mould.

During the Julio-Claudian period a group of single-coloured vessels of bright hues appeared; their shape was identical to that of their contemporary Arretine clay and silver vessels.[52] It is believed that they were also formed through casting,[53] however recent research suggests that they were actually produced through a variation of rotary pressing.[54]

### *Slumping*

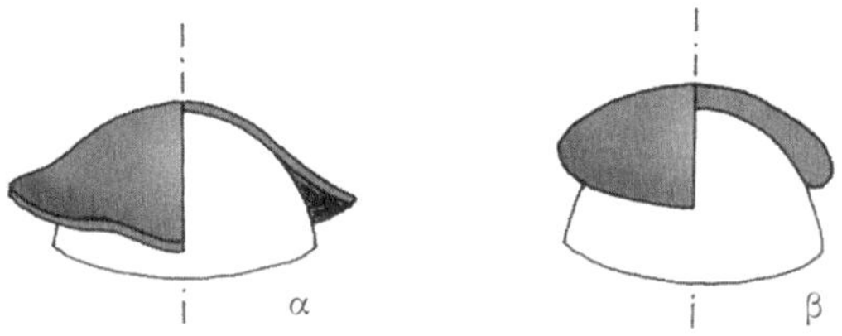


Figure 11. Sagging-slumping technique Lierke 1999, p. 24. fig. 44. Lierke 2009, 41

Slumping is a technique for forming open-shaped vessels by 'slumping' a disk of molten glass heated in the furnace over a convex former mould, or in a concave, open mould. Through gravity, and with appropriate handling from the glassworker, the glass gradually 'slumps' and assumes the shape of the mould.[55] This technique is encountered relatively frequently in the years before the invention of blowing. What vessels formed by slumping have in common is the fact that human breath was not used in their production. Such vessels were formed by heating the glass and using either a simple, open mould, or a bipartite, closed one.

Mosaic vessels, known as *murrina* or *Mo(ύρῤῤ)ινα*, but otherwise left undefined in historical sources,[56] were made of bands and/or horizontal bar or cane sections of glass. Masses of hot glass of adequate colours were put together, tooled and adhered, creating a wide and thick plaque with the desired motif, in great scale, within. This plaque was reheated and pulled out for several metres, thus producing a rod of much smaller diameter with the design all the way through. Thin discoid sections of these rods or canes were cut and used to form mosaic vessels. The bands/cane sections used were sometimes simpler monochrome or polychrome pieces, and at other times composite ones, forming intricate motifs, such as spires, concentric circles, rosettes, etc. These pieces were heated on a flat surface and upon coming into contact with each other they fused and created a disk, whose perimeter was defined by a twisted polychrome – usually white and blue – glass thread. In order to give the mosaic vessel its final form, glassworkers then used the slumping technique on a convex former mould.[57] It is possible that for some types of vessels, glassworkers used a bipartite mould filled with discoid tesserae cut from cylindrical canes. The mould was sealed and heated so that the tesserae would fuse. After the vessel was annealed it was internally polished on a lathe.

*Millefiori* vessels are recognized as a special type of mosaic vessels, even though their forming process was not different technically. *Millefiori*-vessel tesserae were pieces of composite mosaic canes that had the shape of a flower in cross-section.

The twisting network vessels of the *reticella* type are another category of mosaic vessels, whose forming technique was, however, different. They are made of a distinctive glass cane, composed of one or two thin threads of coloured

[50] For a recent overview of earlier glass-forming techniques, see Antonaras 2012, 15–22.

[51] Stern, Schlick-Nolte 1994, 50-52; Lierke 2001, 183-184.

[52] Grose 1991, 2; Grose 1989, 254.

[53] Stern, Schlick-Nolte 1994, 64-65.

[54] Lierke 1999, 55-58.

[55] Stern, Schlick-Nolte 1994, 68-71.

[56] For an overview of Greek and Latin sources where the term appears, see Trowbridge 1930, 83-94.

[57] Stern, Schlick-Nolte 1994, 64-65, 68-69. Very enlightening on the matter are the illustrations in Tait (ed.) 1991, 219-221, where glassmaker Bill Gudenrath recreates the process using modern equipment.

glass twisted onto clear molten glass.[58] The vessels were formed by firing and fusing together short canes, forming a disk, or by winding a long, hot cane around a convex former mould with the help of a long, wide, flat wooden tool (paddle), while the rim was formed by attaching a twisted, bichrome thread.[59]

Marbled mosaic vessels and gold-band mosaic vessels are formed in a similar way.[60] A composite glass cane, with the desired motif formed along its length, was wound spirally on a flat surface. At regular intervals it was additionally pressed so that an undulating motion would complement the spiral decorative motif. The vessel received its final shape through slumping on a convex former mould, a process that further distorted the decoration and rendered it even more intricate.

Some closed-shaped vessels were also formed by firing a blank disk composed of bands of glass. The disk was heated and slumped on an oval convex former mould made of plaster; gravity allowed its bottom part to acquire a conical shape. This part was then compressed into a narrow cylindrical neck with the help of a paddle. At the end of the process the former mould was crushed and its remains were removed from inside the vessel.[61]

In the case of gold-band mosaic vessels, a band made of a sheet of gold leaf encased between two sheets of transparent glass was used together with other composite bands, each made of two or three different colours of opaque glass. Prefabricated elements that resembled the veining of semi-precious stones were also used to form vessels by means of the slumping technique; such vessels imitated their natural stone models very successfully.[62]

***Rotary-pressing***

This technique is almost the same as that used for mould pressing, i.e. the technique of forming an object by pressing viscous glass into an open mould made of plaster or clay; the only difference is that the convex former mould is placed on a potter's wheel so that it can be rotated. According to Pliny,[63] rotary pressing, 'torno teritur', was one of the three techniques that made Sidon a famous glassworking centre.[64]

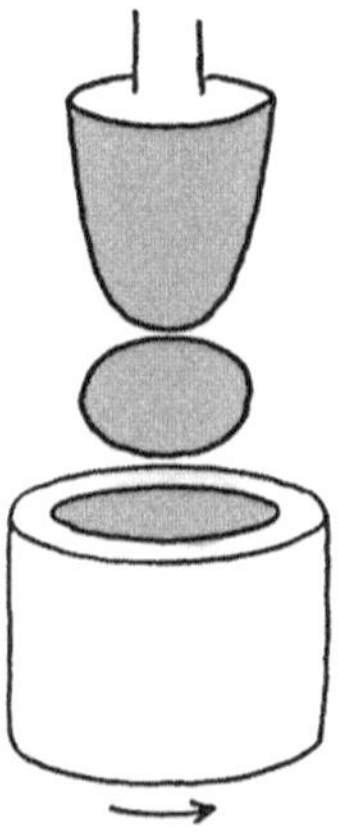

Figure 12. Rotary pressing technique Lierke 1999, p. 46. fig. 102

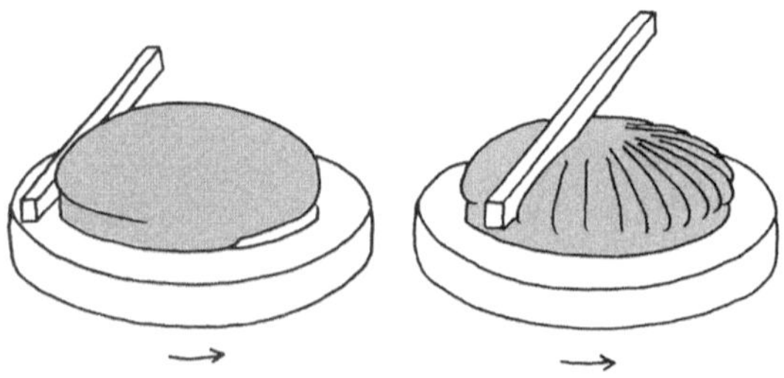

Figure 13. Rotary-pressing techniche of a ribbed bowl Lierke 2009 p. 30

The desired decoration is generated in intaglio on the inside of a plaster concave mould or convex former; the former mould is then placed on the potter's wheel and preheated viscous glass is pressed onto its walls with a plunger, thus shaping the interior of the vessel. The mould is shattered to release the finished vessel.[65] It is quite likely that cameo glass vessels were, at least, partly formed using this technique.[66]

A very common vessel form that was produced through a variation of this technique is the ribbed bowl. Even though there are different theories as to the technique used to form such bowls, the prevailing one today is the following: a preheated disk of glass is placed on a convex former to receive its final shape, while at the same time it is pressed radially with a rod to create the relief ribs to be created.[67] After this the vessel is momentarily reinserted in the furnace and the heat to which it is exposed polishes its exterior. The vessel's interior, as well as the exterior of its rim, are cold-polished on the lathe, a process which leaves visible traces in the form of fine lines.

[58] Stern, Schlick-Nolte 1994, 54.
[59] Auth 1976, 54, col. pl. 51; Stern, Schlick-Nolte 1994, 65-66, 274-275; Lierke 1999, 39-41.
[60] Lierke 1999, 61-66.
[61] Lierke 1999, 64-66.
[62] Weinberg, McClellan 1992, 56-57, no. 48.
[63] Pliny, *Nat. Hist.* 36.193.
[64] Detailed commentary of this passage in Stern 2007, 359.
[65] Lierke 1999, 102-103.
[66] Lierke, Lindig 1997, passim; Lierke 1999, 67-96.
[67] Stern, Schlick-Nolte 1994, 72-78.

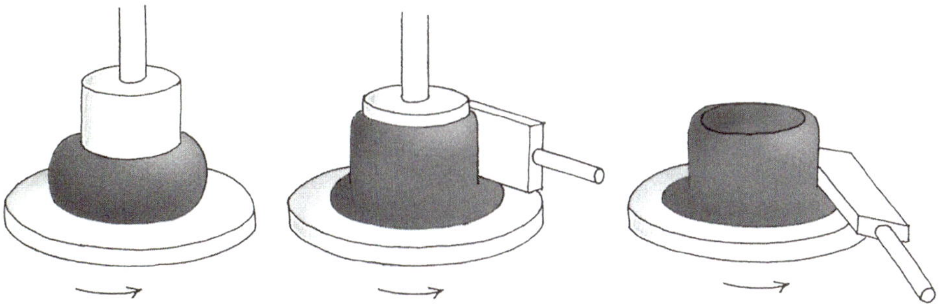

Figure 14. Rotary pressing. Formation of a Cretan pyxis (3rd-1st c. B.C.)
Lierke 2009, p. 32.

Finally, another simple vessel-forming technique, which was heavily influenced by pottery, consists of placing a heated mass of glass on a potter's wheel and forming the vertical walls of the vessel by simultaneously pressing a plunger in the centre of the glass and using a paddle on the exterior: these two tools substitute for the potter's hands. It is thought that this technique was invented in Crete in the early 2nd century BC; it was used in the prodon of particular typetypes of pyxides.[68]

### *Free-blowing*

The technique of free-blowing involves the inflation and then further formation of molten glass through the use of the human breath, channelled through a heat-insulated pipe. According to Pliny,[69] glass blowing, or 'flatu figurare' ('shaping by breath') was one of the three techniques that made Sidon a famous glassworking centre.[70]

Deformed blown glass vessels, wasters from a workshop in Jerusalem, show that this property of glass was discovered in the first half of the 1st century BC, somewhere in the Syro-Palestinian area, where there was already a centuries-long tradition in glassworking.[71] It was initially invented – possibly after observations of and experiments with bead production[72] – as a way of forming small glass tubes from pieces of flat glass folded in a roll and smoothened during reheating. Sometimes a thread of different coloured glass was wrapped around them as well. After one end of the tube was heated up it was sealed and the glassworker blew into its open end, inflating it and forming a simple bottle. After the formed vessel was cracked off from the rest of the tube its rim was shaped.[73]

It took about a century to perfect this initial invention. Around the middle of the 1st century AD free-blowing began to spread throughout the Roman Empire. It appears that, outside of their Syro-Palestinian birthplace, such workshops initially existed only in Italy (Milan, Aquileia, Campania), Switzerland (the

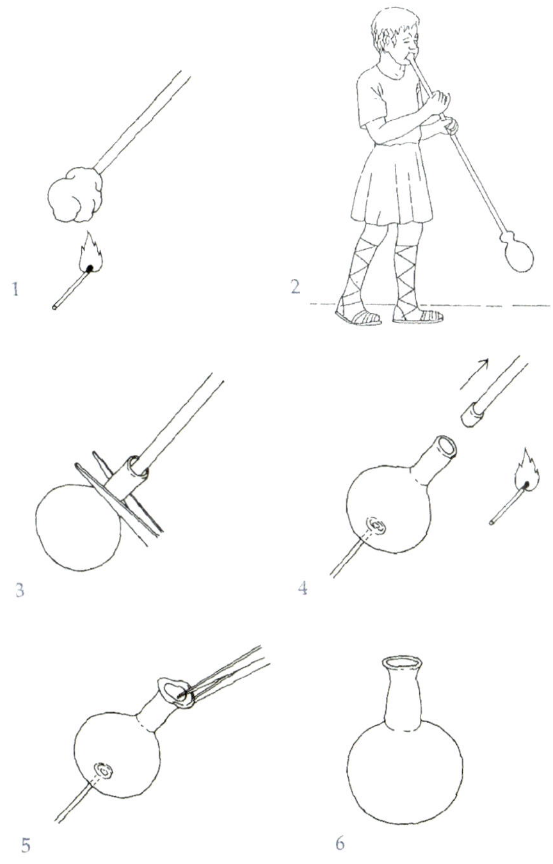


Figure 15. Free blowing technique
Ignatiadou, Antonaras 2008, p. 43. Adapted from Meconcelli Notarianni and Ferrari 1998, p. 13.

68 Lierke 1999, 37-39; Stern, Schlick-Nolte 1994, 79-81.
69 Pliny, *Nat. Hist.* 36.193.
70 For a detailed commentary on the passage, see Stern 2007, 358-359.
71 Israeli 1991, 53.
72 Stern 2004, 87.
73 Israeli 1991, 54; Israeli 2003a, 54-58.

Ticino Valley) and Dalmatia.[74] The period of the technique's spread and popularization coincided with the Augustan age – and the resulting political calm and economic growth. Moreover, this political situation made possible the easy and fast communication between the different provinces. The economic boom in Italy attracted tradesmen and workmen from all over the Empire, and especially from the eastern provinces. The quality and quantity of early blown vessels preserved in the West, in contrast to the relative scarcity of such finds in the East, would seem to support this hypothesis. It appears that the blowing technique was perfected in Italy, and, as also documented in historical sources,[75] particularly in Rome, where glassworkers from the East (Sidonians especially) relocated for financial reasons.[76]

The development and perfection of the free-blowing technique involved at least three points:

First it appears that the form of the furnaces changed. Until this point glass vessels and objects were most probably formed over an open fire, similar to that used by ironworkers. By the third quarter of the 1st century AD, at the latest,[77] the closed, vaulted furnace, in which it is possible to melt glass in a clay crucible, had been developed. According to one interpretation of some ancient Greek passages, it appears that, at first, blowing invole use of preheated glass chunkslumps, known by the term *βῶλος*, which were attached to the tip of a hot ceramic blowpipe.[78] The object being formed could be inserted, with the help of a blowpipe, as many times as necessary in a furnace of this new type; it could thus be adequately reheated until its forming was complete. In contrast to what used to be common practice, and what was required by earlier techniques, the formation of the object now no longer took place over the flame but outside the furnace. This was easier from a technical perspective, and, moreover, it meant that the object was heated uniformly, which is necessary for the production of an even, well-formed vessel.[79]

The second development in the blowing technique was that glassworkers stopped blowing directly at the tip of the glass tube; instead, they began to use a hollow rod – possibly at first made of clay and later of metal – known as the blowpipe. The initial use of a clay blowpipe is considered to be one of the factors facilitating the quick spread of the technique, as such a blowpipe could be produced inexpensively and often even by the glassworkers themselves.[80] Judging from the size of vessels produced, and given that adhesion to the clay blowpipe was limited, it is thought that around AD 70 the use of an iron blowpipe became popular.[81] The iron blowpipe allowed glassworkers to handle bigger pieces of molten glass, and thus to produce the large transport and storage vessels typical of the end of the 1st century. This hypothesis is confirmed by the discovery of blowpipe fragments in waste from workshops of that time. Judging from marks left on the vessels, blowpipes must have had an internal diameter of about 0.5cm and an external one of 1.5-2cm. They must have been about 1m long, judging from comparable blowpipes in use in traditional glassworking workshops of the Eastern Mediterranean.[82]

A third important development in the evolution of glass blowing concerns the use of the blowpipe, or of some other, solid iron rod, onto which the half-finished vessel is transferred for the formation of the vessel rim while it is still hot. The metal adheres to the base of the l with the help of a small chunkamount of molten glass attached to the tip of the blowpipe/rod.

In the early years of blowing, the rims of those open-shaped vessels with very thin walls were simply cracked off, while rims of vessels with thicker walls were subsequently cold-polished. Some closed-shaped vessels with thin walls, whose rim was presumably fire-polished, as there is no pontil mark on their base, were probably treated after being annealed.[83] The next step towards the use of the blowpipe or rod was the decision to form the base of the vessel through the use of a second glass paraison, which was attached to the bottom of mostly open-shaped vessels. In this way the glassworker, by picking up the vessel with the blowpipe/rod again in order to form the base, could also reheat it to form the rim as well.[84] After this development, it was only a matter of time until glassworkers

[74] Stern 1999, 443.

[75] Strabo, *Geog.* 16.2.25.9.

[76] Stern 2004, 82-83.

[77] Baldoni 1987, 22-29.

[78] The term is found in a poem preserved in the Papyrus Poxy 3536; see Stern 2007, 349-353. We also encounter this same term in a poem by Mesomedes of Crete, who flourished in the middle of the 2nd century (*Anthologia Graeca* XVI. 323); see Stern 2007, 353-354.

[79] Stern, Schlick-Nolte 1994, 24-25.

[80] Stern 1999, 446-47.

[81] Stern 1999, 447.

[82] Stern 1999, 447.

[83] Stern 1999, 448.

[84] Grose 1977, 20, figs 3, 4. On the evolution of the technique, see Stern 1999, 448.

understood – before the end of the 1st century AD, and possibly around its middle[85] – that instead of the whole paraison that formed the base of open-shaped vessels, a small quantity of molten glass on the tip of the blowpipe/rod was actually enough to allow the vessel to be picked up and for the rim to be formed by folding or fire-polishing. This technique causes a circular or semi-circular mark to be formed on the base of the vessel; sometimes it is simply a pontil scar, shaped like the tip of the blowpipe/rod, but at other times a piece of the glass covering the blowpipe/rod's tip also remains, resulting in the vessel's partial or complete lack of balance.

The revolutionary invention of blowing greatly facilitated the production of vessels; smaller quantities of glass were needed and the process required much less time compared to earlier techniques. As a result, the price of glass products fell due to inflation caused by over-supply. Strabo illustrates this shift clearly when he writes that in the 2nd half of the 1st century one could buy in Rome a glass vessel in exchange for a copper coin.[86] The devaluation of glass objects in Roman society's economic and aesthetic value system also ensued. At the same time, however, the use of glass became very widespread geographically, reaching even the most remote or under-developed provinces within the Empire and beyond; it spread, furthermore, throughout all social strata. Glass vessels were also put to new uses, such as for the transport and storage of liquids and solid products, in large quantities. Gradually, everyday and utilitarian glass objects began to be sold by weight.[87] Blowing and technical processing in general did not significantly add to the price of glass as raw material. This clearly points to the simplification of the forming process, as well as to the very widespread circulation of glass products, even though it is also known that they were ten times more expensive than clay vessels of equal size.[88]

### Decorative techniques associated with the vessels under study

#### *Mould-blowing*

By the term mould-blowing we mean the blowing of a heated paraison of glass in a previously manufactured container, on the inside of which there is an intaglio decoration that becomes imprinted on the exterior surface of the vessel being formed.

The number of vessels produced through mould-blowing was, even during the time of the technique's greatest popularity in the 1st century AD, only a small fraction of total glassworking production. It was an important technique from an aesthetic perspective, but quantitatively contributed only very little to the total production.[89] The use of this technique with glass, rather than with clay, presents special difficulties. Clay shrinks when it dries and as a result it can be easily removed from the mould, regardless of how finely it is decorated. Glass, by contrast, does not shrink, but inserts itself in the most detailed points of the decoration, where it remains and solidifies. Because of this property of glass, if a vessel were allowed to anneal in the mould it would end up trapped inside it. It was therefore necessary to develop and use bi- and tri-partite moulds.[90] These moulds were made of clay or plaster and sometimes were casts of metal or even glass originals. They were repaired or fully replaced relatively frequently since upon coming into contact with the hot glass they were subjected to strong thermal shocks that wore them out after a short period of use.

The mould-blowing technique, and the use of multi-partite moulds in particular, was invented and developed along the Syro-Palestinian coast, probably from a desire to form glass products that imitated metal prototypes with forged decorations.[91] Mould-blowing, 'argenti modo caelere' is, according to Pliny,[92] another of the three techniques that made Sidon a famous glassworking centre.[93] Mould-blowing starts to be used commercially during the time of Augustus; its development in the western Empire begins probably around the middle of the 1st century AD and is focused on the production of tableware.[94] It remained in vogue until the end of the Flavian era (69-96), when it was superseded by the facet-cut technique. It is believed that because of the wide circulation of mould-blown

85 Stern 1999, 450.
86 Strabo, *Geog.*, XVI, 2, 25. 9; Stern 2007, 362-363.
87 Barag 1985b, 113-116.
88 Stern 1999, 460-466.
89 Fleming 1999, 42. For a general overview of findings from different areas, as well as of specific techniques and products, see Fontaine (ed.) 2010, passim. A concise overview of what is known about this technique can be found in Stern 2010, 25-37. For an example from a large Mediterranean city, see Antonaras 2010g, 241-252.
90 Stern 1995, 46-47.
91 Isings 1957, 45, type 31.
92 Pliny, *Nat. Hist.* 36.193.
93 For a detailed commentary on this passage, see Stern 2007, 359-362.
94 Stern 2001, 41.

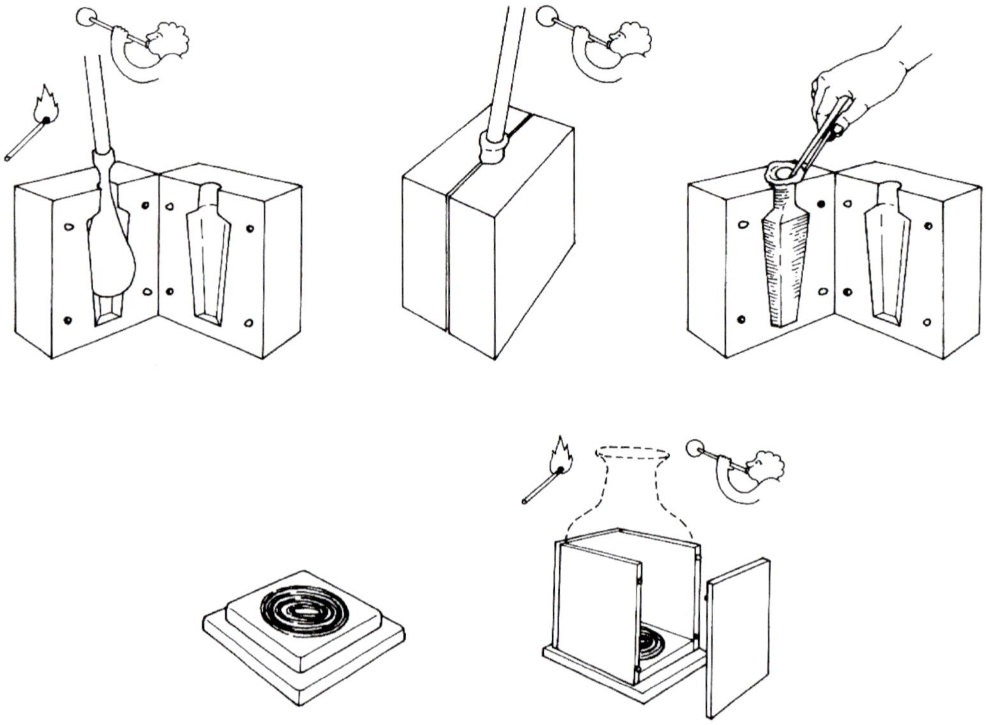

Figure 16. Mold blowing technique
Ignatiadou, Antonaras 2008, p. 185. Adapted from Meconcelli Notarianni and Ferrari 1998, p. 14.

Figure 17. Mold blown, janiform unguentarium, (Cat. No. 701) Thessaloniki, 4th c.

transport vessels in the West, the public there realized earlier how easily they could be produced; as a result, they depreciated in value. By contrast, in the eastern part of the Empire, where there was no mass vessel production employing this technique, such vessels were still produced in the 2nd century AD. Generally speaking, mould-blown vessels continued to be formed until the 4th century, using a very restricted set of decorative themes that mostly depicted human heads.[95] Polygonal *eulogia* vessels, meant to hold sanctified liquid, water or oil from the Holy Land, appeared in the Eastern Mediterranean in the late 6th and early 7th centuries, bearing religious decorations consisting of Christian and Judaic motifs.[96]

In addition to the vessels that were entirely formed in a mould, with only details such as the rim or handle being shaped freely by the glassworker, there are also a considerable number of vessels that were produced by free-blowing, but whose decorations were originally formed in a mould. Such vessels were placed in a mould, or in a ring, with a relief interior, in the early stage of blowing, thus acquiring their relief decoration. They were then given their final shape through free-blowing. As a result, the decoration expanded over a greater surface and thus extruded less. This technique, of dip-mould-blowing, was used at various times during the Roman period, especially during the Late

[95] Stern 1995, 201-246. On the production of mould-blown and dip-mould-blown vessels in a large Mediterranean centre, such as Thessaloniki, throughout the Roman period, and on the relation and juxtaposition of this technique with free-blowing, see Antonaras 2010g, 241-252, esp. 252.

[96] Barag 1970b, 35-63; Barag 1971, 45-63; Stern 1995, 247-269; Newby 2008, passim; for an enlightening introduction, see 12-17.

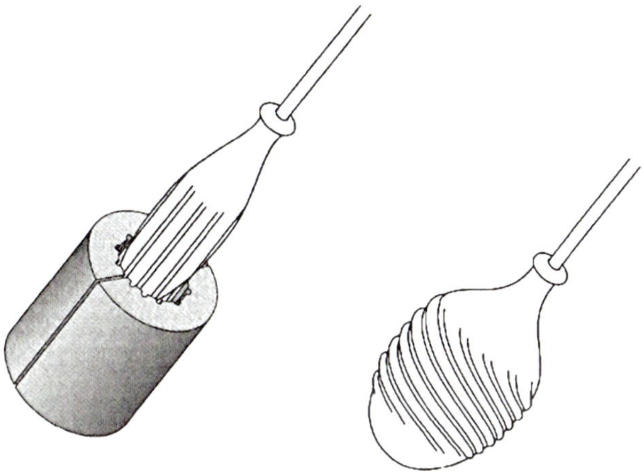

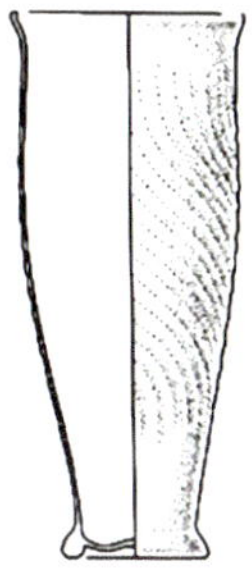

Figure 18. Dip-mould blowing technique Fleming 1999, p. 199, fig. 51-52.

Figure 19. Dip-mould blown bottle, (Cat. No. 259), Thessaloniki, 4th c. MBC BY 321/2 Cat. no. 259. Photo M. Skiadaresis. Copyright Hellenic Ministry of Culture and Tourism.

Imperial period.[97] It has been demonstrated, moreover, that in some cases at least, engraved decorations were first indicated roughly by blowing the vessel in an appropriate mould and then these blanks were completed by cutting.[98] This technique was used at various times during the Roman period.[99]

***Wheel-cutting engraving***

This technique involves the removal of part of the mass of glass from a vessel with the help of a bow drill. It was used to partially form some vessels during Classical Greek and Hellenistic times.[100] During the Roman period all the ways of creating engraved decorations known today were in use. These engraved vessels are the vessels known in the historical sources as *toreumata/τορεύματα* (works in relief);[101] their decorations were formed using a rotating stone or metal wheel and pointed, metal tools.[102] The use of this technique linked glass and glass decorating with small-object manufacture in general and with the production and decoration of silver and bone vessels and objects in particular.[103]

*Geometric motifs*

During the 1st century AD it becomes fashionable to cut the surface of clear, decolourized glass vessels (using a lathe) into contiguous rows of continuous oval- or rice-grain-shaped incisions or facets: the technique known as faceting. Since the facets are relatively deep, such vessels are accordingly made with thick walls. This trend emerges around the third quarter of the 1st century and dies out around the end of the first quarter of the 2nd century.[104] The same technique is used, nonetheless,

Figure 20. Multifaceted beaker, late 1st - early 2nd c. Corning Museum of Glass 59.1.129. Published in *Glass of Caesars*, p. 195

[97] Price, Cottam 1998a, 13; Stern 2001, 27, 133-134.

[98] Sorokina 1977b, 111-122; this technique has been traced to workshops of the 2nd and/or 3rd century in the area of the Black Sea.

[99] Price, Cottam 1998a, 13; Stern 2001, 27, 133-134.

[100] Ignatiadou 2013, 52-54.

[101] Martial, *Epigr.* II.2.1 and 12.74; Clement of Alexandria, *Paedagogus.* 2.3.35; Trowbridge 1930, 109, 166.

[102] This technique has been studied in detail by Paolucci 1997, 17-20 and passim. On techniques that survived through the Middle Ages and into more recent times, see Charleston 1964, 83-100 and Charleston 1965, 41-54.

[103] Paolucci 1997, 63-80.

[104] Oliver 1984, 35-58; Stern 2001, 137.

during the 2nd and 3rd centuries to decorate cheaper vessels, usually hemispherical bowls. The decoration is likewise diminished on such vessels – rather than covering the entire surface, the geometric pattern is sparsely used (to form bands) only on certain parts of the vessel.[105]

During this period it appears that, at least in some areas, the same decorative results are obtained by blowing the vessel in a relief mould and then cutting the imprinted decoration.[106] This technique remains in use in areas of the Rhine until the 4th century. The fact that already in the 1st century and, it seems, up until the 4th, potters tried to reproduce this decorative pattern on clay objects – either by cutting their surfaces before firing or by forming them in moulds – also attests to its wide circulation and popularity.[107]

Figure 21. Beaker with cut decoration (Cat. No. 63), Thessaloniki, 3rd-4th c. AMTH MTH 5474 Cat no 63. Photo A. Antonaras. Copyright Hellenic Ministry of Culture and Tourism.

Horizontal grooves that run across a vessel's interior or exterior surface constituted a common decorative element for glass vessels since Hellenistic times, and this pattern remained in use throughout the Roman period.

During the 3rd and 4th centuries, vessels are often decorated with clusters of thin, shallow, horizontal engravings; these form bands that organize the vessel surface into geometric sections. Sometimes they are combined with wider engraved bands or with other geometric motifs that cover the whole or part of the vessel. The rhomboid network pattern that was quite popular among such motifs is thought to represent the straw covering used to protect vessels.[108]

Wheel engraving was used in combination with the production technique of mould-pressing for the formation of richly decorated vessels.

*Figural scenes*

Figure 22. Engraved dish, (Cat. No. 74), Thessaloniki, 4th c. MBC BY 327/1 Cat no. 74. Photo A. Antonaras. Copyright Hellenic Ministry of Culture and Tourism.

In some cases, by the end of the 2nd century and mainly during the 3rd and 4th centuries, the engraved vessel decoration went beyond the simpler geometric shapes. Glassworkers had at their disposal wheels of varying thickness, either with an oblique or flat edge, as well as pointed tools that could be used to engrave details by hand. These technical advances were sufficient to allow them to create figural decorations, depicting on the one hand either mythological or religious themes and, on the other, scenes from everyday life, such as hunting. It is believed that most engravers of glass and copper objects had, and used, handbooks of decorative motifs from which they copied their designs. Examples do exist, however, in which one can easily recognize that the craftsman was really rendering an existing model.[109] Different workshops have been identified – in Rome, Egypt and along the Rhine – that specialized in particular motifs or that used distinctive techniques for forming silhouettes or for rendering figures.[110] During the 1st century it was used to correct small faults and perfect the decoration of vessels with high relief.[111]

[105] Stern 2001, no. 57, 159.

[106] Sorokina 1977b, 118-119, pl. 4.2; this technique has been traced to workshops of the 2nd and/or 3rd century in the area of the Black Sea.

[107] Stern 1998, 191-193.

[108] Stern 2001, 162-163, nos. 59-60.

[109] Stern 2001, 136.

[110] Stern 2001, 136-138.

[111] Lierke 1999, 100-109.

### *Application of plastic elements*

Decoration with applied plastic threads of varying diameters constitutes the oldest decorative technique in glassworking, already known since Pharaonic, Classical Greek and Hellenistic times, and still in use during Roman times as well. In cases where the thread is attached to the vessel and marvered before the latter acquires its final dimensions, it is fully incorporated into the vessel surface and, provided that it is thick enough and appropriately placed, it may resemble the veining in stone vessels. In the 1st century, moreover, marvered and flattened threads, usually white ones on dark-coloured vessels, are combined with pressed ribs to create the very particular vessels known by the German term *zarte Rippenschalen.*[112] Usually, however, the thread is attached after the vessel has acquired its final dimensions and therefore remains a relief decorative element.

*Thread-like elements*

Throughout the Pharaonic and ancient Greek periods multicoloured threads were used to decorate rod- and core-formed vessels. In Roman times, since the 1st century AD, a different coloured thread begins to be used to decorate blown vessels: often an opaque white thread wrapped around a dark-coloured, often blue, vessel, to striking aesthetic effect.[113] The decoration of glass vessels with attached relief threads of varying diameters becomes more widespread between the 2nd and 4th centuries. It usually involves a thread of glass of the same colour as the vessel body that is wound around either the entire vessel, or only certain parts of it, such as the neck or shoulders.[114]

A special technique known as *snake thread* appears in the 2nd century and develops during the 3rd century, spreading equally in the western and the eastern provinces of the Empire. Snake threading involves the application of thick,

Figure 23. Juglet with applied thread, (Cat. No. 688), Thessaloniki, 4th c. MBC BY 217/32 Cat no. 688. Photo M. Skiadaresis. Copyright Hellenic Ministry of Culture and Tourism.

Figure 24. *Snake-thread* flask, decorated with birds, first half of the 3rd c. Corning Museum of Glass, N.Y., 66.1.20. Published in Whitehouse 2001, p. 224, no. 794.

[112] Stern 2001, 81-82.

[113] Stern 2001, no. 1, 57, where there are also many examples.

[114] Stern 2001, 199-207, nos. 90-97, where there are also many examples.

flattened threads that bear a simple geometric motif in relief on their surface. The threads are usually the same colour as the body of the vessel and form a serpentine pattern on its surface.[115]

During the 4th and 5th centuries we also see vessels decorated with even thicker and wider threads, which, with the help of appropriate jacks, become flattened or folded.[116] This technique, especially as applied to kohl-tube unguentaria, almost reaches a Baroque excess along the Syro-Palestinian coast during this time.

*Circular elements*

Blobs

Already in the 1st century chips of different coloured glass are attached to glass vessels, while they are still hot, through marvering. Initially they are simply set in relief, but later on they are incorporated, through further blowing and marvering, into vessel walls as decoration. In Late Antiquity this type of decoration is back in fashion. Although it is used for different aesthetic effects, the process of producing it is technically identical. Clusters of small glass chips, of one or more colours, which usually form triangular motifs are attached around the bodies of translucent open-shaped vessels.[117] Tesserae, usually blue, are known to have been used to add colour to the decoration of vessels, as excavation evidence from several western sites indicates.[118]

Figure 25. Bowl with applied blobs (Cat. No. 66), Thessaloniki, 4th c.
MBC BY 227/18 CAT NO 66. Photo M. Skiadaresis.
Copyright Hellenic Ministry of Culture and Tourism.

*Appliqué medallions in relief*

During the 1st century, and again in the 4th, it becomes fashionable to decorate vessels by attaching relief elements in the form of medallions depicting Medusa, Bacchus, theatre masks and lion heads. For the most part, these attached medallions (appliqué) were in high relief, although occasionally they could also be flat. They usually decorated the point at which handles were attached to the body of jugs or pitchers, at the base of the handle, but examples of

Figure 26. Bowl with applied blobs (Cat. No. 66), Thessaloniki, 4th c.
MBC BY 227/18 CAT NO 66. Photo M. Skiadaresis.
Copyright Hellenic Ministry of Culture and Tourism.

Figure 27. Appliquee in the form of a hare, Dion, Pieria, 3rd c.
AMTH MTH 183. Photo O. Kourakis.
Copyright Hellenic Ministry of Culture and Tourism.

During the 3rd century and first half of the 4th, yet another variation of the decoration with complex figural elements appears,

[115] Harden *et al.* 1987, 127-139, no. 57-67; Stern 2001, 139, 165-166, nos. 60-61.

[116] Stern 2001, 278-80, 306, nos. 144-145, 169, with many examples.

[117] See Stern 2001, 293-296, nos. 157-179, with an annotated bibliography.

[118] Foy, Nenna 2001, 41.

known as 'flask in a flask'. This time the elements are formed to resemble miniature vessels or animals, such as rabbits, crocodiles, fish or whales. This type of decoration is encountered both along the Syro-Palestinian coast in the East, as well as in the Rhineland provinces of the West. It is believed that this technique was invented in the Syro-Palestinian region in the early 3rd century,[119] and later spread into the West,[120] remaining in use until the first half of the 4th century.[121]

### *Indentation*

While the vessel is still hot it is decorated around its exterior with indentations probably intended to imitate the look of the more valuable, forged, metal prototypes of glass vessels; the indentations are formed through simple pressure with the help of jacks. The same technique was also occasionally applied on clay vessels of the same period.[122] The vessel's body thus becomes sometimes almost cuboid and other times corrugated in shape. This technique, already used on free-blown vessels since the second half of the 1st century,[123] becomes most widespread between the 2nd and 4th centuries.[124]

Figure 28. Indented flask (Cat. No. 671), Thessaloniki, 4th c. MBC BY 255/1 cat. no 671. Photo M. Skiadaresis. Copyright Hellenic Ministry of Culture and Tourism.

Figure 29. Pinched flask, 3rd-5th c. CMOG 59,1,147, Published in Whitehouse 2001, p. 200 no. 349

### *Pinching*

With the help of pincers (small jacks), smaller or bigger protrusions are created on the surface of the vessel; they are usually spread all around the lower part of the vessel body and sometimes form a ring that acts as a substitute for a base.

These protrusions, called fins, are usually flat; they have a small indent in their middle that is sometimes perforated from the pressure exerted by the pincers.[125] This decorative technique appears in the Eastern Mediterranean at the end of the 2nd century[126] and becomes more widespread from the 5th century, remaining in use until the 8th century, at least in Jordan and the Syro-Palestinian area.[127]

A technique that produces similar results, but is nonetheless different in terms of production, was used on vessels with thicker fins that were more regularly formed and arranged. In all likelihood the fins in this case are created through mould-blowing. This technique appears around the 3rd century[128] and remains in use also during the 4th century.[129]

---

[119] Barag 1970a, 8.

[120] Stern 1999, 482.

[121] Fremersdorf 1959, 60-61, pl. 76-79; Abdul Hak 1965, 33, fig. 20; Barag 1970a, 8, figs. 1-4 on page 75, pl. XXII 1-5; Antonaras 2003b, 77-80; Antonaras 2006d.

[122] E.g. a 4th-century vessel found in Thessaloniki, reg. BK 4467/186, held in the Museum of Byzantine Culture, Thessaloniki, rendering in clay a glass vessel of Isings typology form 103, with indentations around the body. See also Antonaras 2010e, 309-319.

[123] Isings 1957, 46, form 32, and 49, form 35.

[124] See Price, Cottam 1998a, 33; Stern 2001, 209-211, 242-243, nos. 99-101, 128-129.

[125] See Stern 2001, 248, no. 134, with relevant bibliography.

[126] See Price, Cottam 1998a, 32-33; Stern 2001, 249-251, nos. 135-137.

[127] See Dussart 1998, 128, BX 111a, table 32/1, page 158, BX 83, table 46/21, page 161, BXII. 1, table 49/1; Stern 2001, 354, no. 201; Gorin-Rosen 2006, 111, note 7.

[128] Weinberg, McLellan 1992, 132, no. 107; Fremersdorf, Polónyi-Fremersdorf 1984, 65, nos. 152-153.

[129] Dussart 1998, 162, BXII 214, table 49/7.

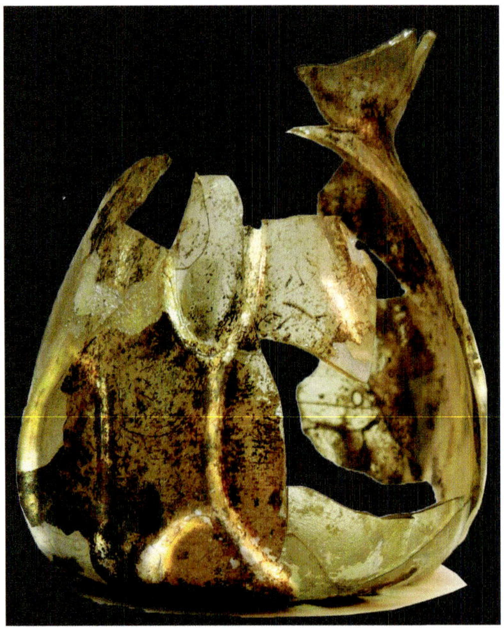

Figure 30. Jar with pinched, birfurcated decoration, (Cat. No. 451), Thessaloniki, late 3rd - 4th c. MTH 3072 CAT. NO. 451. Photo A. Antonaras. Copyright Hellenic Ministry of Culture and Tourism.

Figure 31. Blue amphoriskos with white handles, (Cat. No. 460), Thessaloniki, 1st c. MMST 160/5 CAT. NO 460. Photo A. Antonaras. Copyright Hellenic Ministry of Culture and Tourism.

Finally, another similar but more intricate decorative pattern, called bifurcated or *fadendekor*, can also be produced by using pincers before the vessel has acquired its final size. In its simpler form the pattern is composed of vertical ribs that decorate the body of the vessel, from base to shoulders. In its more complex version, the ribs are compressed at regular intervals, thus creating a rhomboid network.[130]

### *Bi-colouring*

Another relatively simple decorative technique, which was employed from Pharaonic and Classical Greek times and throughout the Roman and Medieval periods, is the use of two or more different colours of glass during the production of a vessel. Usually the handles, base, and/or even attached relief decoration, were formed from a bright glass colour that differed from the colour used for the vessel body.[131]

[130] Weinberg 1988, 3, 80-81 cat. nos. 351-353, drawing in table 4-39, photo in table 4-15/351, 352, where there is also an older bibliography, as well as technical and production details.

[131] For typical examples from various periods and places of origin, see Stern, Schlick-Nolte 1994, 137, 205, 237, 279, nos. 8, 43, 59, 76; Stern 2001, 70, 119, 193, 235, 238, nos. 13, 49, 84, 121, 124.

# 2. Secondary Glass-Workshops

## A. The location of secondary glass-workshops in the Roman world and in Roman Thessaloniki

Glassworking workshops, just like all other artisanal installations requiring the use of open fire, had to be located outside of, or in very close proximity to the city walls. On the one hand there was open, available space there, and, on the other, this arrangement reduced the risk of fire spreading in case of accidents, as well as the everyday disturbance to local residents. In the 14th century, Konstantinos Armenopoulos,[1] copying verbatim passages from Julian of Ascalon,[2] who wrote (c. AD 531–533)[3] and documented the Byzantine legislation in force at that time,[4] mentions that glass-workshops must not be located inside cities. If, however, it becomes necessary for workshops to operate within the city walls, they must be located in uninhabited, remote areas because of the great risk of fire and to avoid harming the health of local residents.[5]

"'Επαρχικὸ. Περὶ ὑελουργῶν καὶ σιδηρουργῶν.

Ὑελουργοὺς καὶ σιδηρουργοὺς τοὺς ἐργαζομένους τὰς ἀξίνας καὶ τὰς ἄμμας καὶ τὰ τοιαῦτα μεγάλα ὄργανα, καὶ ἀνδριαντοπλάστας, οὐ χρὴ ἐν αὐταῖς ταῖς πόλεσι τὰ τοιαῦτα ἐργάζεσθαι. εἰ δὲ ἀνάγκη τις γένηται τούτοις κατοικεῖν τὰς πόλεις, καὶ τὰς τούτων ἐργασίας ἐν αὐταῖς γίνεσθαι, εἰς τοὺς ἀπῳκισμένους καὶ ἰδιάζοντας τῶν πόλεων τόπους δεῖν ταύτας ἐργάζεσθαι. πολὺς γὰρ ὁ ὑπὸ τοῦ πυρὸς πέλων κίνδυνος τοῖς οἰκήμασι καὶ τοῖς σώμασι συνεχὴς λοιμὸς γίνεται"

*On glassworkers and blacksmiths. Glassworkers and blacksmiths who produce hoes and sickles and other large tools, and statue makers should not be working within cities. If there is a need for them to live in the cities and their works to be done in them (the cities), they should work in uninhabited and specific areas, given that there is a great danger of houses catching fire, and of humans catching diseases...*

In addition to archaeological evidence,[6] surviving inscriptions also attest to the activity of glassworkers in various city areas. Such inscriptions reveal that streets, gates and city quarters were named after the group of glassworkers that was active in their immediate vicinity. Inscriptions in Rome, for example, point to the street or quarter of the glassworkers, *Vicus vitrarius*, as well as to a portico 'between the glassworkers', *Porticus inter vitrarios.*[7] In addition, another inscription shows that in Pozzuoli (Puteoli) there existed a quarter called the 'Hill' or 'Slope of the

[1] Konstantinos Armenopoulos was a 14th-century Byzantine jurist who held the post of universal judge of Thessalonica, one of the highest judicial offices in the Byzantine Empire. He is best known for his *Hexabiblos* (1344-45), a legal text in six volumes, in which he compiles a wide range of Byzantine legal sources. See Pitsakis 1971.

[2] Pitsakis 1971, λη'-λθ'; Pitsakis 1992, 246, note 54.

[3] Hakim 2001, 4, with further bibliography at note 21.

[4] Lieberman 1971, 409-17, according to Hakim 2001, 7, note 38, where it is demonstrated that Julian did not record Judaic law, but instead Roman-Byzantine law. This was the prevailing law – on matters of urban planning as well – at least on the Palestinian coast, as is shown by the example of the founding of the glassworking workshops of Tiberias outside the city walls.

[5] *Hexabiblos*, Book 2, title Δ' 'Peri kainotomiōn' (On Innovations), paragraph 19, 'Peri hyelourgōn kai sidērourgōn' (On glassworkers and ironworkers). See Pitsakis 1971, 117-18.

[6] For a detailed presentation of glass-workshops and glass products that circulated, and were produced in a large Mediterranean centre, such as Thessaloniki, see Antonaras 2010b, 93-105; Antonaras 2010c, 301-334. For an overview of the same topic, see also Antonaras 2012, 3-6, 10-15.

[7] *Corpus Inscriptionum Latinarum* 6, 29844 (=093) and Trowbridge 1930, 131, notes 38-40.

Glassworkers', *clivus vitriarius*.[8] Literary sources, such as Strabo,[9] attest to the presence of glassworkers in Augustan Rome, while Martial[10] is witness to the recycling of broken glass in the same city.

Remains that can be connected to artisanal production in Thessaloniki have been discovered in specific areas,[11] i.e. in certain urban public areas that lost their original function and were occupied by craftsmen, e.g. in the ancient agora[12] or in the ruins of the large bathhouse under the Acheropoietos basilica,[13] or along the eastern[14] and western[15] city walls, outside the city walls, and in the eastern[16] and western[17] necropolis.

Workshops that were not impacting on the surrounding areas or presented little in the way of risk to the health and safety of their neighbours, because of the obvious fire risk, were active in several parts of the city.[18]

Glass-workshop buildings, just like other workshops, were usually plain and untidy constructions, made of cheap materials. They often re-used standing walls of older, abandoned buildings, presenting thus only partly new constructions. This fact makes their identification quite difficult. Usually, what identifies the nature of the use of these constructions are either movable find-remains of production, or immovable built constructions such as furnaces or reservoirs. However, as glass was recycled, only small quantities of it were left on site and offering little evidence of the workshop's activities. This is in marked contrast to what is usually found in the vicinity of pottery and metal workshops, where large quantities of deformed vessels and slag are usually encountered. Moreover, secondary glass furnaces, which were small constructions meant to be used only for a short period of time, are often damaged by long exposures to the high temperatures necessary for glassworking. Most of these structures are not well enough preserved to offer clear information on their shape, size and function.

So far, the remains of only one glass-workshop structure have been found in Thessaloniki and its surroundings, at least only one with indisputable evidence, i.e. the presence of a glass furnace, and large quantities of glassworking waste in a shop in the very centre of the city, dated to the late 6th century. Nevertheless, there is presumptive evidence that there were perhaps at least three more glass-workshops within or near the city between the 4th and 6th centuries. These movable finds have been unearthed in different parts of the city, in the necropolises, where

---

[8] *ILS* I, 1224b and Trowbridge 1930, 132, note 42.

[9] Strabo, *Geog.*, 16.2.25.

[10] Martial, *Epigr.*, 1.41.1-5.

[11] On this topic, see Antonaras 2016a, passim.

[12] Pottery workshops were active from the end of the 3rd century BC until the 1st century AD at the site where later the Roman agora was erected. See Valavanidou 2001, 119-123; Boli, Skiadaresis 2001, 91; Adam-Veleni 2003, 146. Several groups of artisans, such as potters, blacksmiths, bone carvers, textile dyers, etc., used the site of the agora for their professional activities during the 5th and 6th centuries AD, when it had already lost its official character. See Vitti 1996, 102-103; also Valavanidou 2001, 123, 127-129. The city's mint operated on the same site during the second half of the 4th century. See Velenis 1996b, 49-60. The society 'syllogos' of purple dyers (porphyrobaphoi), the one referred to as the 'eighteenth' (oktokedakatis) must have been in the proximity of the agora. It was so named as to distinguish it from other similar societies in the city and also in order to specify the exact location of its professional activities. See IG X. 2 no. 291. Discussion on the inscription in Robert 1937, 535, note 3. Vitti 1996, 78-79.

[13] Bakirtzis 1983, 310-329, with bibliography.

[14] An extended complex of workshops was active during the 2nd century BC and 1st century AD under the North Stoa of Galerius' palatial complex, where metallurgical remains, clay matrices, and evidence of purple-dying were documented. See Karamperi, Christodoulidou, Kaïafa 1996, 534. What was possibly a potter's workshop, dated to the 3rd century AD, was found at 18, K. Palaiologou Street. See Kourkoutidou-Nikolaidou, Mavropoulou-Tsioumi 1983, 281-289.

[15] Several furnaces connected with potters' activities were unearthed in rescue excavations: a Roman pottery furnace was found on Eirinis Street, just by the western city wall. See Vavritsas 1972, 505 and Despoini 1982, 81-83, figs. 15-16, pl. 14.

[16] Roman Imperial-period furnaces, possibly used for the production of clay products, were found on the plot of the Museum of Byzantine Culture. See Nalpantis 1988, 381; Nalpantis 1989, 346. Remains of a workshop with moulds for clay lamps were found in the excavation of the Theological Department of the Aristotle University, dated possibly to the 3rd-4th century AD. See Petsas 1966, 334-339, pls. 345-359.In the eastern necropolis, possibly at the Theological Department excavation site, was also found a clay lamp mould dated to the end of the 4th century. See *Thessaloniki Istoria kai Techni* 1986, 22, no. 7.

[17] Furnaces, possibly used for the production of pottery during the Roman Imperial period, were unearthed at plots located on the junction of the Koloniari, Galanaki and Papathanasiou Streets. See Nalpantis 1987, 403-406. A workshop for clay products, possibly for bricks, was unearthed at 93, Monastiriou Street. It is quite possibly one of the workshops which developed west of the city walls on the 'Keramēsios Kampos', toponym known from the 5th Miracle in the second book of St Demetrius' Miracles, dated to the 7th century. See Theocharidou 1985-86, 99; and for the excavation Makropoulou 1988, 379-380. It has been suggested that more than one brick workshop operated simultaneously during the 4th and 5th centuries around Thessaloniki, due to the diversity of brick stamps found in the city's monuments. See Vickers 1973, 285-294, also Theocharidou 1985-86, 109. Two furnaces of indeterminable use were uncovered in the western necropolis, near the Axios complex, next to Langada Street, in 2004. I thank the excavator, E. Lambrothanasi, for providing this information.

[18] Unpublished excavations from several parts of the city. The movable finds are kept in the Museum of Byzantine Culture in Thessaloniki. More than 112 workshops dated from the Hellenistic era to the 15th century, and the artisanal production in ancient and Byzantine Thessaloniki in general feature in a separate study by the author (Antonaras 2016a).

several craftsmen were active, and in urban public spaces following their abandonment or change of their original use, i.e. the large bathhouse under the Acheiropoietos basilica and the site of the ancient agora.[19]

In several cities of the Roman Empire, just as in Thessaloniki, excavations have shown that glassworkers – like other craftsmen – were occupying and using every free city location where they could undertake their professional activities.

Glassworkers in Late Antiquity often housed their workshops in large abandoned (and usually formerly public) buildings in the centre of cities, such as the bathhouse below the Acheiropoietos basilica. Such workshops are located in many cities throughout the Empire[20] in this era: Rome,[21] Solin (Salona),[22] Ashmunein (Hermopolis Magna),[23] Lyon (Lugdunum),[24] Augst/KaiserAugst (Augusta Raurica) in Switzerland,[25] Gamzigrad (Romuliana Felix) in Serbia,[26] Svishtov (Novae)[27] in Bulgaria; Greek sites include (as well as Thessaloniki):[28] Philippi,[29] Delphi,[30] Olympia,[31] Rhodes,[32] Samos[33] and Kos.[34]

Glass-workshops on commercial streets, such as the one at 45 Vasileos Irakeliu Street, and even along the main street of the city, among others, have been unearthed in Edessa[35] and Philippi[36] in Greece and in Bet She'an in Israel.[37]

Moreover a dedicated district or quarter of workshops, similar to the one created in the abandoned area of the ancient agora in Thessaloniki, was, it seems, a quite usual occurrence and similar districts existed in Late Roman times in Soli, Cyprus,[38] Villeneuve-lès-Maguelone,[39] Cologne,[40] Leicester (Ratae Corieltauvorum),[41] Corinth,[42] Samareia,[43] Ljubljana (Emona),[44] and Ptuj (Poetovio).[45] Artisanal clusters had already been created earlier in Aix-en-Provence (Aquae Sextiae) and Autun (Augostodunum)[46] in France, as well as in Britain.[47]

In addition, glassworkings appear in the areas of the city walls (within and beyond) of Marseilles (Massalia),[48] Cologne,[49] and Trier (Treviri).[50]

[19] Detailed presentation of the topic is given in Antonaras 2002b, 237-260.
[20] Van Geesbergen 1999, 107.
[21] Sternini 1989, 105-114.
[22] Auth 1975, 147. For an in-depth view of glassworking residues from Croatia, see Fadić 2004, 95-106.
[23] Bimson, Freestone 1987; Sternini 1995b, 144.
[24] *A travers le verre* 1989, 47-49; Foy *et al.* 1989, 58-59; Sternini 1995b, 181-182.
[25] Sternini 1995b, 193-194.
[26] Ljubinković 1983, 181-182. Under publication by the author.
[27] Four different installations were excavated in the Roman ruins of the city. See Biernacki *et al.* 1976, 158, fig. 41, 42; Olczak 1978, 127-137; Biernacki *et al.* 1991, 110; Olczak 1998, 13; Dyczek 1998, 99-104.
[28] Antonaras 2002b, 237-260; Antonaras 2014a, 95-113; Antonaras 2014d, 263-272.
[29] Gounaris, Velenis 1996, 730-731; Gounaris 1995-2000, 323-331, 351, figs. 2-4, 6; Skordara, Gounaris, Maniatis 2002, 321-326; Gounaris 2002, 28.
[30] Petridis 2002, 90.
[31] Schauer 2001-2002, 42. A Roman glass furnace was located near Thermes Leonidiou, Olympia.
[32] Triantafyllidis 1997, 49, 50, fig. 12.
[33] Isler 1969, 228, pl. 98. 6, a find from the Heraion of Samos, dated to around the middle of the 6th century.
[34] Brouskari 2002, 135-141.
[35] Antonaras, Chrysostomou 2012, 293-303.
[36] Gounaris 1995-2000, 323-356; Gounaris 2002, 28; Skordara 2014, passim.
[37] Gorin-Rosen 2000, 59-60.
[38] Young 1993, 40, note 7.
[39] Foy *et al.* 1989, 67. Sternini 1995b, 155.
[40] Sternini 1995b, 158.
[41] Price, Cool 1989, 24.
[42] Wiseman 1969, 105-106, pl. 32/j, k, m, glassworking residues traced in a layer dated to the late 5th-6th century in the area of the Gymnasion of ancient Corinth.
[43] Sternini 1995b, 179-180.
[44] Plesničar-Gec 1983, 73-74.
[45] Lazar 2003, 221; Korošec 2004, 67-70; Sagadin 2004, 107-114.
[46] Sternini 1995b, 146-148.
[47] Price 1998, 341-342.
[48] Sternini 1995b, 152.
[49] Follmann-Schulz 1989b, 35; Sternini 1995b, 158.
[50] Goethert-Polaschek 1977, 6, note 44; Sternini 1995b, 163.

Finally, glass-workshops also appear in small hamlets and isolated farmhouses across a very wide area, as finds reveal from, *inter alia*, Vienne (Vienna), several in France,[51] Hambach in Germany,[52] Jalame[53] in Israel, Louloudies Kitrous,[54] Palatiano (Ioron),[55] Vrysi Pontokomis,[56] Knossos,[57] and, probably, in Tarrha[58] and Chios.[59]

Furthermore, at least in the northwestern provinces of the Roman Empire, glassworking workshops operated in yet another setting – in the manufacturing zones that developed around military camps, to cater to the needs of the army.[60] It has been revealed that the army was engaged in the distribution and also the production of glass. Archaeological evidence reveals that glass fragments were collected meticulously, obviously with the intention of recycling. Even though it has not yet been ascertained whether the soldiers themselves were involved in glassworking, or whether they only supervised citizens employed by the army, it is certain that the Roman army was active in this enterprise as well.[61]

### B. Glass-workshops in Roman Thessaloniki[62]

#### 1. *East of the Acheiropoietos basilica (S. Georgiadis plot)*

Remains of glassworking activity were unearthed to the east of the Acheropoietos basilica (mid 5th century) on the site of the ruined Roman public bathhouse[63] that covered the area below and around the basilica.[64] Three consequent phases of its floors have been traced[65] and which are dated between the second half of the 2nd and first half of the 4th century.[66] The eastern and northern part of the bathhouse continued to operate after the basilica was erected.[67] The ruins of the bathhouse were unearthed during the excavation of the plot east of the apse in 1966.[68] Movable finds from that excavation included glassworking residues, parts of the glass furnace and a clay melting pot, which are described below and indicate the operation of a workshop in the area.

Excavation data: *11-6-1966. At the level of the mosaics*

1. Mass (16x9x6cm) of yellow plaster covered by dark layer of vitrified clay on which are preserved molten and partly fused fragments of greenish glass (1.5cm thick).
2. Iron scraper with an almost square body (6x5cm) and a handle square in section (1x1cm). The scraper is not commonly found among the iron tools used exclusively or mainly by glassworkers (which include blowpipe, jacks and, probably, scissors), yet the fact that it was found together with the other glassworking items suggests that it was also used in that workshop.[69]
   Excavation data: *13-6-1966. Trench A', Spot O-B/3-4. Depth 0.20-0.40cm*
3. Mass (16x5x6cm) of clay with extensive signs of intensive exposure in fire (its lower part is totally black with blisters); on the sides are preserved traces of vitrification and of greenish glass. It presents part of a circular construction (estimated diam. c. 30-40cm), probably part of the interior rim or shelf of a furnace, or part of the base of the grill, or of the coating of a melting pot. It is part of the same construction as No. 6. With that was found a grey clay shard from the periphery of a discoid object, which has preserved traces of molten greenish glass on its surface (possibly part of a base or supporting system).

[51] Foy *et al.* 1989, 64; Sternini 1995b, 155.
[52] Gaitzsch 1999, 125-149. On the typology of produced vessels, see Follmann-Schulz 2001. On the composition of the glass produced there, see Wedepohl, Gaitzsch, Follmann-Schulz 2001. Final publication of the workshops is in Gaitzsch *et al.* 2000.
[53] Weinberg 1988.
[54] Marki 1996, esp. 239-243; Marki 2002, 65-66; Angelkou 2002, 65-67.
[55] Antonaras, Anagnostopoulou-Chatzipolychroni 2002, 118, fig. 14.
[56] Karamitrou-Mentesidi 2000, 620; Karamitrou-Mentesidi 1982, 298.
[57] Price 1992, 458-459; Carington-Smith 1982, 269.
[58] Buechner 1960.
[59] Ballance *et al.* 1989, 124-125.
[60] Price 2005, 173-174.
[61] Price 1998, 333, where several British examples are cited.
[62] For an overview of the glass-workshops unearthed in Thessaloniki, see Antonaras 2002b, 237-260; Antonaras 2006c, 247-253; Antonaras 2010b, 93-105.
[63] Bakirtzis 1983, 325-326.
[64] Petsas 1967b, 393-396, pl. 299-300; Pelekanidis 1961/62a, 252-253; Bakirtzis 1983, 310-329, esp. 321; Xyngopoulos 1941/52, 472-487.
[65] Bakirtzis 1983, 323.
[66] Asimakopoulou-Atzaka 1998, 127, 227-228, note 129.
[67] Bakirtzis 1983, 324.
[68] Petsas 1967b, 393-396, pl. 299-300.
[69] On the various metal and wooden tools of a glass-workshop, see Sternini 1995b, 83-87.

Figures 32-33. East of the Acheiropoietos basilica (S. Georgiadis plot), Glassworking refuse( the author).

4. Fragments of a roughly semicircular melting pot, reinforced at least twice with layers of clay, covered with molten glass (diam. 13cm, pres. h. 3.5cm, th. 0.6cm). It was coated with a thick layer of reddish clay (1.5cm near the base and 3.5cm on the rim). A thin layer (0.1cm) of greenish glass covers the vessel and the clay coating. Later, another thin grey (0.7cm) layer of clay was applied over the glass in order to reinforce again the pot. A thin layer of glass covers also this outer layer of clay. This reinforcement of the pot reached quite a high level and splayed all around the rim, except for a flat part, which was probably turned towards the glory hole in order to facilitate the movements of the glassblower when taking viscous glass from the kiln. Household vessels were also often used as melting pots, as well as the crude, hand-made vessels fabricated for the purpose.[70]

Excavation data: *13-6-1966. Trench A', Spot O-B/1. 5-3. 5. Depth 0.40-0.65m.*

5. Mass of gray clay with vitrified surfaces (10x6x5cm).

6. Excavation data: *13-6-1966. Trench B', Spot O-B/4-10. Depth 0.40-0.70m.*

7. Mass (6x4x3.5cm) of clay with traces of exposure to intensive fire. The lower part of the mass is totally black with blisters. On the sides there are preserved traces of vitrification and greenish glass. It is part of the same construction as No 1.

8. Circular mass (13x8x7cm) of clay with extensive traces of vitrification, covered almost entirely with blisters.

Figures 34-35. Melting pot fragment (the author).

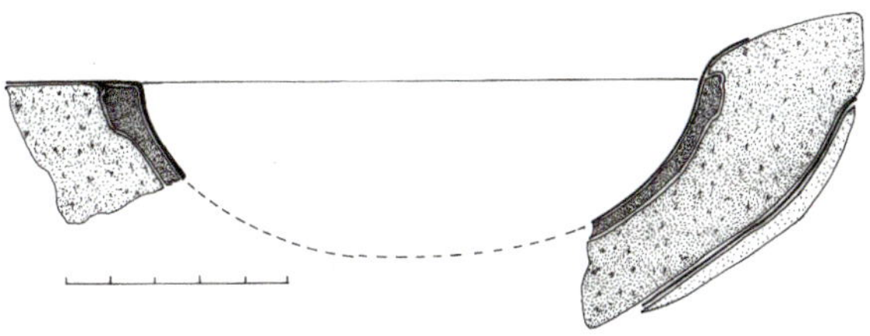

Figure 36. Drawing of melting pot. (Chrysoula Mallia).

[70] On relevant examples, see Price, Cool 1989, 24-25; Sternini 1995b, 73-82; Van Geesbergen 1999, 112-113.

9. Circular clay shard, triangular in section, made of dark-coloured clay (ext. diam. 24cm, h. 3.5cm). Possibly a ring made to hold the melting pot or some permanent similar device, e.g. a wedge.
10. Mass comprised of iron slag and a clay sherd fused together.
11. Reeded strap-handle of a greenish vessel (possibly a large amphora or a single-handled transport vessel) (w. 6cm, pres. l. 4cm).
12. Mass (8.5x6x4cm) of dark-coloured clay with vitrified surfaces.
    Excavation data: *15-6-1966. Trench A', Spot O-B/3½ -4. Depth 0.20-1m.*
13. Fragment of the same melting pot as No. 2. (5x4x2cm). The outer rim created by the red clay layer is not higher than the vessel, as with fragment No. 2; on the contrary it is flat. This part was probably turned towards the glory hole to facilitate the movements of the glassblower when taking viscous glass from the furnace.
14. A small part of what seems to be a glass moil crack-off.
15. A small part of what seems to be a glass moil crack-off; diam. c. 2.5cm.
16. A conical base of a 4th- or 5th-century vessel (Form 70).

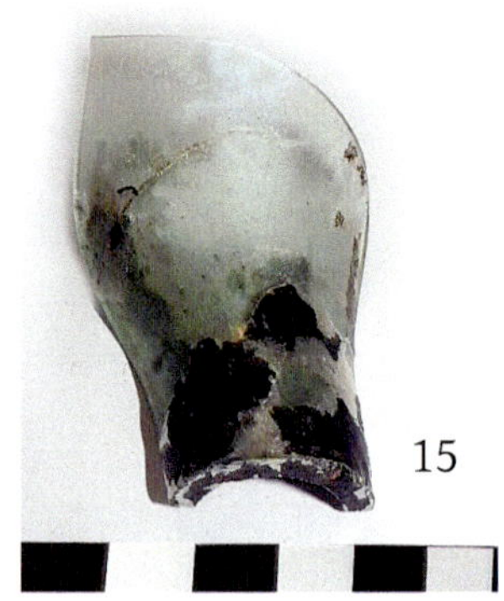


Figures 37-38. Glass moils (the author).

It is not possible to date precisely the operation of the workshop as the reports of the excavation are not fully published. It seems only logical, however, to place it somewhere in the 5th century, presuming that the ruins of the abandoned bathhouse were used before the erection of the basilica, or even during the construction of the monument.

It is also possible that the finds were merely discarded on that site and were originally used in a nearby plot.

### *2. The Ancient Agora*

Excavations for the foundations of the courthouse in Thessaloniki uncovered the city's ancient agora.[71] It appears that when the complex lost its official public character in the late 4th, 5th and 6th centuries, it was occupied by craftsmen of various trades.[72] The remains cannot be accurately dated; it can only be assumed that they belong to the late 4th to early 6th century period, when several craftsmen were active on the site of the agora. The first movable finds related to glassworking activities were unearthed first in the 1962 campaign.

1. Mass (5x3.5x2cm) of three, greenish lumps of glass of irregular shape. These are attached to a mass of red clay (with some tiny gravel inclusions) that was apparently exposed for prolonged periods to high temperatures.
2. A triangular trimming of glass (1.8x1.0x0.5cm), probably a test droplet, of the same tinge and texture as No. 1, which was found with it.
   During the 1966 excavations more glassworking remains were unearthed:
   Excavation data: *Stoa B' (15) depth (in length) 2.60m.*

1a. A mass (8x4x3cm) of olive-green glass adhering to a dark-coloured mass of soil. This mass is covered by a layer of glass (th. 0.5cm), which melted and dripped onto it. On the upper surface of the mass, as it appears, glass had dripped consequently, forming a rough surface with spiky protuberances (th. 1.5cm). The sides are totally covered by a continuous, smooth, thinner (th. 0.5cm) layer of glass. Glass has also dripped onto the periphery of the lower surface,

Figure 39. The Ancient Agora, Glassworking refuse (the author).

[71] Petsas 1968a, 156-161; Petsas 1966/67, 295 ff, nos. 55-57; Adam-Veleni *et al.* 1996, 501, with detailed bibliography.

[72] See Vitti 1996, 102-103; Valavanidou 2001, 123, 127-129.

while the rest of that side remains free of adhesions, probably due to the fact that it was attached to a stone, as its convex shape indicates.

1b-c. Two smaller fragments (3x2x2cm and 4x3x2cm, respectively) of the aforementioned mass.

2. Two tear-shaped test droplets made of the same olive-green glass (3x1.5x1cm and 2x1x1cm, respectively).

3. Flat, ovular mass of olive-green glass (9x6x0.5cm). On its rough surfaces are relief imprints of the surface of the ground where it landed while still hot and malleable.

4. Amorphous, dark-coloured, clay mass (7x6x1.5cm) bearing intensive traces of vitrification.

5a-e In the same area/room were also found deformed shards of decolourized and olive-green glass with no stratigraphic indication; it seems more probable that they were deformed due to exposure to high temperatures during a fire, a quite common phenomenon in Thessaloniki up until the early 20th century. Vessels which had been in fires secondarily often have a characteristic dulling to their surfaces. These can be rough and retain the anomalous shape of the surface on which they lay during the fire. Conversely, vessels deformed during production have smooth and transparent walls. Furthermore, they are only partly unfinished and usually the reason they have been discarded is particularly obvious, e.g. a misshaped rim or base.

Figure 40. The Ancient Agora, Glassworking refuse (the author)

An object once considered to be connected with glassworking was also found in the excavations of the agora in the 1960s, although its exact context is unavailable. It is an almost square (8x8.5x2cm) stone mould. On its summarily curved surface, which is framed by an overhanging rim, the head of an animal *en face*, possibly a wolf or fox, has been presented schematically. The head is on the same level as the rim, raised approximately 0.2cm from the field that surrounds it. The side-surfaces, on the areas where they are not broken, are rough and bevelled, the decorated side being larger than its flat reverse side. The shape and size of the mould are comparable to relevant finds from workshops in the (mainly) western[73] and eastern[74] Empire. It has to be noted, however, that a relevant theme has not been preserved among the extant bases, while the manner in which the theme is rendered *per se*, due to the way it is curved, is not at all suitable for reproduction on a glass vessel.[75]

There are no particular indications for a precise dating of the glassworking waste other than the general chronological time-frame (late 4th to early 6th centuries), i.e. the period in which artisanal activity on the site was thriving.

Finally, further glassworking waste was unearthed in a 4th- or 5th-century context during recent excavations at the same site.[76]

### 3. *Theological Department, Aristotle University*

A building identified as a 'warehouse' was found, between and over graves, on the site of the Theological Department of the Aristotle University of Thessaloniki during salvage excavations conducted on the site in 1965. Several consequent phases of use were identified in the building. It was dated, at least partly, to the Early Byzantine period, probably to the 6th century,[77] as its walls were founded on graves mainly dated to the Tetrarchic period, with a few of them even later, in the reign of Justinian I.[78] Lamp moulds were found within the walls,[79] dating between the

[73] On relevant examples of clay and stone moulds, see France: Foy, Nenna 2001, 51, 82-83; Switzerland: Rütti 1991, 150-152, fig. 103; Germany: Sternini 1995b, 90-91, figs. 129-130, 134.

[74] On relevant examples of clay and stone moulds, see: Romania: Baluta 1981, 111-14, figs. 1-2; Turkey: Akat, Firatli, Kocabaş 1984, 55, pl. 16, fig. 45.

[75] On examples with simpler geometrical motifs and other more elaborate ones representing human figures or animals and inscriptions, see Foy, Nenna 2001, 118-120, with detailed bibliography. Furthermore, the most complete corpus of base marks and moulds appears (three volumes) in Foy, Nenna (eds) 2006a; Foy, Nenna (eds) 2006b; Foy, Nenna (eds) 2011.

[76] *Glass Cosmos 2010*, nos. 264, 295.

[77] Petsas 1966, 334-339, pls. 345-359 (esp. 338-339, pl. 347β).

[78] Petsas 1966, 337.

[79] Petsas 1966, pl. 356α, γ, ε. The decorative motifs on the moulds have parallels in products of Athenian workshops. On moulds with Eros playing double-flute facing to the left, see Perlzwieg 1961, 115, nos. 715-718, esp. 717, pl. 16. Dated to the second half of the 3rd century. The motif survives until the late 4th and early 5th century. On moulds with bucranium, see Perlzwieg 1961, 132-133, nos. 1044-1046, pl. 22. Dated to the first half and to the middle of the 4th century. On moulds with a rosette surrounded by a wine scroll, see Perlzweig 1961, 152, nos. 1788-1809, pl. 31. Dated to the middle of the 4th to the early 5th century. The base mould with three concentric circles appears quite often among Attic lamps

second half of the 3rd and the end of the 4th century. Lamp moulds and other clay finds – vessels and water pipes – found on the site led to the conclusion that it was a functional building of some kind near an older, clay lamp workshop.

Movable finds taken to be the remains of glassworking activities were discovered in the same area, comprising:

1a-c. Three clay masses with layers of colourless greenish glass. The upper surface is covered by consequent layers of glass with spiky protuberances. The side-surfaces are totally covered with a continuous, smooth, layer of glass. Glass has also dripped around the lower surface, while the rest of that side remains clean, probably due to the fact that it was attached to a stone, as its convex shape indicates. Possibly this is an indication of a tank-furnace.

2. Test droplets and trimming sherds of greenish glass, deformed by their contact with the ground while still hot.
3. Decolourized, slightly greenish moil crack-offs. These represent the extreme parts of vessels in production in contact with the blowpipe. This part is cracked off when the rim of the vessel is formed. It is a short part with thick walls. One end is cylindrical, with the dimensions of the blowpipe, while the other side, where the rim of the vessel was eventually cut, widens and forms a cone.
4. Many fragments of glass vessels were found in the same area; these were collected there, probably for melting. Many of these are recognizable and belong to late 4th- and/or 5th-century forms. A large proportion of these fragments are small, dark-green body sherds of unidentifiable vessel forms. The recognizable forms are the following:

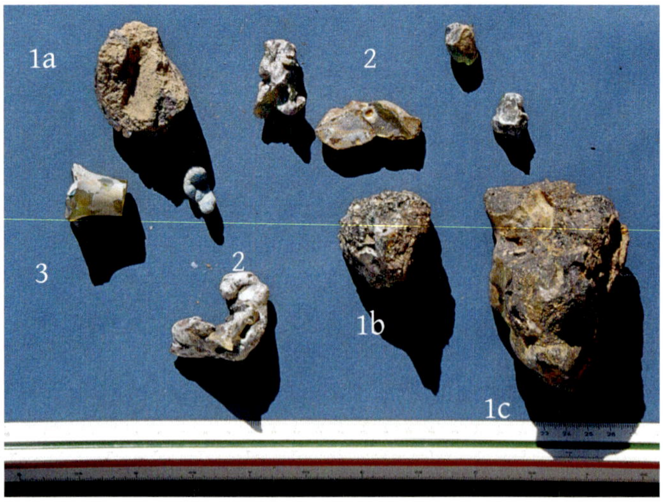


Figure 41. Theological Department, Aristotle University, Glassworking refuse (the author)

   a. Conical bases, found in their dozens, probably from stemmed beakers, in different sizes and in shades of greenish or bluish glass. In a very few cases, small handles have been found on small body sherds, possibly connected with vessels of this form, or with bowls, described as Form 'b'. (Forms 35, 36).
   b. Sherds of hemispherical and deeper bowls, with sharp cracked-off rims in equally large numbers such as 'a' (Forms 12, 13).
   c. Several flat or rounded bases of narrow, cylindrical vessels, stemmed lamps, known in Greek sources as '*ουρίαχοι*'[80] (Form 38).

of the 3rd-4th century; cf. Perlzweig 1961, 160, no. 2047, pl. 34, dated to the 4th century. The fragment of a mould with a wine scroll appears on various types of 3rd- and 4th-century Attic lamps, see Perlzweig 1961, pl. 51, rim form no. 22. In the collections of the Museum of Byzantine Culture there is unpublished material from the same excavation. Included is a mould also similar to the products of 3rd- and 4th-century Attic workshops; cf. Perlzweig 1961, 136-137, nos. 1181-1182, dated to the second half of the 3rd and first half of the 4th century. Also there is a lamp mould published in *Thessaloniki Istoria kai Techni 1986*, 22, no. 7, dated to the end of the 4th century, which is noted as a find from the eastern necropolis, possibly from the same excavation.

[80] Paulus Silentiarius, *Descriptio Sanctae sophiae*, 825, 39); Suida, s.v. οὐρίαχος; Trowbridge 1930, 174, 190, 191.

d. Solid, olive-green base knops, probably parts of conical vessels, most probably lamps (Form 34).
e. Sherds of olive-green bases, most probably parts of ovoid, short-necked, funnel-mouthed vessels with fire-polished rims[81] (Form 70).
f. Fragments of funnel-mouths with cracked-off rim, from vessels which were probably spherical (Forms 62 and 63).

It should be noted that vessels of groups 'a-c' and probably 'd' as well, were used as lamps, a fact that can be reasonably combined with the moulds of clay lamps also found on site, as well as with the fact that the workshop was founded in the extended eastern necropolis of the city, where there was clearly a wide use of lamps. The dating of the workshop is insecure due to the lack of exact excavation data. There are however several clues that can may be indicative:

* The graves in that location are mainly dated to the Tetrarchic period and up to the reign of Justinian.
* The moulds for clay lamps found in the walls of the building are dated between the second half of the 3rd and up to the end of the 4th century.
* The fragments of vessels found on site, which were probably collected for recycling, are dated after the 4th century, and especially in the 5th and 6th centuries.
* The find of a follis of Anastasius I with the glass cullet.

It may be concluded, therefore, that the workshop, if it were operating simultaneously with the workshop making clay lamps, was in use between the second half of the 3rd and the end of the 4th century. However, all the indications, including the glass waste and numismatic evidence, point to a period of probable use as after the Tetrarchy, when the site was used intensively for burials, and even more probably the late 5th, and especially the first half of the 6th century. The types of vessels found on the site reinforce the assumption that mainly lamps were made there, i.e. perfectly normal products for a workshop built in a necropolis.

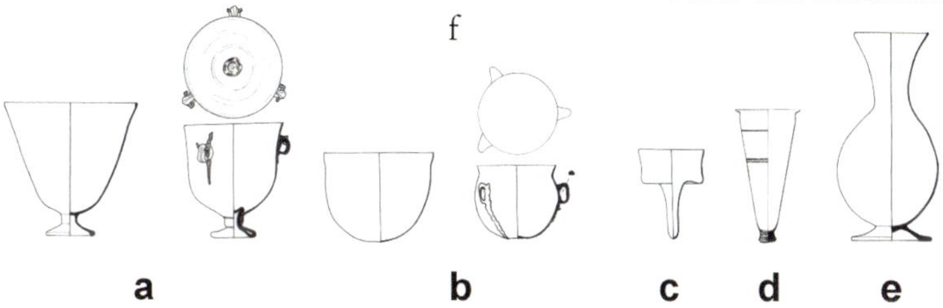


Figure 42a-f. Theological Department, Aristotle University, vessel fragments and forms identified among them (the author).

***4. The site at No. 45 Vassileos Irakleiou Street***

The first firmly identified glass-workshop in Thessaloniki, dated to the late 6th century was excavated in the corner shop of a well built, two-storeyed *insula* at the junction of two streets. Several phases of glass furnaces were preserved in it, along with much glass waste. It was rectangular (9x11m), built with cut stone and rubble with mortar, and had two entrances in its western wall, leading to the street.[82] It probably extended its activities on the pavement in front, presenting a generally quite usual picture of glass-workshops of this period. Namely, during the salvage excavation on the site a range of architectural remains were unearthed, including parts of two intersecting ancient streets and parts of two blocks with high-status 3rd- or 4th-century-AD buildings that underwent several changes in the Early Christian period. The corner shop, at the south end of the eastern block, was formed in a later, 5th- or 6th-century construction phase, and was used as a glass-workshop sometime between the 6th and 7th

[81] *Thessaloniki Istoria kai Techni* 1986, 59, no. 4.
[82] On the glass-workshop, see Antonaras 2014a, 95-113; Antonaras 2014d, 263-272.

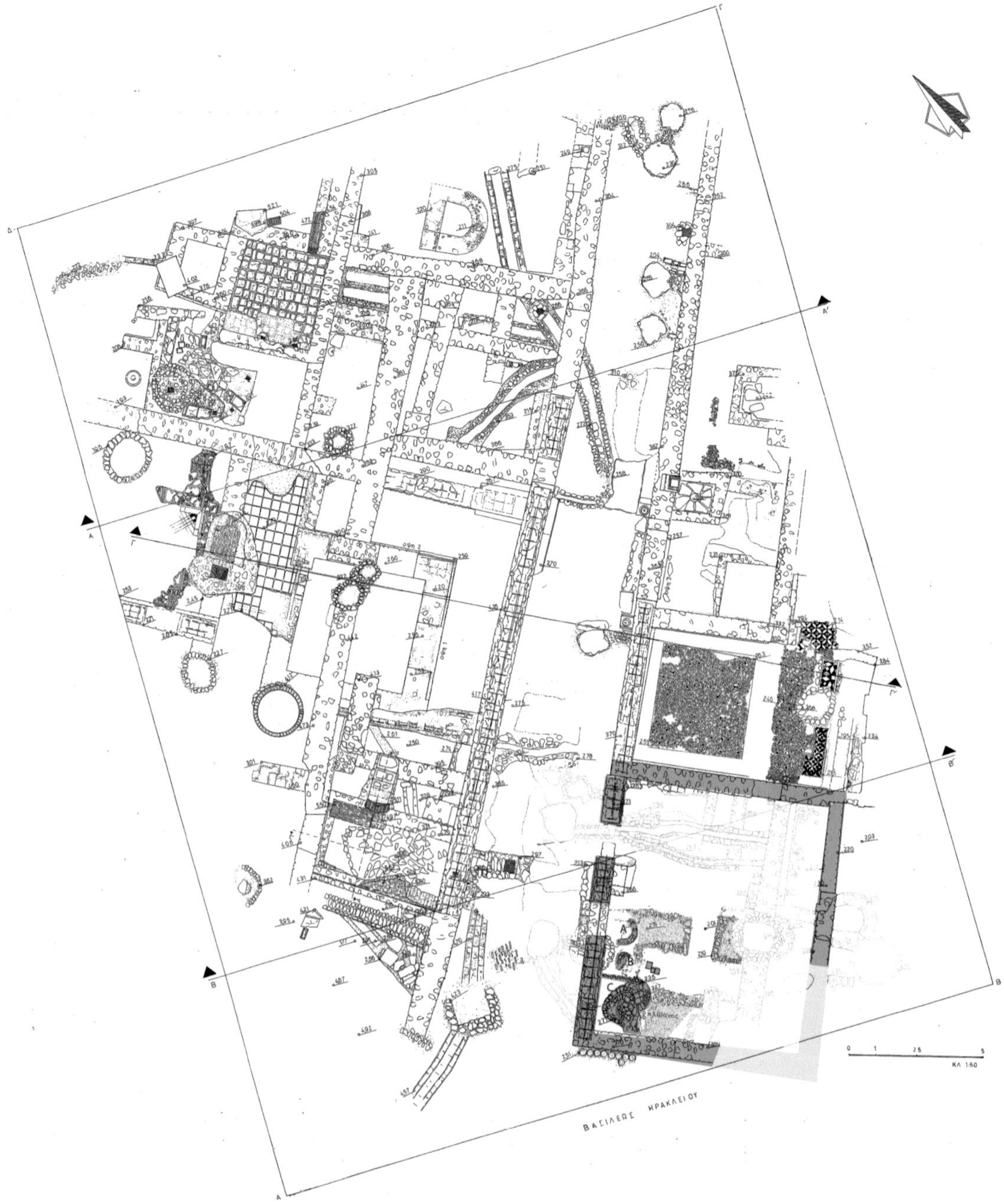

Figure 43. General plan of the excavation at 45, Vasileos Irakleiou street, with the glass workshop highlighted (courtesy of 9th Ephorate of Byzantine Antiquities).

century. On the site several water pipes and reservoirs, datable to the Byzantine and Ottoman periods, were also unearthed, together with some indications for a pottery workshop and lime kiln.[83]

Three circular glass kilns, about 1m wide, were found in the shop that housed the workshop. One of these displayed evidence of four consequent phases of use and repair, while the other two (older) kilns featured only a single, very partly preserved, phase. It seems that it was a tank furnace and no signs of crucibles were unearthed. No clear physical remains of an annealing chamber or furnace have been found on the site.

The quantity of wasters and other glass working material from this excavation is substantial and offers important information about the production processes and the products of the site. Lumps of colourless greenish and dark-

[83] For the excavation, see Akrivopoulou 2014, 255-262.

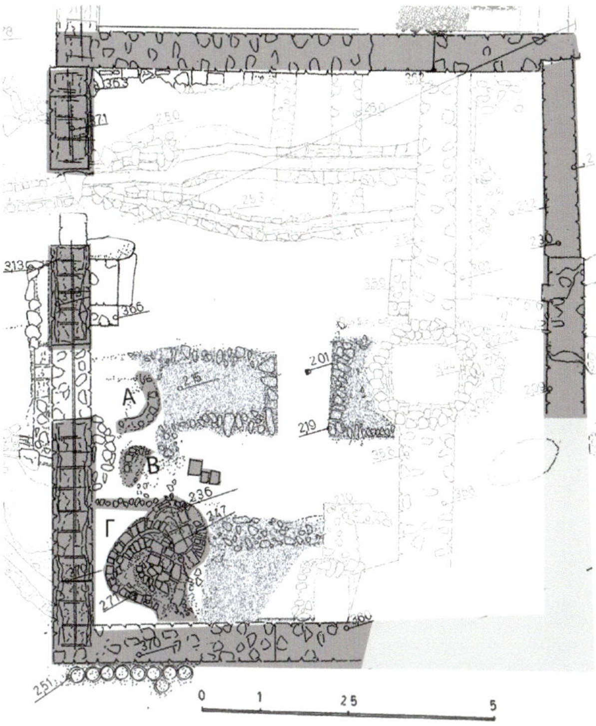


Figure 44. Plan of the workshop at 45, Vasileos Irakleiou street, (courtesy of 9th Ephorate of Byzantine Antiquities).

Figure 45. General view of the workshop (Sofia Akrivopoulou).

Figure 46. Glass kilns B and C. The five phases of furnace C are outlined (Sofia Akrivopoulou).

Figure 47. Raw glass of different colors (the author).

Figure 48. Raw glass and deformed masses (the author).

Figure 49. Testing droplets and deformed masses of glass (the author).

Figure 50. Moils (the author).

green raw glass (2682g) and dark-blue glass (90g) have been found in the Vasileos Irakleiou workshop. Testing droplets are another common residue from glass production and in total a little less than 2.5kg of these were found, all of a greenish tinge. Ring-shaped moil crack-offs (the tubular endings from the blown glass gathers) present another equally important group of finds. Almost all of them (c. 0.5kg) were greenish, which was obviously the principal glass colour used in the workshop, while only a very few were bluish (23g). From their texture and size it may be concluded that the glass-blower used an iron pipe with a gathering end that was 2.5cm wide.

Pieces of vessels have been located in the remains of the workshop, the greenish ones weighing 326g and the bluish ones 35g. Among them the following forms have been identified:

* Several dozen stemmed beakers were recorded in this excavation, mainly with smooth, long or short stems, some with beaded stems, and several examples which were severely distorted. Most were free-blown and just a few bore vertical, dip-mould ribbing. All the pieces were made from a single gather of glass folded and tooled to create the body, stem, and conical base.

* Five base fragments from stemmed beakers found among other workshop waste are unique because they feature stamped inscriptions. On three of them a capital 'K' is visible, on a fourth part of the sunken field is preserved, while the fifth is stamped 'Ms'.

Figure 51. Distorted bases of stemmed beakers (the author).

Figure 52. Dip mould blown stemmed beakers (the author).

Figure 53. Base of a free blown stemmed beaker-like lamp with a long handle and a rim of a flask decorated with dark-colored trail (the author).

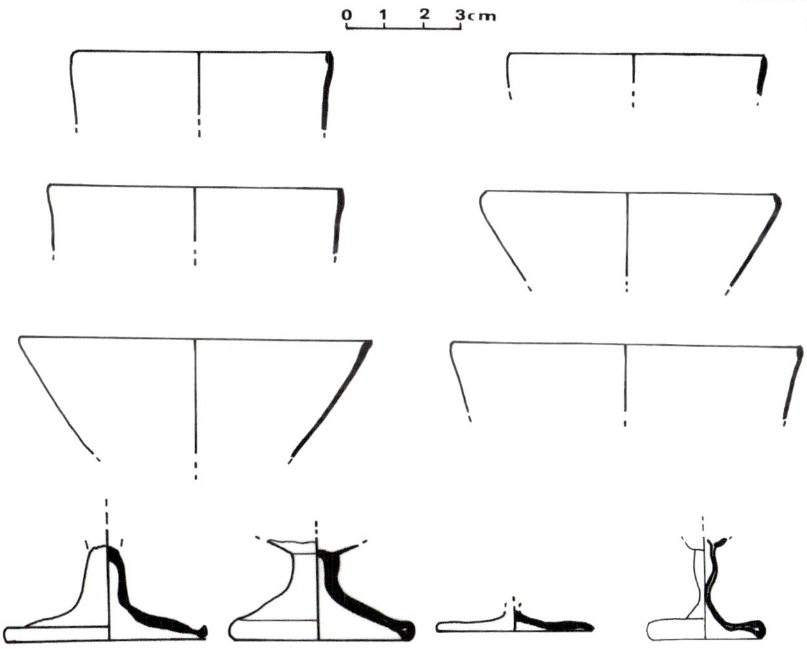


Figure 54. Drawings of bowls and stems of stemmed beakers (Chrysoula Mallia).

Figure 55. Base of a stemmed beaker-like lamp, with long handles (the author).

Figure 56. Bases of stamped stemmed beakers (the author).

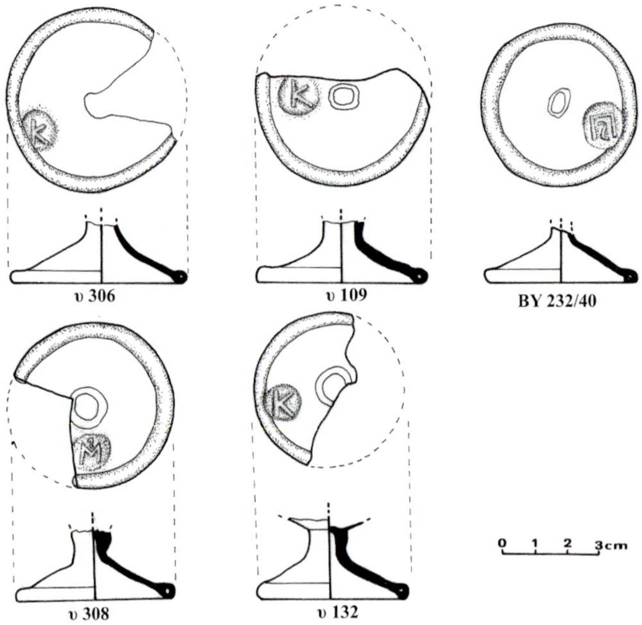


Figure 57. Drawings of stamped stemmed beakers (Chrysoula Mallia).

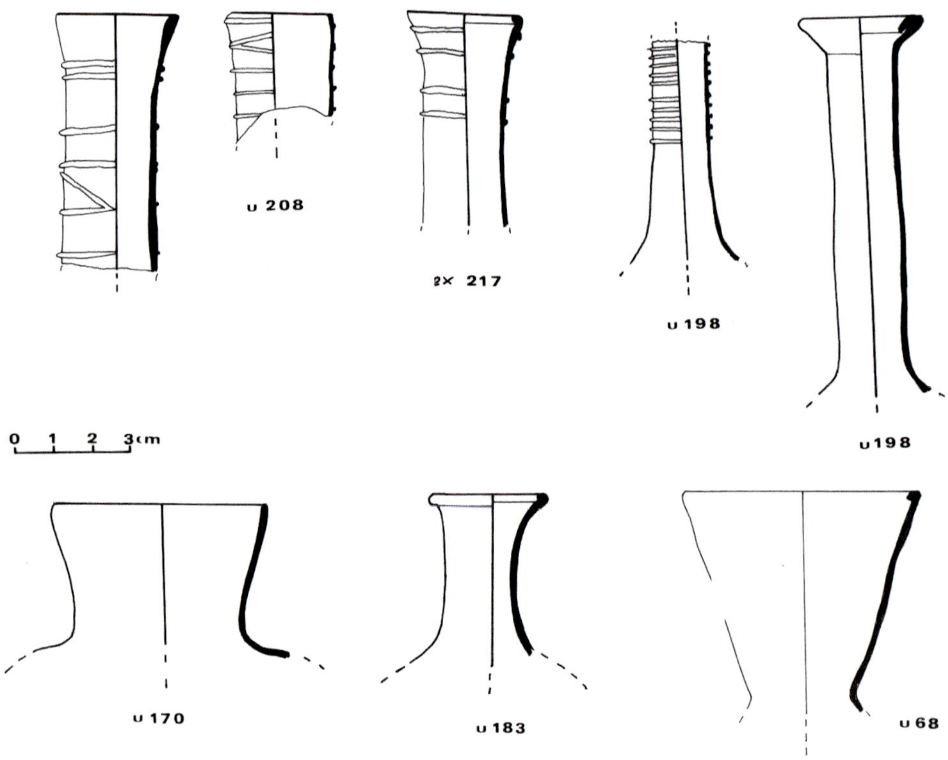


Figure 58. Drawings of necks of close-shaped vessels (Chrysoula Mallia).

Figure 59. Close-shaped vessels/flasks and stemmed beakers (the author).

Figure 60. Bottom of a deformed stemmed lamp (the author).

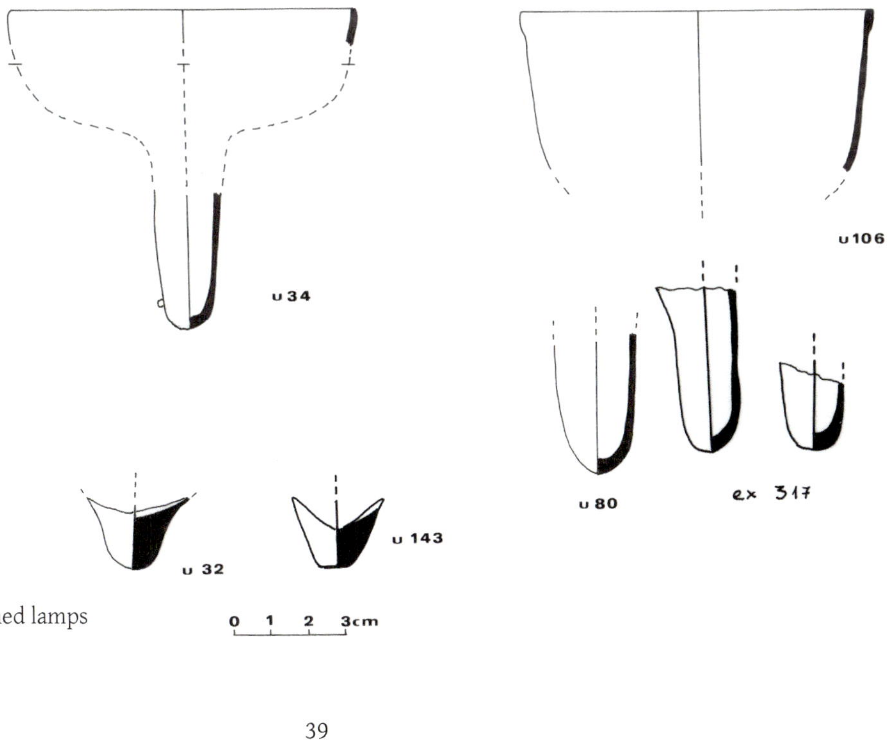


Figure 61. Drawings of stemmed lamps (Chrysoula Mallia).

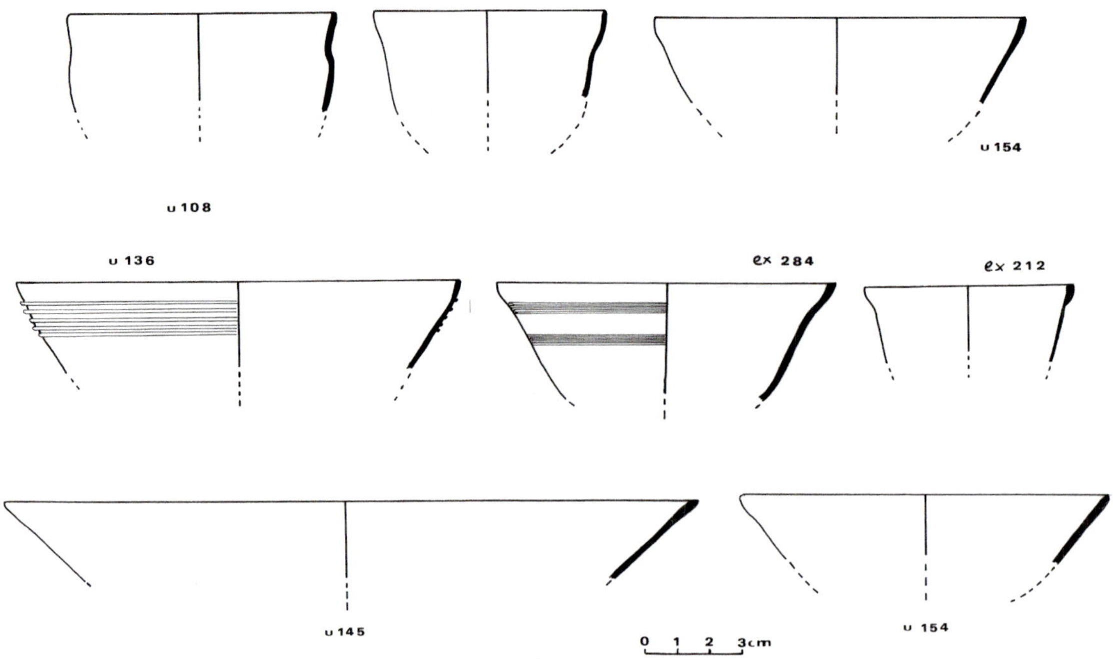


Figure 62. Drawings of bowls (Chrysoula Mallia).

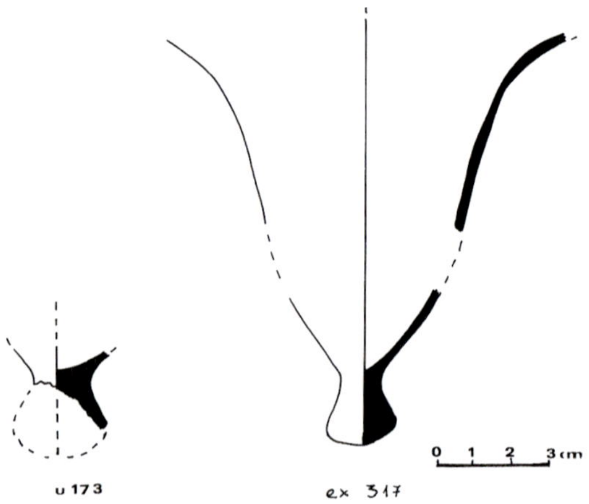


Figure 63. Conical or cylindrical lamps with a massive knop (Chrysoula Mallia).

Figure 64. Deformed fragments of window panes (the author).

Figure 65. Glass gems (the author)

* Funnel-mouthed, closed vessels were the second largest group of vessels identified among the workshop wasters. These have concave bases and funnel-shaped mouths with fire-polished rims, or simple necks with infolded tubular rims, and are all made in the colourless greenish glass used for the vast majority of the locally produced vessels.

* Stemmed lamps were also probably produced in the workshop. At least ten examples were found, one with a quite distorted base. All of them were made in the same greenish glass as the two forms previously noted as possible local products.

* Other forms included bowls, a few hemispherical, wide, conical lamps ending in a large knop, a flask

base and a few small handles of lamps. These were identified among the Early Christian finds from the excavation but they were not among the workshop waste.

* Several fragments of window panes, weighing 235g and made by the cylinder technique, were also found among the workshop waste.

* At least three over-sized, rectangular glass gems in green glass.

***C. Probable Local Products***

The products of Thessaloniki's glass-workshops cannot be determined precisely since deformed or unfinished vessels have been found on sites only occasionally, and only these can offer true evidence of the local production of these forms. However, there are other indications that do allow us to assume with a great degree of certainty which forms of vessels were produced locally, e.g. the fact that certain forms, otherwise quite rare, appear in large numbers in several excavations all over the city, and also that the vast majority of them were made of the same kind of glass in terms of quality and tinge.[84]

There are no excavations indicating the presence of glass-workshops in 1st-century Thessaloniki. In addition, it is generally believed that only a few glass-workshops were active during this period, mainly on the Syro-Palestinian coast and in Italy. However the extremely large numbers in which two forms of 1st-century unguentaria appear, equal to or even larger than in their acknowledged production areas in northern Italy and Ticino, Switzerland, give rise to the conjecture that they were also made in Thessaloniki.

*1. Spherical unguentaria with irregular cracked-off rim, 'Globes' (Form 116)*

More than 29 examples are found in Thessaloniki's necropolises. It must be noted that, with the exception of two examples, all others are completely different from their small-sized, intensely coloured western counterparts, since they are larger (diam. c. 8cm) and made of thin, colourless glass.[85]

*2. Bird-shaped unguentaria (Form 120)*

More than 36 examples are found in Thessaloniki's necropolises. They can be divided into three groups according to the form of head and beak. The most numerous variant has a schematically modelled head and beak, and it seems that it is not present among the western finds, which are generally smaller.[86]

Stronger indications for the local production of glass vessels, almost always indirect, can be traced from the 4th century onwards, a period to which all the following forms can be ascribed.

*3. Hemispherical bowls (Forms 12a-e)*

Hemispherical bowls can with certainty be ascribed to local production. With 19 examples available, they represent one of the most numerous groups of drinking vessels or lamps. Most of them are made from the same glass, or from a glass that is very similar in colour and quality, and they have the same swallow incisions around their unworked rim. Their height ranges between 5 and 10cm. They are widespread during the late 4th and mainly in the 5th century, a period when it seems quite improbable that such an ordinary, utilitarian type of vessel would have been imported from a remote production centre. The assumption that they were locally produced is strengthened by the fact that many sherds of this form were traced among the finds of the glass-workshop that was located on the site of the Theological Department of the university, in the eastern necropolis of the city.

*4. Stemmed beakers (Form 35)*

Stemmed beakers are quite probably another local product. Vessels of this form usually have a fire-polished, or, less frequently, an unworked, cracked-off rim, an inverted conical, calyx-shaped or spherical body, standing on an inherent, conical base with a tubular or, occasionally, beaded stem. Their heights range between 9 and 11cm. They are the most popular vessel form of 5th and 6th centuries, with more than 32 examples registered, but with a far larger number

[84] For an overview of the probable local glass products of Thessaloniki, see Antonaras 2010b, 93-105.

[85] Antonaras 2005, 241-248; Antonaras 2010a, 27-33.

[86] Antonaras 2005, 241-248; Antonaras 2010a, 27-33.

unearthed in general in the city, which also represent a quite homogeneous stylistic group. They are not a demanding product in terms of craftsmanship, and it seems unlikely that they could have been brought to the city from a remote area. Many fragments of this form were found among the remains of the glass-workshop on the site of the Theological Department of the university in the eastern necropolis of the city. They were even more present in the waste of the workshop at 45 Vasileos Irakleiou Street. In the latter there were also five base fragments from stemmed beakers, which bear stamped inscriptions.[87] On three of them a capital 'K' is visible; on a fourth find a part of the sunken field is preserved; the fifth is stamped with 'Ms' and the sixth with 'Π'. These letters may possibly refer to the capacity of the vessel, i.e. a 20th, 40th or 46th, and an 80th or 86th of an unknown unit of liquid measurement.

*5. Stemmed lamps (Form 38)*

Stemmed lamps, which were widely distributed in this region between the 5th and 7th centuries, are another vessel form probably produced in Thessaloniki and more specifically in the workshop at 45 Vassileos Irakleiou Street. There, in the same pit, among the remaining workshop wasters, at least ten examples were found, one with a quite distorted base. All these lamps were made in the same greenish glass as the other products of the workshop.[88]

*6-7. Spherical (Forms 51a, 51b) and pear-shaped (Forms 52a, 52b) bottles*

Bottles with infolded rim, flaring, almost horizontal mouth, cylindrical neck, spherical and pear-shaped, smooth or dip-mould, blown obliquely-ribbed body, dated to the 4th or 5th century, quite possibly represent local products. This form is among the most widely represented throughout the Empire in the late 3rd to early 5th centuries.

The larger ones, their height ranging between 15 and 25cm, with 59 spherical and 27 pear-shaped examples, represent the most numerous group of tableware vessels of this period, outnumbering all other contemporaneous forms combined. Also, smaller ones (Form 139, see infra) with a height of 7-15cm, numbering 31 examples, represent the most numerous group of 4th-5th century unguentaria. Both groups share several common features, both in terms of their shaping, as well as in the glass used for their production, which makes it quite obvious that they were products of the same workshop. For such plain, utilitarian vessels found in such large numbers, it is quite possible to assume that they were locally produced and not imported. In addition, some of the examples are so badly executed, unstable, or with minor deformations on the neck, rim, or even on the body, that it is quite evident that they could not have been brought to the city from a remote production centre. Furthermore, it seems that they were poorly finished creations intended for burial use, possibly sold at lower prices than the ordinary household products, an assumption confirmed by the fact that no scratches are found on any of them, indicating that they were not ordinary vessels taken from the houses of the deceased. The same also applies to ceramic vessels found in the graves of this period.

*8. Spherical vessels with funnel-mouth and plain base (Form 63)*

Unworked rim, funnel-mouth, long cylindrical neck, globular body, slightly concave base. All examples are decorated with shallow, horizontal, incised bands. Their height ranges between 16 and 22cm. This form represents a local, more slender variant of Isings Form 104b, which is assumed to be a local product. Two reasons lead us to assume that this is a locally produced form, i.e. there are seven surviving examples, presenting one of the most numerous group of flasks in the city, and also there is the example of a locally preserved *non-finito* vessel of the type. The *non-finito* vessel is in fact an Isings 104 that had never been fully blown to its normal dimensions.

The rim of the vessel is still uncut, retaining even a short cylindrical ending, part of the crack-off of the moil.[89] Its unfinished character is also suggested by the lack of a pontil mark on its base, while all other examples of this form bear a characteristic annular scar at the centre of their base. The fact that it is a plain, utilitarian vessel is an additional argument for the hypothesis that it is probably not an import to 4th-century Thessaloniki. On the contrary, it appears to be a local product that was left unfinished, either because it was meant to be used as a cheap burial offering, or by pure chance, due to reasons that are obscure today. There are published similar *non-finito* vessels found in the western parts of the Empire.[90] Their single common characteristic is that they all, like our example, have a wheel-cut rim that was finished when the vessel had cooled. When the vessel had been annealed, the glassworker would cut the moil, i.e. the truncated conical part between the blowpipe and the final rim of the vessel, and subsequently finish the rim by polishing it.

[87] Antonaras 2011a, 126-128; Antonaras 2011b, 239-242.

[88] Antonaras 2014a, 108, 110.

[89] Papanikola-Bakirtzi 2002 (ed.), 318, no. 349 (A. Antonaras).

[90] Price, Cool, 1989, 25; Goethert-Polaschek 1977, 153, no. 917 pls. 59 and 252, no. 1458, pl. 16/176d.

*9. Ovular, funnel-mouthed flasks with inherent base (Form 70)*

Thick flasks, with fire-polished rim, funnel-mouth, short cylindrical neck, ovoid body and tall conical base. Their height ranges between 18 and 22cm. They appear to be partly local products, although they bear features of Egyptian products, such as tooling marks on their base, and they are often made of dark-green glass. The fact that this form is not found elsewhere suggests their local production: there are 11 examples in Thessaloniki. (Their only parallels come from Britain.) Some of them should perhaps be considered Egyptian imports, which served as prototypes for a series of local and brighter copies, that represent the form under discussion. It is unlikely that the finds were imports from a remote centre.

*10. Ovular, funnel-mouthed flasks with plain base (Form 71)*

Unworked rim, fairly long, cylindrical neck flaring towards the rim, ovular body, slightly concave base. Almost all examples are decorated with shallow, horizontal, incised bands. Their height ranges between 14 and 18cm. There are 7 examples of these flasks from Thessaloniki and a few more from its hinterland (Edessa, and Marvinci) and no other examples are known from other regions. They must have been local products since they are found only in the Thessaloniki region and they were plain, utilitarian objects, quite unlikely to have been imported from elsewhere. Also, in terms of texture and thickness, they are made of the same glass, have the same type of rim-shaping and same incised decorative bands as the above-mentioned spherical flasks, which are also assumed to be local (Form 63). Both forms share so many characteristics that it is probable that they may even be products of the same workshop.

Thick flasks, with fire-polished rim, funnel-mouth, short cylindrical neck, ovoid body and tall conical base. Their height ranges between 18 and 22cm. They appear to be partly local products, although they bear features of Egyptian products, such as tooling marks on their base, and they are often made of dark-green glass. The fact that this form is not found elsewhere suggests their local production: there are 11 examples in Thessaloniki. (Their only parallels come from Britain.) Some of them should perhaps be considered Egyptian imports, which served as prototypes for a series of local and brighter copies, that represent the form under discussion. It is unlikely that the finds were imports from a remote centre.

*Jugs*

*11. Hemispherical jugs (Form 93)*

Vessels of this type have flaring mouth, infolded rim, a cylindrical neck, narrower towards the rim, hemispherical body and concave base. A curvilinear strap-handle begins on the shoulder and ends at the rim. A thick trail of glass is wound around the rim. Their height ranges between 14 and 16cm. Three examples are known from Thessaloniki, dated to the 4th century, and just one similar jug was found not far away in Edessa. It should be noted that jugs from Syria and Palestine are quite similar.

*12. Eight-sided jugs (Form 94)*

A vessel of this type has a mould-blown, eight-sided body. The shape of its rim, neck and handle are identical to those of the hemispherical jugs. It is 16cm tall. The base bears a mould-made, decorative motif in relief. Three lines intersecting at their central point divide the base into six almost equal areas. The letters 'O' and 'S' can be read, respectively, in two consecutive fields, probably representing the ending of a male name.

The vessel is made from plain, greenish glass, with no impurities or pinprick bubbles, just like the hemispherical jugs. Similar examples have not been traced elsewhere. Judging by stylistic elements, such as the forming of the lip and handle, its general proportions and the texture of the glass, it seems to have been produced by the same workshop as the hemispherical jugs and was probably found in the same grave as one of these. Similar vessels are found in private collections and museums. The form, thanks to technical and stylistic elements, such as the thick trail around the rim and the texture of the glass, is dated to the 4th century.

Several parts of the vessels are identical, which leads to the assumption that hemispherical and eight-sided jugs were products of the same workshop. It is quite probable that they were produced in Thessaloniki since their common features, such as rim and handle shaping, are not traced in other regions. Furthermore, among the finds from Thessaloniki they are the only group of jugs of which five examples have been found, whilst one or, less frequently, two examples have been found of all the other groups. Another feature indicating local production, especially of the eight-sided examples, is their shape itself, and it appears that eight-sided bottles were favoured in 4th-century Thessaloniki, probably imitating products of the workshops of Cologne that came to the city during the period of the Tetrarchy. In addition, the same aesthetics and technological tradition are echoed by the eight-sided jugs for transportation (Form 97), that will be presented later, and these are also considered to be local products.

*13. Eight-sided bottles (Form 97)*

Mould-blown, eight-sided bottles, around 8cm in body diameter, are a relatively rare form of vessel that appears to have been locally produced around the late 4th century. Their lip is horizontal, infolded and flattened and their neck is cylindrical, flaring towards the body. A broad, band-like handle starts at the shoulder of the vessel and ends at the upper end of the neck. They appear in two variants, a taller one 17cm high and a shorter one 12cm high. There are two examples of the taller variant, a smooth one and an obliquely ribbed one. On their base they bear the same mould-blown relief of an ivy leaf.

The two examples of the shorter variant (h. 12 and 13cm respectively) are identical to the taller ones, except for their height and the mark printed on their bases. It is encircled by a relief ring, and consists of four ellipsoidal motifs arranged in a cross-like pattern, probably a four-petal rosette. One of the examples bears a second concentric ring that goes over the middle of the ellipsoid motifs. Since they were blown practically in identical moulds, each of the two variants is obviously the product of the same workshop. Their common characteristics make it almost certain that all four examples were produced in the same workshop. Although no parts of the mould itself were found in Thessaloniki, it is highly likely that they are local products. The only known examples thus far of these vessels, with the specific shape and decorative motifs on their bases, were found in the necropolises of Thessaloniki, and they represent ordinary utilitarian vessels that were quite unlikely to have been imported from another remote glass-making centres. The vessels from Thessaloniki are dated by their context, i.e. generally in the Late Roman period, with some of them placed more specifically in the 4th-5th century (see *infra*).[91] It has to be noted, however, that the modelling technique that involves the use of two moulds is considered to be typical for workshops of northwestern provinces, probably Cologne, in the second half of the 4th century.[92]

*Unguentaria*

*14. Spherical unguentaria (Form 139)*

Unguentaria with infolded, flaring rim, cylindrical neck and spherical body are a miniature version of the bottles already presented and, with 31 examples, are the most widely represented form of unguentaria among the finds from Thessaloniki, dated to the late 3rd to early 5th centuries. They are considered local products firstly because they are simple creations, quite unsuitable for long distance trade and also because of the homogeneity of their group, in terms of material and form. The numerous larger counterparts (Forms 51, 52), which have already been identified as local products, should be added to their already large number, thus reinforcing their local character.

*15. Unguentaria with indented cubical body (Form 141)*

Unguentaria with infolded, slightly flaring rim, tall cylindrical neck, indented cubical body and slightly concave base, ranging between 8 and 11cm in height, are also possible local products. In Thessaloniki they represent the third most numerous group of 4th-century unguentaria, with more than 13 examples registered. They represent a very homogeneous group as far as the forming of their neck and rim is concerned, as well as in the quality and colour of their glass. As they are very easily formed, it appears very probable that they were not imported from elsewhere, rather that they were made locally.

*16. Fusiform unguentaria (Form 147a)*

Fusiform unguentaria might also represent a local, 4th- or 5th-century product. They have an infolded, slightly flaring rim, often quite crudely formed, and a very long (45-55cm), slender, biconical body ending in a solid base knop. These vessels are found throughout the Empire in relatively large numbers and several places have been considered as their production sites. With more than 31 examples, they are the second most numerous type of unguentaria in Thessaloniki of this period; almost all other types are represented by just one or two examples. This group also appears quite homogeneous in terms of size and colour and quality of the glass used for their production. The majority of them are poorly crafted with a quite unsuccessful forming of the lip, which would make their use relatively difficult and the hypothesis of their import from a remote centre quite unlikely.

---

91 Papanikola-Bakirtzi 2002 (ed.), 317-318, no. 348 (A. Antonaras), no. 350 (E. Marki).

92 Stern 2001, 142, 144, 184. The use of two moulds, a smaller ribbed one and a second full-size one, smooth and prismatic, represent a characteristic of the production of the Rhine area, especially Cologne. It first appears in hexagonal vessels with small dolphin-like handles which appear in the first half of the 4th century. Pirling 1966, 79, pl. 40/9α-β, 100/5. Isings 1957, form 100. For further examples see Stern 2001, note 166.

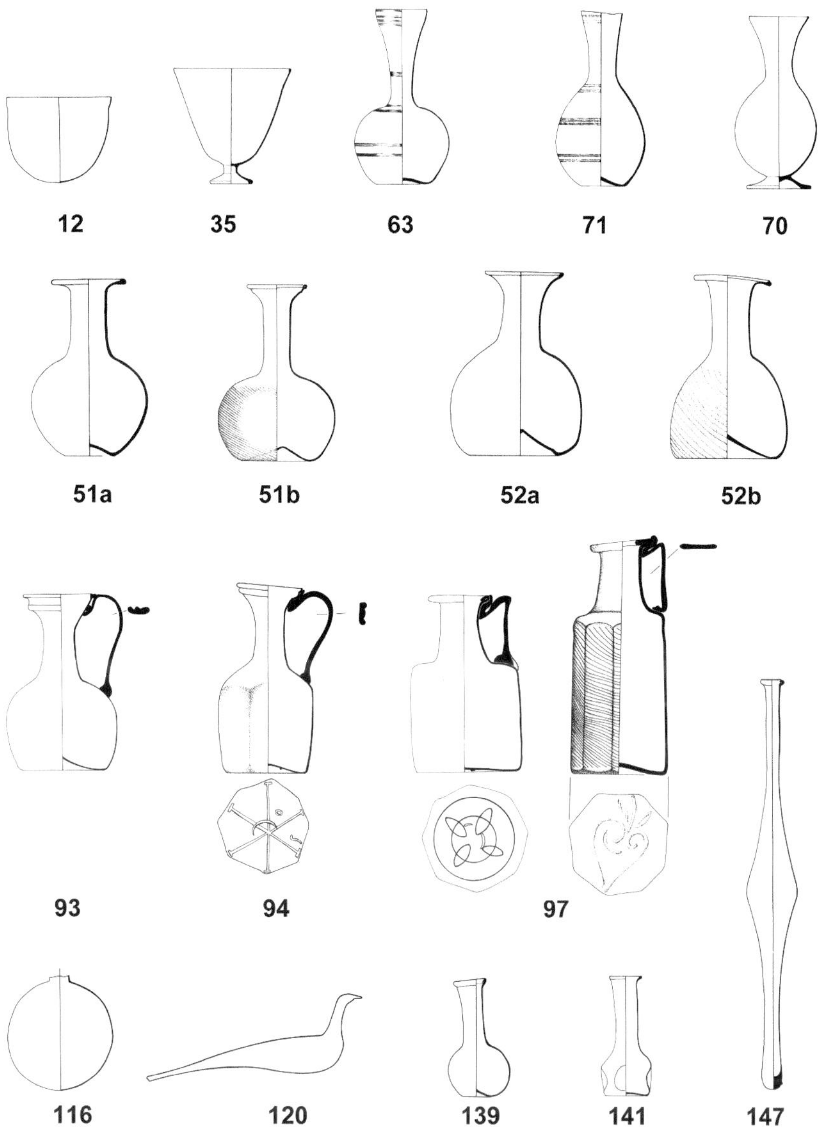


Figure 66. Probable Local Products

### Conclusions

The nature of the local products in Thessaloniki can be determined, by virtue of the nature and scale of glass-workshop remains, only from the excavations of the workshops at the Theological Department site at the university, and the plot at 45 Vasileos Irakleiou Street. From the large number of similar vessels found together, we may conjecture that several types of lamps were produced there: stemmed beakers, stemmed lamps, hemispherical bowls, with or without handles, and cylindrical beakers with conical or cylindrical bases.

The entire range of vessels that can be ascribed, with some certainty, to local workshops comprises plain, utilitarian vessels, occasionally incised or indented, which were intended for the tables of Thessalonians, for lighting their public and private buildings, for transporting and preserving fragrances, cosmetics, pharmaceutical or liturgical substances, dated – with the exception of two forms of 1st-century unguentaria – between the 4th and 6th centuries.

As a general conclusion, we may say that, according to recent finds, remains of glassworking activities in Thessaloniki are exclusively Late Antique, dated to the period after the 4th century. This conclusion agrees with the finds of glass objects found in excavations conducted in the city. From these finds it becomes clear that, with one or two exceptions, the vast majority of 1st and early 2nd century AD glass vessels were imports from well-known glassworking centres primarily in Italy and secondarily in the Eastern Mediterranean region. The resulting picture of the distribution and local production of glass vessels in Thessaloniki agrees with what we already know about

1. East of the Acheiropoietos basilica (S. Georgiadis plot)
2. The Ancient Agora
3. Theological Department, Aristotle University
4. The site at No. 45 Vassileos Irakleiou Street

glass history, where it is generally believed that local glass production spread as early as the late 1st century AD. This is the period when Thessaloniki gradually enters the political scene and strengthens its position in commerce. Its population increases and becomes more cosmopolitan.[93] The communities of Roman *negotiatores* are well attested, as are those of Jews and other foreigners, mainly originating from Asia Minor.[94] This international character can be seen in the buildings and products circulating in the city. Hence, the fact that no 1st- to 3rd-century workshops are found is possibly a result of the fragmentary and virtually accidental nature of salvage excavations, which offer us all the information presented here. Finally, from the late 3rd century onwards, the absolute number of glass vessels increases drastically, while the number of types of plain utilitarian vessels, unlikely to be imported from remote areas, is simultaneously multiplied. From this period onwards, the first indications of local glassworking activities appear, a fact that can be connected with the growth of the population and the economic development and prosperity of Thessaloniki, which then occupied a prominent position in the political and economical life of the Empire. All these factors presented the social and economical background, which permitted and eventually resulted in the founding of glass-workshops and their possibly simultaneous activities in 4th- to 6th-century Thessaloniki.

[93] Voutyras 1997, 81; Sverkos 2001, 99.

[94] On the Roman *negotiatores*, see Rizakis 1983; Velenis 1996a. On Jews, see Nigdelis 1994 and on other foreigners, see Robert 1974, mainly from Asia Minor in Thessaloniki; Sverkos 2001, 99 note 23.

# 3. Glassworkers: People, Artisans and Traders

Glassworkers, like other artisans, are rarely mentioned in historical and literary sources as individuals. This is certainly true for the Classical Greek period, as well as for even earlier times. In contrast, during the Roman Imperial period there are some mentions of glassworkers in connection with taxation[1] or exemptions from earlier taxation,[2] but this data does not really shed much light on the history of glassworking. Very little is known specifically about the economic and social status of glassworkers in Roman society; a few things can only be inferred on the basis of what is known of similar professions. Generally speaking, Roman craftsmen had a very low social status, and furnace-workers, known in ancient Greek as 'banausoi', like glassworkers, belonged to the lowest classes of ancient society. The landed aristocracy considered manual labour as menial; it was the patrons of work who were extolled.[3] It is known, however, that in some cities glassworkers constituted a populous and thriving community, and their associated areas and workshops were identified with their activities and named 'of the glassworkers'.[4]

References to glassworkers are usually vague and concern the profession rather than specific individuals, who always remain unnamed.[5] Seneca in his letters to his friend Lucilius, refers generically to the glassworker and his skill 'to shape glass by his breath into manifold shapes, which could scarcely be fashioned by the most skilful hand', without preserving his name.[6] All the known names of glassworkers, and sometimes also their places of origin, are known from funerary inscriptions or relief inscriptions printed on the vessels themselves. The funerary inscriptions that have survived are written both in Greek, i.e. certain inscriptions from Sparta,[7] Athens,[8] Cilicia[9] and Tyre)[10] and in Latin (inscriptions from Lyon (Lugdunum),[11] Mauritania[12] and Solin (Salona).[13]

During the 1st century, glassworkers or, more probably, glass-workshop owners, primarily Sidonians and a single Cypriot, imprinted their names and places of origin on their products to take advantage of the high reputation of their homelands. In some cases these elements were incorporated into the decoration of the mould, while in others, as in the case of *skyphos* vessels, they were imprinted on the vessels' handles with the help of jacks, i.e. a pair of large pincers, with appropriately shaped jaws. The information was recorded in both Greek and Latin characters, accompanied by *ἐποίει* or *fecit* (made by) respectively.[14]

Names of glass-workers were also imprinted in relief on the bottom of unguentaria. By stamping their names, possibly after the example of some Imperial workshops, these individuals set their products apart and guaranteed the quality of the contents of their vessels.[15] Glassworkers' names were also stamped on the bottom of larger, cylindrical, square and prismatic transport vessels during the blowing process, through the use of a mould that had

---

[1] It is known (Lamprid. *Alex. Severus*, 24.5) that in the 3rd century Alexander Severus imposed a tax on glassworkers. Moreover, Aurelian, Vopiscus (*Historia Augusta, Aurel.* 45.I) taxed Egyptian glass imported to Rome; Trowbridge 1930, 118, note 32.

[2] Constantine the Great exempted *vitrearii* and *diatretarii* from taxation so that they could improve their craft and teach it to their sons. *Codex Theodosianus* 13.4.2, *De excusationibus artificum;* Trowbridge 1930, 119, note 34.

[3] Generally on Roman craftsmen, see Morel 1993, 214-244. For an overview on Roman glassworkers see Stern 2004, passim and esp. 78, 120.

[4] 4th-century catalogues of monuments of Rome and inscriptions from Rome, for example, point to the street or quarter of the glassworkers, *Vicus vitrarius* (*Curiosum Urbis Romae*), as well as to a portico 'between the glassworkers', (*porticus inter vitrarios*). See *Corpus Inscriptionum Latinarum* VI.29844 (=93), Trowbridge 1930, 131, notes 38-40. Another 4th-century inscription shows that in Puteoli, Campania, there was also a quarter called the 'Hill' or 'Slope of the Glassworkers' (*clivus vitriarius*). *ILS*, I, no. 1224b, 269.

[5] Extensive analysis of the bibliography on this topic by Trowbridge 1930, 112-128. Also, detailed references to vessel formation by glass-blowing are made in at least two poems of the Late Roman era: *Anthologia Graeca* XVI. 323 and P. Oxy. 3536, see Deroy 1985, 183-184 and Deroy 1989, 178-184. For the most comprehensive analysis of, and commentary on these texts, regarding the history of glassworking, see Stern 2007, 349-354.

[6] Seneca, *Epistulae morales* 15.90.31.

[7] *Bulletino del'Instituto di Corrispondenza Archeologica* 1844, 147, 1. 24. (non vidi); Foy, Nenna 2001, 67.

[8] Two funerary inscriptions are known, that of Euphrasius (IG III.2, no. 3436) and of Andreas (Sironen 1990-91, 83-85, fig. 19).

[9] Anonymous: (Ου)ιτράριος ((V)itrarios) and Υ(ε)λάριος (Y[e]larios), see Stern 1989, 121-22, with an annotated bibliography.

[10] *Bulletin du Musée de Beyrouth* 29, 1977, no. 117 (non vidi); Foy, Nenna 2001, 67, note 4.

[11] Iulius Alexander Africanus who died in Lyon in the 3rd century, CIL 15.2.1, 7001, see Foy, Nenna 2001, 68, no. 34.

[12] Antas, CIL 8, 9430; Trowbridge 1930, 114, note 11.

[13] Paschasius or Pascasius: *CIL* 3, Suppl. 1-2, 9542; Buljević 2009, 41, fig. 2.14.

[14] Some of the names that have survived are: Ennion, Aristeas, Iason, Meges, Anios, Artas, Nikon, Ariston, Eirenaios, Filippos, Asinius. Glassworker names are listed in detail in Trowbridge 1930, 120-128. For a more recent bibliography, see also Price 1991, 56-75; Stern 1995, 69-74; Foy, Nenna 2001, 176.

[15] De Tommaso 1990, 91-101; Foy, Nenna 2001, 116-118, where there is also an annotated bibliography. For an overview of Croatian finds, see Buljević 2009, 42, fig. 4.17-19.

been carved accordingly.[16] Names written in both Latin and Greek have survived, often in abbreviated form.[17] It is known from an inscription imprinted on vessel bases that a woman, Sentia Secunda,[18] was an active glassworker in Aquileia, while it is also believed that the name Nikais[19] also refers to a woman. Finally, the glassworker Amaranthus stands out because he imprinted his name with the decorative elements that adorned his products.[20]

It is known that some glassworkers were freedmen,[21] while others were free citizens.[22] However the social position of the majority of glassworkers remains unknown. Nevertheless, judging from the distribution and number of surviving signed fragments from particular workshops, it would be safe to assume that at least some glassworkers were involved in large-scale projects and had flourishing businesses; two examples are Ennion,[23] who was active in the East during the 1st century, and Frontinus,[24] active in the West in the 3rd and 4th centuries.

On the basis of the existence of large landed properties and by analogy with what is known for pottery, brick/tile, and metal workshops, it has been hypothesized that glass-workshops in the western provinces of the Roman Empire probably belonged to officials who owned large estates. It is also thought that some flourishing businesses had set up branches in remote areas.[25] In the eastern Roman Empire, by contrast, it is considered more probable that glass-workshops were small-scale, individual businesses that could collaborate freely and form partnerships according to their interests, as demonstrated by examples of vessels that have been blown in identical moulds bearing the names of different glassworkers.[26] According to papyrological evidence, during the 4th century glassworkers, at least in Egypt, were organized within local guilds.[27] Similarly, it has been suggested that Roman *collegia* might also have functioned as guilds, championing the interests of their members.[28]

The price of glass vessels is generally not known. Apart from Strabo's single vague reference to the fact that in Rome one could procure a glass cup for a bronze coin,[29] no other such figures are known. It is, however, obvious that decorated tableware cost more than simple, undecorated, everyday, utilitarian vessels. Emperor Diocletian's *Edict on Prices* (*Edictum De Pretiis Rerum Venalium*),[30] sheds considerable light on the economic aspects of glassworking. The *Edict* sets the maximum sale prices allowed for various products, including those made of glass. The products are separated on the basis of the quality of the glass into more expensive Alexandrian products and cheaper Judean ones. It is believed that Judean refers to the naturally coloured green-blue glass, while Alexandrian refers to the deliberately decolourized glass; these terms are thought to have been in use since the 1st century to distinguish between the two qualities of glass.[31] Furthermore, the *Edict* mentions the prices of glass intended for use in window glazing, likewise distinguishing between the two qualities. This fact, together with relevant archaeological evidence

[16] Foy, Nenna 2001, 118-120, where there is also an annotated bibliography. Tek 2001, 82-87.

[17] For a detailed listing of glassworker names and name abbreviations, see Trowbridge 1930, 120-128. Almost all the vessels with stamped bottoms, arranged by country of origin, are presented for the first time or republished in Danièle Foy and Marie-Dominique Nenna's monumental edition *Corpus des signatures et marques sur verres antiques*, which is at three volumes and growing. In addition, papers have been published on the following names: Miscenius Ampliatus: Buljević 2009, 40, fig. 3.15. P. Gessius Ampliatus: Scatozza-Höricht 1986, 76, fig. 12. C. Caesius Bugaddus: Price 1974, 291-92. Frontinus: Sennequier 1985, 169-82. Rufinus, Antonius Volumnius Ianuarius, Quintus Danus Euhelpistus, L. Aemilius Blasius, C. Salvius Gratus, Cn. Pompeius Cassianus: Buljević 2009, 42, fig. 4. 17-27, where all prior bibliography on these names is assorted. Titianus Hyacinthus: Whitehouse 2003b. Gn Asinius Martialis: Price 2005, 177, note 17. Alexandros: Auth 1976, 111, no. 129. Chresimus: *Roman Inscriptions of Britain* II.2, no. 2419.101, Price 2005, 177, fig. 10.11. Dora, Doros, Euphrasios, Euporianos, Tryphon, Zosimos: Triantafyllidis 2006, passim, esp. 373-374, with annotated bibliography; Stern 2006, esp. 403-404; Weinberg, Stern 2009, 156-158 esp. no. 386; Antonaras 2006a, 414, no. 6. Germanos, Paulinos Antiocheus: Nenna 2006, 432-433, nos. 57, 54, 55. Magnos: Lehrer-Jacobson 1992, 35-38, 40, nos. 8, 9, figs. 1-2. Priscos: Auth 1973, 26, Auth 1976, 111, no. 129. Theodoros: Recent Important Acquisitions, *JGS* 15, 1973, 188, no. 8, new acquisition by the Museum of Art and Archaeology, University of Missouri at Columbia. A second example was published by Lehrer-Jacobson 1986, 46-49. Tiverinos: Auth 1976, 111, no. 129. Trellus and Athenios: Buljević 2009, 40, fig. 3.16, where all prior bibliography can be found.

[18] Harden 1969, 49, 73, pl. IVB. Stern 1999, 456-457, figs. 23-25.

[19] It has been proposed that this was a female glassworker of the 1st century. See Stern 1997, 129-132. Lately Michel Sève has supposed that Neikais could present a form of the male name Νεικαῖος, see Sève, REG 123, 2010, 672-73 no. 76.

[20] Amaranthus is an artisan who was active in Burgundy during the 1st century, and who imprinted his name around the relief Medusa masks with which he decorated his vessels. See Sennequier 1986, 11-18.

[21] Stern 1999, 457.

[22] Foy, Nenna 2001, 67.

[23] Lehrer 1979, passim; Stern 1995, 69-72, where there is also an annotated bibliography.

[24] Arveiller-Dulong, Arveiller 1985, 95-98, 169-172, 241-242, where the older bibliography can be found.

[25] Stern 1999, 459, where there is also an annotated bibliography.

[26] For example, vessels by Jason, Meges and Nikai. See Stern 1995, 68-69, 73-74, for a detailed presentation of this phenomenon.

[27] P. Oxy, 45, 3265 and 54, 3742; Stern 1999, 459, 465-466.

[28] Stern 1999, 459, for earlier bibliography as well.

[29] Strabo, *Geogr*, 16.2.25.

[30] The Edict was issued at the end of 301 in Alexandria, but had already been prepared during the Emperor's stay in Antioch. See Giacchero (ed.) 1974. For a complete commentary on the sections relating to glassworking (16. 1-6), see Stern 2007, 374-388.

[31] Barag 1985b, 113-116.

recently uncovered, seems to overturn the until now dominant hypothesis that window glazing did not appear in the eastern part of the Roman Empire until the 4th century.[32] Finally, a fragmentary section of the *Edict* that has been filled in is thought to refer to the price of glass *tesserae*, which was dependent on its colour.[33]

According to the prices set by the *Edict* a glass vessel, e.g. a Frontinus bottle,[34] cost about ten to twenty times more than a ceramic vessel of comparable size.[35] The weight of such a transport vessel equals approximately one *libra* (327.45gr) and respectively it would cost 20 or 30 denarii, depending on whether it was made of Judean or Alexandrian glass. Generally speaking, and on average, later Roman glass vessels weighed between 150 and 350gr, i.e. approximately between half and one *libra*.[36] Among the most popular forms of glass vessels in the region of Thessaloniki in that period were the plain, globular bottles with tubular rim (Forms 51a and 51b) that weighed between 80 and 120gr, and ovular, funnel-mouth flasks with ring base (Form 70) weighing between 130-150gr; also popular were fusiform unguentaria (Form 147a) weighing between 80 and 100gr. Their prices must have been therefore between 5 and 10 denarii. The same *Edict* gives the daily wage for an unskilled worker as 25 denarii (plus c. 5-10 denarii for his food), while a skilled worker received c. 50-60 denarii. It is therefore estimated that a storage vessel cost roughly as much as an unskilled worker's daily wage, while a tableware vessel or an unguentarium would cost a little more than a worker's food, or around one third or one half of a skilled worker's daily wage.[37] The price comparisons demonstrate that even though a glass vessel was ten to twenty times more expensive than a comparable ceramic one,[38] it was still a cheap, utilitarian object that almost anyone could easily acquire.

The labour costs for producing and forming glass products were low and accounted for a very small fraction of their final price. It is telling, moreover, that plain vessels, much like raw glass, were sold according to their weight. In contrast, engraved vessels or vessels with relief decorations were sold as individual pieces. The practice of selling glass vessels according to weight continued throughout the Middle Ages and is still in use in traditional workshops in the East.[39]

It has been assumed that the low prices in the *Edict* forced certain products out of the market. The reason for this was the extremely low prices at which the merchants were forced to sell and distribute them. More specifically, raw glass cost 13 to 24 denarii per *libra*/pound, depending on the quality of the glass, and one pound of products of the same quality cost 20 to 30 denarii respectively, allowing a 35% and 20% profit. If the great loss of material during the process of production, approximately 45%,[40] is taken into account then the profit is almost half. If, in addition, one deducts from the final price the operating expenses of the workshop, the probable rent for the workshop, the price of fuel, staff wages and taxes, then the result must surely end in a loss.[41] In addition, this financial data explains perfectly the small number of glass fragments found in the excavations. Efficient recycling must have been a high priority for glassworkers, collecting material discarded during their own work and also collecting glass fragments from the locality, if they really intended to stay in business, selling at the prices stipulated by the *Edict*. Given the lack of other sources relating directly to Roman glassworkers, and the organization of their trade, further clues as to their activities may be sought among the potters and glassworkers of Fustat (Cairo) in the 10th to 13th centuries. Documents from there reveal that, at least in some cases, the artisans did not own their workshops, depots, or even tools: these were leased for a certain period (a few months to one year). Glassworkers operated individually but also in partnerships – either for a specific period or project, e.g. a large order. These partnerships could also vary, depending on the monetary value of the work in hand, and could also be structured in unequal shareholdings, with the volume of work offsetting a lower financial contribution, etc.[42]

The long-distance trade of glass and glass products was fairly widespread, as the archaeological evidence now reveals. Excavations of Hellenistic and Roman shipwrecks, both in the western and Eastern Mediterranean areas, have produced valuable data.[43] It has been proven that glass, initially in ingot form and later in lumps, or even as

32 Whitehouse 2001a, 31-43; Antonaras 2010f, 112-113, with several find-spots from Roman Macedonia.

33 Stern 1999, 466.

34 This particular form of vessel is used because it can easily be compared with relevant clay vessels, as they were produced in certain volumes/capacities ranging between 0.5 and 1.5 *libra*. See Stern 1999, 462, note 114; Sennequier 1985, 169.

35 Stern 1999, 462, note 114; Sennequier 1985, 169; Antonaras 2009, 90-91.

36 Stern 1999, 462.

37 Stern 1999, 462.

38 Stern 1999, 462, and note 114; *Edict of Prices*, 15. 98. Giacchero (ed.) 1974, 170.

39 Barag 1985b, 116; Stern 1999, 462-463; Fontaine, Foy 2007, 237-243.

40 Henein 1974, 20, according to Stern 1999, 463 note 121.

41 Stern 1999, 464.

42 Goitein 1967, 87-88. For a synthesis of this evidence, see Stern 1999, 459-460.

43 Radić, Jurišic 1993, 113-138; Foy, Nenna 2001, 101-112, where there is also an annotated bibliography; Foy 2003, 138-145; Fontaine, Foy 2007,

cullet, was transferred from the glassmaking centres in the Eastern Mediterranean to various ports of the Empire for resale to local glassworkers.[44] Maritime transportation was preferred because it was safer and far cheaper, up to ten times cheaper, compared to land transportation, according to the *Edict*.[45] Glass was never the sole, or even the bulk of the transported cargo, but constituted only a small part of it. In some cases, moreover, it is not even clear whether the glass found was part of the cargo or the personal property of crew members or passengers.[46] As well as in its raw state, expensive and decorated glass vessels were also transported (there are only a very few documented cases where simple, utilitarian vessels were involved).[47]

There is almost no archaeological evidence regarding the trade and transportation of glass by land. It is, however, believed that small-scale distribution[48] did take place by means of pack animals.[49] According to written sources, glass was transported in straw baskets and/or was wrapped in layers of hay. This is also confirmed by ethnographic research. In some cases each vessel was protected by its own straw or papyrus wrapping or was placed in an appropriately sized wooden box.[50]

It seems that during the 1st century glass vessels were transported over great distances to the cities where they were finally sold. Moreover, it appears that the large urban centres received products originating from different production sites, both in the East and the West. From the 2nd century onwards, however, because of the wide popularization of glass-blowing, many more vessels were manufactured in situ. As a result, imported products – now coming only from nearby glassworking centres – represent a smaller percentage of the demand.[51] The place of origin of transported glass vessels is difficult to determine because of the restricted archaeological evidence. It is obvious that the great Syro-Palestinian glassworking centres exported their products to areas in Italy, the Balkans and around the Black Sea. Yet western products undoubtedly also made their way to the East, where they were sold, albeit in relatively small numbers.[52]

Regarding the centres where glass objects were sold, it appears they were retailed in shops, together with similar ceramic products, at sites such as Delos, Cosa, Colchester (Camulodunum), and possibly at Herculaneum and Kom Aushim (Karanis).[53] Glass vessels are also known to have been sold from shops where their contents (perfume, medicinal products, etc.) were also produced.[54] In cases where the glassworking workshops were located in the centre of the city, it is very probable that the glassworkers themselves sold their products on site. Regarding how they were sold, it appears that early on glass vessels were sold individually, which makes sense given their high unit prices at the time. Already by the 1st century, however, they were sold in sets, as demonstrated by archaeological evidence at Herculaneum and by 1st-century funerary findings in various parts of the Empire.[55] Relevant testimonies regarding the sale and purchase of vessels in sets of six, or in other such assemblages, survive also in papyri.[56]

The fragmentary nature of the historical sources sheds very little light on glassworkers and their profession. Only careful study and in-depth analysis of archaeological evidence can illuminate aspects of their life and daily routine, such as, for example, details regarding primary and secondary glass production, local and long-distance trade, as well as pricing. Such study will provide a more meaningful and substantive approach to glassworking, allowing for the comparison of glassworkers with other artisans and for their integration into the social fabric of the Roman Empire. We have even less historical evidence regarding glassworking in the Byzantine Empire[57] and in the Arab

237-243.

44 Stern 1999, 467-78; Price 2005, 168-169.

45 Stern 1999, 472-473.

46 Stern 1999, 473.

47 Stern 1999, 472.

48 Stern 1999, 473.

49 Weinberg 1988, 25, note 2, for a 3rd-century Talmudic reference; Stern 1999, 473; Price 2005, 168.

50 Foy, Nenna 2001, 114, with annotated bibliography.

51 Stern 1999, 469-470.

52 Stern 1999, 474; Price, Cottam 1998b, 60.

53 Foy, Nenna 2001, 113-114, with further bibliography; Price 2005, 179-182.

54 Stern 1999, 470-1. This process is known until the Late Middle Ages, see Han 1981, 94-95.

55 Stern 1999, 471, note 175, with relevant bibliography; Price 2005, 179-182.

56 Stern 1999, 472, with relevant bibliography.

57 The fundamental study of the topic is Philippe 1970. For a thorough overview of the forms and the uses of glass vessels in Byzantium, see Antonaras 2010d, 383-430.

world;[58] it is, moreover, only thanks to the extensive Venetian archives that we can gather enough information about the organization of glass production and trade from the 13th century onwards.[59]

Speaking particularly about Thessaloniki, no name or gender, or for that matter social status, of any of the glassworkers operating in the city, or any specific site where they had their workshop, is known. It is certain that the workshops were placed in the abandoned public spaces in the centre of the city and in its outskirts (and even in ordinary shops along the streets in the city centre). Only in one case are their archaeological clues regarding the size of a workshop and its production; however no scientific, archaeometric studies have been carried out concerning the provenance of the raw material they used, which might shed further light on the trade networks and other financial aspects of the lives of Thessalonian glassworkers.

[58] Whitehouse, Carboni 2001, passim; Carboni 2001, passim; Goldstein 2004, passim.

[59] The work of Zecchin (1987), where the relevant historical sources are examined in great detail, is central to the understanding of glass and glassworkers in Murano.

# 4. Typological Classification of the Material Under Study

**1. Mosaic plates**

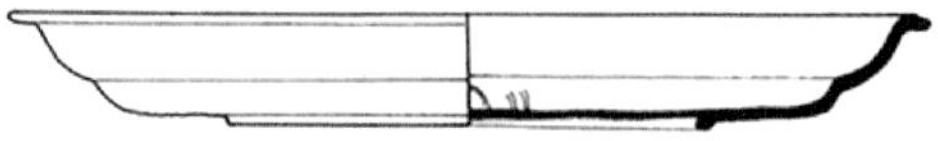

(Cat. no. 1)
(Plates 1, I)

Vessels of this form have horizontal rim, rounded at its tip; concave, flaring walls and flat bottom. Their height is estimated at c. 3cm.

They are made by slumping. Small discoid or oblong pieces of composite rods of glass (i.e. presenting decorative motifs in their cross-section) were placed/arranged one adjoining the other, on a fire-resisting surface and they were heated and fused together. Then this disc was placed on a, probably rotating, convex former mould of the desired shape and by reheating and applying pressure with a concave mould, which formed the lower surface of the final vessel, the originally flat disc took the appropriate shape. At the end of the process the surfaces of the vessel were polished to erase traces of the production procedure.[1]

The disc from which the vessel was made comprised of brownish or olive-green rods, and milky-white ones, which formed motifs resembling veins in marble. Marble-like decoration appears on large vessels, such flat plates and bowls, in a way characteristic of the example from Thessaloniki as well, probably as a result of large distortions occurring during the forming of the vessel.[2]

This example belongs to family IV of David Grose's typology, which consists of vessels made of composite mosaic canes.[3] Flat plates appear only among vessels of this family.[4] They are western products dated to the end of the 1st century BC or to the first decades of the 1st century AD.[5] Examples have been found in Rome, while the vast majority of the extant examples lack provenances.[6] The example from Thessaloniki cannot be dated independently by its archaeological context.

**2. Free-blown dish, plain base (Isings Form 46a, c)**

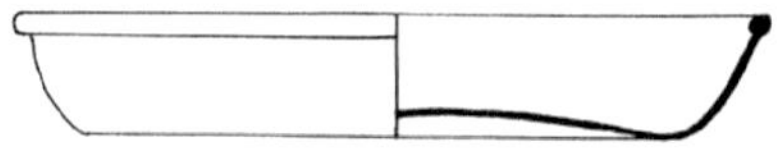

(Cat. nos. 2-3)
(Plates 1, I)

Vessels of this form have infolded tubular rim, conical body, plain base and slightly concave bottom. Their height is estimated at c. 3cm.

They are free-blown. The vessel started as a close-shaped globular vase, then the rim was released from the moil while another pipe was attached to the bottom. The vessel was repeatedly introduced into the furnace, reheated and spun, so as to acquire the desired rim diameter, and the rim was shaped by further tooling.

It is a well-known form, considered to be an Italian product and dated to the 1st century,[7] as are also the finds from Thessaloniki, which are dated to the reign of Vespasian, the destruction time of the bath complex in whose ruins they were found.[8]

**3. Free-blown dish, ring base**

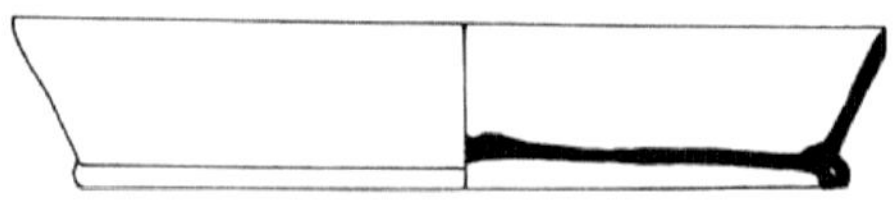

'A' Conical plates with out-turned, pushed-in ring base
(a) handleless and (b) with crimped pseudo-handles
Isings Form 23/49 and Isings Form 43
(Cat. nos. 4, 6)
(Plates 1, I)

Vessels of this form have fire-polished rim, conical body, pushed-in tubular ring base and slightly concave bottom. Their height is estimated at c. 3cm.

[1] Grose 1989, 257.
[2] Grose 1989, 260. Examples of different shape, bearing the same decorative motif are presented in Grose 1989, fig. 159 and 160, from the British Museum and Kunsthistorisches Museum, Wien.
[3] Grose 1989, 256 ff.
[4] Grose 1989, 256-261.
[5] Grose 1989, 257.
[6] Grose 1989, 241-244, with a comprehensive list of mosaic glass vessels in general, found in controlled excavations.
[7] Isings 1957, 61-62, form 46a, c, with relevant, dated examples.
[8] Adam-Veleni *et al.* 1998, 85-102; Mavromichali 2001, 141-142.

They are free-blown, although plates of the same shape were also made by slumping.[9] The vessel started as a close-shaped globular vase, the base probably was shaped at that stage and then the rim was released from the moil, while another blowpipe had been attached to the bottom. Finally the vessel was repeatedly introduced to the furnace, reheated and spun to acquire the desired rim diameter.

It is a well-known form, probably an Italian product, dated to the 1st century.[10] To the same period are dated the finds from Thessaloniki, and more precisely to the reign of Vespasian, the destruction time of the bathhouse in which they were excavated.[11]

To the same form also are ascribed plates with infolded tubular rim, same body shape and two crimped, strap pseudo-handles, attached on opposite sides of their rim.[12] They are considered to be Italian products and are dated to the 1st century,[13] as is dated the find from the bathhouse of Thessaloniki's ancient agora.

'B' Plates with vertical pushed-in ring base
Isings Form 47

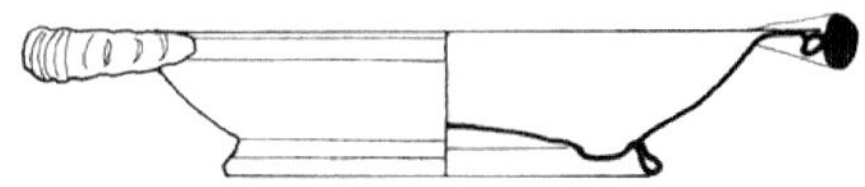

(Cat. no. 5)
(Plates 1, I)

Vessels of this form have fire-polished rim, slightly conical body, pushed-in tubular base and slightly concave bottom. Their height is estimated at c. 3cm.

They are free-blown. The vessel started as a close-shaped globular vase, the base probably was shaped at that stage and then the rim was released from the moil while another blowpipe was attached to the centre of the bottom. Finally the vessel was repeatedly introduced into the furnace, reheated and spun to acquire the desired rim diameter.

This is a relatively rare form, probably an Italian product dated to the 1st century,[14] as are also dated the finds from Thessaloniki, which are dated to the reign of Vespasian, the time of the destruction of the bathhouse in whose ruins they were found.[15]

**4. Rectangular trays, cast decolourized**

(Cat. no. 7)
(Plates 1, I)

Shallow rectangular disks comprise this form. They have horizontal rim that ends also at a horizontal edge, polished by grinding. The body walls are oblique and a gentle curve leads to the flat bottom. A flat, elaborate handle is formed at the centre of each narrow side.[16] Their height is estimated at c. 3cm. The example from Thessaloniki is made of thick, fully decolourized glass of very good quality, without inclusions or pinprick bubbles.

These trays were formed by slumping or by pressure in a double mould; the entire vessel was polished.[17]

The vessel belongs to a small group of elite 1st-century vessels. Fully decolourized vessels appeared as part of a new trend that occurred around AD 60.[18] The shapes are similar to the strongly-coloured vessels produced during the first half of the century[19] and they probably represented examples of the most highest-quality production of

[9] Isings 1957, 39, form 23, with relevant, well-dated examples.
[10] Isings 1957, 63, form 49, with relevant, dated examples; Whitehouse 1997, 69, no. 79, with more recent bibliography, with finds from excavations, dated to the first half of the 1st century.
[11] Adam-Veleni *et al.* 1998, 85-102; Mavromichali 2001, 141-142.
[12] Isings 1957, 59, form 43, with relevant examples dated to the 1st century.
[13] Stern 1977, 57-58, with exhaustive bibliography on similar 1st-century finds, found in controlled excavations and in the market. Harter 1999, 46, form A10, where a more recent bibliography on 1st-century finds from controlled excavations can be found. On more recent information about the origin and the dating of this form, see Weinberg, Stern, 2009, no. 209.
[14] Isings 1957, 62, form 47, with relevant, dated examples; Whitehouse 1997, 68, no. 77, with more recent bibliography, and with finds from excavations dated to the first half of the 1st century.
[15] Adam-Veleni *et al.* 1998, 85-102; Mavromichali 2001, 141-142.
[16] The most relevant example is a tray made of emerald-green glass, dated to the first half of the 1st century, see Stern, Schlick-Nolte 1994, 326-327, no. 98, with dimensions 23.5x11.5x1.7cm, where a large number of relevant, fragmentary finds are listed. Another, much smaller example from the Louvre was published in Arveiller-Dulong, Nenna 2000, 204, no. 256, which is also dated to the first half of the 1st century.
[17] Grose 1991, 9. On the rectangular trays horizontal, not circular, traces of polishing are visible.
[18] Pliny, *Nat. Hist.*, 36.192, where it is mentioned that the transparent and decolourized glass vessels were more appreciated in his era, as they resembled vessels made of rock crystal.
[19] Grose 1989, 254-256. A group of cast vessels that appeared and were used during the first half of the 1st century; most were found at sites of

that period. According to firmly dated finds it appears that decolourized vessels were in use in the late Flavian era, during the reign of Trajan and Hadrian (69-138).[20] The find from Thessaloniki cannot be dated independently by its archaeological context.

Such forms appear in equally small numbers at several sites in the West and East and thus the production centre(s) of this group cannot be identified.[21] In any event, the distribution of the finds renders less feasible the older hypothesis[22] that they were Alexandrian products.[23]

### 5. Bowls smooth, mosaic

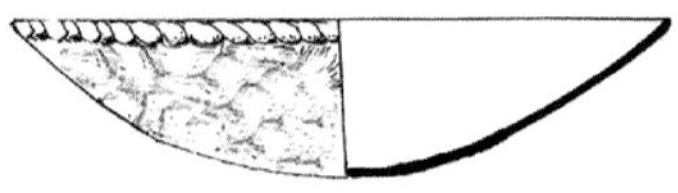

(Cat. no. 8)
(Plates 1, I)

The vessels of this form are (shallow) lens-shaped. Their rim consists of an added twisted coil of two-coloured glass. Their height is estimated at c. 3cm.

They are made by slumping. Small segments, longitudinal in this case, of composite glass rods were placed adjoining one another on a fire-resistant surface and were heated up until they fused together into a solid disk. A twisted, two-coloured rod of glass was added, delineating the rim of the disk. The disk was then placed on a convex former mould of the desired shape and was reheated to obtain the appropriate shape. Finally the internal surface of the vessel was polished to erase the marks of the mould.

The form belongs to family II of David Grose's typology,[24] which consists of vessels with applied rim, made mainly of longitudinal parts of composite glass rods. It is assumed that[25] the motifs of this group derive from the Hellenistic artistic canon,[26] especially connected to the vessels of the Antikythera group. The Roman production, however, was considerably altered both in terms of colours and the combinations in which they appeared, as well as in the shapes of the vessels. More specifically, as regards its shape the vessel belongs to the group of 'striped mosaic lens-shaped' dishes,[27] while on the basis of the way its decoration was rendered it belongs to the group of 'short-strip and meandering-strip' mosaic vessels, of which the relatively short, longitudinal parts are typical, and which are often bent in such a way that they form meanders within the mass of the vessel body.[28] The vessels are western, probably Italian, products and are dated to the Augustan or Julian-Claudian periods.[29] The find from Thessaloniki, which cannot be dated independently by its archaeological context, is ascribed to the same period on the basis of other well-dated parallels. Some examples have been found in Rome, while the exact provenance of the vast majority of the extant examples is not known.[30]

The single published parallel from a Greek site is a *millefiori* fragment from Thebes.[31]

### 6a. Ribbed bowls, rotary-pressed, single-coloured

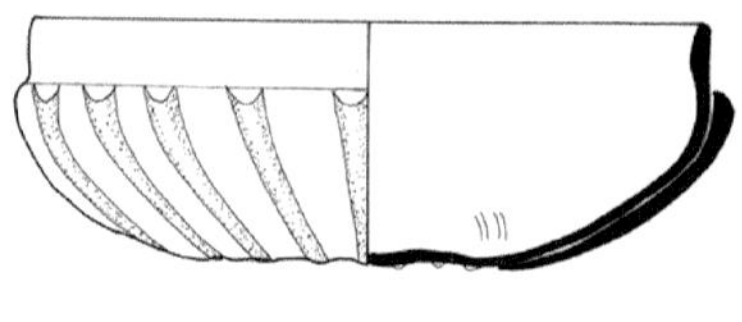

(Cat. nos. 9-24)
(Plates 1, I)

The vessels of this form[32] are characterized by the vertical ribs that decorate their body, beginning almost under the rim and ending at the centre of their bottom. Two main subgroups are discerned: the wider and shallow examples (height c. 4-7cm) and the more narrow, deeper-hemispherical forms (height c. 10-12cm) that occasionally stand on a relatively tall, conical base and have a slightly flaring rim. The rim of the vessels is vertical, fire-polished, and the curved body walls end at a flat base.

The vessels of this group are made by rotary pressing, i.e. by pressing a hot and malleable disk-shaped mass of glass onto a convex former mould. This is attached to a pottery wheel which shapes the glass to the desired shape, hemispherical or more shallow. The ribs that decorate the vessel are formed by pressing the outer surface with

the western provinces of the Roman Empire.

20 Grose 1991, 15.
21 Grose 1991, 15-18, with detailed bibliography of known examples.
22 Grose 1991, 12.
23 Harden 1936, 50.
24 Grose 1989, 249 ff.
25 Grose 1989, 250.
26 Grose 1989, 189-193.
27 Grose 1989, 252.
28 Grose 1989, 252-253.
29 Grose 1989, 252.
30 Grose 1989, 241-244, where a detailed list of mosaic glass vessels in general is presented.
31 Haevernick 1981, 118-119, pl. 44. 4 where a vessel fragment is presented, but with no chronological indication based on the excavation.
32 Isings 1957, 17-21, form 3.

a metal (?) rod while the vessel rotates slowly on the wheel. At the same time the edge of the vessel was pressed, forming the flat rim. The rim was often polished later, after which the exterior and interior of the vessel was also polished. The interior was occasionally decorated with one or two deep horizontal incised grooves placed on the perimeter and the centre of the bottom.[33]

They represent one of the most widespread forms of glass vessel of the 1st century BC to 1st century AD. The production of these vessels began in the early 1st century BC, probably in the workshops of the Syro-Palestinian, and ceased around the middle of the 1st century AD. In the western parts of the Empire, however, and more specifically probably in Italy, they continued to be produced up to the end of the 1st century AD.[34]

Apart from other published sites, extending from the Eastern Mediterranean to Britain,[35] particularly in Greece vessels of this form have been published from several sites: Philippi,[36] Samothrace,[37] Samos,[38] Mytilene (acropolis),[39] Palatiano (Ioron),[40] Pydna,[41] Ierissos (Acanthos),[42] Thebes,[43] Demetrias,[44] Eretria,[45] Athens,[46] Corinth,[47] Argos,[48] Olympia,[49] Amorgos,[50] Delos,[51] Sifnos,[52] Tinos,[53] Perissa (Thera),[54] Tarrha,[55] Knossos[56] and Matala.[57]

Single-coloured examples from Thessaloniki are colourless bluish, dark-blue[58] and amber-coloured. Examples of mosaic and *millefiori* glass have also been found in Thessaloniki.

Vessels found in the area of Thessaloniki's ancient agora (Cat. nos. 9-17) are precisely dated within the last twenty years of the 1st century, the time period for the destruction of the bathhouse in which the finds were discovered.59 Some other vessels from this group cannot be dated precisely. The find from Lete (Cat. no. 21), although it remained in use until the 2nd-3rd century, probably as an heirloom of a wealthy rural family, has been dated to the 1st century

[33] This production technique was discovered and experimentally proved first by R. Lierke, see Lierke 1991, 313-317 and Stern, Schlick-Nolte 1994, fig. 138-144. On the topic see Chapter 1 (Glass-forming techniques: rotary-pressing).

[34] Grose 1989, 244-249 and esp. 247; Stern, Schlick-Nolte 1994, 72-79, where all information concerning this form is listed.

[35] For detailed lists of published examples of this form, see Isings 1957, 17-21; Grose 1989, 244-249; Stern, Schlick-Nolte 1994, 308-309.

[36] Koukouli-Chrysanthaki 1975, 285, pl. 193δ, from a graveyard of the 1st-2nd century. Also, for unpublished fragments from the excavations of the Agora of Philippi, see Nenna 1998, 110, note 101.

[37] Dusenbery 1998, 1072, no. S 148-11, a vessel dated c. 25 BC by the archaeological context in which it was found.

[38] Tölle-Kastenbein 1974, 164, pls. 275-276, finds from Kastro-Tigani, which is dated to the Roman period.

[39] Price 2002, 124, fig. 1/4-6, finds generally dated to the 1st century.

[40] Antonaras, Anagnostopoulou-Chatzipolychroni 2002, 114, fig. 2, finds with no special chronological indications, dated according to similar finds, between the 1st century BC and the 1st century AD.

[41] Personal communication with the researcher of this (under publication) material, Dr Despoina Ignatiadou.

[42] Trakosopoulou 2002, 84, fig. 12, example dated by a coin to 45 BC, and another example from a 1st-century AD grave, possibly from the first half of the century, according to its published contents.

[43] Haevernick 1981, 118-119, pl. 44.7, vessel fragments (22) with no firm dating.

[44] Eiwanger 1981, 99, no. III. 583-584, pl. 88. From a stratum with Late Antique finds (with evidence of contamination).

[45] Andreiomenou 1973/74, 465, pl. 301ζ, dated according to excavation results to the Early Roman period. Also, a fragment from the Athenian Agora: Weinberg and Stern 2009, 33-36, nos. 38-58, figs. 4-5, from contexts ranging from the 1st century BC to the late 1st century AD.

[46] Weinberg, McClellan 1992, 102-103, a vessel of unknown provenance kept in the National Museum, dated on stylistic grounds to the 1st century.

[47] Davidson 1952, 79-80, 95-96, nos. 595-602, figs. 6, 7, from 1st-century layers; Warner Slane 1986, 301, no. 147, pl. 69, a find from a layer with the majority of its content dated to the second and third quarter of the 1st century, although there are also clay fragments dated to the early 2nd century (see 308-309).

[48] Sève 1980, 317-318, no. 39, fig. 52, from a well in use in the second half of the 1st century.

[49] Schauer 2001-2002, fig. 6, fragments of several vases: one can be ascribed to the early production, 1st century BC, of Syro-Palestinian workshops, and the remaining are 1st-century AD products of western workshops.

[50] Triantafyllidis 1998, 96. The deep bowls are dated according to excavation results from the second quarter of the 1st century BC until the early 1st century AD, while the shallow bowls are dated to the Augustan period.

[51] Nenna 1998, 109-111, nos. 301-304, pls. 33-34, group of vessels which seems to disappear in the last quarter of the 1st century.

[52] Mackworth-Young 1949, 88, pl. 31. 2, find from grave no. 14, dated by a coin of Vespasian of AD 75.

[53] Unpublished fragments from the temple of Poseidon and Amphitrite, according to Nenna 1998, 110, note 96.

[54] Gerousi 2002, 134, fig. 1. From a grave with a coin of Augustus dated between 2 BC and AD 11.

[55] Buechner 1960, 110, pl. 36/1, dated according to excavation results to the 1st century or even earlier.

[56] Hayes 1971, 253, nos. 30-31, fig. 6 RR/K/60, finds dated to the last two decades of the 1st century BC; 261, nos. 54-57, fig. 12 MW/58, finds dated to late 1st century BC; 272, nos. 27, 39, fig. 19 RT/60, finds from layers dated to the 1st century BC to first half of the 1st century AD. Price 1992, 421, nos. 69-72, pls. 339-340, finds dated to the Augustan and Neronian eras (27 BC to AD 68).

[57] Kanaki 2002, 104-105, fig. 2, four examples dated on the basis of analogous finds to the 1st century BC.

[58] The example was found in an unpublished excavation of the *fossa*, the external ditch of the western city wall of Thessaloniki, from a plot that provided no firm dating.

[59] Adam-Veleni *et al.* 1998, 85-102; Mavromichali 2001, 141.

BC to the 1st century AD,60 as have the remaining glass finds from that site,61 according to the destruction horizon of the villa rustica in the ruins of which the vessel was found.[62]

### 6b. Ribbed bowls, rotary-pressed, mosaic

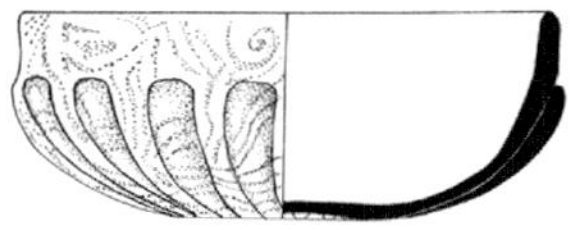

(Cat. nos. 25-29)
(Plates 1, I)

The vessels of this form[63] are characterized by vertical ribs that decorate the body, beginning almost under the rim and ending centrally underneath. Two major subgroups are discerned: wider and shallow types (height c. 4-7cm) and the more narrow and deeper-hemispherical ones, which sometimes stand on a relatively tall base and have a slightly flaring rim (height c. 10-12cm). The rim of these vessels is vertical, fire-polished and the curved body walls end in a flat bottom.

Among the fragments of this form[64] are included those of a vase deformed by fire secondarily, the exact shape of which is now impossible to discern. Nevertheless, the motif that appears on its mosaic body is mainly connected to ribbed bowls and thus it is ascribed to this form, although in the preserved part no signs of ribs are detected. It is equally possible that these fragments belong to a shallow conical bowl with smooth walls.[65] The decorative motif expands on a large area and consists of parallel alternately dark-blue and white layers which form angular spirals.[66]

These vessels are made by rotary pressing. Discoid segments of composite rods of glass, fairly large in diameter in this particular case, were placed adjoining to each other on a fire-resisting surface, then heated until they were fused together and formed a uniform disk. The resulting disk was placed on a convex former mould of the desired shape which was attached to a potter's wheel and reheated. With gravity and the high viscosity of the material while hot it was given the appropriate shape. The outer surface of the disc was pressed by an oblong tool, forming the radiating ribs which are typical of this type of vessels. At the end the surfaces of the vessel were polished to remove traces of the production procedure.

The form belongs to family I of David Grose's typology.[67] These vessels present a western, probably Italian, product and are dated to the Augustan or early Julio-Claudian era[68] (27 BC - AD 68). Examples have been found in Rome, but the vast majority of extant examples are without provenance.[69] From Greece one find was unearthed at Gomati, Chalkidiki.[70]

The vessels from the ancient agora of Thessaloniki are firmly dated to the last twenty years of the 1st century, the destruction date of the bathhouse in which they were found.[71]

### 7. Ribbed bowls, mould-blown

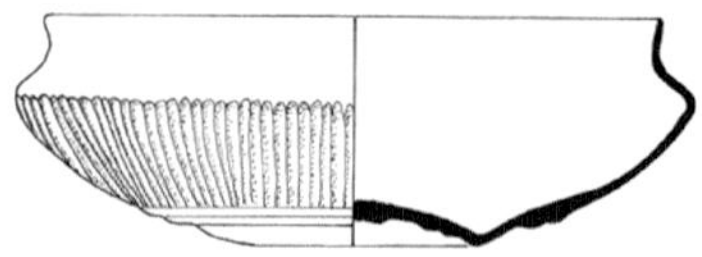

(Cat. nos. 30-36)
(Plate 1, I)

Vessels of this form are short and wide, the rim is slightly everted/calyx-shaped and is finely polished by grinding. Their body is convex and the bottom is mildly concave, forming though a short open ring base. They are decorated by vertical ribs and multiple concentric raised rings on the bottom. Their height is estimated between 4-7cm.

To the same group are ascribed vessels with a considerably more pronounced and concave bottom, and a series of horizontal raised rings delineating the transition to the base/resting surface, although is not clear if they

60 Adam-Veleni, Poulaki, Tzanavari 2003, 214, no. 209 (K. Tzanavari).

61 To mention only the examples illustrated: funnel-mouth vessel (Form 54), dated generally to the late 1st to early 2nd century, and a square, single-handled bottle (Form 75) (Isings 50), with a rim typical of the 1st-2nd centuries. For the shape and date of the rim, see Charlesworth 1966, 26-27, form 1c.

62 Tzanavari, Filis 2000, 160-162; Adam-Veleni, Poulaki, Tzanavari 2003, 81-82, 212, no. 193 (K. Tzanavari). The find comes from a *villa rustica*/ farmhouse built in the middle of the 2nd century that was abandoned, temporarily, in the middle of the 3rd century following raids by Goths.

63 Isings 1957, 17-21, form 3.

64 Apart from the vessels included in the catalogue, a fragment of another vessel was found in a (still unpublished) salvage excavation in the *fossa*, the external ditch of the western city wall of Thessaloniki, with no firm dating.

65 Intact example in Antikesamlung, Berlin, see Platz-Horster 1976, 31, no. 38. A gold glass example in Grose 1989, nos. 599, 337, colour plate on page 237, drawings on page 431.

66 Grose 1989, 269 no. 251, colour illustration on page 214, fragment of a composite, mosaic, ribbed bowl with a ring base, bearing the same motif in different colours.

67 Grose 1989, 247-249.

68 Grose 1989, 249.

69 Grose 1989, 241-244, with a detailed list of all mosaic glass vessels from controlled excavations.

70 *Glass Cosmos* 2010, 383, no. 446, dated to the 1st century.

71 Adam-Veleni *et al.* 1998, 85-102; Mavromichali 2001, 141.

appear among the fragments from Thessaloniki.[72] Another related group comprises vessels shallower than the aforementioned, with vertical rim and shoulders only very mildly pronounced; it is also unclear whether they are included among the Thessaloniki finds.[73]

The vessels of this group are mould-blown in a short, probably bipartite, mould, where the body was shaped; the rim is the result of free-blowing above the mould. Despite the fact that they have been ascribed to Sidonian production,[74] their distribution in both the eastern and Western Mediterranean regions makes it more likely that they were produced contemporaneously in several centres,[75] probably following the migration of the glassworkers from the Syro-Palestinian coastal regions to Italy.[76] Among the other sites[77] with published finds of this group, securely dated examples have been excavated in the region of the Black Sea,[78] all of them dated between the second quarter and the middle of the 1st century. There are also finds from Olympia, dated to the middle of the 1st century,[79] and Edessa, dated to the second half of the 1st century.[80]

The find from Thessaloniki is firmly dated to the last 20 years of the 1st century, the destruction date of the bathhouse in which they were found.[81]

### 8. Ribbed bowls, free-blown (Isings Form 17)

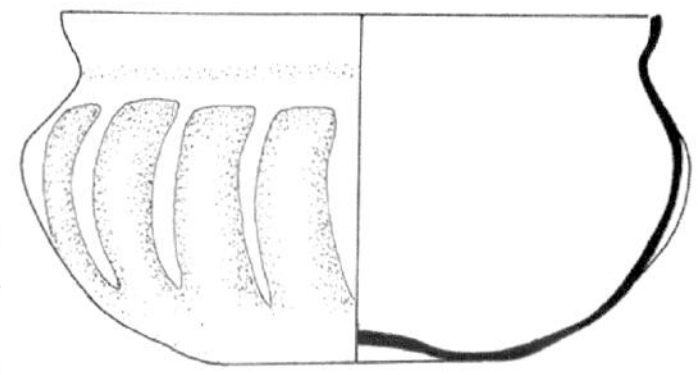

(Cat. nos. 37-38)
(Plates 1, I)

Vessels of this form have unworked, everted rim, globular body, plain flat base and slightly concave bottom. Their body is decorated with vertical ribs. Very often they are additionally decorated with a fine thread, usually white, which is marvered and incorporated into the mass of the body, which is spirally wound from the bottom to the shoulder. Their height is estimated between 5-7cm.

The examples from Thessaloniki are made of amber and colourless green glass.

The vessels of this form are free-blown. The technique of forming the ribbed decoration of the body is still under some debate, and while some researchers[82] consider, convincingly, that they were pinched, others[83] believe they were dip mould-blown, i.e. they were first blown in a small ribbed mould and further inflated with free-blowing until they acquired their final dimensions. Finally, it seems that the decorative thread that was wound around the body was flattened by marvering before the vessel's expansion – either with free-blowing or mould-blowing – to its final dimensions.

Vessels of this form appear more often in the following areas: northern Italy, Switzerland and Slovenia.[84] Other finds are known from Denmark,[85] Britain,[86] Spain,[87] France,[88] Netherlands,[89] Austria (Magdalensberg (Virinum)),[90] Libya,[91] Italy (Cosa and Morgantina (Morgantion)),[92] Greece (Ladi/Didymoteichon,[93] Samothrace,[94] Gomati/Chalkidiki,[95]

[72] For an intact example of this group, see Hayes 1975, 47-48, no. 82, fig. 2. A detailed bibliography is in Stern 1995, 112, note 3.
[73] Stern 1995, 113, no. 14, with further bibliography.
[74] Hayes 1975, 31-33.
[75] Stern 1995, 112.
[76] On the conditions and reasons for the phenomenon, see Stern 1999, 443 and passim.
[77] Stern 1995, 111-113, nos. 13-14, with further bibliography.
[78] Kerch (Panticapaeum): Kunina, Sorokina 1972, 158, 159, 166, fig. 6, no. 12 and fig. 10, no. 16; Kunina 1973, 130-133, fig. 22, no. 1; Kunina 1997, 276, nos. 120-122, with coloured illustrations.Kepoi on the Taman Peninsula: Kunina, Sorokina 1972, 158, 159, fig. 6, no. 42.
[79] Schauer 2001-2002, 36-37, finds dated according to excavation results.
[80] Chrysostomou 2000, 279, 284, fragments from a burial dated according to its contents to the second half of the 1st century.
[81] Adam-Veleni *et al.* 1998, 85-102; Mavromichali 2001, 141.
[82] Harden 1947, 295, no. 21; Berger 1960, 30-31; Czurda-Ruth 1979, 43-44; Haevernick, Pfeffer 1981, XIV.
[83] Fremersdorf 1958b, 13, pls. 20-23; Fünfschilling 1986, 117; Biaggio-Simona 1991, Vol. 1, 72-73.
[84] Isings 1957, 35-36, form 17; Haevernick, Pfeffer 1958; Haevernick 1967.
[85] Müller 1911, 1-46.
[86] Cool, Price 1995, 57-58. Dozens of examples from Britain, dated by excavation results to first half of 1st century.
[87] Price 1981, 408-413 (non vidi).
[88] Price 1988c, 31, fig. 29.
[89] Van Lith 1977, 29-38. Fragments of dozens of examples from Velsen, dated by excavation results to the period of Tiberius, Claudius and Nero (14-68).
[90] Czurda-Ruth 1979, 43-47, nos. 412-467. Finds from layers of Augustan period and first-half of 1st century.
[91] Cool, Price 1995, 57.
[92] Grose 1983, 45. Finds from Augustan layers.
[93] Terzopoulou 1998, 21-22, fig. 10β. Find from a 1st-century cremation.
[94] Dusenbery 1967, 45, figs. 33-34; Dusenbery 1998, 1076-1078. Six examples from graves, some of which are accurately dated to c. 25 BC, and others are dated approximately to the end of the 1st century BC to early 1st century AD.
[95] Trakosopoulou-Salakidou 2000, 363, fig. 7. Find from a grave dated to the 1st century.

Palatiano (Ioron),[96] Edessa,[97] Thebes,[98] Megara,[99] Athens,[100] Corinth,[101] Patras,[102] Olympia,[103] Amorgos,[104] Karditsa,[105] Nisyros,[106] Knossos,[107] Sfakaki/Pangalochorion),[108] Cyprus,[109] Turkey,[110] and the Black Sea region.[111] According to their archaeological contexts, it appears that this group of vessels, which are a development of rotary pressed ribbed bowls, appeared in the Augustan era and its production ceased after the middle of the 1st century.[112] The finds from Thessaloniki cannot be dated independently.

As to the production area of vessels of this form, although they are considered to be products of northern Italy their wide distribution suggests they were made at several sites.[113]

**9. Bowls, everted conical, smooth**

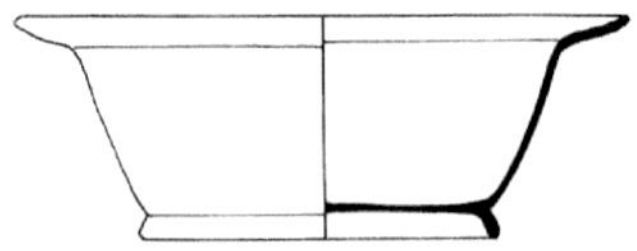

(Cat. no. 39)
(Plates 1, I)

Vessels of this form have ground flaring rim, everted conical body and flaring band base. Their height is estimated between 4-6cm.

These vessels were cast in a double mould and further polished by grinding; they are often decorated with incised decoration.

The vessel belongs to a small group of luxurious vessels dated to the 1st and 2nd centuries.[114] Completely decolourized, cast vessels appeared within the repertory of aesthetic changes occurring around AD 60,[115] succeeding similar vessels made of glass in striking colours that circulated in the first half of the 1st century.[116] It appears, according to the dates of the finds, that decolourized vessels were in use during the late Flavian era and in the periods of Trajan and Hadrian (69-138).[117]

Their production centre(s) cannot be defined as they appeared in equally small numbers at many sites, both in the East and West.[118] In any event the distribution makes the former hypothesis that they were Alexandrian[119] products highly improbable.[120]

The find from Thessaloniki is dated to the last 20 years of the 1st, the date of the destruction of the bathhouse in which it was found.[121]

---

[96] Antonaras, Anagnostopoulou-Chatzipolychroni 2002, 115. Finds from ancient Ioron, dated on the basis of analogous finds to 1st century.
[97] Chrysostomou A. 2000, 283, pl. 320. Vessel from a burial dated on the basis of its context to the second half of the 1st century.
[98] Haevernick 1981, 118, pl. 44. 2. Dated to the 1st century on the basis of analogous finds from other sites.
[99] Weinberg, McClellan 1992, 121, no. 90. Vessel from Megara, dated on the basis of analogous finds to the early 1st century.
[100] Robinson 1959, 86, no. M 23, pl. 54. The vessel probably belongs to the group of ribbed bowls made by rotary pressing.
[101] Davidson 1952, 96, nos. 605-9, fig. 7, pl. 54. Finds, some of which are dated by excavation results to 1st century.
[102] Agallopoulou 1973-1974, 369, pl. 235δ. Find dated by excavation results to first half of 1st century. On the same find, see also, Kolonas 2002, 112, no. 7.
[103] Schauer 2001-2002, 37.
[104] Triantafyllidis 1998, Vol. 2, 32, fig. 23. Five examples from Minoa, dated by excavation results to first half of 1st century.
[105] Intzesiloglou 1980, 269, pl. 117γ. From a clay sarcophagus in Mitropoli (Karditsa) but with no chronological indications.
[106] Jacopi 1939, grave XLI, 533-534, fig. 69. With no precise dating but redated by Isings (1957, 36) as from the Hellenistic period to the time of Nero or Vespasianus (54-79).
[107] Price 1992, decorated examples 57, 428-429, 444-445, nos. 121-4, pl. 342, and plain examples, i.e. the find from Thessaloniki, 448 nos. 171-3, pl. 344. Some of the finds are from layers of the first half of the 1st century, and others are from layers which contained finds of the 1st and 2nd centuries.
[108] Gavriilaki 1989, 457-460, pl. 253β. From a cemetery dated before the 2nd century.
[109] Two vessels, part of the Cesnola collection, now at the Metropolitan Museum of Art, 74. 51. 264 (CG-668), 74. 51. 280 (CG-712), according to Dusenbery 1998, 1076, note 16.
[110] Saldern 1980a, 12-13, nos. 42-44, pls. 2.43, 20.42. One decorated and one plain example from a layer with Hellenistic and Roman finds.
[111] Kunina 1997, 160-161, 295-296, nos. 204, 210-211. Finds from Kerch (Panticapaeum) and Olvia Pontica. Also, nos. 205-209, objects from the antiquities market of unknown provenance, among which no. 208 is undecorated and made of amber-coloured glass, identical to the Thessaloniki find.
[112] Van Lith 1977, 33-35; Grose 1982, 28; Stern 2001, 47.
[113] Whitehouse 2001, 202.
[114] Grose 1991, 12-18; Cool Price 1995, 35-39; Price, Cottam 1998a, 56-57; Rütti 1991, 40, form AR 21, pl. 39; Massabò 2001, 48-49.
[115] Pliny, *Nat. Hist.*, 36.192, who notes that transparent and decolourized glass vessels were more appreciated in his time as they resembled rock crystal vessels.
[116] Grose 1989, 254-256. A group of cast vessels, dated to the first half of the 1st century, most of which were found in the western provinces of the Roman Empire.
[117] Grose 1991, 15.
[118] Grose 1991, 15-18, with exhaustive bibliography of published examples.
[119] Harden 1936, 50.
[120] Grose 1991, 12.
[121] Adam-Veleni *et al.* 1998, 85-102; Mavromichali 2001, 141.

### 10. Bowls, hemispherical with horizontal rim

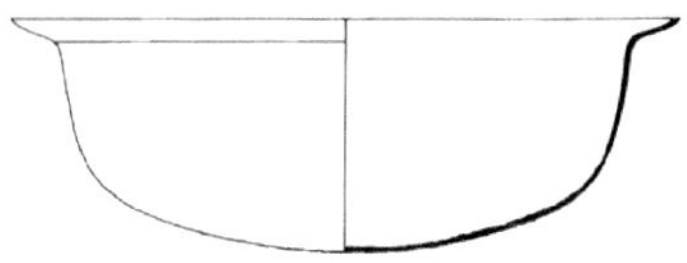

(Cat. nos. 40-41)
(Plates 1, II)

Vessels of this form have pronounced, everted, horizontal rim, hemispherical body and possibly a ring base.[122] They are made of thick, fully decolourized, high-quality glass, with no inclusions and/or pinprick bubbles. Their height is estimated at c. 5cm.

These vessels were cast, made by slumping in a double mould, with a further polishing of the entire vessel.[123]

The vessel belongs to a small group of luxurious vessels dated to the 1st century. Completely decolourized, cast vessels appeared as part of a trend occurring around AD 60,[124] succeeding similar vessels made of glass in striking colours that were circulating in the first half of the 1st century,[125] which quite probably represented the most costly glass of the era. It appears, according to the dates of the finds, that decolourized vessels were in use during the late Flavian era and in the periods of Trajan and Hadrian (69-138).[126] The finds from Thessaloniki cannot be dated independently.

Their production centre(s) cannot be defined due to the fact that they appear in equally small numbers at many sites, both in the East and West.[127] In any event the distribution makes the former hypothesis that they were Alexandrian[128] products highly improbable.[129]

### 11. Bowls/Beakers, hemispherical with crimped handles (Isings Form 43)

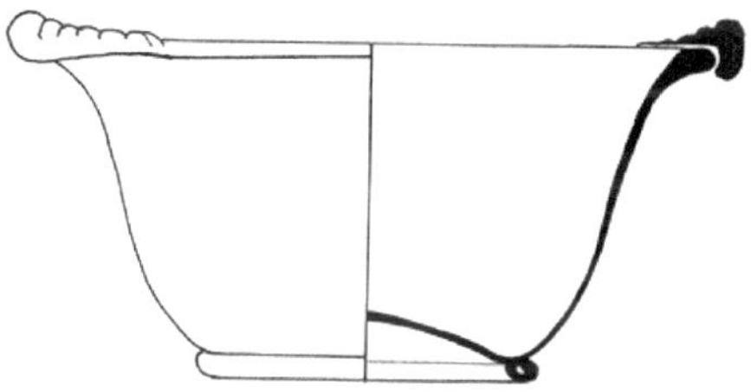

(Cat. no. 42)
(Plates 3, II)

Vessels of this form have flaring fire-polished rim, hemispherical body and pushed-in tubular ring base. Two horizontal, crimped pseudo-handles are placed at opposite sides on the edge of the rim. Their height is estimated at c. 6cm.

These are free-blown vessels, the base is formed by folding part of the lower body mass, and the handles are made of strips of glass attached on the rim of the already-finished vessel and notched (if they were not already notched) on the vessel rim.

This form of vessel is known in both the East and West.[130] Vessels of Form 3b of our classification actually present a shallow variant of Form 11. They were produced during the 1st century and are considered to be Italian products,[131] while there are also known examples that remained in use during the 2nd or 3rd centuries. Furthermore, examples have been reported at Ticino,[132] Tipasa (Algeria),[133] Cave of Scriptures (Israel),[134] Cyprus,[135] Knossos, (Unexplored Mansion),[136] Tarrha,[137] Corinth,[138] Sardis,[139] Hebron,[140] and Newark.[141]

The find from Thessaloniki is dated quite precisely to the last 20 years of the 1st century AD, the destruction date of the bathhouse in which it was found.[142]

---

122 For an illustration of an intact vessel, see Grose 1991, pl. IVa and fig. 2a, diam. 11.5cm.
123 Grose 1991, 9.
124 See note 174, above (Pliny, *Nat. Hist.*, 36.192).
125 Grose 1989, 254-256. Group of cast vessels that appears and remains in use only during the first half of the 1st century, and most of which have been found in sites of the western provinces of the Roman Empire.
126 Grose 1991, 15.
127 Grose 1991, 15-18, with exhaustive bibliography on all known examples.
128 Harden 1936, 50.
129 Grose 1991, 12.
130 Isings 1957, 59, Form 43, with finds dated to the 1st century.
131 Stern 1977, 57-58, with a comprehensive bibliography of analogous finds, including ones from control excavations.
132 Finds from several sites, dated by excavation results to the middle of the 1st century, Isings 1957, 59, and to the second half of 1st century, Biaggio-Simona 1991, Vol. 2, no. 176. 2. 092.
133 Lancel 1967, 9, 94, no. 195. Objects found together with a coin of Marcus Aurelius (161-180) or Lucius Verus (161-169).
134 Barag 1963, 104-105, no. 7. Dated by excavation results between AD 132 and 135.
135 Vessberg 1952, 116. Deep bowl of group B.I.β.3, with no specific information on the excavation site where they were found.
136 Price 1992, 431, 450, αρ. 224-225. Dated by excavation results to the Hadrianic period (117-138).
137 Buechner 1960, 112-113, no. 16, fig. 36. Dated by excavation results to the 2nd or 3rd century.
138 Davidson 1952, 80, nos. 612-613, pl. 54. Dated by excavation results to the 2nd century.
139 Saldern 1980a, 21, nos. 94-99. Finds that cannot be dated by excavation results, but on the basis of analogous finds, between the 1st and 3rd century.
140 Hayes 1975, 64, no. 179, fig. 6. 194. Dated to the 2nd or early 3rd century, according to analogous finds dated mainly to the 1st century.
141 Auth 1976, 95, no. 103. Dated to the 2nd or early 3rd century, according to analogous finds dated mainly to the 1st century.
142 Adam-Veleni *et al.* 1998, 85-102; Mavromichali 2001, 141.

### 12. Bowls, hemispherical (Isings Form 96)

i. Undecorated
(Cat. nos. 43-48)
(Plates 1, II)

ii. Decorated
a. Incised decoration
1. Shallow bands
(Cat. nos. 49-61)
(Plates 1-2, II)

2. Rice-grain-shaped strokes
(Cat. nos. 62-64)
(Plates 2, II)

3. Geometrical or figural decoration
b. Applied elements
1. Blobs
(Cat. nos. 65-66)
(Plates 2, II-III)

2. Threads
(Cat. no. 67)
(Plates 2, III)

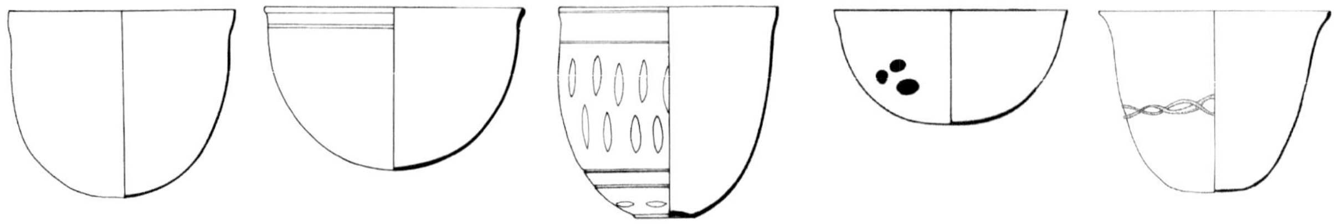

Vessels of this form[143] have unworked rim, slightly everted, hemispherical or even deeper body, flat base, and flat, occasionally slightly concave bottom. Their height is estimated between c. 5-10cm.
These vessels were made by free-blowing and the rim finished after annealing. Because the rims are uneven and the bands of incisions, with which the desired height of the rim was marked, are often discontinued/cut. The majority are plain, undecorated vessels, although decorated examples are also known.

Although attempts have been made to organize them into groups according to whether they have a distinct rim or not,[144] it appears that this was not really intentional and furthermore no chronological or geographical differentiation was noted. It appears, however, that the presence or absence of decoration, and the specific techniques used to form the decoration, does represent a criterion on the basis of which they may be divided into the following subgroups:

i. Undecorated[145]

Undecorated examples represent the most numerous group among hemispherical bowls, known from the late 3rd century, which were most widely distributed during the 4th century, surviving into the 5th century. Dated examples, among others, come from Britain,[146] France (Strasbourg/Augusta Trebecorum, Argentoratum),[147] Germany (Trier/Treviri,[148] Bonn/Bonna,[149] Cologne),[150] Hungary (Csákvár/Floriana),[151] Slovenia,[152] Croatia (Popov Dol),[153] Bosnia (Samobor),[154] Serbia (Belgrade/Singidunum),[155] Gračanica (Ulpiana),[156] Duklja (Doclea),[157] Ravna (Timacum Minus),[158]

143 Isings 1957, 113-114, 131-133, form 96.
144 Ružić 1994, groups IV/6a and IV/6b.
145 It is a subform of Isings form 96a, see Isings 1957, 113-114.
146 Cool, Price 1995, 88-90, pl. 5. 16. Several examples from the entire country, dated by excavation results to the 4th century.
147 Arveiller-Dulong, Arveiller 1985, 102-105, nos. 185-200. Dated by excavation results to the second half of the 3rd or to the 4th century.
148 Goethert-Polaschek 1977, form 49a, 50-59, nos. 156-220, pls. 38-39. Dated by excavation results to the 4th century.
149 Follmann-Schulz 1988, 103-105, nos. 379-387, pls. 44-45. Dated by excavation results to the last third of the 3rd century and mainly to the 4th century.
150 La Baume 1961, fig. 2, nos. 4-6. Dated by excavation results to the first half of the 4th century.
151 Barkóczi 1988, form 37b, 75-76, nos. 76-80, pls. VII, LXXII. Dated by excavation results to the 4th or early 5th century.
152 Petru 1976, pl. IX/1, 5. Dated by excavation results to the 4th century; Šubic 1976, 46, nos. 56-57, pl. VII/56-57, dated end of 4th century according to similar finds; Damevski 1976, pl. 15/1-3; Lazar 2003, 120, form 3.10.1, fig. 35. Dated by excavation results to the second half of the 3rd and/or 4th century.
153 Damevski 1976, 66, pl. XV/2. Find without provenance but dated on the basis of analogous finds to the 3rd or 4th century.
154 Paškvalin 1976, 119, pl. IX/2, dated on the basis of analogous finds to the 4th century.
155 Nikolić-Đorđević 1990, form 4b, 42, pl. II/17, dated by excavation results to the end of the 3rd century.
156 Ružić 1994, 38, nos. 552-554, pl. XXV, dated by excavation results to the second half of the 4th century.
157 Cermanović-Kuzmanović 1976, 179, pl. IV/8, form dated by excavation results to the 4th century.
158 Petković, Ružić 2005, 106, pl. IX, G. 87, one dated by excavation results to between AD 350 and 380.

Bulgaria (Krivina/Iatrus),[159] Greece (Veroia,[160] Thasos,[161] Edessa,[162] Liknades/Kozani).[163] (Only catalogue numbers 44[164] and 48[165] can be dated on the basis of their archaeological context.)

ii. Decorated
a. Incised decoration[166]

Hemispherical bowls decorated with engraving/wheel-cutting, according to motif and complexity of decoration, can be organized into three main groups:

1. The simplest in terms of their decoration examples, which represent the most numerous group, bear a mere rudimentary decoration, usually in the form of shallow bands under their rim or/and on the upper part of the body. These vessels are contemporary with the undecorated ones and to the vessels decorated with applied blobs and were probably produced in the same production centres.

Sites with well-dated finds include the following: United Kingdom,[167] Germany: Trier (Treviri),[168] Bonn (Bonna),[169] Strasbourg (Augusta Trebecorum, Argentoratum),[170] Hungary: Ságvár (Tricciana),[171] Serbia: Gradac,[172] Beočin,[173] and Greece: Patras.[174]

Only Cat. nos. 54-59[175] and 60-61[176] among the finds from Thessaloniki can be dated independently.

2. The vessels with decoration comprising of bands of ovular or rice-grain-shaped (ellipsoid) strokes, present a group well-defined aesthetically and chronologically. Their decoration presents a simplified variant of the 1st-century vessels with multifaceted engraving, and they are dated to the 3rd-4th century. This group was very widely distributed and it is thought that it was produced in at least four centres.[177]

Well-dated find-spots include the following – France: Rouen (Rotomagus),[178] Strasbourg (Augusta Trebecorum, Argentoratum);[179] Germany: Cologne,[180] Black Sea;[181] Slovenia: Ptuj (Poetovio);[182] Croatia: Sremska Mitrovica (Sirmium),[183] Zagreb;[184] Serbia: Belgrade (Singidunum),[185] Ravna (Timacum Minus),[186] Gračanica (Ulpiana);[187]

[159] Gomolka 1978, 26, 28-29, fig. 1/8, dated to the middle of the 4th and the early 5th century.
[160] Allamani-Souri 1991b, 302-303, pl. 114ε, example from a grave dated to the 4th century, according to its context.
[161] Tsouris 1996-98, 61-62, fig. 12, finds dated from excavation results to the 5th and end of 6th century.
[162] Chrysostomou A. 2000, 360, pl. 400B, vessel from a burial dated by a coin of emperor Theodosius I (379-395).
[163] Petsas 1961/62, 215, pl. 253δ, without chronological indication regarding the find.
[164] Pelekanidou 1993, 376, photo 6, find from a grave with coins of the 2nd-3rd century.
[165] See Chapter on Find-Spots: 'Sindos, Thessaloniki Prefecture, Industrial Zone of Thessaloniki 1999-2000'. Graveyard with 4th-century finds.
[166] It is a subgroup of Isings typology, form 96b1, see Isings 1957, 114-116. For illustrated, intact examples from Trier (Treviri), see Goethert-Polaschek 1977, 50-62, form 49, nos. 174, 182, 185, pl. 38-39, dated from excavation results to the 4th century.
[167] Cool, Price 1995, 88-90, pl. 5. 16, with many examples from the entire country, dated from excavation results to the 4th century.
[168] Goethert-Polaschek 1977, 59-61, forms 49b, nos. 222-232, figs. 20-21, pl. 38-39, dated from excavation results to the 4th century.
[169] Follmann-Schulz 1988, 103-105, nos. 376-378, 385, pls. 43-44, dated from excavation results to the last third of the 3rd century and mainly to the 4th century.
[170] Arveiller-Dulong, Arveiller 1985, 105, nos. 201-203, 204, dated from excavation results to the second half of the 3rd or to the 4th century.
[171] Barkóczi 1988, 75, form 37a, no. 75, pl. VII, dated from excavation results to the 4th and to second half of 4th century respectively.
[172] Ružić 1994, 38, no. 535, pl. XXV, dated from excavation results to the 3rd-4th century.
[173] Šaranović-Svetek 1986, 58, no. 10, dated from excavation results to the 4th century.
[174] Dekoulakou 1975a, 113-114, pl. 63δ, from a grave, whose more precise chronological indications are not mentioned. This part of the cemetery was in use from the Neronian era (54-68) up to the reign of Justinian II (565-578).
[175] See Chapter 9 (Find-Spots), '47-49 and 58-62, Sokratous Street, 1968 (today Mirarchou Koufitsa St.)'. From the excavation results the finds were buried after the second half of the 4th and more probably into the 5th century.
[176] See Chapter 9 (Find-Spots), 'Agiou Dimitriou and Agias Sofias Street, at their junction, 1969'. The burial date – not production date – is connected to the destruction of the church in the 7th century.
[177] See Chapter on *Decoration of the Material Under Study* on the engraved decoration that appears on vessels from Thessaloniki.
[178] Sennequier 1985, 54-55, no. 26, dated on the basis of analogous finds to the 2nd-3rd century.
[179] Arveiller-Dulong, Arveiller 1985, 106-107, nos. 205-06, dated from excavation results to the second half of the 3rd or to the 4th century.
[180] Fremersdorf 1967, 68-70, 73, 90-93, pls. 32-35, 40, 72, 74-78, 80, with no firm chronological indications.
[181] Sorokina 1967, 72, fig. 3, nos. 24-26, fig. 5. no. 12.
[182] Šubic 1976, 44, nos. 47, 51, pl. VI, dated to the 1st and to the end of the 4th century, respectively, according to analogous finds.
[183] Šaranović-Svetek 1986, 58, nos. 9, 11-12, dated from excavation results to the 3rd-4th century.
[184] Damevski 1976, 66, pl. XV/1, a find that is not dated by excavation results but according to analogous finds to the 3rd-4th century.
[185] Nikolić-Đorđević 1990, 42, form 4c, pl. II/18-19, III/20, 24, dated from excavation results to the end of the 3rd-early 4th century; Ivanišević, Nikolić-Djordjević 1997, 128-132, fig. 52, finds from layers of mid 2nd/mid 3rd century; Bikić 1997, 161, pl. 5. 1, from a layer dated to the 3rd-4th century.
[186] Ružić 1994, 38, no. 539, pl. XXV, dated from excavation results to the first half of the 4th century.
[187] Ružić 1994, 38, no. 540, pl. XXV, dated from excavation results to the mid 4th century.

Gamzigrad(Romuliana Felix),[188] Turkey: Üçtepe,[189] Sardis;[190] Egypt: Kom Aushim (Karanis);[191] and Greece: Corinth,[192] and Athens.[193]

3. The third group comprises of vessels with elaborate geometrical and figural representations, along with *diatreta* vessels, which are also dated to the 3rd-4th century.[194] No examples of this group have been published from the region of Thessaloniki, although excavations in the city's necropolises have brought to light (unpublished) examples with relatively complicated geometrical decoration known from vessels in museums in the West.[195]

b. Applied decoration

1. Blobs[196]

This type of decoration includes the *potoria gemata*, expensive metal vessels decorated with enamel or with semiprecious stones and/or with colourful glass gems.[197]

The decoration is made while the vessel was still attached to the blowpipe and still adequately warm; it was marvered on a marble surface where lumps of glass were arranged in such a way that once they were attached to the vessel they would form the desired motif. After that the vessel was reheated to polish the surface and render the blobs shiny and smooth. Another probable method of applying blobs is by touching the heated tip of a rod of glass briefly onto the vessel and then swiftly removing it. After repeating this procedure as many times as necessary to create the desired motifs the vessel was reheated and the blobs smoothened.[198]

Mainly dark-blue blobs were used for the decoration, supplemented in some cases by red and green ones. They were arranged around the body of the vessel in a single or a double row, or in triangular formations, occasionally alternating with large single blobs.[199]

Well-dated examples have been found in the United Kingdom;[200] Germany: Trier (Treviri),[201] Cologne;[202] France: Rouen (Rotomagus),[203] Strasbourg (Augusta Trebecorum, Argentoratum);[204] Italy: Aquileia;[205] Hungary: Ságvár (Tricciana), Dunaújváros (Intercisa), Szőny (Brigetio), Sopron (Scarbantia), Regöly, Csákvár (Floriana),[206] Slovenia;[207] Serbia: Belgrade (Singidunum);[208] Croatia: Sisak (Siscium);[209] Bulgaria: Krivina (Iatrus);[210] Romania: Topraichioi;[211] Black Sea region;[212] Egypt: Kom Aushim (Karanis).[213] In Greece examples of this form have been published from

188 Živić 2003,p.100, cat. no. 101 dated to the 4th century.
189 Lightfoot 1991, 90-95, pl. 1, with bibliography on the relevant finds from Turkish sites.
190 Saldern 1980a,15-18, no. 64, pl. 3, with no special chronological indications.
191 Harden 1936, 120, no. 317, pl. 14, dated to the 2nd-3rd century.
192 Davidson 1952, 95, no. 595, pl. 6, from a grave dated from excavation results to the 2nd century.
193 Weinberg, Stern 2009, 94-96, nos. 195-200, fig. 13, from a 3rd-century context.
194 Fremersdorf 1951, fig. 3, 4/4, 8/1-2, pl. 1-5, 8-9, 11/2-3; Fremersdorf 1967, 21-22, 26, pl. 61-63, 65, 85.
195 For an illustrated example, see Whitehouse 2001, 264-265, no. 452, with references to similar finds from Europe.
196 It is a subgroup of Isings form 96b2, see Isings 1957, 133. For illustrated, intact examples from Trier (Treviri), see Goethert-Polaschek 1977, 50-62, form 49, nos. 188, 210, 233-234, pl. 39, dated from excavation results to the 4th century.
197 Fremersdorf 1962, 11.
198 The present author wishes to thank Marianne Stern for this information.
199 On this see the *Chapter on the decoration.*
200 Cool, Price 1995, 88-90, pl. 5. 16, with examples from the entire country dated from excavation results to the 4th century.
201 Goethert-Polaschek 1977, 50-62, forms 49a-d, nos. 188, 210, 233-234, 268, pls. 38-39, 41, dated from excavation results to the 4th century.
202 Fremersdorf 1962, 7 and passim, where it is noted that this decoration culminates at the end of the 3rd/beginning of the 4th century.
203 Sennequier 1985, 64-65, no. 41, dated on the basis of analogous finds to the 4th century.
204 Arveiller-Dulong, Arveiller 1985, 112-113, nos. 209-210, dated from excavation results to the second half of the 3rd or to the 4th century.
205 Calvi 1968, pl. 26. 4-5, two vessels dated to the 4th century.
206 Barkóczi 1988, 96-100, form 67a, b, nos. 144-156, dated from excavation results to the second half of the 4th and to the end of the 4th- early 5th century, respectively.
207 Petru 1976, pl. VIII/6-7, dated from excavation results to the second half of the 4th century; Lazar 2003, 120-121, form 3. 10. 2, fig. 35 dated from excavation results to the second half of the 3rd and to the 4th century.
208 Nikolić-Đorđević 1990, 42, 45, form 4c, pl. III/25-26, dated from excavation results to the end of the 3rd and into the 4th century. Ivanišević, Kazanski 2009, 121, 126-127, fig. 5.19; Ivanišević 2009, 36-37, cat. No. 7, dated to the second quarter of the 5th century.
209 Damevski 1976, 66, pl. XV/3, find with no further excavation information, dated on the basis of analogous finds to the 3rd-4th century.
210 Gomolka 1978, 26, 29, fig. 1/5, dated to the middle of the 4th or early 5th century.
211 Opait 1985, 250-252, pl. 1.
212 More than 70 vessels from various excavations in the NE Black Sea region, which were dated as a whole mainly to the 4th century, and to the first half of the 5th. Part of the finds were considered to be local productions, see Sorokina 1970, 71-79; Sorokina 1971, 85-101. The excavations in the Black Sea region were later redated, see Sazanov 1995, 332-333, moving the glass finds as well, half a century or a little more, and dating this technique to the end of the 4th, mainly to the 5th and up to the early 6th century.
213 Harden 1936, no. 331, dated from excavation results between the 3rd and 5th century.

Diocletianopolis (modern Armenochori, Kastoria),[214] Thasos,[215] Athens,[216] Corinth,[217] and Olympia.[218] As a whole the vessels are dated between the end of the 3rd and the early 6th centuries, but they appear mostly in the 4th and in the first half of the 5th century.

2. Applied thread

In some rather rare cases in these regions a (usually) dark-blue thread of glass is used, either single or in a twisted pair, to decorate the rim or the body at the height of its maximum diameter.

The decoration is applied while the vessel is still attached to the blowpipe and it can be rotated easily, with swift moves so that the thread will not cool and cause the breakage of the vessel. This technical detail explains the sloppiness that almost is almost invariably obvious in the rendering of this type of decoration.

The fragment from Thessaloniki, which might also belong to a conical vessel (Form 31), finds its closest parallels in vessels excavated in the Black Sea region,[219] in Syria[220] and in Egypt,[221] and they are dated to the late 4th to early 5th century. Similar vessels have been also found in Britain,[222] in Strasbourg (Augusta Trebecorum, Argentoratum),[223] and in Hungary (Szőny (Brigetio), Várdomb, Dunaújváros (Intercisa)).[224]

The hemispherical bowls, as well as the taller conical beakers, represent the prevailing form of 4th-century drinking vessels, both in the West and East. It is quite probable that they were also used as lamps, as is indicated, among other things, by their totally unworked, rough rim.[225]

It appears that they were produced in many centres, which for now cannot be defined more precisely. One of these was probably Thessaloniki and more precisely the workshop found in the eastern Necropolis, in the plot of the Theological Department of the University.[226] As to the origin of the decorated examples, generally it is believed that those bearing polychrome blobs are German products, while the single-coloured, dark-blue decoration is indicative of an Eastern, Syro-Palestinian and Egyptian production site.[227]

### 13. Bowls, hemispherical with three handles

(Cat. no. 68)
(Plates 2, III)

This form appears in two main variants. The examples of the first variant have a truncated conical body and are ascribed to Uboldi group I.1. and Isings Form 134. They have a rim bigger or equal to the height of the vessel, three vertical handles placed at the rim, flat base and slightly concave bottom. Known examples have been found both in the West and the East, while so far in the region of Thessaloniki none have yet been found.

The examples of the second variant have hemispherical or ovular bodies and they comprise group I.2. of Uboldi. Vessels of this form have unworked rim, hemispherical body, three small suspension handles around their rim, flat base and slightly concave bottom. In some cases the handles continue down on the body, as smooth or undulating ribbons, almost reaching the base of the vessel.[228] Their height is estimated at c. 8cm.

[214] Petkos 2000, 586, fig. 10, finds from the Armenochori at Kastoria (Diocletianopolis), dated from excavation results to the 4th or to the second half of the 4th century.
[215] Tsouris 1996-98, 61-62, fig. 12, finds dated from excavation results between the 5th and to the end of the 6th century.
[216] Robinson 1959, 147, M 248, pl. 55, dated by excavation results to the early 4th century.
[217] Davidson 1952, 97-98, no. 615-616, pl. 7, which are dated from excavation results to the 4th century.
[218] Schauer 2001-2002, 41, finds ascribed to the 4th-5th century.
[219] See Sorokina 1967, 72, fig. 3 no. 27; Sorokina 1970, 71-79, esp. pl. 2 and page 74. Decoration type IF5 or type III3. The vessels as a group are dated mainly to the 4th century or beginning of the 5th. See Sorokina 1970, 71-79. Redated by Sazanov (1995, 332-333) to the second half of the 5th and to the first half of the 6th century.
[220] Philippe 1964, 58, fig. 69.
[221] Edgar 1905, pl. III 32. 487, find from Hawara.
[222] Cool, Price 1995, 89, pl. 5. 16, nos. 553-557 finds from York, dated from excavation results to the 4th century, and numerous relevant finds from several sites in Britain.
[223] Arveiller-Dulong, Arveiller 1985, 112-113, no. 210, dated from excavation results to the 3rd or first half of the 4th century.
[224] Barkóczi 1988, 93, form 62, nos. 134-136, dated from excavation results to the last third of the 4th or early 5th century.
[225] Cf. comments on conical beakers (Form 31).
[226] See Chapter 2.B.
[227] Fremersdorf 1962, 9, the polychrome nature of the blobs is considered to be the result of aesthetic choices of those German tribes who started to spread c. AD 260. For oriental products of this kind, see Israeli 2003b, 193.
[228] Antonaras 2007, 51-54, finds from the museum's basilica at Philippi, dated by excavation to the Justinian period.

Similar vessels to the find from Rentina (Cat. no. 68) are reported in Chios,229 Kourion,230 Gerasa,231 Sardis,232 Italy,233 in Dalmatia at Gata,234 Galovac235 and Plutaj, around the peninsulas of Taman and Tiritan in the Azov Sea,236 and in Svishtov (Novae)237 in Bulgaria. The rim of these examples is infolded tubular, and the shape of their body is less deep. An example with analogous shape, elongated handles and forming of the rim, has been reported in Istria, Romania,238 and another example now in the RGM, Mainz.239 The closest parallels to this find are two lamps kept in the Corning Museum of Glass240 and in the British Museum,241 dated to 4th-5th and 6th centuries respectively, dates that coincide with the archaeological contexts of the aforementioned finds. The find from Rentina should be ascribed to this same period, which cannot independently be dated more precisely.

### 14. Bowls, hemispherical honeycomb (Isings Form 107a)

(Cat. nos. 69-70)
(Plates 3, III)

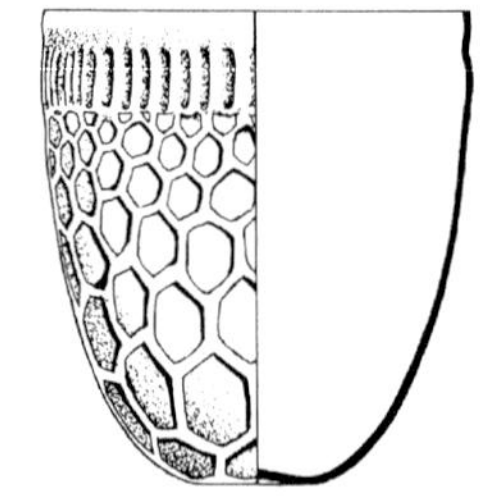

The vessels of this form[242] have vertical, slightly everted, unworked rims, body hemispherical or deeper than the hemispherical flat base, and flat or occasionally slightly concave bottom. Their height is estimated between 7-10cm.

These bowls are dip-mould-blown, i.e. first blown in a small and shallow mould and then further expanded by free-blowing.[243]

The relief decoration of these vessels usually consists of vertical ribs which begin directly under the slightly flaring rim and covering the entire circumference of the vessel. Lower down, the body and base are covered by a web of ovular or hexagonal cells resembling honeycomb. In some cases the web is supplemented by concentric circles,[244] with applied blobs[245] or with a rosette at the bottom.[246] The cell motif is distorted from the dip-mould-blowing; the cells closer to the base are considerably larger than those closer to the rim.

The examples unearthed in the region of Thessaloniki are made of deep blue, colourless green and decolourized glass.

Analogous and firmly dated objects from sites (and in museums) include France: Marseilles (Massalia),[247] Arles (Arelate),[248] Saint-Blaise,[249] Port-Vendres;[250] Germany: Cologne,[251] Trier (Treviri),[252]

[229] Ballance *et al.* 1989, 125, 127, F 17a, b, e, p, fig. 50, finds which most probably belong to the mid-7th century destruction phase. Apart from the small handles (17a-c), long smooth ones (17d), and undulating ones – extended and decorated with tooled or pinched horizontal projections – just like our examples (17e, f), also appear, with the difference that the vessel walls are not bending.
[230] Young 1993, 43, fig. 5, finds dated from excavation results to the end of the 7th century.
[231] Crowfoot, Harden 1931, pl. XXX/41 form C2.
[232] Saldern 1980a, 46-49.
[233] Uboldi 1995, 108, note 68, 69.
[234] Fadić 1994a, 217-219, where it is assumed that their use begins already from the 4th century, and continues into the 6th.
[235] Fadić 1994a, 217, note 58, finds dated from excavation results to the 5th-6th century.
[236] Ivachtchenko 1995, 321, pl. 10, lamps with long, smooth handles, dated from excavation results to the 5th and 6th century.
[237] Turno 1989, 166-167, fig. 2/4, form 4, dated to the end of the 4th and to the 5th century.
[238] Bâjenaru, Bâltâc 2000-2001, 478-483, fig. 4. 1, pl. X. 8-10, finds from a basilica, dated to the end of the 5th century, which was destroyed in the early 7th century.
[239] Fremersdorf 1975, 95, pl. 6, the vessel was acquired by the museum from the market, probably from Paris, or from the German antiquities market, see Uboldi 1995, 109, note 74.
[240] Whitehouse 2001, 227-228, no. 799, the long ends of the handles of the lamp are also undulating, while the lower body is decorated with three, applied, theatrical masks. Dated to the 4th-5th century according to stylistic reasons, e.g. the presence of the theatrical masks.
[241] Buckton (ed.) 1984, 106, no. 115, the object was bought in Tyre and is considered to be a 6th-century Syrian product. The handles have relatively long, flat ends.
[242] Isings 1957, 133, form 107a, with many parallel finds dated mainly to the 4th and 5th centuries.
[243] Haberey 1966, 208-212, assumes that the mould is shallow, almost flat, and the ribs are on its vertical walls and the net pattern on its base.
[244] Istria (Istros): Minchev 1988, 48, 57, form XXIII pl. VIII/39, dated from excavation results between the first and third quarter of the 4th century. Museum of Fine Arts, Boston: Saldern 1968, no. 41, dated on the basis of analogous finds to the 4th century.
[245] Hayes 1975, 147, 154, 226, no. 643, dated on the basis of analogous finds to the beginning or to the middle of the 4th century. The single analogous find was located in an unpublished salvage excavation at the *fossa*, the external ditch of the western city wall of Thessaloniki. The excavation revealed no precise dating of the finds.
[246] Doppelfeld 1966, 46 and photos 54-55.
[247] Foy 1995, 200, form 13e, no. 87-88, pl. 10, finds dated from excavation results to the end of the 5th or early 6th century.
[248] Foy 1995, 200, form 13e, no. 86, pl. 10, find dated from excavation results to the middle or to the second half of the 5th century.
[249] Foy 1995, 200, form 13e, nos. 84-85, pl. 10, finds dated from excavation results to the 5th century.
[250] Foy 1995, 200, form 13e, nos. 89-91, pl. 10, finds dated from excavation results to the 5th century.
[251] Fremersdorf 1961, 57-58, pl. 113, found in a sealed grave with coins of Constantine II, struck c. AD 335; Doppelfeld 1966, 46, photo 56.
[252] Goethert-Polaschek 1977, 62, form 50, cat. no. 237, pl. 25/309a, 40/237, dated from excavation results to the 4th century.

Hamburg,[253] Düsseldorf;[254] Serbia: Rtkovo,;[255] Hungary: Sopron (Scarbantia),[256] Csákvár (Floriana),[257] Százhalombatta (Matrica),[258] Budakalász,[259] Szob,[260] Pusztaszentpál,[261] Nógrádverőce;[262] Slovenia: Ljubljana (Emona),[263] Praetorium Latobicorum,[264] Trebnje;[265] Croatia: Gata;[266] Bulgaria: Krivina (Iatrus),[267] Varna (Odessos),[268] Kerch (Panticapaeum),[269] Cherson, (Chersonesus Taurica),[270] Kom Aushim (Karanis);[271] Israel: Horbat Rimmon,[272] Ha-Gosherim,[273] Akhziv,[274] Khirbet esh-Shubeika,[275] Jalame,[276] Mezad Tamar,[277] Newark,[278] Boston,[279] Los Angeles,[280] Corning,[281] and Wheaton[282] (all of them dated to the 4th century). In Greece published examples are reported from Samothrace,[283] the Early Christian basilica of Solinos in Cassandra Chalkidiki,[284] and Olympia.[285] The finds from Thessaloniki cannot be dated by their excavation data.

Honeycomb bowls, although similar motifs on are known on glass vessels from the 1st century,[286] are the expression of a decorative movement of the 4th century, with a clear preference for vessels with relief decoration and which led to the revival of the mould-blowing technique. Their distribution over the entire area of the Roman Empire makes it difficult locating their production centre. The western part of the Empire,[287] the region of the Black Sea,[288] Syria or Egypt[289] have all been proposed as potential production sites, yet none of these assumptions can exclude the other two candidates, or clearly prevailing as the production area of these peculiar vessels.

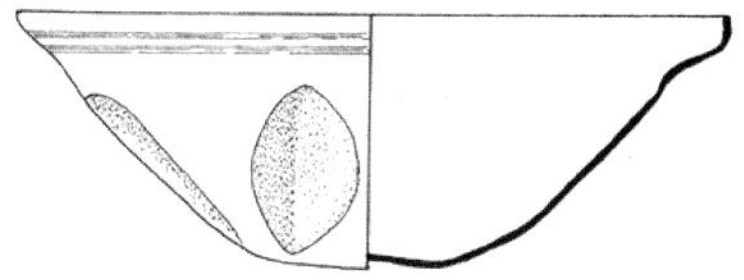

### 15. Bowls, inverted conical, indented (Isings Form 117)

(Cat. nos. 71-73)
(Plates 3, III)

Vessels of this form have unworked rim, possibly slightly ground, everted conical body and flat bottom. The body walls are decorated with vertical indentations. Their height is estimated between 7-8cm.

[253] Saldern *et al.* 1974, 176, cat. no. 487, this example is the only one of this form which does not have vertical grooving around the rim. Dated on the basis of analogous finds to the 4th century.
[254] Saldern 1974, 85, no. 85, dated to the 4th-5th century according to relevant finds.
[255] Han 1986, 92-93, drawings 25.11-1, the site of 'Glamija', a fortress now submerged in Danube with two phases, the earlier dated to the 4th and the second to the 6th century.
[256] Barkóczi 1971, 72, fig. 3, find dated according to analogous finds to the end of the 4th-early 5th century.
[257] Barkóczi 1971, 72, fig. 4, from a grave dated according to the results of the excavation to the end of the 4th-early 5th century.
[258] Barkóczi 1971, 72, fig. 5, find dated according to analogous finds to the end of the 4th-early 5th century.
[259] Barkóczi 1971, 72, fig. 6, find dated according to analogous finds to the end of the 4th- early 5th century.
[260] Barkóczi 1971, 72, fig. 7, find dated according to analogous finds to the end of the 4th-early 5th century.
[261] Barkóczi 1971, 72, fig. 8, find dated according to analogous finds to the end of the 4th-early 5th century.
[262] Barkóczi 1971, 73, fig. 10, find dated according to analogous finds to the end of the 4th-early 5th century.
[263] Barkóczi 1971, 71, fig. 2, find dated according to analogous finds to the end of the 4th-early 5th century.
[264] Petru 1975, 8, fig. 1, pl. 10, find dated according to excavation results at the end of the 4th century.
[265] Lazar 2003, 121, form 3. 10. 3, fig. 35 dated according to excavation results to the 4th century.
[266] Fadić 1994a, 220, where, although 4th-century examples are mentioned, it is assumed that the vessel found in this particular Justinian basilica of St. Cyprianus, even if it were indeed produced in the 4th century, remained in use until the 6th.
[267] Gomolka 1978, 26, fig. 1/3, find dated from excavation results between the early 4th and the early 5th century.
[268] Minchev 1988, 48, 57, form XXIII pl. VIII/40, find dated from excavation results to the 4th-5th century.
[269] Sorokina 1962, 231, figs. 11.5, 13.5, from excavation layers dated to the 3rd-4th century.
[270] Golofast 2001, 126-127, drawings 81:11, 15-20, where many examples dated from excavation results to the late 5th and the early 6th century are noted.
[271] Harden 1936, 165 no. 472, pl. XVI.
[272] Gorin-Rosen 2004, 113-114, find from a graveyard of late 4th or early 5th century. Also published are all the known finds from Israel.
[273] Gorin-Rosen 2004, 113-114, find dated generally to the late 4th-early 5th century.
[274] Syon 1998, fig. 15:6, find dated generally to the late 4th-early 5th century.
[275] Gorin-Rosen 2002b, 314, 316, fig. 7:35, find dated generally to the late 4th-early 5th century.
[276] Weinberg 1988, 79-80, fig. 4-38: 350, find dated by excavation results to the late 4th-early 5th century.
[277] Erdmann 1977, 110, 114, no. 921-922, pl. 8:921, 922, find dated generally to the late 4th-early 5th century.
[278] Auth 1976, 85, find which is not dated directly, yet it is mentioned among other analogous, 4th-century vessels.
[279] Saldern 1968, no. 42, find dated on the basis of analogous finds to the 4th century.
[280] Saldern 1980b, 62, cat. no. 55, find dated on the basis of analogous finds to the 4th-5th century.
[281] Whitehouse 2001, 110-113, nos. 606, 610 objects dated to the 4th-5th century on the basis of analogous finds.
[282] Dusenbery 1971, 15-16, cat. no. 15, fig. 12, find dated on the basis of analogous finds to the 4th century.
[283] Lehmann, Lehmann (ed.) 1969, 196, note 20, fragment found in an older, disturbed layer and dated to the 4th century on the basis of analogous finds.
[284] A Christian basilica in use between the 4th and 6th centuries. A blue fragment bearing a wide relief stripe decorated with vertical, parallel indentations. I thank the excavator of the site, I. Papangelos, for his permission to study the glass finds of the excavation. On the excavation, see Papangelos 2005. On the glass finds, see Antonaras 2014e, 97-110.
[285] Schauer 2001-2002, 41, fig. 14, stylistically ascribed to the production of the 4th-5th century.
[286] Isings 1957, 45-46, form 31.
[287] Hayes 1975, 147, where he considers more probable a western origin for these vessels, although he acknowledges that there are several vessels from the eastern provinces as well.
[288] Barkóczi 1971, 83.
[289] Dusenbery 1971, 16.

These vessels are free-blown and the vessel was probably left with the moil during the annealing process and the rim was cracked off later. The indentations were made while the vessel was still hot and attached to the blowpipe with the use of a pointed tool, probably the glassblower's jacks.

Vessels of this form have been mainly found in the western provinces of the Roman Empire[290] and in the Balkans. Dated examples are known from come Britain,[291] Normandy,[292] Trier (Treviri),[293] Krefeld-Gellep (Gelduba),[294] Worms (Borbetomagus),[295] Strasbourg (Augusta Trebecorum, Argentoratum),[296] Martigny (Octodurum), Switzerland,[297] Bordeaux (Burdigala),[298] Marseilles (Massalia) and its hinterland,[299] and Rome and other Italian sites.[300] In the Balkans they have been found at Banoštor (Bononia),[301] Sremska Mitrovica (Sirmium),[302] Kostol (Pontes),[303] Mihajlovac, the site of Mora Vagei, Serbia,[304] Ravna (Timacum Minus),[305] Gamzigrad (Romuliana Felix),[306] Gračanica (Ulpiana),[307] and Rtkovo.[308]

The finds from Thessaloniki cannot be dated by their archaeological context with precision, yet, in accordance with parallel finds from other regions and sites, they can be placed in the production of the second half of the 4th or the 5th century.

### 16. Bowls, smooth free-blown, hemispherical or shallower (Isings Form 116)

(Cat. nos. 74-83)
(Plates 3-4, III)

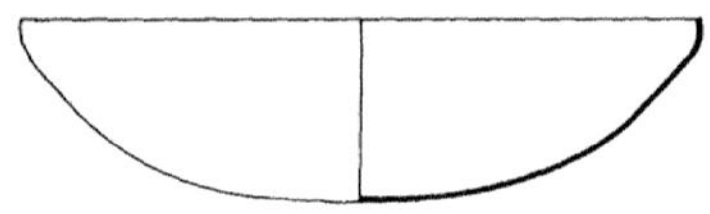

Vessels of this form have unworked rim, hemispherical or shallower body and their bottom is relatively flat. Their height is estimated between 5-7cm.

They are free-blown, annealed retaining the moil, and the rim was later cracked off. On the rim are usually visible shallow horizontal incisions which are seemingly carelessly disturbed in the final cutting of the rim. Almost all were made of clear, relatively colourless glass, without inclusions and pinprick bubbles. Most of the preserved examples retain parts of incised decoration rendering complex motifs and scenes.

This form is mainly known from the western provinces of the Roman Empire and is dated to the 4th and early 5th century.[309] Judging by the incised decoration which is quite often found on them, they must had been made in various sites, both in Italy as well as in the Eastern Mediterranean. This hypothesis is corroborated first of all by the generally accepted fact that in this later period a large number of local glass-workshops were operating throughout the Empire. Furthermore, it is corroborated by the diversity of technical ways for rendering different decorative themes, indicating different workshops, and also by the use of both the Greek and Latin alphabet on them.[310] Among others, dated examples are known from Britain;[311] France: Le clos de la Lombarde, Narbonne,[312]

---

290 Isings 1957, 147-149, form 117, where a large number of finds from sites in Germany and the Low Countries are presented, all of them dated to the second half of the 4th or to the early 5th century.

291 Cool, Price 1995, 104-105, where a large number of examples dated from excavation results to the 4th century, and especially to the second half of the 4th century. At Colchester (Camulodunum) a fragment was found in a layer dated to AD 150-300 (no. 730).

292 Sennequier 1994, 60, form HN. 3. 20B, pl. 2.

293 Goethert-Polaschek 1977, 39, form 28, nos. 105-112, pl. 33, dated from excavation results to the second half of the 4th century.

294 Pirling 1959, 218-219, 221, grave no. 1007a, fig. 3. 2 and grave 1248, fig. 6. 1, finds dated from excavation results to the early 5th century.

295 Bernhard 1979, 20, fig. 14, vessel from a grave dated to the second half of the 4th century.

296 Arveiller-Dulong, Arveiller 1985, 157, no. 346, dated on the basis of analogous finds to the middle of the 3rd- middle of the 4th century.

297 Martin 1995, 100, fig. 1/7, find dated from excavation results to the second half of 4th century.

298 Foy, Hochuli-Gysel 1995, 155-156, fig. 13, nos. 7-8, finds dated to the 5th century.

299 Foy 1995, 201, form 15, finds dated from excavation results to the 4th and to the first half of the 5th century.

300 Sternini 1995a, 258, finds dated from excavation results to the late 4th and into the 5th century.

301 Šaranović-Svetek 1986, 60, no. 35, find dated from excavation results to the second half of the 4th to first half of the 5th century.

302 Šaranović-Svetek 1986, 60, no. 36, find dated from excavation results to the second half of the 4th to first half of the 5th century.

303 Ružić 1994, 49, nos. 949-950, finds dated from excavation results to the second half of the 4th to first half of the 5th century.

304 Ružić 1994, 49, no. 951, find dated from excavation results to the second half of the 4th to first half of the 5th century.

305 Ružić 1994, 49, no. 952, find dated from excavation results to the second half of the 4th to first half of the 5th century.

306 Ružić 1994, 49, nos. 953-4, find dated from excavation results to the second half of the 4th to first half of the 5th century.

307 Ružić 1994, 4, nos. 955-958, finds dated from excavation results to the second half of the 4th to first half of the 5th century.

308 Han 1986, 92-93, drawings 25. 1-6. From 'Glamija', now a submerged Danube fortress that operated in two periods, in the 4th and 6th centuries. Verena Han considered them as local, Mysian products, referring to Gomolka (1978, 28) and Barkóczi (1968, 80-81).

309 Isings 1957, 144-145, form 116.

310 See Chapter 5 (Engraved Decoration).

311 Cool 1995, 13, fig. 3/9, a form of vessel which appears in several sites, and which, according to secure excavation results, appears in the second third of the 4th century, and remains in use even after the end of the century. Detailed presentation of British finds, see Price, Cottam 1998a, 124-125, all of them dated from excavation results to the 4th century.

312 Foy 1995, 201, form 15, no. 98, pl. 11, find dated from excavation results to the early 5th century.

Marseilles (Massalia);[313] Germany: Trier (Treviri),[314] Bonn (Bonna),[315] Mainz (Moguntiacum);[316] Switzerland: Augst;[317] Italy: Ventimiglia (Albium Intemelium),[318] Luni (Luna),[319] Descenzano,[320] Portoguardo,[321] Ravenna-Classe,[322] Ravenna (Ravenna),[323] Turin (Augusta Iulia Taurinorum),;[324] Slovenia: Šmarata;[325] Hungary: Plismarót,[326] Tác (Gorsium);[327] Serbia: Gradac,[328] Karataš (Diana),[329] Kostol (Pontes),[330] Belgrade (Singidunum),[331] Gamzigrad (Romuliana Felix),[332] Ravna (Timacum Minus),[333] Čačak,[334] and Gračanica (Ulpiana).[335]

Only Cat. nos. 82 and 83 can be dated on the basis of the excavation clues and it appears that they were interred in the 7th century; Cat. nos. 76 and 78 are dated second half of 4th or into the 5th century.[336]

### 17. Calyx-shaped bowls

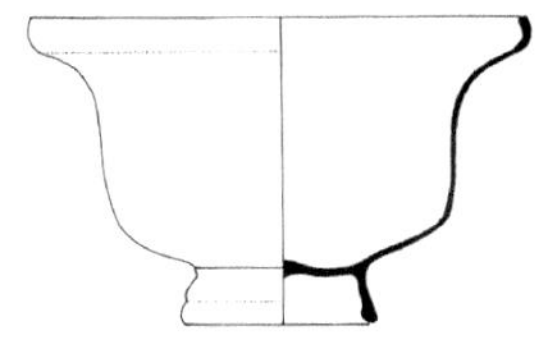

(Cat. nos. 84)
(Plates 4, IV)

The single vessel of this form has fire-polished, out-turned rim, calyx-shaped body and slightly concave bottom. On the relatively high, applied, conical base are visible tooling marks, traces of the production procedure. Their height is estimated at c. 7cm.

The vessel is free-blown. The body was formed first and then a second flat mass of glass was attached under it, which was shaped by further tooling into the vessel's conical base. Finally a metal rod or a blowpipe was attached to the under surface, the moil was cut and the rim was everted and fire-polished.

No exact parallels for this form have been located. The only vessels considered adequately related to this shape were found in the collection of the Corning Museum of Glass and are dated to the 4th-6th centuries.[337] The excavation did not yield a precise dating for the find and thus the form's dating is based on the colour and the texture of its glass. This type of dark green-coloured glass appears mainly among vessels found at Egyptian sites.[338] Furthermore, bases which are formed by pincers, whose acute tooling marks are left visible on the finished vessels, are also mainly found in Egypt. In addition, the dull glass, covered in pinprick bubbles, appears among Late Imperial vessels, and to this period the find from Kalochori is also ascribed.

### 18. Bowls with scalloped rim (Isings Form 42d)

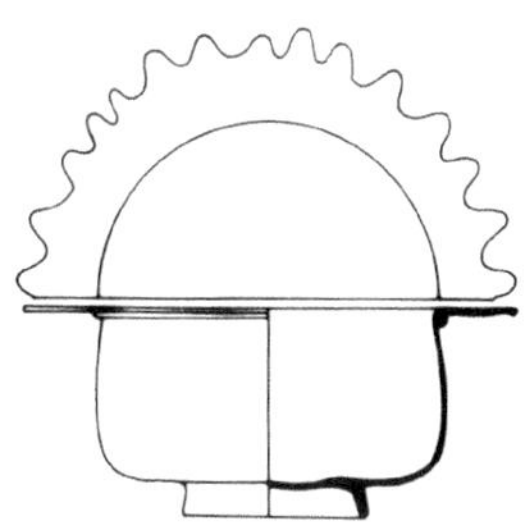

(Cat. no. 85)
(Plates 4IV)

The vessels of this form[339] have horizontal rim with fire-polished, scalloped edge; vertical body walls, flat bottom and low, pushed-in, slightly conical ring base. The edge of the rim

---

313 Foy 1995, 201, form 15, no. 99, pl. 11, find dated from excavation results to the middle of the 5th century.
314 Goethert-Polaschek 1977, 28-32, form 15a, pls. 30-32, finds dated to the 4th century, after AD 324.
315 Follmann-Schulz 1988, 119-121, nos. 457-464, pls. 51, 52/460-464, finds dated from excavation results to the 4th century.
316 Harter 1999, 50-51, 187, form A17, subform 3rd, no. 213, pl. 9, find dated to the 4th century.
317 Rütti 1991, 65, form AR 59. 1, pl. 56/11270-1274, finds dated from excavation results to the second quarter of the 4th and into the early 5th century.
318 Paolucci 1997, 127, dated on stylistic grounds to the end of the 3rd century.
319 Paolucci 1997, 140-141, dated on stylistic grounds to the 4th century.
320 Paolucci 1997, 150-151, dated on stylistic grounds to the second half of 4th century.
321 Paolucci 1997, 157-158, dated on stylistic grounds to the end of the 4th century.
322 Paolucci 1997, 161-162, dated on stylistic grounds to the beginning of the 4th century.
323 Paolucci 1997, 168-169, dated on stylistic grounds to the mid 4th century.
324 Paolucci 1997, 178-179, dated on stylistic grounds to the 4th century.
325 Lazar 2003, 85, form 2. 6. 4, fig. 30, find dated from excavation results to the second half of the 4th to early 5th century.
326 Barkóczi 1988, 66-67, form 27, no. 46, pl. V, find from a grave dated from excavation results to the second half of the 4th century.
327 Barkóczi 1988, 66-67, form 27, nos. 47-48, pl. V, finds dated from excavation results to the second half of the 4th century.
328 Ružić 1994, 39, nos. 563-565, finds from Saldum, dated from excavation results to the second half of the 4th to the early 5th century.
329 Ružić 1994, 39, nos. 566-567, pl. XXVII/1, finds dated from excavation results to the early 6th century.
330 Ružić 1994, 39, nos. 568-570, pl. XXVII/2, finds dated from excavation results to the second half of the 4th to the early 5th century.
331 Nikolić-Đorđević 1990, pl. II/8, finds dated from excavation results to the second half of the 4th to the early 5th century.
332 Ružić 1994, 39, nos. 572-575, pl. XXVII/4, finds dated from excavation results to the second half of the 4th century.
333 Ružić 1994, 39, nos. 576-581, pl. XXVII/5, finds dated from excavation results to the second half of the 4th century.
334 Ružić 1994, 39, no. 582, pl. XXVII/6, find from the *Rimske Terme*, dated from excavation results to the second half of the 4th to the early 5th century.
335 Ružić 1994, 39, nos. 583-584, pl. XXVII/7, 8, finds dated from excavation results to the second half of the 4th to the early 5th century.
336 See Chapter on Find-Spots: 'Agiou Dimitriou and Agias Sofias Street, at their junction, 1969', and 'Sokratous Street, 1968 (today Moirarchou Koufitsa St.)'.
337 Whitehouse 1997, 79-80, no. 104, find dated on stylistic grounds to the 4th century. Less similar but also comparable is Whitehouse 2001, 110-111, no. 604, find dated, also on stylistic grounds, to the 4th-6th centuries.
338 It is the fourth type of glass that Harden identified among the finds from Kom Aushim (Karanis) in Egypt in 1931. Also personal inspection of the same material held at the Kelsey Museum, Ann Arbor and also in the Metropolitan Museum, New York. For a colour illustration of a relevant example, see Whitehouse 1997, no. 112.
339 Isings 1957, 58, form 42d.

and the transition from the body to the rim are occasionally emphasized with a fine thread of strongly-coloured glass, wound around it. The catalogued example from Thessaloniki is colourless greenish and is decorated on the lip with a colourless bluish thread, while a fragment of a second example is not decorated with an additional thread.[340] Their height is estimated at c. 6cm.

The vessel was made by free-blowing. The base was formed while it was still attached on the blowpipe from the rim. After the attachment of another blowpipe on the vessel's bottom and the removal of the first blowpipe from the top the rim was shaped: first the moil was cracked off and a strongly coloured thread was applied at the edge, then the rim was reheated and bent outwards and finely pushed in to form its scalloping edge.

The majority of the published examples are kept in public and private collections, i.e. Toledo, Ohio,[341] California,[342] Corning,[343] Berlin,[344] Bonn (Bonna),[345] Cologne,[346] Düsseldorf,[347] Switzerland,[348] and Rome.[349] Fewer examples are found in controlled excavations: Albenga (Albingaunum) Italy,[350] Solin (Salona), Croatia,[351] Kom Aushim (Karanis), Egypt,[352] Carthage,[353] and Esch, Holland.[354] The example from Thessaloniki appears to be the only one that has this particular shaping of the base and the decoration with a strongly-coloured glass.

The vessels of this form are mainly dated to the 4th century. The fact, however, that vessels of the same shape appear in pottery in the 5th[355] and 6th[356] centuries, and in silverware also in the 6th century,[357] makes it quite likely that glass vessels continued to be produced at least during the 5th century, as is also indicated by glass finds from Rome,[358] Carthage,[359] Marseilles (Massalia), and Saint-Blaise.[360] Although the vessel is ascribed to the category of bowls, the particular rendering of the rim makes it highly probable that it was a vessel for presenting solid food and not for drinking.

The exact archaeological whereabouts of the find from Thessaloniki are not known and it is dated to the 4th-5th century by comparison to analogous finds from other sites.

### 19. Calyx-shaped beakers (Harden Type G1ii)

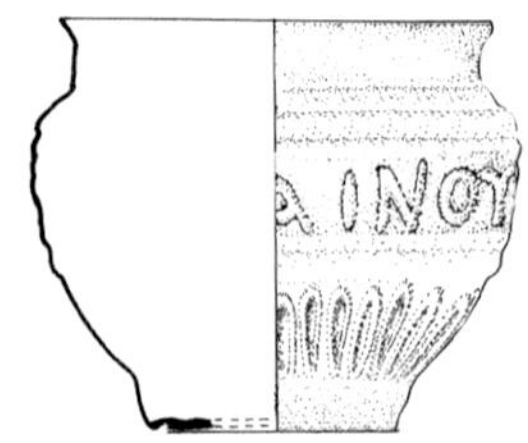

(Cat. nos. 86-88)
(Plate IV )

Vessels of this form have an unworked slightly everted rim, short and wide neck, calyx-shaped body and flat bottom. This form is represented in Thessaloniki by fragmentarily preserved examples. These fragments are colourless bluish with greenish striations. Their height is estimated at c. 7cm.

340 The find comes from an unpublished excavation of the *fossa*, the external ditch of the western city wall of Thessaloniki, from a site that did not reveal a precise dating.
341 The two examples that are kept in the Museum of Toledo, Ohio are provenanced from Syria and are ascribed to Egyptian production. See Harden 1936, 111, note 2, photo 2g.
342 Saldern 1980b, 108, no. 107.
343 Whitehouse 2001, 141, no. 650.
344 Platz-Horster 1976, 88, no. 176.
345 *Antiken aus Rheinischem Privatbesitz* (Bonn 1973), 206, no. 340, pl. 151, according to Platz-Horster (1976, 88, no. 176).
346 La Baume, Salomonson 1976, 57, no. 183, pl. 25, 4.
347 Saldern 1974, 125, no. 164.
348 Dörig 1975, no. 351.
349 Fremersdorf 1975, 63, no. 576, pl. 25, find from the Catacomb of St. Sebastian.
350 Pastorino 1998, 111, no. 16. Dated to the 5th century according to analogous finds from Rome.
351 Auth 1975, 160, nos. 66, 67, pl. 32.
352 Harden 1936, 97 and 111, pl. XIV.
353 Tatton-Brown 1984, 195, no. 10; Duncan-Jones 1994, 11, dated to the 4th-5th century, according to analogous finds. The same researcher mentions three more, unpublished, at least then, fragments from the Jordanian site of El-Lejjun which were found in layers that cannot be dated before the 6th century.
354 Isings 1957, 58.
355 A corresponding form appears in North African red-slip pottery, dated to the mid- or late 4th century, see Hayes 1972, 82-83, drawings 54. Clay examples, made of red and Macedonian grey clay, found in graves at Sindos are also dated to the 4th century, see Keramaris, Papagianni 2000, 147-148, fig. 12. From the city of Thessaloniki are known clay examples made of Macedonian grey clay, dated to the end of the 4th-early 5th century, see Hayes 1972, 405-06; Papanikola-Bakirtzi 2002 (ed.), 313, no. 340, (P. Kampanis), dated to the 4th-5th century; Makropoulou 2007, 76, pl. 9, 64, vessel BK 4501/8, dated to the second half of the 4th century.
356 Robinson 1959, 116, no. 349, pl. 71, late 6th century; Hayes 1972, 150-51; Hayes 1976, 27, no. 128, pl. 17. Analogous find from the ancient agora of Thessaloniki see Karagianni 2001, 151, fig. 10.
357 Boyd 1992, 29, S44. 1, S44. 2, silver lamp from an ecclesiastical treasure of the mid-6th century.
358 Sternini 1995a, 248, 274, pl. 6/33-34, in the temple of Magna Mater at Rome, found in a layer dated between AD 390 and 480; Sternini 2001, 26, 42, nos. 50-52, fig. 8.
359 Tatton-Brown 1984, 195, no. 10, fig. 65.
360 Foy 1995, 201, form 16, nos. 101-105, pl. 11, some of these finds are dated from excavation results to the 6th century.

These vessels are mould-blown in a tripartite mould. The upper part of the body was shaped by two vertical parts while the lower body and the bottom were shaped by a curved, almost hemispherical part.

Only one of the three fragments from Thessaloniki can be almost fully graphically reconstructed, with the exception of its bottom part, on which is preserved only the first of the seemingly three concentric raised rings that were decorating it. The bulbous body of this vessel is fully preserved as well as the band of vertical, adjoining parallel petals on the lower body, the lower double band that delineated the field of the inscription, and the upper triple band that delineated it towards the rim. In addition the unworked, slightly flaring rim is also preserved.

On the first example, part of the inscription field is preserved and it reveals parts of the letters 'Y' and 'Φ'. On the second example are preserved the letters 'ΦΩΠΑ'. According to other fully preserved examples, the inscriptions on both vessels can be restored as: 'ΕΥΦΡΑΙΝΟΥ ΕΦΩ ΠΑΡΕΙ'. This inscription has been interpreted in different ways but the most convincing is the one that considers it an abbreviated form of 'ΕΥΦΡΑΙΝΟΥ ΕΠΙ ΤΟΥΤΩΙ ΕΦΩΙ ΠΑΡΕΙ'. This phrase, which partly appears in the New Testament,[361] translates as: 'Rejoice with what you are present in'.[362] On the third example are preserved the letters 'ΡΔΟΣ' and the phrase has been restored as 'ΚΕΡΔΟΣ ΚΑΙ ΕΥΦΡΟΣΥΝΗ ΕΝ ΟΙΝΟΠΟΣΙΑΙ', and it has been considered to have been intended to stimulate potential buyers' interest.[363]

It is a form that is present in the entire ancient world. It has been systematically studied by Donald B. Harden as long ago as 1935 (Form G1i and ii) and again in 1943/4,[364] while E. Marianne Stern (1995)[365] filled in the gaps by presenting all the finds that had been published in the meanwhile. Vessels identical to the ones under discussion, i.e. Harden form G1ii, have been found on sites in Cyprus,[366] Lebanon,[367] Berezan (Borysthenes Island),[368] Olbia Pontica,[369] Italy,[370] Tunisia,[371] and Windisch (Vindonissa) in Switzerland.[372] Moreover, analogous vessels are housed in several museums (Berlin Antikenmuseum,[373] Oppenländer Collection,[374] and the Corning Museum of Glass).[375] In Greece, a vessel of this form was excavated at Galaxidi (Chaleion),[376] while fragments found in the Athenian Agora[377] and at Delphi[378] could equally belong to form G1i of the Harden typology.

Examples of Harden G1i, which are almost identical to the aforementioned vessels, except that they are slightly shorter and the field of the inscription is delineated on both sides by a double band instead of the triple band at the top that differentiates form G1ii, have been uncovered at sites[379] such as Smyrna,[380] Kerch (Panticapaeum),[381] Loo (near Sochi in southern Russia),[382] Italy[383] and North Africa,[384] while other finds may be seen in the Constable-Maxwell Collection[385] and the Carnegie Museum of Natural History.[386]

It is probable that they present products from a Sidonian workshop,[387] as their distribution corresponds to that of other Sidonian products, or even from a workshop imitating Sidonians, although for the time being this assumption cannot be proved. The shape of the lower part of the mould looks like the Sidonian examples, and it is probable that it was invented in the workshop of the famous glassworker Ennion.[388]

361 Matthew 26:50: ''Εταῖρε ἐφ' ᾧ πάρει;'
362 Stern 1995, 97. Pseudo-Zonarae, lexicon, s.v. ''Εφ' ᾧ πάρει', 928, line 14. For the possibility that there is an Epicurean connotation in the inscription, see Fontaine, Margos 2010, 80-83.
363 Mavromichali 2001, 133.
364 Harden 1935, Harden 1944-1945.
365 Stern 1995, 97. Also see Weinberg and Stern 2009, 65-66, no. 146, fig. 8, from a context dated between the late 1st and late 2nd century.
366 Harden 1935, 174.
367 Harden 1944-1945, 92.
368 Sorokina 1962, 230-231, fig. 15 (lower), the vessel was found in a 3rd-century layer.
369 Harden 1935, 174.
370 Harden 1935, 174.
371 Harden 1944-45, 292.
372 Berger 1960, 49-50, no. 124, pl. 8.
373 Harden 1935, 174; Platz-Horster 1976, no. 62.
374 Saldern *et al.* 1974, 159, no. 447.
375 *JGS* 15 (1979), 187, no. 5; Whitehouse 2001, 24-25, nos. 487-488.
376 Threpsiadis 1972, pl. 72α; Zafeiropoulou 1975, pl. 4.
377 McClellan 1983, 74-75, fig. 5. Unidentified, possibly Harden form G1ii, found at the bottom of a 2nd-century well
378 McClellan 1983, 77, from Antikyra in Phokis, the exact form is not mentioned.
379 As summarized in Stern 1995, 98, note 6.
380 Harden 1935, 174.
381 Sorokina 1962, 230-231.
382 Ivanova, Goloubev 1961, 284 ff, fig. 4.
383 Grose 1973-76, 179.
384 Braat 1963, 106-107, no. 10, pl. 35/1.
385 Maxwell 1979, 174-175, no. 312.
386 Oliver 1980, 70, no. 64.
387 Stern 1995, 97.
388 Stern 1995, 97.

It is dated around the middle of the 1st century, as is this entire group of products, according to the find from Siphnos ascribed to Harden form G1v,[389] and dates the entire G1 form before the third quarter of the 1st century.[390] An additional indication for such a date is the particular resemblance they bear with the clay vessels dated to the middle of the century.[391] Some of the vessels have been found in strata of the 2nd and 3rd centuries[392] and thus it is believed[393] that because of their relatively high price they were protected by their owners and remained in use for a longer period of time.

The destruction date of the two finds is firmly fixed to the last twenty years of the 1st century.[394] The third example cannot be dated independently by its archaeological context and is therefore placed in the middle or second half of the 1st century according to analogous finds from other sites.

**20. Bowls, hemispherical (Isings Form 12)**

(Cat. nos. 89-90)
(Plate IV )

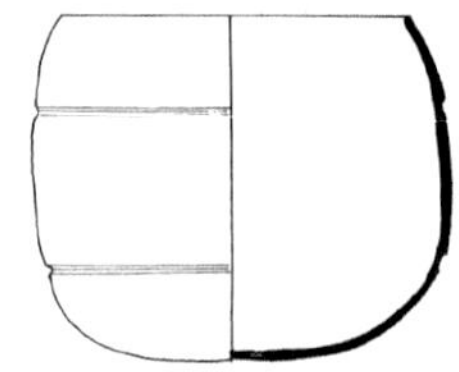

Vessels of this form have unworked rim, and depending on the curve of the upper body they are grouped into cylindrical shapes with vertical walls, and hemispherical ones with curved body walls. The lower part of the body tapers towards the bottom with what is known as a high *kick*, i.e. a tall, acutely conical protuberance at its centre. Their height is estimated between 6 and 8cm.

They are often made of dark, colourless or colourless-greenish glass, decorated with usually two deeply engraved horizontal grooves, one under the rim and another at the mid-body, which divide it into two almost equal parts.

They are free-blown and decorated by wheel-cut engraving.

It is a well-known form[395] usually referred to as a 'Hofheim' cup from the name of the site where the first published examples were found. They are considered to be Italian products imitating the Italian eggshell clay vessels of the Augustan era.[396] In addition, they are obviously related to the eggshell clay vessels which appear already in the second quarter of the 1st century BC in the Athenian Agora.[397]

Examples of this form appear in large numbers in sites of the Western Mediterranean, while in the Eastern Mediterranean they do occur, but only in far smaller numbers. These glass vessels appear already in the Augustan period,[398] however the majority of the extant examples is dated to the second and third quarter of the 1st century.[399] Well-dated examples have been reported from Colchester (Camulodunum),[400] Trier (Treviri),[401] Cologne,[402] Bonn (Bonna),[403] Strasbourg (Augusta Trebecorum, Argentoratum),[404] Magdalensberg (Virinum),[405] Augst,[406] Aquileia,[407] Zadar (Iader),[408] Solin (Salona),[409] Bodrum (Alikarnassos),[410] Kalınharman in Manissa (Magnesia),[411] Kepoi, on the

[389] Harden 1944-45, 92, pl. IX 1-3; Mackworth-Young 1949, 86, pl. 28/3-5.
[390] Harden 1944-45, 88; Stern 1995, 97.
[391] Braat 1963, 10, with reference to a form from Hofheim, pl. VI, 16.
[392] Sorokina 1962, 231.
[393] Stern 1995, 97, note 14; McClellan 1983, 74-75.
[394] Adam-Veleni *et al.* 1998, 85-102.
[395] Isings 1957, 27-29, form 12.
[396] Stern 2001, 45, with further bibliography.
[397] Robinson 1959, 11-13, nos. F18-19, pl. 1. Group F is dated between 75 BC and the end of the 1st century BC.
[398] Czurda-Ruth 1979, 37, nos. 324-418.
[399] Cool, Price 1995, 65; Stern 2001, 73, no. 16, with further bibliography.
[400] Cool, Price 1995, 64-68, with a thorough bibliography on NW provinces of the Roman Empire, and on some Eastern ones also.
[401] Goethert-Polaschek 1977, 41-42, form 30, dated to the reigns of Claudius (41-54) and Nero (54-68).
[402] Fremersdorf 1958b, 38-39, pls. 75-76, dated from excavation results to the first half of the 1st century. Fremersdorf 1967, 54, pl. 6, lower part, dated from excavation results to the first half of the 1st century.
[403] Follmann-Schulz 1988, 99, nos. 358-375, pl. 42, without strict dating according to the excavation results, except for no. 360, which is dated to the second half of the 1st century.
[404] Arveiller-Dulong, Arveiller 1985, 52, no. 68, dated according to similar finds to the 1st century.
[405] Czurda-Ruth 1979, 37-43, nos. 324-409, fragments dated from excavation results to the period between Augustus and Caudian (27-54 BC).
[406] Rütti 1991, vol. 2, 51-53, form AR 34, dated from excavation results between late Augustan times and the period of Traianus (27-117 BC).
[407] Calvi 1968, 67, 72-73, nos. 169-172, pl. C. 5, dated to the 1st century.
[408] Ravagnan 1994, 180, 185, 187-188, nos. 351, 361, 367, 369, dated to the 1st century according to similar finds.
[409] Auth 1975, 158, no. 45 pl. 29/45, dated according to analogous finds to c. the middle of the 1st century.
[410] In the city museum, according to Czurda-Ruth 1979, 39.
[411] Lightfoot 1989, 26-27, no. 10: pl. 9/3, no. 11: fig. 1/2, pl. 2/1, dated to the 1st century according to analogous finds.

Taman Peninsula,[412] Kerch (Panticapaeum),[413] Samothrace,[414] Krenides-Philippi,[415] Corinth,[416] Tarrha,[417] Knossos,[418] Agios Ermogenis at Kourion,[419] Idalion,[420] Episkopi at Kourion, and Cyprus.[421]

The vessels from Thessaloniki are dated to the last twenty years of the 1st century, in accordance with the destruction date of the baths from which they were excavated.[422]

### 21. Conical beakers with straight walls and raised ovular protuberances, 'Lotus Beaker' (Isings Form 31)

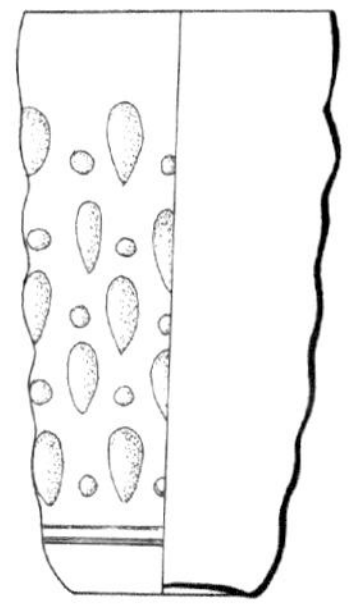

(Cat. nos. 91-94)
(Plate IV)

The vessels of this form have an inverted, truncated conical shape. Their vertical rim is left unworked and they stand on a flat bottom. The walls of the body are decorated with rows of raised, stepped ovular motifs, turned almost always towards the base, and which alternate with smaller, hemispherical protuberances. In addition there are also known examples with additional decorative motifs, such as wine scrolls or theatrical masks.[423] The finds from Thessaloniki are made of colourless greenish glass. Their height is estimated at c. 12cm.

The vessels of this form are blown in a mould which consisted of three vertical parts and sometimes a fourth, discoid, part which shaped the bottom of the beaker. The rim was not cracked off immediately after the extraction from the mould, it was cut after the vessel's annealing and was either left unworked or was slightly ground at a later time. The vast majority of the preserved examples with this decoration are inverted truncated conical beakers. However there are also a few examples known that have the same shaped body but are finished as rhyta[424] or close-shaped vessels, handleless[425] or single-handled.[426]

It is a quite widely spread form of luxury vessel.[427] Analogous finds have been excavated in Britain, Spain, Holland, Italy (mainly southern), northern Switzerland, Egypt, Israel and the Arab Peninsula.[428] In Greece there are published finds from Samothrace.[429] Vessels with identical decoration with the finds from Thessaloniki have been excavated in Surduk (Rittium),[430] Transdierna (Tekija),[431] Pontes (Kostol),[432] Mihajlovac (Mora Vagei),[433] Ravna (Timacum Minus),[434] Keszthely,[435] Costanza (Tomis),[436] Kerch (Panticapaeum),[437] and at Golgi in Cyprus;[438] furthermore, analogous finds are kept at Yale University Art Gallery,[439] in Leiden at the Rijksmuseum van Oudheden,[440] and in the Toledo Museum of Art.[441]

From finds of this form from Pompeii and Ercolano (Herculaneum)[442] it is evident that their production began before AD 79; what remains unclear however is the period they ceased being produced.

412 Sorokina 1977a, 136, fig. 6/1, 2, from graves dated to the middle of the 2nd century.
413 Kunina 1997, 313, no. 293, dated from excavation results to the second half of the 1st century.
414 Dusenbery 1967, 45, no. 36; Dusenbery 1998, 1081-1082, no. S 200-5, from a grave of the late 1st century BC or the early 1st century AD.
415 *Glass Cosmos* 2010, 397, no. 479, from a grave dated to the first half of the 1st century.
416 Davidson 1952, 79, 81, 93, 101, no. 591, fig. 6, no. 638, fig. 8, finds dated from excavation results to the 1st-2nd and the 1st century, respectively.
417 Buechner 1960, 110, no. 6, fig. 1, pl. 36, fragment probably dated to the 1st century.
418 Price 1992, 426-427, nos. 138-146, pl. 343, fragments from layers dated to the Claudian (41-54), Flavian (61-96), or even later eras.
419 McFadden 1946, 486, no. 112, pl. 45, from a burial dated to the 1st century according to its context.
420 Isings 1957, from a grave of the 1st century.
421 Oliver 1983, 252, 255, no. 51, fig. 4/51, pl. 41/10 from a burial of the first half of the 1st century.
422 Adam-Veleni *et al.* 1998, 85-102; Mavromichali 2001, 141.
423 Fadić 1988, 51, fig. 4. For a thorough analysis of the variants of this form, see Stern 1995, 103-105.
424 Tatton-Brown 1991, 74-75, fig. 94.
425 Hayes 1975, 48-49, no. 86, pl. 7.
426 Harden *et al.* 1987, 162-163, no. 84.
427 Isings 1957, 45-46, form 31.
428 Detailed catalogue of finds is listed by Stern (1995, 105-107) and Whitehouse (2001, 13-14).
429 Dusenbery 1998, 1079, find dated from excavation results to the second half of the 1st century.
430 Damevski 1976, 65, find dated to the 1st-2nd century.
431 Ružić 1994, 41, no. 600, find dated to the end of the 1st-2nd century.
432 Ružić 1994, 41, nos. 601-602, finds dated to the end of the 1st-2nd century.
433 Ružić 1994, 41, no. 603, find dated to the end of the 1st-2nd century.
434 Ružić 1994, 41, no. 606, find dated from excavation results to the middle of the 2nd century.
435 Barkóczi 1988, 94, no. 139, dated to the 1st century.
436 Bucovala 1968, 52, no. 57, find from Neptun and dated according to similar finds.
437 Sorokina 1962, 217, fig. 4.
438 Stern 1995, 107, note g.
439 Matheson 1980, 49, no. 126.
440 Stern 1995, 106 note 8.c.
441 Stern 1995, 103-104, no. 8.
442 Scatozza-Höricht 1986, 19, pl. 1 (upper row centre); Scatozza-Höricht 1991, fig. 16a (upper left).

It is thought that vessels of this form are products of at least two production centres, one situated in Western Europe and the other in the Eastern Mediterranean region, and mainly for two reasons: first, due to technical differences, such as the number of parts that comprised the mould and the size of the vessels, and, second, due to the several variants in their decoration. In the Eastern Mediterranean region simpler vessels with two-stepped protuberances prevail, and which were occasionally combined with smaller hemispherical ones, such as the examples from Thessaloniki.[443]

All finds from Thessaloniki were found in excavations in the area of the ancient agora. Most of them are dated, according to the destruction time of the baths in which they were found, to the reign of Domitian,[444] while the fragment found in the agora, but probably in the Odeion, is dated, on the basis of the general dating of this form, between the second half of the 1st and the early 2nd century.

Several interpretations have been suggested for the decoration of these vessels, e.g. lotus buds,[445] almonds[446] or (olive) tree knots.[447] The most convincing interpretation[448] is that they are connected to Hercules' club, and the characteristic knots that covered it, because the characteristic two-stepped profile of the knots and their overall picture is typical of representations of the club in glass,[449] as well as in other media, such as metal and marble.[450]

**22. Conical beakers with straight walls, 'Mythological Beaker'**

(Cat. no. 95)
(Plate IV)

The vessels of this form have an inverted truncated conical shape. Their rim is unworked and s body walls are straight and they stand on a flat bottom. The walls of the body are decorated blown motifs. Their height is estimated at c. 12cm.

They are blown in a mould that consisted of four vertical parts and a fifth (discoid) part that sh some cases vessels with the same body shape have been finished as funnel-mouth flasks.[451]

The decoration unravels on a wide band which covers almost the entire surface of the body standing human figures, each framed on both sides by columns on a stepped pedestal, and a capi style. The base is decorated by two or three raised concentric circles.

Four different groups of beakers have been discerned on the basis of their decoration.[452] The find from Thessaloniki belongs to group III.[453] In this group, high above the standing figures a garland connects the capitals, and a rosette appears in the centre of each compartment and above the garland. A band of well-designed branches covers the lower part of the vessel under the figures. According to analogous intact examples, one female and three male figures were represented around the beaker.

On the extant fragment the figure of the goddess Tyche is preserved almost intact, standing on a pedestal; she is in *contraposto* with the right leg 'engaged' and the left 'free'. The goddess, unlike many other examples where she is presented wearing a short tunic which leaves the upper part of her body naked, here we have her in a tunic and himation. She is rendered frontally but her head, according to other preserved examples, was turned to the right. With her left hand she holds a cornucopia and with the right hand a rudder.

An also frontally rendered naked male figure is partly preserved to the right of Tyche. In the right hand of this figure an object is visible which has been identified as an aryballos hanging from a chain.[454] The male figures have been identified with athletes and the branches on the lower part of the vessel with victory wreaths.[455] A newer and more convincing study, however, identifies the three men with Apollo, Dionysus and Hermes, based on a careful

---

443 Stern 1995, 104.
444 Adam-Veleni *et al.* 1998, 85-102; Mavromichali 2001, 141.
445 Eisen 1927, vol. 1, 290.
446 Harden *et al.* 1968, 55.
447 Isings 1976, 353.
448 Isings 1976, 353.
449 Stern 1995, 105, note 38.
450 Stern 1995, 104-105, where are also illustrated relevant works in marble and metal.
451 Whitehouse 2001, 57, no. 528.
452 Davidson-Weinberg 1972, 29-30.
453 Davidson-Weinberg 1972, 44-46; Wight 1994, 47-51.
454 Mavromichali 2001, 133.
455 Davidson-Weinberg 1972, 46.

examination of the objects that each one of them holds.[456] According to the conclusions of this study, the object held by the figure in the second compartment should be interpreted as a *phiale* for libations.[457]

Beakers of this form are held in museums and collections in Europe and America,[458] including the National Museum in Athens.[459] Firmly dated examples have been excavated in Cyzicus,[460] Slovenia,[461] Windisch (Vindonissa)[462] and Masada,[463] and they are dated to the second half of the 1st or the early 2nd century.

Vessels identical to the find from Thessaloniki are kept in the Sackler Gallery,[464] in the Oppenländer Collection,[465] in a private collection in London,[466] the Haeretz Museum,[467] and in the Rolin Museum, Autun (Augostodunum).[468] None of these vessels was found in systematic excavations and thus they cannot be dated independently. The find from Thessaloniki is the first one of this form that was found within a systematic excavation and its destruction can be dated precisely in the last twenty years of the 1st century.[469]

### 23. Conical beakers with straight walls, multifaceted engraved decoration

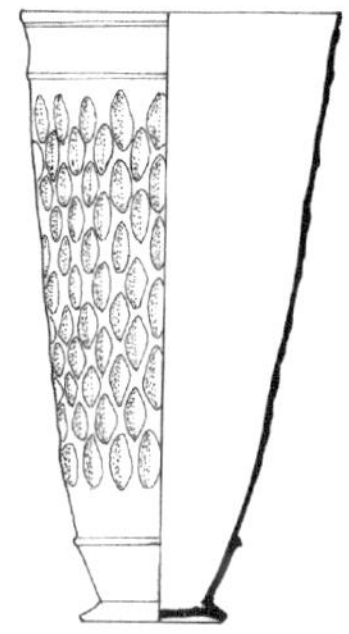

(Cat. no. 96)
(Plates 4, IV)

The fragment belongs to the upper part of the body of a conical vessel. Almost all vessels bearing multifaceted decoration belong to two subgroups of conical beakers: slender or squat. Due to the inclination of the rim, it is not easy to ascribe it to either of the two groups of conical beakers, because neither of them have a flaring rim.[470] It could also belong to a kantharos perhaps, like the example in the Saint Louis Art Museum, and more specifically to the genesis of the handle.[471] Their height is estimated at c. 15cm.

These vessels were made by free-blowing and then they were cold-polished inside and out, and the base was shaped. Then, parts of the surface layer were wheel-cut and the characteristic decorative motif of the adjoining ovular cuts/incisions was made.

Examples of vessels decorated with this technique have been found throughout the Roman Empire and even beyond its borders, to the North and to the East. They are considered Oriental, probably Egyptian, products and they are dated, according to archaeological data, to the second half of the 1st century.[472] To the same period is ascribed the find from Thessaloniki, although it cannot be dated independently by the archaeological context in which it was found.

The fully decolourized fragment from Thessaloniki presents the only extant find of this group from the city and its wider region. Other published finds from Greece are reported in Thebes, Greece[473] and Corinth,[474] as well as in the Benaki Museum.[475]

---

456 Wight 1994, 47-48.
457 Wight 1994, 47.
458 Whitehouse 2001, 54-56, no. 527.
459 Weinberg, McClellan 1992, 130, no. 103, a find of unknown provenance.
460 Bailey 1992, 29-33, the grave was dated according to a clay lamp and other finds to the end of the 1st or early 2nd century.
461 Ložar 1935, 97-105, from a grave with a glass bowl (Form 8) and a coin of Domitian (81-96).
462 Berger 1960, 51-52, no. 126, from a layer dated between AD 60-85.
463 Barag 1991, 137, dated from excavation results between AD 74-115.
464 Wight 1994, 50.
465 Saldern *et al.* 1974, 166, 167 fig. 457, no. 457.
466 Wight 1994, 50.
467 Wight 1994, 50. A find of unknown provenance.
468 Wight 1994, 50. A find of unknown provenance.
469 Adam-Veleni *et al.* 1998, 86.
470 Oliver 1984, 35-5. The definitive publication in which more than 100 examples are listed, of all known shapes and decorative motifs.
471 Oliver 1984, 45, fig. 17.
472 Oliver 1984, 38-39. For more recent finds from Belgrade, see Bikić, Nikolić-Djordjević, Simić 2001, 197-198, fig. 12, finds from 2nd-century layers.
473 Haevernick 1981, 120, pl. 44. 1, fragments of four vessels dated on the basis of analogous finds from other sites to the 1st century.
474 Davidson 1952, 101, no. 641, fig. 8, pl. 54, finds dated from excavation results to the 2nd century.
475 Clairmont 1977, 9, no. 28, pl. 1.

### 24. Conical beakers with curved walls, pointed

(Cat. nos. 97-98)
(Plates 4, IV)

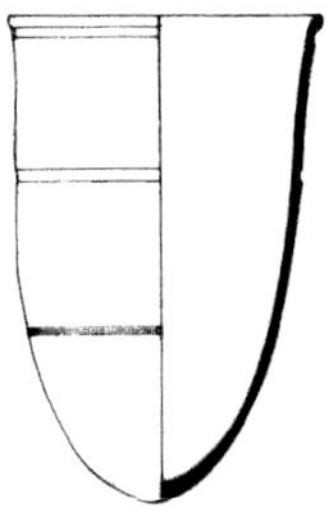

### 25. Conical beakers with curved walls, flat-bottomed (Isings Form 29)

(Cat. nos. 99-100)
(Plates 4, IV)

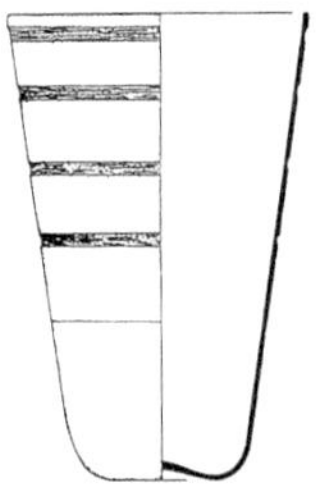

Vessels of this form have fire-polished rim, cylindrical or everted truncated conical body, usually with flat bottom, or very seldom with a pointed bottom. They are made of thick glass decorated with three deep, incised horizontal grooves, one under the rim and two more in equal distances, dividing the body into three equal parts. Their height is estimated between 10-11cm.

The vessels of this form are free-blown and decorated by wheel-cutting.

They present a well-known form,[476] dated to the 1st century. They appear in the Augustan period (27 BC - AD 14), but the majority of the extant examples are dated to the reigns of Tiberius or Claudius (14-54), and they continue to be produced into the third quarter of the 1st century.[477] It seems that this form imitates fine egg-shell pottery vessels which appear already from the second quarter of the 1st century BC.[478]

Dated examples have been published from excavations in sites at Italy,[479] where it is believed that they were produced,[480] Britain,[481] Switzerland,[482] Austria,[483] Slovenia,[484] Samothrace,[485] and Vize (Vizye).[486] In addition, they have been found in Cyprus, Amathous[487] and Limassol,[488] where it seems that they were kept as heirlooms for quite some time after their production had ceased.

The examples with flat bottom from Thessaloniki are dated to the last twenty years of the 1st century, according to the destruction time of the baths where they were found.489 The greenish, pointed example (Cat. no. 97) lacks any information about the whereabouts of its find-spot and is dated to the 1st century according to analogous finds from other sites.

### 26. Conical beakers with curved walls, pushed-in ring base (Isings Form 34)

(Cat. nos. 101-104)
(Plate IV)

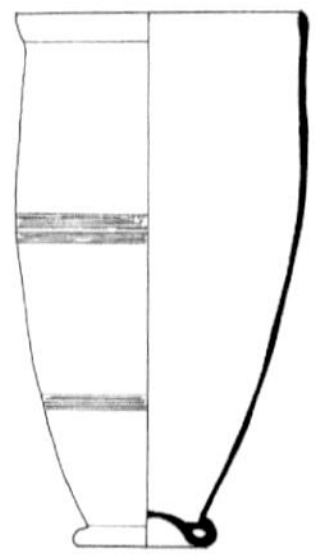

Vessels of this form have slightly everted fire-polished rim, cylindrical body and pushed-in tubular ring base, which is often quite massive. They are made of thick glass decorated usually with three, incised horizontal grooves: one under the rim and two more in equal distances on the body, which divide it into three equal parts. Their height is estimated at c. 10-12cm.

They are free-blown vessels. The base is shaped by folding and further tooling of the lower part of the body mass. The decoration was wheel-cut. This form is also known in silver examples, predating AD 79.[490]

[476] Isings 1957, 44, form 29.
[477] Cool, Price 1995, 68; Stern 2001, 46, 79-80, with thorough bibliography.
[478] Robinson 1959, 11-13, no. F18, F19, pl. 1. Group F is dated between 75 BC and the end of the 1st century BC.
[479] Isings 1957, 44.
[480] Stern 2001, 46.
[481] Colchester (Camulodunum): Cool, Price 1995, 68-69, with thorough bibliography on finds from the NW provinces of the Roman Empire and selectively on the oriental provinces.
[482] Locarno (Solduno): Biaggio-Simona 1991, vol. 1, 100, pl. 10, from graves dated to the first half of the 2nd century.
[483] Magdalensberg (Virinum): Czurda-Ruth 1979, 99, nos. 753-761, fragments dated prior to AD 45.
[484] Ljubljana (Emona): Petru 1972, 163, pl. 70, from a grave dated by coins of Augustus and Nero (27 BC - AD 14; AD 54 - 68).
[485] Dusenbery 1967, 46, no. 39; Dusenbery 1998, 1086, 1088-1089, no. S 219-19, from a grave dated to c. AD 25.
[486] Mansel 1940, 133, fig. 49b, dated from excavation results, probably, to the first half of the 1st century.
[487] Vessberg 1952, 120, find dated to the Antonine period (138-192).
[488] Vessberg 1953, 166, pl. 58, find dated to c. AD 220.
[489] Adam-Veleni *et al.* 1998, 85-102; Mavromichali 2001, 141. Two more examples, one with incised grooves and another with marvered threads have been unearthed in a 1st-century grave in the western necropolis of Thessaloniki, under publication by the author, see Antonaras, Lamprothanasi, in press (1).
[490] Whitehouse 1997, 227, comments on catalogue entry no. 384, finds from Pompeii.

It is a well-known form[491] dated to the 1st century. It appears already in the first quarter of the century, continues to be produced until the third quarter, and is considered to be a product of north-Italian workshops.[492] Dated examples have been reported, among others, from many excavations, including Colchester (Camulodunum),[493] Moers-Asberg (Asciburgium),[494] Windisch (Vindonissa),[495] Locarno (Solduno),[496] Magdalensberg (Virinum),[497] Aquileia,[498] Moghegno,[499] Solin (Salona),[500] Kerch (Panticapaeum),[501] Corinth,[502] Samothrace,[503] Mytilene,[504] and Knossos.[505]

The example from Thessaloniki is dated to the last twenty years of the 1st century, according to the destruction date of the baths in which it was found.[506]

### 27. Conical beakers, straight walls with indentations, plain base, Isings Form 32

(Cat. no. 105)
(Plates 4, IV)

Vessels of this form have slightly everted, unworked rim, conical body and stand on a flat, slightly concave bottom. The vessel from Thessaloniki is made of colourless slightly greenish glass. Their height is estimated at c. 10cm.

These vessels are free-blown. They were marvered to obtain the conical body shape. Gentle pressure with a pointed tool around the body formed the vertical oval indentations, creating the impression of a folded vessel. The rim, after the cutting-off of the moil, was left unworked or was lightly abraded and polished and then was decorated with incisions.

Vessels of this form appear already from the second half of the 1st century and remain in use probably until the 3rd century,[507] although it appears that after the 2nd century their rims are often fire-polished.[508]

Dated analogous finds are published from Cologne,[509] Trier (Treviri),[510] Düsseldorf,[511] Windisch (Vindonissa),[512] Ticino, Switzerland,[513] Slovenia,[514] Tomis,[515] Cyprus,[516] Amathous,[517] Tipasa,[518] Tripolis,[519] Kom Aushim (Karanis),[520]

[491] Isings 1957, 48-49, form 34.
[492] Stern 2001, 81.
[493] Cool, Price 1995, 68-69, nos. 348-351, finds dated to the 1st century. With full bibliography on finds from the NW provinces of the Roman Empire and selectively on the Eastern provinces.
[494] Van Lith 1987, 48-50, nos. 180-186, dated to the reign of Tiberius (14-37) and Nero (54-68).
[495] Berger 1960, 46, nos. 105-106, dated a little before and during the Flavian period (69-96).
[496] Biaggio-Simona 1991, vol. 1, 115-116, dated to the end of the 1st or to the early 2nd century.
[497] Czurda-Ruth 1979, 102, nos. 762-764, dated to the reign of Tiberius (14-37) and Claudius (41-54).
[498] Calvi 1968, 56, no. 135.
[499] Biaggi-Simona 2002, 297-299, 14 examples, dated from excavation results between the late 1st century and the early 2nd century.
[500] Auth 1975, 161, no. 74 pl. 30/74, dated according to analogous finds, c. the Flavian period (69-96).
[501] Kunina 1997, 313 nos. 295-296, 299, dated from excavation results to the third quarter of the 1st century, and to the late 1st to the early 2nd century, respectively.
[502] Davidson 1952, 101, no. 637, dated from excavation results to the 1st century.
[503] Dusenbery 1967, 46, no. 41 redated by the same researcher (1998, 1084-1085, no. S212-B), with a new date for the grave, c. first quarter of the 1st century.
[504] Price 2002, 125, fig. 3, finds that appear also at Benghazi and Knossos.
[505] Price 1992, 430, nos. 203-204, pl. 345, fragments from layers dated to the first half or to the middle of the 1st century.
[506] Adam-Veleni *et al.* 1998, 85-102; Mavromichali 2001, 141.
[507] Isings 1957, 46-47, form 32; Harden 1936, nos. 376-82, 391-401.
[508] Cool, Price 1995, 88; Lazar 2003, 101.
[509] Fremersdorf, Polónyi-Fremersdorf 1984, 21-22, nos. 53, 57, finds dated to the first half of the 3rd century.
[510] Goethert-Polaschek 1977, 74, form 53c, nos. 311-313, pl. 43, dated from excavation results to the second half of the 3rd century.
[511] Saldern 1974, 106, no. 120, a vessel with unknown provenance, considered to be a 2nd-century Syrian product.
[512] Berger 1960, 46, nos. 107-109, pl. 7, dated to the last third of the 1st century. The bases of these vessels are not preserved, and it remains uncertain if they had a ring base or other means of standing.
[513] Biaggio-Simona 1991, vol. 2, no. 176. 2. 222, dated from excavation results to the second half of the 1st century.
[514] Lazar 2003, 98-102, form 3. 5. 1-6, fig. 32, finds from several sites in Slovenia, dated from excavation results to the second half of the 1st and the 2nd century.
[515] Bucovală 1968, 46, no. 43, from a 2nd-3rd-century context.
[516] Vessberg 1956, pl. 44, 25 examples in the Museum of Nicosia.
[517] Vessberg 1953, 164 fig. 1. β, dated by a coin of Antoninus Pius (138-161) to the second half of the 2nd century. The vessel is slightly different, with a cylindrical and not conical body.
[518] Lancel 1967, 18, 83-84, nos. 163-164, form 20, pl. VIII.1, 8, dated according to numerous shorter but still analogous finds from the same cemeteries, to the 1st-3rd centuries.
[519] Price 1985, 73, fig. 6:2, nos. 14-6. The archaeological whereabouts of the vessels are not known. It is noted that similar vessels were produced in the late 1st and 2nd centuries and later in the Eastern provinces.
[520] Harden 1936, 145-148, form A.II.1, no. 376-82, form A.III.1, nos. 391-401, pl. IV, XV, finds, which according to the internal dating of the site, were destroyed and discarded after the 3rd century.

Samothrace,[521] Palatiano (Ioron),[522] Thebes, Greece,[523] Athens,[524] Oropos,[525] and Corinth.[526]

The find from Thessaloniki cannot be dated independently.

### 28. Conical Beakers with straight walls, indentations, and pushed-in ring base (Isings Form 35)

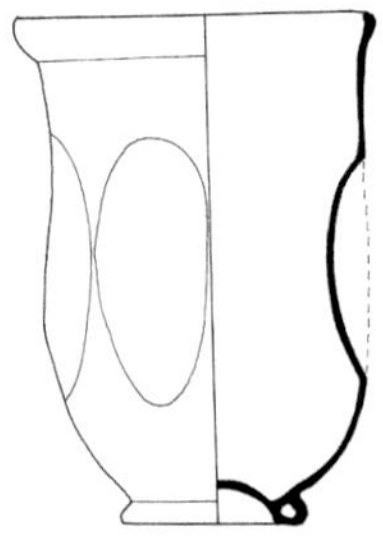

(Cat. nos. 106-109)
(Plates 4, V)

Vessels of this form have slightly everted, unworked rim, conical body, pushed-in tubular ring base and slightly conical bottom. The vessel from Thessaloniki is made of colourless slightly greenish glass. Their height is estimated between 10-15cm.

These vessels are free-blown. They were marvered to achieve the conical body shape, and the pushed-in base was formed by folding and tooling the lower body mass. Gentle pressure with a pointed tool created the slightly conical bottom and further pressure around the body formed the indentations that create the impression of a folded vessel. The rim, after the cutting-off of the moil, was left unworked or was lightly abraded and polished.

This form of beaker appears in the second half of the 1st century and continued to be in use into the 2nd century.[527] Published examples are known from Colchester (Camulodunum),[528] Cologne,[529] Trier (Treviri),[530] Windisch (Vindonissa),[531] Logatec (Longaticum) in Slovenia,[532] Tripolis,[533] Nymphaion in Crimea,[534] Tomis,[535] Ercolano (Herculanaeum),[536] Cyprus,[537] Thera,[538] Antikyra,[539] Thebes, Greece,[540] Oropos,[541] Athens,[542] and Corinth.[543]

The find from Thessaloniki cannot be dated independently and on the basis of analogous finds is ascribed to the second half of the 1st or 2nd century.

### 29. Beakers, cylindrical

(Cat. nos. 110-111)
(Plates 4, V)

The vessels of this form have unworked, cracked-off rim, vertical walls which form a cylindrical body and a flat bottom. Their height is estimated at c. 8cm.

[521] Dusenbery 1998, 1089, nos. S220-7, a similar find from a grave dated to c. AD 25.
[522] Antonaras, Anagnostopoulou-Chatzipolychroni 2002, 116, finds from ancient Ioron, dated on the basis of analogous finds to the 1st-2nd century.
[523] Haevernick 1981, 123, 126-128, pl. 40. 8, fragments of vessels dated to the 1st to 3rd centuries. The archaeological whereabouts of the vessels are not known.
[524] Whitehouse 1997, 111, no. 169, dated on the basis of analogous finds to the middle or end of the 1st century.
[525] Pologiorgi 1998, 44, no. 2, vessel fragments from a 1st-century burial, dated according to the glass finds of the grave.
[526] Davidson 1952, 101-103, fig. 9, no. 644, from a grave dated to the 1st or 2nd century.
[527] Isings 1957, 49-50, form 35, where it is noted that it is found more often in the Mediterranean provinces and less in the NW provinces.
[528] Cool, Price 1995, 69-70, finds dated from excavation results to the second half of the 1st and into the beginning of the 2nd century. The full bibliography of dated finds from Britain and other NW provinces which are also datable within this time-frame.
[529] Fremersdorf, Polónyi-Fremersdorf 1984, 24-25, nos. 62, 63, finds dated to the first half of the 3rd century.
[530] Goethert-Polaschek 1977, 81-82, form 58c, nos. 359-361, fig. 29, pl. 45, dated to the 4th century.
[531] Berger 1960, 46, nos. 107-109, pl. 7, dated to the last third of the 1st century. The bases of the vessels are not preserved and it remains uncertain if they stood on the bottom or on a tubular ring base.
[532] Lazar 2004, 60-61, nos. 12, 13, from a site with finds dated to the second half of the 1st, up to the 3rd century. The vessels *per se* were dated, on the basis of analogous finds from Slovenia, which are presented in detail in the article, to late 1st or 2nd century.
[533] Price 1985, 73, fig. 6:2 no. 17, according to analogous finds is considered to be a product of the 1st century.
[534] Kunina 1997, 315, no. 303, which is considered to be a Cypriot product, and another from a private collection in Russia (Kunina 1997, 315, no. 304), which is also considered to be from Cyprus, and are dated according to analogous finds to the 1st century.
[535] Bucovală 1968, 50, no. 52, from a 1st or 2nd-century context. Bucovală 1984, 62, fig. 7, two more finds dated with greater precision to the second half of the 1st century.
[536] Scatozza-Höricht, 40-41 no. 69, pl. IX, XXVIII. 1249, a form that appears also in Pompeii among the finds of AD 79.
[537] Vessberg 1952, 123, pl. IV. 13-16, with no archaeological or chronological indication.
[538] Dragendorff 1903, 286-288, fig. 481t, u, from graves dated to the 2nd century.
[539] Baziopoulou-Valavani 1980, 262-263, pl. 114, vessel from a grave dated to the second half of the 1st or 2nd century, and a second example from a grave dated by a coin of Domitian (81-96).
[540] Haevernick 1981, 123, 126-128, pl. 40. 7 16, fragments dated between the 1st and 3rd centuries, with no archaeological or chronological indication.
[541] Pologiorgi 1998, 42, no. 7, fragments of a vessel from a burial dated to the second half of the 1st century, dated according to the glass vessels tit contained.
[542] Weinberg and Stern 2009, 47 ff, nos. 94-96, fig 6, from contexts ranging from 1st to mid-2nd century.
[543] Davidson 1952, 101-103, fig. 9, nos. 645-648, dated from excavation results to the 1st and 2nd century. Warner Slane 1986, 302, 308-309, no. 148, from a layer with the majority of its finds dated to the second and third quarter of the 1st century, although it contained a few clay sherds dated to the early 2nd century.

It is made by free-blowing, marvering to get its cylindrical shape, and pressure by a blunt tool to flatten the bottom. The incised decoration was done after the annealing, and possibly even in a different workshop.

The two examples from Thessaloniki are made of two totally different qualities of glass, probably indicating different production centres as well. One is made of greenish glass, of not particularly good quality, and the second is made of good quality, totally decolourized glass.

There is no excavation data for the finds to provide a firm dating independently. Vessels of similar shape are known from Britain,[544] and the decoration of the vessels, consisting of shallow grooves carelessly executed, is widely used on beakers in the same regions, mainly in the 2nd century.[545] Several similar finds from Naxos[546] and Cyprus are also ascribed to the same period. These were decorated with vertical indentations and have been regarded as local products.[547] Finally, the closest, probably, finds to this form were excavated in the Black Sea region, at Tanais,[548] Kerch (Panticapaeum),[549] and Tomis,[550] and are dated to the second half of the 2nd and to the early 3rd century.

Therefore, although it is not clear if two or even more production centres of this form were operating, e.g. in the West and in the Black Sea region, it is quite clear that they can be dated to the middle of the 2nd or early 3rd century.

### 30. Beakers, ovular

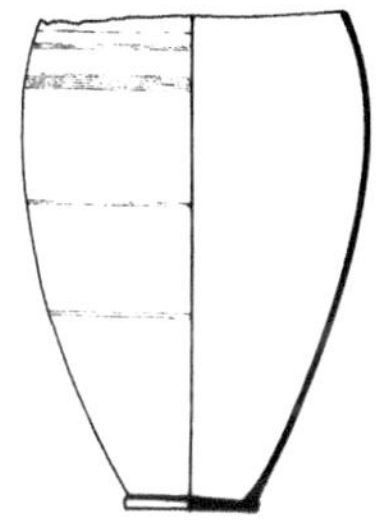

(Cat. no. 112)
(Plates 4, V)

Vessels of this form have unworked rim, ovular body, pushed-in tubular ring base and slightly concave bottom. Their height is estimated at c. 8cm.

The vessels were free-blown and their base was shaped with further tooling with jacks. The moil was left on the vessel during the annealing and only after the incision of the decoration was the rim carelessly cracked off at the desired height, destroying some of the incisions on the rim in a way quite typical for the Late Roman period.

The example from Thessaloniki is made of colourless greenish glass.

Only one exact parallel to the vessel has been traced, in Serbia, at Kolovrat.[551] Furthermore, this form of vessel is directly related with, if not identical to, vessels known also in Trier (Treviri).[552] Generally however the closest related groups are the conical and cylindrical beakers,[553] which, just like the aforementioned parallels, ascribe our find to a production date of between the late 3rd and the early 5th century. The other glass vessels found with it may also be assumed to have the same date.[554]

### 31. Conical beakers with straight, smooth walls, plain base (Isings Form 106a, b, c)

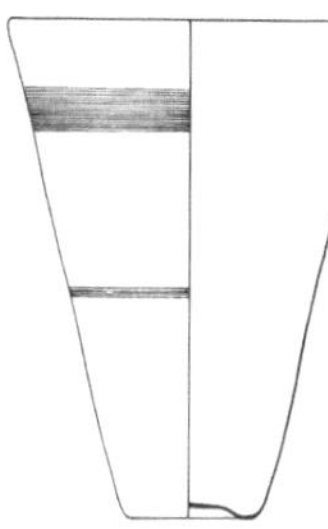

(Cat. nos. 113-119)
(Plates 4-5, V)

The vessels from Thessaloniki have unworked rim, inverted truncated conical body, straight body walls and narrow, flat bottom.[555] The majority have flat bottoms,[556] and only one has a

[544] Price, Cottam 1998a, 94-95, fig. 34. This form is very widespread, mainly in excavations of houses, and is dated to the second and third quarter of the 2nd century.
[545] Cool, Price 1995, 79-82. This decorative technique, although known in cylindrical beakers already from the end of the 1st century, becomes very popular mainly in the second and third quarter of the 2nd century; during the 3rd century it goes out of fashion.
[546] Zafeiropoulou 1999, pl. 88ε, dated according to analogous finds.
[547] Vessberg 1956, 142, form A.IV.α.2, fig. 44/30-31. See also, Hayes 1975, 64, nos. 188-190, vessels dated on the basis of analogous finds to the second half of the 2nd- early 3rd century.
[548] Sorokina 1965, 206, pl. 1/2-8, finds dated from excavation results to the second half of the 2nd- early 3rd century; some are dated more precisely by a firmly dated layer of destruction at c. AD 240.
[549] Sorokina 1965, 205, with thorough bibliography in footnote 9. They are findings from graves dated by coins and their content in general to the end of the 2nd - early 3rd century; Kunina 1997, 319, no. 327, find dated from excavation results and analogous finds from other sites to the second half of the 2nd- first half of the 3rd century.
[550] Bucovală 1968, 47, no. 46, similar vessel, with slightly out-splayed rim, find dated from excavation results to the 2nd-3rd century.
[551] Ružić 1994, 50, no. 967, dated from excavation results to the 4th century.
[552] Goethert-Polaschek 1977, 80-81, form 58b, nos. 352-357, pl. 45, fig. 28, dated from excavation results to the first half of the 4th century; idem, 87-88, form 64, nos. 384-386, fig. 33, pl. 46, it seems that it is a shorter and wider variant of the same form and is also dated to the middle of the 4th century.
[553] Form 32.
[554] Two Janiform unguentaria (Form 146), dated at least to the 3rd-4th century.
[555] Isings 1957, 129, forms 106a, b, c. These vessels were divided into four main groups depending on the shape of their base and on whether they were decorated or not.
[556] Form 106a of Isings 1957, 126-129. Cat. nos. 114-118 are classified as this form.

slightly concave bottom.[557] Their height is estimated at c. 14cm.

The finds from Thessaloniki are made of colourless greenish glass.

These vessels were made by free-blowing and further marvering to acquire their conical shape. In a few cases they were blown in a conical mould. The fact that none of the examples bears a pontil mark, in conjunction with their unworked rim, shows that the moil was cracked off later, after the annealing of the vessel. The cracking at the desired height was carried out by a sudden thermal shock, and after the incision decoration was done.[558]

Their standard decoration features bands of horizontal incisions, or, less often, wheel-cut motifs and representations. Quite often they are decorated around the body with blobs of strongly-coloured glass in one or more colours, applied isolated or in groups.

Vessels of this form are dated to the 4th and into the early 5th century. Examples of this form have been found in various sites in Britain;[559] France:[560] Strasbourg (Augusta Trebecorum, Argentoratum);[561] Germany: Trier (Treviri);[562] Switzerland: Augst;[563] Hungary:[564] Dunaújváros (Intercisa),[565] Szőny (Brigetio),[566] Budapest (Aquincum),[567] Ságvár (Tricciana);[568] Romania: Alba Iulia (Apulum), Sîntana de Mureş, Bologa (Resculum) and Gornea;[569] Slovenia: Ravno Brdo;[570] Bosnia and Herzegovina: Stoca;[571] Croatia: Sisak (Siscium), Virobica, Senj (Senia), Solin (Salona), Popov Dol, Radovanci,[572] Sremska Mitrovica (Sirmium),[573] Sviloš; Serbia: Komini (Municipium 'S...'),[574] Kruševlje,[575] Gradac,[576] Čezava (Novae),[577] Karataš (Diana),[578] Ravna (Timacum Minus),[579] Belgrade (Singidunum),[580] Bela Palanka (Remesiana),[581] Kostolac (Viminacium);[582] FYROM: Isar-Marvinci;[583] Bulgaria: Krivina (Iatrus),[584] Sozopol (Apollonia Pontica),[585] Varna (Odessos), Balchik (Dionysopolis) and Nesebar (Mesembria).[586] In addition, many vessels of this form, decorated with blue blobs, have been found in the region of the Black Sea, southern Russia,[587] but their

[557] Form 106b of Isings 1957, 126-129. Cat. no. 113 is classified as this form.
[558] Stern 2001, 293 no. 157. For a different shaping of rims, see Stern 2001, 294, no 158.
[559] Cool, Price 1995, 88-92. They are all dated to the 4th century onwards, and along with the hemispherical bowls, they present the most numerous groups of 4th-century drinking vessels in the NW provinces of the Roman Empire.
[560] Morin-Jean 1922-23, 140-141 fig. 262, for conical vessel with flat base (Form 106); Morin-Jean 1922-23, 142, fig. 189 for conical vessel with pointed base (Form 107).
[561] Arveiller-Dulong, Arveiller 1985, 145-152, nos. 321, 324, 327, 331, 333, 335, 338. These vessels are dated to the 4th and early 5th century. There is a full bibliography on finds from NW provinces.
[562] Goethert-Polaschek 1977, pl. 43/314.
[563] Rütti 1991, vol. 2, 66-67, 75, nos. 1434-50, form AR 68. Finds dated from excavation results to the end of 4th - early 5th century.
[564] Barkóczi 1988, 82-95 nos. 98-110, pls. IX-X, LXXIII, finds from various sites, dated to the second half of the 4th or into the early 5th century, most of them dated from excavation results. Especially nos. 105-108, which are decorated with incised grooves, are assumed to be products of north-Italian workshops.
[565] Radnóti 1957, 153, pl. XXXIII/12.
[566] Barkóczi 1968, 78, no. 47, pl. 38/1. From a 4th- or early 5th-century cemetery.
[567] Kaba 1958, 447, pl. II/4. Benkő 1962, T. XXXIV/6.
[568] Burger 1966, 137, form 7. Many examples from graves dated to the middle of the 4th century. It was noticed that they were used both as drinking vessels and as lamps. Also, in note 433, are mentioned many other similar finds from other sites excavated by the same researcher.
[569] Baluta 1978, 105, fig. 5/5, it seems that the general dating to the 4th century was concluded according to analogous finds from other sites, as no reference is given to the excavation data concerning the vessel.
[570] Petru 1976, 17, pl. VIII/6. Dated to the second half of the 4th century according to the other objects found in the same grave.
[571] Paškvalin 1976, 135, pl. VIII/2. Dated to the 4th century according to analogous finds from other sites, and also based on characteristic structural elements of the grave in which it was found.
[572] Damevski 1976, 66, pl. XV/4. The method of dating is not mentioned, but is probably based on analogous finds.
[573] Cermanović-Kuzmanović 1981a, 11-12, form III/2A, 18, pl. III, photo 10. From a 3rd-4th-century horizon of graves. Another example is noted in Šaranović-Svetek 1986, 61, no. 48, pl. IV, 2, which was found with 4th-century coins.
[574] Cermanović-Kuzmanović 1976, 180, pl. V/3.
[575] Šaranović-Svetek 1986, 61, no. 52. From a sealed grave with a coin of Constantius II (337-361). The vessel bears around its lip the inscribed phrase 'ACCIPE CALICE PIE ZES'.
[576] Ružić 1994, 51, nos. 975-976. Dated, probably according to excavation results, to the second half of the 4th- early 5th century.
[577] Ružić 1994, 51, no. 977, pl. XXXVIII/7. Dated, probably, according to excavation results, to the 4th century.
[578] Ružić 1994, 51, no. 978, pl. XXXVIII/8. Dated, probably, according to excavation results, to the second half of the 4th century.
[579] Ružić 1994, 51, nos. 1002-1007, pl. XXXVIII/11 dated, probably, according to excavation results, to the second half of the 4th- early 5th century; Petković, Ružić 2005, 44, no. 56, pl. XII, G. 134, 2, dated from excavation results between AD 350 and 380.
[580] Nikolić-Đorđević 1990, 48, pl. III/31, IV/32-33, from a stratum of the first half of the 4th century.
[581] Pejić 1982, fig. 1.
[582] Zotović, Jordović 1990, 61, pl. XVI: 6, dated by a coin of Trajan (98-117). The same vessel is also presented in Ružić 1994, 50, note 106, which, due to its remarkably early dating, is considered as an archetype of 4th-century production.
[583] An example decorated with blue dots: Sokolovska 1986, 127, pl. 48/4, from a grave dated by coins in the reign of Constantius (305-306). Undecorated example: Sokolovska 1986, 114, pl. 45/15, from a grave dated by coins in the middle of the 4th century.
[584] Gomolka 1979, 149, 157-158, nos. 80-81, 84, 91, pl. 58, 60.
[585] Minchev 1988, 49, form XXIV, no. 41, pl. VIII, 41. It is a completely conical vessel Isings Form 106c, dated to the 4th century according to analogous finds.
[586] Minchev 1992, 75, form II. 4, drawing 6.
[587] Sorokina 1970, 70-79; Sorokina 1971, pl. 3/2, pl. 4/3.

production site has not been located. Many conical vessels have been found in Kom Aushim (Karanis),[588] and from the same site must originate many more examples that are now housed in Western collections, provenanced from Fayum (Arsinoe). The closest geographical parallel is the hitherto single published item from a Greek site, a find from the Pieria region.[589]

One of the ascertained production sites of the conical vessels is Jalame in Israel.[590] The workshop, according to the numismatic evidence, was operating between AD 351 and 383.[591] According to the excavator of the site, Gladys D. Weinberg, the Western products are slightly different from Eastern ones. The Syro-Palestinian examples are colourless greenish, almost completely colourless, and usually they are thick-walled. The Egyptian finds are olive-green, with thinner walls and decorated often with cobalt blue blobs, just like Syro-Palestinian products. Because of the large number of examples found in Western Europe, it is probable that they were produced at several sites, at least one of them situated in that part of the Empire. Some of the Western artefacts are decorated, in addition to the blue blobs, with blobs of other colours as well. This polychromy has not been noticed in Eastern examples and it has been assumed that it presents a peculiarity of western European production.[592] Vessels of the same group with thick greenish walls, were probably also produced in Pannonia.[593]

Vessels of this form have been identified as lamps as well as drinking beakers. Most probably they fulfilled both uses. Oily remains have been preserved on the walls of vessels from Kom Aushim (Karanis) and inside one from northern France.[594] They were probably used standing in wooden tripods, similar to those found at Karanis,[595] regardless of whether they were used as lamps or beakers. Their use as lamps[596] is illustrated in a 4th-century mosaic from the synagogue at Hamat in Tiberias, where conical glass lamps are represented lit on the branches of a menorah.[597] Their use as beakers is proved, *inter alia*, by an incised inscription 'ACCIPE CALICE PIE ZES' on the rim of an example from Vojvodina,[598] as well as the wall-paintings on two 4th-century tombs from Thessaloniki[599] and Osenovo in Varva[600] respectively. A different, and quite probably temporary, use of these vessels for gambling purposes is demonstrated by the find of four dice in one conical vessel from Kom Aushim (Karanis).[601]

None of the finds from Thessaloniki can be firmly dated by their context and their ascription to the 4th and early 5th century is based on analogous finds from other sites.

### 32. Conical beakers with straight smooth walls, ring base (Isings Form 109c)

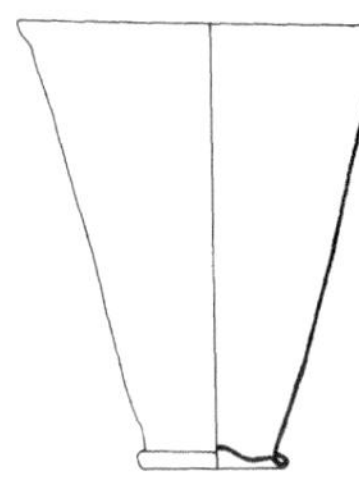

a. Slender:
(Cat. nos. 120-121)
(Plates 5, V)

b. Wide-mouthed:
(Cat. no. 122)
(Plates 5, V)

Vessels of this form[602] have unworked, slightly everted rim, cylindrical or conical body, pushed-in ring base and slightly concave bottom. Their height is estimated between 9-11cm.

They were made by free-blowing, marvering, and further tooling with jacks to form the base.

The finds from Thessaloniki are made of colourless greenish glass.

Examples of this form can be divided into two groups:

[588] Harden 1936, 155-166, nos. 457-463, pl. 5 and 16.
[589] Find from the *Source of Athena* site in Platamonas, dated on the basis of analogous finds to the Late Imperial period. Adam-Veleni, Poulaki, Tzanavari 2003, 212, no. 195 (P. Adam Veleni).
[590] Weinberg 1988, 87-94, figs. 4-46, 4-47.
[591] Merker 1988, 103.
[592] Fremersdorf 1962, 9, where the polychromy of the glass blobs expresses the aesthetics of the German tribes and appears c. AD 260. Israeli 2003b, 193.
[593] Barkóczi 1988, 84-85, no. 109f, pl. 10.
[594] Weinberg 1988, 89; Harden 1936, 155, note 2.
[595] Dauterman-Maguire, Maguire, Duncan-Flowers 1989, 62.
[596] Crowfoot, Harden 1931, 197-200.
[597] Gorin-Rosen 1998, 24.
[598] Šaranović-Svetek 1986, 61, no. 52, pl. IV, 1.
[599] Colour illustration in Papanikola-Bakirtzi 2002 (ed.), 532-533.
[600] Minchev, Georgiev 1981, 11-12, pl. VIII.
[601] Harden 1936, 155.
[602] Isings 1957, 137-138, form 109c.

a. The more slender forms. These finds, judging by their shape, appear to have evolved from 1st- and 2nd-century beakers.[603] Well-dated, analogous examples were reported on sites in Germany:[604] Bonn (Bonna),[605] Zülpich-Enzen,[606] Mainz (Moguntiacum),[607] Trier (Treviri),[608] Cologne;[609] Hungary: Szőny (Brigetio);[610] Croatia: Sremska Mitrovica (Sirmium);[611] Serbia: Belgrade (Singidunum),[612] Novi Sad,[613] Kolovrat;[614] and Cyprus,[615] which are as a whole dated between the second half of the 3rd and the second half of the 4th century. The finds from Thessaloniki cannot be dated according to their archaeological context, only by analogous finds dated to the same period.

b. Wide-mouthed forms. Analogous, well-dated finds come from sites in Germany: Mainz (Moguntiacum);[616] Belgium: Samson;[617] Hungary: Kisàrpàs,[618] Ságvár (Tricciana),[619] Szombathely (Savaria);[620] Bulgaria: Svishtov (Novae);[621] Serbia: Kolovrat;[622] and Cyprus.[623] As a rule these are dated between the 4th and the early 5th century. The find from Thessaloniki is dated by its archaeological context to the first half of the 4th century.[624]

### 33. Cylindrical beakers with small knop

(Cat. nos. 123-129)
(Plates 5, V)

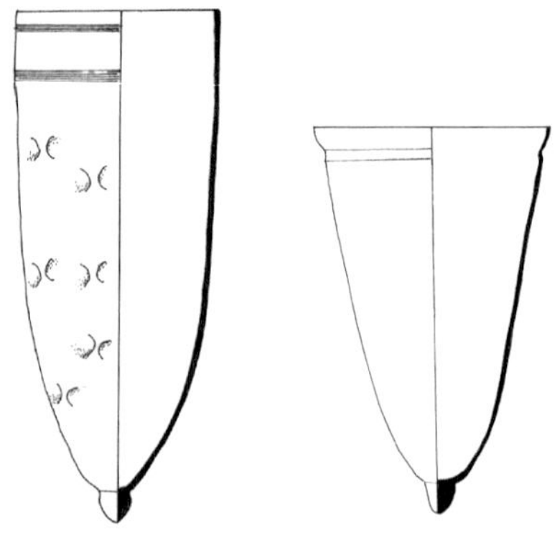

Fragments of ovular bases with a small central knop belong to another form of, probably wide-mouthed, cylindrical or slightly conical vessels, used also as lamps. Their height is estimated at c. 10cm.

Examples of this form are occasionally decorated with engraving, or an applied fine thread of glass wound around the body, and also pinching.

This form appears from the early 5th century and survives throughout the 6th century.[625] Examples have been unearthed in Marseilles (Massalia) and Saint-Blaise,[626] Rome,[627] Cornus Sardinia,[628] Fiesole (Faesulae),[629] Ravenna,[630] Invillino.[631] There are also finds from northern Lombardy, Monte Barro[632] and S. Giulia,[633] as well as Dalmatia, Split

603 Isings 1957, forms 26 and 34.
604 Harter 1999, 77, with a full bibliography of the Germanic world.
605 Follmann-Schulz 1988 82-83, nos. 281-284, pl. 35, finds dated to the second half of the 3rd century.
606 Follmann-Schulz 1989a, 49, 51, no. 4, fig. 1/4, find from a sarcophagus with a burial dated to c. AD 360.
607 Harter 1999, 77, form B22b, dated to the second half of the 3rd and into the second quarter of the 4th century.
608 Goethert-Polaschek 1977, 76-80, nos. 323-358, pls. 16/175α, 16/176b, 21/230d, 23/249a, 25/295a, 44-45, form 58a, dated from excavation results between the early- and the second half of the 4th century.
609 Fremersdorf 1959, 74, pls. 116-117, two examples decorated with an applied web of glass, which are dated from excavation results to the second half of the 3rd century.
610 Barkóczi 1988, 78, nos. 84-85, form 42, finds dated from excavation results to the end of the 3rd - early 4th century.
611 Šaranović-Svetek 1986, 60, no. 41, dated from excavation results to the 4th century. Cermanović-Kuzmanović 1981a, 18, nos. 53, 54, finds dated generally to the 4th century.
612 Ružić 1994, 50, no. 964, dated from excavation results to the second half of the 4th century.
613 Šaranović-Svetek 1986, 60, no. 44, dated on the basis of analogous finds to the 4th century.
614 Ružić 1994, 50, no. 965, dated from excavation results to the 4th century.
615 Vessberg 1952, 123, form B. II. β. 2, pl. IV/2-5, with no special chronological indications.
616 Harter 1999, 78, form B24, vessel with applied thread, dated to the second half of the 4th to early 5th century.
617 Harter 1999, 78, note 353, with prior bibliography, find dated to the end of the 4th to early 5th century.
618 Barkóczi 1988, 80-82, nos. 92, 96, pl. VIII, IX, form 47a, finds dated from excavation results to the end of the 4th to early 5th century.
619 Barkóczi 1988, 80-82, nos. 93-94, 97, pl. LXXIII, IX, form 47a, finds dated from excavation results to the second half of the 4th century.
620 Barkóczi 1988, 80-82, no. 95, pl. LXXIII, form 47a, finds dated from excavation results to the second half of the 4th century.
621 Billewicz 1971, 136-137.
622 Ružić 1994, 50, no. 965, dated from excavation results to the 4th century.
623 Vessberg 1952, 123, form B. II. β. 2, pl. IV/7, 9, with no chronological indications.
624 The burial is dated, according to a clay vessel found in it, and by which it can be accurately dated, to the first half of the 4th century. See Makropoulou 2007, 323, pl. 27, 69.
625 Uboldi form III.2, see Uboldi 1995, 119-120, fig. 4. 22-27, for a detailed discussion of the type.
626 Decorated with an applied thread, Foy 1995, form 205, 21c, pl. 14, finds dated from excavation results to the 5th-6th century. For plain examples see Foy 1995, 206, form 22c, pl. 14, finds dated from excavation results to the 6th-7th century.
627 Sternini 1989, 36, pl. 6. 41, from a layer dated to the late 4th- mid 5th century; Sternini 1993 figs. 20-22f and 24; Sternini 2001, 28, no. 120, fig. 12. 120.
628 Uboldi 1995, 119, note 126.
629 Uboldi 1995, 119, note 126.
630 Curina 1983, nos. 11, 14-15.
631 Uboldi 1995, 119, note 126.
632 Uboldi 1998, 208, fig. 2. 17, dated from excavation results to the end of the 5th to the first half of the 6th century.
633 Uboldi 1996, 186, nos. 6-8; Uboldi 1998, 209, fig. 3. 16-18, dated from excavation results, and similar finds, to the 6th-7th century.

(Spalatum)[634] and Gata,[635] and Serbia, Karataš (Diana),[636] Kostol (Pontes),[637] Belgrade (Singidunum),[638] Mihajlovac, the site of 'Mora Vagei',[639] Rtkovo,[640] Ćupria (Horeum Margi),[641] Gračanica (Ulpiana),[642] and Obrenovac.[643] Still further examples are known from Kom Aushim (Karanis) and Sidon,[644] Sardis,[645] Asprovalta,[646] and Thasos.[647] The sole intact example of this form was found in an Early Christian basilica at Turbe-Varošluk in Bosnia.[648]

The rim of this form, which is not preserved on any of the examples, based on the preserved find from Bosnia, must had been vertical and fire-polished. The cylindrical body in some cases is decorated with irregular rows of indentations formed by pointed pincers. This type of decoration appears on vessels from the late 2nd[649] and mainly between the 3rd-6th centuries,[650] especially in the Syro-Palestinian region.[651] Pinching survives as a technique until the 7th century in Palestine.[652] In Jordan it appears from the 4th and remains in use until the 8th century.[653] The presence of pinching suggests a possible Syro-Palestinian origin for these vessels; this hypothesis is corroborated by an intact example, bearing a typically Syrian incised decoration, now in the Metropolitan Museum, New York.[654]

Only Cat. nos. 127-129 can be independently dated. Although there are no special stratigraphic clues, they were found in a Christian church dated to the reign of Justinian (527-565), which was ruined in the early 7th century.[655] These dates present the *termini ante* and *post quem*, respectively, for the dating of the fragments.

### 34. Cylindrical beakers with massive knop

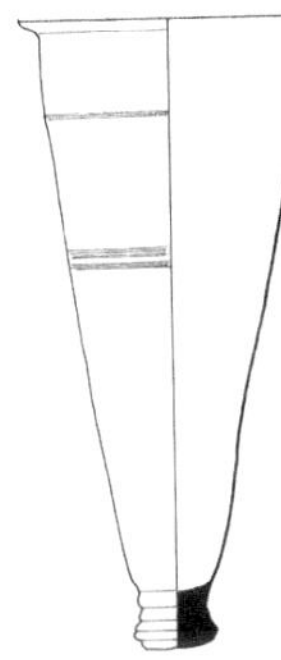

(Cat. nos. 130-132)
(Plates 5, V)

The conical and hemispherical lower body parts of vessels terminating in a massive squat knop most probably belong to lamp forms. Their height is estimated at c. 18cm. This form corresponds to Uboldi group III.1,[656] dated between the 5th and 7th centuries.[657]

Examples of this form have been found in southern France,[658] northern Lombardy,[659] Rome (Palatine

[634] From the palace of Diocletian. De Maine 1979, 64, 116, pl. 11, R 57, from layers dated generally between the 3rd and 6th centuries.
[635] Fadić 1994a, 220-221, photo page 221. The finds are dated to the 6th-7th century, the period when the basilica was functioning, but they are not recognized as lamps. On the contrary, they are considered as unguentaria, with unidentified body shape, and as their presence in a church is considered to be curious, it is assumed that they were used for the transportation and/or holding of holy oil or myrrh.
[636] Ružić 1994, 55, no. 1149, pl. XLIII/5, find dated from excavation results to the 4th-6th century.
[637] Ružić 1994, 55-56, nos. 1150-1169, 1170-1171, pl. XLIII/6, finds dated from excavation results to the end of the 5th - early 6th century.
[638] Pop-Lazić 2002, 36, 58-59, from a grave dated to the end of the 4th century. The find is dated between the 4th and 6th century according to analogous finds.
[639] Ružić 1994, 56, nos. 1172-1179, pl. XLIII/8-9, finds dated from excavation results to the end of the 5th-6th century.
[640] Ružić 1994, 56, no. 1180, pl. XLIII/7, find dated from excavation results to the 4th-6th century.
[641] Ružić 1994, 56, nos. 1181-1184, pl. XLIII/8-9, finds dated from excavation results to the 4th-6th century.
[642] Ružić 1994, 56, nos. 1185-1188, pl. XLIII/10, find dated from excavation results to the 4th-6th century.
[643] Ružić 1994, 56, no. 1189, pl. XLIII/11, find dated from excavation results to the 4th-6th century.
[644] Harden 1936, 158, nos. 466-67, pl. XVI.
[645] Saldern 1980a, 52-53, form 4, pl. 23, no. 298, find dated from excavation results between AD 400-616.
[646] *Glass Cosmos* 2010, 350, no. 377, dated to the 5th-6th century.
[647] Abadie-Reynal, Sodini 1992, 91-92, fig. 38, V6, this form appears in layers dated at c. AD 600.
[648] Paškvalin 1976, 112, 118, no. 38, pl. VIII/4. The researcher dated the find to the second half of the 4th century. It seems however more secure to ascribe it to production of the 5th century, especially after the recent redating of the excavation, see Marijanović 1990, 126-126, fig. 15.
[649] Čremošnik 1965, 197, finds dated from excavation results generally between the 2nd and 4th century; Rütti 1991, 13/2, 72-73, form 64. 2, nos. 1400-42, pl. 65.
[650] Vessberg 1952, 146, pl. IV. 22 (vessel type A.I.); Saldern *et al.* 1974, 243-46, no. 714; Bulat 1976, 89-90, pl. II/1, VII/2, find from a sarcophagus with a 4th-century burial; Doppelfeld 1959, 156-157, fig. 3. 6, find from Cologne, dated from excavation results to the early 4th century; Saldern 1974, 98-99, nos. 107-108; Auth 1976, 129, 138, nos. 160, 175; Oliver 1980, 112, no. 182; Barkóczi 1988, 153, no. 341, pls. XXXI, LXXXVII, find from Szőny (Brigetio) dated from excavation results to the middle of the 3rd century.
[651] Delougaz, Haines 1960, pl. 50. 9; Whitehouse 1997, 199-200, no. 349; Whitehouse 2001, 124, 148, nos. 628, 663; Stern 2001, 247-251, nos. 133-137, with thorough bibliography.
[652] Stern 2001, 354, no. 201.
[653] Baur 1938, 536, no. 67 (244), fig. 18; Dussart 1998, 128, 158, 161, respectively BX. 111a, pl. 32/1; BX 83, pl. 46/21; BXII. 1, pl. 49/1.
[654] Caron 1997, 42-44, no. 11, example dated, according to its characteristic decoration with bunches of grapes under an inscription 'ΠΙΕ ΖΗΣΗΣ', to the late 4th or 5th century.
[655] Marki 1985, 159-188, esp. 171.
[656] Uboldi 1995, 116-119, 118, fig. 4. 16-21. For this form see also, Sternini 1995a, 259, 261, 264-265.
[657] The fragments with acutely pointed bases could equally probably belong to 6th-century amphorae, such as the one in the National Archaeological Museum in Athens, see Weinberg, McClellan 1992, 135, no. 111.
[658] Foy 1995, 197, 226, pl. 8, nos. 59-63, finds from Le clos de la Lombarde, Narbonne, Port-Vendres and Saint-Blaise, dated from excavation results to the 5th-6th century.
[659] Brescia (Brixia), S. Giulia: Uboldi 1996, 186, nos. 4-5; Uboldi 1998, 209, fig. 3. 14, 15, finds dated from excavation results and analogous finds to the 6th-7th century.

Hill[660] and Crypta Balbi),[661] Carthage,[662] Salamis, Cyprus,[663] Gortyna,[664] Sardis,[665] and Egypt.[666] An intact example of this form, probably a lamp, can be seen in the Corning Museum of Glass.[667]

Only Cat. no. 132 can be independently dated. Although there are no special stratigraphic clues, the building in which they were found was a Justinian-era church destroyed in the early 7th century.[668] These dates present the *termini ante* and *post quem*, respectively, for the dating of the fragments.

### 35. Stemmed beakers, inherent stem (Isings Form 111.1)

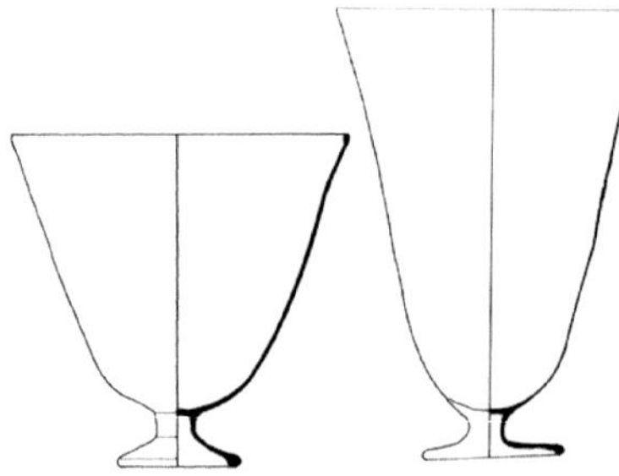

i. Tubular stem
(Cat. nos. 133a, 134a, 135d, 136, 138, 141, 142a, 145, 149, 151, 152, 159, 160, 163, 164)

ii. Beaded stem
(Cat. nos. 133b, 134b, 135a, b , c , 140, 142b, 144, 147, 148, 150, 154, 155, 156, 157, 161)

iii. Unidentified
(Cat. no. 143)
(Plates 5, VI)

Vessels of this form have either fire-polished or even unworked rim; body inverted truncated conical, calyx-shaped or globular, which stands on a tubular or beaded stem that in turn is based on a discoid, slightly conical base. Their height is estimated between 9-15cm.

This form of vessel was apparently created as tableware, which, according to the very often unworked rim and the huge number of extant examples from Early Christian churches, were very widely used as lamps too.

All the examples from Thessaloniki are made of greenish glass. Only one example decorated with strongly-coloured, applied blobs around its body has been found.[669]

This form is widely distributed throughout the Mediterranean. North of the Alps only very few examples have been found.[670] It was considered that they appeared from the 4th century,[671] although it would seem that there are no firmly dated examples predating the 5th century,[672] and they continue to be produced until at least the 7th century. It is not possible to discern the production of the 5th century from that of the 7th by the macroscopic study of glass alone. The examples from the region of Thessaloniki are loosely dated by their archaeological contexts and mainly according to the following parallels in the period between the 5th and 7th centuries.

According to the technique employed to form their base, they are divided into the following groups:

1) Beakers with inherent, hollow stem (Form 35).

The inherent base is created by folding the same mass of glass from which the body of the vessel is created. The base is either flat or slightly raised towards the centre, forming a low cone. There is a local, late 6$^{th}$-century variant of this form which features a small stamped numeral on the base.[673] It is not easy to discern chronologically the finds of this group because essentially they do not change much, apart probably from the colour and texture of the glass, which gradually gets thicker and darker-green compared to the earlier examples.

As to the shape of the stem they can be classified into the following subgroups:

[660] Sternini 2001, 28, nos. 115-119, from Late Antique layers, dated between the 5th and 7th century according to similar finds.
[661] Saguì 1996, fig. 8. 10, finds from a layer of the second half of the 7th century.
[662] Sternini 1998, 137, figs. 8-9; Sternini 1999, 94, fig. 5. 58-59.
[663] Chavane 1975, 65, pl. 63/172, find dated from excavation results to the 7th century.
[664] Sternini 1997, pl. XLVII. 1. De Matteis, De Tommaso 2001, passim, from late-7th century layers.
[665] Saldern 1980a, 52-53, nos. 287-297, pl. 23, finds dated from excavation results between the 5th and the early 7th century.
[666] Edgar 1905, 19, nos. 32483-32485; Mossakowska-Gaubert 2000, 348, nos. 18-21; Foy 2001, 466-467, no. 9; Keller 2008, fig. 22. 2.
[667] Whitehouse 1997, 250, no. 428, dated on the basis of analogous finds to the 4th-5th century.
[668] Marki 1985, 159-188, esp. page 171.
[669] Finds from Rentina (Cat. no. 145). For a vessel with the same shape and decoration from Israel, see Israeli 2003b, 197, no. 236, dated according to similar finds to the 4th-6th century.
[670] Harden 1975, 22-23.
[671] Slane 1994, 165-166, no. 111. From a pit dated to the period of the Tetrarchy, with 6th-century intrusions. The author describing the find mentions that beakers with globular (beaded) stems are the rule for 6th-century finds in Corinthia. Also from Corinth are published finds dated to the 4th century; see Davidson 1952, 81, 103, nos. 654-655.
[672] The finds from Kom Aushim (Karanis), which were originally dated to the 4th century, now, due to the more precise dating of the clay finds from the same contexts, have been re-dated to the end of the 5th century; see Hayes 1975, 2-3, 84, pl. 11, 25.
[673] Antonaras 2011a, 126-128; Antonaras 2011b, 239-242; Antonaras 2014a, 106, 110.

i) Those forms with tubular stem, comprising the most numerous group.

Examples of this group are known from southern France,[674] where local production is attested,[675] Torcello,[676] which also boasts local production,[677] Caričin Grad (Iustiniana Prima),[678] and Ras.[679] There are numerous find-sites in Dalmatia and other locations,[680] including Putalj,[681] Majsan,[682] Srima on Šibenik,[683] Galovac near Zadar (Iader),[684] Gata,[685] Svishtov (Novae) in Bulgaria,[686] Argos,[687] Athens,[688] Thasos,[689] Demetrias,[690] Aegina,[691] Olympia,[692] Chios,[693] Samos,[694] Delos,[695] Kos,[696] Asprovalta,[697] Agios Philon[698] and Salamis, Cyprus,[699] Sagalassos,[700] Iasos Caria,[701] Latakia (Laodikeia Paralos),[702] and Qal'at al Madiq (Apameia on the Orontes).[703]

ii) Those forms with beaded stem, probably shaped with the help of jacks. This subgroup appears to be less widespread. Examples have been found at Port-Vendres,[704] Svishtov (Novae) in Bulgaria,[705] Constantinople,[706] Anamur (Anemurium),[707] Iasos, Caria,[708] Salamis, Cyprus,[709] Nea Anchialos (Phthiotides, Thebes),[710] Athens,[711] Aegina,[712] Corinth,[713] Argos,[714] Chios,[715] Samos,[716] Delos,[717] and Kos.[718]

[674] Foy 1995, 208, form 23a, pl. 15. 176-177, 179-188. This form appears in southern France at the end of the 5th- early 6th century and dominated in the second half of the 6th.

[675] Foy 1995, 208-209, form 23. A workshop producing them was excavated in Maguelonne and is dated to the end of the 6th or in the 7th century.

[676] Sternini 1995b, 257, with a thorough bibliography.

[677] The workshop was operating in the middle of the 7th century. Sternini 1995b, 257, with thorough bibliography.

[678] Bavant, Kondić, Spieser (ed.), 1990, 208-211, nos. 90-105, pl. XXXII, from layers dated to the end of the 6th to early 7th century.

[679] Nikolić-Đorđević 1999, 108-109, form 5/1, finds from a Christian basilica and private houses, in layers of the late 5th and 6th century.

[680] Fadić 1994b, 243-246.

[681] Fadić 2001, 133-134, fig. 31b, finds dated from excavation results to the 4th-5th century.

[682] Fadić 1994a, 214, note 18.

[683] Fadić 1994a, 214, note 19.

[684] Fadić 1994a, 214; Fadić 2001, 134, with relevant bibliography.

[685] Fadić 1994a, 213-215. Dozens of finds dated from excavation results to the 6th or the early 7th century.

[686] Turno 1989, 163-165, forms 1, 2, 3, 5, finds dated between the 3rd and 6th century according to analogous finds.

[687] Aupert 1980, 403, 449, nos. 410-418, pl. 50. 410-418, finds dated to the 6th and mainly to the second half of the same century.

[688] Weinberg and Stern 2009, 149, nos. 345-55, fig. 20, from 5th- or 6th-century contexts.

[689] Abadie-Reynal, Sodini 1992, 91-92, fig. 38 V7, 8. This form appears in all layers dated between the early 5th and late 7th or even 8th century.

[690] Eiwanger 1981, 96, nos. II. 658-II. 665, pl. 86, with no archaeological or chronological indications.

[691] Felten 1975, 73, pl. 26 no. 135, find dated from excavation results to the period after the reign of Justinian I (527-565).

[692] Schauer 2001-2002, 42, fig. 13.

[693] Ballance *et al.* 1989, 127, F 14c-d, fig. 50, finds which, quite probably, belong to the mid 7th-century destruction phase.

[694] Tölle-Kastenbein 1974, 164, pl. 277 b, find from Kastro-Tigani, dated to the Roman period.

[695] Nenna 1998, 121 no. D54, pl. 37, Early Christian find, dated to the post 4th-century period, according to analogous finds.

[696] Brouskari 2002, 137, 139, finds dated from excavation results to the 5th and first half of 6th century.

[697] *Glass Cosmos* 2010, 349-350, nos. 372-5, dated to the 4th/5th and 5th/6th centuries.

[698] Du Plat Taylor, Megaw 1981, 227, nos. 26, 28, 30, figs. 46: 26, 28, 30, finds dated before AD 666.

[699] Chavane 1975, 57-59, no. 165, finds dated from excavation results to the 7th century.

[700] Lightfoot 1993, 175-176, nos. 17-30, fig. 122, from 4th-century layers.

[701] Baldoni, Berti 1996, 76, 78-79, fig. 13, find with no special reference to its dating.

[702] Salibi 1979, 139-142, figs. 5. 3, 8, from the excavation of the settlement at Ras Ibn Hani, dated from excavation results between the 4th and 6th century.

[703] Pirling 1977, 142-143, 147, pls. 2, 3, finds dated from excavation results to the 4th-5th and, probably, the 6th century.

[704] Foy 1995, 208, no. 178, find dated from excavation results to the second half of the 6th or into the 7th century, and is considered to be an imported product from the Orient.

[705] Turno 1989, 163-165, form 4, finds dated from excavation results to the 5th century.

[706] Saraçhane: Hayes 1993, nos. 17-18, 20, finds dated from excavation results to the 6th or 7th century.

[707] Stern 1985, fig. 3.

[708] Baldoni, Berti 1996, 76, 78-79, figs. 11-12, finds from two Early Christian basilicas, with no special reference to their dating.

[709] Chavane 1975, 57-59, nos. 165, 166, finds dated from excavation results to the 7th century.

[710] Sotiriou 1929, 105-106, fig. 137, centre of upper row, finds dated probably to the destruction period of the basilica.

[711] From the Athenian Agora, Robinson 1959, 116, nos. M341-346, finds dated to the early 6th century. The two examples illustrated in pl. 54, M341-342, belong to subform ii; Weinberg and Stern 2009, 149, nos. 338-45, fig. 20, from 5th- and first half of 6th-century contexts. These vessels were made of two or three masses of glass, one for the bowl, one for the base and the stem, or one for the stem and another for the base.

[712] Felten 1975, 73, pl. 26, no. 136, upper left, finds dated from excavation results to the period after the reign of Justinian I.

[713] Slane 1994, 165-166, no. 111, from a pit dated to the period of the Tetrarchy, with 6th-century intrusions. K. Slane in the catalogue entry mentions that beakers with globular (i.e. beaded) stem are the rule for 6th-century finds in Corinthia.

[714] Aupert 1980, 403, 449, no. 412, pl. 50. 412, find dated to the 6th or to the second half of the 6th century.

[715] Ballance *et al.* 1989, 125, 127, F 14a, b, e, p, fig. 50, finds which, quite probably, belong to the mid 7th- century destruction phase.

[716] Isler 1969, 228, figs. 59-61, pl. 98. 8, finds from the Heraion on Samos, dated approximately to the middle of the 6th century; Tölle-Kastenbein 1974, 164, pl. 277 a, c finds from Kastro-Tigani dated to the Roman period.

[717] Nenna 1998, 121, no. D53, pl. 37, Early Christian find, dated post 4th century according to analogous finds.

[718] Brouskari 2002, 137, 139, finds dated from excavation results to the 5th and to the first half of the 6th century.

### 36. Stemmed beakers, inherent foot, three handles

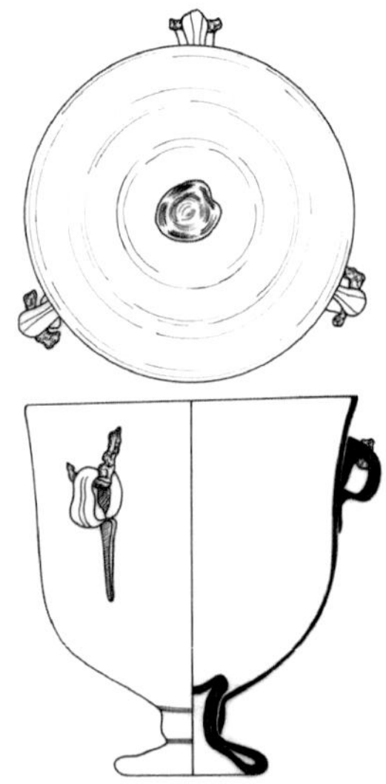

(Cat. nos. 165-172)
(Plates 6, VI)

Vessels of this form have usually unworked rim or in rare cases a fire-polished one, inverted truncated conical, calyx-shaped or hemispherical body, standing on an inherent tubular or beaded stem, based on a discoid or slightly conical base. The vessel is supplemented with three small handles arranged almost equidistantly high on the body or around the rim. Their height is estimated at c. 10cm.

The inherent base is formed by the folding of the lower part of the mass of glass, from which the body was made. In some cases it is totally flat and in other cases it rises steadily towards the centre, forming a low cone.

The only example which can be dated independently is Cat. no. 165, which is made of olive-green glass and is dated by its archaeological context to the 5th century.[719]

This form is more widespread in the East Mediterranean.[720] Very rarely are there sufficient numbers of body fragments preserved to allow even graphically the restoration of the exact shape of the body. Parallels mentioned below are almost always in reality handles that were found detached from the body. Handles of this form have been found in southern France,[721] northern Lombardy at Monte Barro,[722] Brescia (Brixia), S. Giulia,[723] Carthage,[724] Caričin Grad (Iustiniana Prima),[725] Ras,[726] Kourion on Cyprus,[727] Gerasa,[728] Medinet Madi (Narmouthis),[729] Bet She'an (Nysa-Scythopolis),[730] Nahariya,[731] Philippi,[732] Thasos,[733] Argos,[734] Olympia,[735] and Corinth.[736] Handles with long ribbon-like ending attached to the body and extending almost to the base of the vessel were found at Argos,[737] Philippi,[738] Chios,[739] Samos,[740] Iasos Caria,[741] Sagalassos,[742] Istria,[743] Cherson (Chersonesus Taurica),[744] and Israel.[745]

It seems that this form became popular from the end of the 5th and continued up to Medieval times. It survived at least until the 9th century in Cairo (Foustat),[746] while 11th-century examples are known from the Corinth agora.[747] It has been assumed that from this form evolved the 11th-century type of the also hanging, but quite larger in size, mosque lamps.[748]

---

719 Kissas 1988, 209, fig. 4; Papanikola-Bakirtzi 2002 (ed.), 285, no. 299 (A. Antonaras), dated to the 5th century.
720 Uboldi 1995, form V, 124-125.
721 Foy 1995, 206-207, form 22c, finds dated from excavation results to the 6th-7th century.
722 Uboldi 1998, 208, fig. 21, 4, conical vessels dated from excavation results to the end of the 5th to the first half of the 6th century.
723 Uboldi 1998, 209, fig. 3, hemispherical and conical vessels dated from excavation results, and according to analogous finds from other sites, to the 6th-7th century.
724 Hayes 1978, 187, lamp handles from layers dated to the 5th-6th century, or even later.
725 Bavant, Kondić, Spieser (ed.), 1990, 208, 211, nos. 108-109, fig. 149, pl. XXXII, from layers dated to the end of the 6th - early 7th century.
726 Nikolić-Đorđević 1999, 108-109, form 4/2, from layers in a basilica and private houses, dated to the late 5th and 6th centuries.
727 Young 1993, 43, from a basilica which was in use between the end of the 5th and the middle of the 7th century.
728 Crowfoot, Harden 1931, 206, pl. XXX, 50; Baur 1938, 524-525, fig. 20, 376, form F.
729 Silvano 1999, 19, 22-25, 35-35, pl. IV, finds from the site of an ecclesiastical building, dated generally between the 4th and 7th centuries; especially for the lamps, on the basis of comparison with analogous finds from other sites, the date is more precise, between the 5th and 7th centuries.
730 Hadad 1998, 63, 64-65, 67, form 1, finds that appear in layers dated between the 5th and 14th centuries.
731 Dauphin, Edelstein 1984, 105, 109, LXIVa, finds from a basilica dated to the first half of the 6th century, which was probably destroyed in the early 7th century.
732 Antonaras 2007, 51-54 finds from museum's basilica, dated from excavation results to the Justinian period.
733 Abadie-Reynal, Sodini 1992, 92-93, fig. 38, pls. XV. c., XVI. a, b, dozens of examples of small-sized handles.
734 Aupert 1980, 403, 447, no. 408, pl. 50. 408, find dated to the 6th or to the second half of the 6th century.
735 Schauer 2001-2002, 41-42, handles ascribed to the truncated conical Isings Form no. 134.
736 Wiseman 1969, 105-106, pl. 32/j, together with glassworking remains which were located in a layer dated to the late 5th-6th century and found in the area of Corinth's ancient gymnasium.
737 Abadie-Reynal, Sodini 1992, 93, fig. 38 V16. Handle of a curved body with long, smooth ending.
738 Antonaras 2007, 51-54, finds from the Museum's Basilica, dated from excavation results to the Justinian period.
739 Ballance *et al.* 1989, 125, 127, F 17a, b, e, p, fig. 50, finds which, quite probably, belong to the mid 7th-century destruction phase. Small handles (17a-c), and others with long, smooth (17d) and notched or undulating ending (17e, f).
740 Tölle-Kastenbein 1974, 164, pl. 278 e-k, small handles and one with long and smooth ending from Kastro-Tigani dated to the Roman period.
741 Baldoni, Berti 1996, 76, 78, figs. 7-10, small handles with smooth, long ending, finds from two Early Christian basilicas, without firm dating.
742 Lightfoot 1993, 176-177, nos. 31-32, fig. 122, small handles, from layers dated to the 5th and to the 5th-6th century, respectively.
743 Bâjenaru, Bâltâc 2000-2001, 474-479, fig. 3, pl. VI-IX, finds from the basilica dated to the late 5th century, which was destroyed in the early 7th.
744 Golofast 2001, 139-141, drawings 85:1, 86, examples dated from excavation results to the end of the 6th century and in the 7th century.
745 Gorin-Rosen 2002a, 119-121, drawing 2, finds dated to the 5th-6th century, mainly on the basis of analogous finds from other sites.
746 Scanlon, Pinder-Wilson 2001, 55, fig. 26, drawing 26.
747 Davidson 1952, 111-112, 726-729, pl. 57, from layers dated to the 11th-12th century.
748 Scanlon, Pinder-Wilson 2001, 53.

**37. Stemmed beakers (Isings Form 111)**

1. Applied, solid cylindrical stem
i. Squat, cylindrical stem
(Cat. no. 164a, b)

ii. Slender, cylindrical stem with a constriction at mid-height
(Cat. no. 164c)

iii. Conical body ending on four ogival arches, on slender stem
(Cat. no. 146)

2. Applied, conical stem
(Cat. nos. 153, 164d)
(Plates 5, VI)

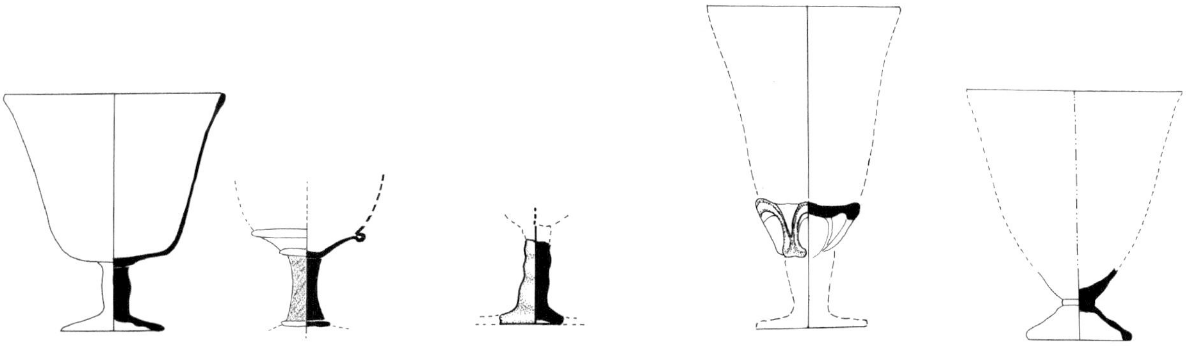

Vessels of this form have either fire-polished or even unworked rim; body inverted truncated conical, calyx-shaped or globular, which stands on an applied tubular or beaded stem, which is based on a discoid, slightly conical base. Their height is estimated between 9 and 15cm.

In the examples of this subgroup the stem and base are made from two different masses of glass, occasionally in glass of different colour.[749] It is considerably less widespread than stemmed beakers with inherent base (Form 35) and probably appears later, at the end of the 6th and throughout the 7th century, although both forms co-exist for a while.

Stemmed beakers were apparently created as tableware, but quite probably they were also used as lamps, according to the large number of extant examples from Early Christian churches. All examples from Thessaloniki are made of greenish and bluish glass. This form is widely distributed throughout the Mediterranean. North of the Alps actually no examples have been recorded.

They are divided as:

1. Beakers with solid cylindrical stem. According to the size and shape of the stem the following two groups have been identified:

i. Beakers with short and squat cylindrical stem (Cat. no. 164a), for which a north African origin has been proposed.750 They are considered to be earlier than the beakers of subgroup ii.[751] Dated finds have been published from Thasos,[752] Aegina,[753] Agios Philon,[754] Salamis on Cyprus,[755] Latakia (Laodikeia Paralos),[756] Cagliari (Caralis)[757] and the northern Black Sea littoral.[758]

ii. Beakers with taller, slender cylindrical stem with a constriction at mid-height (Cat. no. 164c), which are quite often made of bluish glass. It has been assumed that they are the younger of the two subgroups and that they do

[749] Examples are mentioned by Foy 1995, 208, nos. 188-189. See also Hayes 1978, 187.
[750] Foy 1995, 209, no. 189. A typical characteristic of these is the short and relatively thick stem that appears quite often in Tunisia during the 6th century. For Carthage, see Hayes 1978, 187, 190, nos. 25-33, fig. 3-4, from layers dated between the 6th and 7th century. Also, at Bir el Knissia in Carthage, Hayes 1990, 289, 293, nos. 33-37, from layers dated between the late 5th and up to the 7th century.
[751] Hayes 1990, 289.
[752] Abadie-Reynal, Sodini 1992, 92-93, fig. 38 V9, 10, these examples are far fewer than those of form 1. This form appears in all layers dated between the early 5th and the end of the 7th or the 8th century.
[753] Felten 1975, 73, pl. 26, no. 136, lower, finds dated from excavation results to the period after the reign of Justinian I (527-565).
[754] Du Plat Taylor, Megaw 1981, 227, nos. 25, 27, fig. 46: 25, 27, finds dated to the period before AD 666.
[755] Chavane 1975, 62, pl. 18/167, find with no chronological indication.
[756] Salibi 1979, 139-142, fig. 5. 6, from the excavation of the settlement at Ras Ibn Hani, dated from excavation results between the 4th and 6th century.
[757] Stiaffini, Borghetti 1994, 82, 143-144, nos. 450-457, pls. 115-118, finds dated to the 4th-6th century, according to analogous finds.
[758] Golofast 2009, 318-319, wherein further bibliography on finds from several sites dated on the basis of their context into the second half of the 6th or 7th century, and in other cases up to the 9th.

not appear before the end of the 6th or the 7th century.759 For these stems it has been proposed760 that they are associated with calyx-shaped bodies.

None of the finds from Thessaloniki can be dated independently.

Several stems of subform 37.1.i. have been unearthed in the ruins of a wealthy Late Antique house in the centre of the city, most of them related to its destruction phase.[761] Dark green examples prevail, although bluish examples also occur, and most appear to be quite clumsily made artifacts.

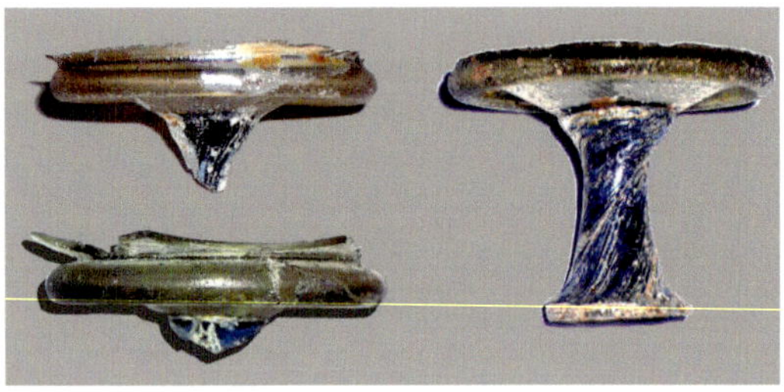

An interesting variant of the form of stemmed beakers made from two different gathers, one for the bowl and the other for the stem and base, appeared in the ruins of the late-6th century glass- workshop at 45 Vasileos Irakleiou Street (Cat. no. 164b). There are three examples of this form. The bowl was made in greenish glass with a characteristic fold at the lower edge, while the long smooth stem was made in dark-blue glass, as, presumably, was the discoid base, now missing.[762] One or two fragmentary examples, unstratified and hitherto unpublished, with the same fold at the lower part of the body have been unearthed in the excavations of the city's agora, conducted in the 1960s and a third one in the excavations at Diikitiriou Square at 1993.763 An identical find from Chersones in the Black Sea region, is dated to the late 6th-early 7th century.[764]

A conical fragment (ΜΒΠ/BY 52) from the hilly suburb of Thessaloniki, Panorama, seems to belong to the group of the beakers with tall stem, and more precisely to a particularly large and heavy subform. It is dated to the 4th or 5th century, most probably before the middle of the 5th century, when a basilica was founded on the site, destroying an older burial building[765] with which the glass find is probably connected. Analogous finds have been found in Carthage,[766] Rome,[767] and Marseilles (Massalia).[768]

iii. A rare decorative variant of the form is preserved in a small fragment from Thessaloniki, which is described as a beaker with inverted truncated conical body, the lower part of which ends in four ogival arches. It probably stood on a cylindrical solid stem and a conical base. The fragment from Thessaloniki is made of bluish glass.

According to analogous finds, the beaker was made of two separate masses of glass. The conical body, with fire-polished rim, was made by blowing a glass mass, which was then marvered to make its conical form, and its lower part was pinched into four converging arcades. The base was made from a second mass of glass. The beaker stood on a flat discoid base and the stem of the foot was slender and cylindrical.

This form is quite rare and the only published parallels are from Italy, Aquileia[769] and other Italian sites,[770] and Slovenian Koper/Capodistria (Capris).[771] It seems quite probable that the form evolved from vessels of the same shape which appear in the 4th-5th century and featured on the lower part of the body some similar applied or inherent decoration.[772] From the exaggerated drawing of the decoration of these early vessels evolved the three-dimensional arcading of the lower part of the body that characterizes this form.

759 Foy 1995, 208-209.
760 Foy 1995, 208, find dated from excavation results to the end of the 7th century.
761 Finds from the excavation of 3 Agapis Street, see Chapter 9 (Find-Spots), '3 Agapis Street, 1996'.
762 Antonaras 2014a, 110, fig. 12.28.
763 Personal observation, material kept in the Archaeological Museum of Thessaloniki and in the Museum of Byzantine Culture.
764 Golofast 2001, 114-115, pl. 50:16. From a complex dated on the basis of pottery finds to the end of the 6th to early 7th century.
765 Tsigaridas 1973b, 498-501, drawing 12, pl. 462α.
766 Sternini 1998, 138-39, no. 10. The fragments are generally dated, according to the relative (among themselves) connection, and to analogous finds, between the 5th and 7th century.
767 Sternini 1995a, 277, fig. 9. 97, finds from the Palatine Hill, dated from excavation results to the middle of the 5th century.
768 Foy, Bonifay 1984, 295, fig. 3. 86, finds dated from excavation results to the 6th century.
769 Calvi 1968, pl. 27; Mandruzzato, Marcante 2005, 68, no. 65, find dated to the 4th century.
770 Uboldi 2009, 221-228, wherein all similar finds are cited.
771 Guštin (ed.) 1991, 14, 46, 60, nos. 26-27, two examples from a layer dated to the 7th-9th century.
772 Platz-Horster 1976, 88, no. 17, which according to analogous finds is dated to the 4th-5th century; Barkóczi 1988, 93, no. 136, find from Dunaújváros (Intercisa) dated to the first half of the 5th century; Whitehouse 1997, 105-106, no. 158, dated on the basis of analogous finds approximately to the 4th-5th century. This decoration continues to be used later, at least in the West. There are known Longobard and Frankish vessels dated to the 6th-7th century, see Stern 2001, 348 no. 195, where a plethora of these are dated from excavation results.

The find cannot be dated independently, because it lacks exact information about the layer and context in which it was found, and thus it is generally dated post 5th century, probably 7th or 9th, according to the date of the Slovenian find and the overall morphological evolution of the form.

2. Beakers with low, conical, applied stem; subform 37.2.

Another variant of stemmed beakers, is also present in Thessaloniki, although very rarely, with just a couple of examples preserved. In general it is a rare product but finds are known from several sites in the East Mediterranean.

The body was bell-shaped or somehow conical. A separate wad of glass, or the squeezed, beaded upper part of the base formed the stem of the vessel. The conical base was formed from a second paraison of glass, free-blown and shaped as a small bowl or plate with a slightly convex bottom, short vertical walls and a fire-polished rim. In some cases a translucent dark-blue fine thread was applied on the upper edge of the circumference of the base. The same form of beaker was occasionally supplemented with three presumably small handles with long endings reaching the bottom of the bowl.[773] The finds from Thessaloniki preserve only a part of the base and thus it is not possible to say whether they were handled or not. Judging by the pontil mark (c. 2cm in diameter) preserved at the centre of the base's under-surface, the rim of the vessel was presumably fire-polished.

At least from one example from Thessaloniki (Cat. no. 164d) it is evident that the vessel was made of two paraisons of different hues, the body of transparent bluish and the base of transparent greenish glass. The other two examples of this subform both have bowl and the base made of colourless greenish glass.[774]

Analogous finds are known from Athens,[775] Corinth,[776] Kourion[777] and Salamis on Cyprus,[778] Sardis,[779] and Beirut.[780]

The finds from Thessaloniki are both dated independently. One was found among the finds from the operation period of a glass-workshop active at the end of the 6th century.[781] The second was unearthed in the ruins of a sumptuous Late Antique urban house.[782]

### 38a, b. Beakers with tall cylindrical base/stemmed lamps

1. With hollow base
2. With solid base

(Cat. nos. 173-175)
(Plates 6, VI- VII)

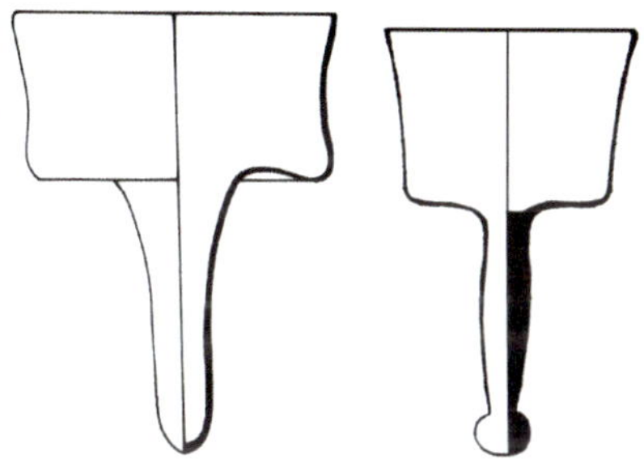

This form comprises vessels with vertical, fire-polished rims, truncated conical, or hemispherical body, long cylindrical base with flat or beaded bottom. Their height is estimated between 9-12cm. The lamps of this form are known in Greek sources as 'οὐρίαχοι', i.e. 'those that have a tail'.[783] Stemmed lamps were designed to be inserted into metal lighting devices, the so called *polykandela*. These devices were introduced in the Late Roman period, probably replacing the multi-nozzled lamps, and remained the prevailing form of chandeliers for the Byzantine, Arab and Ottoman eras. They were suspended by three or four chains, and were usually in the shape of a ring or a pierced, open-worked disc with circular perforations around its periphery, and into which stemmed lamps were inserted. *Polykandela* are known in more innovative forms, e.g. openwork crown, tree-, church-, or cross-shaped with brackets, occasionally dolphin-shaped, ending in small rings/ holders of stemmed lamps. They were widely used for the illumination of religious and secular buildings, which

[773] Vessels from Corinth under study by the present author.

[774] The first example is Cat. no. 153, from the western necropolis, without firm independent dating. For the second example see Antonaras 2014a, fig. 12.30.

[775] Weinberg, Stern 2009, 160-61, nos. 339, 344-45, fig. 20, handled examples dated to the second half of the 5th and into the 6th century; Weinberg, Stern 2009, 160, nos. 338, 344, handleless examples (dated to the 6th and mid- to second half of the 5th century); idem, 165, no. 381, for a lamp supplemented with a wick holder (dated to the 6th century).

[776] Davidson 1952, nos. 654-55, dated to 4th century; several more examples from the area east of the theater and gymnasium, all of them dated to the late 5th and 6th centuries are currently under publication by the present author.

[777] Young 2007, 495, no. 45, pl. 16d, from a lot dated to the late (?) 7th century.

[778] Chavane 1975, 65, no. 169, pls. 18 and 63.

[779] Saldern 1980a, nos. 37-376, pls. 12. 24.

[780] Jennings, 2004-2005, figs. 6.1:1, 11, 6.3:1.

[781] Antonaras 2014a, 110, fig. 12.29; Antonaras 2014d, 263-272.

[782] For the excavation of the site at No. 3 Agapis Street, see Kanonidis 1996b, 490-493.

[783] Paulus Silentiarius, *Descriptio Sanctae sophiae*, 825. Suida, s.v. οὐρίαχος; Trowbridge 1930, 174, 190, 191.

required dozens or even hundreds of stemmed lamps for the larger edifices.[784] The form also appears in a variant with an internal, narrow cylinder stemming from the centre of its bottom, which was used as a wick holder.[785] Depending on whether the base of the vessel is solid or hollow, two groups are discerned:

1) With a narrow, hollow, slightly conical, usually practically cylindrical base, ending in a flat bottom.

This is type IV.2 in the Uboldi typology,[786] which is dated to the 6th-7th century. Examples have been found at Gardanne,[787] Marseilles (Massalia),[788] Saint-Blaise,[789] northern Lombardy, Monte Barro,[790] Brescia (Brixia), S. Giulia,[791] Ravenna,[792] Torcello,[793] in Rome,[794] Crypta Balbi[795] and Palatine Hill, near the temple of Magna Mater,[796] in Ostia,[797] in Carthage,[798] in several Dalmatian sites,[799] Putalj,[800] Gata,[801] Galovac,[802] Split (Spalatum),[803] in Serbia at Caričin Grad (Iustiniana Prima)[804] and Ras,[805] in Romania at Galata[806] and Istria,[807] and at Svishtov (Novae)[808] Bulgaria, Ukrainian Cherson (Chersonesus Taurica),[809] Constantinople,[810] Sardis,[811] on Cyprus at Kourion,[812] Agios Philon,[813] and Salamis,[814] in Iasos Caria,[815] Beirut,[816] Gerasa,[817] Petra,[818] Qal'at al Madiq (Apameia on the Orontes),[819] in Nahariya,[820] in Ashqelon,[821] Bet She'an (Nysa-Scythopolis),[822] in Samaria,[823] in Cairo (Foustat),[824] in the area of Fayum (Arsinoe)

[784] Xanthopoulou 2010, 46-53, 281-316; Boura and Parani 2008, 58; Motsianos 2011, 336-346.
[785] For the Gortyna finds, see Sternini 1988, figs. 24-29, from the excavation of basilicas dated to the 6th and second half of the 7th century.
[786] Uboldi 1995, 121-124.
[787] Foy 1995, 206, form 22b, finds dated from excavation results to the early 6th century.
[788] Foy 1995, 206, form 22b, finds dated from excavation results to the 7th century.
[789] Foy 1995, 206, form 22b, finds dated from excavation results to the 6th century.
[790] Uboldi 1998, 208, fig. 2. 15, 16, dated from excavation results to the late 5th to the first half of the 6th century.
[791] Uboldi 1998, 209, fig. 3. 11-13, dated from excavation results, but also according to analogous finds, to the 6th-7th century.
[792] Curina 1983, nos. 11, 16.
[793] Gasparetto 1979, fig. 16.
[794] For finds from older excavations, see Uboldi 1995, 121, note 145.
[795] Saguì 1998, 206, fig. 8. 9, in a layer of the second half of the 7th century.
[796] Sternini 1995b, 260-262, 265; Sternini 2001, 28, no. 121, finds dated to the 6th-7th century.
[797] Sternini 2001, 28, note 49, finds dated from excavation results to the 6th-7th century.
[798] Sternini 1998, 137, 139, fig. 7, dated on the basis of analogous finds between the 5th and 7th century; Tatton-Brown 1984, 199, 203, figs. 65/59, 67/91, dated from excavation results to the 6th century.
[799] Fadić 1994b, 247-248.
[800] Fadić 2001, 136-137, fig. 31d, find dated from excavation results to the end of the 5th or to the 6th century.
[801] Fadić 1994a, 215-217, three-handled examples, with handles around their neck a quite rare form, are also mentioned. It is noted among the finds from Svishtov (Novae) in Bulgaria (Turno 1989, 166). It must be underlined however that this form of lamp is made with a funnel-shaped, or cylindrical base precisely in order to be placed in a *polykandelon* and is not intended to be used independently hanging from metal hooks. The finds are not dated from excavation results, nor on the basis of analogous finds.
[802] Fadić 2001, 136, with further bibliography.
[803] From Diocletian's palace, De Maine 1979, 64, 116, pl. 11, R 58, 62, from layers dated generally between the 3rd and 6th centuries.
[804] Bavant, Kondić, Spieser (ed.) 1990, 208-209, 211, nos. 110-112, pl. XXXII, from layers dated to the late 6th - early 7th century.
[805] Nikolić-Đorđević 1999, 108-109, form 4/1, in layers dated to the late 5th and 6th centuries, from the excavation of a basilica and private houses.
[806] Mirchev 1953, 8, fig. 10, find from a 5th-century basilica, which was destroyed in the 6th century.
[807] Bâjenaru, Bâltâc 2000-2001, 471-74, fig. 2, pl. I-V, finds from a basilica erected at the end of the 5th century, and destroyed in the early 7th.
[808] Turno 1989, 166-167, fig. 2/1, 2, 3. Three subforms are noted, depending on whether their base is cylindrical or conical and if they have suspension handles. All three are dated 5th-6th century.
[809] Golofast 2001, 138-139, drawings 85:2-12, a plethora of examples dated from excavation results to the late 5th century and up to the 7th.
[810] Hayes 1993, no. 14, from a layer dated to the second half of the 6th century; and ibid, 403 nos. 31, 37, 38, figs. 150-151, along with several, uncatalogued, similar fragments from layers dated to the 7th century.
[811] Saldern 1980a, 23, 49-52, pl. 11, nos. 274-80, finds dated between the 5th and 7th centuries.
[812] Young 1993, 43, from a basilica operating between the end of the 5th and middle of the 7th centuries.
[813] Du Plat Taylor, Megaw 1981, 227, nos. 35-46, fig. 46:35-46, finds dated to the period before AD 666.
[814] Chavane 1975, 62-64, 66-68, pl. 63/174-180, finds dated from excavation results to the second half of the 7th century.
[815] Baldoni, Berti 1996, 76, 78, figs. 8-9, finds from two Early Christian basilicas, with no firm dating.
[816] Foy 2000, 245-247, finds from a glass-workshop dated to the early or middle 7th century.
[817] Crowfoot, Harden 1931, 198, form 2B2, pl. XXIX/24-26, dated from excavation results to the 7th-8th century. Uboldi 1995, 121, note 142.
[818] Keller 2006, 225, form VII.51a, from contexts dated to the 6th to the 8th centuries.
[819] Pirling 1977, 145-147, pl. 5, finds dated from excavation results to the 4th, 5th and, probably, 6th centuries.
[820] Dauphin, Edelstein 1984, 105, 109, LXIVa, finds from a basilica dated to the first half of the 6th century which was probably destroyed in the early 7th century.
[821] Katsnelson, Jackson-Tal 2004, 106, fig. 3:3, 4, finds dated from excavation results to the 6th-7th century.
[822] Hadad 1998, 63, 69-72, form 4, finds appearing in layers dated to the 5th and up to the mid 8th century.
[823] Crowfoot 1957, 405, fig. 96. 5, 995. Their dating is archaeologically set between the 6th and 15th century.
[824] Pinder-Wilson, Scanlon 1973, 22, figs. 18-19.

Medinet Madi (Narmouthis),[825] at Gortyna,[826] Argos,[827] Larisa,[828] Nea Anchialos (Phthiotides, Thebes),[829] Thasos,[830] Philippi,[831] Samos,[832] Chios,[833] Delos,[834] and Kos.[835] Even more examples are kept in private collections.[836]

This form remained in use both in the East and the West, at least until the 9th century.[837] The fragments from Thessaloniki were found in two churches, which were destroyed around the 7th century.[838]

2) Vessels with inverted truncated conical body; applied, thin, solid, cylindrical base, usually ending in a bead. This form is Uboldy type IV.1.[839] Only very few examples are found in the West, while the majority are found in the East, where they are dated from the 5th century onwards. Analogous finds are reported from Rome,[840] Karadirek (Turkey),[841] Beirut,[842] Gerasa,[843] Samaria,[844] Madaba,[845] Shavei Zion,[846] Bet She'an (Nysa-Scythopolis),[847] Khan El-Ahmar,[848] Agios Philon (Cyprus),[849] Salamis (Cyprus),[850] and Gortyna.[851]

This form survives for quite a long time, at least up to the 11th century, as is proved by finds from Torcello,[852] Cairo (Foustat),[853] and Corinth.[854] The finds from Thessaloniki[855] very probably fall into this later period.

### 39. Ladles

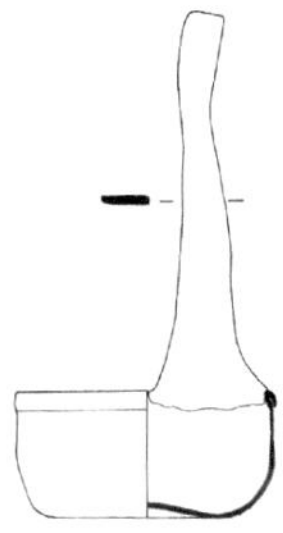

(Cat. nos. 176-177)
(Plates 6, VII)

The vessels of this form are characterized by their long, vertical strap-handle. The body is wide cylindrical, or truncated conical with flat bottom and the rim is either plain fire-polished or tubular.[856] Their height is estimated between 10-12cm.

The body is free-blown and the handle was shaped by the flattening and drawing of a separate mass of glass.

[825] Silvano 1999, 17-20, 34-35, pl. V, finds from the site of an ecclesiastical building, dated generally between the 4th and 7th centuries, while the lamps especially are dated, according to analogous finds, between the 5th-7th centuries.
[826] Sternini 1988, nos. 27-30, from the excavation of basilicas dated to the 6th and second half of the 7th century.
[827] Bommelaer, Grandjean 1972, 221, 223, fig. 118, finds from a building which was destroyed at the end of the 6th century.
[828] Papanikola-Bakirtzi 2002 (ed.), 285, no. 298 (A. Ntina), dated to the 5th-6th century according to analogous finds.
[829] Sotiriou 1929, 105-106, fig. 137, at the centre of the upper row, a find which is probably dated to the period of the basilica's destruction.
[830] Abadie-Reynal, Sodini 1992, 91-92, fig. 38 V1-4, dozens of examples of the form appear in layers dated to the period between AD 600 and the 8th century.
[831] Antonaras 2007, 51-52, fig. 4, finds from the museum's basilica, archaeologically dated to the Justinian period.
[832] Isler 1969, 228, fig. 57, pl. 98. 7, find from the Heraion on Samos, dated around the middle of the 6th century. Tölle-Kastenbein 1974, 164, pl. 277e, find from Kastro-Tigani dated to the Roman period.
[833] Ballance *et al.* 1989, 125, 127, F 16a, b, fig. 50, finds which, quite probably, belong to the mid 7th-century, destruction phase.
[834] Nenna 1998, 122, no. D59, pl. 37, Early Christian find, dated post 5th century, according to analogous finds.
[835] Brouskari 2002, 137, 139, finds dated from excavation results to the 5th and to the first half of the 6th century.
[836] Saldern *et al.* 1974, 251, no. 731.
[837] Fragments of 36 examples were found at 'Saranda Kolones' in Paphos, in layers dated to the 9th century, see Megaw 1971, 131, 145-146, fig. 5. 18. For examples used in the West during the 9th and probably until the 14th century, see Uboldi 1995, 124, note 159.
[838] See Chapter 9 (Find-Spots), 'Agiou Dimitriou and Agias Sofias Street, at their junction, 1969', for Cat. nos. 173-175, and '1 Pringipos Nikolaou, 1961', regarding lamps MBΠ, BY 25/1-3 (Cat. nos. 175a-c).
[839] Uboldi 1995, 120-121. This form is dated to the 6th-7th century.
[840] Fremersdorf 1975, 96, pl. 60, no. 876, find without firm archaeological date.
[841] Lightfoot 1989, 52, no. 80, photo on pl. 8, 4, dated to the 6th-7th century according to analogous finds.
[842] Foy 2000, 243-245, finds from a glass-workshop dated to the early or middle 7th century.
[843] Crowfoot, Harden 1931, 198, form 2B1, pl. XXIX/21-23, finds dated from excavation results to the 7th-8th century; Uboldi 1995, 121, note 142; Meyer 1987, 205, 212-213, figs. 11m, 13a-d.
[844] Crowfoot 1957, 405, fig. 99, archaeologically dated between the 6th and 15th centuries.
[845] Uboldi 1995, 121 note 142.
[846] Barag 1967c, 68, fig. 16, 24.
[847] Hadad 1998, 63, 68-69, form 2, finds from layers dated to the 5th and up to the 14th century.
[848] Meimaris 1990, 424, fig. 164. Dozens of bases of this form were found in a pit, dated by context to the 8th century.
[849] Du Plat Taylor, Megaw 1981, 227, no. 34, fig. 46:34, find dated to the period before 666 AD.
[850] Chavane 1975, 62-66, pl. 63/168-170, finds dated from excavation results to the 7th century.
[851] Sternini 1988, 19, nos. 24-25, from the excavation of basilicas dated to the 6th and second half of the 7th century.
[852] Uboldi 1995, 121, notes 139-140.
[853] Scanlon, Pinder-Wilson 2001, 53-54, form 25a, 25b, drawing 25a, b, finds dated 9th-10th century.
[854] Davidson 1952, 121, no. 801, pl. 60. From a layer dated 11th-12th century.
[855] Unpublished excavation in the area north of the Saint Demetrios basilica, reg. no. MBΠ, BY 212.
[856] Lazar 2003, 123, forms 4. 2. 1 and 4. 2. 2.

Relevant finds were unearthed at Herculaneum,[857] Albingaunum (Albenga),[858] Ljubljana (Emona),[859] and Tipasa.[860]

The finds from Thessaloniki are precisely dated to the second half of the 1st century, because the baths, in whose ruins they were found, were destroyed in the last two decades of the 1st century.[861]

This form imitates metal prototypes,[862] and appears in the early 1st, and survives until the first half of the 2nd century. Although the largest number of preserved examples from Ljubljana (Emona) led to the hypothesis that they represent a local product,[863] they more probably seem to be a north Italian product imported in large quantities to that region by newcomers, the Italian colonists.[864]

Their use in Ljubljana (Emona) was originally thought to have been restricted solely to libations.[865] This approach was later confuted in favour of the vessel's probable use as tableware, but still connected to burial habits.[866] The finds from Thessaloniki were discovered in public baths, and this sheds further light on the question of their use, corroborating in favour of the vessels use as banquet tableware.

**40. Pans, 'trullae'**

(Cat. nos. 178-184)

(Plate VII)

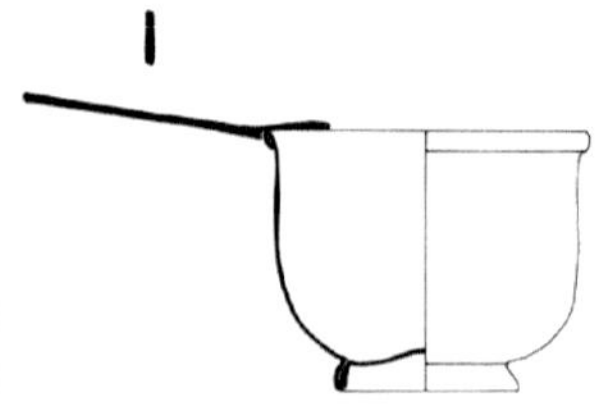

Vessels of this form are characterized by their long, horizontal, strap-handle. They are discerned into two variants: those with shallow and those with deep body. They stand either on a pushed-in ring base or directly on their flat bottom.[867] Their height is estimated at c. 6cm.

The body is free-blown and the handle was shaped by the flattening and drawing of a separate mass of glass.

Examples of this form appear in the 1st century and they are present up to the early 2nd. The form survived up to the 3rd century, shaped and decorated differently.[868] Analogous finds are reported from Pompeii,[869] Padova (Patavium),[870] Ticino,[871] Slovenia,[872] Algeria,[873] and Cyprus.[874]

The finds from Thessaloniki are precisely dated to the second half of the 1st century from their discovery in a baths destroyed in the last two decades of the century.[875]

These vessels imitate metal prototypes[876] and clay examples are also known.[877] They were used in baths, at banquets, and for libations.[878]

[857] De Franciscis 1963, 138, fig. 1, five ladles with strap-handles found in a shop in Ercolano (Herculaneum) packed and ready for selling, witnessing that their production time was very close to AD 79, the time of the Vesuvius explosion and their ultimate destruction.
[858] Massabò 2001, 130-131, no. 76, find dated from excavation results to the end of the 1st- early 2nd century.
[859] Petru 1972; Plesničar-Gec 1972, passim; De Maine 1985, 135-160; De Maine 1988, 129-144, where the dozens of ladles found in the excavations of the northern necropolis of Ljubljana (Emona) are examined; Lazar 2003, 123, forms 4.2.1 and 4.2.2, finds dated from excavation results to the second half of the 1st and first half of the 2nd century.
[860] Tipasa: Lancel 1967, 10, 21, finds from graves dated to the 1st, and 1st-2nd century.
[861] Adam-Veleni *et al.* 1998, 85-102; Mavromichali 2001, 141-142.
[862] Künzl 1997, 21, fig. 10, silver example from the hoard from Xanten (Colonia Ulpia Traiana, Vetera); Massabò 2001, 131, on the left, bronze example from Pompeii.
[863] Plesničar-Gec 1976, 35-37.
[864] Lazar 2003, 126.
[865] Plesničar-Gec 1976, 37.
[866] De Maine 1988, 143.
[867] Isings 1957, 92, form 75a, b; Lazar 2003, 123, forms 4. 1. 1 and 4. 1. 2.
[868] Fremersdorf 1958b, pl. 133; Fremersdorf 1959, pls. 50-53; Whitehouse 1997, 198, for comments on no. 346.
[869] Scatozza-Höricht 1986, 37-38, forms 15-16, pl. XXVII.
[870] Zampieri 1998, 154-155, with no precise date provided from the excavation.
[871] Biaggio-Simona 1991, pl. 8, no. 139. 2. 036; Carazzetti, Biaggio-Simona 1988, 107, fig. 127, find from a grave dated between AD 80-130.
[872] Šubic 1976, pl. 6:52; Lazar 2003, 123, forms 4. 1. 1 and 4. 1. 2, finds dated by the archaeological context to the second half of 1st, first half of the 2nd century.
[873] Lancel 1967, 94, no. 196, find from Tipasa, with no precise date provided from the excavation.
[874] De Ridder 1909, no. 493, pl. XXIV, find from Idalion, with no precise date provided from the excavation.
[875] Adam-Veleni *et al.* 1998, 85-102; Mavromichali 2001, 141-142.
[876] Künzl 1997, 18-19, 24, fig. 7; 13 silver examples from the hoards of Boscoreale and Manching (Vallatum). Later, 4th- and 6th-century examples are also known, see *Wealth of the Roman World 1977*, 54-55 (For no. 106), 95 (For no. 158); Papanikola-Bakirtzi 2002 (eds), 221-222, no. 256 (E. Meramveliotaki), find dated to the 6th century, possibly for bath use.
[877] Künzl 1997a, 475, pls. 8-9, 13d, 14-15, 21a, with detailed bibliography on analogous clay and metal finds dated to the end of the 1st- early 2nd century.
[878] Whitehouse 1997, 198, with relevant bibliography.

### 41. Calyx bowl with three handles

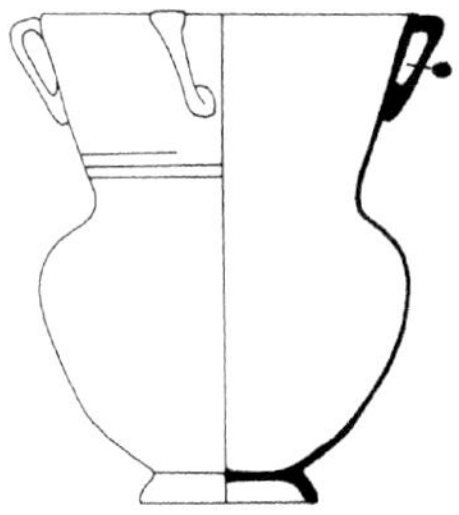

(Cat. no. 185)
(Plates 6, VII)

The vessel of this form has fire-polished rim, wide funnel-mouth, calyx-shaped body and flat bottom. The conical base is short, probably pushed-in. Three vertical coil handles are arranged around the mouth. A fine thread of glass is spirally wound around the lower part of the mouth/neck below the handles.[879] Their height is estimated at c. 14cm.

The vessel is made by free-blowing, the base shaped by pressure with jacks and folding of the lower part of the body. Then the vessel was transferred to a blowpipe attached on the bottom, the moil cracked off, the rim fire-polished, and the handles then added. While the vessel was rotating slowly horizontally, the fine thread of glass was wound around the lower part of the mouth.

The vessel is made of clear, colourless, greenish glass.

No firmly dated parallels have been found,[880] however the dating of the Thessalonean find was made possible by its own archaeological context: it was found in a sarcophagus along with a glass bracelet[881] and a spherical vessel of form 51a (Isings Form 101), which date their deposition time around the late 3rd or into the 4th century.[882]

This form probably derived from the bell-shaped vessels which are usually standing on a foot with beaded stem, or, less often, on a pushed-in or pad base. They appear already in the second half of the 1st or beginning of the 2nd century.[883] The most closely related forms are those of the typology of Clasina Isings 36b and 36c, which are dated to the 2nd and 3rd-4th century, which are also occasionally decorated with an applied thread around the neck.[884] It seems that this form of vessel evolved later to a different type of lamp with similar body, but now the handles were placed around the neck and the base was quite different,[885] and finally it entered the glassware of the Arab world and survived until the 10th century.[886]

The example from Thessaloniki, however, seems to belong to a different, contemporaneous line of evolution, which is probably connected to the silver lamps[887] and which, in all probability, later evolved into the Islamic glass lamps with handles on the shoulders, and no longer on the rim, that appeared at the end of the 1st millennium[888] and became very popular during the 13th and 14th centuries.[889]

### 42. Kantharoi, free-blown (Isings Form 38a)

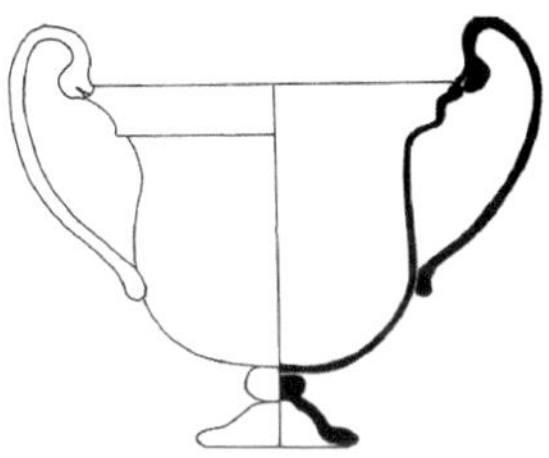

(Cat. nos. 186-187)
(Plate VII)

The vessels of this form have slightly everted, fire-polished rim, deep calyx-shaped body, and foot with a beaded stem and conical base. Two long, vertical coil handles start at mid-height of the body, and, following the curve of the body and rising up well beyond the rim, bend and end applied to it. Their height is estimated between 8-10cm.

They are free-blown, and the stem of foot and the base are separate masses of glass which are attached to the body while still hot and then formed by further tooling. At the end the coil handles, which were shaped while they being applied, were added.

The form derived from the ancient Greek shape of the vessel, which was produced in that era as well, probably more often in metal and mainly of silver.[890] At least three variants of 1st-century glass kantharoi are known, discerned according to the

[879] The vessel and all finds from this excavation were discussed in detail by the present author in Antonaras (in press 1 and 2).
[880] The only analogous find was located in an unpublished excavation of the *fossa*, the external ditch of the western city wall of Thessaloniki, from a site which did not render any more specific dating.
[881] Glass bracelets in the Mediterranean region became popular from around the 3rd century onwards, see Spaer 1988, 51; Antonaras 2003a; Antonaras 2006b.
[882] Antonaras in press 1 and 2.
[883] Isings 1957, 50-52, form 36.
[884] Whitehouse 2001, nos. 674, 782 of the 3rd-4th century; Harden *et al.* 1987, 123, no. 55, vessels dated to the 3rd century, on the basis of the archaeological context in which they were found, but also on stylistic grounds.
[885] Steckner 1988, 257-270, examples found on Samos, dated to the early 7th century.
[886] Carboni, Whitehouse 2001, 76, no. 6, objects dated to around 9th or 10th century.
[887] Similar shapes with identical, low base but with bigger rim, appear in silver lamps dated to the 6th century, and which have been found in various hoards from the Eastern Mediterranean. Newman, Lie 1992, 82-83, figs. 37. 1, 37. 2; Mundell-Mango 1992, 137, fig. 18, lamp of *argyroprates* Sergios; Snow 1992, 202, fig. 2, lamp from a hoard at Hama (Epiphania).
[888] Kröger 1995, 182, no. 235, find from Nishapur, dated by the archaeological context in which it was found to the 10th or 11th century.
[889] Carboni, Whitehouse 2001, 227-238, nos. 113-118, objects dated to the 13th-14th century.
[890] Examples are known from Boscoreale, La Baume 1964, 117, fig. 100, and from a hoard at Hildesheim, Gehrig 1967, 24 , figs. 19-20.

size of their handles and type of base. Due to the fragmentary preservation of the example it is not exactly clear to which variant it belonged, but most probably it can be associated with the previously mentioned shape.[891] Published examples from the Western provinces include Windisch (Vindonissa),[892] Valkenburg,[893] Nijmegen (Noviomagus Batavodurum), Xanten (Colonia Ulpia Traiana, Vetera), Vechten (Fectio), Avignon (Avennio), Hofheim, Bonn (Bonna), Vervoz (Vervigum), and Pompeii; they can be dated to the second half of the 1st century.[894] They are considered to be north Italian products which were mainly distributed in the northwestern provinces of the Roman Empire.[895]

The handles from Thessaloniki are made of colourless greenish glass, as was the body too, judging by the very small fragment preserved.

The example from Thessaloniki is dated to the last two decades of the 1st century, according to the destruction time of the baths in which it was found.[896]

### 43. Kantharoi, mould-blown

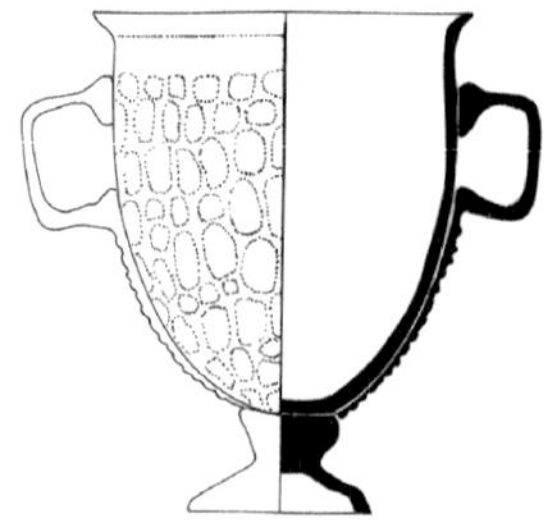

(Cat. nos. 188-189)
(Plate VII)

The vessels of this form[897] have unworked, slightly everted rim, hemispherical or deeper than hemispherical body, and bottom usually flat or occasionally slightly concave. They stand on an applied conical base which is connected to the body by a beaded stem. Two rectangular, almost square, strap, or small loop-handles, both variants with long undulating endings that reach the base of the vessel, are placed one opposite the other under the rim. Their height is estimated between 10-12cm, although much larger examples are also known.[898]

The body of the vessels of this form are dip-mould-blown, i.e. they are initially blown in a small and shallow mould and they are further inflated and modelled by free-blowing. While the vessel was still attached to the blowpipe, a mass of glass was applied under the bottom that was shaped by further tooling into the base. The, (probably partly prefabricated) handles were added at the end.

The relief decoration of the vessels usually consists of vertical ribbing directly from the slightly flaring rim, covering the entire circumference. Lower down, the body and bottom are covered by a web/network of ovular and hexagonal cells which form a honeycomb pattern.[899] This form evolved from the handleless bowls with identically decorated body which appear in the 4th century (Form 14).

The examples from the region of Thessaloniki are made of dark-blue and colourless greenish glass.

Identical examples have been found in Israel,[900] Egypt[901] and Syria,[902] and their production site should be expected in one of these regions. It is quite probable that they are indeed products of one large workshop of that region, because of their dispersed geographical distribution, but also because of their relatively complicated production. Another example of this form has also been traced in the unpublished material of the Early Christian basilica at Solinas in Cassandra, Chalkidiki.[903]

---

[891] Isings 1957, 53-54, form 38a; Van Lith 1978/79, 66, with thorough bibliography.
[892] Berger 1960, 38-39.
[893] Van Lith 1978/79, 64-66.
[894] Isings 1957, 53-54.
[895] Harden *et al.* 1987, 109.
[896] Adam-Veleni *et al.* 1998, 85-102; Mavromichali 2001, 141.
[897] These vessels evolved from Form 14, dated mainly to the 4th and 5th centuries.
[898] Unpublished, blue example with a height that exceeds 20cm. From the collection of the Museum für Islamische Kunst, reg. no. I. 2856, bought in Halepo in 1918 and exhibited in the Bode Museum, Berlin.
[899] See Form 14.
[900] Two finds have been published: a blue example from the church at Horbat Kenes, Karmiel, dated to the 4th-7th century; and the second from the baptistery at Nir Galim, on the southern coast of Israel, also dated to the 4th-7th century. See Gorin-Rosen 2000a, 237; Gorin-Rosen 2002a, 122.
[901] Example from a Christian place of worship in south Sinai at 'Ein Najila (4th-7th century), see Gorin-Rosen 2000a, 237. Another, intact example is known from Kom Aushim (Karanis), held at the Kelsey Museum (No 5361) and dated to the Late Roman period. This example consists of a colourless green bowl on which has been attached a base, comprised of a low, bead-like stem and conical base. Two opposed, blue, 'Pi'-shaped, strap-handles with long undulating endings that reach down to the base of the vessel are placed below the rim, at a similar height to the one that has been traced on the find from Rentina. It should be noted that among the unregistered glass fragments from the excavation of Rentina there are traces of fragments also of a blue, long and undulating handle.
[902] Blue example from Raqqa (Kallinikos), dated to the 7th-8th century by Lamm 1929-30, 54, pl. 12:3.
[903] Blue body and handle fragments, see Antonaras 2014e, 102, fig. 5, dated from excavation results between the 4th and 6th centuries, the period when the basilica was in use. I thank the excavator, I. Papangelos, for allowing me to publish the glass finds from this excavation.

The finds from Rentina cannot be firmly dated independently, although they are finds from an Early Christian basilica and its cemetery and thus are ascribed to the period between the 4th and 6th centuries.

### 44. Flasks with collar-shaped rim (Isings Form 70)

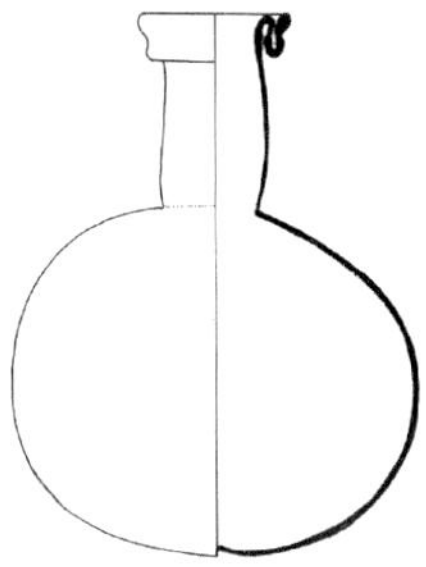

(Cat. nos. 190-192)
(Plate VII)

Vessels of this form have a rim which is turned out, down, up, and out in such a way as to form a band-like collar from which only a small edge protrudes at the lip. The neck is cylindrical, slightly splaying above the constriction at its base; the body is globular and the bottom is almost flat, slightly concave. Their height is estimated between 13-19cm.

These vessels were made by free-blowing, and they are often decorated with simple incised decoration.[904]

The extant examples of this form were mainly found at sites in the East Mediterranean.[905] It seems that they started to be produced, probably in Asia Minor,[906] already by the middle of the 1st century; they were produced up to the early 2nd century.

Examples have been found at Ercolano (Herculaneum),[907] Kerch (Panticapaeum),[908] Costanza (Tomis),[909] Anıtkaya,[910] Cyprus,[911] Akko (Acre, Antiocheia Ptolemais),[912] Castra,[913] Thera,[914] Naxos,[915] and Sifnos.[916] Furthermore, a jug with such a collar-shaped rim is published from Nea Kerdyllia.[917]

In addition, examples are kept also in Western museums such as Royal Ontario Museum, Toronto,[918] Toledo Museum of Art, Ohio,[919] and the Wolf collection.[920]

The vessels from Thessaloniki are dated, on the basis of the destruction time of the baths in which they were found to the last twenty years of the 1st century.[921]

### 45. Bottles squat conical body with curved walls, with constriction (Isings Form 82A2)

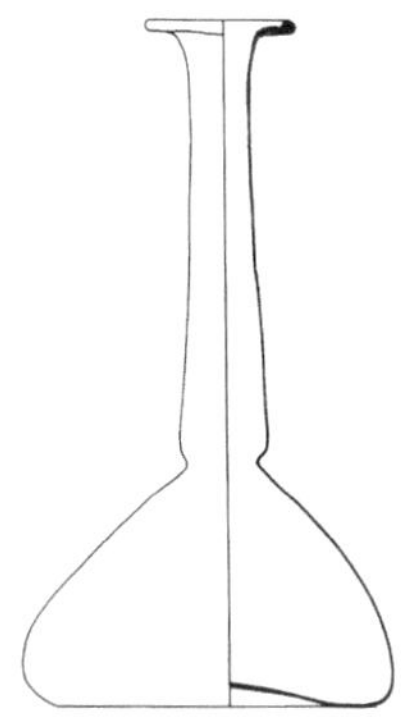

(Cat. nos. 193-196)
(Plates 7, VII)

Vessels of this form have everted, infolded tubular rim. Long cylindrical neck, double the height of the body, with a constriction at its base. The body is squat hemispherical, and the bottom is intensely concave almost hemispherical. Their height is estimated at c. 17cm.

The vessels of this form are free-blown and their rim formed while they were held by a metal rod attached on their bottom.

The vessels are ascribed to a variant of a well-known form which appears at the end of the 1st century and survives until the early 3rd century.[922]

---

904 Deeper incisions indicate an early date, prior to AD 79, while on later examples they are shallow, almost invisible; see Stern 2001, 102.

905 Isings 1957, 90, form 70, dated to the 1st century; Barag 1970c, vol. 2, pl. 41, form 15.3.

906 This special forming of the rim is considered to be characteristic of the production of an Asian Minor workshop, probably one from Pergamon, because clay and metal vessels of Pergamon origin indeed have similar rims. This workshop was operating in the 1st century; see Sorokina 1988, 55-60.

907 Scatozza-Höricht 1986, 56, nos. 125, 126, pl. 8, 18, 35, finds dated before the explosion of Vesuvius in AD 79.

908 Kunina 1997, 297, nos. 214-216, finds dated to the second half of 1st century based on its context and analogous finds from other sites.

909 Bucovală 1968, 86-87, no. 144, find from a grave dated from excavation results to the first half of the 1st century.

910 Lightfoot 1989, 28, no. 15, drawing 1:4, fig. 1:6, colour illustration 9:4, dated to the 2nd century according to analogous finds.

911 One example without precise provenance, Vessberg 1952, pl. 7:10, 16:2, and a second from the Oasis of Lemesos, Vessberg 1952, pl. 23:20, 24:110, which has been dated around AD 220.

912 Flavian find, around AD 69-79, according to Stern 2001, 102, in the comments of Cat. no. 37.

913 Gorin-Rosen 1998, 7, 8, finds from the site at Castra near Haifa, the single reference to their date is made in the captions of the illustrations, placing them in the Early Roman period.

914 Dragendorff 1903, 286-288, fig. 481 l, o, from graves dated from the 1st to the 4th centuries. Glass finds from the graves, at least those illustrated, as a whole can be ascribed to the production of the 1st and 2nd century.

915 Zafeiropoulou 1999, pl. 86β, dated according to analogous finds.

916 Mackworth-Young 1949, 88, pl. 32. 4 14 (13), find from grave 14, dated by a coin of Vespasian at AD 75.

917 Malama, Ntarakis 2008, 84, pl. 77, MA 6370, from a grave dated, on the basis of the glass vessels, to the second half of the 1st century.

918 ROM: Hayes 1975, 37-38, 58-59, nos. 144-146, fig. 4, pl. 11, finds dated to the second half of the 1st, or to the early 2nd century, according to analogous finds.

919 Stern 2001, 102, mentioned in the discussion of no. 37.

920 Stern 2001, nos. 37, 38, finds dated to the second half of the 1st or to the early 2nd century according to analogous finds.

921 Adam-Veleni *et al.* 1998, 85-102; Mavromichali 2001, 141.

922 Isings 1957, 97-98, form 82A2, with several examples from Oriental sites. Detailed bibliography can be found also in Dussart 1998, 165, who

Identical vessels to Cat. no. 196 are known from Ravenna[923] and Kassel (Castellum Cattorum),[924] and which are dated to the 2nd-3rd century, while there is a vessel identical to Cat. no. 194 in the National Museum of Budapest.[925] Other vessels of this form were found at Sremska Mitrovica (Sirmium),[926] Kostolac (Viminacium),[927] and probably also in Berlin.[928] In Greece some published examples identical to Cat. no. 196 were found at the Antimacheia, Kos,[929] Thera,[930] and Oropos.[931]

Identical vessels to Cat. no. 195 were unearthed on Samothrace[932] and Tripiti Melos.[933]

### 46. Bottles with conical body, curved walls, squat, without constriction (Isings Form 82B2)

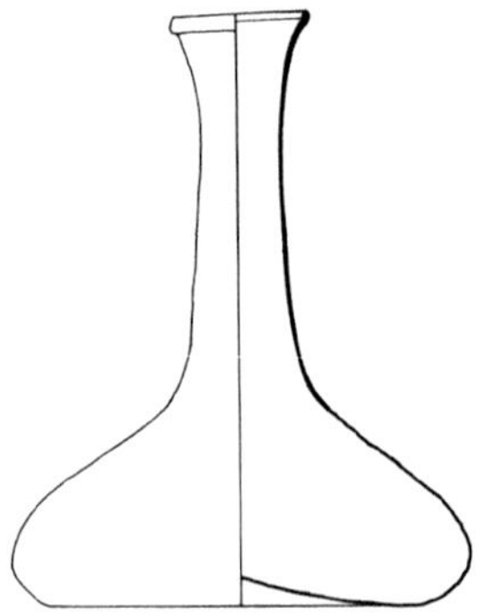

(Cat. nos. 197-198)
(Plates 7, VII)

Vessels of this form have inverted, almost horizontal, tubular infolded rim, and a long cylindrical neck three times taller than the body. The body is squat, conical with curved walls and flat slightly concave bottom. heir height is estimated at c. 11-20cm.

Vessels of this form are made by free-blowing, and since usually they have no pontil marks on the bottom, it seems that the shaping of the rim was done while they were held by jacks from the base of the neck.

The examples from Thessaloniki are made of colourless, bluish glass.

The form, usually with a body smaller than the examples from Thessaloniki, appears mainly in the West[934], i.e. in Britain,[935] Germany,[936] France,[937] Switzerland,[938] Italy,[939] and Dalmatia.[940] Further finds are known from Hungary,[941] Serbia (Kostolac/Viminacium),[942] FYROM (Skopje/Scupi),[943] and Bulgaria (Chatalka).[944] In the East, finds are detailed from Tipasa,[945] Dura-Europos,[946] Limassol,[947] and Amathous.[948] Greek find-spots are Nea Anchialos (Phthiotides

comes to the same conclusion about the chronology in the discussion on form BXIII 1211a, pl. 50.

923 De Tommaso 1990, 62, form 37.

924 Boosen 1984, 60, no. 110, dated on the basis of analogous finds to the 1st-2nd century, and is considered an Oriental product.

925 Barkóczi 1988, 125, no. 241, pl. XIX, LXXXI, object of unknown provenance, which for stylistic reasons is dated to the early 2nd century.

926 Ružić 1994, 31, pl. XX/1, find dated between the 2nd and 4th century.

927 Ružić 1994, 31, pl. XX/5, find dated from excavation results to the end of the 2nd century.

928 Platz-Horster 1976, 67, no. 128, dated on the basis of analogous finds to the 2nd-3rd century, and is considered an Oriental product.

929 Zervoudaki 1971, 545-549, pls. 556, 557, vessels from graves dated to the 2nd-3rd, as well as the second half of the 3rd century.

930 Dragendorff 1903, 286-288, fig. 481 h, from graves dated between the 1st and 4th century. The glass finds, at least those illustrated, can be ascribed to the production of the 1st and 2nd century.

931 Pologiorgi 1998, 41, no. 2 MΩ 265, 268 pl. 12β, finds from a grave dated according to the glass vessels to the second half of the 1st century.

932 Dusenbery 1998, nos. W12-12, 13, XS-527, examples dated from excavation results to the second third, and to the middle of the 2nd century.

933 Zafeiropoulou 1972, 605, pl. 558, nos. 534-535, from a destroyed Roman grave, dated on the basis of analogous finds to the 1st-2nd century.

934 Isings 1957, 99, form 82B2, with many examples from France, Holland, and mainly Germany.

935 Price, Cottam 1998a, 175-177, with detailed bibliography on dated examples from Britain, several with stamped base, dated from excavation results to the second and third quarters of the 2nd century.

936 Example from Trier (Treviri) in Goethert-Polaschek 1977, no. 613. Example from Cologne, Fremersdorf 1958b, 42 pls. 88-89, seven examples dated from excavation results to the early 2nd and to the 2nd century. Fremersdorf 1961, 49-50, pl. 97, example dated at around AD 100.

937 Arveiller-Dulong, Arveiller 1985, 90-91, find from Strasbourg (Augusta Trebecorum, Argentoratum) with stamped base dated to the late 1st-2nd century according to analogous finds.

938 Biaggio Simona 1991, vol. 1. 156-156, example dated to the 3rd century from Locarno (Solduno).

939 De Tommaso 1990, 61-62, form 35, with a large number of finds from Italy dated from excavation results to the 2nd century. In the western part of the Empire, the bottom of these unguentaria is quite often decorated with an impressed medallion.

940 *Trasparenze imperiali 1997*, 138, nos. 89-90, finds from Zadar (Iader), Nona or Benkovac (Asseria), dated according to analogous finds to the 2nd-3rd century.

941 Barkóczi 1988, 117-118, nos. 211 and 213, pls. XVII, XVIII, LXXX, finds from Szombathely (Savaria), and of unknown provenance in the National Museum of Budapest, dated to the early 2nd century and which have been considered by the researcher to be Italian products.

942 Zotović, Jordović 1990, 94, 101, pls. CXXXIX:3, CXXVIII: 8, finds from graves dated from excavation results to the years before Antoninus Pious (138-161) and between the reign of Domitian (81-96) and Traian (98-117), respectively; Spasić-Djurić 2006, 301-302, figs. 3/7-10, 4/1, 4, finds from a grave dated from excavation results to the first half of 2nd century.

943 Mikulčić 1976, pl. V, no. 375, similar find from a grave dated between the reign of Domitian (81-96) and Hadrian (117-138).

944 Bujukliev 1986, 76, pl. 13, from a cremation dated from excavation results to the second half of the 1st century.

945 Lancel 1967, no. 136 (late 1st- early 2nd) and no. 137 (late 1st-3rd century), dated according to analogous finds.

946 Isings 1957, 99, an example with different, funnel-like mouth.

947 Vessberg 1953, 167, dated from excavation results around AD 220.

948 Vessberg 1953, 163, dated from excavation results to the Antonine era (138-192).

Thebes),[949] Delphi,[950] Eleusis,[951] Athens,[952] Patras,[953] Elliniko in Kynouria (Eva of Thyraiatis),[954] Antimacheia,[955] and Marmaroto, Kos.[956] Some of the examples of this form are dated to the end of the 1st century, while the majority are dated to the 2nd, and some even to the early 3rd century.

Among the finds from Thessaloniki only Cat. no. 197 can be independently dated, by its archaeological context, to the first half of the 2nd century.[957]

### 47. Bottles with conical body, straight walls, flat shoulder (De Tommaso Form 53)

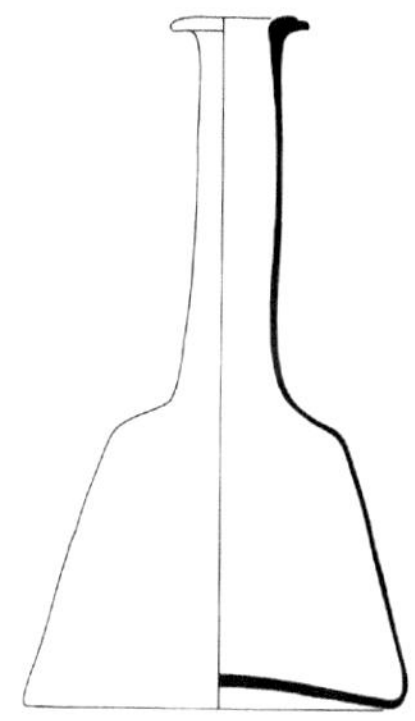

(Cat. no. 199)
(Plates 7, VII)

This form comprises a colourless, greenish flask/unguentarium with inverted, infolded tubular rim, long cylindrical neck almost of the same height as the body; the body is truncated conical, flat shoulder, and flat bottom. This practically colourless bell-shaped vessel is the only preserved example from Thessaloniki of a well-known form of flask.[958] Their height is estimated at c. 16cm.

In Greece analogous finds include two from Samothrace,[959] three from Ierissos (Acanthos),[960] four at Nea Anchialos (Phthiotides Thebes),[961] two at Perissa, Thera,[962] three from Antimacheia, Kos,[963] and one each from Stavros,[964] Ierissos, Chalkidiki,[965] Athens,[966] Tripiti, Melos,[967] Naxos,[968] Paros,[969] Marmaroto, Kos[970] and Alexandroupoli.[971]

A vessel of this form was found in Cyprus (Amathous)[972] and is dated to the 4th century, according to another find from Beirut dated by a coin to the early 4th century.[973]

72 of the 106 known vessels of this form have been found along the Croatian/Adriatic coast, and 54 of them in Zadar (Iader).[974] Parallel examples have been found in north Italy, Milan (Mediolanum),[975] Genoa[976] and Albenga

[949] Sotiriou 1939, 64, figs. 16, 17, 25, finds from graves without firm dating.

[950] Perdrizet 1908, 216, fig. 948, finds from a grave in the eastern necropolis.

[951] Kyparissis 1926, 71, fig. 16, from a grave which is probably dated to the years after AD 267; Mylonas 1931-32, 44-45, fig. 5, finds from a grave with a coin dated to the second half of the 1st century BC; Mylonas 1975, 12, grave A9 vessel 7, dated according to the glass vessel to the second half of the 2nd century; Papangeli 1987, 32, pl. 29, find from a sarcophagus dated from excavation results to the 2nd century.

[952] Threpsiadis, Platonos 1963, 35, pl. 35β, two vessels from a grave; Stavropoulos 1964, 60, pl. 57α, β, two vessels from an ancient Greek grave which was reused in the Roman period; Alexandri 1967b, 98, pl. 90στ, two vessels from a Roman grave; Alexandri 1969, 70, pl. 59β, vessel from a Roman grave.

[953] Agallopoulou 1973-1974, 369, pl. 235, from a grave dated by the glass finds to the first half of 1st century.

[954] Abadie, Spyropoulos 1985, 416, fig. 72, vessel from a grave dated to the 2nd century by the glass vessel and idem 437-438, figs. 125-128, from a grave dated to the 2nd or to the early 3rd century, also by the glass vessels found in it.

[955] Zervoudaki 1971, 545-549, pls. 556, 557, vessels from graves dated to the 2nd-3rd and the second half of the 3rd century.

[956] Touchais, Huber, Philippa-Touchais 2001, 1002, fig. 249, from a grave with glass context dated to the 1st or early 2nd century.

[957] The grave is dated by its glass context (vessels BY 227/13-15, form 129) and by a clay lamp to the first half of the 2nd century, see Perlzweig 1961, 94, no. 259, pl. 8.

[958] Kisa 1908, 390, formentafel A 23, where the form is dated to the late 3rd century.

[959] Dusenbery 1967, 43-44, no. 34, 2nd century; Dusenbery 1998, 1120-21, the vessel, which was also presented in 1967, and another are both are dated to the end of the 1st - early 2nd century.

[960] Trakosopoulou 2002, 86, figs. 17, 18. The graves are dated by coins of Septimius Severus (193-211) and Caracallas (198-211). The dimensions and the proportions appear to be similar to the ones in the production of Zadar (Iader).

[961] Sotiriou 1939, 53-72, fig. 16, reports a 'cemetery of the Late Roman times... totally abandoned during the period when the Christian city was flourishing'; Weinberg 1962, 130, 131, pl. 27, fig. 10, which includes a passage noting that 'Professor Sotiriou was dating the vessels from the graves of Nea Anghialos, generally and on the basis of their context, in the late 2nd and to the 3rd century AD'.

[962] Gerousi 2002, 138-139, fig. 13, from a secondary burial of a sealed grave dated to the 5th century. The excavator considers that the 2nd century is a very early date for this grave and mentions comparable material of the 4th century.

[963] Zervoudaki 1971, 545-549, pls. 556, 557, vessels from graves dated to the 2nd-3rd, and to the second half of the 3rd century.

[964] *Glass Cosmos* 2010, 343-44, nos. 359, 361, from two graves dated to the first half of the 2nd century.

[965] *Glass Cosmos* 2010, 379, no. 440, from a grave dated between AD 190-211.

[966] Alexandri 1970, 43, pl. 48β, at the junction of 'Achilleos' and 'Salaminos' Streets, grave V.

[967] Zafeiropoulou 1972, 605, pl. 558, no. 536, dated on the basis of analogous finds to the 2nd century.

[968] Zafeiropoulou 1999, pl. 89β, dated on the basis of analogous finds.

[969] Zafeiropoulou 1999, pl. 84β, dated on the basis of analogous finds.

[970] Touchais, Huber, Philippa-Touchais 2001, 1002, fig. 249, from a grave with glass context dated to the 1st or early 2nd century.

[971] Skarlatidou 1981, 351-353, pl. 242γ, vessel from a grave dated, by the glass vessel, between the 2nd and 4th century.

[972] Vessberg 1952, 139, pl. VIII. 24.

[973] *Syria* 16 (1935), 62 ff (non vidi).

[974] Fadić 1989, 11, 21; Gluščević 1998, 182-190.

[975] Gluščević 1998, 186.

[976] Pastorino 1989, 40, no. 156, photo on page 63, drawing on pl. VI. The vessel, which has dimensions corresponding to the products of Zadar (Iader), is of unknown provenance, deriving from the collection of Prince Oddone.

(Albingaunum).[977] In France two more vessels in Picardie.[978] Other known examples of this form were found in Romania (Costanza/Tomis),[979] Serbia (Niš/Naissus),[980] Montenegro (Kolovrat near Prijepolje[981] and Komini/ Municipium 'S...'),[982] and in Albania (Apollonia, Illyria).[983] Three more examples were found in Hungary: one is of unknown provenance,[984] one was found in Szőny (Brigetio)[985] and the third in Budapest (Aquincum).[986] Probably vessels from Bulgarian Plovdiv (Philippopolis),[987] dated to the 2nd-3rd century, belong to this form, but the oversimplified drawings and the generic nature of their publication does not permit secure conclusions.

The distribution and production sites of these vessels were thoroughly studied by Smiljan Gluščević.[988] He agrees with De Tommaso[989] and, based on new material from his excavations, he separated the vessels of this form (with height of c. 16-17cm, and a body: neck ratio of 1:1) from the more general form of Isings 82A1. Due to the great number of finds in his region and their uniformity, in terms of size and texture, he ascribed them to a local workshop from Zadar (Iader). They appear in the second half of the 2nd century and were produced throughout the 3rd century.

The find from Thessaloniki cannot be independently dated.[990]

**48. Bottles conical body, curved walls, slender, tubular rim (Isings Form 82A2 var.)**

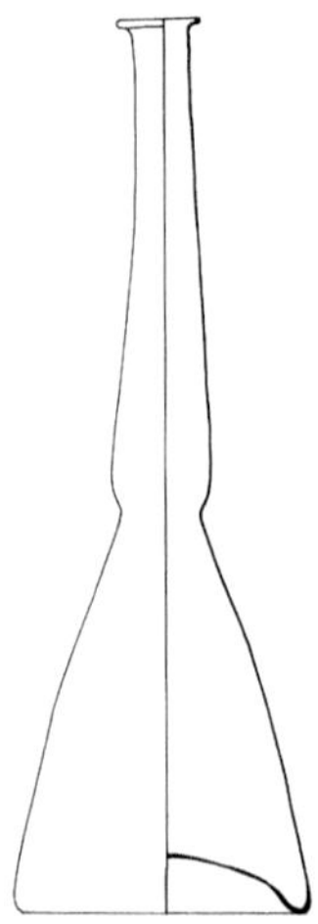

(Cat. nos. 200-204)
(Plates 8, VII)

The vessels of this form have everted rim, either tubular in-folded, or simple, fire-polished ones. They feature a relatively wide, cylindrical neck, double in height compared to the body, mildly splaying above the constriction at its base. The body is elongated, conical, standing on the periphery of the under-surface, which is highly concave, almost hemispherical or domed. Their height is estimated between 16-24cm.

Vessels of this form are free-blown and the shaping of their rim was performed while they were held by a blowpipe attached to their under-surface.

The examples from Thessaloniki are made of colourless greenish glass.

The vessels are ascribed to a variant of a well-known form which appears at the end of the 1st century and survived until the early 3rd.[991] Identical vessels were found in the necropolis of Brescello ( Brixellum),[992] at Skopje (Scupi)[993] and Chatalka,[994] while similar finds from Asia Minor, apart from those in the museums at Afyon[995] and Bodrum (Halicarnassus),[996] are housed in Geneva.[997] In Greece published examples of this form were found at Dikaia,[998] Veroia,[999] Pythagorion, Samos,[1000] Antimacheia, Kos[1001] and Thera.[1002]

[977] De Tommaso 2001, 160-161, the dimensions correspond to the products of Zadar and are dated to the end of the 2nd - early 3rd century.
[978] Dilly, Mahéo 1997, 97-98, nos. 181-182, objects of unknown provenance, dated to the 2nd-3rd century according to analogous finds.
[979] Bucovală 1968, 110-113, nos. 222-233; Ružić 1996, 192; Gluščević (1998, 185, 188) noticed that the vessels from Constanza (Tomis) present common characteristics (body:neck ratio) and that possibly they are products of another, probably local, workshop which appears to operate in the 2nd-3rd century, contemporaneous to the workshop of Zadar (Iader).
[980] Drča 2000, 216-217, 221, no. 47, pl. VI/47, vessels dated to the 3rd-4th century.
[981] Cermanović-Kuzmanović 1992, 389, pl. II/15; Ružić 1996, 192, photo 2.
[982] Cermanović-Kuzmanović 1971, 294, 300-301, pl. 4. 17, dated to the 3rd century.
[983] Mano 1975, 208, pl. III. 1, 3, two 2nd-3rd-century examples.
[984] Barkóczi 1988, 124-125, no. 241, pl. XIX. 241, dated to the early 2nd century.
[985] Barkóczi 1988, 124-125, no. 242, pl. XIX. 242, dated to the early 2nd century.
[986] Kaba 1958, 430, fig. 11. 23.
[987] Cherneva-Tilkian 1997, 49-50, forms IVB and VB, nos. 24-27 (drawings pages 54-55).
[988] Gluščević 1998, 188.
[989] De Tommaso 1990, 73, form 53.
[990] It was found with the glass vessels Cat. nos. 270, 698, 736-737, dated to late 3rd-4th century. Yet it is not known if they were found in one or more burials.
[991] Isings 1957, 97-98, form 82A2, with many examples from Oriental sites. Thorough bibliography at Dussart 1998, 165, who in the discussion of form BXIII 1211a, pl. 50, concludes with the same chronology.
[992] De Tommaso 1990, 86, form 75, find from a grave dated to the Antonine period (138-192).
[993] Mikulčić 1976, 195, pl. VII, no. 132, dated to the reign of Hadrian (117-138) on the basis of a coin and the ceramic finds which were found together.
[994] Bujukliev 1986, 95, nos. 437, 438, pl. 31, finds from a grave dated by coins in the reign of Alexander Severus (222-235).
[995] Lightfoot 1989, 27, no. 13, pl. 3/2, find dated to the middle of the 1st or to the 2nd century on stylistic reasons.
[996] Özet 2000, 134-135, nos. 89, 90, finds dated to the 2nd-3rd century on stylistic reasons.
[997] De Tommaso 1990, 86, form 75.
[998] Triantafyllos, Karadima 1991, 336, pl. 131β, from a tumulus with a cremation dated by the numismatic evidence to the 2nd century.
[999] Exhibited in the Archaeological Museum of Veroia (exh. no. 937).
[1000] Tsakos 1980, 463, pl. 272, vessel from ruined Roman graves.
[1001] Zervoudaki 1971, 545-549, pl. 556, vessel from a grave of the 2nd or 3rd century.
[1002] Dragendorff 1903, 286-288, fig. 481 f, g, from graves dated to the 1st to the 4th century. The glass finds from the graves, at least those

The finds from Thessaloniki cannot be dated independently, on the basis of archaeological information, and only Cat. no. 200 was found in a grave, probably dated to the 3rd century.[1003]

**49. Bottles with cracked-off rim, biconical with indentations**

(Cat. no. 205)
(Plates 8, VII)

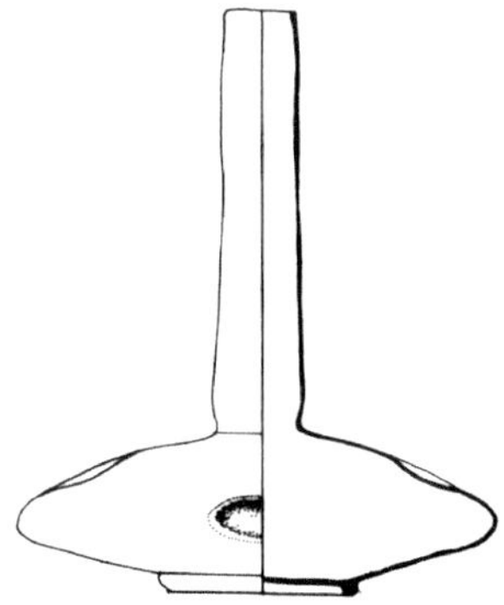

The vessels of this form have cylindrical neck ending in a cracked-off rim that is not particularly shaped; at the transition to the body the neck is strongly constricted. The body is biconical and bears four indentations at its upper part. The vessel stands on a pushed-in ring base. The find from Thessaloniki is made of colourless, greenish glass. Their height is estimated at c. 15cm.

The vessel is free-blown. The body was originally globular and which was squeezed while still hot to arrive at its biconical form. Then the pushed-in base was formed by folding the mass of the lower part of the body. Indentations were created by simple, momentary pressure and these decorate the upper part of the body. The rim of the vessel was left unworked.

Analogous finds were found in Veroia,[1004] Meniida,[1005] Nea Kerdyllia,[1006] and Cilicia;[1007] examples of unknown provenance are kept in Western museums, i.e. the Corning Museum of Glass[1008] and the Römisch-Germanisches Museum in Cologne.[1009]

Similar vessels, with taller, less squat bodies, are also found in the same[1010] and other[1011] excavations, while similar examples of unknown provenance are kept in several museums,[1012] and indicate a general trend in glass production during the 2nd and 3rd centuries to decorate vessels with simple indentations, which were quite probably rendered as hammered motifs on metal prototypes.

The find from Thessaloniki cannot be dated independently by its archaeological context and it can only be ascribed to the general time period of the late 2nd or 3rd century. But if the sarcophagus in which it was found was used only once, then the vessel should be dated to the very end of the 3rd century, because in the same sarcophagus a fusiform flask of Form 147a was found, which appear from the end of the 3rd century onwards.

**50. Bottles spherical with cracked-off rim (Isings Form 103)**

(Cat. nos. 206-214)
(Plates 8-9, VII-VIII)

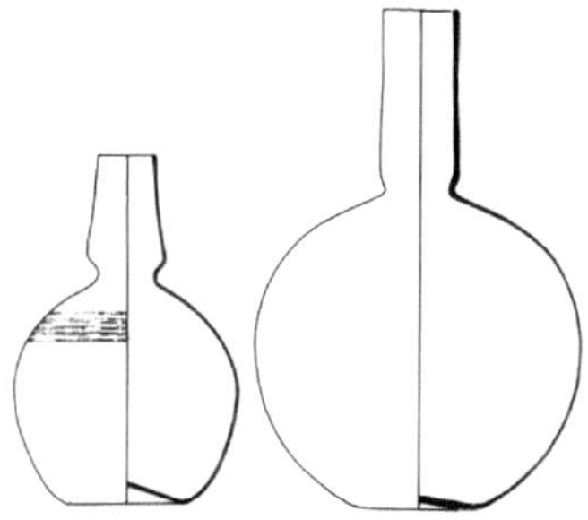

The vessels of this form do not have a particularly shaped rim; their neck is relatively short and cylindrical and splays towards its base, where it bears a strong constriction at the transition to the spherical body; the plain base is flat and the bottom is slightly concave. Their height is estimated between 15-17cm.

These vessels often bear incised decoration, which is usually quite simple and consisting of parallel, horizontal, shallow grooves or bands of fine incisions. Less often they bear more complex geometrical patterns, e.g. concentric intersecting circles,[1013] and only rarely are they decorated with figural incised and/or painted decoration.[1014]

illustrated in the publication, as a whole can be attributed to the production of the 1st and 2nd century.

1003 Makropoulou 2007, 234, pl. 27, from a 3rd-century grave, found together with a vessel of Form 132.

1004 Petsas 1964, 353, pl. 414δ, find from a grave dated to an 'advanced Roman period'.

1005 Chrysostomou 2000, 455-467, fig. 19, from a grave dated by a coin of Gallian to the years after AD 261. On the same find, see also, Chrysostomou 1999-2001, 195-200, fig. 16.

1006 Malama, Ntarakis 2008, 346, pl. 78, MA 7076, from a grave dated to the first half of the 3rd century.

1007 Stern 1984, 136, fig. 4 (centre), from the site of Yumurtalık/Ayas (Aigai Cilicia), dated to the 2nd-3rd century.

1008 Whitehouse 2001, 181, no. 319, object of unknown provenance dated to the 2nd-3rd century according to analogous finds.

1009 La Baume, Salomonson 1976, 48, no. 130, pl. 16. 4, object of unknown provenance dated on the basis of its decoration to the 2nd-3rd century.

1010 Stern 1984, 136, fig. 4 (left), from the site of Yumurtalık/Ayas (Aigai Cilicia); Mikulčić 1976, VI. 334, find from Skopje (Scupi) from a grave dated by a coin and a clay lamp to the late 2nd century.

1011 Find from Naxos, in Zafeiropoulou 1999, pl. 87γ, dated on the basis of similar finds.

1012 For Nicosia, see Vessberg 1952, 136, form A.I.α.I., pl. VIII. 7; for Cologne, see La Baume 1973, 48, no. 129; for the Oppenländer Collection, see Saldern *et al.* 1974, 216, no. 627, dated to the 3rd century.

1013 A vessel from Cherson (Chersonesus Taurica), see Kunina 1997, 166, 299, no. 226, from a grave dated to the second half of the 4th century. For the find from Dunaújváros (Intercisa), see Barkóczi 1988, 146, no. 314, pl. XXVIII, find dated from excavation results to the late 3rd- early 4th century. Fora vessel from Bonn (Bonna), see Follmann-Schulz 1988, 38, pl. 7. 88, find dated from excavation results to the last third of the 3rd century. For the three examples from Duklja (Doclea) in Montenegro, see Cermanović *et al.* 1975, 154-155.

1014 Whitehouse 2001, 271-272, no. 865.

They are free-blown and the constriction at the base of the neck is created by jacks used to shape the cylindrical neck. The rim remains unworked or slightly ground.

It is a widely distributed form[1015] which appears mainly in the West, but also in the East, where the examples with fire-polished rim prevail.[1016] They appear around the end of the 3rd century and they remain in use into the 4th. This form appears at the same sites in the same period in smaller size, under 15cm in height, and these examples are discussed among the unguentaria (Form 145).

Among the extremely numerous published examples of this form, there are finds from Colchester (Camulodunum),[1017] Strasbourg,[1018] Trier (Treviri),[1019] Mainz am Rhein (Moguntiacum),[1020] Bonn (Bonna),[1021] Cologne,[1022] Augst,[1023] Ptuj (Poetovio),[1024] Celje (Celeia),[1025] Szőny (Brigetio),[1026] Dunaújváros (Intercisa),[1027] Aquileia,[1028] Kostol (Pontes),[1029] Gračanica (Ulpiana),[1030] Sremska Mitrovica (Sirmium),[1031] Morović,[1032] Niš (Naissus),[1033] Mala Kopašica,[1034] Duklja (Doclea),[1035] Komini (Municipium 'S...'),[1036] Dion,[1037] Strovolos,[1038] Oropos,[1039] Argos,[1040] Dura-Europos,[1041] Ashdod (Azotos?),[1042] and Bosra (Nova Trajana Bostra).[1043]

The examples from Thessaloniki are dated to the 3rd-4th century, according to analogous finds from other sites. More precise chronological clues exist for the vessels, i.e. Cat. nos. 206,[1044] 207,[1045] 208,[1046] 210-211,[1047] 214,[1048] which are independentdated to the same period by as the analogous finds from other sites.

---

[1015] Isings 1957, 122-123, form 103. This form also has identical clay vessels, see La Baume, Salomonson 1976, no. 589.

[1016] On the matter see Dussart 1998, 151, where analogous finds from the Syro-Palestinian region are listed.

[1017] Cool, Price 1995, 153-154, finds dated from excavation results to the second half of the 3rd and to the first half of the 4th century.

[1018] Arveiller-Dulong, Arveiller 1985, 127-128, nos. 276-278, the vessel, dated from excavation results, is ascribed to the second half of the 3rd or the beginning of the 4th century.

[1019] Goethert-Polaschek 1977, nos. 902-914, form 93, pls. 58-59, the finds, dated from excavation results, are ascribed to the second half of the 3rd and the first half of the 4th century.

[1020] Harter 1999, 140-141, form G21, among the remaining examples in the museum, 7 can be dated more precisely between the middle of the 3rd and the middle of the 4th century.

[1021] Follmann-Schulz 1988, 37-38, pl. 7. 85-87, dated from excavation results to the second half of the 3rd century.

[1022] Fremersdorf 1958a, 43-44, pls. 74-77, finds dated from excavation results to the end of the 3rd century; Fremersdorf 1967, 56, 110-112, pls. 11, 112-113, 116-117.

[1023] Rütti 1991, I 54 AR 154.1, 130-131, nos. 2599-2602, pl. 110, finds decorated with incisions, dated from excavation results to mid 3rd to mid 4th century.

[1024] Lazar 2003, 145-146, forms 6. 2. 7, finds from a grave dated first half to end of 3rd century.

[1025] Lazar 2003, 145-146, forms 6. 2. 6, finds from a grave dated to the 4th century.

[1026] Barkóczi 1968, 63, 67, no. 12a, b. Two undecorated examples from Szőny (Brigetio), dated from excavation results around middle of 3rd century. For another example, see Barkóczi 1988, 145, no. 315, pls. XXVII, LXXXVIII, from a grave dated to the last third of the 3rd century or beginning of 4th. A fourth example is published in Barkóczi 1988, 145, no. 317, pl. LXXXVIII, from a grave with a coin of Aurelian (270-275), dated to the last third of the 3rd century.

[1027] Radnóti 1957, 146, no. 132, from a grave dated to the end of the 3rd- early 4th century which contained a coin of Diocletian (284-305). Also, a vessel decorated with incisions in Barkóczi 1988, 146, no. 314, pl. XXVIII, a find dated from excavation results to the end of the 3rd- early 4th century.

[1028] Calvi 1968, pl. 22. 6, vessel dated to the 3rd century.

[1029] Ružić 1994, 15, no. 68, dated, probably from excavation results, to the middle of the 4th century.

[1030] Ružić 1994, 15, no. 69, dated, probably from excavation results, to the middle of the 4th century.

[1031] Cermanović-Kuzmanović 1981a, 8, 17, nos. 12-13, pl. II, photo 3. Two examples without firm dating, from a phase of the cemetery dated to the 3rd-4th century. Based on analogous finds from other sites they were later dated to the second half of the 4th century, see Šaranović-Svetek 1986, 64, 99.

[1032] Milošević 1979, 115-116, find from a grave dated according to its context to the middle of the 4th century.

[1033] Drča 2000, 212, no. 11, pl. II, find dated from excavation results to the end of the 3rd and to the 4th century.

[1034] Stamenković, Ivanišević, Pešić 2016, p. 28-30, ill. 16:1, from a cremation dated between the 2nd and the late 3rd century.

[1035] Cermanović *et al.* 1975, 154-155, formI/2c, photos 12, 13, 15. Although none of the examples can be dated independently, the introduction of this form in Damaltia is estimated in the first half of the 3rd century, as some of the examples were found in cremations.

[1036] Srejović 1975, 182, pl. IV/4, find from a burial dated to the last quarter of the 3rd century.

[1037] Pantermalis 2001, 351-352, figs. 2-3, 5.

[1038] Malama, Christopoulou, Ntarakis 2000, 78, fig. 15, find from a grave dated to the 3rd century.

[1039] Pologiorgi 1998, 47, pl. 14γ, find dated on the basis of analogous finds to the 3rd-4th century.

[1040] Morou 1981, 109, pl. 50α, vessel from a Roman grave, with no other chronological indication.

[1041] Clairmont 1963, nos. 543-562, dated from excavation results to the second half of the 2nd - the first half of the 3rd century. This is the earliest published example of this form.

[1042] Barag 1967a, 37, no. 13.

[1043] Dussart 1998, 151, form BX.511, vessel with fire-polished rim; Coscarella 1994, 393, example dated according to analogous find from Samaria to the 3rd century.

[1044] Nalpantis 2003, 80, 129-130, drawing pl. 43, from a grave dated to the end of the 3rd century.

[1045] Makropoulou 2007, 214, pls. 26, 68, from a grave used at the end of the 3rd century by an Antonianus of Probus, dated between AD 276-282.

[1046] Makropoulou 2007, 284, pl. 25, from a *cubiculum* used from the end of the 3rd to the end of the 5th century.

[1047] From a sarcophagus with glass finds dated to the 3rd-4th century belonging to Forms 50, 35 and 147a.

[1048] From the excavation of a house with mosaic pavements dated to the second half of the 4th and the 5th century. See Chapter 9 (Find-Spots), '47-49, 58-62 Sokratous Street (today Moirarchou Koufitsa St.), 1968'.

### 51. Bottles with tubular rim, spherical

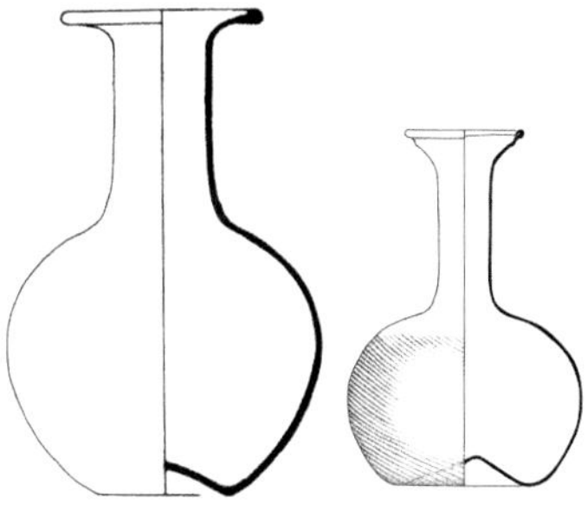

a. Smooth
(Cat. nos. 215-253)
(Plates 9-12, VIII-IX)

b. Ribbed
(Cat. nos. 254-273)
(Plates 12-14, IX)

Vessels of this form are handleless, have everted, infolded tubular rim, cylindrical neck, spherical body and they stand on a slightly concave bottom. As a whole they are made of colourless greenish or colourless bluish, often bubbly glass. Their height is estimated between 15-22cm.

The smooth types of these vessels are free-blown, and the ribbed examples are dip-mould-blown, i.e. they were initially blown in a small ribbed mould and then expanded to their final dimensions by further free-blowing and rotating. The rim of both groups is shaped by jacks and continuous rotation, while the vessel was attached at its bottom on a blowpipe, which left a characteristic mark, known as pontil mark or scar.

It is a very widely spread form[1049] which appears mainly in the West, but also in the East. It appears around the end of the 3rd century and remains in use throughout the 4th and up to the early 5th century. This form also appears at the same sites, and in the same contexts and strata, in a smaller format, under 15cm in height. These smaller examples are treated as a different form, and will be presented among the flasks/unguentaria (Form 139).

The two canonical groups are:

a. Spherical smooth
b. Spherical with oblique ribbing

Among the extremely numerous preserved finds the following are particularly noted because they can be independently and relatively firmly dated:

a) Smooth: Mainz am Rhein (Moguntiacum),[1050] Strasbourg (Augusta Trebecorum, Argentoratum),[1051] Trier (Treviri),[1052] Cologne,[1053] Szőny (Brigetio),[1054] Ságvár (Tricciana),[1055] Dunaújváros (Intercisa),[1056] Ptuj (Poetovio),[1057] Duklja (Doclea),[1058] Sremska Mitrovica (Sirmium),[1059] Kostol (Pontes),[1060] Ravna (Timacum Minus),[1061] Hrtkovci ('Gomolava'),[1062] 'Kruševlje' (Sviloš),[1063] Kostolac (Viminacium),[1064] Niš (Naissus),[1065] and in Romania, Veţel (Micia).[1066] The only published parallel from Greece was found at Nea Anchialos (Phthiotides Thebes).[1067]

b) Ribbed variant: Mainz am Rhein (Moguntiacum),[1068] Krefeld-Gellep (Gelduba),[1069] in Hungary, Csákvár (Floriana),[1070]and Pécs (Sopianae);[1071] in Croatia, Bakar (Volcera), Zadar (Iader), Solin (Salona), Makarska (Inaronia),

[1049] Isings 1957, 119-120, form 101, group of taller examples, taller than 15cm.
[1050] Harter 1999, form G27a, 146-148, 286-288. nos. 1037-1049, pls. 50, 104, 90, 102, some of which, at least, are dated from excavation results between the end of the 3rd and middle of the 4th century.
[1051] Arveiller-Dulong, Arveiller 1985, 119-126, nos. 218-272, finds dated from excavation results to the 3rd and to the first half of the 4th century. This form also includes small-sized finds, probably unguentaria.
[1052] Goethert-Polaschek 1977, form 79a, figs. 15/6, 23/1, finds dated to the 4th century. In terms of height they can be ascribed to form 98 in Goethert-Polaschek (1977, 154-155, fig. 59) and dated from excavation results to the second half of the 4th century.
[1053] Fremersdorf 1958b, pl. 86.
[1054] Barkóczi 1988, 137 nos. 288-289, first half of the 4th century, and idem, 139, no. 294, find dated from excavation results to the end of the 4th century.
[1055] Barkóczi 1988, 138 no. 290, find dated from excavation results to the second half of the 4th century.
[1056] Barkóczi 1988, 138 no. 291, find dated from excavation results to the second half, or to the end of the 4th century.
[1057] Šubic 1976, 58, pl. IV/36, 37 finds dated according to analogous finds to the 1st-2nd, and 4th centuries.
[1058] Cermanović *et al.* 1975, 153-154, form I/2a, examples dated from excavation results to the second half of the 3rd and into the first half of the 4th century.
[1059] Cermanović-Kuzmanović 1981a, 17, 11; Šaranović-Svetek 1986, 63, nos. 90, 91, 94, examples dated from excavation results to the 4th century and to the second half of the 4th century.
[1060] Ružić 1994, 14, no. 55, dated, probably from excavation results, to the second half of 4th century.
[1061] Ružić 1994, 14, no. 56, dated, probably from excavation results, to the second half of 4th century.
[1062] Šaranović-Svetek 1986, 63, no. 87; Ružić 1994, 14, no. 57, dated, probably from excavation results, to the end of the 3rd- mid 4th century.
[1063] Šaranović-Svetek 1986, 63, nos. 86, 88, 89, dated on the basis of analogous finds to the end of the 3rd- mid 4th century.
[1064] Ružić 1994, 15, no. 62, pl. V/7, dated to the end of the 2nd century by a coin of Antonine Pious (138-161).
[1065] Drča 2000, 212, 218, no. 10, pl. II/ 10, dated from excavation results to around the 4th century.
[1066] Baluta 1978, 103, find dated, probably from excavation results, to the 3rd century.
[1067] Sotiriou 1939, 64 fig. 18, examples from graves without firm dating.
[1068] Harter 1999, 147-148, 286-288, form G27b, nos. 1050-1051, pls. 51, 104. 82. 4, 104. 83. 2, dated to the end of the 4th or early 5th century, as with other analogous finds from Germanic areas.
[1069] Pirling 1959, 225-226, grave 530, fig. 11. 17, find dated from excavation results to the late 4th century.
[1070] Barkóczi 1988, 139, no. 297, find dated from excavation results to the end of the 4th- early 5th century.
[1071] Füllep 1969, pl. 48/29, vessel with oblique ribbing at the upper part of the body, dated from excavation results to the second half of the 4th-

Štrbinci,[1072] Sremska Mitrovica (Sirmium),[1073] Beška ('Brest');[1074] in Serbia, Ravna (Timacum Minus),[1075] Niš (Naissus);[1076] in Slovenia, Ptuj (Poetovio);[1077] in Romania, Alba Iulia (Apulum).[1078] The only published parallel from Greece was found in an excavation at Patras.[1079]

The examples from Thessaloniki are dated between the 3rd and 4th centuries, which is set by analogous finds from other sites.

The following examples can be independently dated with relevant precision:

Smooth forms: Cat. nos. 216,[1080] 217,[1081] 218,[1082] 219,[1083] 220,[1084] 224,[1085] 225,[1086] 226,[1087] 227,[1088] 228,[1089] 230,[1090] 231,[1091] 232,[1092] 235,[1093] 239,[1094] 253.[1095]

Ribbed forms: Cat. nos. 255,[1096] 258,[1097] 270.[1098]

It is a very common vessel form that was quite probably locally produced, as is additionally indicated by the fact that several of them are poorly constructed, unstable, or with minor deformations at the neck, rim or even the body. It therefore appears probable that either from the beginning they were made hastily to be sold more cheaply for burial use – most of the preserved examples were found in graves in the cemeteries of the city – or, as is more probable, they were simply items of poor workmanship from local workshops sold cheaply and mainly used in burials. In addition, although they must have been quite a common form of household vessel, none of the examples shows traces of use on the bottom, a fact that seems to corroborate the assumption that they were indeed made especially for this use, and at the same time excludes the possibility that these particular objects were brought to the cemetery from the homes of the deceased.[1099]

**52. Bottles with tubular rim, pear-shaped**

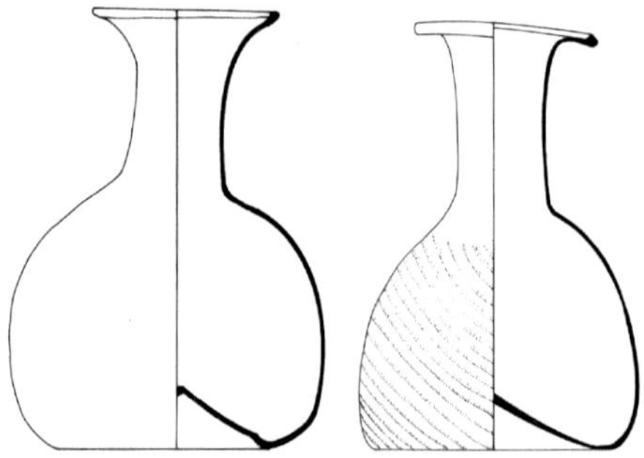

a. Smooth
(Cat. nos. 274-296)
(Plates 14-17, IX-X)

b. Ribbed
(Cat. nos. 297-300)
(Plates 17, X)

Vessels of this form are handleless, have everted, infolded tubular rim, cylindrical neck, pear-shaped body, and stand on a slightly concave

early 5th century, which is an identical copy of vessel Cat. no. 260.

1072 Damevski 1976, 66, 85, pl. XIII/4, this form is dated in Croatia, according to finds from various other sites, to the 3rd and 4th century.

1073 Cermanović-Kuzmanović 1981a, 17, 10; Šaranović-Svetek 1986, 63, no. 93, from a cemetery operating in the 3rd-4th century.

1074 Šaranović-Svetek 1986, 63, 92, dated from excavation results to the late 3rd- mid 4th century.

1075 Petković, Ružić 2005, 44, no. 56, pl. XII, G. 140.1 dated from excavation results between AD 350 and 380.

1076 Drča 2000, 212, 218, nos. 9, 12, 13, pl. II/9, 12, 13, dated from excavation results to the late 3rd- early 4th century.

1077 Lazar 2003, 141-143, form 6. 2. 3, the examples that can be dated from excavation results are dated to the 4th and to the second half of the 4th century.

1078 Baluta 1978, 103, find dated probably from excavation results to the 3rd century.

1079 Dekoulakou 1976, 103-104, pl. 81β, γ, vessel with vertical ribbing from a Christian cemetery.

1080 From a grave with a bowl with indentations, Form 15, dated to the 4th century.

1081 From a grave dated to the 4th century according to the pottery it contained; Makropoulou 2007, 50, pl. 68.

1082 From a grave dated, according to the pottery it contained, to the first half of the 4th century. Makropoulou 2007, 45, pl. 24.

1083 From a grave with a coin of the 2nd-3rd century.

1084 Makropoulou 2007, 27, from the construction technique of the grave the burial is dated to the early 4th century.

1085 Nalpantis 2003, 130, pl. 43, from a grave with coins of AD 334-348.

1086 Nalpantis 2003, 34, 130, drawing 8, from a grave with coins of AD 317-348.

1087 Karydas 1994, 501, from graves dated by their construction technique to the 4th century.

1088 Karydas 1994, 501, from graves dated by their construction technique to the 4th century.

1089 Karydas 1994, 501, from graves dated by their construction technique to the 4th century.

1090 Makropoulou 2007, 300, pl. 24, from a grave dated to the 3rd-4th century.

1091 Makropoulou 2007, 312, pl. 24, from a grave dated to the 4th century.

1092 Makropoulou 2007, 138, pl. 24, from a grave dated by its construction technique to the end of the 3rd or the 4th century.

1093 From graves dated to the 4th-5th century which remained in use until the 7th-8th century, see 9th EBA 2002, 19.

1094 From a sarcophagus, found together with Cat. no. 185, Form 45, and a glass bracelet (reg. no. MΘ 11548). For more on the dating of this assemblage, see the discussion of Form 45.

1095 From a grave with a coin of Constantine I.

1096 Makropoulou 2007, 342, pl. 24, from a grave dated to the end of the 3rd - early 4th century.

1097 Makropoulou 2007, 310, pl. 24, from a grave dated according to the clay finds to the first half of the 4th century.

1098 Globular unguentaria with cracked-off rim, Form 145 and fusiform unguentaria form F47a dated to the late 3rd-4th century, and conical bottles, Form 47, dated to the 3rd century.

1099 These remarks apply also for the clay vessels found in cemetery burials in Thessaloniki.

bottom. As a whole, they are made of colourless greenish or colourless bluish, often bubbly glass. Their height is estimated between 15-25cm.

The smooth variants are free-blown and the ribbed ones are dip-mould-blown, i.e. they are initially blown in a small ribbed mould and then expanded to their final dimensions by further free-blowing and rotating. The rim of both groups is shaped by jacks and continuous rotation, while the vessel is attached at its bottom to a blowpipe, leaving a characteristic mark known as pontil scar.

It is a very widely spread form[1100] which appears mainly in the West, but also in the East. It appears around the end of the 3rd century and remains in use throughout the 4th and up to the early 5th centuries. This form also appears at the same sites and in the same contexts and strata in a smaller sized form (under 15cm in height). These smaller examples are treated as a different form and will be presented among the flasks/unguentaria.

The two canonical groups are:

a) Pear-shaped smooth.
b) Pear-shaped with oblique ribbing.

Among the extremely numerous preserved finds the following are listed as they can be independently and relatively firmly dated:

a) Smooth form: Bonn (Bonna),[1101] Colchester (Camulodunum),[1102] Duklja (Doclea),[1103] Transdierna (Tekija),[1104] Costanza (Tomis).[1105] Two published parallels from Greece were unearthed in excavations in Edessa[1106] and Nea Kerdyllia.[1107]
b) Ribbed variants: Gračanica (Ulpiana),[1108] Sremska Mitrovica (Sirmium)[1109] and Szőny (Brigetio).[1110]

The examples from Thessaloniki are dated between the 3rd and 4th centuries, as determined by analogous finds from other sites.

The following examples can be relatively firmly dated independently:

Smooth forms: Cat. nos. 274,[1111] 275,[1112] 277,[1113] 278,[1114] 279,[1115] 280,[1116] 281,[1117] 282,[1118] 294.[1119]
Ribbed variants: Cat. nos. 297,[1120] 298.[1121]

It is a very common vessel form which was quite probably locally produced, as is additionally indicated by the fact that several of them are poorly constructed, unstable, or with minor deformations at the neck, rim or even to the body. It therefore appears probable that either they were from the offset made to be sold for burial use – and most of the preserved examples were found in graves in the city's cemeteries – or they were the results of poor workmanship from local workshops and sold at lower prices and mainly used in burials. In addition, although they must had been quite a common form of household vessel, none of the examples shows traces of use on the bottom, a fact that corroborates the assumption that they were indeed made especially for this use, and at the same

[1100] Isings 1957, 119-120, form 101, group of tall, slender examples, taller than 15cm.
[1101] Follmann-Schulz 1988, 28-29, nos. 46-49, pl. 4, finds dated between the first half of the 3rd and the 4th century.
[1102] Cool, Price 1995, 150-151, find from a grave dated to the 3rd-4th century.
[1103] Cermanović *et al.* 1975, 155-156, form I/4, examples dated from excavation results to the first half of the 4th century.
[1104] Ružić 1994, 13, no. 32, dated, probably, from excavation results to the 3rd-4th century.
[1105] Bucovală 1968, 71-76, nos. 105-120, finds dated to the 2nd-3rd century, some of them at least from their context.
[1106] Chrysostomou A. 2000, 373, pl. 418B, vessel from a burial dated according to its context to the middle of the 4th century.
[1107] Malama, Ntarakis 2008, 332, pl. 77, MA 7047, from a grave dated to the first half of the 4th century.
[1108] Ružić 1994, 13, no. 34, pl. III/2, dated, probably from excavation results to the 3rd- first half of the 4th century.
[1109] Cermanović-Kuzmanović 1981a, 14, 17; Ružić 1994, 13, no. 33, pl. III/1, vessel dated to the 3rd-4th century.
[1110] Barkóczi 1968, 71, 76, fig. 37/1, find with almost cylindrical body dated to the second half of the 4th century.
[1111] Makropoulou 1991, 257-270, photo 19, the grave is dated from excavation results to the end of the 3rd- early 4th century.
[1112] Makropoulou 2007, 54, pl. 25, the grave, which was used as an ossuary, is dated to the first half of the 4th century according to the clay vessels found inside.
[1113] From a grave with coins of Constantine I.
[1114] Nalpantis 2003, 67, 128, drawings 36, pl. 43, from a grave with coins of Diocletian (minted 293-295) and Licinius (minted 312-317), probably dated to the second decade of the 4th century or even later.
[1115] Nalpantis 2003, 128, probably dating to the 3rd-4th century according to an analogous find from the excavation at the plot at the conjunction of Koloniari-Galanaki-Papathanasiou Streets; the find was accompanied by a coin of AD 335-337, see Nalpantis 2003, 129, note 267.
[1116] Nalpantis 2003, 56, 128, drawings 24, pl. 43, probably dated to the 3rd-4th century.
[1117] Makropoulou 2007, 241, pls. 23, 68, the grave and the burial in it are dated to the first half of the 4th century, according to the clay context.
[1118] Makropoulou 2007, 303, pl. 23, from a grave dated to the first half of the 4th century.
[1119] From a grave with a coin of Constantine I.
[1120] Nalpantis 2003, 48, 128, drawing 17, pl. 43, from a grave with coins of AD 330-340, probably dated to the second quarter of the 4th century.
[1121] Nalpantis 2003, 48, 128, drawing 18, pl. 43, from a grave with coins of AD 330-340, probably dated to the second quarter of the 4th century.

time eliminates the possibility that these particular objects were brought to the cemetery from the homes of the deceased.[1122]

### 53a, b. Baby-feeders or lamp-fillers

a. Inherent spout
(Cat. nos. 301-303)
(Plates 17, X)

b. Applied spout
(Cat. nos. 304-312)
(Plates 17-18, XI)

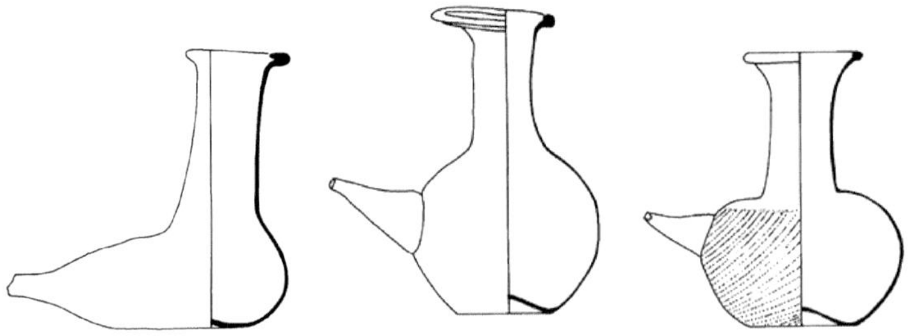

Baby feeders, or lamp fillers, are relatively small vessels characterized by a small spout at the maximum diameter of the body, which is either inherent or applied. The vessels with applied spout are funnel-mouthed with tubular infolded rim or fire-polished rim, long cylindrical neck equal or taller than the body, body spherical, and mildly concave bottom. Their height is estimated between 8-13cm.

In Thessaloniki only handleless examples are known, while in other regions there are many examples recorded which are supplemented with a vertical strap-handle that begins on the shoulder and ends under the rim.

The known examples so far from Thessaloniki are made of naturally coloured greenish or bluish glass, and in two cases (Cat. nos. 307, 311) blue glass has been used especially for the creation of the spout.

The vessels are almost always free-blown, although there is one example (Cat. no. 307) which is mould-blown and its body is covered by oblique ribbing. The rim of the vessels was shaped after the removal of the moil and while the vessel was held by a metal rod attached to its bottom. The small and slender spout created on their body represents the main characteristic of the group, giving it the name that they are generally known by. Technically they can be divided into two groups according to the method used to create the spout, which can be either inherent or applied.

In the first case (Cat. no. 301) the spout is created by reheating the vessel and, especially, heating it up locally at the desired part of the body, which is then pierced by a fine metal rod, and by drawing it out during the extraction of the rod from the rims of the pierced area. The applied spout (Cat. nos. 307-310) is formed by the jab of a metal rod covered with hot, viscous glass. A hot blob of glass covering the tip of a pointed metal rod is attached on the still hot body of the vessel. The moment that the body walls are heated enough at the desired spot, they are pierced with an abrupt, violent thrust of the rod, which is equally swiftly removed, transforming the blob into the conical spout.[1123]

The probable uses of this form remain uncertain. There is a testimony in the sources that they were indeed real baby feeders,[1124] but it has also been proposed that they were used for the filling of oil lamps.[1125] Although the remark in the ancient written source should represent adequate evidnce, the contents of the vessel remains puzzling, bearing in mind its size. (This would seem to apply no matter the contents – milk, water, or other liquid. Moreover, the fragile spout would surely be dangerous when teething starts.[1126] Lately it has been proposed by some researchers that the device was used for pumping human milk.[1127] Another hypothesis suggests some sort of medical use, and specifically as a drip feeder for medicinal liquids.[1128] The more probable hypothesis seems to be that 'baby feeders' were used for filling oil lamps, especially clay lamps: their very small filling holes would seem to necessitate the use of a funnel or a spout. However, the obvious discrepancy between the large number of lamps found in excavations and the small number of glass 'lamp fillers' does not support this hypothesis unreservedly.

---

1122 These remarks also apply for the clay vessels found in cemetery burials in Thessaloniki.
1123 Stern 2001, 198.
1124 Hilgers 1969, s.v. *titina or Ubuppa*, 80; Kisa 1908, t. II, 332.
1125 Isings 1957, 118; Calvi 1968, 77.
1126 Generally on the discussion about their possible uses, see Arveiller 1985, 116-117.
1127 Rouquet 2003, 164-170.
1128 Welker 1974, 95-98.

Examples of baby feeders have been published from both the western and eastern provinces of the Roman Empire, i.e. Aquileia,[1129] Trier (Treviri),[1130] Heddernheim (Nida),[1131] Cologne,[1132] Bonn (Bonna),[1133] Strasbourg (Augusta Trebecorum, Argentoratum),[1134] Rouen (Rotomagus),[1135] Hungary,[1136] Nicosia,[1137] Agios Hermogenis at Kourion,[1138] Tyre,[1139] Jalame,[1140] and other Palestinian sites.[1141] Private and public collections in Europe[1142] and America[1143] have also been reported to hold finds of this form. In Greece examples have been published from Samothrace,[1144] Nea Anchialos (Phthiotides Thebes),[1145] Nicopolis,[1146] and Rhodes.[1147]

This form is generally dated from the 2nd century up to the end of the 4th,[1148] and one example is published with decorations typical of the 5th-6th century.[1149] Although it cannot be proved yet by archaeological data, it seems that the variant with the inherent spout is the earlier, and the one with the applied spout appears later and also remains widely used in the 4th century.

Only Cat. nos. 302,[1150] 303[1151] and 311[1152] can be independently; they are ascribed to the same general time-frame as finds from other sites.

### 54. Funnel-mouth flasks, globular, with long curved neck and very small body spouts

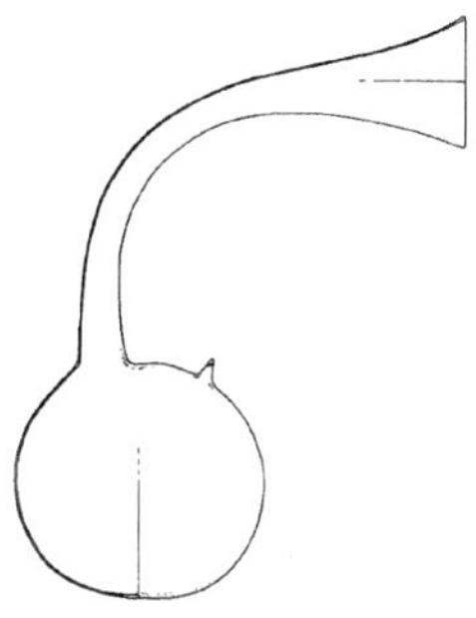

(Cat. no. 313)
(Plate XI)

Vessels of this form have unworked rim, funnel-mouth, narrow, long cylindrical neck, globular body and flat, slightly concave bottom. The neck is bent, giving a deviation of the mouth some 30° from the vertical axis. At the upper part of the body, on the side under the oblique rim, there are usually two small applied lugs, or actual spouts, and there are known examples with just one spout. Their height is estimated between 15-19cm.

The vessels of this form are free-blown. The lugs or spouts were created by the application of a hot blob of glass from the tip of a pointed metal rod on the still hot body of the vessel. The moment that the body walls are heated enough at the desired spot, they are pierced with an abrupt, violent movement of the rod, which is equally swiftly removed transforming the blob into the conical spout. These lugs were probably used to control the flow of the contents, which due to the obliqueness of the neck could not flow if extra air was prevented from entering the vessel body from another point. Thus, the user could stop or allow the flow of the vessel's contents

[1129] Calvi 1968, pls. 12:1-4, the two handless, bag-shaped examples with inherent spout are dated to the 2nd-3rd century, while the other two examples, with applied spout, one handleless and one single-handled, are dated to the 1st-2nd century.

[1130] Goethert-Polaschek 1977, form 116b, 197-198, pl. 66, dated to the 2nd century. Goethert-Polaschek 1977, form 123b, 208-209, pl. 69, dated to the first half of the 4th century. All examples are single-handled vessels, and they are divided into groups according to whether they have an applied or pushed-in base.

[1131] Welker 1974, 95-98, form 15g. Single-handled vessel dated to the end of the 2nd century. In the discussion of this form analogous vessels found in Essen, Cologne, and Worms (Borbetomagus) are mentioned.

[1132] Doppelfeld 1966, 48, photo 80 (below). Single-handled vessel with braided handle, of unknown exact provenance, dated on stylistic grounds to the 2nd-3rd century; Fremersdorf, Polónyi-Fremersdorf 1984, 86, no. 194, vessel dated from excavation results to the early 3rd century.

[1133] Follmann-Schulz 1988, 61-62, nos. 189-191, pl. 22, vessel no. 191 is dated to the first half of the 2nd century.

[1134] Arveiller-Dulong, Arveiller 1985, 116-118. One vessel dated to the 2nd century and two more dated to the end of the 3rd or to the first half of the 4th century, from excavation results.

[1135] Sennequier 1985, 193-195, nos. 298-300, pl. on page 192, dated by their context between the late 2nd and 4th century.

[1136] Barkóczi 1988, 127, no. 248, pl. XX, find dated from excavation results to the 2nd century.

[1137] Vessberg 1952, 148-149, pls. X:1-2, XX:4-5, they are vessels of the early, bird-shaped form with coherent spout. Dated to the second half of 2nd century according to the pottery with which it was found. 19 Cypriot examples are listed. Examples both with plain and tubular rim appear.

[1138] McFadden 1946, 475, no. 32, pl. 39, from a burial, which, according to its context, is dated to the 1st century.

[1139] Dussart 1998, 177, form BXIV. 11, two examples from graves with coins which imply that they were used during the 3rd and 4th century.

[1140] Weinberg 1988, 79, cat. nos. 341-342, figs. 4-37, spouts of two examples, which, probably, are connected to early 2nd-century graves.

[1141] Barag 1970c, form 23. 1, generally dated to the 2nd century.

[1142] Stern 2001, 22, 146 and 198, no. 89, dated on the basis of analogous finds between the 2nd and 4th century.

[1143] Hayes 1975, 81, no. 299, fig. 299 (page 207). Dated to the end of the 3rd century as it is ascribed to the later Syro-Palestinian products, i.e. class XII, main Syro-Palestinian series, which is assumed to have evolved after AD 256 as it is totally absent from finds from Dura-Europos; Auth 1976, 152, no. 199.

[1144] Dusenbery 1998, no. K3-1, a find that has been assumed to date to the end of the 1st century BC.

[1145] Sotiriou 1939, 64, fig. 26, an example with inherent spout (published also in Weinberg 1962, 132, pl. 28).

[1146] Vokotopoulou 1971, 335, pl. 314, example with inherent spout from a grave with 2nd-century coins.

[1147] Konstantinopoulos 1969, 472, pl. 472δ, example with inherent spout dated to the second half of the 1st century, see ASA 50, 1946, pl. XXXIV, 32.

[1148] Isings 1957, form 99.

[1149] Auth 1976, 152, no. 199, with relevant bibliography.

[1150] From a cemetery dated to the 2nd-3rd century. See Chapter 9 (Find-Spots), 'Sindos, Pref. Thessaloniki. Good Year Plant, 1981'.

[1151] Find from a necropolis dated to the 1st-3rd century. See Chapter 9 (Find-Spots), 'Melenikou and Chatziandreou Street, 1994'.

[1152] In the same grave was found Cat. no. 343. It is from a Late Roman or Early Christian cemetery, See Chapter 9 (Find-Spots), 'International Fair of Thessaloniki, 1979'.

by sealing or opening the hole of the lug/spout with his/her finger. This flow-control system must have been particularly useful at events such as banquets and for libation offerings.

Analogous finds are known from Damascus,[1153] Jerusalem,[1154] Cologne,[1155] Arles (Arelate),[1156] Albenga (Albingaunum),[1157] Crete,[1158] and Amphipolis;[1159] they are dated to the late 1st and early 2nd centuries, a date which seems to be confirmed also by the archaeological context of those that have them.

The find from Lete (Cat. no. 313) has been dated1160 to the 2nd-3rd century, according to the destruction time of the villa rustica in the ruins of which it was found. Nevertheless, the parallels for this find from other sites, and the rest of the glass finds from the same stratum belong to forms known partly from the 1st century BC, and mainly from the 1st-2nd centuries AD.1161 This observation combined with the rural character of the site allows the hypothesis that the glass vessels were heirlooms that were in use by the members of the same family for quite a long time – at least from the middle of the 2nd century, when the villa was founded, until the middle of the 3rd century, when the building was destroyed.

### 55. Funnel-mouth flasks, bulbous spear-shaped, slender on plain base, wide mouth and unworked rim

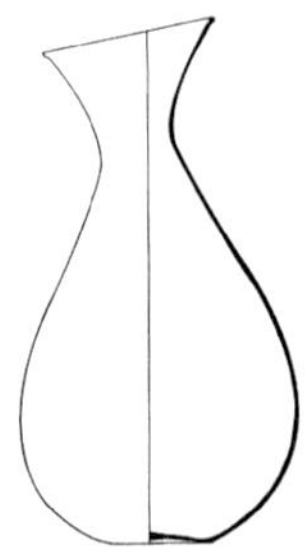

(Cat. nos. 314-315)
(Plates 18, XI)

Vessels of this form have a funnel-mouth ending in a cracked-off rim, pear-shaped body, and flat, slightly concave bottom. Their height is estimated between 15-17cm.

These vessels were free-blown; the rim was cracked-off most probably after the vessel's annealing.

The examples are made of clear, colourless, slightly greenish glass.

The only analogous find is from Duklja (Doclea)[1162] and is also decorated with shallow, incised bands. It is securely dated to the 2nd century on the basis of an associated coin. However, placing the find within the general evolution context of glass vessels, a later dating, at least to the 3rd century, seems more plausible.

Cat. no. 315 cannot be dated independently, nor can Cat. no. 314 be directly dated; the latter was found in an area of the necropolis mainly used in Late Imperial times.[1163]

### 56. Funnel-mouth flasks, pear-shaped, slender, on plain base, wide mouth and fire-polished rim

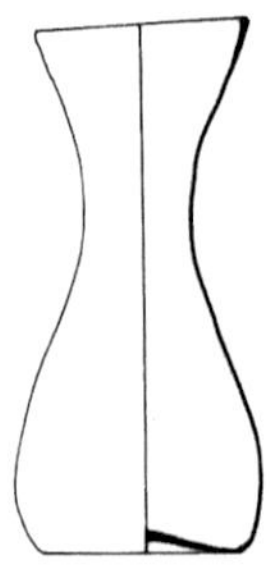

(Cat. nos. 316-317)
(Plates 18, XI)

Vessels of this form have fire-polished funnel-mouth, short neck, pear-shaped body and flat, mildly concave bottom. Their height is estimated between 9-16cm.

These vessels were free-blown and, as no pontil mark is visible underneath, it seems that in order to polish the rim the vessel was held by jacks, probably by the neck.

The examples from Thessaloniki are made of colourless, greenish glass.

Analogous finds are published from Cyprus[1164] and in museum collections in North America, at Corning Museum of Glass[1165] and Royal Ontario Museum, Toronto.[1166] Cat. no. 316 was found in a grave dated to the first half of the

[1153] Al 'Ush 1964, 55, no. 1, fig. 44, vessel of unknown provenance, erroneously considered to be 10th-century, Islamic product.
[1154] Brosh 2003, 34, 37, no. 458, find of unknown provenance erroneously published as Islamic product.
[1155] La Baume, Salomonson 1976, 44, no. 104, pl. 13/2, find of unknown provenance dated for stylistic reasons to the 1st-2nd century.
[1156] Morin-Jean 1922-23, 181-182, fig. 239; Foy (ed.) 2003, 132, 135, dated to the 1st century for stylistic reasons.
[1157] Massabò 2001, 122-123, no. 71, find dated from excavation results to the end of the 1st to early 2nd century.
[1158] A vessel with reg. no. EAM 11694, with only one spout, exhibited in the temporary exhibition *Efthrafsti Polyteleia*; it is dated to the late 1st to early 2nd century.
[1159] Two unpublished finds with one spout. The vessels are exhibited in the Amphipolis museum.
[1160] Tzanavari, Filis 2000, 160-162; Adam-Veleni, Poulaki, Tzanavari 2003, 212, no. 193 (K. Tzanavari). The find from a farmhouse/*villa rustica* built in the middle of the 2nd century and temporarily abandoned in the middle of the 3rd century following raids by Goths (idem, pages 81-82).
[1161] The illustrated finds are listed, i.e. for ribbed bowl, see Isings 1957, 19-20 form 3b, dated to the 1st century BC to 1st century AD. Square, single-handled bottle, Isings Form 50, with a characteristic rim of the 1st-2nd. See also Charlesworth 1966, 26-27, form 1c.
[1162] Cermanović *et al.* 1975, 153, form I/1A, vessel from a grave dated by a coin of Phaustina Minor to the 2nd century.
[1163] See Chapter 9 (Find-Spots), 'Aristotle University of Thessaloniki, Law School 1965, 1966'.
[1164] Vessberg 1952, 134, form III.β.1, pl. VII/32-33.
[1165] Whitehouse 1997, 161, no. 278, vessel dated for stylistic reasons probably to the 4th century.
[1166] Hayes 1975, 93, nos. 316-317, pl. 21, vessels dated to the first half of the 4th century on the basis of analogous finds from the Syro-Palestinian

4th century.[1167] Generally, funnel-mouth vessels became very popular during the 4th and 5th centuries.[1168] It is very probable that the examples from Thessaloniki were made in the Syro-Palestinian region, where funnel-mouth vessels were in general produced.

**57. Funnel-mouth flasks, pear-shaped, slender plain base and narrow mouth**

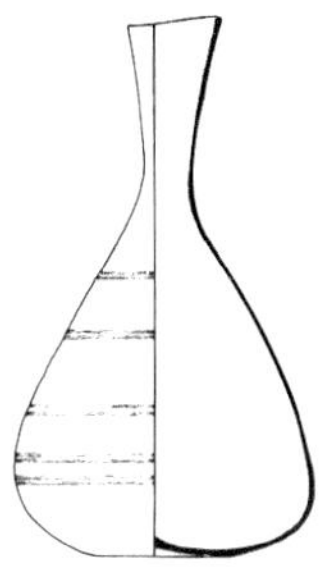

(Cat. no. 318)
(Plate 18, XI)

Vessels of this form have narrow, relatively conical neck, ending to a fire-polished rim, pear-shaped body, and standing on a flat, slightly concave bottom. Their height is estimated at c. 14cm.

These vessels were free-blown and no pontil mark is visible underneath, thus it is thought that the vessel was held by jacks, probably from the base of the neck, so as to crack off the moil and polish the rim. Examples like the one from Thessaloniki were decorated with incisions of shallow horizontal bands after the annealing process was completed.

The example from Thessaloniki is made of colourless, greenish glass.

Similar examples were found in Gerasa,[1169] Jerusalem,[1170] Hadithe,[1171] and Kefar'Ara[1172] dated to the end of the 3rd, but mainly in the 4th century.

The fact that analogous finds were traced exclusively in the Syro-Palestinian region, in addition to the colour and the texture of the glass of our find, and to some technical characteristics, i.e. the fire-polished rim, the absence of a pontil mark, and the finishing of the vessel while held by jacks from the neck, make it quite feasible that the origin of this form was also Syro-Palestinian.

The find from Thessaloniki cannot be dated by its archaeological context, rather it is dated to the 4th century according to the dating of analogous vessels and the dating of analogous incised decoration.

**58. Funnel-mouth flasks, pear-shaped, ribbed (Isings Form 133 min)**

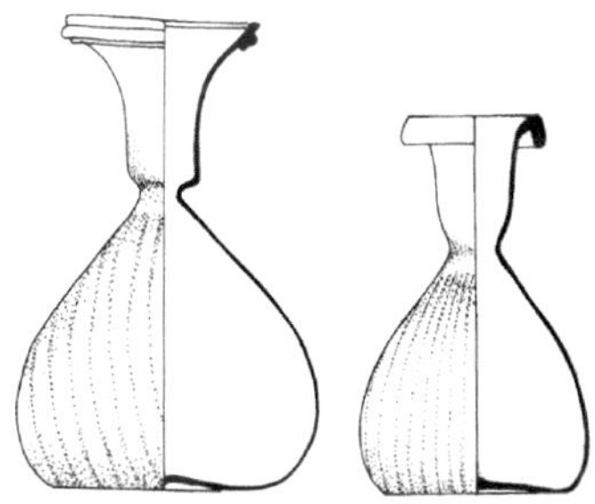

(Cat. nos. 319-320)
(Plates 18-19, XI)

Vessels of this form have tubular rim, sometimes band-like and sometimes accentuated by an applied thread, funnel-mouth constricted at the transition to the pear-shaped body. The resting surface is flat and the bottom is slightly concave. The body is decorated with faint vertical ribbing. Their height is estimated between 12-13cm.

These are dip-mould-blown vessels and the ribbing on their body is formed during mould-blowing. They were made in a two-part mould, open at the bottom, as indicated by the anomalies imprinted underneath. It was then inflated by free-blowing to its final shape and while held by jacks – most probably at the constricted area at the base of the neck – the final shaping of the rim was done. As a final step the relief thread was applied around the rim.

Vessels of this form were produced in Egypt, as is indicated by the two parallels from Fayum (Arsinoe), and also by the colour and texture of the glass used.

Cat. no. 319 was found in a 4th- or 5th-century cemetery.[1173] Parallels from Egypt, one of them identical[1174] and one quite similar,[1175] are not firmly dated, while another from Tomis[1176] was dated too early. Similar Western vessels, of bigger dimensions, are dated to the 4th century,[1177] while other tableware parallels from Egypt are also dated to the

region.
[1167] Makropoulou 2007, 46, pl. 25.
[1168] Stern 2001, 149, 269, 217, no. 106, vessel of the same form with vertical ribbing dated to the first half of the 4th century. This work cites a large number of analogous finds dated to the late 3rd and to the 4th centuries.
[1169] Dussart 1998, 128, pl. 32/7, no. BX 1112, find dated from excavation results to the end of the 6th century.
[1170] Finds from three necropolises dated from excavation results to the end of the 3rd and to the 4th century, according to Dussart 1998, 132, who cites Barag 1976, 203, 205, no. 35, fig. 98:3.
[1171] Dussart 1998, 132, pl. 34/1, no. BX 1122, find dated according to analogous finds to the end of the 3rd to the early 4th century.
[1172] According to Dussart 1998, 132, who cites Sussman 1976, 99, fig. 4/4, find dated from excavation results between the 4th and 6th centuries.
[1173] See Chapter 9 (Find-Spots), '4 N. Kapatou Street, 2001'.
[1174] Edgar 1905, 42, no. 32588, pl. VI, from Fayum (Arsinoe).
[1175] Edgar 1905, 42, no. 3259, pl. VI, from Fayum (Arsinoe).
[1176] Bucovală, Paşca 1992, 253, 259, 264, pls. 5 and 9, from a grave dated to the late 2nd to early 3rd century.
[1177] Isings 1957, 161, form 133; Arveiller-Dulong, Arveiller 1985, no. 226, the vessel, according to the researcher, is ascribed to Isings Form 101,

4th,[1178] or, according to the re-dating of the relevant archaeological layers, into the 5th or 6th centuries.[1179] Finally, the use of a thick trail around the rim, the overall impression of the vessel shape, and the use of dip-mould-blowing, make it plausible to date this form to the 4th or 5th century.

### 59. Funnel-mouth flasks, slender biconical body, horizontal rim

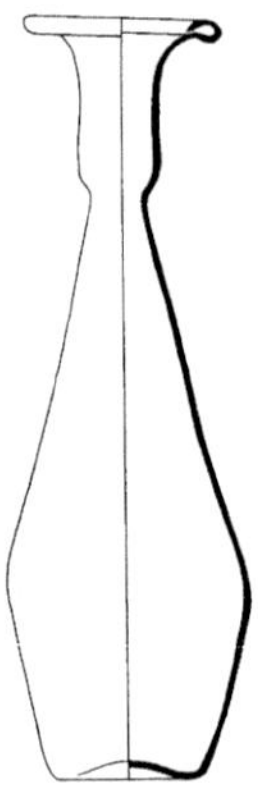

(Cat. nos. 321-322)
(Plates 19, XI)

Vessels of this form have wide, horizontal, infolded tubular rim, short cylindrical neck with intense constriction at the transition to the biconical body. The resting surface is flat and the bottom is slightly concave. Their height is estimated between 12-15cm.

The vessel is free-blown and the rim was probably shaped while the vessels was held by jacks from the base of the neck, at the point of the constriction.

No analogous finds have been traced, but Cat. no. 321 is dated according to excavation results to the end of the 4th century.[1180] It appears that the form derives from ovular vessels with fairly tall, slightly conical neck, which appear around AD 200,[1181] and which by the first half of the 3rd century evolve into slender, pear-shaped vessels with similar but shorter necks, constricted at its base, but without, however, the characteristic horizontal rim of the find from Thessaloniki, which is an element of further evolution of the form.[1182] The production site of the vessels is impossible to trace, although the forming of the neck and the rim are strongly reminiscent of Egyptian artefacts.[1183] Nevertheless they should not be considered Egyptian imports as the glass does not have the typical Egyptian dark, olive-green hue.

The examples from Thessaloniki are made of colourless, greenish glass.

### 60. Funnel-mouth flasks, pear-shaped, slender with pushed-in base

(Cat. no. 323)
(Plates 19, XI)

Vessels of this form have a funnel-mouth ending in a ground, or sometimes fire-polished, rim; the body is pear-shaped and stands on an inherent pushed-in conical base. Their height is estimated at c. 16cm.

These vessels were free-blown, with the base shaped by the constriction and folding of the lower part of the body. Due to the bad state of preservation of the example from Thessaloniki it is not clear whether jacks were used to hold it from the base of the mouth, or if the vessel was attached from underneath to a blowpipe to finish the rim.

The example is made of greenish glass.

It is a relatively widespread form, produced on the Syro-Palestinian coast, and consequently it is believed that it appeared in the middle of the 5th century, representing an evolution of funnel-mouth vessels with plain base, which were produced in the region during the 4th century.[1184] Undecorated analogous finds are published from Jordanian Beit Ras (Capitolias),[1185] Jerusalem,[1186] and Edessa, Greece;[1187] furthermore a group of vessels of this form are known that all display the same vegetal decoration.[1188]

On the basis of the aforementioned parallels the example from Thessaloniki is dated to the 5th century.

regardless of its slightly different shape.

[1178] Harden 1936, 186, from houses dated later than AD 300.

[1179] Hayes 1975, 2-3; Whitehouse 1997, 81-82.

[1180] I thank my colleague Maria Polychronaki for the identification of the grave coins: a bronze coin of the second half of the 4th century (BN 2021/1), which dates the burial with the glass unguentarium, and a small, bronze denomination (minimus), dated to the 5th-6th century (BN 2021/2), from a secondary use of the grave. The vessel was accompanied by a glass unguentarium (Form 139).

[1181] Fremersdorf, Polónyi-Fremersdorf 1984, nos. 113-114.

[1182] Fremersdorf, Polónyi-Fremersdorf 1984, nos. 115-116.

[1183] See Form 58 (Cat. nos. 319-320).

[1184] Stern 2001, 262, 268-269, 297, with relevant bibliography. Her dating seems quite secure, since she suggests the second half of 5th century based on technological details, i.e. the method of forming the base, which appears in independently dated finds, only from the 5th century onwards.

[1185] Harden 1966, 51-52, fig. 10 (upper row, second from right), an undecorated find dated to the 4th-5th century.

[1186] Israeli 2003b, 169, no. 183, find dated according to analogous finds to the 4th-5th century.

[1187] Chrysostomou A. 2000, 384, 387, pl. 437, vessel from a grave with no other chronological indications, from a graveyard dated to the 4th-5th century.

[1188] Auth 1996, 103-105, with detailed bibliography. This form is dated to the 4th century by her, while previous researchers had suggested a 4th- or 5th-century date, e.g. Harden 1959, 3-20; Weinberg 1963, 25-28.

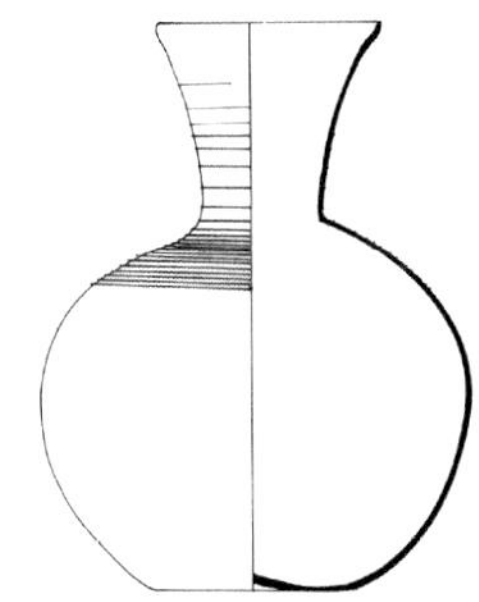

**61. Funnel-mouth flasks, spherical with short neck (Isings Form 104b var.)**

(Cat. nos. 324-336)
(Plates 19-20, XI-XII)

Vessels of this form have funnel-mouth with unworked or slightly ground rim; short, practically non-existent neck, spherical body and flat, slightly concave bottom. Their height is estimated between 12-15cm.

These vessels were free-blown and the decorated examples, at least, were held by the blowpipe from the bottom to finish them off, by spiralling a thread of glass around the neck.

The examples from Thessaloniki are made of colourless, greenish glass.

This form appears in the late 3rd century, and remains in fashion mainly during the 4th and into the early 5th century.[1189] Examples of this form have been excavated mainly in the NW provinces of the Empire. It deserves noting, however, that the examples from the West and the Balkans are decorated with shallow incised bands and never with an applied thread around the neck, as is the case with the majority of examples from Thessaloniki. This type of decoration appears more often on funnel-mouth vessels from the eastern provinces,[1190] and it probably offers an indication for the origin of the vessels of this form, or at least for the place of the original inspiration of their decoration.

Analogous finds come from Germany: Trier (Treviri),[1191] Bonn (Bonna),[1192] Cologne,[1193] Mainz (Moguntiacum);[1194] Croatia: Dalj near Osijek (Mursa),[1195] Sremska Mitrovica (Sirmium);[1196] Serbia: Belgrade (Singidunum),[1197] Gamzigrad (Romuliana Felix),[1198] Gračanica (Ulpiana),[1199] and Kostolac (Viminacium).[1200] Similar vessels are known from Greece, i.e. Agios Ioannis Theologos[1201] and Elis.[1202]

Only Cat. no. 327, can be independently dated by its archaeological context to the end of the 3rd century.[1203]

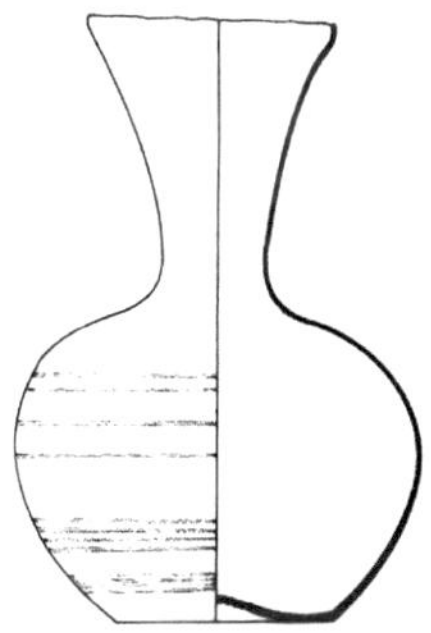

**62. Funnel-mouth flasks, spherical with tall neck and unworked rim, squat (Isings Form 104b)**

(Cat. no. 337)
(Plates 20, XII)

Vessels of this form have funnel-mouth with unworked or slightly ground rim. The conical mouth continues uninterrupted towards the body, forming an almost cylindrical part like a neck at its lower part. The body is spherical, and the bottom is slightly concave; their height is estimated at c. 13cm.

These vessels were made by free-blowing and they are often decorated with shallow bands of incisions. The rim was cracked off and finished after the vessel was annealed and incised, as indicated by the fact that the incisions around the rim are often partly destroyed by the cutting-off of the rim.

The examples from Thessaloniki are made of colourless, greenish glass.

It is a form that appears at the end of the 3rd century, spreading mainly during the 4th and survived into the early 5th century.[1204] Large numbers of vessels of this form have been mainly discovered in the NW provinces of the Empire,

[1189] Isings 1957, 123-125, form 104b.
[1190] Weinberg 1988, 70-71.
[1191] Goethert-Polaschek 1977, 163-179, some of the vessels ascribed to form 101b, esp. nos. 988, 998, 1057, pls. 61-62, can be divided from the rest of the vessels, due to the different shape and size of mouth, and be ascribed to the form under discussion. The finds from excavation results are dated between the early and second half of the 4th century.
[1192] Follmann-Schulz 1988, 74-75, nos. 243-244, pl. 29, finds dated from excavation results to the second half of the 3rd century.
[1193] Doppelfeld 1959, 156-157, fig. 3. 5, find dated from excavation results to the early 4th century.
[1194] Harter 1999, 142-143, form G23a, dated to the end of the 3rd and to the 4th century, esp. nos. 976, (idem, 331, pl. 47, 94/62), dated from excavation results around the 4th century and no. 998 (idem, 331, pl. 95/65), dated from excavation results to the middle of the 4th century.
[1195] Bulat 1976, 90, no. 15, pl. II/4, V/2, donated to the museum (without archaeological data).
[1196] Cermanović-Kuzmanović 1981a, 16, no. 4, 8, examples dated from excavation results to the 4th century.
[1197] Nikolić-Đorđević 1990, pl. V/40, example dated from excavation results to the end of the 3rd-4th century.
[1198] Ružić 1994, 13, no. 42, example dated from excavation results to the 4th century.
[1199] Ružić 1994, 13, no. 43, example dated from excavation results to the end of the 3rd to the first half of the 4th century.
[1200] Zotović, Jordović 1990, 95, pl. CXLIII: 4, dated between the end of the 1st and the 2nd century. Probably the same vessel appears also in Ružić 1994, 14 no. 49, an example generally dated between the 1st and 2nd century.
[1201] Dakoronia 1987, 228-231, pl. 134, from a jar burial with no other chronological indication.
[1202] Karagiorga 1971, 142, pl. 121ε, from an excavation dated probably to the 2nd century.
[1203] Makropoulou 2007, 214, pl. 24, from a grave dated to the end of the 3rd century by a coin of Antonianus Provus (276-282).
[1204] Isings 1957, 123-125, form 104b.

and less often in Eastern sites. It is quite probable that they were produced in several glassworking centres. Sites with firmly dated parallels include Germany: Strasbourg (Augusta Trebecorum, Argentoratum),[1205] Bonn (Bonna),[1206] Trier (Treviri),[1207] Cologne,[1208] Mainz (Moguntiacum);[1209] Italy: Aquileia;[1210] Hungary: Dunaújváros (Intercisa),[1211] Budapest (Aquincum),[1212] Szőny (Brigetio);[1213] Austria: Lorch (Lauriacum);[1214] Croatia: Osijek (Mursa),[1215] Sremska Mitrovica (Sirmium);[1216] Montenegro: Duklja (Doclea);[1217] Serbia: Brza Palanka (Egeta),[1218] Belgrade (Singidunum);[1219] Bulgaria;[1220] Greece: Corinth[1221] and Dion;[1222] Cyprus;[1223] Jordan: Beit Ras (Capitolias),[1224] Ajlun;[1225] and Israel: Samaria-Sebaste.[1226] The find from Thessaloniki cannot be dated independently.

### 63. Funnel-mouth flasks, spherical with tall neck and unworked rim, slender (Isings Form 104b local)

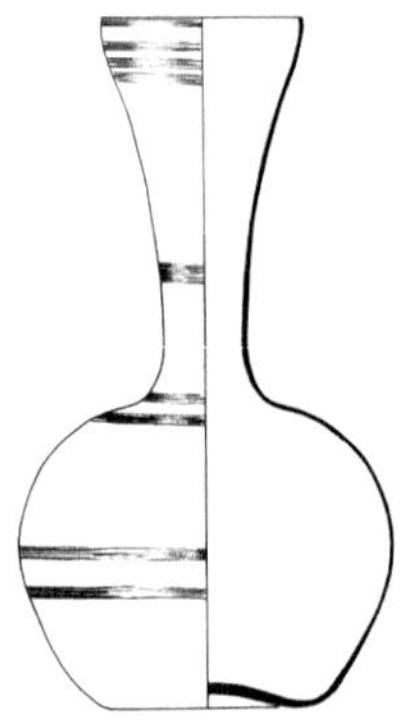

(Cat. nos. 338-344)
(Plates 20-21, XII)

Vessels of this form have funnel-mouth with unworked or slightly ground rim. The conical mouth continues uninterrupted towards the body forming an almost cylindrical neck section at its lower part; the body is ical, and the bottom slightly concave. This form presents a variant of form Form 62, but different in terms of its more slender proportions, by the fact that the body represents a smaller part of the entire vessel, and by the decorative incised bands which are quite characteristic, consisting of actual bands of shallow incisions. As no true parallels were traced in other regions, and because of the relatively large numbers of finds from Thessaloniki, it is considered a local product. This hypothesis is corroborated by the fact that among the local finds a blank, unfinished vessel, is preserved (Cat. no. 338), which was inexplicably left in that state and placed in a grave as a grave good in this condition. Their height is estimated between 16-22cm.

The only vessels with which this form presents a real connection are those from Trier (Treviri)[1227] and Bonn (Bonna),[1228] with which they have in common all other characteristics and elements except for the pushed-in tubular base which distinguishes Germanic cts. They are dated to the 4th century based on the dating of the general form Form 62 (Isings Form 104b), with which they are associated, as none of them has preserved any archaeological clues that enables an independent dating.

1205 Arveiller-Dulong, Arveiller 1985, 131-133, form 60, nos. 284-297, finds dated from excavation results to the end of the 3rd and mainly to the 4th century.
1206 Follmann-Schulz 1988, 74, no. 247, pl. 29, find dated from excavation results to the 4th century.
1207 Goethert-Polaschek 1977, 163-179, form 101b, nos. 978-1101, pls. 61-63, finds dated from excavation results between the early and second half of the 4th century.
1208 Fremersdorf 1959, 78-79, pls. 128-129, finds dated to the second half of the 4th century; Doppelfeld 1959, 157, fig. 3.2, grave 5, dated from excavation results to the early 4th century; Fremersdorf 1961, 62-63, pl. 122, find dated to the 4th century.
1209 Harter 1999, 142-143, form G23a, dated from excavation results to the end of the 3rd and the 4th century.
1210 Calvi 1968, 148, nos. 296-300, finds dated to the 4th century.
1211 Radnóti 1957, 146, no. 45, pl. XXIX/7. In addition, the same reference mentions another example from a grave dated to the end of the 3rd century, which, according to the researcher, is probably of western origin.
1212 Three examples from excavations dated to the end of the 3rd and early 4th century, see Barkóczi 1988, 143, where there is an earlier bibliography on the matter.
1213 Barkóczi 1988, no. 308 a, b, pls. XXVI, LXXXVIII, finds dated from excavation results to the end of the 3rd or to the early 4th century.
1214 Thomas 1962, 91, 103, pl. 4.
1215 Bulat 1976, 89, no. 2, pl. I. 1, V/1, with no excavation data.
1216 Cermanović-Kuzmanović 1981a, 16, nos. 1-3, 5-6, examples dated from excavation results to the second half of the 3rd and first half of the 4th century.
1217 Cermanović-Kuzmanović 1976, 179, dated to the late 3rd to early 4th century, pl. IV. 14.
1218 Ružić 1994, 13, no. 39, example dated from excavation results to the late 3rd to early 4th century.
1219 Ružić 1994, 13, no. 39, 50, examples dated from excavation results to the late 3rd-4th century.
1220 Djingov 1964, 106-108, fig. 33 (First vessel from the left, upper row).
1221 Wiseman 1969, 105-106, pl. 32/j, along with remains of glassworking activities found in a layer dated to the late 5th-6th century in the area of the ancient gymnasium at Corinth.
1222 Pantermalis 2001, 351, fig. 4.
1223 Vessberg 1952, 133, flask A.I.y.I, pl. VII:14.
1224 Harden 1966, fig. 10 (First from right, upper row), from a grave dated to the 4th-5th century.
1225 Harden 1966, 52-53, fig. 12 (First from left, upper row), dated to the 4th century.
1226 Crowfoot 1957, fig. 95, no. 10.
1227 Goethert-Polaschek 1977, 156-163, some of the vessels form 101a, esp. nos. 941, 955, 956, pl. 60. The finds are dated from excavation results to between the first half and late 4th century.
1228 Follmann-Schulz 1988, 74-75, no. 250, pl. 30, find dated on the basis of analogous finds from the same city to the 4th century.

### 64. Funnel-mouth flasks, spherical with tall neck and fire-polished rim, squat

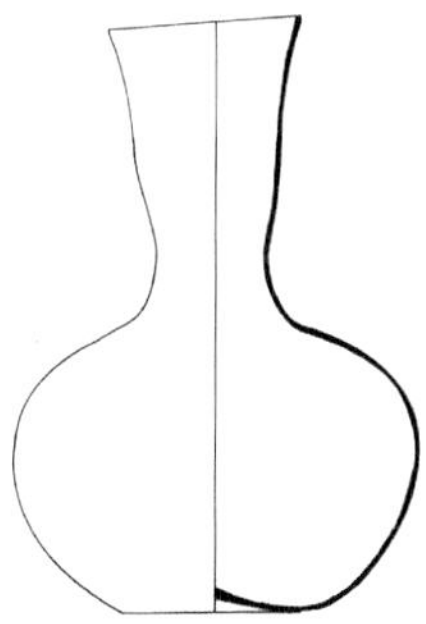

(Cat. nos. 345-348)
(Plates 21, XII)

Vessels of this form have long funnel-mouth with fire-polished rim, short cylindrical neck, globular body and flat, slightly concave bottom. Their height is estimated between 15-17cm.

These vessels were free-blown, then, and with the vessel held by some form of jacks, quite probably from the neck, the funnel-mouth was shaped, the rim fire-polished, and the bottom was pushed and raised forming a 'kick', in order to make the vessel more resistant to stresses and fractures.

The examples from Thessaloniki are made of olive-green glass.

It is a form that appears at the end of the 3rd century, becoming more popular during the 4th, and surviving into the early 5th century.[1229] Vessels of the same shape, occasionally decorated with vertical ribs, are popular in Jordan[1230] and in Palestine particularly,[1231] where it is believed that they were produced. Later examples have been found in Sardis.[1232]

Among the examples from Thessaloniki only one (Cat. no. 345) can be dated independently to the 4th century.1233

### 65. Funnel-mouth flasks, globular, with tall neck and fire-polished rim, slender

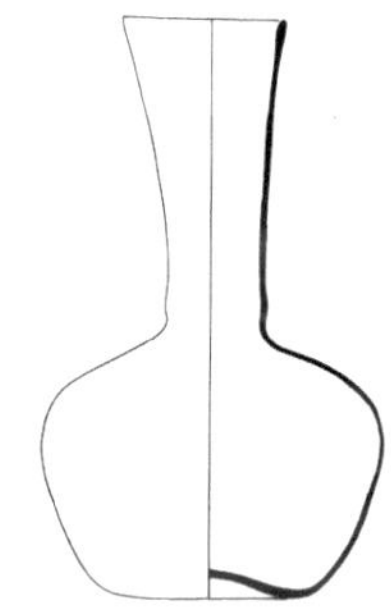

(Cat. no. 349)
(Plates 21, XII)

Vessels of this form have long, relatively narrow conical neck, fire-polished rim, globular body and flat, slightly concave bottom. Their height is estimated at c. 16cm.

The example from Thessaloniki is made of colourless, greenish glass.

These vessels were free-blown and after the inflation of the paraison and the modelling of the body, the vessel was held by jacks, quite probably from the neck, and the funnel-mouth shaped; the rim was fired, and thus polished, and the bottom was pushed and raised, forming a small 'kick' to make it more resistant to stresses and fractures.

It is a form that appears at the end of the 3rd and mainly during the 4th and 5th centuries. The find from Thessaloniki is also placed in this same period, based on its archaeological context.[1234] As a result of the texture of the glass and the way the rim and underneath were shaped, it should be ascribed to Syro-Palestinian production.[1235] Analogous finds have bshed in Hungary[1236] and others are housed in the Royal Ontario Museum, Torontoto Museum.[1237]

### 66. Funnel-mouth flasks, globular, with thread wound around the mouth

(Cat. no. 350)
(Plates 21, XII)

The vessel has a funnel-mouth ending to a fire-polished rim. The neck was probably long and cylindrical. The body is globular and the bottom is mildly concave. Around the mouth is spirally wound 13 times a blue thread of glass. Their height is estimated between 15-20cm.

[1229] Barag 1970c, pl. 43, form 24, 28.

[1230] Harden 1966, fig. 10 (First from right, upper row), from Beit Ras (Capitolias), a grave dated to the 4th-5th century.

[1231] Stern 2001, 215, with exhaustive bibliography on relevant finds. These vessels are identified as products of a Palestinian workshop dated to the first half of the 4th century. See also Israeli 2003b, 168, no. 179, with references to analogous finds from the region, which are ascribed to the production of the 4th century.

[1232] Saldern 1980a, 70-71, nos. 477-488, pls. 14, 26, from early Byzantine layers.

[1233] Makropoulou 2007, 162, from a burial quite probably dated to the 4th century, according to a worn coin which was found in it. This group of graves is dated to the period after the early 4th century, because of the bricks bearing stamped Christograms and crosses that were used in their construction; the period of their use probably continues into the 6th century, since the only coin found in them was a decaexanummium of Justinian I. For structural and stylistic reasons the wall-painted, barrel-vaulted grave of this complex is dated to the 6th century (see Marki 2006, 247). Furthermore, the graves are dated after the middle or to the second half of the 5th century, as when they were dug the mosaic floors of the building (probably a martyrium) were destroyed; these mosaics are securely dated to this period, see Asimakopoulou-Atzaka 1998, 151, 264-65, pls. 190-91.

[1234] See Chapter 9 (Find-Spots), '4, N. Kapatou Street, 2001'.

[1235] Hayes 1975, 80-81, a vessel as per Barag 1970c, form XV 12.

[1236] Barkóczi 1988, 142, 143, no. 310, pls. XXVI, LXXXVI, find of unknown provenance dated by analogous examples to the 3rd-4th century and assumed to be of eastern origin.

[1237] Hayes 1975, 93, no. 312, pl. 22, dated due to its shape and decoration with applied blue blobs, to the early or middle of the 4th century; it is considered to be a Syro-Palestinian product, see ibid, 80-81.

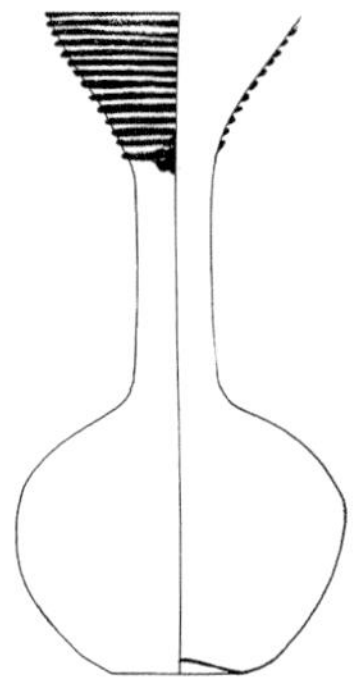

The vessel is made by free-blowing, then, after the vessel was held by some kind of jacks, quite probably from the neck, the funnel-mouth was shaped, the rim fired and thus polished, and the bottom was pushed and raised in order to make it more resistant to stresses and fractures. Finally the thread of blue glass was wound around the mouth.

The vessel is made of colourless, greenish glass and decorated with a blue thread. The blue thread wound around the mouth associates our find with Syro-Palestinian products.[1238] This type of decoration mostly appears on vessels with a tall cylindrical neck ending in a funnel-mouth, and they are the closest parallels for the Thessalonean find. The mouth of the vessel, however, is far wider than usual for these products, and nothing has been preserved from the supposedly long cylindrical neck. Examples of this form have been found in Khirbet esh-Shubeika,[1239] Bet She'an (Nysa-Scythopolis), Gerasa, Khirbet el-Kerak/Beth Yerah (Philoteria), and elsewhere, and they are dated to the late 4th and into the 5th century.[1240]

The vessel was found in a grave dated to the 4th century.[1241]

**Undecorated version**
(Cat. no. 351)
(Plates 21, XII)

The vessels of this form have tubular infolded rim and funnel-mouth; the body was probably globular, with a concave bottom. It could represent, however, an ill-conceived reconstruction from different vessels that might belong to Syro-Palestinian production of the 3rd -4th centuries.[1242]

### 67. Funnel-mouth flasks, globular, with indentations

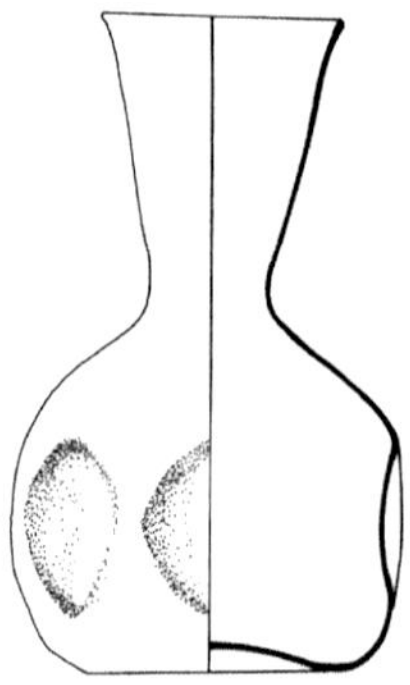

(Cat. no. 352)
(Plates 21, XII)

The vessels of this form have a funnel-mouth and fire-polished rim. The transition from the mouth to the body is constricted. The body is globular and decorated all around at its maximum diameter with six indentations. The bottom is slightly concave. Their height is estimated at c. 12cm.

These vessels were free-blown. After the inflation of the paraison and the modelling of the body, and while the vessel was held by some form of jacks from the base of the mouth, which created the constriction at that point, the conical shape of the mouth was formed, the rim fire-polished, and the bottom pressed and raised in order to make the vessel more durable. Finally the indentations on the body were created by gentle thrusts with a blunt tion.

The example from Thessaloniki is made of greenish glass with pinprick bubbles and inclusions; it cannot be dated independently as its excavation records are unavailable.

Analogous vessels have been published from Jordan: Umm Qeis (Gadara, Antiochia Semiramis, Seleukeia),[1243] Gerasa;[1244] and Israel:[1245] Bet She'an (Nysa-Scythopolis).[1246] Vessels of the same shape, but a little bigger in size, and decorated with ribs instead of indentations, are quite common in Palestine.[1247] It seems that the form of these indented vessels, like the example from Thessaloniki, which cannot be dated independently, presents yet another example of the art of Syro-Palestinian glassworkers in the 4th century.

---

1238 Weinberg 1988, 70-72, with finds dated to the 4th century and a plethora of references to analogous finds dated to the 4th century onwards from the Syro-Palestinian region. Stern 2001, 263, 269, nos. 162-165, with a thorough discussion on the characteristics of this late Syro-Palestinian production, and with examples dated to the 6th-7th century. Israeli 2003b, 164, with discussion on the matter, which concludes by dating the relevant finds to the 6th-7th century.

1239 Gorin-Rosen 2002b, 186, 317, fig. 8, from a burial cave dated by numismatic finds to the 4th century.

1240 Barag 1970c, form 15. 29-1, pl. 29 (First and second from right).

1241 Makropoulou 2007, 162, pl. 25, dated by a worn 4th-century coin found inside, but also on typological reasons based on the burial architecture.

1242 Hayes 1975, 60, 68, nos. 154, 211, pls. 12, 13, finds dated to the 3rd century. They are different in the rendering of the lip and base, although the overall picture is similar.

1243 Dussart 1998, 91, form BVII. 2412, pl. 18/6, find dated from excavation results to the end of the 3rd century.

1244 Dussart 1998, 91, form BVII. 2412, pl. 18/6, find dated from excavation results to the 4th century.

1245 Barag 1970c, pl. 40, form XV 12-2.

1246 Fitzgerald 1931, pl. V, fig. 10 (second row, last object on right), find from a grave dated to the 4th century.

1247 Stern 2001, 215, with detailed bibliography on relevant finds. The vessels are identified as products of a workshop in Palestine dated to the first half of the 4th century; Israeli 2003b, 168 no. 179, with references to analogous finds from the region, ascribed to the 4th century.

### 68. Funnel-mouth flasks, globular, with vertical ribs

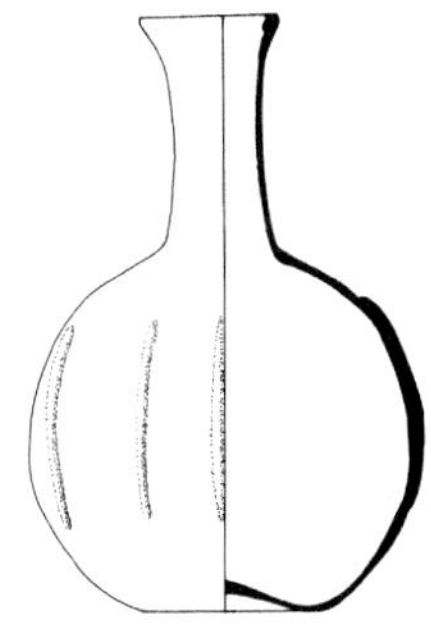

(Cat. no. 353)
(Plates 21, XII)

Vessels of this form have funnel-mouth ending to a tubular infolded rim, short cylindrical neck, globular body and slightly concave bottom. Their height is estimated at c. 13cm.

These vessels were free-blown and the vertical ribs that decorate the entire body were formed by pinching, while the vessel was still hot and malleable. The shaping of the mouth and the rim was carried out after the removal of the blowpipe from the rim, and when a second pipe was attached to the bottom of the vessel allowing its further unobstructed rotation.

Flasks with vertical ribs appear already in the 1st century, probably in its second half.[1248] The find from Thessaloniki, however, as the excavation data affirm,[1249] in addition to the characteristics of the vessel itself (e.g. the texture of the glass, the thickness of the material, the way the rim was shaped and its considerably larger size compared to 1st-century flasks), is probably much more likely to be of Late Roman production, i.e. 4th-5th century, to which are ascribed similar finds from the Syro-Palestinian region,[1250] Egypt,[1251] and a private collection as well.[1252]

### 69. Funnel-mouth flasks, pear-shaped, bulbous

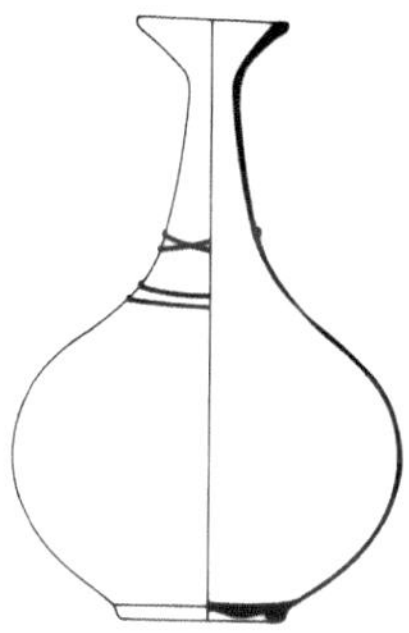

(Cat. no. 354)
(Plates 21, XII)

Vessels of this form have fire-polished rim, funnel-mouth, conical neck, with a smooth transition to the bulbous body, pushed-in ring base, and slightly concave bottom. A fine thread of glass is carelessly wound around the neck and the shoulders four times. Their height is estimated at c. 15cm.

The vessel is made by free-blowing and the base formed while the vessel was still attached to the blowpipe from its rim. Then a blowpipe was attached underneath, the moil cracked off, and the rim formed and fire-polished.

Vessels of this form appear in the late 2nd and mainly during the 3rd century; they can vary considerably in the shape of the body, i.e. to the degree of squatness, but they share all other characteristics. Examples are known from several sites and collections, including Cologne,[1253] Kassel (Castellum Cattorum),[1254] Trier (Treviri),[1255] Slovenia,[1256] Hungary,[1257] Costanza (Tomis),[1258] Tipasa,[1259] Cyprus;[1260] collections: Wolf,[1261] Oppenländer,[1262] Hentrich,[1263] and in Newark Museum.[1264] In Greece analogous finds, otherwise identical but with narrower mouth, have been reported

[1248] Isings 1957, 40-41, form 26b, finds dated to the 1st century. Trier (Treviri): Goethert-Polaschek 1977, 114-115, form 70c, nos. 618-620. Unguentaria with vertical ribs dated from excavation results to the Tiberian-Flavian (14-96) period; Black Sea: Kunina 1997, 293, no. 199, dated to the 1st century according to analogous finds; Siphnos: Mackworth-Young 1949, 39, pl. 32, from a grave dated to the 1st century; Cyprus: Vessberg 1952, pl. VII. 16.

[1249] Makropoulou 2007, 308, from a 4th-century grave.

[1250] Barag 1970c, pl. 13:19.

[1251] Harden 1936, 186, 200, pear-shaped examples, nos. 542 and 544, pl. XVII, finds from the phases of the settlement dated post AD 300, which were re-dated later to the 5th-6th century; for the re-dating, see Hayes 1976, 2-3.

[1252] Saldern 1974, no. 116, dated to the 4th century according to similar finds.

[1253] Loeschcke, Willers 1911, pl. XXXI, no. 528, and esp. no. 529; Fremersdorf, Polónyi-Fremersdorf 1984, 44, no. 111, find dated from excavation results to the middle of the 3rd century.

[1254] Boosen 1984, nos. 158, 159, finds from Olvia dated to the 2nd-3rd century.

[1255] Goethert-Polaschek 1977, 149, a squat version of the Form is form 91, pl. 14, 157d. 59, dated to the second half of the 2nd century.

[1256] Šubic 1976, 42, no. 24, pl. 1/4, find dated on the basis of analogous finds to the 3rd century; Lazar 2003, 143, form 6. 2. 4, from various sites, dated from excavation results to the second half of the 2nd and to the 3rd century.

[1257] Barkóczi 1988, 141, form 120, nos. 303-305, pl. 25, finds from various sites dated from excavation results, some of them to the 2nd and others to the second half of the 4th century.

[1258] Bucovală 1968, 64-65, 3rd-century finds.

[1259] Lansel 1967, 52-53, form 10-11, nos 65, 67, it is a slender and taller form, compared to the find from Thessaloniki, dated from excavation results between the end of the 1st and the 3rd century; idem, 53, nos. 68-69, for a more squat form, compared to the find from Thessaloniki, dated from excavation results around the middle of the 2nd century.

[1260] Vessberg 1952, 135, form VIII.B.I, pl. 8.2.

[1261] Stern 2001, 212, no. 102, squat vessel which is dated to the 2nd or more probably to the 3rd century and is assumed to be a product from Asia Minor.

[1262] Saldern *et al.* 1974, no. 592, vessel dated to the 3rd century and is considered to be a product of the Eastern Mediterranean.

[1263] Saldern 1974, 150, no. 221, dated probably to the 2nd century and is considered to be an oriental product.

[1264] Auth 1976, 124, no. 154, dated on the basis of analogous finds to the 3rd century.

from Argos,[1265] Athens,[1266] Ierissos (Acanthos),[1267] and Nea Anchialos (Phthiotides Thebes),[1268] and the most closely related examples were found at Strovolos/Nea Kerdilia.[1269] The find from Thessaloniki cannot be dated independently.

### 70. Funnel-mouth flasks, ovular with applied coil base

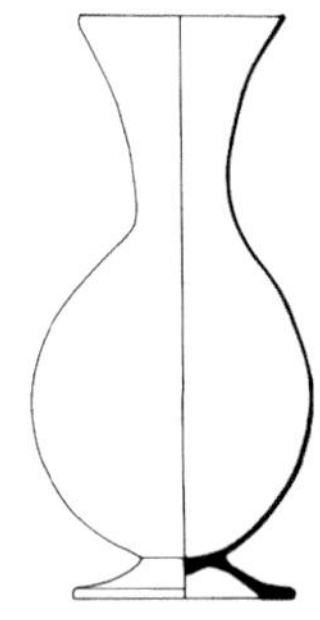

(Cat. nos. 355-365)
(Plates 22-23, XII- XIII)

The vessels of this form have funnel-mouth ending into fire-polished or infolded tubular rim; the body is ovular and stands on an applied tall, conical base. Their height is estimated between 18-22cm.

These vessels were free-blown, the base is formed by a second mass of glass applied under the body and shaped into its conical form with jacks. The traces of this tooling are almost always visible on the extant examples of this form. The rim was finished while the vessel was held by jacks, probably from the base.

The examples are made of thick, dark-green or bluish glass.

This form, because of the dark-green colour of the examples and the obvious tooling marks on the base, are clearly connected to Egyptian workshops. However, due to the relatively large number of finds from Thessaloniki and the absence of the form among the finds from other regions, the example probably represents a local variant of the pear-shaped funnel-mouthed flask with inherent base (Form 60). A variant particularly related to an ovular vessel was found in Britain,[1270] and thus accordingly dated to the 4th or 5th century. The examples from Thessaloniki, which can be independently dated,[1271] support a dating around the 4th century.

### 71. Funnel-mouth flasks, ovular, plain base

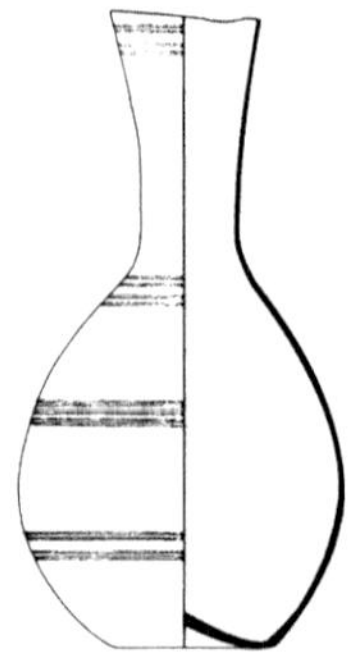

(Cat. nos. 366-372)
(Plates 23, XIII)

Vessels of this form have conical neck ending in an unworked rim, with ovular body standing on a slightly concave bottom. Their height is estimated between 14-18cm.

These vessels were free-blown. The rim was finished by cracking-off after the annealing and the vessel decorated with shallow incised bands.

The examples are made of clear, slightly greenish glass.

Analogous finds have been reported from Edessa, Greece,[1272] and Isar-Marvinci.[1273] Vessels of this form share several technological and sylistic characteristics with Fform 63, which are also funnel-mouthed and which are considered to be local products from the texture and the thickness of the glass used for their production, the way the rim was cracked-off and left unworked, and from the shallow incised bands that decorate them. These characteristics are strongly suggestive of local production.

The examples from Thessaloniki are firmly or loosely dated by their archaeological contexts to around the 4th century.[1274]

1265 Alexandri 1963, 61-62, pl. 78α, β, vessel from a Roman grave with no other chronological indications.
1266 Weinberg, McClellan 1992, 133-134, no. 109, vessel of unknown provenance, dated to the 3rd century, according to analogous finds. Weinberg assumed that they were produced in Tomis.
1267 Trakosopoulou 2002, 86, fig. 18, an example dated by a coin of Caracalas (198-211).
1268 Sotiriou 1939, fig. 16 (upper row, second from right), an example from a grave with no firm chronological indications.
1269 Malama, Christopoulou, Ntarakis 2000, 78, find from a grave dated to the 3rd century; Malama, Ntarakis 2008, 92, pl. 78, MA 6391 (From a grave dated to the first quarter of the 3rd century), idem, 95, pl. 78, MA 6396, from a grave dated to the second half of the 2nd century or later.
1270 Harden 1959, 3-20.
1271 Cat. no. 356: Makropoulou 2007, 57-58, from a grave dated to the 4th century according to the pottery found in it; Cat. no. 358: Makropoulou 2007, 108, pls. 26, 68, from a grave dated to the end of the 3rd or to the 4th century; Cat. no. 357: Makropoulou 2007, 242, pls. 26, 68, from a grave probably dated to the second half of the 4th century; Cat. no. 359: Makropoulou 2007, 253, pl. 26, from a grave dated to the end of the 3rd or to the 4th century.
1272 Chrysostomou A. 2000, 201, pl. 205, a vessel of the same shape, decorated with warts or pinches, from a burial dated to the 3rd century by a coin.
1273 Sokolovska 1986, 127, pl. 45/1, from a grave dated to the late 3rd century according to the pottery found in it.
1274 Cat. nos. 367, 368 were found in an excavation with graves predominantly dated to the period of the Tetrarchy. See Chapter 9 (Find-Spots), 'Aristotle University of Thessaloniki, Theological Department Plot, 1965'. Cat. no. 369 was found in the Ramona area in the western necropolis of Thessaloniki, along with a single-handled eight-sided vessel, Form 97, also dated to the 4th century. Cat. no. 366 was also found in the western necropolis and is dated from excavation results to the 4th century, see Papanikola-Bakirtzi 2002 (ed.), 470, no. 643 (E. Marki).

### 72. Funnel-mouth flasks, cylindrical, with short body

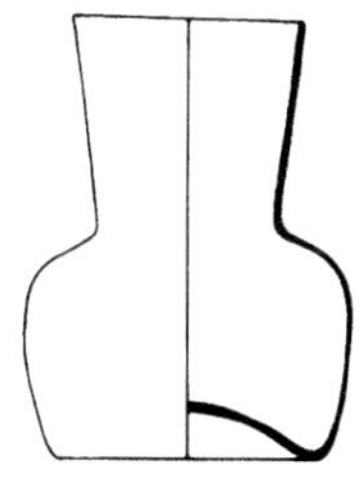

(Cat. no. 373)
(Plates 23, XIII)

The vessels of this form do not have a particularly shaped rim. The end of the wide funnel-shaped neck is fire-polished. The body, which is approximately the same height as the neck, is almost cylindrical, or inverted truncated conical, and stands on a domed bottom. The example from Thessaloniki is made of yellowish glass. Their height is estimated at c. 10cm.

The vessel is free-blown, and the body was formed probably by marvering. The rim was fire-polished while the vessel was held, probably by jacks, from the neck. The bottom has the shape of a low dome, a rare feature among vessels from Thessaloniki, but very common among Syro-Palestinian products.

Vessels of this group exist among the Syro-Palestinian products which appear at the end of the 3rd and early 4th centuries, and remain popular until the 6th century. This form is inherited almost identically from Islamic glassware and survived at least until the 8th or 9th century.[1275] Examples are known from several sites in Israel,[1276] e.g. Horbat Rimmon,[1277] Bet She'arim (Besara),[1278] Jerusalem[1279] and Kefar Sava,[1280] and to the same regions are also ascribed vessels of unknown provenance held in private collections and museums.[1281]

The example from Thessaloniki cannot be independently dated because the excavation and chronological data have not been preserved.

### 73. Funnel-mouth flasks, cylindrical, tall body (Isings Form 132)

(Cat. no. 374)
(Plates 23, XIII)

Vessels of this form have unworked rim, funnel-mouth, narrow cylindrical neck, cylindrical body tapering at the shoulders and flat, slightly concave bottom. Their height is estimated at c. 17cm.

These vessels are free-blown and the body marvered to make them cylindrical; then the mouth was released from the pipe, which was then attached underneath, reheated and widened to the desired diameter. The rim was cracked off and left unworked, or in other cases it was mildly ground after the annealing.

The vessel belongs to a subgroup of funnel-mouth vessels with cylindrical, quadrilateral or slightly conical bodies, the latter assumed to be mould-blown.[1282]

Published examples are reported from Germany: Mayen,[1283] Cologne-Rodenkirchen,[1284] Cologne,[1285] Trier (Treviri);[1286] France: Riceys;[1287] Italy: Ravenna[1288] and Rome;[1289] and Syria: Gerasa.[1290] According to the dating of the examples, this

---

Slightly different examples (Cat. nos. 371-372) were found in a building dated to the Justinian period. See Chapter 9 (Find-Spots), 'Agiou Dimitriou and Agias Sofias, the junction of, 1969'; in the same building other 4th-century glass vessels were also found. For their early date witness also an identical find from the eastern necropolis, dated according to a pottery vessel found together, to the 4th or 5th century, see Panti 2010, 466-485.

[1275] Israeli 2003b, 353, no. 470, with relevant bibliography.

[1276] Barag 1970c, 43, pl. 5:15, 7. 4e, 6, form XV:25, 27-1.

[1277] Gorin-Rosen 2004, 115-116, nos. 7, 8, fig. 2, finds dated by archaeological context to the late 4th- early 5th century.

[1278] Roffia 1993, 160, 162-163, nos. 373-374, dated on the basis of analogous finds to the 3rd or first half of the 4th century; Israeli 2003b, 171, nos. 188, 187, decorated with a fine thread around the neck, which are dated on the basis of analogous finds to the 6th century.

[1279] Bagati, Milik 1958, 144, nos. 2-4, fig. 34:2-4.

[1280] Ayalon 1994, 32-33, pls. 4. 2, 5. 5. 4, 5, from a grave used from the 3rd-4th until the 5th-6th centuries.

[1281] Hayes 1975, 91, no. 300 (undecorated example) and no. 301 (decorated with a fine thread around the neck). The vessels are dated to the first half of the 4th and later into the 4th century, respectively, according to analogous archaeological finds. Unpublished examples are also cited from the Museum of Istanbul and the Fitzwilliam Museum, Cambridge, which were found in Tyre and Bethlehem respectively.

[1282] Isings 1957, 160-161, form 132.

[1283] At least five examples have been found, three of them with oblique ribs, one with incised grooves and one mould-blown, in the form of a bunch of grapes. The vessels are dated after AD 370 and up to the early 5th century, see Isings 1957, 160-161, form 132, with older bibliography.

[1284] Vessel with oblique ribbing, probably mould-blown, dated from excavation results to the last quarter of the 4th century, see Isings 1957, 160-161, form 132, with older bibliography.

[1285] Vessel decorated with indentations on the body and dated probably from excavation results to the 3rd-4th century, see Isings 1957, 160-161, form 132, with older bibliography.

[1286] Goethert-Polaschek 1977, 317, no. 245, form 111, pl. 22, find dated from excavation results to the second half of the 4th century.

[1287] A small-sized vessel of this form, from a 4th-century layer, see Isings 1957, 160-161, form 132, with older bibliography.

[1288] At least four examples have been found, dated to the 4th-5th century, see Isings 1957, 160-161, form 132, with older bibliography.

[1289] Four-sided vessel with identical neck and rim, dated from excavation results to the 4th century, see Isings 1957, 160-161, form 132, with older bibliography.

[1290] Baur 1938, 531, no. 385, fragment dated to the 4th-5th century.

form appears in the last quarter of the 4th century and remained in use in the 5th century as well.

Usually the vessels are left undecorated or occasionally they bear fine, shallow incised horizontal grooves, like the example from Thessaloniki. Less often they bear indentations on the body, and mould-blown examples decorated with oblique ribbing on the body are also known. Finally, there is one example whose body is shaped like a grape bunch.

The example from Thessaloniki is made of olive-green glass, just like most of the other published examples of the form. The majority of known examples are found in the western provinces and despite the fact that not a large number of them are known, it seems more probable that they were made in that part of the Empire. The find from Thessaloniki is dated by its archaeological context most probably to the 4th or 5th century.[1291]

### 74. Amphorisks

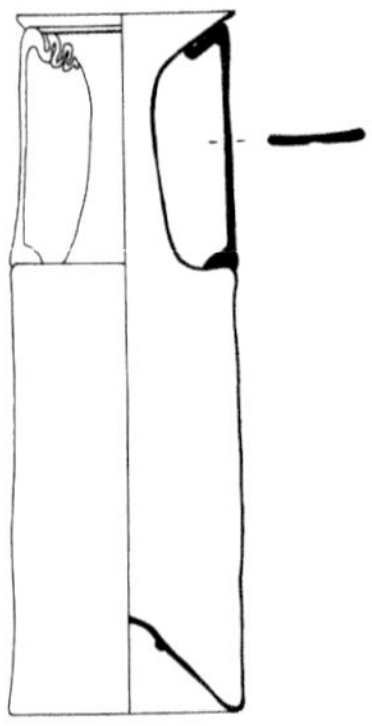

a. Amphorisks with flat base
(Cat. nos. 375-381)
(Plates 23-24, XIII)

The vessels of this form have everted, almost horizontal, fire-polished rim, which forms a funnel-mouth, short cylindrical neck, almost horizontal shoulders, cylindrical body and concave bottom. Two vertical strap-handles start from the shoulders, bend at the rim and then fold and end under it. A thick trail of glass is attached under the rim. The examples from Thessaloniki are made of greenish glass. Their height is estimated between 16-25cm.

The vessels are mould-blown, except for the neck and the rim, which are formed with free-hand tooling. It appears that their size was predefined, and probably precisely for that reason they were blown in a mould, and their size corresponded to certain volume units, i.e. two and four sextarii.[1292]

This form appears around the middle of the 3rd century and continues in use mainly during the 4th, and surviving into the early 5th century. Vessels of this form have been found both in the West and East, including[1293] Britain: Colchester (Camulodunum);[1294] Germany: Bonn (Bonna),[1295] Trier (Treviri),[1296] Mainz am Rhein (Moguntiacum),[1297] Cologne;[1298] Serbia: Belgrade (Singidunum),[1299] Čezava (Novae),[1300] Karataš (Diana),[1301] Vlastonice,[1302] Čačak;[1303] Hungary: Mucsfa,[1304] Budapest (Aquincum),[1305] Dunaújváros (Intercisa);[1306] Black Sea region: Kerch (Panticapaeum);[1307] Egypt: Behnasa (Oxyrhynchos),[1308] Kom Aushim (Karanis);[1309] Greece: Veroia,[1310] Vrysi at Pontokomi/Fordaia,[1311] the Athenian Agora,[1312] and Frangonisi in Eleia.[1313]

Although their shape equates with the most typical shape of transport vessels, the fact that they were often decorated strongly suggests that they were mainly used as tableware vessels, and this might well apply also for pointed amphorisks.

[1291] For the dating of the graves, see Chapter 9 (Find-Spots), 'Koloniari-Galanaki-Papathanasiou, 1987'.

[1292] 1 sextarius = 0.543l. See Whitehouse 1997, 266-267, and the discussion of vessel no. 455, with relevant bibliography.

[1293] Isings 1957, 157-158, form 127, finds dated from excavation results between the end of the 3rd and end of the 4th century.

[1294] Cool, Price 1995, 201, 218, they appear in Colchester (Camulodunum) around AD 230, and remain in use throughout the 4th century.

[1295] Follmann-Schulz 1988, 52-53, nos. 150, 151, pl. 18, 19, dated from excavation results to the late 3rd and to the 4th century.

[1296] Goethert-Polaschek 1977, 233-234, form 141, nos. 1395-1398, finds dated to the 4th century.

[1297] Harter 1999, form F5, 258-259, 343, nos. 826-829, finds dated from excavation results to the middle, or to the second half of the 4th century.

[1298] Fremersdorf 1967, 131-135, pls. 156-162, finds, some of which are dated generally to the 4th, and others dated from excavation results to the end of the 4th century, see Isings 1957, 158, for relevant bibliography.

[1299] Ružić 1994, 22, no. 148, find dated from excavation results to the 3rd-4th century.

[1300] Ružić 1994, 16, no. 77, find dated generally to the end of the 3rd-4th century.

[1301] Ružić 1994, 16, nos. 78-79, finds dated generally to the end of the 3rd-4th century.

[1302] Ružić 1994, 16, no. 80, find dated generally to the end of the 3rd-4th century.

[1303] Ružić 1994, 16, no. 81, find dated probably from excavation results to the second half of 4th century.

[1304] Barkóczi 1988, 202, no. 504, pl. LVIII, find dated from excavation results to the last quarter of the 4th- early 5th century.

[1305] Barkóczi 1988, 203, no. 505, pl. LIX, find dated from excavation results to the late 3rd- early 4th century.

[1306] Radnóti 1956, 151, nos. 9, 84, 129, 134, pl. XXXII; Barkóczi 1988, 203-204, nos. 506-510, pl. LIX, finds dated from excavation results to the 4th and to the second half of the 4th century.

[1307] Kunina 1997, 332-333, no. 400, dated according to analogous finds and context to the 4th century; it is ascribed to production of the Rhine region.

[1308] Petrie 1925, 16, pl. xlv. 9, dated to the 3rd-4th century.

[1309] Harden 1936, class XI.B.3β.

[1310] Allamani-Souri 1991a, 301-302, pl. 114δ; Allamani-Souri 1991c, 39, fig. 1, example from a grave dated to the 4th century according to its context; Allamani-Souri 1991b, 302-303, pl. 114ε, example from a grave dated to the 4th century on the basis of its context.

[1311] *Glass Cosmos* 2010, 192, no. 35, dated for no specified reason to the 1st-2nd century.

[1312] Thompson 1949, 217, pl. 41. 2, find dated prior to AD 267; Weinberg and Stern 2009, 146, nos. 330-31, fig. 19, from 4th-century contexts.

[1313] Gialouris, 1966, 171, pl. 182ς, without a specific reference to the dating of the grave.

The finds from Thessaloniki, except for Cat. nos. 375[1314] and 376,[1315] are not independently dated and thus they are generally ascribed –according to parallels – to the period between the end of the 3rd and the early 5th century.

b. Pointed Amphorisks
(Cat. nos. 382-384)
(Plates 24, XIII)

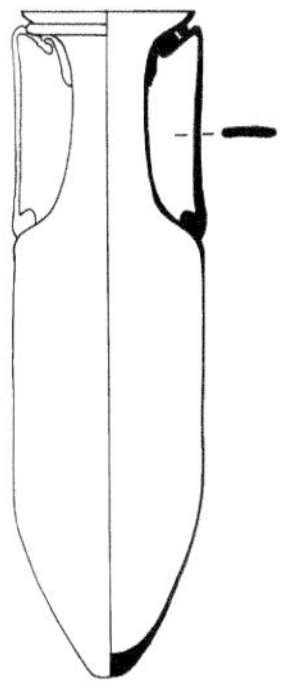

Vessels of this form have an everted, almost horizontal, fire-polished rim, which forms a funnel-mouth, short cylindrical neck, almost horizontal shoulders, cylindrical body and pointed bottom that occasionally ends into a small knop. Two vertical strap-handles start from the shoulders, bend at rim height, and then are folded and attached under the rim. A thick trail of glass is wound around the underside of the rim. The examples from Thessaloniki are made of greenish glass. Their height is estimated between 18-30cm.

The vessels are mould-blown, except for their neck and rim, which are shaped with free-hand tooling; very often they bear incised geometrical decoration.

This form probably appears already from the 3rd century, but in any event they continued into the 4th century, as the patterns of their decoration and the texture of the glass reveal;into the early 5th century. Examples are kept in museums in the East and West, including Toronto,[1316] Newark,[1317] New Haven,[1318] and Corning.[1319] Amphorisks of this shape have been found in excavations of the following sites: Ravenna-Classe;[1320] Hungary: Nagybarátfalu,[1321] Dunaújváros (Intercisa),[1322] Ságvár (Tricciana);[1323] Serbia: Sremska Mitrovica (Sirmium),[1324] Bela Palanka (Remesiana),[1325] Jagodin Mala-Niš (Naissus),[1326] Glavnik (Vindenis),[1327] Kuršumlja (Ad Fines);[1328] FYROM: Zajas,[1329] Dračevo,[1330] Ochrid,[1331] Živojno,[1332] Isar-Marvinci,[1333] Lopate;[1334] in Egypt at Kom Aushim (Karanis),[1335] and Homs (Emesa) in Syria.[1336] In the Volos museum there is an example from an excavation in Magnesia,[1337] and a variant with angular body has been found at Patras.[1338]

Although their shape represents that of the most typical transport vessel, the fact that they are quite often decorated strongly suggests that they were mainly used as tableware vessels, which was probably the case also for the amphorisks with flat bases.

The only one among the finds from Thessaloniki which can be dated by its archaeological context is Cat. no. 382,[1339] while the remaining ones are dated on the basis of parallels from other sites between the 4th and 5th century.

### 75. Square bottles (Isings Form 50)

(Cat. nos. 385-396)
(Plates 25, XIII-XIV)

The vessels of this form have infolded, horizontal rim, short cylindrical neck, four-sided or square body and flat

[1314] Makropoulou 2007, 197, pl. 28, find from a grave dated to the 3rd-4th century.
[1315] Makropoulou 2007, 269, pls. 28, 69, from a grave dated to the 3rd-4th century.
[1316] Hayes 1975, 102, nos. 365-366, dated according to analogous finds.
[1317] Auth 1976, 130, no. 162, dated according to analogous finds.
[1318] Matheson 1980, 86, no. 235, dated according to analogous finds.
[1319] Whitehouse 1997, 260-261, no. 445, dated on the basis of analogous finds to the 4th-5th century.
[1320] Isings 1957, 78, from a 4th- or 5th-century cemetery.
[1321] Benkő 1962, 5, no. 6b, pl. XXII, 1, 9. The same vessel is also presented in Barkóczi 1988, no. 535, pl. LXII, CXIII/533, a find dated to the second half of the 4th century.
[1322] Radnóti 1957, 151, 158, no. 58, pl. XXXI/3. The same appears also in Barkóczi 1988, no. 533, pl. LXI/533, a find dated to the first half of the 4th century.
[1323] Burger 1966, 138, fig. 125, pl. XCVIII/1, find dated by its archaeological context between AD 355-363.
[1324] Šaranović-Svetek 1986, 62, no. 73, pl. VII, 1, find dated generally to the 4th century. Cf. another example, also dated generally to the 4th century, in Cermanović-Kuzmanović 1981a, 19, no. 98.
[1325] Ružić 1994, 22, nos. 141-142; Drča 2000, 219, 226, nos. 18-19, figs. 18-19, finds dated to the second half of 4th century.
[1326] Ružić 1994, 22, no. 144; Drča 2000, 219, 226, no. 20, fig. 20, find dated to the 4th century.
[1327] Ružić 1994, 16, no. 80, find dated by archaeological context to the 4th century.
[1328] Kondić 1992, fig. 4, dated to the 4th century.
[1329] Bitrakova 1966, 96, photo 2, two examples, slightly different, from a grave dated to the end of the 4th century.
[1330] Josifofska 1956, 287, find from a grave dated to the late 4th- early 5th century.
[1331] Bitrakova 1966, 96, without a reference to the dating of the find.
[1332] Mikulčić 1976, 192, 196, pl. VIII, A-B, dated first half of the 4th century, according to context.
[1333] Sokolovska 1986, 127, pl. 46/5, from a grave dated by coins of Constantius to the middle of the 4th century.
[1334] Nikolovski 2006, 247-248, fig. VII/6, finds from Late Roman graves, without a more precise dating.
[1335] Harden 1936, 258, 264, pl. XX/793, dated from excavation results to the end of the 2nd to early 3rd century.
[1336] Abdul Hak 1965, 32-33, fig. 19, four examples from a burial complex dated between the middle of the 2nd to the middle of the 3rd century, or to the middle of the 4th century, according to Stern 1977, 83-84.
[1337] Hourmouziadis, Asimakopoulou-Atzaka, Makris 1982, 62-63, fig. 33, without more information concerning the excavation or its dating.
[1338] Dekoulakou 1976, 103-104, pl. 81β, γ, vessel from a cemetery dated to the Christian era.
[1339] Makropoulou 2007, 265, pl. 28, from a grave dated to the first half of the 4th century, according to its pottery context.

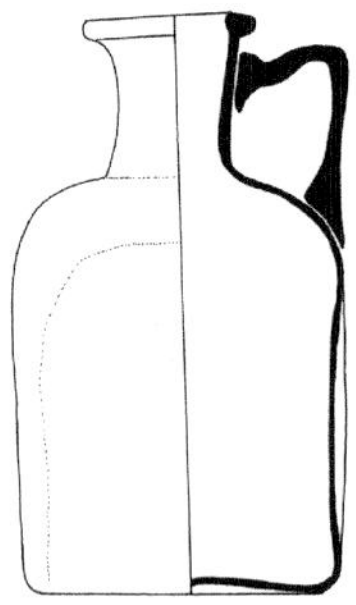

bottom. A wide, vertical, strap-handle, usually reeded, starts on the shoulder and ends on the rim. Their height is estimated between 12-35cm.

They are mostly mould-blown, except for the neck and the rim which were shaped with free-hand tooling. The mould-blown examples often feature underneath a range of decorative relief motifs, which were carved on the mould and then stamped during the mould-blowing process.[1340]

The decorative motifs on the undersides of the examples from the baths of the Roman agora are only partly preserved. The following can be identified:

a) Greek cross which divides the bottom into four equal quadrants, each with a dot at its centre and 6 or 7 concentric semicircles centred at the corner.

b) Indiscernible symbol identified as the Greek letter Π, but it seems perhaps part of a very fragmentarily preserved ellipsoid motif.

c) A base, delineated all over with large dots that were probably intended to act as a firmer support and more secure stand for the vessel; in the centre is a 'pelte' shield and the Latin letter 'L'.

d) Square bottom of a vessel with relief decoration which is organized in concentric ring bands. Part of the exterior band is preserved, which was decorated with a wine scroll and 'pelte' shields. The corners have some indiscernible, partly preserved, geometrical motifs.

Although various *corpora* of base marks have been created, it is still not possible to draw conclusions regarding the provenance or the content of these vessels and no exact parallels have been located either.[1341]

Less often, vessels of this form were free-blown and their sides flattened with a flat tool while they were still hot and malleable. This technique appears to be more common among finds from Eastern provinces, while it is almost unknown among Western finds.[1342]

Vessels of this form as a whole are made of thick, greenish or bluish glass.

They have been divided into two almost contemporaneous groups, on the basis of whether their height is greater than the width of their side. The shorter examples appear already in the first half of the 1st century, while the taller examples appear in the Flavian period.[1343] It is a form known in the East, but mainly appearing in the West.[1344] The earliest examples are dated to the first half of the 1st century.[1345] They get their widest distribution in the second half of the 1st century and it was thought that they met with their greatest popularity in the period between AD 70 - 120/130.[1346] It has been proved now, however, that at least in the NW provinces they continued to circulate throughout the 2nd, and even probably into the early 3rd century,[1347] a fact that, according to the parallels noted in the present work, seems to be confirmed for the Balkans as well.

Detailed presentations of all known examples have been published in Isings,[1348] Charlesworth,[1349] and Cool and Price.[1350] From the Balkan regions, the published examples which were not noted in these publications include finds from Starigrad (Argyruntum),[1351] Zaton,[1352] Istra (Istria),[1353] Zadar (Iader),[1354] Bakar (Volcera), Croatia, Sisak

[1340] Antonaras 2006a, 413-420.
[1341] Foy, Nenna (ed.) 2006; Foy, Nenna (ed.) 2006a; Foy, Nenna (ed.) 2011.
[1342] Charlesworth 1966, 33.
[1343] Isings 1957, 64, 66.
[1344] Isings 1957, 63-66, form 50a and 50b, where many examples are cited.
[1345] Czurda-Ruth 1979, 135-136, nos. 1020-1022, finds dated by archaeological context between the late years of Augustus and up to Tiberius (10-37).
[1346] Charlesworth 1966, 26-40.
[1347] Cool, Price 1995, 179-199, esp. 184, where large quantities of vessels of this form from various sites are analyzed, and which are dated by coins or by clay finds from their own contexts.
[1348] Isings 1957, 63-66, form 50a and 50b.
[1349] Charlesworth 1966, 26-40.
[1350] Cool, Price 1995, 179-199.
[1351] Fadić 1989, 11, nos. 53-52, vessels dated from excavation results to the 1st or into the early 2nd century, according to analogous finds.
[1352] Gluščević 1991, mentions 12 bases of vessels with no particular chronological indications; reference is also made to various other finds from Dalmatia, some of them previously unpublished, dated mainly to the 1st-2nd century.
[1353] Fadić 1986, 103-109, form 2, finds dated for stylistic reasons to the 1st-2nd century, and the author also assumes that they present part of local Dalmatian production.
[1354] Fadić 1999-2000, 185-189, with examples of bases decorated with the Christogram, dated to the 3rd-4th century, and including an older bibliography on the Dalmatian area and Zadar (Iader).

(Siscium),[1355] Budva (Buthoe),[1356] Kostol (Pontes),[1357] Čezava (Novae),[1358] Transdierna (Tekija),[1359] Karataš (Diana),[1360] Brza Palanka (Egeta),[1361] Belgrade (Singidunum),[1362] Ravna (Timacum Minus),[1363] Gračanica (Ulpiana),[1364] Sremska Mitrovica (Sirmium),[1365] Szombathely (Savaria),[1366] Győr (Arrabona),[1367] Dunaújváros (Intercisa), Budapest (Aquincum), Iványc, Szőny (Brigetio),[1368] Costanza (Tomis),[1369] Chatalka,[1370] Asenovgrad (Stenimahos),[1371] Radiuvene,[1372] Varna (Odessos),[1373] Balchik (Dionysopolis),[1374] Mangalia (Callatis).[1375] Published examples in Greece were found at Samothrace,[1376] Palatiano (Ioron),[1377] Lete,[1378] Amphipolis,[1379] Naxos,[1380] Perissa, Thera,[1381] Corinth,[1382] Kallio,[1383] Athens,[1384] Oropos,[1385] Knossos,[1386] and Olympia.[1387]

Cat. nos. 385-393 from Thessaloniki were buried in the last two decades of the 1st century, the time of the destruction of the bath complex in which they were found.[1388]

### 76. Jugs, globular, trefoil-mouth

(Cat. no. 397)
(Plates 25, XIV)

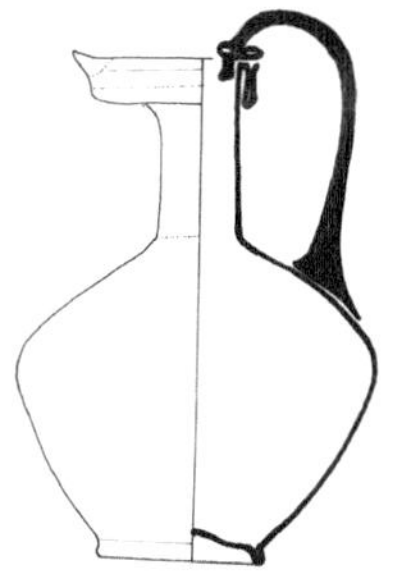

The vessels of this form have infolded, trefoil-mouth, cylindrical neck, globular body and inherent pad base. The vertical, coil handle, which starts on the shoulder, exceeds the height of the vessel and bends and finishes on the rim. On the preserved lower part of the vessel an incised horizontal groove is visible, and, judging by analogous finds from the Black Sea region, the upper part of the vessel must have had another one on it. Their height is estimated at c. 15cm.

The vessel is free-blown, the base is shaped by further tooling of the same mass of glass with the body. The rim has been folded out, down and up, and then again drawn up and out in order to get the desired spouted form. The handle was formed by applying with the help of a metal rod a lump of hot glass on the vessel's shoulder and by further drawing it upwards and bending it towards the rim.

1355 Damevski 1976, 80, pl. VIII/1, vessels dated by analogous finds between the 1st and 2nd century.
1356 Cermanović-Kuzmanović 1976, 176, pl. II/4, vessels dated by analogous finds between the 1st and early 3rd century.
1357 Ružić 1994, 11-12, no. 1, 11-12, finds dated by their archaeological context to the 3rd and 3rd-4th century, respectively.
1358 Ružić 1994, 11, nos. 3-6, finds dated by their archaeological context to 3rd-4th century.
1359 Ružić 1994, 11, nos. 7-8, finds dated by their archaeological context to the 1st- early 2nd century.
1360 Ružić 1994, 11-12, nos. 9-10, finds dated by their archaeological context to the 3rd-4th century.
1361 Ružić 1994, 12, no. 13, find dated by archaeological context to the 3rd century.
1362 Nikolić-Đorđević 1990, 50, nos. 35, 36, finds dated by their archaeological context, one of them to the late 1st- early 2nd century, and the other between the 1st and 3rd century.
1363 Ružić 1994, 12, nos. 18-24, finds dated by their archaeological context, some of them in the late 1st- early 2nd, and others to the 3rd-4th century.
1364 Ružić 1994, 12, nos. 25-30, finds dated by their archaeological context to the late 1st- early 2nd century.
1365 Ružić 1994, 12, no. 31, find dated by archaeological context to the 3rd-4th century.
1366 Barkóczi 1988, 173, nos. 407, 416, pl. XXXVII, XCV, finds dated by their archaeological context to the end of the 1st and to the 2nd century.
1367 Barkóczi 1988, 177, no. 419, finds dated to the early 2nd century.
1368 Barkóczi 1988, 172-178, nos. 403-427, finds dated to the end of the 1st and to the 2nd century.
1369 Bucovală 1968, 31-32, form V, nos. 17-18, which are dated on the basis of analogous finds to the 2nd century.
1370 Bujukliev 1986, 76, pl. 13, finds from a cremation dated by archaeological context to the second half of the 1st century.
1371 Milchev, Petkov 1965, 51, note 6.
1372 Miatev 1928, 313, fig. 202, find dated by a coin of Antoninus Pius (138-161) (non vidi), according to Milchev, Petkov 1965, 51, note 11.
1373 Milchev, Petkov 1965, 51, fig. 7, dated to the 2nd century according to analogous finds. It was also found together with a cylindrical glass beaker, Isings form 34, which appears in the second half of the 1st century and survives until the 4th century. Minchev 1990, 59-60, 71-73, dated from excavation results to the middle 2nd - early 3rd century.
1374 Minchev 1990, 59-60, 71-73, dated to the end of the 1st- into the 2nd century.
1375 Minchev 1990, 59-60, 71-73, dated to the end of the 1st- into the 2nd century.
1376 Dusenbery 1998, 1125-1126, nos. S199-5, find dated from excavation results to the end of 1st century BC, or early in the 1st century AD; also S208-3, S219-21 and XS-528, finds dated from excavation results to the first quarter of the 1st century.
1377 Antonaras, Anagnostopoulou-Chatzipolychroni 2002, 115, fig. 5, dated between the 1st and 3rd century according to analogous finds.
1378 Adam-Veleni, Poulaki, Tzanavari 2003, 211, no. 191 (K. Tzanavari). The find was dated to the 2nd-3rd century from its location in the destruction layer of a *villa rustica* dated to this time. However the glass find, as is the case with the other glass objects with which it was found, is datable approximately to the 1st century.
1379 Malama 2000, 65, fig. 21, without a reference to its dating, judging though by the illustrated content of the grave the burial is dated to the 1st century.
1380 Zafeiropoulou 1999, pl. 86δ, dated according to analogous finds.
1381 Gerousi 2002, 135, fig. 3, find from a grave dated by its archaeological context to the second half of 1st century.
1382 Davidson 1952, 103, nos. 656-662, fig. 9, finds dated by archaeological context to 1st and 2nd century.
1383 Zafeiropoulou 1982, 6, 8, pl. ΔΘ' (page 8, left), ascribed to the period between the 1st and 4th centuries, according to other examples known from the bibliography. According to the illustrated co-finds it rather seems to be ascribed to the 1st century.
1384 Weinberg, McClellan 1992, 119-120, no. 89, find dated to the 1st-2nd century according to similar finds.
1385 Pologiorgi 1998, 45-46, pl. 13 ΜΩ 256, find dated according to its context and analogous find to the 1st-2nd century.
1386 Price 1992, 424, nos. 111-120, pl. 341, finds dated to the 2nd century and later, except for one dated to the Flavian period (69-96).
1387 Schauer 2001/2002, 40, fig. 9, finds dated between the 1st and early 3rd century, probably according to similar finds.
1388 Adam-Veleni *et al.* 1998, 85-102; Mavromichali 2001, 141.

This form of vessel is dated to the 1st century, a period that coincides with the destruction time of the bath complex where the example from Thessaloniki was found. Identical jugs have been found at Pompeii,[1389] Olvia Sardinia,[1390] Ljubljana (Emona),[1391] Tomis,[1392] Kerch (Panticapaeum),[1393] Olbia Pontica,[1394] and Herakleion, Crete.[1395] A similar vessel was also found at Avenches (Aventicum).[1396]

Although they have been considered to be products of Italian[1397] and Eastern Mediterranean[1398] workshops, for the time being there is no particular clue which might ascribe them to either region.

### 77. Jugs, conical with undulating handle

(Cat. nos. 398-402)
(Plate XIV)

The vessels of this group are not sufficiently preserved to exactly identify the shape of their body, which appears to be conical. In some cases they stand on their under surface and in some other cases they stand on a ring base. What is mainly preserved is the undulating ending of their handle. Their height is estimated between 9-13cm.

These vessels were made by free-blowing. Judging by other, better preserved examples, after the removal of the blowpipe from the rim, and the attachment of the vase onto another pipe attached to the centre of the bottom, the moil was removed and the rim drawn into its collar-like shaped; the ribbed handle was added afterwards. The long, wavy end of the handle extends to the lower part of the body.

It is quite possible that they are part of a relatively well-known form of vessels, with conical body, which appears in the middle of the 1st century; their distribution culminates in the third quarter of the same century and they continue to be used into the early 2nd century.[1399]

Although they are considered to represent a product of a workshop, either Western, from north Italy, or from the Rhine region,[1400] or from Asia Minor,[1401] it seems equally convincingly that they can be ascribed to the SouthEastern Mediterranean, and more precisely a Cypriot workshop.[1402]

The finds from Thessaloniki are dated, on the basis of the destruction time of the bath complex where they were found, to the last twenty years of the 1st century.[1403]

Analogous finds are published from Cologne,[1404] the Black Sea region,[1405] Kerch (Panticapaeum),[1406] Ljubljana (Emona),[1407] Pompeii,[1408] Zadar (Iader),[1409] Budva (Buthoe),[1410] Cyprus,[1411] Idalion Cyprus,[1412] Samothrace,[1413] and Galaxidi (Chaleion).[1414]

---

1389 Isings 1957, 75-76, form 56b.
1390 Stiaffini, Borghetti 1994, 68, 128-129, no. 336, pl. 54-55, find dated to the 1st century according to analogous finds.
1391 Lazar 2004a, 67, no. 59.
1392 Bucovală 1984, 60, fig. 2, finds from a cremation dated from excavation results to the second half of the 1st century. Furthermore these jugs are considered to be Italian products based on the similarity with those depicted on the wall-painting in the grave of Vestorius Priscus, Pompeii. More careful observation, however, reveals that the similarities of the vessels are actually the high handles, while their bodies and bases are totally different.
1393 Kunina 1997, 303, no. 246, fig. 135, example from a grave dated to the 1st century.
1394 Kunina 1997, 303, no. 247, example from a grave dated to the 1st century, which was found with a *trulla*, a pan with horizontal handle, see Kunina 1997, 197, no. 202.
1395 Alexiou 1966, 408, pl. 437δ', στ', find from a grave with several burials dated to the 1st and early 2nd century.
1396 Bonnet-Borel 1997, 54, form AV V 148, dated between AD 40-80 from excavation results.
1397 Bucovală 1984, 60-61, fig. 3, these are considered to be Italian products, based on the similarity with those depicted on the wall-painting in the grave of Vestorius Priscus, Pompeii. More careful observation, however, reveals that the similarities of the vessels are actually the high handles, while their bodies and bases are totally different.
1398 Kunina 1997, 303, nos. 246-247.
1399 It is related with the much taller Isings Form 50, see Isings 1957, 73-74.
1400 Ravagnan 1994, 172, no. 334.
1401 Sorokina 1988, 57, fig. 2:24.
1402 Stern 2001, 50, 95, no. 32, with a detailed catalogue of all internationally published analogous finds.
1403 Adam-Veleni *et al.* 1998, 85-102; Mavromichali 2001, 141.
1404 Fremersdorf 1958b, pl. 47.
1405 Sorokina 1988, 57, fig. 2:24, finds dated generally to the 1st-2nd century.
1406 Kunina 1997, 302-303, nos. 242-244, they are considered to be Cypriot products dated to the second half of the 1st century, according to analogous finds.
1407 Petru 1976, 28, pl. 5:4, find from a cremation dated to the second half of the 1st century.
1408 Scatozza-Höricht 1988, 45, fig. 1:12489.
1409 Ravagnan 1994, 172, no. 334, vessel dated on the basis of analogous finds to the second half of the 1st- early 2nd century; *Trasparenze imperiali 1997*, 166 no. 140, from a grave dated to the second half of the 1st century.
1410 Veličković 1976, 168, pl. II:1, find dated on the basis of analogous finds to the 1st-2nd century, whereby it is assumed that it was made in Syria and arrived in Budva (Buthoe) via Aquileia.
1411 Vessberg 1952, pl. 5:15, 18.
1412 Cesnola 1903, pls. 95:5, 95:1-4, 6.
1413 Dusenbery 1998, 1128, no. XS-529, with no precise dating, probably first half of the 1st century.
1414 Threpsiadis 1972, 197, pl. 76α. Published for the second time by Zafeiropoulou 1975, 54, pl. 4:5, where the vessel is dated on the basis of

### 78. Globular jugs, without rim. 'Dilekythoi'

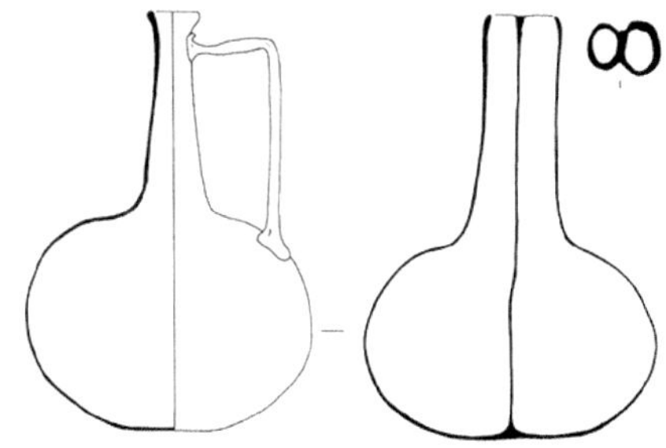

(Cat. no. 403)
(Plates 25, XIV)

This form includes tableware vessels with one handle that in some cases were made as a single, united vessel which was internally divided into two or three compartments by the skill of the glassblower.[1415] The rim is vertical or slightly flaring or infolded. There are published examples with a globular body and others with a mould-blown quatrilinear one.[1416] They stand on a flat or slightly concave bottom and feature a usually wide, ridged strap-handle. The find from Thessaloniki is made of colourless, bluish glass. Their height is estimated between 15-20cm.

This form is relatively rare and probably is not dated very accurately. Vessels with bipartite and tripartite body were produced already from the first half of the 1st century[1417] and were still in use in the 2nd, and up to the 4th century. Most of the extant examples were found in the western provinces of the Roman Empire, where they were probably produced, i.e. Britain,[1418] Spain,[1419] Switzerland,[1420] France,[1421] Germany,[1422] and Italy.[1423] There are also published examples from the East, such as from Kerch (Panticapaeum),[1424] Athens,[1425] Herakleion,[1426] Palatiano (Ioron),[1427] Amphipolis,[1428] and Dura-Europos.[1429]

The find from Thessaloniki cannot be dated independently because there are no archaeological data connected with it, and it is loosely dated to between the 1st and 4th centuries, in accordance with analogous finds from elsewhere.

### 79. Globular, aryballoid jugs, without rim

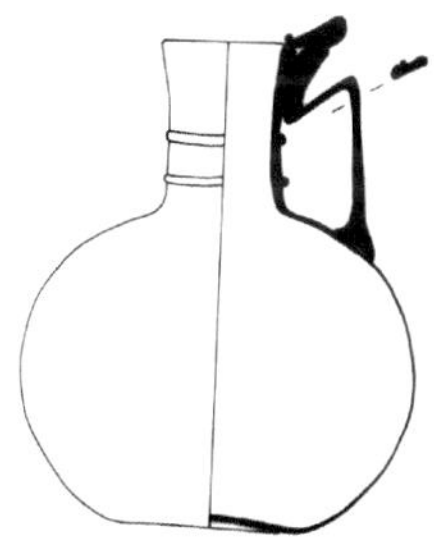

(Cat. no. 404)
(Plates 25, XIV)

Vessels of this form have vertical fire-polished rim, cylindrical neck, globular body, and they stand on a flat, slightly concave bottom. A vertical strap-handle starts on the shoulder, bends at acute angle, touches the neck and continues upwards ending on the rim where it forms a thumb-rest pad. A fine thread of glass is wound spirally twice around the neck. Their height is estimated at c. 13cm.

The body of these vessels was modelled by free-blowing, then a pipe was attached underneath to further shape the rim; the rim was then released from the blowpipe and the moil, heated up for polishing, and the thread wound before the handle was applied.

The find from Thessaloniki is made of decolourized glass.

No analogous find has been traced, yet the find itself was unearthed in a grave used from the 2nd and up to the first half of the 4th century,[1430] and thus is dated to this period. On artistic grounds it should not be ascribed to the production of the late 3rd or to the 4th century; it looks rather like a 2nd- or early 3rd-century product.

---

analogous finds to the end of the 1st-2nd century.

[1415] Cool, Price 1995, 203, with a discussion about their being formed as a single, united vessel.

[1416] Antonaras, Anagnostopoulou-Chatzipolychroni 2002, 117, fig. 13.

[1417] Isings 1957, 31-32, form 14.

[1418] Cool, Price 1995, 203-204, with examples from various British sites dated from excavation results to the middle and second half of the 1st century.

[1419] Cool, Price 1995, 203, find dated from excavation results to the second quarter of the 1st century.

[1420] Avenches(Aventicum): Bonnet-Borel 1997, 54, form AV V 146, pl. 27, dated from excavation results between the 1st and 3rd centuries, most probably from the second half of the 2nd century; Augst: Rütti 1991, form 168, nos. 4067-4072, pl. 152, dated from excavation results between the first half of the 1st and 4th centuries.

[1421] Strasbourg (Augusta Trebecorum, Argentoratum): Arveiller-Dulong, Arveiller 1985, 54, no. 70, find from a cemetery in use after AD 150; Seine-Maritime: Sennequier 1985, 189, no. 290, pl. 4, dated from excavation results to the second half of the 2nd century.

[1422] Goethert-Polaschek 1977, form 131, dated, without archaeological indications, to the 4th century.

[1423] Meconcelli-Notarianni 1979, 64, no. 56, generally dated between the 1st and 3rd century.

[1424] Kunina 1997, 287, no. 169 (colour pl. page 132), square, mould-blown vessel, dated probably from excavation results to the 1st-2nd century, and assumed to be a product of the area of modern France or Germany.

[1425] Dilekythos, National Archaeological Museum, A.E. 28243, dated to the 1st-2nd century, presented in the exhibition *Euthrafsti polyteleia. Gyalina angeia tou Ethnikou Archaeologikou Mouseiou*, National Archaeological Museum, Athens, 16 April - 30 July 2002.

[1426] Harden *et al.* 1968, 62, no. 78.

[1427] Antonaras, Anagnostopoulou-Chatzipolychroni 2002, 117, fig. 13, generally dated according to analogous finds from other sites to the 1st-2nd century.

[1428] *Glass Cosmos* 2010, 387, no. 457, from a grave dated to the 1st century.

[1429] Clairmont 1963, 128-129, no. 647, pl. XXXV, dated to the period before AD 256.

[1430] See Chapter 9 (Find-Spots), 'Leoforos Stratou. Camp of the 3rd Army Corps. Building of YENEΔ / EPT3 TV Station, 1969'.

### 80. Jugs, spherical, funnel-mouth, plain base

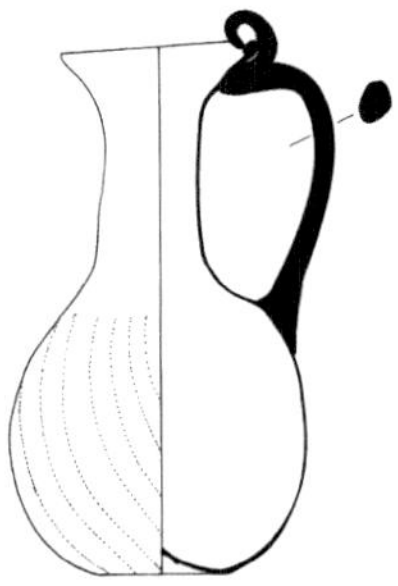

(Cat. no. 405)
(Plates 25, XIV)

Vessels of this form have flaring, infolded tubular rim, cylindrical neck, and has a globular body decorated with vertical ribs that are more visible on the lower part of it, and mildly concave bottom. A coil handle starts on the upper part of the body, rises up, drawing away from the neck, bends, adheres under the rim, folds and ends over it. Their height is estimated at c. 18cm.

The vessels were dip-mould-blown, i.e. the paraison was initially blown in a small ribbed dip-mould and then expanded to its final size by free-blowing. The shaping of the rim and the base, as well as the application of the handle, were carried out while the vessel was held and rotated by a metal rod temporarily attached underneath.

The example from Thessaloniki is made of colourless, greenish glass.

A vessel of similar shape, without ribbing, was recorded in Hungary[1431] and is dated to the 3rd century. The technically defective, almost sloppy modelling of our example, the colour and texture of the glass in combination with the modelling technique used, suggest a date of around the 4th century.

The glass vessel found with it also points to the same period,[1432] along with the general dating of that part of the cemetery where they were discovered.[1433]

### 81. Jugs, globular, funnel-mouth, with pushed-in ring base

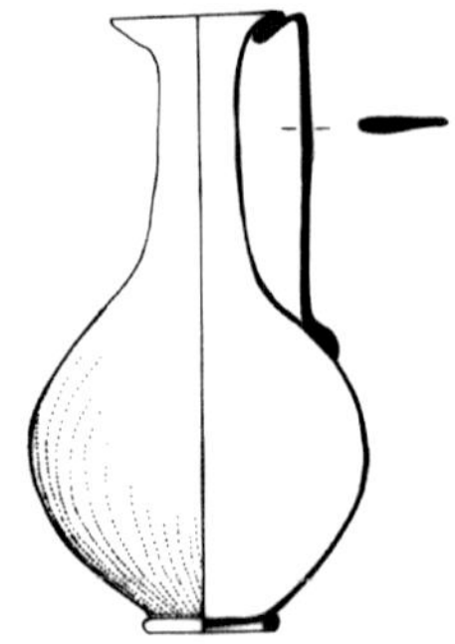

(Cat. no. 406)
(Plates 26, XIV)

Vessels of this form have flaring, almost horizontal fire-polished rim, cylindrical neck tapering towards the rim, globular body with vertical ribs, more strongly visible on its lower part, pushed-in tubular ring base and flat bottom. A smooth strap-handle which starts on the upper part of the body, rises vertically, bends in an acute angle and adheres under the rim. Their height is estimated at c. 15cm.

The vessels were dip-mould-blown, i.e. the paraison was initially blown in a small ribbed dip-mould and then was free-blown to its final dimensions. The shaping of the rim and the base, as well as the application of the handle, were carried out while the vessel was held and rotated by a metal rod temporarily attached underneath.

The vessel from Thessaloniki is made of bluish-green glass.

The colour and texture of the glass from which it was made, in combination with the modelling technique used, ascribe the vessel to production of the 4th century. Similar vessels were found in Budapest,[1434] Kisàrpàs,[1435] and Budapest (Aquincum),[1436] and are dated to the 4th century. The find from Thessaloniki cannot be independently dated.

### 82. Jugs, globular, with tubular rim

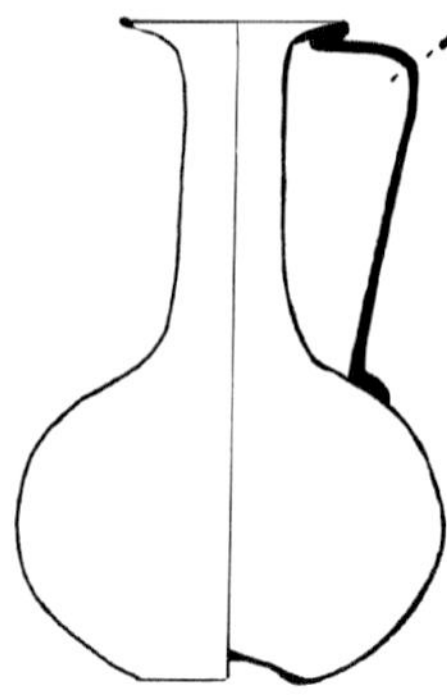

(Cat. no. 407)
(Plates 26, XIV)

Vessels of this form have flaring, almost horizontal, infolded tubular rim, a long cylindrical neck, globular body, and slightly concave bottom. The vertical, strap-handle begins on the shoulder, bends acutely and ends on the rim. Their height is estimated at c. 15cm.

These vessels were made by free-blowing, the rim was shaped after the removal of the blowpipe and the cutting off of the moil, while the vessel was held by another pipe attached centrally

[1431] Barkóczi 1988, 68, no. 395, pls. XXXVI, XCIV, unprovenanced vessel dated according to analogous finds.
[1432] Form 70, Cat. no. 362 (MΘ 3760).
[1433] See Chapter 9 (Find-Spots), 'Agiou Dimitriou Street, extension to the East, 1960'.
[1434] Barkóczi 1996, 97, no. 307, pls. XXIX, LXVII, unprovenanced vessel from the Museum of Plastic Arts, Budapest, dated on the basis of analogous finds to the 4th century.
[1435] Barkóczi 1988, 189, no. 462, pl. XLVI, CI, vessel dated from excavation results to the end of the 4th- early 5th century.
[1436] Barkóczi 1988, 189, no. 464, pl. XLVI, vessel dated from excavation results to the second half of the 4th century.

underneath. Finally the handle was applied; it was formed by adhering a flatten band-like blob of glass onto the shoulder and drawing it upwards and inwards to bend it and adhere it to the rim.

The find from Thessaloniki is made of colourless, greenish glass.

No exact parallel has been found, but since the vessel was found in grave used from the end of the 3rd until the end of the 4th century it is dated to this same period.[1437]

**83. Jugs, ovular, smooth (Isings Form 120a)**

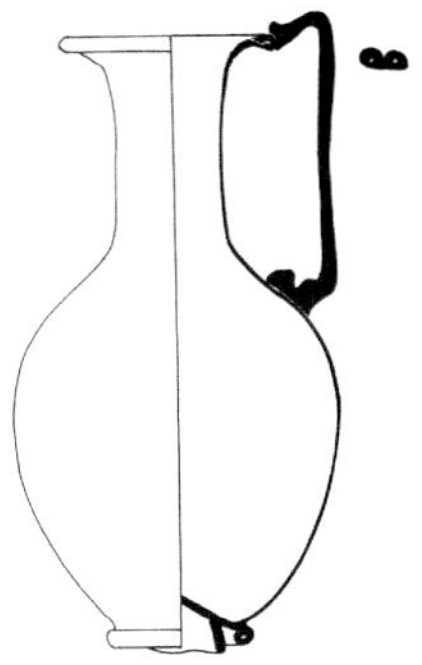

(Cat. nos. 408-409)
(Plates 26, XIV)

Vessels of this form have flaring, infolded tubular rim, cylindrical neck, ovular body and applied ring base. A vertical strap-handle with a median rib begins on the shoulder, bends at an acute angle and ends on the rim. Their height is estimated at c. 14cm.

The body of these vessels was made by free-blowing, and then the base was applied to the body; after that it was transferred from the blowpipe to the pontil, which was attached underneath the vase and finally the rim was formed and the handle attached to the body.

The vessels from Thessaloniki are made of colourless, greenish and decolourized glass.

These vessels are a variant of a well-known Western form which appears already in the late 2nd and 3rd centuries.[1438] During the 4th century such vessels were in use in the West[1439] and in the Balkans.[1440]

According to its archaeological context Cat. no. 408 is dated to the 4th century[1441] and 409 to the first half of the 4th century.[1442]

**84. Jugs, ovular, with oblique ribbing (Isings Form 120b)**

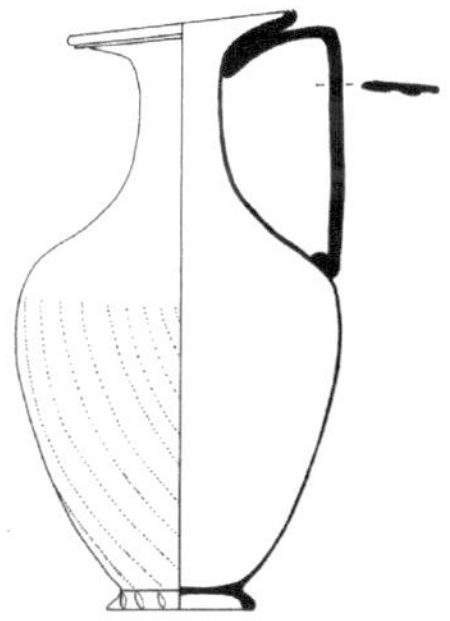

(Cat. nos. 410-411)
(Plates 26, XIV)

Vessels of this form have flaring, almost horizontal, infolded tubular rim, with cylindrical neck and a tall, ovular body; they feature a pushed-in ring base with slanting tooling marks, and a flat bottom. A strap-handle with median rib starts on the shoulder, draws upwards vertically, bends acutely and adheres under the rim. A thick trail of glass is wound under the rim. The body retains fine oblique ribs up to the shoulder.

Their height is estimated at c. 19cm.

The vessels of this form were dip-mould-blown, i.e. they were initially blown in a small ribbed dip-mould and then were free-blown, reaching their final dimensions. The shaping of the rim and the base, as well as the winding of the trail under the rim, and the application of the handle, were carried out while the vessel was held and rotated by a metal rod temporarily attached centrally underneath.

The vessel from Thessaloniki is made of blue-greenish glass.

The colour and texture of the vessel's glass, in combination with the thick trail under the rim,[1443] and the modelling technique of the base,[1444] ascribe it to production of the 4th century. This form is part of a well-known western type dated to the 4th century.[1445] Exact parallels have notbeen traced except for some quite similar, but not identical,

[1437] Makropoulou 2007, 304, pl. 69, from a tile grave dated by its pottery context between the end of the 3rd and the end of the 4th century.
[1438] Price, Cottam 1998a, 161-162, fig. 71a, a form dated to the late 2nd or 3rd century.
[1439] Isings 1957, 149-150, form 120a, where analogous finds are cited.
[1440] Ružić 1994, 18, no. 101, pl. X/4, find from Belgrade (Singidunum), dated from excavation results to the second half of the 3rd - middle of the 4th century.
[1441] Makropoulou 2007, 309, pl. 29, based on the grave's typology.
[1442] Makropoulou 2007, 320, pls. 29, 68, based on a clay vessel found with it.
[1443] Goethert-Polaschek 1977, 214, form 124b, nos. 1303-1304, pl. 70, a form dated from excavation results to the 4th century; Harden *et al.* 1987, 142-144, nos. 70-72, jugs dated from excavation results to the end of the 3rd and the 4th century.
[1444] Goethert-Polaschek 1977, 214, form 124b, no. 1303, pl. 70, a form dated from excavation results to the 4th century.
[1445] Isings 1957, 150-151, form 120b, with analogous finds dated to the 4th century. Among them appear several examples decorated with oblique ribbing and some of them feature the same technique of base forming and decoration.

vessels from Makriygialos in Pieria,[1446] Beška,[1447] Bonn (Bonna),[1448] Mainz (Moguntiacum),[1449] and those from Trier (Treviri) especially,[1450] which are dated on the basis of their archaeological contexts to the end of the 3rd, and mainly to the first half of the 4th century. The find from Thessaloniki cannot be dated independently more firmly, and thus it is placed to the 4th century according to the analogous finds from other sites.

**Trefoil variant**
(MΘ 542)

The example has a trefoil, fire-polished rim, short neck, ovular body, pushed-in tubular ring base and slightly concave bottom. The ribbon handle starts on the upper part of the body and ends on the rim. The body is covered by fine oblique ribs.

Vessels of this form were dip-mould-blown, i.e. they were initially blown in a small ribbed mould and then were free-blown, reaching their final dimensions. The base was formed by folding the lower part of the body by jacks while the vessel rotated. A metal rod was attached centrally underneath, the rim area was released from the blowpipe, the moil was cracked off, the actual rim was fire-polished, and then the handle was applied.

No exact parallels were found; it belongs to a variant of trefoil jugs known in the West, with very short, almost non-existent neck, as with our find, which are occasionally decorated with oblique ribs on the body[1451] and is dated to the late 4th century.[1452] In addition, the relatively dark glass of the vessel and the modelling technique, i.e. dip-mould-blowing, corroborate a Late Roman production date in the 4th century. No further information on the excavation was preserved and thus the vessel cannot be dated more firmly.

## 85. Jugs, fusiform

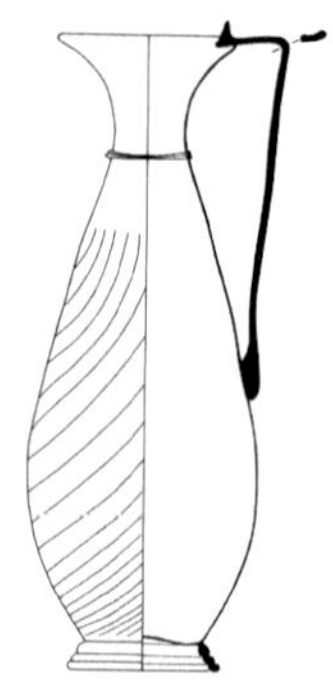

(Cat. no. 412)
(Plates 27, XIV)

The vessel of this form has fire-polished rim, funnel-mouth, short neck accentuated by an applied thread wound around it, fusiform body, and slightly concave bottom. The conical coil base is formed by a trail wound three times forming a high cone. The strap-handle has a rib along its centre, starts from the middle of the body, rises vertically to rim height, where it bends at a right angle and adheres to it. The handle widens on the rim enclosing one third of it, and rises over the rim forming a triangular crowning. At the base of the handle there is an applied medallion bearing a lion head in relief. On the body of the vessel faint oblique ribs are visible. The height of the vessel of this form is 28cm.

The vessel is made of olive-green glass by dip-mould-blowing, i.e. the paraison was initially blown into a small ribbed mould in order to stamp the ribs on it and then it was inflated by free-blowing to obtain its final dimensions. Furthermore the trail of the base was wound three times under the body and was given its conical shape. Finally a blowpipe was attached underneath, the pipe was removed from the rim, the moil was cracked off from the rim, which was reheated and shaped; then the thread was wound around the neck and the lion head medallion, which was stamped from a blob of glass on a smooth surface, e.g. the marver, was applied under the base of the handle.

Vessels which share all characteristics of the form have been traced at such locations as Kerch (Panticapaeum),[1453] Kolovrat,[1454] and certain sites in Israel.[1455]

[1446] Despoini 1976, 249-250, pl. 198, from a grave dated by coins of Constantius II (355-361).
[1447] Ružić 1994, 18, no. 100, pl. X/3, find dated from excavation results to the first half of the 4th century.
[1448] Follmann-Schulz 1988, 63-64, no. 199, pl. 23, find dated from excavation results to the first half of the 4th century.
[1449] Harter 1999, 139, 273, form G20b, esp. no. 924, dated to the end of the 3rd and to the 4th century.
[1450] Goethert-Polaschek 1977, 214, form 124b, nos. 1303-1304, pl. 70, form dated from excavation results to the 4th century; Harden *et al.* 1987, 142-144, nos. 70-72, jugs dated from excavation results to the end of the 3rd and 4th century.
[1451] Goethert-Polaschek 1977, 224, no. 1345, pl. 72, find dated according to analogous finds to the last third of the 4th century.
[1452] Isings 1957, 154-155, form 124c, where examples dated to the second half of the 4th and the 5th century are listed; Goethert-Polaschek 1977, form 132, dated to the last third of the 4th century; Harrter 1999, 149, form G30a, find dated to the second half of the 4th century.
[1453] Kunina 1997, 208, fig. 103, 175a, 6, identical to the vessel dated to the 4th century, based on its shape and decoration, but also based on its context which is securely dated to the 4th century.
[1454] Ružić 1994, 17-18, no. 96, free-blown vessel of the same shape, with incised geometrical decoration, dated from excavation results to the 4th century.
[1455] Israeli 2003b, 190, no. 224, free-blown vessel dated to the 4th century, according to analogous finds. Based on its colour and the comical mask, which is attached on the base of the handle, it has been ascribed to Egyptian production.

Several features of the vessel are quite widespread in the 4th and 5th centuries, including dip-mould-blowing in a ribbed mould, the thread around the neck and the triangular protuberance of the handle; all are known in the glassware of the 4th century, especially in Syro-Palestinian production.[1456]

The peculiar base is known among finds of the 4th and 5th centuries in Rome, Sardis, Israel and Egypt.[1457]

The medallions with theatrical masks or lion heads,[1458] attached on the base of the handle or on the body, reappear in the 4th century, and offer another clue for the dating of the vessel. Medallions with lion head appear below the bases of 4th-century handles in finds from: Germany,[1459] Kerch (Panticapaeum),[1460] Ptuj (Poetovio) Slovenia,[1461] Budapest,[1462] Palestine,[1463] Fayum (Arsinoe),[1464] Cairo,[1465] and Corinth.[1466] Also from the same period are known an ovular cup from Reims (Durocortorum),[1467] now kept in the British Museum, and a hemispherical bowl from Højvang[1468] in Denmark, both of them decorated around the body with lion head medallions.[1469]

Although the exact grave in which the vessel was discovered is uncertain, it is known however that it was unearthed in a cemetery whose entire inventory, according to the preserved photographs, was dated to the 4th or 5th century.[1470]

### 86. Jugs, pear-shaped, oblique ribs

(Cat. no. 413)
(Plates 27, XIV)

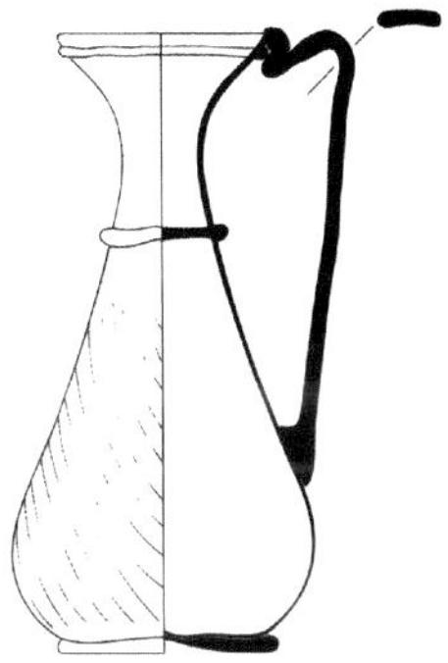

The vessel of this form has tubular, infolded rim, funnel-mouth, pear-shaped body and true pad base. It is decorated around the rim and at the base of the neck by two thick coils of glass. The strap-handle starts from the middle of the body, rises almost vertically to rim height, bends, and adheres to it. The body is decorated with faint dip-mould-blown oblique ribs. Their height is estimated at 20cm.

The vessel is dip-mould-blown, i.e. initially blown in a small ribbed mould and free-blown to its final dimensions. Then the base was applied and formed and the vessel transferred to a rod attached at the centre of the base; the rim was then shaped and the decorative trails and handle added.

The vessel and the base are made of yellowish glass, while the handle and the two decorative trails are made of olive-green glass.

---

1456 Harden *et al.* 1987, 147-148, nos. 75, 76; Israeli 2003b, 185, 186-189, nos. 220-223, with relevant bibliography.

1457 Harden 1936, 217, no. 658, pl. XIX, listed as 'solid coil base'; Davidson 1952, 99, no. 633, pl. 8, generally dated to the Roman period; Sternini 2001, 31, 68-70, nos. 204-219, where the author, presenting the material found under the temple of Magna Mater, mentions a large number of similar artefacts and their analogous finds both from the West and East.

1458 A similar find was located in an unpublished excavation of the *fossa*, the external ditch of the western city wall of Thessaloniki, from a site which did not offer a more specific dating.

1459 Kisa 1908, Vol. I, 187, fig. 93, from the vom Rath Collection. The vessel is identical in shape and decoration to the one under examination, and, according to Hayes, is probably of oriental origin.

1460 Kunina 1997, 208, fig. 103, 175a, б, the vessel is dated to the 4th century on the basis of the glass vessel with which it was found, and also because from the oblique ribbing that decorates it.

1461 Šubic 1976, 48, no. 69, pl. I, 7, the vessel is dated erroneously by the author to the 1st century, comparing it with medallions depicting female faces.

1462 Barkóczi 1996, 104-105, cat. no. 335, pls. XXXIII, LXX, unprovenanced, donated find, made of colourless, greenish glass. Erroneously citing the mask from Stein, the researcher dates the find to the 1st-2nd century.

1463 Hayes 1975, 144, no. 606, pl. 37, medallion with lion head, probably from Palestine.

1464 Hayes 1975, 144, no. 605, pl. 37, medallion with lion head from a vessel with faint oblique ribbing on the body, which was compared by Hayes with our find.

1465 Edgar 1905, 77-78, nos. 32762-70, pl. X. The Cairo Museum has three medallions with theatrical masks (32762-32765), one with a lion head (32768), and one with a female face (32766); all are made of colourless, greenish glass. The same museum also has an opaque white medallion (32767) with a face, which is attached on a dark subtract. It is not clear whether this object was originally attached to a vessel, or if it was made for independent use, e.g. as a pendant. It seems more probable that the objects are not contemporary among themselves, and that they represent productions of both the 1st and 4th centuries.

1466 Davidson 1952, 98, no. 618, pl. 54, medallion with lion head found in a layer dated generally to the Roman period.

1467 Isings 1964, 60, photo 2.

1468 Isings 1964, 61, photos 3-4. In this particular example the lions have almost entirely closed mouths, like the one in Budapest, as opposed to all other known examples, which are represented with widely open mouth.

1469 Isings 1957, 160, form 131; Isings 1964, 59-63. The decoration around the body with lion heads, quite probably shares the same prototypes with the decoration of clay, *terra sigillata* and *terra nigra* vessels, dated to the 2nd-3rd century. Also, the spouts of the clay mortars (Dragendorff Form 45), are often formed like lion heads. For metal artefacts, lion-head spouts are known at least from the Hellenistic era. To this group belongs a beaker from Dalj (Teutoburgium) assumed to date to the 3rd century, see Šaranović-Svetek 1986, 60, no. 32, pl. II, 2; Fadić 1997, 88, 182, no. 172.

1470 See Chapter 9 (Find-Spots), 'Veroia, Tsirakoglou plot, 1940'.

Exact parallels have not been traced. It is very similar to Isings Form 121b, dated to the late 3rd-early 4th century.[1471] In addition, a vessel from Kerch (Panticapaeum),[1472] as well amples of our form Form 87, have the same pear-shaped body, but with a different base, and they are dated to the 4th-5th century. In addition the dip-mould-blown ribbing, the decorative trails around the rim and the neck,[1473] the colour and texture of the glass[1474] from which it was made, are all consistent with a dating to the 4th-5th century.

The vessel was found in a 4th-century cluster of graves.[1475] The grave in which it was found belongs to a later phase of burials and it also contained a fusiform clay jug known from several 4th-century graves in the western necropolis of Thessaloniki.[1476] All these clues justify the dating of the vessel to, probably, the late 4th century.

### 87. Jugs, pear-shaped, smooth

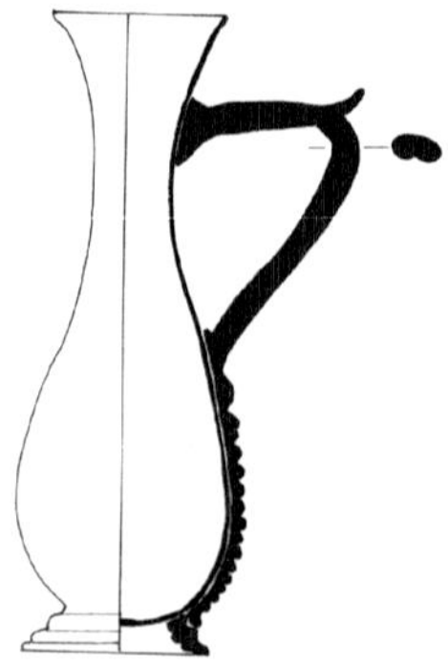

(Cat. no. 414)
(Plates 27, XIV)

Only one example of this form was found in the excavations of the city. The flaring rim is ground, the short neck is conical, like a funnel-mouth, and the body is pear-shaped. A coil handle starts on the middle of the neck as a horizontal coil, bends in a right angle to form a thumb-rest pad, and then curves to adhere onto the upper part of the body, continuing as an undulating coil to the base of the vessel. The conical, coil base consists of a thick trail wound four times. The body of the vessel is made of dark-blue glass and the base and the handle are made of dark-green glass. Their height is estimated at c. 21cm.

The vessel is made by free-blowing, the base and the handle have been added while the still hot mass of the body was on the blowpipe. The rim was cracked off and mildly ground after the annealing.

The jug is identical in shape and dimensions with one in the Corning Museum of Glass,[1477] one in the Israel Museum in Jerusalem,[1478] and the example in the Miho Museum outside Kyoto, Japan,[1479] which, with a few more vessels, are products of the same workshop. Apart from the vessel at Corning, identical examples can be seen at Newark,[1480] Toledo, Ohio,[1481] and Kassel (Castellum Cattorum).[1482] Another jug, with an applied medallion with a theatrical mask on the body, is kept in Boston.[1483] Other identical vessels with vertical ribs on the body are in Berlin,[1484] Cologne,[1485] and Corning.[1486]

The globular vessel of this group, kept in Jerusalem, is decorated with oblique ribbing.[1487] In addition, vessels of this production group in the shape of a human head are to be seen in Berlin[1488] and Corning.[1489]

The use of glass of one colour for the body and another for the base and the handle appears, apart from the example from Thessaloniki, in the finds from Kom Aushim (Karanis)[1490] and Kassel (Castellum Cattorum).[1491]

The vessels of this group, although they have been found in far distant regions, share several common features. Most are made of dark-blue glass, although there are known purple examples as well.[1492] The base is formed by a thick trail of glass wound three or four times around the bottom, while held on the blowpipe, forming a relatively tall and embossed base. In addition, the rim of the vessels is unworked or slightly ground and the neck is conical.

---

1471 Isings 1957, 152, form 121b, with several examples from western cemeteries dated between the end of the 3rd and the end of the 4th century.
1472 Kunina 1997, 321-322, no. 399, dated from excavation results to the early 5th century and ascribed to a Syrian workshop.
1473 Harden *et al.* 1987, 142-144, nos. 70-72, jugs dated from excavation results to the end of the 3rd and to the 4th century.
1474 Harden *et al.* 1987, 146, no. 74, jug dated to the 4th century, of different shape, but with the same combination of colours.
1475 See Chapter 9 (Find-Spots), 'Efxeinou and Panagopoulou Streets, 1994'.
1476 Unpublished material presented in Thessaloniki in the permanent exhibition of the Museum of Byzantine Culture, *From the Elysian Fields to the Christian Paradise:* two examples in a cist grave dated to the 4th century, and other examples in the showcases with clay vessel typology.
1477 Whitehouse 2001, 182, no. 724.
1478 Israeli 2003b, 191, no. 225, colour plate on page 153.
1479 Goldstein 2001, 114, 207, cat. no. 165.
1480 Auth 1976, 109, no. 127.
1481 *TMA, Art in Glass* 1969, plate on page 31.
1482 Boosen 1984, no. 40, colour pl. III.
1483 Saldern 1968, 93, pl. 55.
1484 Platz-Horster 1976, 48, no. 79.
1485 La Baume 1973, pl. 39, D95.
1486 Whitehouse 2001, 121, no. 623.
1487 Israeli 2003b, 192, no. 225, colour plate on page 153.
1488 Greifenhagen 1962, 64, no. 9; Platz-Horster 1976, 45, no. 70.
1489 Whitehouse 2001, 74-75, no. 548.
1490 Harden 1936, no. 658, yellow body with colourless, greenish base.
1491 Boosen 1984, no. 40, dark-blue body, with colourless, bluish handle and base.
1492 Stern 1995, 241-242, no. 162; Whitehouse 2001, 127-129, no. 633.

The handle bends in a right angle with the bend spot accentuated, and its lower part, adhered to the body, is undulating and goes all the way to the base. Most of the examples are formed by mould-blowing.

This group has been ascribed in the past to the production of some 4th-century Eastern workshop on no particular archaeological evidence.[1493] Archaeological finds of similar bases, which are considered a peculiarity of this workshop, are known from Corinth[1494] and Egypt,[1495] but with no firm dating however. The suggested date was finally confirmed by finds of the 4th and 5th centuries from Carthage,[1496] Rome, and elsewhere.[1497] This late date is further confirmed by new finds from an Early Christian basilica at Solinas (Kassandra, Chalkidiki).[1498]

The find from Thessaloniki is dated to the 4th century on the basis of the clay vessel found with it, and also on the basis of the typology of the tomb in which it was found.[1499]

### 88. Jugs, inverted truncated conical, with horizontal shoulder

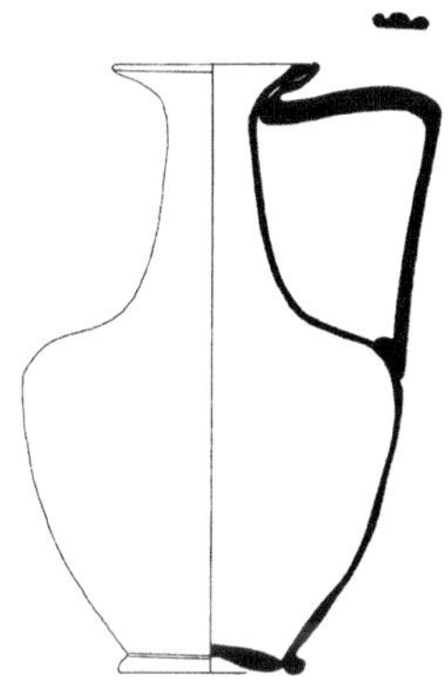

(Cat. no. 415)
(Plates 27, XIV)

The vessel of this form has flaring, tubular infolded rim, cylindrical neck, tapering toward the rim. The body is inverted, truncated conical with horizontal shoulder; applied coil ring base; vertical, slightly flaring strap-handle with central rib that starts on the shoulder, rises vertically, and at rim height bends in a right angle and adheres under the rim. Their height is estimated at c. 16cm.

The vessel is made of colourless, greenish glass and is free-blown. The base was attached and shaped while the vessel was held by the blowpipe attached on the rim. Then the vessel was transferred onto a pipe attached underneath and the rim was shaped and the handle applied.

This vessel might be considered as a large variant of a form mainly known by western, 4th-century examples.[1500] The single real analogous find is reported in Pannonia[1501] and is dated to the first half of the 4th century. The find from Thessaloniki is independently dated generally to the 3rd-4th century.[1502]

### 89. Jugs, verted truncated conical, slender, with shoulder

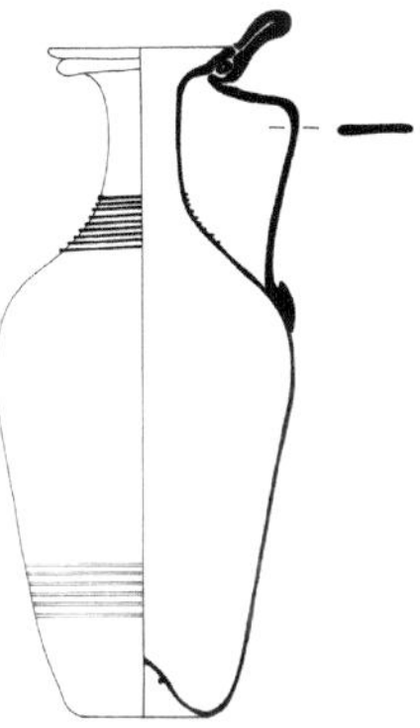

(Cat. nos. 416-417)
(Plates 27, XIV)

The vessels of this form have flaring fire-polished rim, cylindrical neck tapering towards the rim, inverted truncated conical body and concave bottom. A strap-handle starts on the shoulder, rises vertically, bends in an obtuse angle and adheres under the rim, forming a pronounced fold which in one case creates a thumb-rest pad. The vessels are decorated with a fine thread wound spirally around the body either on the shoulders or lower towards the bottom, and occasionally also under the rim. Their height is estimated at c. 18cm.

These vessels were made by free-blowing and marvering of the paraison with the appropriate angle. The shaping of the rim, the winding of the threads and the addition of the handle were carried out after the vessel was transferred to a pipe attached to the underneath.

The vessels from Thessaloniki are made of colourless, greenish glass.

---

1493 *Glass from the Ancient World 1957*, 141; Whitehouse 1997b, 367-375.
1494 Davidson 1952, 99, no. 633 pl. 8, dated generally to the Roman period.
1495 Harden 1936, 217, no. 658, pl. XIX, referred to as 'solid coil base'.
1496 Sternini 1998, 140-141, nos. 58-62. The fragments are generally dated, according to their stratigraphical relation and their analogous finds, between the middle of the 4th and 5th century.
1497 Sternini 2001, 31, 68-70, nos. 204-219. Where, in presenting the material found under the temple of Magna Mater, the author mentions many analogous finds both from the West and the East.
1498 A church functioning between the 4th and 6th century. Fragment of cracked-off rim, and body with vertical ribs and another fragment of an undulating handle, both made of blue glass. I thank the excavator of the basilica, I. Papangelos, for his permission to publish the glass finds of the excavation. For the excavation, see Papangelos 2005; on the glass finds see, Antonaras 2014e, 97-110.
1499 Makropoulou 2007, 162, pl. 29.
1500 Isings 1957, 152, form 121a.
1501 Šaranović-Svetek 1986, 21, 61, no. 55, pl. VI/1, XVIII/5, a find from the site of Brest (Beška), dated from excavation results to the first half of 4th century.
1502 Nalpantis 2003, 61-62, 127-128, drawing 30, pl. 43, dated probably to the 4th century.

Exactly analogous finds to the jugs of this form have not been traced. The decoration of the vessels with fine threads, and especially in the area under the rim, is more usual during the 3rd and mostly during the 4th century, i.e. the period when glass jugs were common burial goods in Thessaloniki. For at least one of the examples there is a real possibility that it was found together with a fusiform unguentarium (Form 147a) that is also dated to the 4th century.

### 90. Jugs, inverted truncated conical, squat, plain base

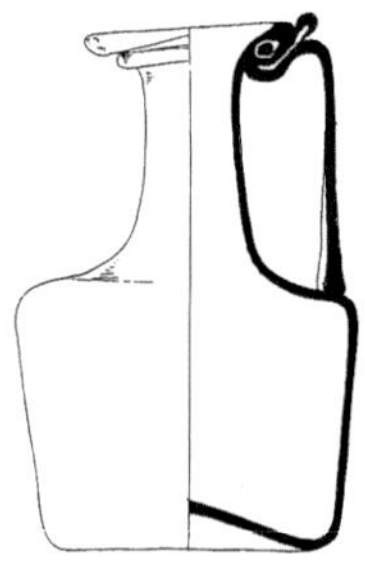

(Cat. no. 418)
(Plates 27, XIV)

Vessels of this form have horizontal tubular infolded rim, short cylindrical neck, inverted truncated conical body, and flat, slightly concave bottom. Their height is estimated at c. 18cm.

The vessels were probably made by mould-blowing, but almost equally probably they were free-blown and shaped by marvering the paraison. After the vessel was attached to a pipe centrally underneath, the rim was shaped and the handle applied.

The vessel from Thessaloniki is made of colourless, greenish glass.

Conical, single-handled vessels that are analogous to this form are known from Kisàrpàs[1503] and Duklja (Doclea),[1504] and they are dated to the 4th-5th century. It is essentially a short variant Isings Form 126.[1505] Furthermore, their shape is related to the cylindrical mould-blown jugs that were in fashion in this period both in the West and East.[1506]

The vessel from Thessaloniki, according to the other glass finds from the grave, must belong to a 4th-century burial.[1507]

### 91. Jugs, inverted, truncated conical, squat, pushed-in tubular base, dip-mould-blown

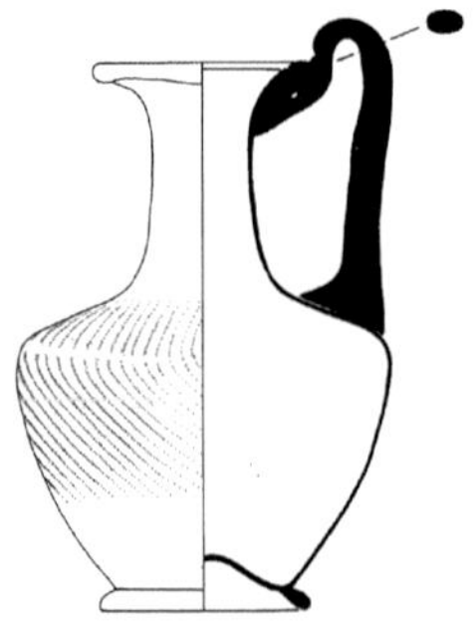

(Cat. no. 419)
(Plates 28, XIV)

The rim of this vessel is flaring, tubular infolded; the neck is cylindrical tapering towards the rim, the body is inverted, truncated conical with almost horizontal shoulder, the base is low, pushed-in and tubular; the bottom is mildly concave. A coil handle starts on the shoulder, rises vertically over the rim, bends acutely and adheres under the rim. The body is decorated with faint oblique ribs. Their height is estimated at c. 16cm.

The vessels are dip-mould-blown, initially blown in a small ribbed mould and then free-blown to its final dimensions. The shaping of the rim and the base, as well as the addition of the handle, were undertaken while the vessel was held and rotated by a pipe attached centrally underneath.

No exact parallels have been located. The modelling technique used, i.e. dip-mould-blowing, was common in the 4th century, however, a period when glass jugs were relatively common grave goods in Thessaloniki anyway. Furthermore, the texture and olive-green colour of the glass does not appear in the first centuries of the Christian era, while it is quite common during the 4th century. The find-spot and the stylistic and structural features of the jug also point to a 4th-century date.[1508]

### 92. Jugs, inverted truncated conical, squeezed with pushed-in tubular ring base, free-blown

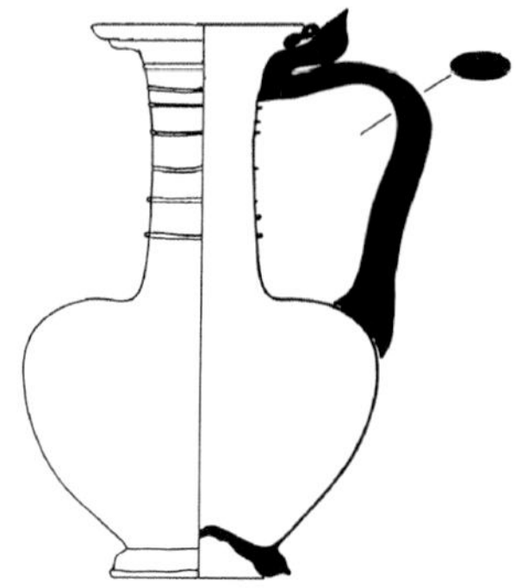

(Cat. no. 420)
(Plates 28, XIV)

This vessel has a flaring tubular infolded rim, cylindrical neck, inverted truncated conical body with horizontal shoulders, relatively tall tubular base and a mildly concave bottom. A coil handle slightly flattened, starts on the shoulder, rises almost vertically, slightly flaring from the body and bends and adheres under the rim. A fine thread of glass is wound around the neck. Their height is estimated at c. 13cm.

[1503] Barkóczi 1988, 153, no. 458, pls. XLV, CI, find dated from excavation results to the end of the 4th- early 5th century.
[1504] Cermanović *et al.* 1975, 157-158, form II/2A, with 4 examples from graves dated to the second quarter of the 4th century.
[1505] Isings 1957, 156-157, form 126, with analogous finds dated to the 4th century.
[1506] Form 95.
[1507] It was found in a sarcophagus with vessels of Forms 12, 52a, 61, 90 and 103.
[1508] Makropoulou 2007, 264, pl. 29. This part of the necropolis was used around the 4th century.

The vessel is made by free-blowing; the base was applied and shaped while the blowpipe and the moil were still on the rim; then the vessel was transferred to a pipe attached underneath and the rim shaped, the thread wound around the neck, and finally the handle was added.

The vessel is made of greenish glass.

The vessel belongs to a form of jugs mainly known from the West and is dated to the 4th century.[1509] The find from Thessaloniki is dated by the results of the excavation to the first half of the 4th century.[1510]

### 93. Jugs, hemispherical

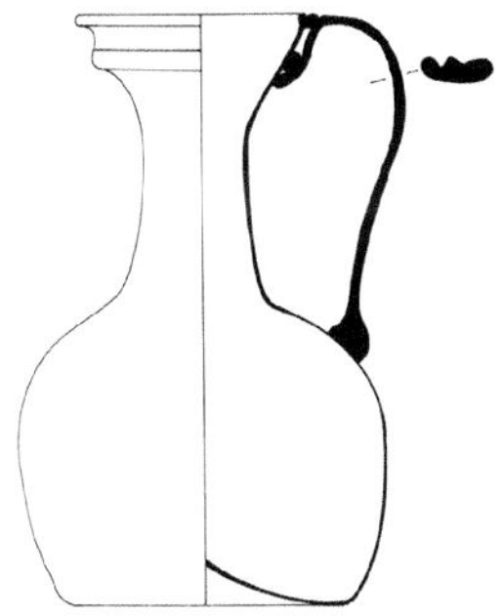

(Cat. nos. 421-423)
(Plates 28, XIV-XV)

Vessels of this form have flaring, tubular infolded rim, cylindrical neck which tapers towards the rim, hemispherical body, and concave bottom. A ridged strap-handle starts on the upper part of the body, rises almost vertically, flaring a little from the body, bends and adheres under the rim. Under the rim a trail of glass is wound. Their height is estimated between 14-16cm.

The vessels are free-blown. The shaping of the rim and the base, the application of the handle and the trail around the neck were carried out while the vessel was held by a pipe attached underneath, thus permitting undisturbed rotation.

The vessels from Thessaloniki are made of olive-green glass.

The colour and the texture of the glass ascribe the vessels approximately to production of the 4th century. Exact published[1511] parallels have not been traced; a fairly similar one from Edessa, Greece, also offers no firm chronological clues,[1512] The vessels, judging by stylistic features, i.e. the shaping of the rim, the handle and the overall proportions, and also the colour and the texture of the glass, must represent works from the same workshop as the eight-sided jugs, which was probably operating in Thessaloniki. Also, they are quite similar to vessels from Syria, dated between the 4th and 6th century,[1513] and from Israel.[1514]

Only one of the examples, Cat. no. 423, can even be vaguely dated by its context, as it was found in a cluster of the necropolis with Late Roman or Early Christian burials.[1515]

### 94. Jugs, eight-sided

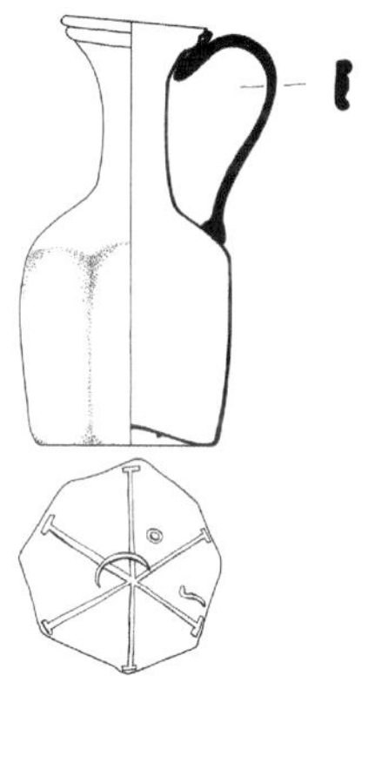

(Cat. nos. 424-425)
(Plates 28, XV)

The vessels of this form are characterized by their eight-sided body. Their rim is tubular infolded; they have funnel-mouth, cylindrical neck, eight-sided body and mildly concave bottom. A strap-handle with a median rib starts on the shoulder and ends at the mouth. Around the mouth a thick trail is wound. Their height is estimated at c. 16cm.

The vessels are mould-blown. The neck, rim and handle were shaped by free-hand tools. On the bottom there is a stamped base mark. This comprises three lines intersecting at the centre; these divide the bottom into six almost equal fields. In two adjacent fields the Latin letters 'O' and 'S' are visible, probably the ending of a (male) name.[1516]

These vessels are made of greenish glass, free of impurities and pinprick bubbles. Vessels identical to the find from Thessaloniki have not been traced. The vessels, judging by stylistic

[1509] Isings 1957, 152, form 121a, with analogous finds dated to the 4th century.

[1510] In the grave were found coins of Constantine I (306-337) and Constantius II (329-333), as well as a clay vessel dated to the first half of the 4th century, see Makropoulou 2007, 249, pl. 65.

[1511] A vessel of this form was found in a tile-grave in the western necropolis, along with 4th-century pottery. In this particular part of the necropolis (see Chapter 9 (Find-Spots), 'Junction of Koloniari-Galanaki-Papathanasiou Streets') only graves dated to the 4th century were found, while some of the built-graves were re-used in the 5th century. I thank the excavator, Mr Dimitris Nalpantis for the relevant information.

[1512] Chrysostomou A. 2000, 321, pl. 362, from a burial with a single glass vessel find inside, which in turn was dated according to the vessels from Thessaloniki to the 4th-6th centuries.

[1513] Dussart 1998, 178, form BXIV. 221, pl. 74, photo 60. 9, finds from the Amman Museum, Kefar'Ara and Tyre.

[1514] Barag 1970c, pl. 19 (First row, first vessel from left), from El Bassa, nowadays known as Betzet.

[1515] See Chapter 9 (Find-Spots), '13 Eleftherias Street, Ampelokipoi, 1977'. In the same grave a Form 139 vessel was also found.

[1516] Auth 1976, 111, no. 129, eight-sided vessel, slightly slenderer, with the name ΠΡΙΣΚΟΣ imprinted underneath, dated for stylistic reasons to the 1st-2nd century; Lehrer-Jacobson 1986, 62; Israeli 2003b, 256, where the author mentions an identical vessel from a museum in Jerusalem.

criteria, i.e. the shape of the rim, the handle and their overall proportions, and also by the colour and texture of the glass, must be works from thesame workshop as the hemispherical jugs of our Fform 93. Similar vessels, however, are kept in Jerusalem[1517] and in a collection in Germany.[1518] To the same late period are also dated similar six-sided vessels from Pannonia, Beška,[1519] Čalma/Sremska Mitrovica (Sirmium),[1520] and an unprovenanced example from the Novi Sad Museum,[1521] and one more from Cologne.[1522] Furthermore, a ten-sided vessel from Dunaújváros (Intercisa) shares all the features with our find except for the number of sides; it is also dated to the 4th century.[1523] The find from Thessaloniki, although it cannot be dated independently by excavation data, on the basis of technological and stylistic features, i.e. the thick trail under the rim and the colour and texture of the glass, is dated to the 4th century. Despite the distribution of analogous finds, mainly in Germany and the Balkans, it represents an Eastern Mediterranean product, probably even from Thessaloniki.[1524]

### 95. Jugs, cylindrical (Isings Form 126)

(Cat. nos. 426-429)
(Plates 29, XV)

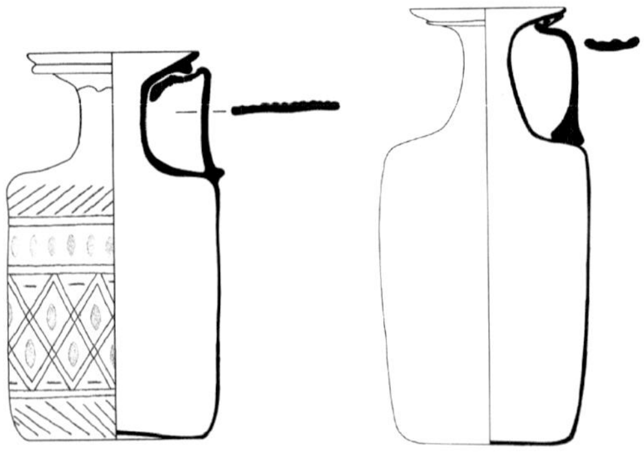

Vessels of this form have almost horizontal fire-polished rim, short cylindrical neck, and flat bottom. A wide, ridged strap vertical handle connects the shoulder with the rim.

A thick trail of glass is wound under the rim. Very often the body is decorated with bands of incised geometrical motifs, combined with engraved and wheel-cut motifs. The body of Cat. no. 427 is divided into two unequal bands, the lower double the size of the upper one, and which are delineated by a double line. The upper is filled with vertical ovular strokes; the lower is filled by a double row of lozenges, each of them formed by a double line. In the centre of each lozenge there is a vertical ellipsoidal stroke, while in the triangular fields, which are formed between the ends of the lozenges at the edge of the band, a thin horizontal incised line is visible. The rest of the body, above and below these bands, is covered by rows of parallel oblique incisions forming an oblique hatching. The examples from Thessaloniki, like the majority of the preserved examples, is made of thick greenish glass. Their height is estimated between 19-24cm.

Sometimes the body is mould-blown, probably in order to create vessels of predefined capacity,[1525] while the neck and rim areas were modelled by free-hand tools.

It is a widely known form both from the East and the West.[1526] Finds dated by their archaeological context are ascribed to the period between the late 3rd and up to the early 5th century. Examples are preserved from Britain;[1527] Germany: Trier (Treviri),[1528] Bonn (Bonna),[1529] Cologne;[1530] Hungary: Dunaújváros (Intercisa),[1531] Pécs,[1532] Szőny

[1517] Israeli 2003b, 256, no. 330, eight-sided, purple vessel with a six-petal rosette inscribed in a circle and a hexagon. Dated to the 2nd-3rd century on the basis of analogous finds and considered to be a product of the Eastern Mediterranean.

[1518] Stern 2001, 142, 144, 184-185, no. 77, eight-sided, purple vessel with green handle, dated for stylistic reasons in second half of 4th century and assumed to be a product of the NW Empire, possibly Cologne.

[1519] Šaranović-Svetek 1986, 62, no. 66, pl. VII, 3, dated by a coin of Constantius II (337-361).

[1520] Milošević 1973, 87, 92, drawing 4/2 pl. IV/1; Šaranović-Svetek 1986, 62, no. 68, pl. XVIII/1, dated by coins of Licinius (308-324), Constantine I (306-337) and Constantius II (337-361).

[1521] Šaranović-Svetek 1986, 62, no. 67, pl. VIII, 3, dated to the 4th century according to analogous finds.

[1522] Loeschcke, Willers 1911, Vol. 1, 40, nos. 393-396, Vol. 2, pls. XXXI/393, 396, vessels without independent chronological indications.

[1523] Barkóczi 1988, 181, no. 437, pl. XCVII, find dated from excavation results to the 4th century and considered an Oriental import.

[1524] This vessel was published in Antonaras 2006a, 414, 416.

[1525] Similar vessels, which were finished as double-handled amphorae, are assumed in some cases, at least, to have been made in sizes fixed as multiples of a sextarium (0.543l). On the topic, see Whitehouse 1997, 266-267, in the discussion on vessel no. 455, where a relevant bibliography is cited.

[1526] Isings 1957, 156-157, form 126.

[1527] Cool, Price 1995, 201-203, finds from various sites dated from excavation results to between the 3rd and t5th centuries.

[1528] Goethert-Polaschek 1977, 216-217, form 125, nos. 1308-1315, finds dated to the 4th century.

[1529] Follmann-Schulz 1988, 49-51, nos. 140, 141, 147-149, pls. 16, 17, dated from excavation results to the second half of 3rd-4th century.

[1530] Loeschcke, Willers 1911, pls. XXIV/319, XXVI/318; Doppelfeld 1966, 65, photo 150, find dated from excavation results to the first half of the 4th century; Fremersdorf 1967, 127-129, pl. 148-155; La Baume, Salomonson 1976, 72, pl. 38/260, find from Rhine region dated to the 4th century, according to analogous finds.

[1531] Radnóti 1957, 150, pl. XXXI/1, which is ascribed to the production of the 4th century.

[1532] Barkóczi 1988, 201, no. 501, pls. LVII, CVIII, find dated from excavation results to the first half of 4th century.

(Brigetio),[1533] Jánosházapuszta;[1534] Croatia: Osijek (Mursa);[1535] FYROM: Dračevo;[1536] Syro-Palestinian region;[1537] Egypt: Kom Aushim (Karanis),[1538] Behnasa (Oxyrhynchos),[1539] Sahara.[1540] In Greece published analogous finds were unearthed at Paliouri,[1541] Dion,[1542] Edessa,[1543] and Corinth.[1544]

Although cylindrical jugs were in use from the 1st century as transport vessels, very early on they were decorated with incisions and equally probably were also used as tableware vessels.[1545] The 4th-century examples as a whole should be ascribed to tableware.[1546] This assumption is corroborated by the fact that more expensive decolourized glass was occasionally used for their production. In addition they were very often decorated with bands of complex incised motifs that echo elaborate casings, originally wicker, which were probably later replaced by gemmed-metal ones; the cutters used their skills to render imitations of these casings on the surfaces of the glass. A further indication that the form was intended for tableware was the fact that vessels of this form appear in mosaic representations in conjunction with drinking vessels.[1547]

More precise chronological clues come from Cat. no. 426, found in a 4th-century grave, and 427, found in a burial of the 3rd or 4th century.[1548]

### 96. Jugs, figural in the shape of human head

(Cat. no. 430)
(Plates 29, XV)

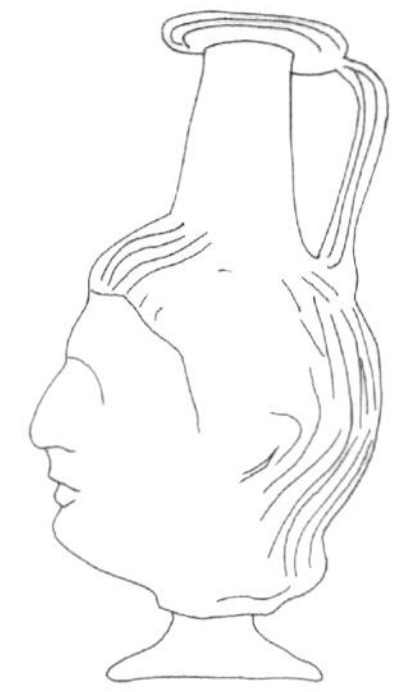

This vessel was intended to render a beardless male head.[1549] In the preserved fragment we can see part of the ending of the hair, made by wide, vertical and parallel ribs. A part of the man's smooth neck is also preserved, as is also the seam between the two sections of the mould, which did not fit perfectly and resulted in a difference in the level of the surface of the neck on the two parts.[1550] The vessel is made of olive-green glass. Their height is estimated at c. 15cm.

The vessel was blown in a bipartite mould. According to other analogous finds, the relatively long, cylindrical neck, ending in a horizontal, tubular infolded rim. Also, according to analogous finds it can be assumed that it was supplemented with a strap-handle that began high on the body – almost at the base of the neck – and ended under the rim.

Analogous finds are known from Germany,[1551] France,[1552] and Luxembourg,[1553] dated by their contexts to the 4th-5th century; they are considered western products, probably from Gaul or Cologne.[1554] The underneath of the vessel is so anomalous and bumpy that almost certainly the vessel could not stand on its own. The traces of crescent-shaped

[1533] Barkóczi 1988, 201, no. 502, pl. LVIII, find dated from excavation results to the first half of 4th century.
[1534] Barkóczi 1988, 201, no. 503, pls. LVII, CVIII, find dated from excavation results to the last third of the 4th or to the early 5th century.
[1535] Damevski 1976, 85, pl. XIII. 3; Bulat 1976, 89, no. 3, pl. VII. 6; Šaranović-Svetek 1986, 20, pls. IX/1, XXI/2, vessel dated according to analogous finds to the 4th century. The vessel was more precisely dated in *Trasparenze imperiali 1997*, page 168, to the first half of the 4th century, or c. the middle of the 4th century.
[1536] Josifofska 1956, 287, fig. 5, only the upper part of the vessel is preserved. The grave is dated by context to the 4th century.
[1537] Hayes 1975, 82, pls. 24/363, nos. 364-365, first half or middle of the 4th century; Israeli 2003b, 257, no. 332, a vessel dated according to analogous finds to the 3rd-4th century.
[1538] Harden 1936, 234 ff, pl. XIX 4739-4740, dated to the latest period of the site, which was originally placed to the 4th and now around the 5th century. On the re-dating, see Hayes 1975, 2-3; Whitehouse 1997, 81-82.
[1539] Isings 1957, 157, where the original bibliography is cited. It is a find from a grave dated to the 4th century.
[1540] Isings 1957, 157, where the original bibliography is cited. It is a find from a burial chapel dated to the 6th century.
[1541] Tsigarida, Vasileiou, Naoum 2009, 396-397, fig. 40, find dated to the later period of use from a Roman building in use up to the 6th century.
[1542] Pantermalis 2001, 352, figs. 7-8.
[1543] Chrysostomou A. 2000, 329, pl. 370B, vessel from a burial dated to the Early Christian period.
[1544] Davidson 1952, 104, no. 665, fig. 9, find dated from excavation results to the 3rd century.
[1545] Stern 2001, 50-51, 138.
[1546] Stern 2001, 138, 162-164.
[1547] Baratte 1989, 145, fig. 4, mosaic dated to the end of the 2nd century from El Jem, now in the Bardo Museum, Tunis.
[1548] Makropoulou 2007, 340, pl. 29.
[1549] I thank E. M. Stern for her suggestions and her help on the identification of this fragment. It seems, however, that the find from Panorama combines features of both Types (A and B) of Stern's typology. Namely, its body renders a single, male, beardless head (Type B of the 4th century; Stern 1995, 214, with further bibliography), which was standing on an applied, conical base (Type A of the 4th century, Stern 1995, 213, with further bibliography). For a 4th-century vase with two faces, standing on a conical base, see Harter, 1999, 150-151, 299, form H2, no. 1182, pl. 53, where other 4th-century examples are cited.
[1550] A similar asymmetry is visible on the vessel from the Corning Museum of Glass, see Whitehouse 2001, 72-74, nos. 545-546, as well as the one from Cologne, see Fremersdorf 1961, pl. 167.
[1551] Fremersdorf 1961, pls. 166, 167, with references to analogous finds from private and public German collections. Kreuznach am Rhein: Stern 1995, 214, note 76.
[1552] Grand Hole at Chassemy (Aisnes): Stern 1995, 214, note 75.
[1553] Steinfort: Stern 1995, 214, fig. 99, note 75. This particular example is identical in colour and dimensions to the one from Thessaloniki.
[1554] Stern 1995, 213-214.

scars on the slightly convex under-surface indicate that the vessel was originally standing on an applied, conical base. No traces of the base were found in the grave. Probably the vessel was already baseless when it was placed in the grave. The same phenomenon is observed in clay vessels used in burials and it seems that lower-quality objects, apparently sold at lower prices, or even partly broken or repaired vessels, were used as grave goods.

The vessel itself is dated to the 4th or 5th century, most probably before the middle of the 5th century, the period when a Christian basilica was founded on the site and which replaced the earlier burial building,[1555] with which probably the glass find should be connected.

### 97. Bottles, eight-sided, single-handled

(Cat. nos. 431-434)
(Plates 29-30, XV)

The vessels of this group are characterized by their eight-sided body. The rim is flaring, tubular, infolded and flattened; the neck is wide and cylindrical, tapering towards the rim. A wide strap-handle starts from the shoulder and ends on the upper part of the neck. This form appears in two variants, a taller one (17 and 18cm) and a shorter one (12 and 13cm), which are both made of the same thick, blue-greenish glass.

The vessels were mould-blown in a smooth eight-sided mould, and in one case the paraison was first blown in a small ribbed mould which imprinted relief ribs on its surface. The ribs of the latter became oblique by the rotation of the vessel before it was placed in the second, eight-sided mould and which gave it its final shape. The bottom of the mould, in which the vessels of the higher variant were blown, was decorated with an ivy leaf, and the bottom of the shorter variant had four cross-wise arranged ovals – which might give a four-petal rosette – inscribed in a circle: a motif that in one case is supplemented by a second concentric circle which runs through the middle of the ovals. The neck, rim and handle were shaped by free-hand tooling while the vessel was rotating with the help of a metal rod/blowpipe, temporarily attached centrally underneath.

Due to the identical base marks and also certain common technological features, i.e. the shape of the handle and the rim, it seems quite secure to assume that all four examples are products of the same workshop, which was probably operating in Thessaloniki.[1556]

However, it must be mentioned that the technique involving the successive use of two moulds, although also appearing in Syro-Palestinian products, is considered characteristi workshops of the northernNorthern provinces, possibly the Cologne area, in the second half of the 4th century.[1557]

Exact parallels could not be traced, except for a 4th-century example from Dračevo,[1558] although this is slightly different in terms of the form of the lips and the shape of the handle. Similar vessels exist in private collections in Germany[1559] and museums in Jerusalem[1560] and in Newark,[1561] Aphrodisias[1562] and Ephesus.[1563] The hexagonal ribbed vessels from Pannonia, Beška,[1564] Čalma[1565] and Novi Sad[1566] are technically similar, and, like the other parallels, have a funnel-shaped mouth. Smooth hexagonal examples are known from Cologne with the same mouth, but which

[1555] Tsigaridas 1973b, 498-501 (esp. 499), drawing 12, pl. 462α.
[1556] See Chapter 2.C, regarding the probable local glass products. See also Antonaras 2006a, 413-414.
[1557] Stern 2001, 142, 144, 184. The use of two moulds, a smaller one with oblique ribs and a second, full-sized, smooth and prismatic, is a characteristic typical of the production of the Rhine region, especially Cologne. It first appears in the hexagonal vessels with tiny dolphin-like handles, the earliest examples of which are dated to the first half of the 4th century. Pirling 1966, 79, pl. 409α-βb, 100/5. Isings Form 100. For further examples see Stern 2001, note 166.
[1558] Josifofska 1956, 287, only the bottom of the vessel is preserved, on which, according to Josifofska, only the ring-shaped pontil scar is visible. The vessel has the same blue-greenish tinge as the examples from Thessaloniki; each of its sides is 4cm long. The grave is dated by contents to the 4th century.
[1559] Stern 2001, 184-185, no. 77, for technological reasons the object is dated to the second half of the 4th century.
[1560] Israeli 2003b, 256, no. 330, a vessel dated on the basis of analogous finds to the 2nd-3rd century and ascribed to East Mediterranean workshops.
[1561] Auth 1976, 111, no. 129, eight-sided vessel, with the name ΠΡΙΣΚΟΣ imprinted around its bottom, which was dated for stylistic reasons to the 1st-2nd century; Lehrer-Jacobson 1986, 62; Israeli 2003b, 256, mentions an identical vessel from a museum in Jerusalem.
[1562] Tek 2001, 86, Appendix no. 16, a vessel dated to the Late Roman period, bearing on its base in relief the inscription ΣΕΥΤΟΣ.
[1563] Tek 2001, 86, a vessel dated to the Late Roman period, bearing on its base in relief the inscription ΤΙΒΕΡΙΝΟΥ.
[1564] Šaranović-Svetek 1986, 62, no. 66, pl. VII, 3, dated by a coin of Constantius II (337-361).
[1565] Site near Sremska Mitrovica (Sirmium), see Milošević 1973, 87, 92, drawing 4/2, pl. IV/1; Šaranović-Svetek 1986, 62, no. 68, pl. XVIII/1, dated by coins of Licinius (308-324), Constantine I (306-337) and Constantius II (337-361).
[1566] Šaranović-Svetek 1986, 62, no. 67, pl. VIII, 3, for some unprovenanced examples dated to the 4th century.

are not strictly dated.[1567] Finally, a smooth eight-sided vessel, which was possibly without a handle, is known from Mainz (Moguntiacum).[1568]

The vessels from Thessaloniki are dated by their context, generally to the Late Roman period and some of them more specifically to the 4th-5th century.[1569] Their dating is confirmed by analogous Balkan and Germanic material.

### 98. Jars, globular, smooth with tubular rim (Isings Form 68)

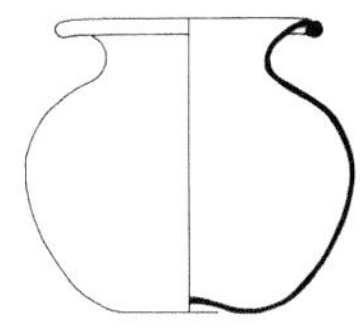

(Cat. no. 435)
(Plates 30, XV)

Vessels of this form have everted, practically horizontal rim with infolded lip, which is lower than horizontal, a short neck, globular body and slightly concave bottom. Their height is estimated between 7-8cm.

These are free-blown vessels, and the rim was shaped while the vessel was held by jacks, most probably from the neck.

The example from Thessaloniki is made of colourless, bluish glass.

It is a miniature[1570] variant of the large[1571] cinerary vessels which were widespread in the Western provinces during the 1st and 2nd centuries. These miniature vessels appear mainly in the western provinces, i.e. France: Barentin;[1572] Italy: Albenga (Albingaunum);[1573] Switzerland: Augst;[1574] Germany: Trier (Treviri),[1575] Heddernheim (Nida);[1576] Hungary: Budapest,[1577] Szőny (Brigetio),[1578] and Szombathely (Savaria).[1579] There are also known examples from the Eastern provinces, i.e. Naxos,[1580] Cyprus,[1581] and Dura-Europos.[1582] Their distribution peaks at the end of the 1st, and during the 2nd century,[1583] although 3rd- and 4th-century examples are known as well.[1584]

They are considered to be north Italian products,[1585] but it has also been suggested that at least some of them were produced in Pannonia.[1586] It has been assumed that they were also produced in the Rhine region in the first half of the 2nd century.[1587]

The find cannot be firmly dated by its context, but it seems more probable that it belongs to the late 1st or 2nd century, a period when the form was most popular.

### 99. Jars, globular, smooth with fire-polished rim (Isings Form 94 var.)

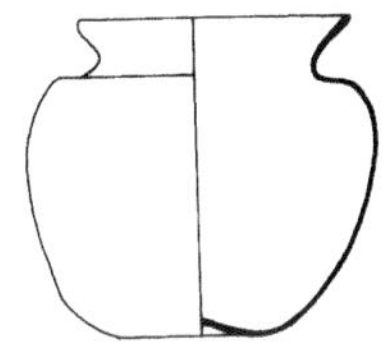

(Cat. no. 436)
(Plates 30, XV)

The vessel of this form has a wide funnel-mouth ending into a fire-polished rim. The body is globular, and the bottom is slightly concave. Their height is estimated at c. 5cm.

The vessel is made by free-blowing, and during the shaping of the mouth a fold was formed there, with the help of a jack, to create a decorative relief ring.

---

1567 Loeschcke, Willers 1911, Vol. 1, 40, nos. 393-396, Dalmatia, Vol. 2, pl. XXXI/393, 396.
1568 Harter 1999, 254, form E2, no. 0118, pl. 58, vessel with the same shape of rim as the examples from Thessaloniki and dated to the end of the 3rd - early 4th century, according to its context.
1569 Cat. nos. 431-432: Papanikola-Bakirtzi 2002 (ed.), 317-318, no. 348 (A. Antonaras), 350 (E. Marki); Marki 2006, 218, (Cat. no. 433) from a grave dated to the first half of the 4th century.
1570 Isings 1957, 88-89, form 68.
1571 Isings 1957, 86-87 form 67a.
1572 Sennequier 1985, 73, no. 62 (drawing on page 74).
1573 Massabò 2001, 116, nos. 64-68.
1574 Rütti 1991, 13/2, 108-109, form AR 114, 115, nos. 2138-2170, pls. 93-94.
1575 Goethert-Polaschek 1977, 242-243, form 147d, nos. 1440-1447, pl. 76.
1576 Welker 1974, 43-44, form 7, nos. 106-108, pl. 7, finds dated from excavation results between the second half of the 1st and first half of the 2nd century.
1577 Barkóczi 1988, 205-206, form 181, cat. nos. 513, 515, 517, 520, pls. LX, CXI.
1578 Barkóczi 1988, 205-206, form 181, cat. nos. 516, 518-519, pls. LX, CXI.
1579 Barkóczi 1988, 205, form 181, cat. no. 514, pls. LX, CXI.
1580 Zafeiropoulou 1999, pl. 88α, dated according to analogous finds.
1581 Vessberg 1952, 145-146, pl. V/20, 21.
1582 Clairmont 1963, 141, no. 749, pls. XVI, XXXVI, dated to the 2nd-3rd century.
1583 Welker 1974, 43-44, form 7, nos. 106-108, pl. 7, 1st century to first half of the 2nd century; Goethert-Polaschek 1977, 242-243, nos. 1440-1447, dated to the second half of the 1st century; Massabò 2001, 116, nos. 64-68, dated to the middle and late 1st century.
1584 Goethert-Polaschek 1977, 241-242, nos. 1436-1439, dated to 4th century; Barkóczi 1988, 206, form 181, cat. nos. 519-520, pls. LX, CXI, examples dated from the 1st to the 4th century; Rütti 1991, 108-109, 13/2, form AR 114, 115, nos. 2138-2170, pls. 93-94, finds dated from the 1st to the 4th centuries.
1585 Probably from the region of Aquileia, according to Czurda-Ruth 1979, 157, no. 1163, or from Campania, according to Scatozza-Höricht 1986, 70. This discussion is presented in detail by Massabò 2001, 116.
1586 Barkóczi 1988, 205.
1587 Barkóczi 1988, 205.

The vessel from Thessaloniki is made of bluish glass.

Although the vessel cannot be dated independently it can be ascribed to a miniature variant of a form known between the 2nd and 4th centuries.[1588] An indication in favour of a date to the 3rd and probably 4th century is the relatively low quality of the glass, quite bubbly and disproportionate to the relative thinness of the walls; the relatively intense natural-coloured type of glass is a further sign. Quite similar is a mid 4th-century find from Belgrade (Singidunum).[1589] The only other parallels that could be traced were found in the Syro-Palestinian region, and are dated to the end of the 3rd or to the 4th century.[1590]

### 100. Jars cylindrical, with short neck, wide (Isings Form 130b)

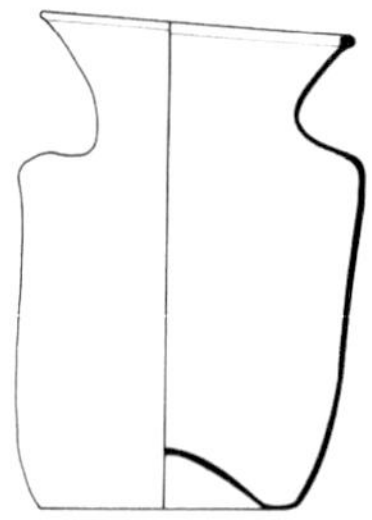

(Cat. nos. 437-438)
(Plates 30, XV)

Vessels of this form have funnel-mouth which ends at a tubular infolded rim; very short and wide neck, and cylindrical body slightly tapering towards the mildly concave bottom. Their height is estimated at c. 12cm.

The vessels of this form are free-blown and marvered, and the mouth and the rim were shaped while a blowpipe was adhered underneath the vessel, and it was tooled through that.

The object probably represents an Oriental variant of Isings Form 130b.[1591] A similar concept, but not quite so cylindrical, applies to a vessel from Jordanian Umm Qeis (Gadara).[1592] Similar vessels were unearthed at Argos,[1593] Kallio,[1594] and Oropos.[1595]

The example from Thessaloniki is made of colourless, greenish glass, with a typical colour and texture for Late Imperial times, but it does not preserve any other excavation or chronological clues. It can be dated to the 3rd-4th century on the basis of the aforementioned analogous finds.

### 101. Jars, quadrate, mould-blown

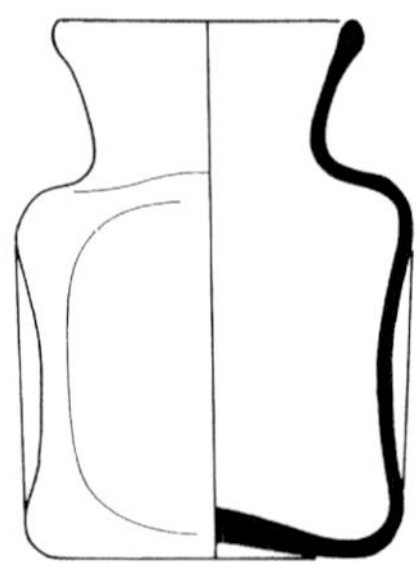

(Cat. nos. 439-441)
(Plates 30, XV)

The vessels of this group are characterized by their quadrate, almost square body. Their flaring rim is usually fire-polished and there is one example with tubular infolded rim. The neck is short and funnel-mouth. The bottom is flat and in one case bears a mould-blown base mark. Their height is estimated at c. 9cm.

The vessels are mould-blown, as the sharp edges of its sides indicate. At least in one case the mould included the bottom, on which was inscribed a simple geometrical pattern of three lines intersecting at the centre.

Vessels with cuboid bodies are known and widespread already from the 1st and 2nd centuries.[1596] Those earlier vessels differ from examples of this form in their vertical rim, which was folded to form some sort of collar. The group of simpler vessels, without a fold on the neck, are known from the 2nd century in Cologne,[1597] with at least three examples whose mouth is identical to the one from Thessaloniki. A vessel now in Warsaw is dated to the 3rd

[1588] Isings 1957, 111, form 94, miniature variant, dated to the 2nd-3rd century, which appears without the ring base which is typical of the large storage examples.

[1589] Pop-Lazić 2002, 58-59, from a grave dated to the middle of the 4th century by the coin of Constantius II (337-361) found in it.

[1590] Barag 1970c, pl. 6/232 on page 30, and pl. 34 (Form of jar 16) on page 154. These vessels were found in a burial on the Mount of Olives dated by its pottery context to the end of the 3rd or to the first half of the 4th century. I extend my thanks to colleague, Yael Gorin-Rosen, of the Israel Antiquities Authority for the translation of the relevant parts and the more general explanations on the matter.

[1591] Isings 1957, 159-160, with examples dated to the 4th century.

[1592] Dussart 1998, 92, form BVII. 2424, fig. 18/14, a find dated from excavation results to the 4th century.

[1593] Morou 1981, 109, pl. 50β, 4 vessels from a Roman grave with no other chronological indications.

[1594] Zafeiropoulou 1982, 6, pl. E.α., from grave no. 13, vessels nos. 1, 2, dated by a coin of Hadrian (117-138).

[1595] Pologiorgi 1998, 37-38, no. 4 ΜΩ 274, pl. 9β, a find from a grave dated by the glass vessels to the first half of the 2nd century.

[1596] Isings 1957, 81, form 62; Goethert-Polaschek 1977, 182, form 103, pl. 63.

[1597] Fremersdorf 1958b, 54, pls. 123-124, decorated on the bottom with concentric circles, and with the letters CCQC on one of them; dated from excavation results to the early and later 2nd century, respectively.

century.[1598] Vessels of this form continued to be produced during the 4th and 5th centuries, according to finds from Ravenna,[1599] Hungary,[1600] Israel,[1601] Syria,[1602] and Turkey.[1603]

The two examples from Thessaloniki are made of greenish, quite bubbly glass, with inclusions; they were found in graves dated to the early 4th century (Cat. no. 440)1604 and to the middle (Cat. no. 441)1605 of the 4th century, a date also indicated by the quality of the glass. The third example, with a base marked bottom, although found in the ruins of a 6th-century church1606 that was ruined in the 7th century, cannot be dated after the 4th or early 5th century.

### 102. Jars, quadrate, indented, with short neck

(Cat. no. 442)
(Plates 30, XV)

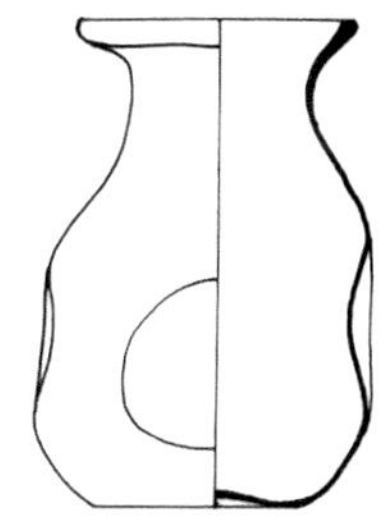

The vessels of this form have tubular infolded rim and a rudimentary (very wide and short) cylindrical neck. The characteristic feature of the form is that the body (almost always) bears four indentations which render the shape square. The bottom is almost flat, slightly concave. Their height is estimated at c. 8cm.

It is a well-known form ascribed to the production of 4th-century Oriental workshops.[1607] Analogous finds have been located in the following sites and museums: Toronto,[1608] Mainz (Moguntiacum),[1609] Sremska Mitrovica (Sirmium),[1610] Sopron (Scarbantia);[1611] Syria: Bosra (Nova Trajana Bostra);[1612] Israel: Samaria;[1613] and Jordan: Mahayy,[1614] dated the same period as the long-necked examples and they are probably also products of Oriental workshops.

Almost all the vessels from Thessaloniki are made of bubbly, olive-green glass with inclusions, while only a few are made of less coloured (partly decolourized?) glass, which is also not free from pinprick bubbles.

The examples from Thessaloniki cannot be dated firmly independently and they are ascribed on the basis of the aforementioned parallels to the late 3rd, but mainly to the 4th century.

[1598] Filarska 1952, 180-181, no. 191, pl. XLI/2, vessel of unknown provenance ascribed to Egyptian production of the 3rd century on stylistic reasons.

[1599] Isings 1957, 81, examples originally published in *Notizie degli Scavi di Antichità* 1904, 185 ff, dated from excavation results to the 4th-5th century.

[1600] Barkóczi 1988, 155-156, no. 349, pls. XXXII, XCI (Find from Dunaújváros (Intercisa) dated from excavation results to the 3rd century, and ascribed to a production that covers also the 4th century), and no. 350, pls. XXXII, XCI (unknown provenance dated to the 3rd-4th century according to analogous finds).

[1601] Parallel for Cat. no. 441: Rectangular vessel, with rim of the Syrian type, with folding that creates the illusion of a thick thread under the rim, bearing the same mark on the bottom as Cat. no. 441, which is described as a 'star with eight arms', and is dated to the 4th-5th century (Israeli 2003b, 240, no. 312). Israeli mentions another, similar, unpublished vessel from a grave at Khirbet Lid dated to the 4th century.
Parallel for Cat. no. 439: Israeli 2003b, 240, no. 311, dated to the 3rd-4th century according to analogous finds and ascribed to East Mediterranean production.

[1602] Hayes 1976, 39, 60, 115-116, nos. 156, 445, pls. 12, 19, find from Syria, dated according to analogous finds to the 3rd or early 4th century; Dussart 1998, 92, form BVII. 2425, pl. 18, find from Bosra (Nova Trajana Bostra), dated to the end of the 3rd o- early 4th century according to analogous Syrian finds; Coscarella 1994, 398, pl. IIIe, dated on the basis of analogous finds to the end of the 3rd o- early 4th century, a period this form was in fashion in Syria; Chéhab 1986, 232-234, pl. XLII, find from Tyre dated from excavation results to the end of the 3rd to early 4th century.

[1603] Lightfoot, Arslan 1992, 198-199, nos. 133-134, dated on stylistic grounds to the 4th-5th century. Vessels with rectangular body, with height twice the size of each side, fire-polished rim, made of glass similar to the type the finds from Thessaloniki.

[1604] Makropoulou 1991, 257-270, photo 22; Papanikola-Bakirtzi 2002 (ed.), 574, no. 810 (D. Makropoulou), dated to the end of the 3rd- early 4th century, with a most probable dating to the early 4th century by a bronze buckle discovered with it.

[1605] Makropoulou 2007, 262, pls. 28, 68, from a grave with a Christogram painted on the wall, and a coin find of the middle of the 4th century.

[1606] See Chapter 9 (Find-Spots), 'Agiou Dimitriou and Agias Sofias Streets, at their junction'.

[1607] Burger 1966, 137; Stern 1977, 100-101, no. 28; Šaranović-Svetek 1986, 29, 65, pls. XIII/1, 2 and XVIII/3; Barkóczi 1988, 151, form 131; Roffia 1993, 108. The only researcher to reach different conclusions is Glusčević (1998, 184-185), who concluded that most of the finds from Zadar (Iader) are securely dated by coins and clay vessels to the 2nd century, with some of them up to the 3rd century; he even assumed that, from their large quantities, they were local products.

[1608] Hayes 1975, 39, 66, pl. 12, no. 153, similar vessel, dated to the 3rd century on the basis of analogous finds.

[1609] Harter 1999, 104, 246, form D21b, no. 726, pl. 30, find from an older excavation, dated between the 3rd, 4th and early 5th centuries according to analogous examples; it is ascribed to Syro-Palestinian production.

[1610] Ružić 1994, 33, no. 458, vessel dated to the 3rd-4th century on the basis of analogous finds.

[1611] Barkóczi 1988, no. 339, pls. 31, 90, dated to the 3rd-4th century on the basis of analogous finds.

[1612] Dussart 1998, 92, form BVII. 2423, pl. 18, find from a grave dated to the 3rd-4th century.

[1613] Crowfoot 1957, 409, no. 5, fig. 94/5, finds from a 3rd-century grave.

[1614] Dussart 1998, 92, note 7, find from a grave dated to the end of the 3rd to early 4th century.

### 103. Jars, globular, with wide funnel-mouth and fire-polished rim

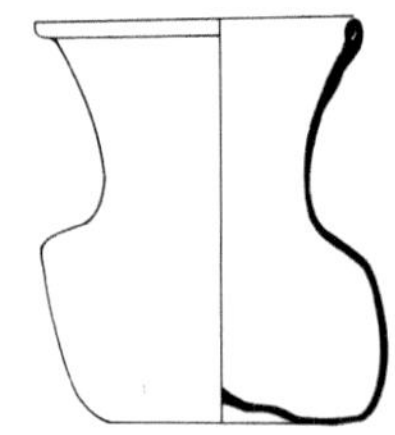

(Cat. no. 443)
(Plates 30, XV)

The vessel of this form has funnel-mouth ending into a tubular rim, a short and wide neck, an almost cylindrical body and slightly concave bottom. Their height is estimated between 12-15cm.

The vessel is free-blown, and the mouth and the rim were finished off while the vessel was held by a blowpipe attached underneath.

The vessel was found in a sarcophagus with contents dated to the late 3rd, and quite probably the 4th century.[1615]

### 104. Jars, squat globular, with horizontal rim

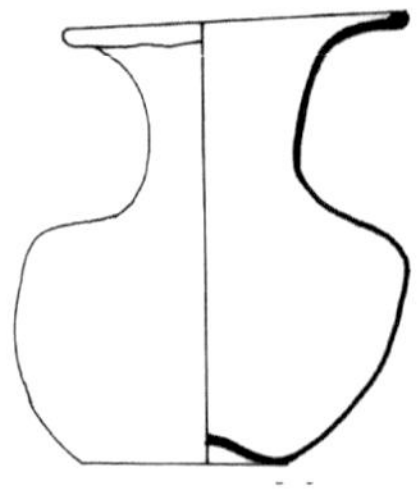

(Cat. nos. 444-445)
(Plates 30, XVI)

Vessels of this form have a wide, open mouth ending into a horizontal, infolded rim. The ne cylindrical, tapering towards the almost globular body that stands on a flat slightly concave is estimated at c. 5cm.

The vessels of this form are free-blown and the shaping of the rim was carried out while the ve: probably from the neck. The examples from Thessaloniki were made of colourless, greenish glass.

It does not seem to be a widespread form and only two parallels have been traced, both in Hungary,1616 while analogous vessels are to be found in the Royal Ontario Museum.1617 The example (Cat. no. 445) from Thessaloniki, whose excavation information we have, was found in a cluster of 4th-century wall-painted graves.[1618]

### 105. Jars globular, smooth, with fire-polished rim and a cut-out

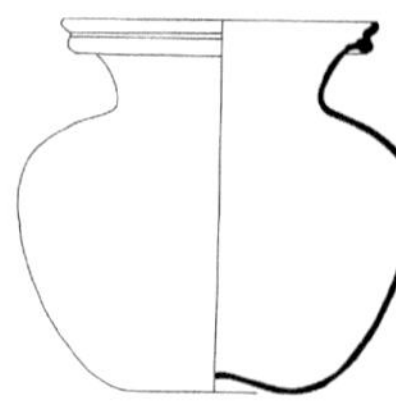

(Cat. no. 446)
(Plates 30, XVI)

The vessel of this form has funnel-mouth ending to a fire-polished rim. The body is globular and the bottom is slightly concave. Under the mouth is a cut-out forming a relief ring. Their height is estimated at c. 9cm.

The vessel is made by free-blowing and during the shaping of the mouth a cut-out was formed by jacks, creating the impression of an applied trail under the mouth.

It is a vessel directly connected to the Syro-Palestinian products of the 4th century,[1619] although the quality of the glass is not typical for that region. Among other dated examples are ones from Homs (Emesa),[1620] Khirbet el-Kerak (Philoteria),[1621] Bosra (Nova Trajana Bostra),[1622] Samaria, and Gerasa.[1623] Analogous finds from the European region are known from Hungary,[1624] as well as from private collections in Budapest[1625] and Cologne.[1626]

1615 Forms 12, 52a, 61, 90, 103, vessels Cat. nos. 63, 287-288, 328-329, 418, 443.
1616 Barkóczi 1988, 206, form 181, cat. no. 520, pls. LX, CXI, a vessel with unknown exact provenance, dated to the 4th century, according to an analogous find from Pécs; Fülep 1969, 12, pl. 48. 8, find from a grave in Pécs, dated from excavation results to the 4th century.
1617 Hayes 1975, 79, no. 292, identical vessel with indentations in the lower body, which is identified as a Syro-Palestinian find, dated probably to the 3rd century.
1618 Marki 2006, 269.
1619 Dussart 1998, 92, form BVII.2431, pl. 19, with examples from various sites dated from the excavation results to the 4th-5th century; Hayes 1975, 94-95, nos. 325, 326, pl. 22, identical, undecorated vessels dated according to analogous finds to the early or middle 4th century. For examples decorated with a coloured thread around their body or/and with handles, see Hayes 1975, 96-97, nos. 330, 332, 385, pls. 22-23, which, according to analogous finds, are dated to the second half of the 4th or to the first half of the 5th century. See also Matheson 1980, 116, nos. 307-308.
1620 Abdul Hak 1965, 30, figs. 9, 11, finds from a necropolis dated to the 2nd-3rd century, although it seems, judging by the illustrated finds, that it continued to be used at least during the 4th century.
1621 Dussart 1998, 92, form BVII. 2431, pl. 19/1, find dated according to analogous finds to the 4th-5th century.
1622 Coscarella 1994, 400, pl. III d, finds dated to the 4th century.
1623 Coscarella 1994, 396, 398, finds dated to the 5th century.
1624 Barkóczi 1988, 152, no. 337, pls. XXXI, XC, find from Sopron (Scarbantia) dated according to analogous finds to the 3rd-4th century, and ascribed by the researcher to production from the East Mediterranean.
1625 Barkóczi 1996, 35, no. 54, pls. E, VI, XLI, unprovenanced find dated according to analogous finds to the 4th century.
1626 Loeschcke, Willers 1911, 64, no. 1079a, b, pl. LII, unprovenanced finds from a private collection.

The find from Thessaloniki cannot be dated independently and it is generally ascribed to the 3rd or 4th century on the basis of the rest of the excavated finds,[1627] a date which fits within the time-frame provided by the parallels mentioned.

**106. Jars, globular, ribbed, with fire-polished rim**

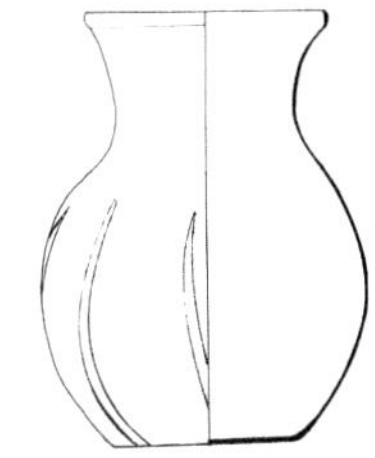

(Cat. no. 447)
(Plates 31, XVI)

This vessel has a funnel-mouth ending into a fire-polished rim, a globular body, with vertical ribs around it, that stands on a flat, slightly concave bottom. Their height is estimated at c. 10cm.

The vessel is made by free-blowing. The ribs were pinched, probably before the vessel was expanded to its final dimensions. The rim was shaped while the vessel was held by a metal rod or blowpipe attached to its base.

The vessel from Thessaloniki is made of blue-greenish glass.

The texture and quality of the glass, the quite asymmetrical ad thus /unbalanced body shape indicate an approximate 4th-century date for the vessel. In addition, the pinched ribs on the body are also indicative of a specific production time and region. According to the distribution of the finds these are Syro-Palestinian products dated to the second half of the 3rd century,[1628] or, more probably, to the 4th or 5th centuries.[1629] Dated examples have been found at Homs (Emesa),[1630] Khirbet el-Kerak (Philoteria),[1631] Bosra (Nova Trajana Bostra),[1632] Pella,[1633] and Samaria.[1634]

**107. Jars, bulbous, smooth**

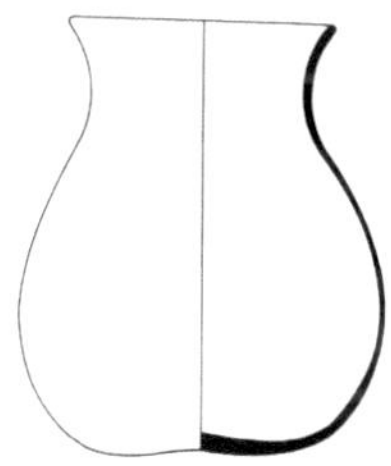

(Cat. nos. 448-449)
(Plates 31, XVI)

Vessels of this form have funnel-mouth and tubular rim, bulbous body and concave bottom. Their height is estimated at c. 10cm.

They are free-blown. The mouth and rim were finished while the vessel was held by a metal rod attached underneath.

Cat. no. 448 was discovered in a grave with a 2nd- or 3rd-century coin and glass vessels dated to the end of the 3rd and, most probably, to the 4th century.[1635] The excessive thickness of the body and the low quality of the glass seem to corroborate such a late dating, as is indicated by the naturally dark hue and the relatively large numbers of pinprick bubbles and sandy inclusions.

For Cat. no. 449, the excessive thickness of the body and the low-quality glass, naturally intensely coloured and quite bubbly, would appear to support a date after the 2nd, and possibly into the 4th century.

**108. Jars, pear-shaped, short**

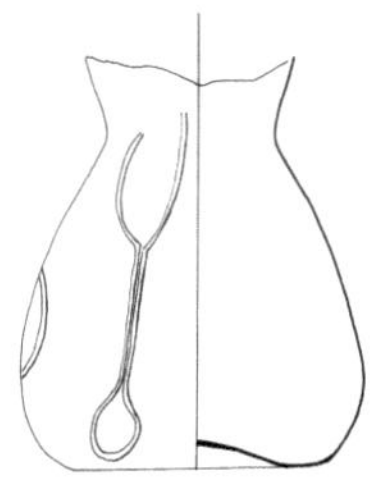

(Cat. nos. 450-451)
(Plates 31, XVI)

The vessels of this form have no particular rim. The end of the funnel-shaped neck is fire-polished. The body is hemispherical and the bottom slightly concave. Their height is estimated at c. 11cm.

These vessels were free-blown. After the moil was cracked off the rim was reheated, widened and fire-polished, while the vessel was held by jacks, probably from the neck. Cat. no. 440 has pinched vetical ribs that were constricted in pairs at their middle, forming a rhomboid net or web.

[1627] Makropoulou 2007, 235, pl. 25.
[1628] Hayes 1975, 160, no. 658, pl. 42, vessel dated on stylistic reasons to the second half of the 3rd century, considered to be a Syro-Palestinian product.
[1629] Dussart 1998, 92, form BVII. 2432a1, with the relevant discussion.
[1630] Abdul Hak 1965, 30, fig. 10, find from a necropolis dated to the 2nd-3rd century, although it seems that, judging by the illustrated finds, it continued to be used at the least during the 4th century.
[1631] Dussart 1998, 92, form BVII. 2432a1, pl. 19/4.
[1632] Coscarella 1994, 400, pl. IV a-b, finds dated to the 4th-5th century.
[1633] McNicol *et al.* 1982, pl. 136, 1, find from a grave dated to the Late Roman period.
[1634] Crowfoot 1957, 413, no. 17, fig. 95/17, dated to the 4th-5th century.
[1635] Vessels Forms 12a and 51.

The examples from Thessaloniki are made of naturally coloured, greenish glass.

Analogous finds have been located, as is the case with all funnel-mouth vessels, in the Syro-Palestinian rgion. Others dated examples have been discovered in Gerasa[1636] and Jericho,[1637] while analogous finds appear also at Kerch (Panticapaeum).[1638]

Cat. no. 451, according to its decoration, is dated to the 3rd-4th century, more probably to the 4th,[1639] whereas Cat. no. 450 was found in a grave dated to the second half, or to the end of the 3rd century.[1640]

### 109. Jars, pear-shaped

(Cat. no. 452)
(Plates 31, XVI)

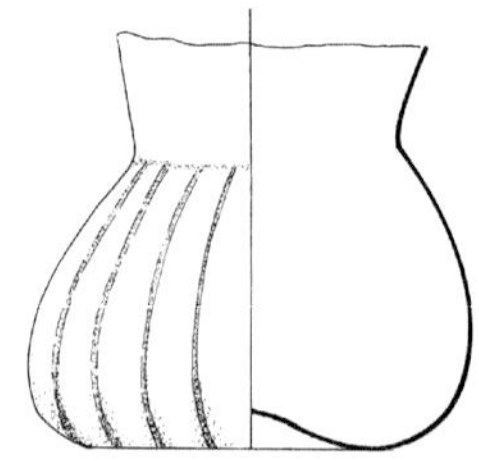

The vessel has a funnel-mouth which ends, at least in the present state of preservation, on an uneven cracked-off rim; the pear-shaped body features 12 vertical ribs that continue at the periphery of the base. The underneath is smooth and slightly concave at its centre. Their height is estimated at c. 6cm.

The vessel is dip-mould-blown, initially in a small ribbed mould and then expanded to its final dimensions by free-blowing. It is not clear if the rim was ever finished or whether it broke later in this peculiar way.

The vessel cannot be dated firmly on the basis of the excavation data, but it is made of olive-green glass, a typical feature for Late Imperial production in the East Mediterranean, probably of the 4th-5th century. Furthermore, the modelling technique, i.e. dip-mould-blowing, appears and becomes popular in the same period, offering yet another indication for defining the date of the form around the 4th or 5th century.

### 110. Jars, bulbous, with pinching

(Cat. no. 453)
(Plates 31, XVI)

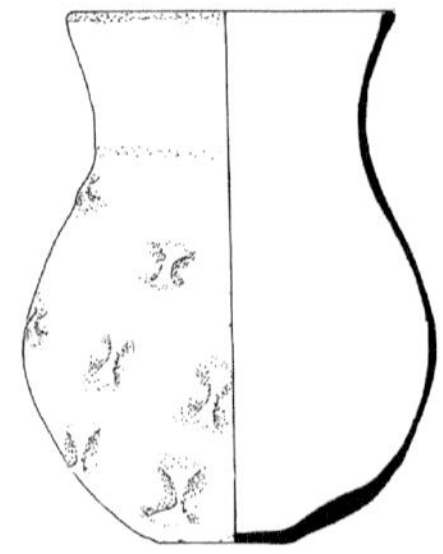

The vessel has short funnel-mouth that ends onto a fire-polished rim; the body is bulbous with a flat, slightly concave bottom. The body is decorated all around with irregular rows of small relief folds, i.e. by pinching. Their height is estimated between 8-9cm.

The vessel is made y free-blowing., After the moil was cracked off, the rim was reheated, widened and fire-polished while the vessel was held by jacks, probably from the neck. At that stage the surface was pinched by a pair of pointed pincers and its relief decoration made.

The vessel is made of clear colourless, greenish glass.

The decoration with analogous pinches in irregular rows appears on vessels from the end of the 2nd,[1641] and mainly in the 3rd-6th centuries,[1642] especially in the Syro-Palestinian region.[1643] This decoration remains in use until the 7th century in Palestine;[1644] in Jordan it appears from the 4th century and remains in use up to the 8th.[1645] In the West this decoration does not become so popular and the only example yet traced is from Cagliari(Caralis).[1646] This decorative technique, therefore, suggests a Syro-Palestinian provenance for the vessel.

---

[1636] Dussart 1998, 132, form BX 1123, pl. 34/2, find from a temple to Jupiter, dated by its context to the Early Christian period, after the 3rd and between the 4th and 6th centuries.

[1637] Baramki 1937, pl. XXI, 1, from the excavation of a synagogue dated to the 8th century, according to Dussart 1998, 132, note 1, form BX 1123.

[1638] Sorokina 1971, 94, 234, drawing in pl. 5/1; Sorokina 1962, photo 7.4, where the photo caption gives a date to the end of the 4th century, whereas on page 234 it is given as in the original publication (*IAK* 25, 1907, 54). These two vessels are identical to the form under discussion.

[1639] Weinberg 1988, 80, note 161; Stern 2001, 341, no. 189. See also Chapter 5 (Decorative Techniques).

[1640] Pazaras 1981, 372-387.

[1641] Čremošnik 1965, 197; Rütti 1991, 70, 13/2, form AR 60.3, nos. 1355-68, pl. 62, dated from excavation results between the middle of the 2nd and the middle of the 3rd century.

[1642] Vessberg 1952, 146, pl. IV. 22 (vessel form A.I.); Saldern *et al.* 1974, 243-246, no. 714; Bulat 1976, 89-90, pl. II/1, VII/2; Auth 1976, 129, 138, nos. 160, 175; Oliver 1980, 112, no. 182; Rütti 1991, 72-73, 13/2, form 64.2, nos. 1400-42, pl. 65, dated from excavation results to the end of the 3rd-4th century.

[1643] Delougaz, Haines 1960, pl. 50. 9; Whitehouse 1997, 199-200, no. 349; Whitehouse 2001, 124, 148, nos. 628, 663; Stern 2001, 247-251, nos. 133-37, with thorough bibliography.

[1644] Stern 2001, 354, no. 201.

[1645] Baur 1938, 536, no. 67 (244), fig. 18; Dussart 1998, page 128, BX.111a, pl. 32/1; page 158, BX83, pl. 46/21; page 161, BXII.1, pl. 49/1.

[1646] A similar vessel, also a jar, whose decoration is supplemented by four large indentations. See Stiaffini, Borghetti 1994, 82-83, 144, no. 458, pl. 119, find dated with reservations to the 3rd century, according to analogous finds.

In Thessaloniki pinching appears on lamps dated to the 6th-7th century,[1647] however it seems more secure to date the vessel to the 4th-5th century, a period which fits better with the other grave goods found with it in the same burial.[1648]

### 111. Aryballoi

(Cat. nos. 454-458)
(Plate XVI)

The vessels of this form[1649] have a tubular rim folded out, down, up, and inwards; they also feature a short cylindrical neck, globular body and flat, slightly convex bottom. Two small loop-handles are attached to the neck under the rim. Their height is estimated at c. 11cm.

These aryballoi were free-blown.

Aryballoi were already known in Classical Greece. They were made of different materials, i.e. clay and metal, and core-formed aryballoi were produced already in the 6th century BC.[1650] These vessels contained fragrant oils applied to the body after bathing. Usually they were held from the wrist by a system of three chains; each one of these chains ended in a loop attached to each handle, and the third secured the vessel's stopper.[1651]

After the invention of free-blowing, aryballoi were made initially with thin walls, which were in fashion already in the first half of the 1st century, and a little while later they were made with thick walls[1652] in a form that survived rather longer, into the early 3rd century.[1653]

The finds from Thessaloniki, due to the characteristic shape of their rim, are ascribed to the group of aryballoi known from sites in the Black Sea region, which have been considered products from the Asia Minor, probably from Pergamon, dated to the second half of the 1st to the first quarter of the 2nd century.[1654] Less probable seems the assumption that the aryballoi were made in the Syro-Palestinian region – where only a very few examples have been found – and intended mainly for export to those regions where public baths played an important role in social life.[1655]

Aryballoi, mainly thick-walled forms, have been found in the Western Europe,[1656] Slovenia,[1657] Croatia,[1658] Bulgaria,[1659] Romania,[1660] Cyprus[1661] and Syria.[1662] In Greece they have been found at Corinth,[1663] Galaxidi (Chaleion),[1664] Samothrace,[1665] Sifnos,[1666] Monasteriaki Kephala, Knossos,[1667] Kastelli Kissamou,[1668] and Herakleion.[1669]

The finds from Thessaloniki are dated securely to the second half of the 1st century, by the time of the destruction of the baths in which they were found.[1670]

---

1647 Form 33, Cat. nos. 128-129.

1648 Form 12β, Cat. nos. 45, 50.

1649 Isings 1957, 78-80, form 61.

1650 Stern, Schlick-Nolte 1994, 214-215, with relevant bibliography.

1651 Davidson 1952, 105, drawing of no. 676.

1652 This form appears in Pompeii already before AD 79, see Scatozza-Höricht 1986, 55-56, form 40, nos. 112-120, pl. XXXIV.

1653 Cool, Price 1995, 156-159; Stern 2001, 50.

1654 Sorokina 1987, 42-43, group B.

1655 Stern 2001, 44-45, numerous finds from the Black Sea region and Dalmatia were perceived, on the basis of this assumption, as Syro-Palestinian imports.

1656 Isings 1957, 78-80, form 61; Cool, Price 1995, 156-159; Bonnet-Borel 1997, 47-49, forms AV V 125-129, dated from excavation results to the 1st-2nd century.

1657 Šubic 1976, 41, no. 13, pl. 3.21, find with thick walls, dated on the basis of analogous finds to the 1st century; Lazar 2003, 171-172, form 8.3.2, finds with thin walls dated from excavation results between the 1st and the 3rd centuries.

1658 Damevski 1976, 65, pl. 7.1, finds with thick walls, dated on the basis of analogous finds between the 1st and 3rd centuries.

1659 Nikolov, Bujukliev 1967, 23, 15, fig. 12b, find with thick walls, dated from excavation results to the second half of the 1st century; Bujukliev 1984, 14, photo 15, find with thick walls, dated from excavation results to the middle of the 2nd century; Belivanova 1999, 43-44, where various finds from Bulgaria are cited, which are dated from excavation results to the end of the 1st - early 2nd century.

1660 Tomis: Bucovala 1968, 81-83, nos. 129-130, 132-134, examples with thick walls, finds dated from excavation results between the 1st and 3rd century; nos. 131 and 135, examples with thin walls, from which no. 135 is dated from excavation results to the 3rd century.

1661 Vessberg 1952, pl. 9.33-36, finds that cannot be dated from excavation results.

1662 Dura-Europos: Clairmont 1963, 115, no. 565, pl. XXXIV, find with thin walls, dated from excavation results between the end of the 1st century and AD 256.

1663 Davidson 1952, 105, no. 676, fig. 11, find with thick walls dated from excavation results to the 1st-2nd century.

1664 Threpsiadis 1972, 197, pl. 76β:γ. Published again by Zafeiropoulou (1975, 55, pl. 4:6), where the vessel is dated on the basis of analogous finds to the 2nd century.

1665 Dusenbery 1998, 1130, nos. S240-1, photo and drawing on page 1131, find with thick walls dated from excavation results to the middle or to the third quarter of the 1st century BC.

1666 Mackworth-Young 1949, 87, grave 6, no. 4, pl. 29:1. 6(4), find with thin walls from a grave with finds dated to the 1st century.

1667 Carington-Smith 1982, 278, pl. 40a, find with thick walls, dated from excavation results to the 1st century.

1668 Tzedakis 1978, 370-372, pl. 192α, from a 2nd-century burial.

1669 Alexiou 1975, 340-341, pl. 246α, β, from a burial of the second half of the 2nd century, dated by a coin which was found in it.

1670 Adam-Veleni *et al.* 1998, 85-102; Mavromichali 2001, 141-142.

### 112. Amphorisks, globular, flat base

(Cat. nos. 459-460)
(Plate 31, XVI)

Vessels of this form have flaring, tubular rim, long cylindrical neck, equal in height to the body, globular body and flat bottom. Two almost vertical coil handles start on the shoulders, bend, and then adhere to the vessel at mid-neck height. Their height is estimated between 5-6cm.

One of the examples is made of dark-blue glass, while the handles are made of milky-white, opaque glass. The body of the second example is purple and the handles of colourless greenish glass. The use of two different colours of glass is quite usual in vessels of this form.

The production of this form begins right after the middle of the 1st century and they were probably made in more than one centre. It is also assumed that, although they were made in the Syro-Palestinian region, they were almost exclusively exported to western provinces, where baths were playing an important role in social life, which was not the case in their assumed production region.[1671] Analogous finds have been traced in Ljubljana (Emona),[1672] Zadar (Iader),[1673] Kerch (Panticapaeum),[1674] Sifnos,[1675] and Jerusalem.[1676]

Information about the excavation and its context are preserved only for Cat. no. 460, which is thus dated to the 1st century.[1677]

### 113. Amphorisks, ovular, with ring base

(Cat. nos. 461-462)
(Plates 31, XVI)

The vessels of this form have flaring, fire-polished rim, short neck, ovular body and pushed-in tubular ring base and flat bottom. Two very fine coil handles, start on the shoulder, rise vertically and adhere at the mid-height of the neck or under the rim. Their height is estimated between 6-7cm.

The vessels of this form are made of fine, transparent, colourless or slightly greenish glass.

They are free-blown. The shaping of the rim and the application of the handles was probably carried out while the vessel was held by jacks from its base.

This form is a miniature variant of its contemporary tableware vessel.[1678] Exact parallels with the finds from Thessaloniki have been found at Heddernheim (Nida),[1679] Este,[1680] Locarno-Muralto,[1681] Aquileia,[1682] Zadar (Iader), Nona or Benkovac (Asseria),[1683] Starigrad (Argyruntum),[1684] Kerch (Panticapaeum),[1685] Dura-Europos,[1686] and Athens.[1687] From the distribution pattern of the finds it was assumed that they were produced in northern Italy or Ticino.[1688]

The finds from Thessaloniki are dated by excavation results to the 1st century.[1689]

[1671] Stern 2001, 44-45.
[1672] Plesničar-Gec 1972, grave 90, pl. 22.
[1673] Ravagnan 1994, 43-44, nos. 47, 50, which are dated on the basis of analogous finds to the second half of the 1st or to the early 2nd century. In particular Ravagnan's vessel no. 50 is illustrated in colour in *Trasparenze imperiali* (1997, 132, no. 79), where it is apparent that the vessel has inverted the colours of our example from the Macedonian Museum of Modern Art excavation, i.e. opaque white body and blue handles.
[1674] Kunina 1997, 321, nos. 340, 342, finds dated from excavation results to the middle or third quarter of the 1st century and to the 1st century, respectively.
[1675] Mackworth-Young 1949, 87, grave no. 6, vessel no. 4, pl. 29:1. 6 (4), dated to the 1st century.
[1676] Israeli 2003b, 259-260, nos. 333-335, 337, finds dated to the 1st century according to analogous finds.
[1677] On the basis of the glass vessels with which it was found, Cat. nos. 460-462, 568-569, 577-578, 595; Forms 112, 113, 126α, 126β, 129 and 130.
[1678] Isings 1957, 32-34, form 15, dated from around the middle of the 1st century to its close.
[1679] Welker 1974, 63-65, form 12, finds from a 1st-century grave.
[1680] According to Welker (1974, 65), finds dated from excavation results to the 1st and to the last third of the 1st century respectively.
[1681] Carazzetti, Biaggio-Simona 1988, 45, no. 19, pl. II, from a grave of the second quarter of the 1st century.
[1682] Calvi 1968, 24, pl. 1. 2-1. 4, finds dated to the 1st century.
[1683] Ravagnan 1994, 36-37, nos. 29, 31, 33, dated on the basis of analogous finds to the second half of the 1st or to the early 2nd century; *Trasparenze imperiali* 1997, no. 77, a find dated from excavation results to the early 1st century; it is made of opaque greenish glass.
[1684] Fadić 1989, 39, no. 136, pl. X, 2, dated on the basis of analogous finds to the 1st century.
[1685] Kunina 1997, 321, no. 339, quite tall, 15cm, find dated from excavation results to the second half of the 1st century.
[1686] Clairmont 1963, 30, no. 123, pl. 20, grave with two phases of use, dated to the 1st and the early 2nd century.
[1687] Alexandri 1972, 115-118, pl. 76δ, vessel made of marbled, blue-white mosaic glass from a group of graves dated to the Roman period, which, judging by the other glass finds, are dated to the 1st century.
[1688] Ravagnan 1994, 36-37.
[1689] As is indicated by the other glass vessels found with it: Cat. nos. 460-462, 568-569, 577-578, 595; Forms 112, 113, 126α, 126β, 129 and 130.

### 114. Amphorisks, ovular, mould-blown on flat base

(Cat. no. 463)
(Plate XVI)

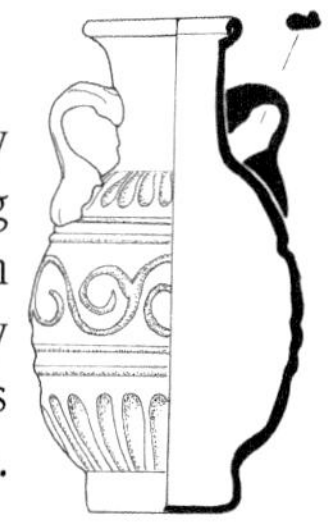

The rim of vessels of this form is flaring, tubular infolded, the neck is short and cylindrical, the body is ovular tapering toward the bottom where it forms a ring creating the impression of a short ring base. The bottom of the vessel is flat. The body bears relief decoration. The body at the maximum diameter zone is surrounded by a band decorated with a vine scroll. This central band is flanked by bands of vertical, tangent parallel petals that face towards it and cover the upper and lower parts of the body. The example from Thessaloniki is made of purple glass. Their height is estimated at c. 7cm.

The body of vessels of this group is made in a mould comprised of two vertical parts, while the neck and rim are free-blown.

Vessels with this body are either finished to resemble juglets, or, more often, amphorisks.[1690] It is considered[1691] that they are miniature renderings of mould-blown amphorisks, like those produced in the workshop of Ennion.[1692] Relevant vessels shaped as amphorisks are to be found, *inter alia*,[1693] in collections at Boston,[1694] Newark,[1695] Leiden,[1696] Berlin,[1697] and the Oppenländer Collection.[1698] Examples of vessels shaped as juglets, which can be dated by their excavation context, were discovered on Samothrace[1699] and in the Crimea,[1700] while vessels shaped as amphorisks are found in burials at Kerch (Panticapaeum).[1701]

Although it is not possible to date precisely the burial in which the vessel was discovered, there are chronological indications however, such we know it was a cremation and that the second deformed glass unguentarium from that burial, on the basis of analogous finds,[1702] is datable to the 1st century, that link our find with the 1st century, a period that coincides with the time that this form was anyway popular.

### 115. Juglets, hexagonal

(Cat. no. 464)
(Plate XVI)

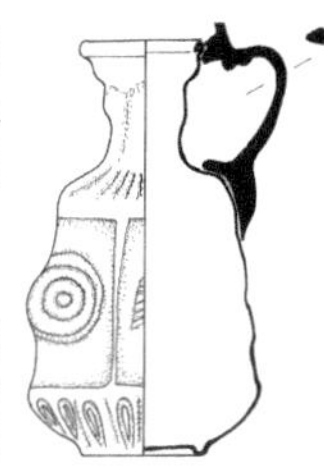

The rims of these vessels are infolded tubular, and the neck is relatively long and cylindrical; the shoulders are curved. The body is hexagonal and curves towards the bottom, which is flat. A strap-handle with a central rib starts on the shoulder and ends under the rim with the tiny fold that forms a thumb-rest pad.[1703] Their height is estimated at c. 9cm.

The shoulders and the lower, curved part of the body are decorated by a band of vertical, adjoining petals. The largest part of the body is covered with six adjoining, rectangular fields featuring Dionysiac motifs: a two-handled vessel, a *syrinx*, a discoid object probably a *phiale*, a krater, crossed *thyrsoi*, and a juglet. The bottom of the vessel is decorated by three raised, concentric circles. The find from Thessaloniki is made of blue transparent glass, like the majority of the preserved examples of this type.[1704]

[1690] Stern 1995, 152, no. 55, with exhaustive bibliography on all published examples.
[1691] Stern 1995, 152, no. 55.
[1692] Kunina 1973, 101-106, figs. 1-6; Kunina 1997, 273, no. 109.
[1693] A full catalogue of vessels of this form, and their place in the evolutionary process of mould-blown vessels, and which justify their dating to the 1st century, can be found in Stern 1995, 152-153. In addition, for an Athenean find, see Weinberg and Stern 2009, 68, no. 151, fig. 8, from a 1st-century context.
[1694] Saldern 1968, no. 33.
[1695] Auth 1976, 71, 199, nos. 66, 329-330.
[1696] Stern 1995, 152, note 3d, e.
[1697] Platz-Horster 1976, 39, nos. 56-57.
[1698] Saldern *et al.* 1974, 145-146, nos. 413-415.
[1699] Dusenbery 1998, 1079-1080, find dated to the third quarter of the 1st century.
[1700] Kunina, Sorokina 1972, 161-162, fig. 7, no. 34, form of unguentaria I.2.b, dated to the 1st and especially to the end of the 1st century, from grave 32; Kunina 1973, 118-119, fig. 16.1.5.
[1701] Kunina 1997, 280, no. 140; Kunina 1973, 119, fig. 16.1.3; Kunina 1997, 280, no. 138; Kunina 1973, 119-120, fig. 16.1.4, from a grave dated by a coin to the middle of the 1st century.
[1702] Unregistered unguentarium, for analogous finds see Goethert-Polaschek 1977, form Trier 66b, 350, dated to the 1st century.
[1703] The description of the vessel is made on the basis of analogous finds held in the Toledo Museum of Art, and presented in detail by Stern (1995, 161-166). One two-handled example of this group is also known.
[1704] Another example, also made of blue glass, was found in an unpublished excavation of the *fossa*, the external ditch of the western city wall of Thessaloniki, from a site which did not provide a more specific dating.

The vessels of this form are blown in a mould comprised of three or four parts: two or three were vertical and convex, forming the body walls, and one was hemispherical – forming the lower part of the body and the bottom. The shaping of the rim and the attaching of the handle were done with free-hand tooling.

Although the decoration of these vessels was considered for a long time of Judaic character, it is now clear that they almost typically express the religious syncretism and domination of the ancient Greek religion of the eastern coastal regions of the Mediterranean.[1705] According to the motifs appearing on the six panels of the body, five different groups can be discerned. The find from Thessaloniki belongs to a group of vessels which try to replicate, by their shape, decoration and unidentified structure (probably an altar), specific features connected to the cult of Dionysus.[1706]

This form was created in an eastern workshop, most probably situated in the region of Sidon,[1707] as indicated by the distribution pattern of the extant finds; it was later copied by glassworkers in the West.[1708]

Analogous vessels were found in Syria,[1709] Armenia,[1710] the Black Sea,[1711] and Greece.[1712]

Although most of the finds cannot be dated by their own archaeological context, it is believed – for structural and stylistic reasons – that these vessels were produced in the first half of the 1st century, and probably in its second quarter. The find from Thessaloniki is dated to the same period, as the destruction period of the bath complex where it was discovered can be dated to the last twenty years of the 1st century.[1713]

### 116. Unguentaria, globular, unworked rim, 'spheres' (Isings Form 10)

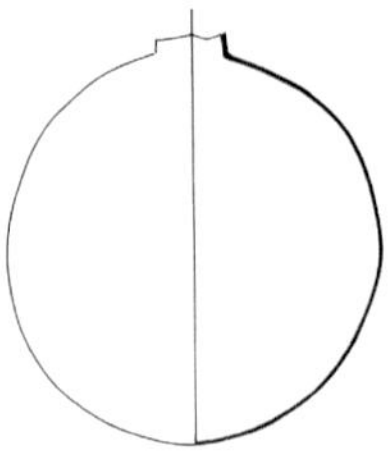

(Cat. nos. 465-493)
(Plates 31-32, XVI)

This form comprises perfectly spherical vessels with extremely short neck and cracked-off rim. The sphere of the body is not deformed in the slightest, not even around the underneath, and unlike what is normally encountered with other glass vessels. In Thessaloniki this form appears mainly made of colourless greenish, and, quite rarely, yellowish and purple glass. None of the preserved examples, however, presents the relatively usual decoration, found at other sites, of the fine, strongly coloured thread wound spirally around the body.

In addition, thus far the Thessaloniki finds comprise exclusively extremely thin-walled examples, and not the thick-walled form so well know from elsewhere.[1714] Their height is estimated between 7-9cm.[1715]

They are free-blown vessels. Because of their extremely short neck, it is believed that after filling they were sealed by reheating the neck end, which was later broken off to unseal the vessel and use the contents. A few examples reveal the remains of a very fine and sophisticated lead sealing system, with a bell-shaped lid and four long 'legs' arranged cross-ways, enclosing the entire body. To date this device has only been unearthed in Thessaloniki and not reported elsewhere.[1716]

1705 Stern 1995, 84-86.
1706 Stern 1995, 85, 163.
1707 Stern 1995, 85.
1708 Stern 1995, 82-83.
1709 According to Stern 1995 (163, note 10, and including the relevant bibliography), such finds may be seen today in the following museums: Metropolitan Museum of Art, British Museum, Carnegie Museum, and the Corning Museum of Glass.
1710 Stern 1995, 163, note 11, wherein the relevant bibliography.
1711 Sorokina 1967, 70, fig. 2, nos. 20, 23.
1712 Chalkis: Agios Stefanos, Choremis 1973, 300-301, pl. 259γ and Choremis 1974, 28, 31, fig. 7; Knossos: Price 1988b, 34, fig. 3 no. 28, the find comes from a layer dated to Nero (54-68) or Flavian (69-96); Aegean region: today in the British Museum, see Stern 1995, 164 note 13, wherein the relevant bibliography; Athens: today in the Metropolitan Museum of Art, see Stern 1995, 164, note 13, wherein the relevant bibliography; Ierissos in Chalkidiki: a handleless, milky-white example dated to the middle of the 1st century, see *Glass Cosmos* 2010, 378-79, no. 438; Philippi, one example dated to the first half of the 1st century, see *Glass Cosmos* 2010, 398, no. 481.
1713 Adam-Veleni *et al.* 1998, 85-102; Mavromichali 2001, 141.
1714 Examples of spheres of thick glass, exhibited in the Archaeological Museum of Patras and in RGM Köln.
1715 On the attribution of these fine examples to a local Thessalonean workshop, see Antonaras 2010a, 27-33.
1716 Antonaras, Keramaris 2012, 94-97.

Numerous examples have been found in Italy,[1717] Switzerland,[1718] Germany,[1719] Holland,[1720] and France;[1721] they are considered products of north Italian workshops and are dated to the 1st century.[1722]

In Greece, apart from the examples from Thessaloniki, analogous finds have been found at Patras,[1723] Thebes,[1724] and Monasteriake Kephala in Knossos.[1725]

Several of the examples from Thessaloniki are dated by their archaeological context to the 1st century.[1726] Due to the fact that they are clearly different from their Western parallels, and at the same time very closely related among themselves, they are considered to be local products.

Remains of their contents, in the form of white, pink and red powder, are relatively often preserved inside them. Chemical analyses of these powders have revealed in one case a mixture of wine and resins, and in another a mixture of fat and cosmetics,[1727] as well as $MnCo_3$ and $MnSo_3$.[1728] Chemical analyses were conducted in vessels from Thessaloniki with similar contents:[1729] the blue powder was Egyptian blue, the red ocher, and the pink madder.

### 117. Unguentaria, figural, date-shaped (Isings Form 78d)

(Cat. nos. 494-495)
(Plate XVII )

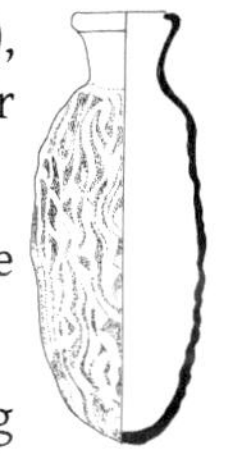

The vessels of this form appear to render naturalistically[1730] the fruit of the palm tree (*Phoenix dactylifera*), i.e. dates. The entire surface of the vessel is covered by 'wrinkles', similar to those of a ripe date, and their size, which is estimated at approximately 7cm, is also identical to the size of many of these fruits.

The examples from Thessaloniki, just like the vast majority of the internationally known examples, are made of amber glass (to resemble the colour of dates). Their height is estimated between 7-9cm.

The vessels are mould-blown, except for the short neck and rim, which were shaped by free-hand tooling after the vessel was extracted from the mould. Usually they were made in a two-part mould, although there are some examples made in a tripartite mould.[1731] No traces of the seams are preserved on the examples from Thessaloniki, and thus this constructional technological detail remains uncertain.

Generally it is a widely known and distributed form of unguentarium[1732] that appears in many regions of the Roman Empire.[1733] Examples, firmly dated by their archaeological context, are reported from Spain,[1734] Germany,[1735]

---

[1717] Isings 1957, 26, finds from Pompeii, dated prior to AD 79; De Tommaso 1990, 40-42, with a thorough bibliography of finds from the Italian Peninsula, with examples dated mainly to the 1st century; Maccabruni 1991, 112, examples dated to the 1st century from excavations in the region of Pavia (Ticinum).

[1718] Biaggio-Simona 1991, vol. I, 120-23, pls. 12, 13 and 51, with a large number of finds from the region of Ticino dated to the 1st century; Carazzetti, Biaggio-Simona 1988, 44, no. 15, from a grave dated to the first half of the 1st century; Rütti 1991, form AR 123, nos. 2247-48, pl. 100, finds from Augst, dated from excavation results between the reign of Tiberius (14-37) and Taijus (98-117).

[1719] Isings 1957, 25-26, form 10, with finds dated from excavation results to the 1st century; Goethert-Polaschek 1977, 155, 157, form 155, nos. 1497, 1504, fig. 79, with finds from Trier (Treviri) dated to the middle of the 1st century; Harter 1999, 87, form D2, finds from Mainz (Moguntiacum), dated to the Claudian era (41-54).

[1720] Isings 1957, 26, finds from Nijmegen (Noviomagus), dated from excavation results to the 1st century.

[1721] Sennequier 1985, 200-201, nos. 310-311, dated to the 1st century according to analogous finds.

[1722] Isings 1957, 27; De Tommaso 1990, 41.

[1723] Agallopoulou 1973/74, 369, pl. 235β; Dekoulakou 1975c, 115-117, pl. 65δ; Kolonas 2002, 113, no. 9.

[1724] Haevernick 1981, 119, fragments of three vessels dated on the basis of analogous finds from other sites to the 1st century.

[1725] Carington-Smith 1982, 270-272, pl. 4/2, 36b, fragments of at least ten spheres, found in a grave dated to the second half of the 1st century.

[1726] Vessels Cat. nos. 468-477, 480-492, 515 were found in graves together with bird-shaped unguentaria dated also to the 1st century. For three (at least) more finds from the eastern necropolis, also dated to the 1st century, see Tsimpidou-Avloniti, Kaïafa 2004, 273, fig. 5.

[1727] De Tommaso 1990, 41.

[1728] Isings 1957, 26.

[1729] The analyses were carried out by Er. Mirtsou of the Archaeological Museum of Thessaloniki, using X-ray defraction (XRD).

[1730] According to Stern (1995, 92) the vessel represents the oblong 'Medjool' date.

[1731] Stern 1995, 92.

[1732] Isings 1957, 94, form 78d.

[1733] A large number of examples, which cannot be dated independently by archaeological results, are cited in Stern 1995, 93, note 186.

[1734] Stern 1995, 92, note 182, from graves securely dated to the middle of the 2nd century.

[1735] Goethert-Polaschek 1977, 96, no. 453, pl. 5, 55e, find from Trier (Treviri), from a grave dated between AD 41-54.

Switzerland, Italy,[1736] Croatia,[1737] FYROM,[1738] Crimea,[1739] Greece (Amphipolis[1740] and Crete),[1741] Israel,[1742] and Syria.[1743] This form appears in the middle of the 1st century and ceases to be produced in the early 2nd century.[1744] Because of their small size and their thick walls it appears that they survived for a relatively long period and thus are found in later graves dated to the 2nd and up to the 4th century.[1745] The finds from Thessaloniki cannot be independently dated by their archaeological context.

Over 60% of the finds published internationally were found in the Middle East, which was most probably their production region.[1746] Phoenicia was a well-known glass producing region with a very long tradition in glassworking already by the 1st century. This area, of course, was a most appropriate region for the production of a vessel that imitated the fruit of the palm tree, and which, in Greek, the current language of that time, gave its name to the entire region – Phoenicia.

As to their contents, apart from the general remark that it must had been a liquid due to the narrow neck, it has been assumed that it was probably palm tree oil, which was not only used for culinary purposes but in medicines as well.[1747]

### 118. Unguentaria, figural, grape-shaped (Isings Form 78e)

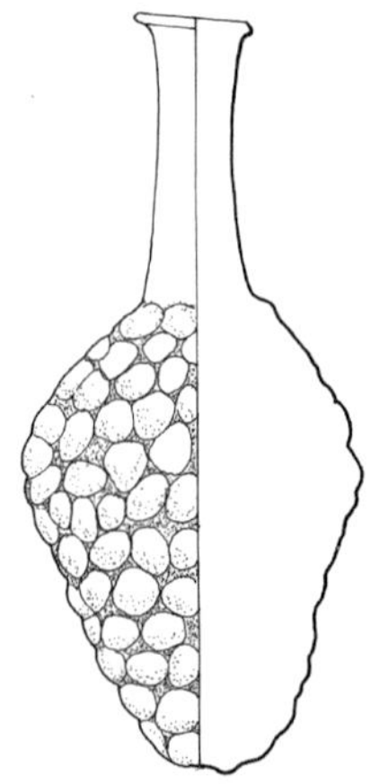

(Cat. no. 496)
(Plate XVII)

The vessels shaped in the form of grapes comprise another form of unguentaria. The rim is flaring, tubular infolded; the long cylindrical neck is almost the same height as the body, giving the characteristic shape from a tripartite assembly. The bottom is slightly concave at its centre. The example from Thessaloniki is made of opaque, blue glass. Their height is estimated between 16-17cm.

The body is mould-blown in a bipartite mould consisting of two vertical parts. The seam of the joint of the two parts is usually invisible, hidden along the sides of the vessel. The neck and the rim were shaped by free-hand tooling.

Vessels rendering wine grapes appear in three distinct periods of Roman times. The oldest examples[1748] reproduce the bunch with greater naturalism, unlike the later ones that render it more schematically. The former are known already from the third quarter of the 1st century in Greece[1749] and Italy,[1750] and from the early 2nd century in French Esvres.[1751] Examples of this form of unknown provenance are held in Western museums[1752] and they are dated on the basis of the aforementioned finds. The production site of the earlier examples is not known, but it is assumed that it might have been the Syro-Palestinian coast.[1753]

---

1736 Scatozza-Höricht 1986, no. 105 and Stern 1995, 93, note 175, 176, for examples from Pompeii and Ercolano (Herculaneum), dated prior to AD 79; Calvi 1968, 106, nos. 251-253, pl. 17/3, for an example dated to the 2nd half of the 2nd century from Aquileia.

1737 Raknić 1968, 214, pl. 2, from graves dated to the end of the 1st or to the early 2nd century; Damevski 1976, 65, 72, pl. IX/5, from Nin, with no indications of its dating.

1738 Mikulčić 1976, 194, 201, nos. 376-377, pl. 12, finds from Skopje (Scupi), dated from excavation results between the reign of Domitian (81-96) and Hadrian (117-138).

1739 Kunina 1997, 280-281, nos. 146-148, with older bibliography. Only no. 146 is dated from excavation results and was found in a grave dated to the 1st century.

1740 Romiopoulou 2002, 76, no. 23, photo on page 73, find from a grave with burials from the 1st century BC and 1st century AD. The same find appeared in *Glass Cosmos* 2010, 388, no. 459, from a grave dated to the second half of the 1st century.

1741 Knossos: Carington-Smith 1982, 280, nos. 65-67, fig. 6, pl. 39f, g, h, from a grave dated to the second half of the 1st century; Matala: Kanaki 2002, 104, fig. 1, dated on the basis of analogous finds to the 1st century.

1742 Stern 1995, 93, note 177, examples from graves dated to the second half of the 1st century; Israeli 2003b, 140, nos. 150-151.

1743 Clairmont 1963, 40, no. 150, pl. 21, example from Dura-Europos dated around AD 100.

1744 The timespan in which this form appears and becomes popular, more or less coincides with what is generally known to happen in glass history, i.e. that mould-blown figural vessels made of intensely coloured glass meet their greatest popularity in the second half of the 1st century, while in the early 2nd century they almost disappear as vessels made of colourless glass came into fashion, and on which it is extremely difficult to achieve a clearly visible/perceivable relief decoration. See Tatton-Brown 1991, 76.

1745 Stern 1995, 93, notes 184, 185, finds from Germany and Israel.

1746 Stern 1995, 92-93.

1747 Stern 1995, 94, notes 182, 183, with detailed references to the historical sources of that era.

1748 Isings 1957, 94, form 78e.

1749 Carington-Smith 1982, 280-281, no. 69, fig. 6, pl. 39j, a grave dated to the third quarter of the 1st century.

1750 Calvi 1968, 103, no. 195, which was found in Pompeii and is dated to the era before AD 79, and idem, 106 no. 254, pl. 2, for an object from Aquileia.

1751 Stern 1995, 208, note 42.

1752 Saldern *et al.* 1974, 174, no. 475; Hayes 1975, 49, no. 90; Stern 1995, 180, nos. 109-110.

1753 Stern 1995, 180.

The second subgroup comprises two-handled and handleless vessels resembling bunches of grape; they are a little larger than those of the 1st century. The two-handled vessels are present in the northwestern provinces of the Empire and they are considered to be local products there, while the handleless variant prevails in the East. Both forms of this subgroup are dated to the end of the 2nd century.[1754]

Finally, the third subgroup comprises vessels with body blown in a tripartite mould that consisted of two vertical parts and one discoid base. The neck and rim of the vessel were free-blown and finished by free-hand tooling. The body is modelled like a grape bunch. It is covered with continuous rows of hemispherical protuberances that are interconnected and interlocked alternately. At the transition from the body to the neck there is a constriction. This subgroup has been mainly found, and was probably made in the Syro-Palestinian region. None of the preserved examples is independently dated on the basis of its archaeological context.[1755] They are ascribed to the 3rd century based on stylistic features[1756] that have to do with the evolution of mould-blowing and their placement in this process.[1757] This form also appears (rarely) in a single-handled variant which was probably dated to the 4th century.[1758]

O example, although it a was a confiscated find[1759] and thus without excavation data, quite probably belongs to the earliest subgroup.[1760] As the grapes are rendered so naturalistically and that the object is made of opaque glass, which ceased to be used after the early 2nd century,[1761] an early dating appears quite justified.

Although grapes are of great symbolic importance for paganism, Judaism and Christianity,1762 it seems that the selection of its shape for the production of unguentaria must be ascribed to the general tendency in glassware, especially in the 1st century, a period of great experimentation in this trade, to render forms of the natural environment in the most naturalistic possible way.

**119. Unguentaria, figural, fish-shaped**

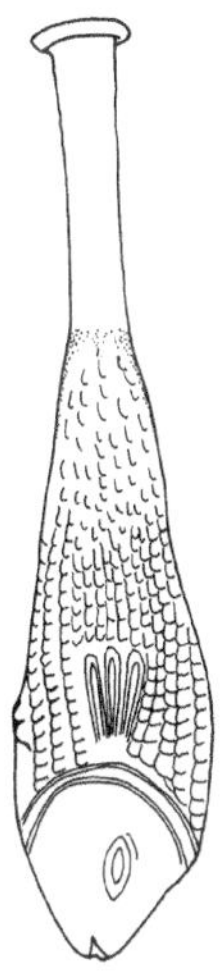

(Cat. no. 497)
(Plates 32, XVII)

Vessels approximately 25-30cm long, rendering fish naturalistically comprise this form. The head of the fish is actually the bottom of the vessel, the body of the fish is the vessel's body and the point where the tail of the fish would begin, a relatively long cylindrical neck starts, ending into a narrow flaring, tubular or band-like rim. The example from Thessaloniki is made of olive-green glass. Their height is estimated at c. 25cm.

They are made in a two-part mould. The seam at the joint of the two parts of the mould is usually visible along the belly and the back of the fish. The long cylindrical neck is free-blown. The rim was modelled after the vessel's removal from the mould and the cracking off of the moil, while it was held by jacks, quite probably from the base of the neck, as the constriction there indicates.

Mold-blown vessels naturalistically rendering the form of a fish are actually quite rare. Apart from the fragment from Thessaloniki a few more examples are known, all of them olive-green and almost identical. Two examples were found in excavations from the eastern Balkans, one at Ladi, near Didymoteichon (Greece)[1763] and the other from Romanian Istria.[1764] They are both dated securely, according to the other finds from the graves in which they were deposited, to the 1st century. The third example was found in Zadar (Iader), Croatia, and was dated generally between the 1st and 3rd centuries.[1765] Another example is in Hermitage, an unprovenanced artefact dated, for stylistic reasons, to the 2nd-3rd century.1766 The Metropolitan Museum in New

1754 Isings 1957, 108-109, form 91a; Stern 1995, 190-191, no. 119, with detailed bibliography.
1755 Stern 1995, 191. It should be noted that this particular find does not have a firm dating (see Bucovală 1968, 116), and was found in a context generally dated to the 2nd century.
1756 Stern 1995, 191-195, nos. 120-128.
1757 Stern 1995, 191.
1758 Stern 1995, 191.
1759 Confiscated in 1973 at Katerini, from two residents of the village of Lofos, in the Pieria region.
1760 Isings 1957, 94, form 78e.
1761 Tatton-Brown 1991, 76.
1762 It is well known that wine is often referred to in the Old and New Testament, and symbolizes the people of Israel. Thus, in Christian art it symbolizes those believers who live abiding by the Law of Christ, see THIE, Vol. 2, s.v. 'Ampelos', cols. 391-392. For Jews it also symbolizes the proof of God's promise to bring them back to Chanaan. It refers to the huge grape bunch that needed two men to carry it to Moses from the brook of Eshcol, and which, along with other fruits, gave proof of their arrival into the lands of Chanaan, see *Numbers*, 13:23 1-27. 5.
1763 Terzopoulou 1998, 21, fig. 11.
1764 Bucovala 1979, 95; Alexandrescu 1966, 219, XXV, 6, and page 513, pl. 81.
1765 Fadić 1997, 90, ascribed to a Syrian workshop, page 211, no. 223.
1766 Kunina 1997, 281, colour photo on page 37, no. 152, and photo 152.

York possesses yet another find,[1767] and another was bought for the Schimmel Collection in 1960.[1768] The Pilkington Museum in Lancaster (England) has a specimen said to have been found at Balıkesir (Adrianou Therai), Turkey, and dated to the 1st century.[1769] The item bought for the Museum of Natural History, Carnegie, is also dated to the the 1st century.1770 Another find is part of the Amatler Collection in Barcelona[1771] and the British Museum has a further fish-shaped vessel found at Arles (Arelate), France, assumed to be a local Gaulish product, dated to the 3rd century.[1772] The dating and attribution to a local workshop was probably based on the alleged similarity of the vessel to the fish-shaped vessels of Isings Form 95a.[1773] A closer and more careful observation shows that, although both forms render the form of a fish, they are completely different in their conception and execution. The 3rd-century vessels lack almost any sign of plausibility, bordering the caricature, in contrast to the naturalistic rendering of the 1st-century vessels. Furthermore, the later examples are free-blown, embellished with applied threads of glass, whereas the earlier ones are mould-blown in a bipartite mould.

The find from Thessaloniki lacks the archaeological data that would facilitate its independent dating. According to analogous finds without firm archaeological dating, it should be ascribed to production of the 1st century.

As regards the potential symbolism that these vessels might incorporate, any relation should be excluded from the corpus of, anyway, much later clay and bronze fish-shaped early-Christian vessels, where the fish bears its own Christian symbolism and connotations;[1774] rather they should be connected to the general tendency of that era for figural vessels rendering forms of deities, humans, animals, and fruit in the mos naturalistic ways possible.

**120a, b, c. Unguentaria, bird-shaped (Isings Form 11)**

a. Naturalistically rendered
(Cat. nos. 498-507)
(Plates 32-33, XVII)

b. With globular head and short beak
(Cat. nos. 508-515)
(Plates 33, XVII)

c. With schematized head
(Cat. nos. 516-537)
(Plates 33, XVII, XVIII)

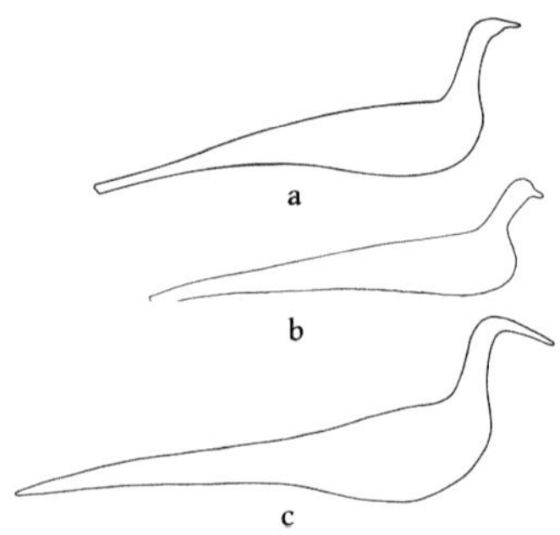


The vessels of this form render the shape of a bird, probably a dove with elongated tail. Their height is estimated between 15-20cm. On the basis of the shape of the bird's head and beak, three subgroups are discerned:

a) Those with globular head and very small beak.
b) Those with ovular head and long beak.
c) Those with head and beak rudimentary, or not modelled at all, but rendered by a simple bending of the tubular neck.

It is a well-known vessel type which mainly appear in Western Europe.[1775] A very large number of finds comes from the necropolises of the Swiss Ticino[1776] and Avenches (Aventicum), near Lausanne.[1777] There are also finds from Germany,[1778] Holland,[1779] northern Italy,[1780] Dalmatia,[1781] Pompeii and Herculaneum,[1782] Patras,[1783] Herakleion,

[1767] Eisen 1927, pl. 78 (middle and centre).
[1768] Recent Important Acquisitions, *J.G.S.* II, 1960, 139, no. 7.
[1769] Recent Important Acquisitions, *J.G.S.* XVI, 1974, 125, no. 4.
[1770] Oliver 1980, 71, no. 66.
[1771] Oliver 1980, 71.
[1772] Tatton-Brown 1991, 83, photo 103; Harden *et al.* 1968, 61, no. 77 (K. Painter).
[1773] Isings 1957, 112. Dated to the 3rd century and ascribed to the production of the NW provinces.
[1774] Galavaris 1970, 57-58; Dauterman-Maguire, Maguire, Duncan-Flowers (ed.) 1989, 22-23.
[1775] Isings 1957, 27, form 11. In detail on finds of this form from Thessaloniki, see Antonaras 2005, 241-248 and Antonaras 2010a, 27-33.
[1776] Biaggio-Simona 1991, Vol. 1, 125-129, where all archaeological indications are listed that date bird-shaped unguentaria between AD 10-30 and 125.
[1777] Biaggio-Simona 1991, Vol. 1, 125-129.
[1778] Fremersdorf 1961, 19, pl. 1, 2, four vessels from Cologne. Another vessel, this time from Krefeld-Gellep (Gelduba), is mentioned on page 19, dated to the 1st century according to analogous finds; Goethert-Polaschek 1977, 257, form 157, nos. 1527-1529, figs. 79-80, finds from Trier (Treviri) dated from excavation results to the middle of the 1st century.
[1779] Isings 1957, 27, from the necropolis of Nijmegen (Noviomagus Batavodurum) dated to c. 30-70.
[1780] Calvi 1968, pl. 18/2, find from Aquileia dated to the 1st century according to analogous finds. Maccabruni 1991, 112, examples dated to the 1st century from excavations in the region of Pavia (Ticinum).
[1781] Buljević 2002β, 395, nos. 12-18, finds from Solin (Salona).
[1782] Scatozza-Höricht 1986, 53-54, no. 106.
[1783] Kolonas 2002, 112-113, no. 8, dated, probably, by analogous finds.

Crete,[1784] and Monasteriaki Kephala, Knossos.[1785]

Bird-shaped vessels were produced in the region of the Swiss Ticino,[1786] as well as in Avenches (Aventicum), near Lausanne,[1787] and the period of their use ranges between AD 10-30 and 125.[1788] The dimensions of the finds from Thessaloniki, associated with subform C, and the very large numbers of them found in Thessaloniki, make it quite probable that they were also produced in Thessaloniki.

The production method remains uncertain. It is quite probable however that they were made by the heating and further shaping of a glass tube, using some kind of lamp-working technique. In addition, these vessels are special in that they present one of the earliest known containers intended to be used once only. The tip of the bird's tail remained open after the shaping of the vessel. It was only after the filling of the vessel with its contents that the tail was sealed by local reheating, and buyers had to break it to access the contents.

Some traces of their contents were preserved due to their closed shape. It appears that they contained generally the same substances s the plain spheres, the globular unguentaria Form 116. Chemical analyses show that[1789] they contained a mixture of $MnCO_3$ and $MnSO_3$, a blend of red ochre, gyps and quartz. It is considered that it was some kind of cosmetic. In one intact vessel a liquid is preserved that has been identified with a perfume, most likely rosewater,[1790] but this vessel was quite probably (re)sealed during the 19th or early 20th century.[1791] Chemical analyses of the powder contents of bird-shaped vessels from Thessaloniki showed that the blue powder is Egyptian blue, the red ochre and the pink madder.[1792]

The finds from Thessaloniki that have archaeological data can be dated by their contexts to the 1st century.[1793] There is no chronological difference noted among the three subforms,[1794] and examples of all of them have been found in the same, or in contemporaneous graves.

### 121. Unguentaria, globular, unworked rim, milky-white

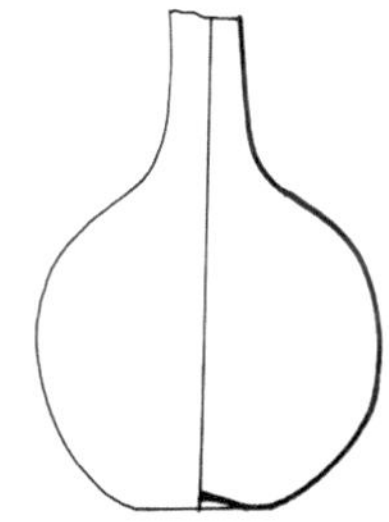

(Cat. no. 538)
(Plate 33, XVIII)

This form includes vessels with narrow cylindrical neck, globular body, flat, slightly concave bottom, made of opaque milky-white glass. Their height is estimated at c. 6cm.

They are very rare vessels with known examples in Israel[1795] and in the region of Ticino.[1796]

As to the modelling technique, these vessels are considered to be products of the earliest blowing technique, i.e. without the intervention of a blowpipe, but rather by blowing directly into a glass tube. By locally heating up one end of it the tube was sealed, inflated and shaped into a small-sized globular or oval vessel.[1797] The vessel from Thessaloniki is the only one among the extant examples that has the neck preserved; it is cylindrical with an unworked end and with no trace of a rim. If the hypothesis about the modelling technique is correct, then the neck is actually part of the glass tube from which the vessel was shaped, and thus probably preserves its original diameter and thickness.

[1784] Alexiou 1966, 408, pl. 437δ, dated from excavation results to late 1st- early 2nd century.

[1785] Carington-Smith 1982, 272, figs. 4/12, 13, 14, 16, pl. 36c, d, e, finds dated from excavation results between the first quarter of the 1st century and up to c. AD 70.

[1786] Biaggio-Simona 1991, Vol. 1, 125-129.

[1787] Morel *et al.* 1992, 9-10. This workshop was operating according to archaeological indications between AD 40-70.

[1788] Biaggio-Simona 1991, Vol. 1, 125-129, where all archaeological indications dating them are listed.

[1789] Isings 1957, 26; Carington-Smith 1982, 293; Pérez-Arantegui, Paz-Peralta and Ortiz-Palomar 1996.

[1790] A vessel found in Rovasenza, Italy, and now in the Museo di Antichità in Turin (Augusta Iulia Taurinorum). Carducci, 1968, 76, agrees with Whitehouse 1997, 121.

[1791] Personal observation during an on-site inspection of the vessel.

[1792] The analyses were carried out by Er. Mirtsou of the Archaeological Museum of Thessaloniki using X-ray diffraction (XRD).

[1793] Cat. nos. 499, 508-11, 514, 518-523, 527-530, 536-537, were found together with spheres (Form 116), or with other unguentaria also dated to the 1st century. On more finds from the eastern necropolis, also dated to the 1st century, see Tsimpidou-Avloniti 2003, 238 and Tsimpidou-Avloniti, Kaïafa 2004, 273, fig. 5.

[1794] Bird-shaped vessels, and especially those similar to the third subgroup, were produced in the western provinces of the Empire even much later. They are dated by the results of excavations to the end of the 3rd century and to the first half of the 4th century, see Arveiller-Dulong, Arveiller 1985, 141-142, nos. 306-308, photos 306-308, with older bibliography.

[1795] Two ovular vessels made of opaque white glass are kept in the Israel Museum in Jerusalem (Israeli 2003b, 113, no. 94-95), and dated on the basis of manufacturing technique to the 1st century BC to the 1st century AD.

[1796] Biaggio-Simona 1991, 129-130, pl. 15, fig. 56, two identical vessels made of white glass, dated from excavation results to the 1st-2nd century.

[1797] Grose 1984, 33-34, fig. 13.

They are considered1798 to be the Eastern counterpart of spheres (Form 116) of Western production, and that they were made for a special use or contents. It must be noted, however, that an example was found in the Ticino region where spheres were very popular and widespread.

The find from Thessaloniki, due to its scant excavation data, cannot be dated independently; only on the basis of the date of analogous finds from other sites it is placed in the period between the 1st century BC and 1st century AD.

**122. Unguentaria, globular, fire-polished, flaring rim (Isings Form 6)**

(Cat. nos. 539-541)

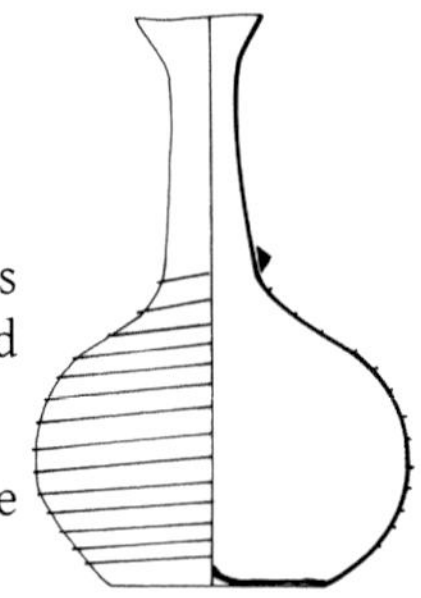

122a. Decorated by a wound thread
(Plates 34, XVIII)

Vessels of this form have small, flaring, fire-polished rim, cylindrical neck widening towards its base. The globular body stands on the flat, slightly concave bottom. The body is often decorated by a thin thread of glass of a different colour. Their height is estimated between 10-12cm.

The examples from Thessaloniki are made of translucent blue and amber glass and they are decorated with an opaque white thread.

They are free-blown. While they were still attached to the pipe, and before the final shaping of the bottom, the thread was spirally wound around the neck, body and the entire bottom. The rim was shaped after the detachment of the blowpipe and the cracking-off of the moil by local reheating of that area and probably also rotation. This procedure leaves no marks of use of a second blowpipe on the bottom or jacks that were holding it from the neck.

Vessels of this form were made both in the West and in the East, with small differentiations in the decoration and the shape.[1799] The vessels from Thessaloniki are probably Western products, due to the shaping of the rim.[1800]

Examples have been found in excavations both in the West and East, e.g. Rome,[1801] Italy,[1802] Kassel (Castellum Cattorum),[1803] and Israel.[1804] In Greece they have been found at Samothrace,[1805] Macedonia,[1806] Athens,[1807] Oropos,[1808] Mytilene,[1809] Mamousia Aigialeia,[1810] Pylos,[1811] Patras,[1812] and Sfakaki (Pangalochorion).[1813] Furthermore, examples of this form are held in museums and private collections, i.e. Corning Museum of Glass,[1814] Yale University Art Gallery,[1815] Newark Museum,[1816] Römisch-Germanischuseums, Cologne;[1817] the Oppenländer[1818] and Wolf[1819] Collections, and the and Fondation Custodia.[1820]

The vessels are dated between the 1st century BC and 1st century AD, mainly to the second half of the 1st century BC and the first half of the 1st century AD, and to this period the Thessalonean finds should be ascribed.

[1798] Israeli 2003b, 113, no. 94.
[1799] Stern 2001, 43, with detailed bibliography.
[1800] Cf. Stern 2001, 58, with many analogous finds.
[1801] Grose 1977, 20-23, no. 9, fig. 6, from a layer dated to the late 1st century BC.
[1802] De Tomasso 1990, 39-40, form 5. The form is dated to the Claudian era (41-54) and several examples from Italy and elsewhere are listed.
[1803] Spartz 1967, no. 8.
[1804] From a grave in Ein-Gedi, dated from excavation results at c. 40 BC, see Stern 2001, 43, 64, where the main bibliography on the matter is found.
[1805] Dusenbery 1998, 1094-96, eight examples from three different graves dated to the last quarter of the 1st century BC.
[1806] Weinberg, McClellan 1992, 112-113, no. 78, dated on the basis of analogous finds to the late 1st century.
[1807] Weinberg, McClellan 1992, 117, no. 83.
[1808] Pologiorgi 1998, 94, nos. 10, 11, vessels of this form dated to the Augustan period (27-14 BC), on the dating see page 102.
[1809] Chatzi 1972, 593, pl. 542ζ, vessels from Roman graves. Chatzi 1973, 509, pl. 476, two vessels from Roman graves.
[1810] Dekoulakou 1975d, 120, pl. 69γ, find without specific chronological clues.
[1811] Kaltsas 1983, 24-25, pl. 32δ, decorated example from a grave dated to the 1st century.
[1812] Dekoulakou 1975c, 115-117, pl. 66γ, finds without specific chronological clues; Kolonas 2002, 122, 124, nos. 35, 43, dated on the basis of analogous finds to the 1st-2nd century.
[1813] Gavriilaki 1989, 457-460, pl. 253β, γ, 254α-δ, find from a Roman cemetery dated prior to the 2nd-4th century period.
[1814] Whitehouse 2001, 169-170, nos. 700-701.
[1815] Matheson 1980, 27-28, no. 74, esp. pages 25-26, no. 66.
[1816] Auth 1976, 90, no. 95.
[1817] La Baume 1973, no. D18.
[1818] Saldern *et al.* 1974, 221-222, nos. 645, 646, 648.
[1819] Stern 2001, 43-45, 57, 64-66, nos. 1, 8-9.
[1820] Stern 1977, 30-31, no. 5.

122b. Undecorated

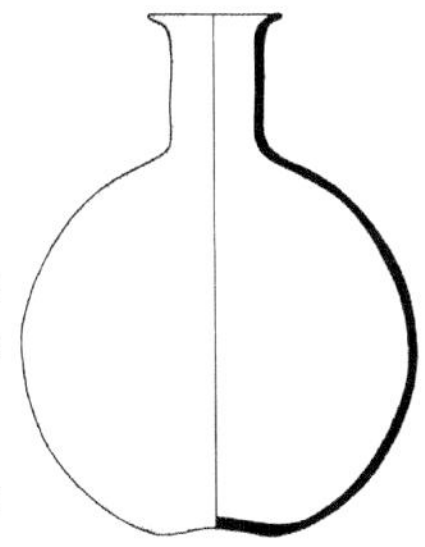

(Cat. nos. 542-550)
(Plates 34, XVIII, XIX)

Vessels of this form have simple, fire-polished, inverted rim, a cylindrical neck of the same height as the body or even shorter. The body is globular, standing on a flat, slightly concave bottom. Their height is estimated at c. 8cm.

The vessels of this form are free-blown and usually the shaping of the rim was carried out while they were held by jacks, probably from the base of the neck.

Exclusively vessels with simple fire-polished rim appear among the finds from Thessaloniki, which is considered an indication of their Western origin.[1821]

The examples from Thessaloniki are made of colourless, bluish, dark-blue and red glass.

It is a very widespread form during the 1st century, mainly in Italy, where they were quite probably produced, and in the Western provinces in general.[1822] In Greece examples have been published from Samothrace,[1823] Amphipolis,[1824] Ierissos (Acanthos),[1825] Gomati (Chalkidiki),[1826] Edessa,[1827] Kato Vatheia (Eubeia),[1828] Athens,[1829] Eleusis,[1830] Megara,[1831] Argos,[1832] Patras,[1833] Samos,[1834] Mytilene,[1835] Thera,[1836] and Marmaroto, (Kos).[1837]

The examples, Cat. nos. 542-543,[1838] 546-547,[1839] from Thessaloniki, which can be independently dated, are dated to the 1st century.

**123. Unguentaria, squat, spherical, with flaring rim**

(Cat. nos. 551-552)
(Plates 34, XIX)

Vessels of this form have fire-polished, slightly inverted rim, cylindrical neck the same height or even taller than the body. The body is squat globular, standing on a flat bottom. Their height is estimated between 5-7cm.

The vessels of this form are free-blown and the shaping of the rim was carried out while they were held by jacks, probably from the base of the neck.

Among the finds from Thessaloniki exclusively vessels with simple, fire-polished rim appear, which is considered to be an indication of Western origin.[1840]

The examples from Thessaloniki are made of opaque, blue and purple glass.

1821 Stern 2001, 44.

1822 Isings 1957, 22-23, form 6; Stern 1977, 47 ff, no. 11 pl. 2, with all bibliography known at the time; De Tommaso 1990, 39-40, 42-43, forms 4, 5, 7, with finds from various sites in Italy, dated from excavation results to the 1st century.

1823 Dusenbery 1998, no. S153-A, S157-4-10 (examples dated from excavation results to the fourth quarter of the 1st century BC), S161-3, 4, S167-11, S168-1, S171-2, S201-10, 11 (examples dated from excavation results to the end of the 1st century BC or in the early 1st century) and S206-4, S207-11, 12, S 210-1, S220-4, S223-1 (examples dated from excavation results to the first quarter of the 1st century).

1824 Romiopoulou 2002, 76, photo on page 73, finds from a grave with burials ranging from the 1st century BC to the 1st century AD.

1825 Trakosopoulou 2002, 84-85, figs. 12-15, examples from graves dated by coins to the second half of the 1st century BC and to the 1st century AD.

1826 Trakosopoulou-Salakidou 2000, 344, photo 9, find from a burial dated by a coin to the 1st century.

1827 Chrysostomou A. 2000, 228, pl. 234, vessel from a burial dated by its context to the 1st century.

1828 Themelis 1969, 204, pl. 213, vessels from one grave, handed over to the Archaeological Service by the finder.

1829 Alexandri 1973-74c, 128-131, pl. 102, from a Roman cemetery; Zachariadou 2003, 16.

1830 Mylonas 1975, 130-131, pl. 195, Γ29-209, 210, from a burial dated to the second half of the 1st century.

1831 Zoridis 1981, 39, pl. 20, finds from a cist grave which contained among others tear-shaped glass unguentaria.

1832 Alexandri 1963, 61-62, pl. 78α, β, vessels from Roman graves; Papachristodoulou 1968, 130, pl. 73δ, five vessels from a Roman grave, which, judging by the illustrated clay vessels found in it, is dated to the 1st century.

1833 Agallopoulou 1973-1974, 369, pls. 235-236, four vessels from two graves, one dated according to its glass contents to the first half of the 1st century and the other, also according to its glass contents, to the 1st century; Papapostolou 1983, 27 no. 23, fig. 10, pl. 21α, find dated from excavation results to the first quarter of the 1st century.

1834 Tsakos 1977, grave of Vakentis V, 377, 415, pl. 138ε, dated to the 1st century.

1835 Chatzi 1973, 509, pl. 476, vessels from Roman graves.

1836 Dragendorff 1903, 286-288, fig. 481i, from graves dated from the 1st to the 4th centuries. The glass finds from these graves, at least those illustrated, as a whole are datable to production of the 1st and 2nd centuries.

1837 Touchais, Huber, Philippa-Touchais 2001, 1002, fig. 249, from a grave with other glass finds dated to the 1st or early 2nd century.

1838 From the same burial with Cat. nos. 551, 597, dated to the 1st century.

1839 From the same burial with Cat. nos. 480-484, 552, 527-528, 624, dated to the 1st century.

1840 Stern 2001, 44.

This form does not seem to have met with great popularity and analogous finds were found only in Ercolano (Herculaneum),[1841] Aquileia,[1842] Magnesia,[1843] Athens,[1844] and Agios Nikolaos, Crete.[1845]

On the basis of these parallels, and also according to the glass vessels found with them,[1846] the finds from Thessaloniki are dated to the 1st century.

### 124. Unguentaria, bulbous (Isings Form 16)

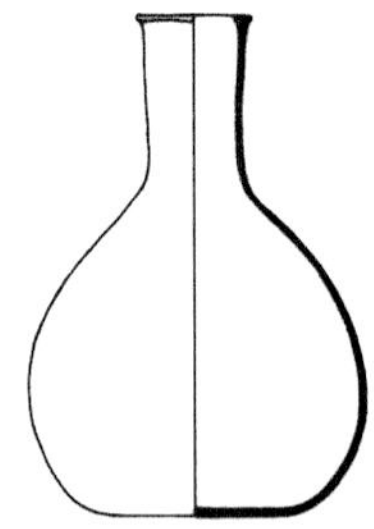

(Cat. no. 553)
(Plates 34, XIX)

Vessels of this form have a characteristic rim, infolded and flattened, a wide cylindrical neck, bulbous body, and a flat, slightly concave bottom. Their height is estimated at c. 9cm.

The example from Thessaloniki is made of thick bluish glass, just like most of the published parallels.

It is free-blown and the shaping of the neck was carried out while the vessel was held by jacks from the base of the neck, where a characteristic constriction is formed by the pressure of the tool. This form is quite widely known, both in the West and East.

It is dated between the second quarter and the end of the 1st century[1847] or in the third quarter of the 1st century.[1848] It is probable that the Eastern products bear a characteristic constriction on the base of the neck, while the Western artefacts, like the example from Thessaloniki, have a smooth and straight base to the neck.[1849] In Greece some analogous vessels have been found at Samothrace,[1850] Sifnos,[1851] Knossos,[1852] Patras,[1853] Athens[1854] and Oropos.[1855] The example from Thessaloniki lacks any special archaeological information that would allow us to date it more precisely.

### 125. Unguentaria, ovular (De Tommaso Form 21)

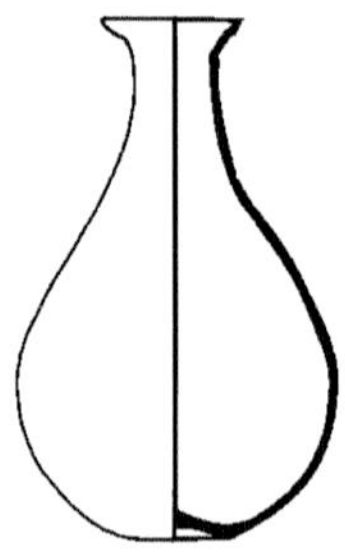

(Cat. nos. 554-558)
(Plates 34, XIX)

Vessels of this form have fire-polished flaring rim, short neck, ovular body and flat bottom. Their height is estimated between 5-7cm.

These vessels were made by free-blowing and the rim was shaped and polished while it was held by jacks, probably from the neck.

The examples from Thessaloniki are made of translucent amber and dark-blue glass.

It is a widely known form of unguentaria which appears from the early 1st century and up to the end of the same century.[1856]

---

1841 Scatozza-Höricht 1986, 63, pl. XX. 127, XXXV; De Tommaso 1990, 38, form 2, find also made of opaque glass, dated prior to AD 79.

1842 Calvi 1968, 33, 44, pl. A, 14F, similar vessels dated to the 1st century.

1843 Hourmouziadis, Asimakopoulou-Atzaka, Makris 1982, 62-63, fig. 33, without references to more specific archaeological or chronological indications.

1844 Alexandri 1972, 115-118, pl. 76, three vessels made of marble-like blue and white glass, from Roman graves, which, judging by the rest of the glass finds, are dated to the 1st century; Zachariadou 2003, 16.

1845 Davaras 1985, 192-193, fig. 23A, one find from a grave dated by a 1st-century coin and a second (idem, 194-195, fig. 25β), from a grave with coins of the reign of Caligula (37-41), dated to the second half of the 1st century.

1846 Cat. no. 551 found together with vessels form 122 and 130. Cat. no. 552 was found in a grave along with vessels of forms 116, 122, 123, 46 and 120c.

1847 Isings 1957, 34-35, form 16, with dated examples from United Kingdom, Germany, Holland, Switzerland, Italy, and Dura-Europos.

1848 Price, Cottam 1998a, 171-172, with exhaustive bibliography on British examples supporting such a tight time-frame.

1849 Stern 1977, 40-41, no. 8, with exhaustive bibliography on finds from the entire Empire, with examples dated between the second quarter and the end of the 1st century. Detailed presentation of Italian examples in De Tommaso 1990, 49-50, form 18.

1850 Dusenbery 1998, 1110-1111, nos. S 207-13, dated from excavation results to the first quarter of the 1st century.

1851 Mackworth-Young 1949, 88, pls. 32, 3, 14(2), dated from excavation results to the fourth quarter of the 1st century.

1852 Carington-Smith 1982, 274, nos. 23-24, fig. 4, pl. 37e-f, from a grave dated to the second half of the 1st century.

1853 Papapostolou 1983, 25, nos. 21α-στ, pls. 18α-δ, 19α-β, fig. 8 and probably also nos. 22α-στ, from a mausoleum dated to the first quarter of the 1st century; Agallopoulou 1973-74, 369, pl. 236γ, from a sarcophagus dated to the 1st century.

1854 Alexandri 1972, 115, pl. 76β, from a grave dated to the 1st century according to the illustrated glass context.

1855 Pologiorgi 1998, 116, no. 7 (ΜΩ 321-323), pl. 54γ left, (ΜΩ 321-323), finds from a burial dated to the second half of the 1st century.

1856 Isings 1957, 22-23, form 6, with numerous dated examples of the 1st century from the western, mainly, provinces.

Vessels with simple, fire-polished rim appear exclusively among the finds from Thessaloniki, which is an indication of a Western production site.[1857] The form was thoroughly studied by De Tommaso,[1858] who assumes that they were produced in northeastern Italy on the basis of the distribution of the finds. In addition, analogous finds are reported from Bakar/Buccari (Volcera), Croatia,[1859] and Zemaljski Muzej, Sarajevo.[1860] In Greece analogous finds are published from Edessa,[1861] Galaxidi (Chaleion),[1862] Athens,[1863] Corinth,[1864] Patras[1865] and Samos.[1866] The finds from Thessaloniki cannot be independently dated.

### 126a. Unguentaria, tear-shaped, with fire-polished rim (Isings Form 8)

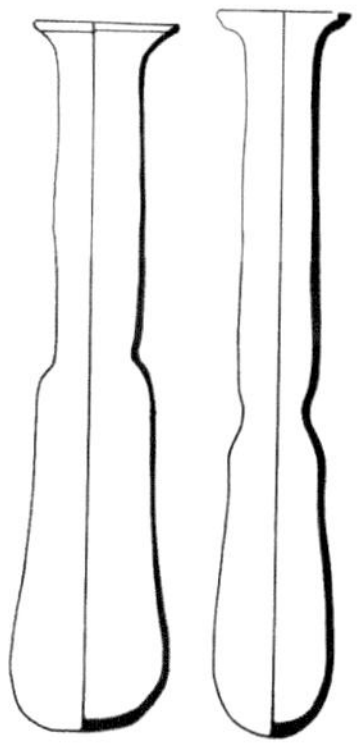

(Cat. nos. 559-571)
(Plates 35, XIX)

### 126b. Unguentaria, tear-shaped, with tubular rim (Isings Form 8)

(Cat. nos. 572-578)
(Plates 35, XIX)

Vessels of this form have inverted rim, either infolded tubular or fire-polished. There is a long, cylindrical neck which is often constricted at its base. The body is not much wider than the neck and the bottom is either left convex, giving a tear- or drop-shaped form to the vessel, or it has been flattened, rendering the body almost conical. Their height is estimated between 6-12cm.

The vessels of this form are free-blown and usually the shaping of the rim is carried out while they are held by jacks, probably from the base of the neck. In one of the preserved examples the rim was squeezed while it was still hot and was rendered almost trefoil in shape.

These unguentaria present one of the most widely spread and numerous groups of 1st-century glass vessels. They appear from the beginning of the century, their distribution culminates around the middle of it and they cease to circulate in the early 2nd century.[1867] Sites with dated examples include Britain,[1868] France,[1869] Germany,[1870] Switzerland,[1871] Italy,[1872] Slovenia,[1873] Hungary,[1874] Croatia,[1875] Serbia,[1876] Bulgaria,[1877] and the Syro-Palestinian

[1857] Stern 2001, 44.

[1858] De Tommaso 1990, 52, form 21, where finds from various Italian sites are cited, dated from excavation results to the first third of the 1st century.

[1859] *Trasparenze imperiali 1997*, 102 no. 27, find dated from excavation results to the 1st century.

[1860] Paškvalin 1976, 111, no. 22, pl. IV, 2, unprovenanced vessel.

[1861] Chrysostomou A. 2000, 217, pl. 219, vessel from a burial dated by the glass finds to the 1st century.

[1862] Zafeiropoulou 1975, pl. 5, no. 8, find from an excavation with no clear chronological indications.

[1863] Alexandri 1972, 115-118, pl. 76δ, vessel from a group of Roman graves, which, judging by the rest of the glass finds, is dated to the 1st century.

[1864] Davidson 1952, 104, no. 666, fig. 9, pl. 55, dated to the 2nd century.

[1865] Agallopoulou 1973-1974, 369, pl. 235, vessel from a grave dated by the glass finds to the 1st century.

[1866] Tsakos 1977, fig. V, no. 16, pl. 138.

[1867] Isings 1957, 24, form 8.

[1868] Price, Cottam 1998a, 169-170, with exhaustive bibliography of dated examples from Britain. The examples with fire-polished rim are dated between AD 43-75/80. On the contrary, the examples with ring-shaped rim appear in the third quarter of the 1st century and probably remained in use during the late 2nd- early 3rd century.

[1869] Arveiller-Dulong, Arveiller 1985, 50-52, nos. 54-67, finds from Strasbourg (Augusta Trebecorum, Argentoratum) dated from excavation results to the 1st century. The work referenced includes a detailed bibliography of analogous finds from the West and East.

[1870] They appear in many sites between the first third of the 1st century and to the 2nd century. See Goethert-Polaschek 1977, 91-95, 97-104, forms 66b, 69a. Harter 1999, 92-94, forms D8-10, where analogous finds from other German sites are listed.

[1871] They appear already in the Augustan era (27-14 BC) and they become popular in the period between the reigns of Tiberius (14-37) and Hadrian (117-138). See Rütti 1991, 114-116, forms AR 128. 1, 129, 130. Biaggio-Simona 1991, 142, pl. 6.

[1872] They appear in great numbers in many sites dated between the Tiberian (14-37) and Hadrianic (117-138) eras. See De Tommaso 1990, 82-83, forms 71-72, with a detailed bibliography on dated finds from Italy.

[1873] Petru 1972, pls. VII, XII-XIV ; Petru 1976, pl. 2/9; Plesničar-Gec 1972, pl. CXXXV/18, finds dated to the 1st and early 2nd century; Lazar 2003, 175-179, forms 8. 6. 1, and 8. 6. 2, dated to the 1st- early 2nd century and 2nd-3rd century, respectively, with examples from various Slovenian sites.

[1874] Barkóczi 1988, 111-112, forms 83, 84, examples appear at various sites, at least in the western part of ancient Pannonia, and are dated to the 1st - early 2nd century.

[1875] *Trasparenze imperiali* 1997, 99-101, finds from various sites dated to the 1st- early 2nd century.

[1876] Ružić 1994, form III/4a, 23-25, this form is dated to the 1st-2nd century, although examples are mentioned from sites dated to the 3rd century; Zotović, Jordović 1990, 93, pl. CXXXVII: 6, find from a grave in Kostolac (Viminacium) with a coin of Antoninus Pius (138-161).

[1877] Bujukliev 1986, 95, pl. 27, finds from a burial dated by coins to the second half of the 1st century.

region.[1878] In Greece examples have been reported from excavations at Sifnos,[1879] Knossos, Samothrace,[1880] Kozani,[1881] Gomati (Chalkidiki),[1882] Ierissos (Acanthos),[1883] Veroia,[1884] Spilia (Eordaia),[1885] Larisa,[1886] Magnesia,[1887] Nea Anchialos (Phthiotides Thebes),[1888] Nicopolis,[1889] Galaxidi (Chaleion),[1890] Athens,[1891] Eleusis,[1892] Megara,[1893] Argos,[1894] Naxos,[1895] Herakleion,[1896] Agios Nikolaos,[1897] Marmaroto, Kos,[1898] Aigion,[1899] Patras,[1900] Corinth,[1901] Thera,[1902] Oropos,[1903] and Edessa.[1904]

Among the finds from Thessaloniki both shapes of rims and bottoms appear. Although they are discovered in the same groups of finds the tubular rim is considered indicative of an Oriental provenance and the fire-polished rim of a western provenance.[1905]

The examples from Thessaloniki are all made of colourless, bluish glass, except for an amber find, and they are dated to the 1st century. Only Cat. nos. 577-578 can be independently dated.[1906]

### 127. Unguentaria, tubular (Isings Form 27)

(Cat. nos. 579-580)
(Plates 35, XIX)

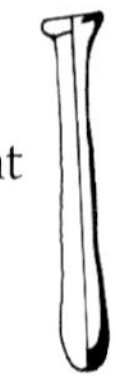

Vessels of this form have a tubular, flaring rim and a cylindrical body ending in a convex bottom. Their height is estimated at c. 6cm.

They are free-blown and the rim was shaped while the vessel was held by jacks, probably, from the neck.

---

1878 Stern 1977, 35-38, no. 7, with detailed bibliography on finds from the Eastern Mediterranean region, according to which the production of the form is located between the first and third quarter of the 1st century; Dussart 1998, 168, form BXIII.1321, pl. 54/10-13, with examples from various sites dated from the 1st century BC to the 3rd century AD, with a higher concentration in the 1st century; Israeli 2003b, 208-209, with analogous examples dated mainly to the 1st century.

1879 Mackworth-Young 1949, 80-81, grave 6 (pl. 29. 6. 6) dated to the 1st century, grave 14 (pl. 32, 6. 14 (3)) with coins of Vespasianus (69-79).

1880 Dusenbery 1998, 1108, nos. S205-10, 14, examples dated from excavation results to the first quarter of the 1st century.

1881 Karamitrou-Mentesidou 1993, 20, fig. 2, illustrated in the middle of the picture, finds from the hill of Agios Athanasios, with no further information about the excavation.

1882 Trakosopoulou-Salakidou 2000, 344, photo 9, find from a burial dated by coin to the 1st century.

1883 Trakosopoulou 2002, 84-85, figs. 11, 12, 15, examples from graves dated precisely to the second half of the 1st century BC and up to the second half of the 1st century AD.

1884 Petsas 1964, 353, pl. 414δ, find from the fill of a Roman cemetery.

1885 Karamitrou-Mentesidou 1985, 274, pl. 131ε, find which according to the glass vessel it was found with is dated to the 1st-2nd century.

1886 Tziafalias 1988, 260-263, pl. 138β, vessel from a destroyed Roman grave.

1887 Hourmouziadis, Asimakopoulou-Atzaka, Makris 1982, 62-63, fig. 33, with no reference to more precise chronological indications.

1888 Sotiriou 1939, 64, fig. 26, examples from graves with no precise chronological indications.

1889 Chrysostomou 1984, 35-36, four vessels from two graves: one is dated by a coin of Augustus (27-14 BC) and the other to the last quarter of the 1st century.

1890 Threpsiadis 1972, 199, pl. 76β (β, δ, ε), finds from Roman graves; Zafeiropoulou 1975, pl. 5, nos. 10-13, finds from excavations with no clear chronological indications.

1891 Alexandri 1967a, 97, pl. 90ζ, two vessels from a Roman grave; Alexandri 1973-74c, 128-131, pl. 102, from a Roman cemetery.

1892 Mylonas 1975, 131, pl. 195 (I30-822), from a grave dated to the 1st century; page 13 (A12-9), from a grave dated to the end of the 1st century; pages 130-131 (Γ29-207), from a burial dated to the second half of the 1st century; page 91 (Θ-749α), find dated to the 1st century according to analogous finds.

1893 Zoridis 1981, 39, pl. 20, finds from a cist grave which contained globular glass unguentaria Form 122.

1894 Alexandri 1963, 61-62, pl. 78α, β, vessels from Roman graves; Bruneau 1970, 490-493, fig. 147; Morou 1981, 109, pl. 49γ, 50 vessels from a Roman grave with no further chronological indications.

1895 Zafeiropoulou 1977, 308-311, pl. 180γ, find from Roman graves.

1896 Sakellarakis 1965, 562, pl. 709, finds from a grave which was in use, according to the numismatic finds, between the middle of the 1st century and the 2nd.

1897 Davaras 1985, 166, pl. 48η, from a grave dated by its contents to the second half of the 1st century; page 168, fig. 18, from a grave dated to the 1st century, according to the other objects found with the glass vessels; page 171, fig. 19, from a grave with a coin dated to AD 37/38.

1898 Touchais, Huber, Philippa-Touchais 2001, 1002, fig. 249, from a grave with glass co-finds of the 1st or the early 2nd century.

1899 Petsas 1971, 177, pl. 164, with 16 vessels from four burials, with no chronological indications.

1900 Agallopoulou 1973-1974, 369, pl. 236, three vessels from a grave dated according to the glass finds to the first half of the 1st century; Papapostolou 1983, 27, no. 24. fig. 11, pl. 21γ, find dated from excavation results to the first quarter of the 1st century.

1901 Davidson 1952, 105, nos. 666, 671, fig. 11, finds dated from excavation results to the 2nd and 3rd century.

1902 Dragendorff 1903, 286-288, fig. 481 a, b, from graves dated from the 1st to the 4th centuries. The glass finds of the graves, at least those illustrated, as a whole are ascribed to production of the 1st and 2nd centuries.

1903 Pologiorgi 1998, 115, nos. 3-4 (ΜΩ 324-325), pl. 54α, β, finds from a burial dated to the second half of the 1st century; page 43-44 (ΜΩ 257) pl. 13α, (ΜΩ 207) pl. 16γ from a burial dated according to the glass finds to the 1st century.

1904 Chrysostomou A. 2000, 216-217, pl. 219, vessels from a burial dated according to its glass finds to the 1st century. A similar one is presented in pl. 255, with no chronological indications. Chrysostomou A. 2000, 283, pl. 321, vessels from a burial dated (page 279) by its content to the second half of the 1st century.

1905 Barag 1970c, 198, according to Stern 1977, 37-38 note 3.

1906 From the same burial with Cat. nos. 461-462, 460, 568-569, 595, 606-607, Forms 112, 113, 126, 129, 131, which are also ascribed to the same time-frame of the 1st or early 2nd century.

The finds from Thessaloniki are made of colourless, greenish glass.

This form is known both in the West and East.[1907] Dated examples are known from the 1st to late 4th century. Sites with well-dated finds include Strasbourg (Augusta Trebecorum, Argentoratum),[1908] Cologne,[1909] Heddernheim (Nida),[1910] Zülpich-Enzen,[1911] Mainz (Moguntiacum),[1912] Augst,[1913] Valkenburg,[1914] Ticino,[1915] several sites in Italy, (Angera (Angleria), Pavia(Ticinum), Aquileia, Ravenna, Bologna),[1916] Serbia: Kostolac (Viminacium),[1917] Costanza (Tomis),[1918] Sardis,[1919] Dura-Europos,[1920] El-Jish (Jiscala),[1921] and Kom Aushim (Karanis).[1922] In Greece, published finds are known from Kozani,[1923] Delphi,[1924] and Knossos.[1925]

One of the examples under research, Cat. no. 580, was found in an Early Christian basilica below the castle at Rentina, making it quite probably date around the 4th century onwards, while the second example, Cat. no. 579, cannot be independently dated.[1926]

**128. Unguentaria, conical body, straight walls, no shoulder, short neck, fire-polished rim**

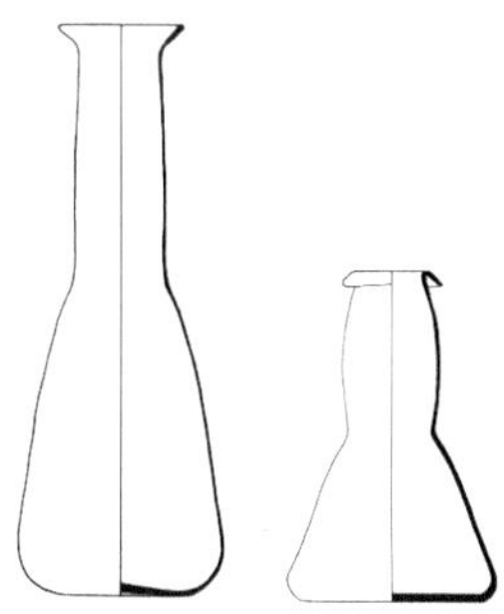

(Cat. nos. 581-585)
(Plates 35, XIX)

**129. Unguentaria, conical body, straight walls, no shoulder, short neck, tubular rim**

(Cat. nos. 586-596)
(Plates 35-36, XIX, XX)

Vessels of this form have everted rim, either tubular or fire-polished, and a wide, cylindrical neck equal in height to the body, often bearing a constriction at its base. The body is conical standing on its flat bottom. Their height is estimated between 7-13cm.

The vessels of this form are free-blown and usually the shaping of the rim was carried out while it was held by jacks, probably from the base of the neck.

Mainly vessels with tubular rim appear among the finds from Thessaloniki, which is considered an indication of Oriental/Eastern origin, while the simpler, fire-polished rim indicates a western product.[1927]

The examples from Thessaloniki are made of colourless, bluish glass, except for the one example with fire-polished rim made of opaque turquoise glass.

This form is dated to the 1st century and it is considered that it predates the vessels with very long neck.[1928] Finds

[1907] Isings 1957, 41, form 27, which appears in the reign or Nero (54-68) and survives until the 5th century. Isings considers it as an evolution of form no. 8.
[1908] Arveiller-Dulong, Arveiller 1985, 63-64, no. 91, find from Strasbourg (Augusta Trebecorum, Argentoratum) dated to the early 3rd or 4th century, according to the archaeological context in which it was found.
[1909] Fremersdorf 1958b, 44 pl. 91 (right), find dated to the 2nd century according to the archaeological contexts in which they were found.
[1910] Welker 1974, nos. 115-117, pl. 7, finds dated from the Flavian era (69-96) to the 2nd century.
[1911] Follmann-Schulz 1989a, 55, no. 15 pl. 2, find dated to the third quarter of the 4th century.
[1912] Harter 1999, 93, form D10a with flaring fire-polished rim from a burial dated to the 1st century; form D10b with tubular rim finds from sarcophagi dated to the first half of the 4th century and to the late 3rd or into the early 4th century.
[1913] Rütti 1991, Vol. I, 53; Vol. II AR138, 120-121, pl. 102, nos. 2381-2399, a great number of finds with tubular rim and dated from excavation results between the Claudian era (41-54) and the 4th century.
[1914] Van Lith 1978/79, 55-56, no. 197, pl. 9, finds from the early years of Claudius' reign (41-54).
[1915] Biaggio-Simona 1991, Vol. I, 143, dated by their archaeological context to the 4th-5th century.
[1916] De Tommaso 1990, 80-81, form 65, who considers these vessels as Oriental imports.
[1917] Spasić-Djurić 2006, 303, fig. 4/2, find from a grave dated from excavation results to the first half of the 2nd century.
[1918] Bucovala 1968, two examples, nos. 250-251, dated from excavation results to the 2nd-3rd and 3rd-4th century, respectively.
[1919] Saldern 1980a, 24, 26, no. 128, pl. 6, finds from graves dated to the 1st-2nd century.
[1920] Isings 1957, 41, finds dated to the 1st-2nd and to the end of the 2nd century.
[1921] Isings 1957, 41, finds dated to the 4th and 5th century.
[1922] Harden 1936, 278, nos. 836-837, pl. XX, this form is considered (page 270) to have appeared before the 3rd century.
[1923] Karamitrou-Mentesidou 1993, 20, fig. 2, where it is illustrated horizontally, between Medieval and modern vessels; it is a find from the hill of Agios Athanasios, but with further information about the excavation.
[1924] Perdrizet 1908, 216, fig. 951, finds from Pylaia.
[1925] Carington-Smith 1982, 277, nos. 39-45, drawings 5 pl. 38f-q, finds dated to the second half or the last third of the 1st century.
[1926] Makropoulou 2007, 152, pl. 68.
[1927] Stern 2001, 44.
[1928] Isings 1957, 42, form 28a, with a great number of dated examples.

are known, inter alia, from Serbia (Kostolac/Viminacium),[1929] Bulgaria (Chatalka),[1930] Episkopi at Kourion, Cyprus,[1931] Macedonia,[1932] Samothrace,[1933] Amphipolis,[1934] Sifnos,[1935] Larisa,[1936] Athens,[1937] Eleusis,[1938] Mamousia Aigialeia,[1939] Herakleion,[1940] Agios Nikolaos (Crete),[1941] Megara,[1942] Argos,[1943] Corinth,[1944] Mytilene,[1945] Thera,[1946] Rhodes,[1947] Antimacheia (Kos),[1948] Oropos,[1949] Kato Vatheia (Eubeia),[1950] and Edessa (Greece).[1951]

The following are firmly dated by their archaeological contexts to the 1st or early 2nd century: Cat. nos. 508,[1952] 588-590,[1953] 595,[1954] 592 and 584.[1955]

### 130. Unguentaria, conical body, straight walls, no shoulder, long neck, fire-polished rim

(Cat. nos. 597-607)
(Plates 36, XX)

### 131. Unguentaria, conical body, straight walls, no shoulder, long neck, tubular rim

(Cat. nos. 608-620)
(Plates 36-37, XX)

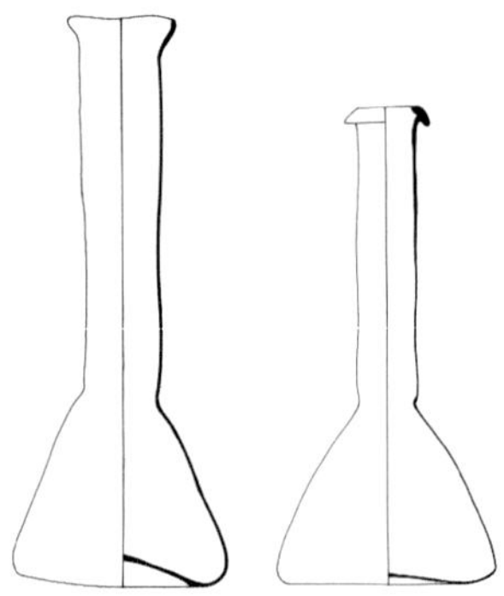

Vessels of this form have inverted rim, either tubular infolded or fire-polished. There is a long, cylindrical neck, three times the body height, often constricted at its base. The body is conical and stands on its flat bottom. The height of the vessels of this form ranges between 3 and 16cm, but usually is c. 10-15cm.

The vessels of this form are free-blown and usually the shaping of the rim was carried out while they were held by jacks, probably from the base of the neck.

Among the finds from Thessaloniki there are equal quantities of vessels with tubular rims, which is considered to be an indication of Oriental origin, and vessels with simple fire-polished rims, indicating their Western provenance.[1956]

The examples from Thessaloniki are made of colourless, bluish and greenish glass.

---

1929 Zotović, Jordović 1990, 61, pl. XVI: 1, dated by a coin of Trajan (98-117).

1930 Nikolov, Bujukliev 1967, 23, 15, fig. 12d, from a cremation dated from excavation results to the second half of the 1st century; Bujukliev 1986, 95, pl. 27, finds from a burial dated by coins to the second half of the 1st century.

1931 Oliver 1983, 252, 255, no. 51, fig. 4/58-59, pl. 41/14-15, from a burial dated to the second half of the 1st century.

1932 Weinberg, McClellan 1992, 115-116, no. 79, five examples from a grave dated by its context to the 1st century.

1933 Dusenbery 1998, nos. S179-4, 5, S180-3, 4, S184-2, S191-G, S210-2, examples dated from excavation results to the last quarter of the 1st century BC, with nos. S200-4 dated to the early 1st century.

1934 Romiopoulou 2002, 76 (photo page 73), finds from a grave with burials ranging from the 1st century BC to the 1st century AD.

1935 Macworth-Young 1949, grave 4 (84, pl. 25, 1. 4 (1, 2) dated to the 1st century; grave 14 (88, pl. 32, 3, 14 (2)) by a coin of Vespasianus (69-79); grave 20 (90, pl. 34, 5) by coins of Vespasianus and Titus (79-81); grave 22 (92, pl. 37, 1. 22 (1-2)), probably, dated to the second half of the 1st century.

1936 Tziafalias 1988, 260-263, pl. 138β, vessels from a destroyed Roman grave.

1937 Alexandri 1972, 115-118, pl. 76, vessel from a group of Roman graves, which, judging by the rest of the glass finds, is dated to the 1st century; Alexandri 1973-74c, 128-131, pl. 102, from a Roman cemetery; Zachariadou 2003, 16.

1938 Mylonas 1975, 130-131, pl. 195 (Γ29-208), from a burial dated to the second half of the 1st century.

1939 Dekoulakou1975d, 120, pl. 69γ, find with no particular chronological indications.

1940 Sakellarakis 1965, 562, pl. 709, finds from a grave that was in use, according to the numismatic evidence, between the middle of the 1st century and the 2nd century.

1941 Davaras 1985, 157-158, from a grave dated by a coin of Caligula (37-41).

1942 Zoridis 1983, 37, pl. 21γ, two vessels from a grave dated from excavation results to the 1st century.

1943 Alexandri 1963, 61-62, pl. 78α, β, vessels from Roman graves; Morou 1981, 109, pl. 49γ, vessels from a Roman grave with no other chronological indications.

1944 Davidson 1952, 105, nos. 668-669, fig. 11, finds dated from excavation results to the 1st and 2nd century.

1945 Archontidou-Argyri 1988, 458-459, pl. 279α, vessels with no chronological indications.

1946 Dragendorff 1903, 286-288, fig. 481 d, e. from graves dated from the 1st to the 4th centuries. The glass finds from the graves, at least those illustrated, as a whole are ascribed to the 1st and 2nd century.

1947 Konstantinopoulos 1965, 598-599, pl. 762, two vessels from a Roman grave.

1948 Zervoudaki 1972, 683-685, pl. 638, two vessels from a grave dated to the 1st century.

1949 Pologiorgi 1998, 41, no. 3, find from a grave dated to the second half of the 1st century (page 55, ΜΩ 258, 259), pl. 16γ, δ), from a burial dated to the 2nd century by the glass finds.

1950 Themelis 1969, 204, pl. 213, vessels from a single grave that were presented to the Archaeological Service by the finder.

1951 Chrysostomou A. 2000, 219, pl. 222, vessels from a burial dated by its context to the 1st century.

1952 It was found together with a bird-shaped unguentarium of Form 120 (Cat. no 583) and thus dated to the 1st century.

1953 From a grave with a clay lamp dated to the first half of the 2nd century, see Perlzweig 1961, 94, no. 259, pl. 8.

1954 From a grave which, according to the rest of the glass finds (Cat. nos. 460-462, 568-569, 577-578, 595, 606-607), is dated to the 1st century.

1955 Both finds are dated to the 1st century, and they were found in one grave which, as it seems, belongs to the oldest phase of use of that part of the necropolis, where burials range from the 1st to the 4th centuries. See Makropoulou 1990a, passim.

1956 Barag 1970, I, 198, according to 1977, 37-38, note 3.

This form is mainly dated to the 1st or early 2nd century, and it is considered to be an evolution of the vessels with necks of equal height as the body.[1957] Dated examples are reported from Britain,[1958] France,[1959] Germany,[1960] Hungary,[1961] Serbia (Kostolac/Viminacium),[1962] FYROM (Skopje/Scupi),[1963] Bulgaria (Chatalka),[1964] Cyprus (Agios Ermogenis),[1965] and Algeria (Tipasa).[1966] In Greece, published finds are reported from Alexandroupoli,[1967] Samothrace,[1968] Abdera,[1969] Palatiano (Ioron),[1970] Mitropoli Karditsas,[1971] Kato Ktimeni,[1972] Larisa,[1973] Nea Anchialos (Phthiotides Thebes),[1974] Corfu,[1975] Antikyra,[1976] Galaxidi (Chaleion),[1977] Kerameikos,[1978] Eleusis,[1979] Athens,[1980] Elliniko Kynouria (Eua Thyreatidos),[1981] Kallio,[1982] Patras,[1983] Argos,[1984] Sifnos,[1985] Samos,[1986] Mytilene,[1987] Marmaroto, Kos,[1988] Herakleion,[1989] Olympia,[1990] Knossos,[1991] Nea Kerdyllia,[1992] Doxipara Nicopolis (Epirus).[1993]

---

1957 Isings 1957, 42-43, form 28b, with a great number of dated finds.

1958 Price, Cottam 1998a, 173-174, with detailed bibliography on dated finds from the Britain, dated from excavation results to the last quarter of the 1st and to the first quarter of the 2nd century.

1959 Arveiller-Dulong, Arveiller 1985, 64-66, nos. 92-101, finds from Strasbourg (Augusta Trebecorum, Argentoratum) dated according to analogous finds to the 1st-2nd century, with a detailed bibliography of analogous finds from the West and East.

1960 Fremersdorf 1958b, 43-44, pl. 90, five examples from Cologne.

1961 Barkóczi 1966-7, no. 16a, find from Szőny (Brigetio) dated to the 2nd century.

1962 Zotović, Jordović 1990, 60-61, 82, 95, analogous to form no. 130), pls. XIV:3, LXXVIII: 3, CXLIII: 5, finds from three graves. The first grave contained a coin of Nerva (96-98), the second a coin of Hadrian (117-138), and the third is dated between the end of the 1st and the 2nd century. In addition (analogous to form no. 131), see page 106, pls. CLXXXVIII: 2 and CCI: 3, from graves dated from excavation results up to the middle of the 2nd century.

1963 Mikulčić 1976, pls. I, II, III, V, VI, VII, from graves dated between the Augustan era (27-14 BC) and the 3rd century.

1964 Nikolov, Bujukliev 1967, 23, 15, fig. 12a, from a cremation dated from excavation results to the second half of the 1st century; Bujukliev 1986, 95, pl. 27, finds from a burial dated by coins to the second half of the 1st century.

1965 McFadden 1946, 475, 495, pls. 39/35-36, 44/98-103 from a burial, which, according to its content is dated to the second half of the 1st century.

1966 Lancel 1967, nos. 127-134, finds dated from excavation results between the end of the 1st and the early 3rd century.

1967 Skarlatidou 1981, 351-353, pl. 242γ, vessels from graves dated between the 1st and 2nd century, according to the glass vessel.

1968 Dusenbery 1998, nos. S205-11, 12, 13, S216-2, S217-3, 4, 5, S218C-3, S219-15, 16, 17, 18, S220-2, 3, S232-2, examples dated from excavation results to the first quarter of the 1st century; W2-2, W3-D, RDS16-2, 3, S245-3, examples dated to the third quarter of the 1st century.

1969 Lazaridis 1965, 453-461, pl. 544, two vessels from Roman graves.

1970 Antonaras, Anagnostopoulou-Chatzipolychroni 2002, 117, fig. 11, finds dated on the basis of analogous finds to the 1st-2nd century.

1971 Tziafalias 1976, 183, pl. 130, two vessels from Roman, stone, sarcophagi.

1972 Intzesiloglou 1987, 272, pl. 160, two vessels handed over by the finder.

1973 Tziafalias 1988, 260-263, pl. 138β, two vessels from destroyed Roman graves.

1974 Sotiriou 1939, 64, figs. 16, 25, examples from graves with no precise chronological indications.

1975 Preka-Alexandri 1992, 333, pl. 100, two confiscated vessels.

1976 Baziopoulou-Valavani1980, 262-263, pl. 114, vessel from a grave dated to the second half of the 1st and up to the 2nd century.

1977 Threpsiadis 1972, 199, pl. 76β (α), find from a Roman grave; Zafeiropoulou 1975, pl. 5, nos. 14, 15, finds from excavation with no special chronological indications.

1978 Stroszeck 2000, pl. 24, finds from a grave dated to the early 2nd century.

1979 Mylonas 1931-1932, 44-45, fig. 5, finds from a grave with a coin dated to the second half of the 1st century BC; Mylonas 1975, 92 (Γ7-124α), pl. 195, from a burial dated by the glass vessel to the second half of the 2nd century; (page 11, A3-5), from a burial in a sarcophagus dated to the second half of the 2nd century dated only by the glass vessel.

1980 Alexandri 1968, 75, pl. 44α, γ, twelve vessels from a pit grave with no dating indications; Alexandri 1973-74c, 128-131, pl. 102, from a Roman cemetery.

1981 Spyropoulos 1980, 149-152, pls. 55α, 56ε, finds from graves with no special date references; Abadie, Spyropoulos 1985, 432, fig. 113, vessel from a grave dated to the second half of the 2nd century, according to a clay lamp.

1982 Zafeiropoulou 1982, from grave 2 (page 5, pl. Δ.α.), grave (page 6, pl. Δ.ζ.), grave 8 (page 6, pl. Δ.η.), grave 9 (page 6, pl. Δ.Θ.) all of which, according to their content, are dated to the second half of the 1st or in the early 2nd century.

1983 Dekoulakou 1975b, 115, pl. 64γ, without precise dating; Dekoulakou 1975a, 113-114, pl. 63δ, from a grave without precise dating.

1984 Alexandri 1963, 61-62, pl. 78α, β, vessel from a Roman grave.

1985 Macworth-Young 1949, grave 7 (pl. 30,1.7 (1)), dated to the 1st century; grave 14 (pl. 32, 3.14 (2)), with coins of Vespasianus (69-79); grave 20 (page 90, pl. 34, 5.20 (3-5,15-16)), with coins of Vespasianus and Titus (79-81).

1986 Tsakos 1977, 377, 377, pl. 138στ, ζ (page 415), grave of Vakentis V, dated to the 1st century.

1987 Archontidou-Argyri 1988, 458-459, pl. 279α, vessels with no chronological indications.

1988 Touchais, Huber, Philippa-Touchais 2001, 1002, fig. 249, from a grave with other glass finds dated to the 1st or to the early 2nd century.

1989 Sakellarakis 1965, 562, pl. 709, finds from a grave which, according to the coins found in it, was in use between the middle of the 1st and the 2nd centuries; Alexiou 1966, 408, pl. 437στ'. Undefined number of finds, three of which are illustrated, from a grave with several burials dated to the 1st and the early 2nd century.

1990 Schauer 2001/2002, 40 fig. 10, finds dated between the late 1st and the 2nd century, probably, according to analogous finds.

1991 Wardle 1972, 275-276, fig. 4/10-11, finds from a grave dated by the clay lamps found in it to the second half of the 1st or to the first half of the 2nd century.

1992 Malama, Ntarakis 2008, 148, pl. 76, MA 6567 (From a grave dated to the Roman imperial period); idem, page 163, pl. 76, MA 6609 (From a grave dated to the early 3rd century); idem, page 199, pl. 76, MA 6725 (From a grave dated to the first half of the 3rd century); idem, page 209, pl. 76, MA 6755 (From a grave dated to the first half of the 3rd century); idem, page 293, pl. 76, MA 6942 (From a grave dated to the transition from the 2nd to the 3rd century); idem, page 346, pl. 76, MA 7074 (From a grave dated to the first half of the 3rd century); idem, page 346, pl. 76, MA 7075 (From a grave dated to the first half of the 3rd century).

1993 Vokotopoulou 1971, 335, pl. 314, miniature example from a grave with coins dated to the 2nd century. Chrysostomou 1984, 36, vessel from a grave with a clay lamp dated to the first half of the 1st century.

The following can be dated by their archaeological contexts to the 1st or early 2nd century (Cat. nos. 593,[1994] 608,[1995] 599,[1996] 606,[1997] 603,[1998] 597,[1999] 600-602).[2000]

### 132. Unguentaria, conical body, curved walls, no constriction (Isings Form 82B2 var.)

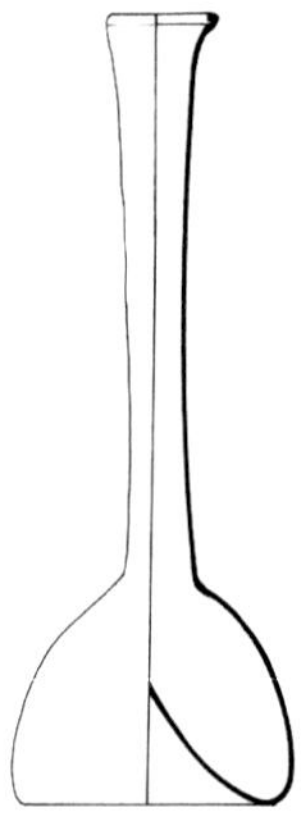

(Cat. nos. 621-624)
(Plates 37, XX)

Vessels of this form have inverted, almost horizontal, tubular infolded rim and a cylindrical neck, some three times higher than the body. The body is conical with convex walls and stands on a conical bottom. Their height is estimated between 11-20cm.

Vessels of this form are free-blown and usually the shaping of the rim was carried out while the vessel was held by jacks, probably from the base of the neck, as is indicated by the lack of a pontil mark underneath.

The examples from Thessaloniki are made of colourless, bluish glass. No exact parallels for the find from Thessaloniki were found, yet this form is directly connected to the (admittedly quite wider at the base) form 46, and thus is generally dated to the period between the late 1st and early 3rd century.

Similar vessels, particularly in Greece, were found at Thebes,[2001] Nea Kerdyllia,[2002] Eleusis,[2003] Athens,[2004] Oropos,[2005] Agios Nikolaos (Crete),[2006] Kasteli Kissamos,[2007] and Marmaroto (Kos).[2008] Some of the examples of this form are dated to the late 1st century, the majority are dated to the 2nd century, and a few even to the early 3rd century.

The following examples from Thessaloniki can be independently dated on the basis of their archaeological context: Cat. no. 624, dated to the 1st century[2009] and Cat. no. 621, dated between the 1st and 3rd centuries.[2010]

### 133. Unguentaria, globular, with fire-polished rim and cut-out

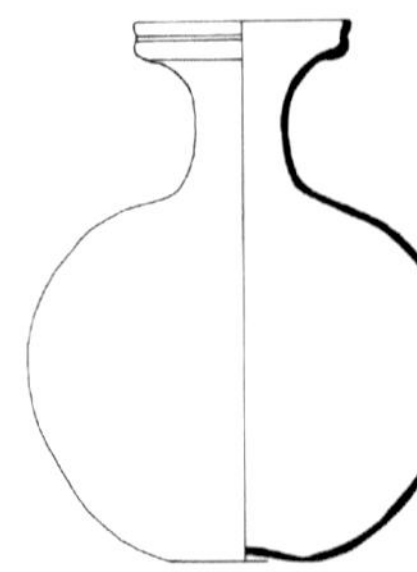

(Cat. no. 625)
(Plates 37, XX)

Vessels of this form have everted rim with vertical fire-polished lip, short cylindrical neck, perfectly globular body with flat, slightly concave bottom: under the rim a seemingly inherent cut-out is formed. Their height is estimated at c. 10cm.

The vessel is made of colourless greenish glass and is free-blown. During the shaping of the rim, a fold was formed by the use of jacks to create the cut-out under the rim.

The way that the rim was formed with a cut-out is typical of Syro-Palestinian products dated to the period from the late 3rd century onwards, although there are higher concentrations in the 4th and 5th centuries.[2011] The find from Thessaloniki, however, was found in a grave dated to the second half of the 3rd or the early 4th century.[2012]

[1994] Found together with a glass vessel dated to the 1st century (Cat. no. 599).

[1995] From a sarcophagus in use from the last quarter of the 2nd century until the early 4th century. See Chapter 9 (Find Spots), '232 Agiou Dimitriou Street, 1962'.

[1996] Found together with a glass vessel dated to the 1st century (Cat. no. 593).

[1997] Found together with glass vessels dated to the 1st century (Cat. nos. 460-462, 568-569, 577-578, 595, 607).

[1998] Together with clay unguentaria (reg. no. ΜΘ 6309, ΜΘ 6310), dated to the 1st century.

[1999] Found together with four glass vessels dated to the 1st century, two of them deformed by fire (Cat. nos. 542-543, 551, 683).

[2000] Finds from an Early Roman cemetery, See Chapter 9 (Find-Spots), 'Egnatias and Melenikou Street, 1969'. In particular, Cat. no. 601 was found together with Cat. no. 544, and Cat. no. 602 with Cat. no. 394, both of which are dated to the 1st-2nd century.

[2001] Keramopoulos 1917, 210-211, fig. 151, vessel from a grave with no clear chronological indications.

[2002] Malama, Ntarakis 2008, 288, pl. 76, MA 6933, from a grave dated after the middle of the 3rd century.

[2003] Kyparissis 1926, 71, fig. 16, from a grave probably dated after AD 267; Mylonas 1931-32, 44-45, fig. 5, finds from a grave with a coin dated to the second half of the 1st century BC; Mylonas 1975, 12, grave A9 vessel 7, dated by the glass vessel to the second half of the 2nd century; Papangeli 1987, 32, pl. 29, find from a sarcophagus dated from excavation results to the 2nd century.

[2004] Alexandri 1973-74c, 128-131, pl. 102, vessel with vertical ribbing from a Roman cemetery.

[2005] Pologiorgi 1998, 51, no. 1 (ΜΩ 255), pl. 15δ, find from a grave dated by its glass vessels to the 2nd-3rd century.

[2006] Davaras 1985, 157-158, from a grave dated by a coin of Caligula (37-41).

[2007] Pologiorgi 1985, 75, fig. 6, finds from a grave in which a glass bowl was also found dating the burial to the 2nd-3rd century.

[2008] Touchais, Huber, Philippa-Touchais 2001, 1002, fig. 249, from a grave dated by other glass finds to the 1st or early in the 2nd century.

[2009] From a grave with glass finds (Cat. nos. 480-484, 552, 527-528, 546-547, Forms 116, 122, 120c, 123, 132) dated to the 1st century.

[2010] Makropoulou 2007, 235, pl. 27.

[2011] Hayes 1975, 82, in the discussion on the characteristics of the author's Syro-Palestinian classes; Barag 1970c, pls. 12/20, 34/7, 7-1, 3 18, 42/17, 43/30, 44/10. Israeli 2003b, 234, in the author's introduction on the jars/vessels of the region. The only globular vessel with a tall, albeit cylindrical, neck and identically shaped rim is now in the museum at Khirbet el-Kerak, but with no chronological indications, see Dussart 1998, 151, form BX513 pl. 43/13.

[2012] Makropoulou 2007, 102, pl. 25, 68.

### 134. Unguentaria, ovoid with indentations, convex base

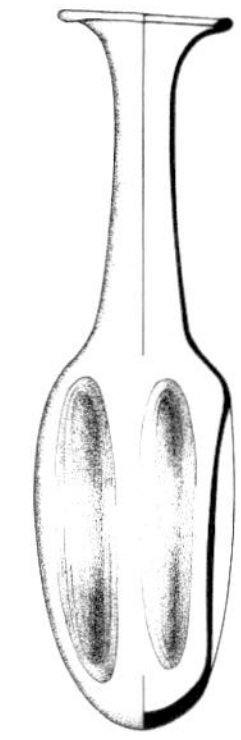

(Cat. nos. 626-627)
(Plates 37, XX)

Vessels of this form have tubular, infolded rim, cylindrical neck, elongated – almost cylindrical – indented body and narrow convex bottom, slightly flattened at its centre. Their height is estimated at c. 17cm.

These vessels are free-blown and the body while still hot and malleable was marvered to elongate it, and was pushed by jacks lengthwise at 6-7 specific areas all around it to achieve its final shape. Finally, the mouth and rim were shaped while the vessel was manipulated, being held by a blowpipe attached underneath.

This form[2013] is mainly known from the Western provinces, although it is not unknown in the East, admittedly in smaller numbers. As is indicated by the finds, it becomes more popular in the late 2nd and mainly during the 3rd century, and remained occasionally in use even in the early 4th century.

Dated finds are reported from Britain,[2014] Germany: Trier (Treviri),[2015] Bonn (Bonna),[2016] Heddernheim (Nida),[2017] Cologne,[2018] and Kassel (Castellum Cattorum);[2019] France: Strasbourg (Augusta Trebecorum, Argentoratum);[2020] Italy: Angera (Angleria),[2021] Aquileia;[2022] Hungary:[2023] Győr (Arrabona),[2024] Budapest (Aquincum),[2025] Dunaújváros (Intercisa);[2026] Romania: Mangalia (Callatis);[2027] Syria: Dura-Europos,[2028] Bosra (Nova Trajana Bostra),[2029] Khirbet Fahil (Pella),[2030] and Cyprus.[2031] The only published find from Greece was reported from Knossos.[2032]

The vessels from Thessaloniki are made of colourless, greenish glass.

Particular clues for the dating of the following examples come from the burials in which they were found: Cat. no. 627 was found in a grave dated to the second half of the 3rd century, particularly to the last two decades;[2033] Cat. no. 626, most probably, was found in a burial monument used from the middle of the 2nd and, at least, until the end of the 3rd century.[2034]

### 135. Unguentaria, cylindrical, with short neck, slender (Isings Form 102b oriental var. similis)

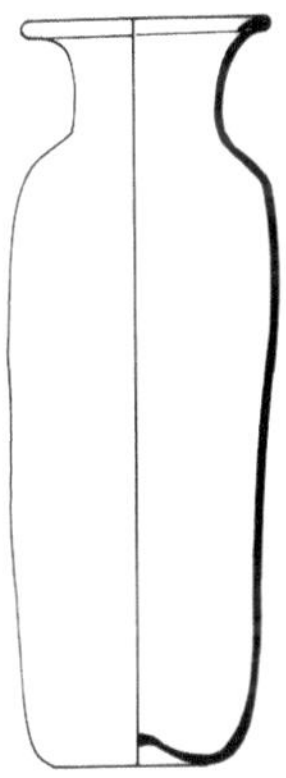

(Cat. nos. 628-629)
(Plates 37-38, XXI)

Vessels of this form have a funnel-mouth which ends to a tubular infolded rim. It has a very short neck and cylindrical body that tapers towards its base. The vessel stands on its flat, slightly concave bottom. Their height is estimated at c. 13cm.

2013 Isings 1957, 99-100, form 83. Although one example from London is dated to the early 2nd century, this has been re-dated to the 3rd-4th century, according to the vessel's use as a cinerary urn, see Cool, Price 1995, 162, where it is mentioned that this form appears in the late 2nd and mainly the 3rd century, while it is not clear if it remains in use also into the 4th century.
2014 Cool, Price 1995, 162-163, nos. 1251-1252, finds dated mainly to the 3rd century, and also a find from London which was earlier dated to the 2nd century, and later redated to the 3rd or 4th century.
2015 Goethert-Polaschek 1977, 122, form 76a, nos. 667-668, pl. 53, form dated to the 2nd century according to analogous finds.
2016 Follmann-Schulz 1988, 19-20, nos. 3-7, pl. 1, vessels dated according to excavation results, between the 2nd and the end of the 3rd century.
2017 Welker 1974, 130-131, nos. 294-295, pl. 18, finds dated according to analogous finds to the late 2nd or early 3rd century.
2018 Fremersdorf 1958a, 26-27, pls. 22-24, finds dated to the 2nd century; Fremersdorf, Polónyi-Fremersdorf 1984, 56 no. 135, which is dated from excavation results to the second half of the 3rd century.
2019 Boosen 1984, 88, no. 183, dated to the 2nd-3rd century according to analogous finds.
2020 Arveiller-Dulong, Arveiller 1985, 92, 240, no. 162, find dated from excavation results to the 3rd or to the early 4th century.
2021 De Tommaso 1990, 80, form 65, find from a grave dated to the first half of the 3rd century.
2022 Calvi 1968, no. 311, find dated to the 3rd-4th century.
2023 Barkóczi 1988, 121-122, nos. 228-233, pl. 18, finds dated from excavation results to the 2nd and 3rd century.
2024 Barkóczi 1966-7, 77, 88, no. 21, pl. 29/4, finds from graves dated by a coin of Hadrian (117-138) and one of Antoninus Pius (138-161), respectively.
2025 Kaba 1958, pl. 11/20, finds dated to the first half of the 2nd century, dated according to excavation results.
2026 Radnóti 1957, 144, no. 88, pl. 28/4, dated to the second half of the 3rd century; Barkóczi 1988, 122, no. 230, find dated from excavation results to the 3rd century.
2027 Preda 1980, 33-34, pl. XIII, dated to the 3rd-4th century, according to its context.
2028 Clairmont 1963, 131, 139-140, form J, nos. 727-735, pl. 16, nine finds dated to the middle Imperial period, (mid 2nd - mid 3rd century).
2029 Dussart 1998, 168, form BXIII. 1312b2, note 1, vessel dated to the second half of the 4th century according to analogous finds.
2030 Dussart 1998, 168, form BXIII. 1, 312b2, note 1, find from a grave dated to the second half of the 4th century.
2031 Vessberg 1956, 165, pl. 50/13.
2032 Price 1992, 437, 453, no. 297, dated from excavation results to the Severan period (193-235). The author does not find local analogous finds and she connects it to Isings Form 83, which actually might not represent a closely related form, since it lacks the indentations which are so typical and characteristic of the shape of this particular vessel.
2033 The grave is dated at around AD 280-290 on the basis of the pottery finds and the style of the wall-paintings of the grave, which are dated by Marki 2006, 220-221.
2034 Marki 2006, 211.

The vessels of this form are free-blown and marvered, the mouth and the rim were shaped by further tooling while the vessel was handled by a metal blowpipe attached underneath.

The examples from Thessaloniki are made of colourless, greenish and translucent, purple glass.

This form appears in the Eastern Mediterranean region, and the purple colour of one of the examples is also indicative of this same region. It is a later, evolved form of cylindrical unguentarium[2035] that probably represents an Oriental variant of Isings Form 102b.[2036] Parallels have been published from Bosra (Nova Trajana Bostra),[2037] Fiq,[2038] Bet She'an (Nysa-Scythopolis),[2039] Samaria-Sebaste,[2040] Homs (Emesa),[2041] and an example in the Royal Ontario Museum, Toronto.[2042]

The examples from Thessaloniki cannot be independently dated and they are placed in the 4th century according to analogous finds from other sites.

**136. Unguentaria, cylindrical, tall neck (Isings Form 102 var. similis)**

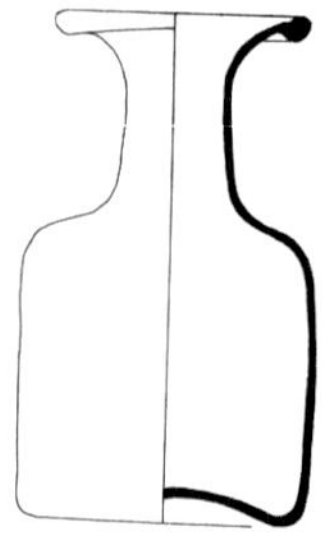

(Cat. no. 630)
(Plates 38, XXI)

The vessel of this form has tubular rim, almost horizontal everted mouth, cylindrical neck and body, and flat, slightly concave bottom. Their height is estimated at c. 9cm.

The vessel could equally plausibly be made either by mould- or free-blowing, and marvering. No pontil mark is visible underneath, an indication that the vessel was held by jacks from the neck while the rim area was shaped.

Although exact parallels have not been traced, the vessel appears to belong to a well-known form[2043] present in the Western provinces, probably from the late 2nd or early 3rd century; it went on to become popular during the 4th cntury.[2044] It is the Western counterpart of Fform 135 – cylindrical flasks with short neck.

The average quality of the glass, quite bubbly with sandy inclusions, naturally intensely coloured, indicates a type of glass present after the end of the 3rd century; this and the shape of the mouth and tubular rim corroborate a dating to around the 4th century. This date is also confirmed by the archaeological context of the find which places it to the first half of the 4th century.[2045]

**137. Kraterisks**

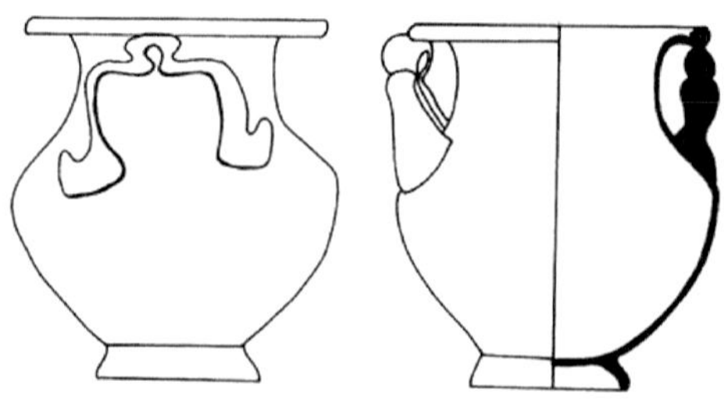

(Cat. no. 631)
(Plates 38, XXI)

The vessel has everted, tubular rim, calyx-shaped body, applied conical base bearing tooling marks. Two Pi-shaped, strap-handles, with a protruding fold at the centre of the horizontal bar, are placed on the shoulders and they almost reach the rim. The body is made of olive-green glass and the handles of dark-blue glass. In Thessaloniki only one example of this very rare form has been unearthed. Their height is estimated at c. 8cm.

Because of its small size it is ascribed to the category of flasks or unguentaria, although its shape accurately represents in miniature a tableware vessel: the krater. Because of the colour and texture of the glass, and the particular way the base is shaped, it can quite probably be considered an Egyptian product.

---

2035 Stern 1977, 79, form 21, where according to the form of the mouth and the rim itself, these cylindrical unguentaria are divided into four groups, dated from the 1st to the 4th century, dating the examples analogous to our form to the 4th century.
2036 Isings 1957, 121, with examples dated to the 3rd and 4th century.
2037 Dussart 1998, 160-161, form BXI.3211b, fig. 48/12, vessel dated on the basis of analogous finds to the 4th century.
2038 Dussart 1998, 160-161, form BXI.3211b, fig. 48/13, vessel dated on the basis of analogous finds to the 4th century.
2039 Fitzgerald 1931, fig. 10, 4th vessel, second row, find from a grave dated to the 4th century.
2040 Hamilton 1939, 68 fig. 2a, find from a grave with a coin of Constantius II (337-361).
2041 Abdul Hak 1965, 29, fig. 8, first from the right, upper row, from a necropolis originally dated to the 2nd-3rd century; the period of its use was re-dated by Stern (1977, 83-84) from the first half of the 2nd to the middle of the 4th century.
2042 Hayes 1975, 62, 110, nos. 170, 410, pls. 12, 25, dated according to analogous finds to the late 3rd-4th century.
2043 Isings 1957, 120-121, form 102a.
2044 Trier (Treviri): Goethert-Polaschek 1977, 185-186, form 10$^{a}$a, pl. 64, dated from excavation results to the first half of 4th century; page 187, form 11$^{a}$a, pls. 64-65, dated from excavation results to the second half of the 4th century. Mainz (Moguntiacum): Harter 1999, 113-114, form F3a, dated to the late 3rd or to the middle of the 4th century.
2045 Makropoulou 1991, 257-270, photo 24; Makropoulou 2007, 46, pls. 27, 68, from a barrel-vaulted grave used in the first half of the 4th century, according to a clay jug found inside.

The single other parallel of this vessel, although rather larger in size, and with differently shaped handles, was found in Bonn (Bonna)[2046] and it is dated to the first half of the 4th century. The structural features of the vessel fit perfectly well with this dating: both the forming of the base with jacks, leaving visible slanting tooling marks, and this particular quality and texture of glass, are quite common for Egyptian vessels, which were originally dated to the 4th-5th century.[2047] By the re-dating of the pottery of the relevant strata, however, it has been proved that their use continued to the 5th-6th century.[2048]

On the basis of the clay vessels[2049] and the glass vessel (Cat. no. 631) found with it, the find from Thessaloniki is ascribed to the first half of the 4th century.

### 138. Amphorisks, cylindrical (Isings Form 127 min.)

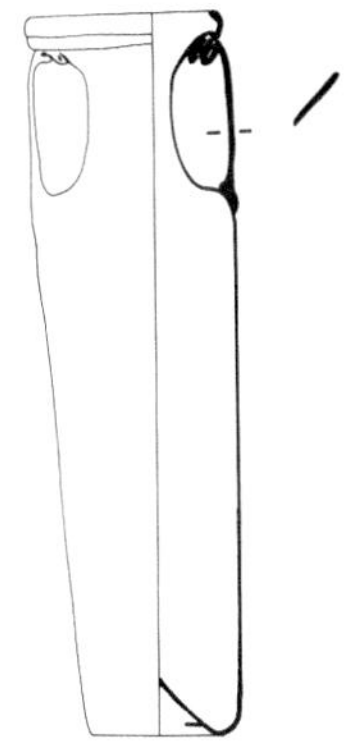

(Cat. nos. 632-635)
(Plates 38, XXI)

Vessels of this form have everted, almost horizontal, fire-polished rim which forms a funnel-mouth, a short, cylindrical neck, almost horizontal shoulders, cylindrical body, and concave, almost conical, bottom. Two vertical strap-handles start on the shoulders and bend under the rim, where they fold and adhere. A thick trail of glass is wound around the underside of the rim. The examples from Thessaloniki are made of greenish glass, and the handles in some cases are made of dark-blue glass. Their height is estimated at c. 15cm.

The essels are actually small-sized examples of Form 74a. This form appears around the middle of the 3rd century and it is mainly popular during the 4th century, but it survives up to the early 5th century. Examples have been found in the West and East,[2050] and they are also dated to the 3rd-4th century. Vessels of similar size have been found, inter alia, at Belgrade (Singidunum),[2051] Šid,[2052] Dion,[2053] and Veroia;[2054] one analogous example is held at the Corning Museum of Glass.[2055]

None of the finds from Thessaloniki can be independently dated and they are ascribed, according to analogous finds from other sites, to the period between the end of the 3rd and the early 5th century.

### 139. Unguentaria, spherical, tubular rim, tall neck (Isings Form 101 min.)

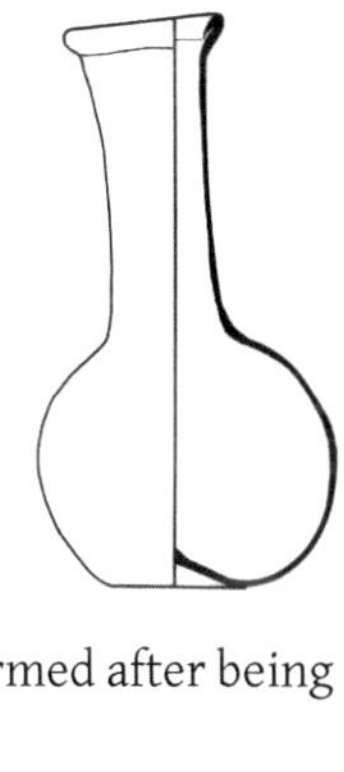

(Cat. nos. 636-666)
(Plates 38-40, XXI-XXII)

This form is one of the most popular type of unguentaria found in the cemeteries of Thessaloniki. They are distinguished from the identically-shaped tableware vessels (Form 51) by their smaller size, which seems, almost arbitrarily, to be presumed not to have exceeded 15cm in height. They are characterized by a globular or pear-shaped body, concave bottom, cylindrical neck, and flaring, vertical or even everted, tubular rim. They appear in natural hues of glass, colourless greenish, olive-green and bluish. Their height is estimated between 7-15cm.

These vessels were made by free-blowing, the body was shaped by free rotation, and the rim was formed after being removed from the blowpipe, which was subsequently attached underneath and the moil cut off.

It is a variant of the miniature examples of Isings Form 101,[2056] which appear already from the early 3rd, become popular in the 4th, and survive into the 5th century.[2057]

---

[2046] Follmann-Schulz 1988, 76, no. 254, pl. 32, dated from excavation results to the first half of the 4th century; Follmann-Schulz 1992, 89-90, no. 48, with colour illustration.

[2047] This technique appears with vessels of different shapes, made of olive-green glass. For examples from Kom Aushim (Karanis), see Harden 1936, nos. 17, 21, 228, 235, 356, 358, 479, 482, 484, 494, 498, 508. For examples from the Cairo Museum, see Edgar 1905, nos. 32.433, 32.423 32.436, 32.426, 32.429, 32.439, 32.443, 32.444.

[2048] Hayes 1975, 2-3, relevant examples in a colour illustration in Whitehouse 1997, 81-82, nos. 107-110.

[2049] These vessels are known and precisely dated from graves of the (mainly) eastern necropolis of Thessaloniki; Nalpantis 2003, 120-121, the jugs correspond to form 1α, dated to the end of the 3rd- first half of the 4th; idem, 123, for the handless unguentaria, form 1α, dated to the first half of the 4th century.

[2050] Isings 1957, 157-158, form 127, finds dated from excavation results between the end of the 3rd and the end of the 4th century.

[2051] Ružić 1994, 22, no. 148, pl. XIV/4, find dated from excavation results to the 3rd-4th century; Pop-Lazić 2002, 58-59, from a grave dated to the middle of the 3rd century by coins of Septimius Severus (193-211) and Gordianus (238) found inside.

[2052] Ružić 1994, 22, no. 147, find dated from excavation results to the 3rd century.

[2053] Pantermalis 2001, 354, fig. 13.

[2054] Allamani-Souri 1991b, 302-303, pl. 114ε, example from a grave dated by its context to the 4th century.

[2055] Whitehouse 2001, 184, no. 729, dated on the basis of analogous finds to the 3rd-4th century.

[2056] Isings 1957, 120.

[2057] Arveiller-Dulong, Arveiller 1985, 120.

Although attempts have been made to group these unguentaria into categories according to whether their rim is vertical or everted,[2058] it does not seem that these elements present a real feature for helping discern them, neither chronologically, nor in terms of their use or provenance: they were a very popular form which was apparently produced in several workshops.

Known examples are reported from France: Strasbourg (Augusta Trebecorum, Argentoratum);[2059] Germany: Mainz (Moguntiacum),[2060] Cologne,[2061] Trier (Treviri),[2062] and Bonn (Bonna);[2063] Switzerland: Augst and KaiserAugst (Augusta Rauricorum and Castrum Rauracense);[2064] Slovenia;[2065] Hungary: Szőny (Brigetio),[2066] Ságvár (Tricciana),[2067] Dunaújváros (Intercisa),[2068] Kisàrpàs;[2069] Montenegro: Komini (Municipium 'S...'),[2070] Duklja (Doclea);[2071] Serbia: region around Belgrade (Singidunum),[2072] Niš (Naissus);[2073] Croatia: Sremska Mitrovica (Sirmium),[2074] Đakovo (Certissa);[2075] FYROM: Skopje (Scupi);[2076] and also Cyprus,[2077] Jordan, Syria,[2078] and Israel.[2079]

The finds from Thessaloniki that can be independently dated are Cat. nos. 637,[2080] 640,[2081] 659,[2082] 664,[2083] and 665.[2084] These are ascribed to the generally known timespan of the form, i.e. 4th-early 5th century.

### 140. Unguentaria, pear-shaped, smooth (Isings Form 101 min. similis)

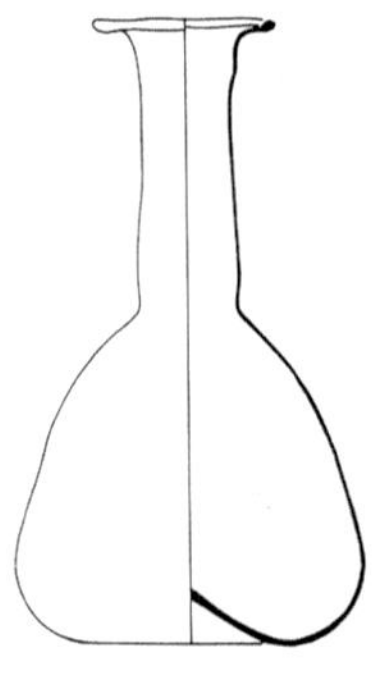

(Cat. no. 667)
(Plates 40, XXII)

The vessel of this form has everted almost horizontal, tubular rim, cylindrical neck, pear-shaped body, and flat, mildly concave bottom. Their height is estimated at c. 13cm.

The vessel is made by free-blowing, the body was further shaped by marvering. The rim was formed after it was released from the blowpipe and the moil, while the vessel was held either from the base of the neck by jacks or by a blowpipe attached temporarily underneath.

2058 Rütti 1991, form AR 146.1 with vertical rims and form AR 148, which is funnel-mouthed; Harter 1999, form with vertical rim D16a, and funnel-mouthed form G27a.

2059 Arveiller-Dulong, Arveiller 1985, 119-126, nos. 218-272 (esp. nos. 260-272, pages 125-126), dated from excavation results to the 3rd and mainly to the 4th century.

2060 Harter 1999, form D16a: 241-243, nos. 681-696, pls. 29, 88, 91-93, 101; idem, 286-288, form G27a, nos. 1037-1049, pls. 50, 90, 102. Vessels of this form can be ascribed both to unguentaria D16a as well as to the group of smaller examples of transportation/tableware G27a, both of which are dated to the 3rd and mainly to the 4th century.

2061 Doppelfeld 1959, 156-157, fig. 3.4, find dated from excavation results to the early 4th century.

2062 Goethert-Polaschek 1977, form 79a, pl. 55, with examples dated from the first half of the 2nd into the 4th century. The somehow larger form 98 (idem, 154-155, pl. 59) is dated from excavation results to the second half of the 4th century.

2063 Follmann-Schulz 1988, 29, no. 49, pl. 4 (pear-shaped example dated from excavation results to the first half of the 4th century); idem, 28-29, nos. 51-61, pl. 5 for globular examples dated to the 4th century.

2064 Rütti 1991, AR 146. 1 (vessels with vertical rims: 54, 123, nos. 2427-2437, pl. 104), AR 148 (Funnel-mouthed vessels: idem, 54, 125, nos. 2453-2457, pl. 105), dated from excavation results from the late 3rd to the early 5th centuries.

2065 Lazar 2003, 192, globular vessels form 8.6.14 and pear-shaped vessels form 8.6.16 (page 193), finds from graves in various sites dated to the 2nd and 3rd centuries.

2066 Barkóczi 1988, no. 287, find dated from excavation results to the first half of the 3rd century.

2067 Burger 1966, 137, pear-shaped examples of unguentaria of form 10, fig. 125/10; globular examples: page 138, forms 12-14, fig. 125/12-14, dated from excavation results to the 4th century, and especially to the middle of the 4th century.

2068 Radnóti 1957, 146, 148, pls. XXVII/1, 7, XXIX/9, vessels from graves dated to the 4th century.

2069 Barkóczi 1988, 119-120, no. 221, pl. XVIII/221, dated from excavation results to the end of the 4th century.

2070 Cermanović-Kuzmanović 1976, 180, pl. V/6, find dated from excavation results to the 4th century.

2071 Cermanović *et al.* 1975, 170, form V/5. From this cemetery five vessels are known, one of which is blue. One of the vessels is precisely dated by a coin to the period of the Tetrarchy and the rest to the 4th century.

2072 Ružić 1994, 29, nos. 381-383, finds dated to the 3rd-4th century on the basis of analogous finds from other sites.

2073 Drča 2000, 216, 220-221, nos. 33, 37-38, 40-42, 45, finds from cemeteries dated to the 3rd-4th century.

2074 Cermanović-Kuzmanović 1981a, 19, 85, 97; Šaranović-Svetek 1986, 66, nos. 134-135, dated on the basis of analogous finds to the 3rd-4th century.

2075 Šaranović-Svetek 1986, 65, nos. 130-131, pear-shaped examples dated by a coin of Licinius, minted AD 320-321.

2076 Mikulčić 1976, 198, 208, pl. VI/106, from a cremation dated to the end of the 2nd- early 3rd century.

2077 Vessberg 1956, 154, fig. 48/8.

2078 Dussart 1998, 165, form BXIII.1121 a-b, pl. 51, globular unguentaria, one of which is dated from excavation results to the 4th century. Form BXIII.1422 a-b, page 169, pl. 55, pear-shaped examples dated to the 4th century according to analogous finds.

2079 Barag 1970c, form XVI 8, dated to the 3rd-4th century; Israeli 2003b, 165, dated to the 3rd-4th century on the basis of analogous finds.

2080 Found with coins dated to the end of the 4th and to the 5th-6th century. Unpublished find and I thank the excavator, mr. Dimitris Nalpantis, for this information.

2081 Makropoulou 2007, 235, pl. 24, from a grave dated to the 3rd-4th century.

2082 It was found together with the glass jug (Cat. no. 423) dated to the 4th century.

2083 With metal finds, among which were amphora-shaped decorative plates, parts of belt fittings. On their dating (mainly) to the 4th century, see Tomić 1996, 197-206.

2084 Petsas 1970a, 347-351 (esp. page 349), from a grave in use for a long period of time and definitely during the 3rd and 4th centuries.

It is a variant of the small-sized vessels of Isings Form 101[2085] which appear already from the early 3rd century and are produced throughout the 4th, although even some 5th-century examples are known.[2086]

The find from Thessaloniki is dated on the basis of coins found with it to AD 325-330.[2087]

### 141. Unguentaria, square, indented, tall neck

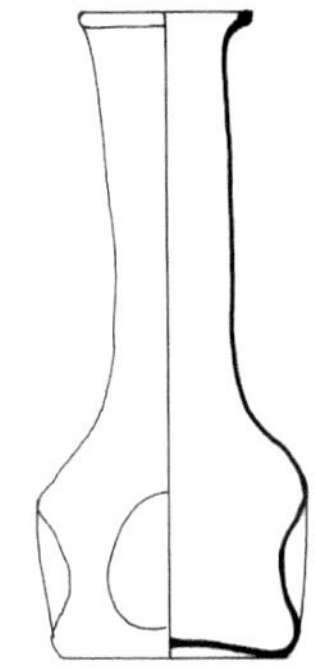

(Cat. nos. 668-682)
(Plates 40-41, XXII)

The vessels of this form have slightly flaring, tubular infolded rim, and long cylindrical neck. Only one example has a truly intensely everted funnel-mouth. They are characterized by the four indentations around their body which render it almost square. The bottom is flat, mildly concave. Their height is estimated between 8-11cm.

It is a well-known, 4th-century form ascribed to the production of Eastern Mediterranean workshops.[2088] Published dated finds include from Germany: Münster,[2089] Düsseldorf,[2090] Mainz (Moguntiacum),[2091] Strasbourg (Augusta Trebecorum, Argentoratum),[2092] Krefeld-Gellep (Gelduba),[2093] Trier (Treviri),[2094] Augst;[2095] Italy: Milan (Mediolanum),[2096] Vatican,[2097] Newark;[2098] France: Paris;[2099] Austria: Lorch (Lauriacum);[2100] Slovenia: Ptuj (Poetovio);[2101] Croatia: Zadar (Iader),[2102] Sremska Mitrovica (Sirmium),[2103] Serbia: Brest (Breska),[2104] Sombor;[2105] FYROM: Skopje (Scupi),[2106] Isar-Marvinci,[2107] Lopate;[2108] Hungary: Szőny (Brigetio),[2109] Budapest,[2110] Ságvár (Tricciana);[2111] Romania: Costanza (Tomis);[2112] Syria: Dura-Europos,[2113] Allepo;[2114] Lebanon: Beirut;[2115] Jordan: Amman-Jebel Jofah;[2116] Israel: Beit Fajjar[2117] and Gezer.[2118] In Greece one example was found at Edessa[2119] and another at Nea Kerdyllia.[2120]

Almost all the vessels from Thessaloniki are made of olive-green glass with pinprick bubbles and inclusions, and only few are made of a naturally less colourful, or more successfully decolourized glass, which in any case is not

2085 Isings 1957, 120.
2086 Arveiller-Dulong, Arveiller 1985, 120.
2087 Makropoulou 2007, 251, pl. 23, from a grave dated by bronze coins of Constantine I and II (306-361).
2088 Burger 1966, 137; Stern 1977, 100-101, no. 28; Šaranović-Svetek 1986, 29, 65, pls. XIII/1, 2, XVIII/3; Barkóczi 1988, 151, form 131; Roffia 1993, 108. The only researcher who came to different conclusions is Gluščević (1998, 184-185), who, according to the excavation results, concluded that the finds from Zadar (Iader) are mainly dated securely by coins and pottery to the 2nd century, some even to the 3rd century, and due to their great numbers he even assumed that they represent a local product.
2089 Reimenschneider 1989, 194, no. 112, pl. 15, 2, unprovenanced object dated to the 3rd-4th century according to analogous finds.
2090 Saldern 1974, 110, no. 129, unprovenanced object considered to be a Syrian product dated to the 3rd-4th century according to analogous finds.
2091 Harter 1999, 145-146, 286, form G26a, nos. 1035-1036, pl. 50, find dated to the 4th century according to analogous finds.
2092 Arveiller-Dulong, Arveiller 1985, 126, no. 271, find from a grave dated by a coin after AD 317.
2093 Pirling 1966, 104, form 200, from graves dated to the first half of the 4th century.
2094 Goethert-Polaschek 1977, 123-124, form 78a, nos. 672-675, drawing 37, pl. 53, finds dated from excavation results to the 4th century.
2095 Rütti 1991, 124, form AR 147.2, nos. 2449-2452, pl. 105, finds dated from excavation results to the 4th and up to the early 5th century.
2096 Roffia 1993, 108, 129, 135, 252, no. 310, unprovenanced objects and dated to the 4th-5th century according to analogous finds.
2097 Fremersdorf 1975, 63, 77, nos. 579, 740, pl. 28e, 34, unprovenanced objects, dated to the 3rd-4th century on stylistic reasons.
2098 Auth 1976, 221, nos. 465-466, examples with no archaeological indications and which have not been dated precisely.
2099 Stern 1977, 100-101, no. 28, pl. 10, unprovenanced find dated according to analogous finds to the end of the 3rd- middle of the 4th century.
2100 Thomas 1962, 105, 107-108, pl. XXVIII/4, from a grave dated to the second half of the 3rd or to the early 4th century.
2101 Šubic 1976, 43, 49, pl. IV, nos. 34-35, dated to the 4th century.
2102 Gluščević 1998, 184-185. 25 examples from 19 graves, 14 cremations and 4 inhumations. 11 of the burials are securely dated by coins and clay finds to the 2nd century, while the rest can be dated up to the 3rd century.
2103 Cermanović-Kuzmanović 1981a, 19, 91-94, dated to the end of the 3rd- early 4th century, as can the entire necropolis in which they were found.
2104 Šaranović-Svetek 1986, 65, dated to the first half of the 4th century by a coin of Constantine I (306-337) and Constantius II (337-361).
2105 Šaranović-Svetek 1986, 65, unprovenanced find, dated to the 4th century according to analogous finds.
2106 Mikulčić 1976, 193, 196, pl. II, VIII, nos. 35, 152, dated to the second quarter of the 4th century by coins of Constantius, Constantine I and II (305-361).
2107 Sokolovska 1986, 126, pl. 45/11, from a grave dated by coins to the period of the Tetrarchy.
2108 Nikolovski 2006, 245, 246, fig. VII/4, finds from two Late Roman graves with no precise date.
2109 Barkóczi 1988, no. 336, pl. 31, from a grave dated from excavation results to the second half of the 3rd century.
2110 Barkóczi 1988, no. 335, pls. 31, 90, object of unknown provenance, dated on the basis of analogous finds to the 4th century.
2111 Burger 1966, 137, fig. 125, form 11, dated by a coin between AD 351-367.
2112 Bucovală 1968, 95-96, nos. 171-172, finds dated to the 2nd and 3rd century and whose exact provenance is not noted.
2113 Clairmont 1963, 142, no. 757, pl. XXXVII/757, the find was dated prior to AD 256, the time of the city's destruction. On the possibility of a later dating, see Dan Barag's review on the work of Clairmont at Dura-Europos (Barag 1967b), which was also accepted by Marianne Stern (Stern 1971, 101, note 1).
2114 Stern 1977, 101, where all prior bibliographies are cited, without a precise dating.
2115 Stern 1977, 101, where all prior bibliographies are cited, without a precise dating.
2116 Barag 1970c, I 96, find dated from excavation results to the 3rd century.
2117 Harden 1966, 50, fig. 8, find dated from excavation results to the Constantinian period, or more generally to the first half of the 4th century.
2118 Barag 1970c, I 94, from a grave in use from the late 1st to the late 4th century.
2119 Chrysostomou A. 2000, 312, pl. 356, vessel from a burial dated according to its context to the 3rd-4th century.
2120 Malama, Ntarakis 2008, 299, pl. 78, MA 6962, from a grave dated to the end of the 2nd to early 3rd century.

free of pinprick bubbles. The large number of finds and the great resemblance they have as a group renders them very probably local products.

Although it has been assumed[2121] that they probably contained holy water or oil and that they had a liturgical use, the fact that they were unearthed in cremations negates the connection of these vessels exclusively with Christians.[2122] On the contrary, the fact that they appear equally in inhumations leads in any event to the natural and anticipated conclusion that their use was universal and not connected to the religion of the deceased. In addition, it has been attempted to connect the place of the vessel in the grave to the origin of the deceased. It was assumed that if it was found near the head, or at the upper part of the body, then an Oriental origin is suggested, and if it was found in the area of the feet, then the person who died was a Westerner.[2123] The finds from Thessaloniki do not permit this hypothesis to be either confirmed or rejected.

Among the finds from Thessaloniki, firm chronological indications are known Cat. nos. 669,[2124] 670,[2125] 671,[2126] 672,[2127] 673,[2128] 674,[2129] and they are thus ascribed to the late 3rd and (mainly) to the 4th century.

### 142a. Unguentaria, funnel-mouthed, spherical, tall conical neck, constriction and fire-polished rim

(Cat. nos. 683-684)
(Plates 41, XXII)

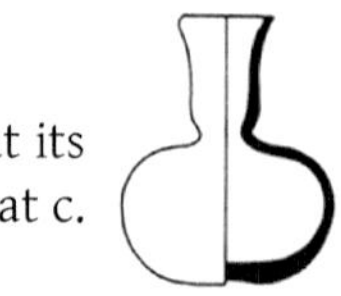

Vessels of this form have a relatively long and narrow conical neck, with an intense constriction at its base, fire-polished rim, globular body and flat, slightly concave bottom. Their height is estimated at c. 12cm.

The body of the vessel is made by free- blowing the paraison. Then, after the vessel was held by jacks, probably by the base of the neck – as the constriction there indicates – the conical shape of the neck was formed and the rim was reheated and fire-polished; finally the bottom was flattened and gently pushed to form the *kick*, so as to provide a standing surface for the vessel and render the body more durable.

This is a miniature version of Form 65, which appears at the end of the 3rd and, mainly, during the 4th and 5th centuries.[2130] Cat. no. 684 can be independently dated to the end of the 3rd or to the 4th century.[2131] From the texture of the glass, the shape of the rim and its base, it seems likely that the artefact is of Syro-Palestinian production.[2132]

### 142b. Unguentaria, funnel-mouthed, spherical, tall conical neck and fire-polished rim

(Cat. nos. 685-687)
(Plates 41, XXII)

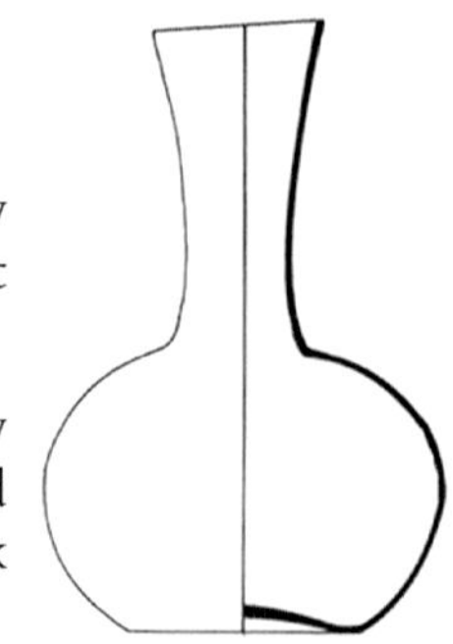

The vessels of this form have a funnel-mouth which ends into a fire-polished rim. They feature a long, cylindrical neck, globular body and flat, slightly concave bottom. Their height is estimated at c. 7cm.

The body of the vessel was formed by free-blowing of the paraison, Then, and while the body was held by jacks – most probably from the base of the neck – the funnel-mouth was formed and the rim was reheated and fire-polished; the bottom was then pushed in to form a mild kick so as to make the body less susceptible to breakage.

Exact parallels have not been traced. Cat. no. 685 was found in a cluster of Early Christian graves, one of which featured wall-paintings.[2133] The dark colour and the inferior quality of the glass, with many pinprick bubbles and larger bubbles and sandy inclusions, are consistent with its dating to the Early Christian period: 4th-6th centuries.

2121 Thomas 1962, 108, form 6; Šaranović-Svetek 1986, 30, this opinion is mentioned by both authors, with no further proof.
2122 As noticed first by Gluščević (1998, 185), referring to the finds of the cemeteries at Zadar (Iader).
2123 Burger 1966, 137, mentioned without special argumentation.
2124 From a group of Early Christian graves. See Chapter 9 (Find-Spots), 'Aristotle University Campus, Sorority Building (ΦΕΘ), 1968'.
2125 Unpublished. From a cemetery dated to the 4th century, in which some of the built-graves were reused in the 5th century. (I thank the excavator of the site, mr. Dimitris Nalpantis, for sharing this information with me).
2126 Makropoulou 2007, 224, pl. 28, from a grave most probably dated to the 4th century.
2127 Makropoulou 2007, 309, pl. 28, from a grave most probably dated to the 3rd century.
2128 Papanikola-Bakirtzi 2002 (ed.), 470, no. 646 (E. Marki), dated to the 4th century.
2129 Papanikola-Bakirtzi 2002 (ed.), 225, no. 261 (E. Marki), dated to the 4th century.
2130 See Form 65 (Cat. no. 349) for analogous finds.
2131 Makropoulou 2007, 335, pl. 69.
2132 Hayes 1975, 80-81, a vessel belonging to form XV12 according to Barag (1970).
2133 See Chapter 9 (Find-Spots) 'Aristotle University Campus, Sorority Building (ΦΕΘ), 1968'.

Cat. no. 686 is dated to the first half of the 4th century.[2134] Cat. no. 687 is firmly dated around the middle of the 4th century.[2135]

### 143. Juglets, trefoil-mouth

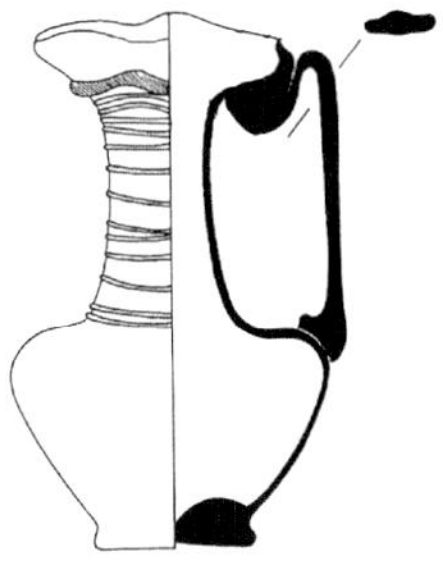

(Cat. no. 688)
(Plates 41, XXII)

Vessels of this form have a flaring trefoil, fire-polished rim, cylindrical neck, horizontal shoulder, inverted truncated conical body, ring base and flat bottom. Strap, with median rib, handle starts from the shoulder and rises vertically, folds, and is attached to the rim. A fine thread is spirally wound around the neck approximately 14 times. A thicker trail is wound under the rim. Their height is estimated at c. 10cm.

The body is made of purple glass, while the handle, the decorative thread around the neck, the trail under the rim, and part of the rim, are made of colourless, greenish glass. The chromatic differentiation on the rim is probably due to some remains of a different colour glass on the tip of the blowpipe when the vessel was blown. On the contrary, the colour of the handle, the thread and the trail all present an aesthetic choice. Purple vessels supplemented with colourless, greenish elements, usually handles, are considered to be Syro-Palestinian products because of the large number of such examples found in the early 20th century southeast of Jerusalem, in the region of Bet Guvrin (Eleutheropolis). The objects of this group are characteristically resistant to weathering, just like the example from Thessaloniki, and among them forms of amforoids[2136] and jars[2137] typical of the 4th century have been identified.[2138]

The example from Thessaloniki is dated by the excavation context to the 4th century.[2139] Although jugs generally are widely distributed amongst 4th-century tableware, and miniature examples with globular, hemispherical, pear-shaped or carrot-shaped body are also known, no examples with the body shape of our find was located.

Analogous finds with trefoil-mouth, some of them rendered also in two colours, are held in Western museums: Corning Museum of Glass,[2140] Yale Art Gallery,[2141] Newark,[2142] Carnegie Museum of Natural History,[2143] Royal Ontario Museum,[2144] and Fondation Custodia.[2145] Eastern repositories include Hama (Epiphaneia)[2146] and Tel Aviv (Ioppi).[2147]

### 144. Amphorisks, spherical, ring base (Isings Form 129)

(Cat. no. 689)
(Plates 41, XXII)

Vessels of this form have a long cylindrical neck, with usually unworked rim. The body is globular and the base is a pushed-in tubular ring. Two strap-handles start from the shoulders, rise vertically, and at the middle of the neck bend in a right angle and adhere to it. Right above the handles a thick trail is wound around the neck forming a raised ring. Their height is estimated at c. 12cm.

They are free-blown, the base is formed by tooling the lower part of the body mass; then the trail is applied around the neck and finally the handles are added.

The example from Thessaloniki is made of colourless, greenish glass.

---

2134 Makropoulou 2007, 286, pl. 24, from a grave used in the first half of the 4th century, according to a clay vessel found inside.
2135 Makropoulou 2007, 258, from a grave with a coin of Constantine I, struck in AD 336-337.
2136 Whitehouse 2001, 184-185, no. 729, find dated to the 3rd-4th century according to analogous finds.
2137 Whitehouse 1997, 169 no. 297, find dated to the 4th-5th century according to analogous finds; there are also cited analogous finds with the same combination of colours.
2138 Israeli 2003b, 258, nos. 344-345.
2139 Papanikola-Bakirtzi 2002 (ed.), 224, no. 259 (D. Nalpantis).
2140 Whitehouse 1997, 190-191, nos. 334-335, examples dated to the 3rd-4th and 4th century, respectively. Whitehouse 2001, nos. 721, 722, 725-727, examples dated to the 4th and 4th-5th century respectively.
2141 Matheson 1980, 76, nos. 197-199. Especially similar is no. 197, with hemispherical, purple body with colourless, greenish handle, dated to the 3rd-4th century according to analogous finds.
2142 Auth 1976, 103, no. 115.
2143 Oliver 1980, 100, no. 156, a somewhat bigger example with pointed base, dated to the 3rd-4th century, according to analogous finds.
2144 Hayes 1975, 79, no. 286, pl. 19, example with pointed base, dated on the basis of analogous finds to the 3rd-4th century.
2145 Stern 1977, 106-107, no. 31, bulbous example with a purple body and a colourless, greenish handle, dated to the first half of the 4th century according to analogous finds.
2146 Stern 1977, 106-107, no. 31, cited as comparable to the above-mentioned bulbous example with purple body and colourless, greenish handle, dated to the first half of the 4th century.
2147 Stern 1977, 106-107, no. 31, cited as comparable to the above-mentioned bulbous example with purple body and colourless, greenish handle, dated to the first half of the 4th century.

The vessels of this form are mainly known from the West, although similar vessels with tubular or fire-polished rim and ribbed body appear also in the East,[2148] where they are dated to the 4th century.[2149] Firmly dated examples have been published from Britain,[2150] Bonn (Bonna),[2151] Trier (Treviri),[2152] Cologne,[2153] and Strasbourg (Augusta Trebecorum, Argentoratum).[2154]

The find from Thessaloniki is dated by the excavation context in which it was found to the 4th century.[2155]

### 145. Unguentaria, spherical, cracked-off rim and tall neck (Isings Form 103 min.)

(Cat. nos. 690-699)
(Plates 41-42, XXII)

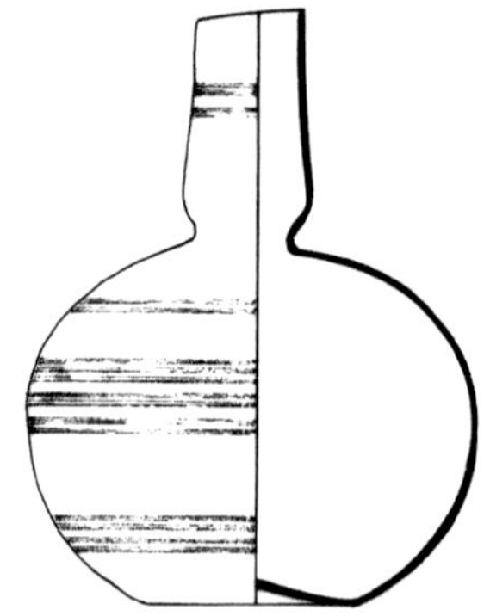

The vessels of this form have no specially formed rim; the rim is relatively short widening towards the body, strongly constricted at its base; the body is globular and the bottom is flat, slightly concave. Their height is estimated between 4-15cm.

The examples of this group often bear incised decorations, usually simple, consisting of parallel shallow grooves or bands of incisions. Less often more complex geometrical motifs appear, i.e. concentric or intersecting circles; quite rarely there are examples decorated with incised figural motifs or with painted themes.

They are free-blown and the constriction at the base of the neck is formed by the jacks used to shape the cylindrical neck. The rim remains unworked or occasionally is mildly ground.

It is a widely distributed form[2156] which mainly appears in the West but also in the East, where a variant with fire-polished rim is also known.[2157] It appears sporadically in the early 3rd century and from the end of the century becomes popular and remains in use in the 4th century as well. This form appears in the same sites and strata as the larger forms – above 16cm – which are included as tableware (Form 50).

Examples of small-sized vessels have been found, inter alia, at Trier (Treviri),[2158] Mainz am Rhein (Moguntiacum),[2159] Augst,[2160] Dunaújváros (Intercisa),[2161] Budapest,[2162] Kerch (Panticapaeum),[2163] Niš (Naissus),[2164] Zaskokin Kosovo,[2165]

---

[2148] *Glassammlungen der DDR 1973-1976*, 35, fig. 18, from Giza, dated to the 4th-6th century; La Baume 1973, no. D94, from Palestine, dated to the 4th century according to analogous finds.

[2149] Isings 1957, 159, form 129, with relevant dated examples.

[2150] Price, Cottam 1998a, 167-168, finds from various sites dated to the late 4th century. The examples from Britain are different from the typical examples of this form having a tubular rim.

[2151] Follmann-Schulz 1988, 39, no. 90, pl. 9, dated from excavation results to the 4th century.

[2152] Goethert-Polaschek 1977, 235-236, form 143, pls. 25, 295b, 75, dated from excavation results to the 4th century.

[2153] Fremersdorf 1933, three examples, pl. 56, nos. 1, 2, 4, dated to the first half of the 4th century; Fremersdorf 1962, 37, pl. 55 dated to the 4th century; Kisa 1908, two examples, fig. 160g, page 325 and fig. 239a, page 523; Loeschcke, Willers 1911, four examples, nos. 133, 466, 467 and 469.

[2154] Arveiller-Dulong, Arveiller 1985, 170-172, nos. 376-382 (drawings on page 223, photos on pages 278-279), from graves dated to the late 3rd or 4th century.

[2155] I thank the excavator, mr. Dimitris Nalpantis, for sharing with me the results of his unpublished excavation.

[2156] Isings 1957, 122-123, form 103.

[2157] On this topic, see Dussart 1998, 151, where analogous finds from the Syro-Palestinian region are cited.

[2158] Goethert-Polaschek 1977, nos. 903-905, 908, 912, form 93, pls. 58-59. The author makes no distinction in terms of vessel use based on their height. The finds are dated from excavation results to the second half of the 3rd and to the first half of the 4th century.

[2159] Harter 1999, 140-141, form G21. No distinction is made by the author as to the size of the vessels, yet to this group can be ascribed Cat. nos 932-934, 938-939, 944, 947-949, 952. From the rest of the museum's examples, 7 preserve chronological indications which ascribe them to the period between the middle of the 3rd and the middle of the 4th century.

[2160] Rütti 1991, I 54 AR 154.1, 130-131, nos. 2599-2600, pl. 110. No distinction is made as to the size and/or the use of the vessels. These finds are dated from excavation results to the middle of the 3rd to the middle of the 4th century.

[2161] Radnóti 1957, 146, no. 132, from a grave dated to the end of the 3rd or early 4th century by a coin of Diocletian (284-305).

[2162] Barkóczi 1988, 145, no. 314, pls. XXVI, LXXXVII, unprovenanced find.

[2163] Kunina 1997, 299, no. 225, from a grave in Kerch (Panticapaeum) dated to the second half of the 3rd century. The same vessel was presented also in Kunina 1984, 156, 158, no. 19, pl. III, 19.

[2164] Drča 2000, 218, no. 11, dated, probably from excavation results to the 3rd-4th century; Ružić 1994, 15, no. 74, dated, probably, from excavation results to the 4th century.

[2165] Ružić 1994, 15, no. 75, dated from excavation results to the middle of the 4th century.

Veroia,[2166] Thermopylae,[2167] Karditsa,[2168] Nea Kerdyllia,[2169] Bosra (Nova Trajana Bostra),[2170] and Hanita.[2171]

The vessels that can be independently dated are consistent with the 3rd- and 4th-century general time-frame of the form (Cat. nos. 692,[2172] 698,[2173] 699[2174]).

### 146. Unguentaria, figural, 'Janiform' (Isings Form 78b)

(Cat. nos. 700-712)
(Plates 42, XXIII, XXIV)

The body of vessels of this form has the shape of two heads arranged back-to-back. For early glass-researchers this shape was reminiscent of the Roman god Janus, and this form was accordingly styled 'Janiform'. The rim is everted, tubular and infolded; the cylindrical neck is relatively long and of the same height as the body. The vessel stands on its flat base. Their height is estimated between 8-9cm.

Most are made in a bipartite mould, consisting of two vertically arranged sections. The seam of the junction of the two parts is concealed in the relief of the hair, usually along the join of the two heads, thus leaving the face area clear of marks. The mould was open at its base and because of this the base of the vessel shows discernible traces of dust, small holes (cavities) and the anomalies of the surface on which the mould was standing when the vessel was formed. The neck and rim were constructed by free-blowing and further tooling.

Janiform vessels belong to a large group of mould-blown vessels that quite naturalistically depict human heads. The body of the vessel features either a whole head, or the frontal part of two heads placed back-to-back. There are also a few examples that comprise more than two heads forming the body. The vessels of the first and second group are occasionally supplemented with a handle. Most of the examples in general are handleless flasks or unguentaria, some are single-handled cups, and very few are double-handled, close-shaped vessels.[2175]

Figural cups in the shape of a human head are already known from the Greek Archaic period and their reappearance, originally in clay, probably is connected with an antiquarianism, a trend for collecting antiquities, in Augustan period.[2176] Glass vessels in the shape of a human head appear towards the end of the 1st century[2177] and continue to be produced until at least the end of the 4th.[2178] Apart from chubby children, which were probably meant to represent Eros or Dionysus glass vessels also vessels depicted other deities, e.g. Medusa and Dionysus-Antinous. In addition, men with negroid facial characteristics or grotesque figures are also rendered. Representations of beardless men, which have been identified with some deity or with the emperor, are known from northwestern Europe in the 3rd and 4th centuries.[2179]

Janiform vessels are typically products of the Eastern Mediterranean, probably from the Syro-Palestinian region, where many of the extant examples are found.[2180] They all belong to Stern Form A4,[2181] dated to the 3rd century, characterized by heavy cheeks and chin, and accentuated curly hair.

[2166] Allamani-Souri 1991b, 302-303, pl. 114ε, example from a grave dated to the 4th century, according to its context.

[2167] Weinberg, McClellan 1992, 74, 131, nos. 105-106, two vessels with applied and incised decoration which are dated to the end of the 2nd century. Although there is no obvious reason to restrict their dating to the early 3rd century, the only other find with relevant decoration to have archaeological data has been re-dated to that period, see von Boeselager 1989, 35. On the excavation, see *BCH* 20, 1896, pl. 29, nos. 2, 9.

[2168] Weinberg, McClellan 1992, 76, 133, no. 108, confiscated object, dated according to analogous finds, to the 3rd century. It is decorated however with an applied decoration in a manner very typical of the 1st century.

[2169] Malama, Ntarakis 2008, 122, pl. 78, MA 6481 (From a grave dated to the first half of the 3rd century); idem, 122, pl. 78, MA 6482 (From a grave dated to the first half of the 3rd century); idem, 145, pl. 78, MA 6555 (From a grave dated on the basis of the glass vessels to the first half of the 3rd century); idem, 346, pl. 78, MA 7077 (From a grave dated to the first half of the 3rd century); idem, 356, pl. 78, MA 7100 (From a grave dated to the first half of the 3rd century); idem, 356, pl. 78, MA 7101 (From a grave dated to the first half of the 3rd century).

[2170] Dussart 1998, 151, form BX. 511, fire-polished rim; Coscarella 1994, 393, another example, dated according to an analogous find from Samaria to the 3rd century.

[2171] Barag 1978, 23, 26-27, no. 45, fig. 12 pl. IV, short vessel with fire-polished rim, dated from excavation results to the 3rd to early 4th century.

[2172] Papanikola-Bakirtzi 2002 (ed.), 470, no. 647 (E. Marki), find from a grave dated to the 3rd century.

[2173] From a grave used during the 3rd century and again in the 4th, according to its glass context (Cat. nos. 199, 270, 736-737).

[2174] Petsas 1970a, 347-351 (esp. 349), from a grave in use for a long time and certainly during the 3rd and 4th centuries.

[2175] Auth 1976, 75, no. 76; Stern 1995, 202.

[2176] Stern 1995, 201, note 5.

[2177] Find from Crete, Carington-Smith 1982, 281, no. 72, fig. 76, pl. 39k, l.

[2178] For examples from Rome, see Sternini 1989, 40-42, pl. 9/54, dated to the end of the 4th- middle of the 5th century. Also, a similar fragment was found in Rome, at Magna Mater temple, dated from excavation results to the first half of the 5th century, see Sternini 1989, 42.

[2179] See Form 96.

[2180] Stern 1995, 203.

[2181] Stern 1995, 210-212, this form is dated to the 3rd century, according to two analogous finds from Egypt loosely dated between the 3rd and

Analogous finds have been found in excavations in Marseilles (Massalia),[2182] Jajce,[2183] Homs (Emesa),[2184] Dura-Europos,[2185] Kom Aushim (Karanis),[2186] and Tarsus;[2187] the dates vary between the end of the 2nd century and the middle of the 5th. Analogous finds from museum collections, without excavation data, are held in Newark,[2188] Düsseldorf,[2189] Toronto,[2190] Berlin,[2191] Cologne,[2192] Warsaw,[2193] and Pittsburgh.[2194]

Finds from Thessaloniki, which can be dated by the archaeological context in which they were found,[2195] are dated to the 3rd and 4th centuries.

**147a Unguentaria, large fusiform, convex base**

(Cat. nos. 713-741)
(Plates 43-45, XXIV)

**147b Unguentaria, small fusiform, flat base**

(Cat. nos. 742-743)
(Plates 45, XXIV)

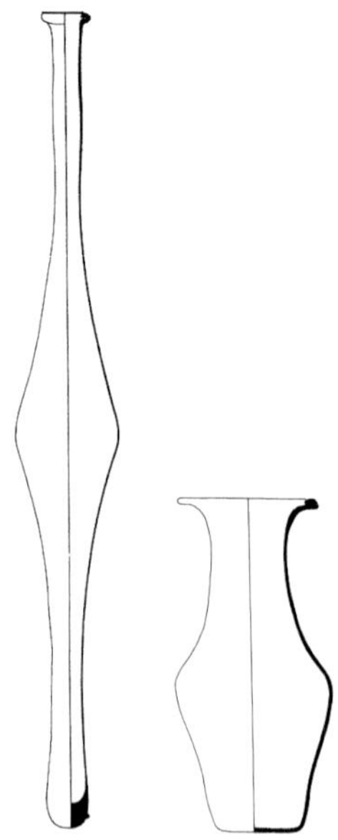

The flasks/unguentaria of this form have a particularly elongated biconical body, which terminates into a cylindrical, or less often into a slightly globular solid toe-base. The neck is cylindrical and the rim is infolded, tubular, slightly everted, and quite often carelessly executed. Their height is estimated between 40-55 and 12-22cm, respectively.

These vessels were made by free-blowing, marvering and drawing out. The rim was shaped by metal jacks, while a blowpipe was attached underneath the vessel, leaving very often a clear pontil mark. These vessels cannot stand on their own and were either kept lying horizontally, or, more probably, held upright within other containers, e.g. kalathoi. The poorly executed rims point to an exclusively burial rather than functional use, or, more probably, to the assumption that they contained some kind of powder, extracted with the help of a small rod.

The vessels from Thessaloniki are made of colourless, greenish glass with many pinprick bubbles and impurities.

It is a form that appears across the entire Empire and is dated to the 4th and 5th centuries.[2196] Dated examples, inter aliea, are known from France: Strasbourg (Augusta Trebecorum, Argentoratum),[2197] Arles (Arelate), Marseilles (Massalia), Verdier (Lunel-Viel);[2198] Germany: Cologne,[2199] Trier (Treviri),[2200] Bonn (Bonna),[2201] Zülpich-Enzen,[2202]

4th centuries.

[2182] Foy, Nenna 2001, 218, no. 389, a vessel found in a sarcophagus dated to the first half of the 5th century.

[2183] Paškvalin 1976, 115, pl. V. 4, a find from a grave dated to the 4th century according to its context.

[2184] Abdul Hak 1965, 30, fig. 11.

[2185] Clairmont 1963, 37, no. 132, pl. XXI, dated to the reign of Commodus (180-192) by the coin found in the grave. In the same city have been found 4 more fragments dated to the period prior to AD 256, see Clairmont 1963, 37-38, nos. 133-144.

[2186] Harden 1936, 214, no. 629, pl. 18, with a Christogram on its base, which was found in a house dated to the 2nd-3rd century. Although this relatively early dating by the excavator was supported more recently (Stern 1995, 212), it seems more probable that it belongs to a later phase (4th-5th century), as has been seconded already by other researchers (e.g. Auth 1976, 74, no. 74).

[2187] Goldman (ed.) 1950, 402, no. 26, fig. 274, according to Stern 1995, no. 151.

[2188] Auth 1976, 74, nos. 74-75, dated according to analogous finds from other sites to the 2nd-3rd century.

[2189] Saldern *et al.* 1974, 170-171, nos. 465-456, generally dated to the 2nd century.

[2190] Hayes 1975, 50, no. 94, pl. 7, dated according to analogous finds from other sites to the late 3rd-4th century.

[2191] Platz-Horster 1976, 47, no. 75, from a private collection, erroneously dated to the 1st-2nd century.

[2192] La Baume 1973, no. C14, pl. 10/4. RGM Cologne, with 7 more examples from the same museum, all from private collections, dated erroneously to the 1st-2nd century.

[2193] Filarska 1952, 112-113, no. 87, pl. XVII/ 3, who dates it to the 2nd-3rd century, according to an analogous find from Dura-Europos placed in the reign of Commodus (180-192) by the coin found in the grave, see Clairmont 1963, 37, no. 132, pl. XXI.

[2194] Oliver 1980, 64, no. 61, dated to the 1st century, although he cites only Hayes (1975), who dated it to the late 3rd-4th century.

[2195] Cat. nos. 700, 712: from a group of Early Christian graves dated to the 4th century, see Chapter 9 (Find-Spots), 'Agiou Dimitriou Street, extension of the street to the East, 1960'; Cat. no. 702: Makropoulou 2007, 131, pls. 26, 68, from a grave which is probably dated to the 3rd century; Cat. no. 703: Makropoulou 2007, 261, pl. 27, from a grave which is probably dated to the 4th century; Cat. nos. 707-708: dated probably to the 3rd-4th century according to the glass vessel with which it was found; Cat. no. 704: from a grave with a bronze belt buckle dated, probably, to the 4th century; Cat. no. 701: from a grave dated to the 4th century, see Papanikola-Bakirtzi 2002 (ed.), 224, no. 258 (E. Marki).

[2196] Morin-Jean 1922-23, 82, form 32, objects dated to the period of the Constantinian dynasty. Isings 1957, 126, form 105, dated to the 4th century.

[2197] Arveiller-Dulong, Arveiller 1985, 143-144, dated from excavation results to the second half of the 3rd or to the first half of the 4th century.

[2198] Foy 1995, 191, pl. 2, finds dated from excavation results to the 4th and to the first half of the 5th century.

[2199] Doppelfeld 1966, pl. 67b, 4th-century finds; Loeschcke, Willers 1911, pl. LII/767.

[2200] Goethert-Polaschek 1977, 142-146, form 85, nos. 797-823, pl. 57, finds dated from excavation results to early or second third of the 4th century.

[2201] Follmann-Schulz 1988, 20-21, pl. 2/9-10, without an independent dating.

[2202] Follmann-Schulz 1989a, 55, no. 14, pl. 2/14, from a sarcophagus dated ca AD 360.

Mainz (Moguntiacum);[2203] Italy: Angera (Angleria),[2204] Vicenza (Vicentia),[2205] Modena (Mutina),[2206] Aquileia;[2207] Hungary: Szőny (Brigetio),[2208] Fenékpuszta (Valcum),[2209] Budapest (Aquincum),[2210] Ságvár (Tricciana),[2211] Dunaújváros (Intercisa);[2212] Romania: Costanza (Tomis),[2213] Mangalia (Kallatis);[2214] Bulgaria: Plovdiv (Philippopolis);[2215] Croatia: Sremska Mitrovica (Sirmium),[2216] Đakovo (Certissa);[2217] Serbia: Belgrade (Singidunum),[2218] Karataš (Diana),[2219] Obrenovac,[2220] Leskovac,[2221] Niš (Naissus);[2222] Montenegro: Komini (Municipium 'S...')[2223] and Duklja (Doclea);[2224] FYROM: Skopje (Scupi),[2225] Lopate;[2226] Syria: Dura-Europos;[2227] Cyprus;[2228] and Israel: Kisra.[2229] In Greece published examples are known from Veroia,[2230] Edessa,[2231] Knossos,[2232] and Nea Kerdyllia.[2233]

Some of the vessels of this form, represented in Thessaloniki by one example (Cat. no. 743), are of an obviously smaller size, with flat, hollow base. According to data from their excavations they are not distinguished chronologically or geographically, and dated also to the 4th-5th century, a period to which the find from Thessaloniki should also be ascribed. Dated examples have been found at several sites, e.g. Bonn (Bonna),[2234] Ságvár (Tricciana),[2235] Dunaújváros (Intercisa),[2236] Costanza (Tomis),[2237] and Duklja (Doclea).[2238]

The vessels of this form are clearly similar to Hellenistic fusiform clay unguentaria.[2239] Although this has been noted quite early on,[2240] it does not seem possible for there to be any kind of connection between the two groups, which appear separated by several centuries. Furthermore, it seems that there was never just one production centre, e.g. Egypt,[2241] or Pannonia,[2242] or the region of the Rhine. On the contrary, it appears far more convincing, from the great number of finds known and their distribution throughout the Empire, that they were produced at several centres contemporaneously. The obvious resemblance among the more than 30 hitherto excavated examples from the cemeteries of Thessaloniki make the city yet another potential production centre for this form.

2203 Harter 1999, form D18, 102-103, no. 719, form dated from excavation results to the late 3rd or mid 4th century.
2204 De Tommaso 1990, 76, form 57, dated to the 3rd-4th century.
2205 De Tommaso 1990, 76 form 57, dated to the 3rd-4th century.
2206 De Tommaso 1990, 76 form 57, dated to the 3rd-4th century.
2207 Calvi 1968, 152, pl. 24/4-6, finds dated from excavation results to the 3rd-4th century.
2208 Barkóczi 1968, 68, 84, no. 25, pl. 34/1; Barkóczi 1988, 128, no. 253, find dated from excavation results to the last third of the 3rd or early 4th century.
2209 Benkő 1962, pl. III/4, found with a coin of Valens (364-378), according to Barkóczi 1988, 128, notes of cat. entry no. 253.
2210 Kaba 1958, pl. VI/2, from a sarcophagus with a burial dated to the last third of the 3rd or early 4th century.
2211 Burger 1966, 106, 134, 139, form 24, figs. 98. 63/3, 122. 335, 125. 24, of the second half of the 4th century and the last third of the 4th century.
2212 Radnóti 1957, 145, no. 52, 119, pl. XXVI/1, 8, find dated from excavation results to the 4th century.
2213 Bucovală 1968, 127-130, figs. 260-270, fig. IX, which are dated to the 4th century according to analogous finds (page 127), and more precisely to the early 4th century, a date probably also supported by other internal indications (see pages 152-153).
2214 Preda 1980, 34, pls. XIII, LVIII, LXXI, dated to the 4th century according to its context.
2215 Cherneva-Tilkian 1997, 51, two examples dated to the 4th century according to analogous finds.
2216 Cermanović-Kuzmanović 1981a, 14, 21, pl. V 87-90, photo 14, from a necropolis dated to the 3rd-4th century.
2217 Šaranović-Svetek 1986, 29, find dated by coins to the first third of the 4th century.
2218 Pop-Lazić 2002, 58-59, from a cremation dated on the basis of analogous finds between the end of the 3rd and the end of the 4th century.
2219 Ružić 1994, 66, find dated from excavation results to the second half of the 4th or early 5th century.
2220 Ružić 1994, 66, find dated from excavation results to the 4th century.
2221 Ružić 1994, 66, find dated from excavation results to the 4th century.
2222 Drča 2000, 214, photos 21-24, examples (4) from the same spot, possibly of the 4th century, dated according to analogous finds to the 4th century.
2223 Cermanović-Kuzmanović 1971, 294, pl. III/15.
2224 Cermanović *et al.* 1975, 169-170, dated according to their context to the end of the 3rd or to the first half of the 4th century.
2225 Mikulčić 1976, 196, pl. VIII. 232, dated by a coin of Constantinus I (324-337) to the second quarter of the 4th century.
2226 Nikolovski 2006, 244, fig. VII/3, find from a grave dated to the first half of the 4th century on typological reasons.
2227 Clairmont 1963, nos. 736, 737, pl. XXXVI/736, finds dated for historical reasons before AD 256, in the period between the 2nd and the middle of the 3rd centuries, and which present the earliest examples of the form.
2228 Vessberg 1956, 162, 166, fig. 50:31, an example from the Cypriot Museum is mentioned and 12 similar ones from the Cesnola collection.
2229 Stern E. 1997, 17, fig. 31, vessel of this form (with indentations around the body) from a cave used for burials during the 4th and 5th centuries.
2230 Allamani-Souri 1991a, 301-302, pl. 114δ, Allamani-Souri 1991c, 39, fig. 1, from a 4th-century grave, according to its context.
2231 Chrysostomou A. 2000, 413, pl. 474, vessels from a burial with no chronological indications.
2232 Frend, Johnston 1962, 186-238 (esp. page 196), figs. 9, 205-210, 231, finds dated to the 4th and 5th centuries according to archaeological contexts.
2233 Malama, Ntarakis 2008, 246, pl. 77, MA 6832, from a grave dated to the middle of the 4th century.
2234 Follmann-Schulz 1988, 20, pl. 2.9.
2235 Barkóczi 1988, 129, nos. 255-266, find dated to the second half of the 4th century.
2236 Radnóti 1957, 154, 155, no. 3, pl. XXVIII/1, a 4th-century find, according to the archaeological context.
2237 Bucovală 1968, 129, 152-153, fig. 265, find dated to the 4th century (more precisely the beginning).
2238 Cermanović *et al.* 1975, 169-170, form V/4B, drawing 40, from a grave dated to the first half of 4th century.
2239 For examples from Thessaloniki, see Adam-Veleni *et al.* 2000, pl. 139γ, 141α.
2240 Vessberg 1956, 205-206.
2241 Radnóti 1957, 145, pl. XXVI/8; Filarska 1952, pl. XXXVIII/7.
2242 Barkóczi 1988, 128.

The finds from Thessaloniki, that can be dated by the archaeological context in which they were found, are dated to the period between the late 3rd and up to the 4th century (Cat. nos. 725,[2243] 727,[2244] 736-737,[2245] 740,[2246] 713,[2247] 714,[2248] 71 718-719,[2249] 721,[2250] 722.[2251]

**148 'Stirring rods' (Isings Form 79)**

(Cat. nos. 744-752)
(Plate XXIV)

Objects comprising this form actually are fragments of twisted rods, most usually twisted in one or two directions, or, more rarely, smooth ones. The ends of the rods were shaped in different ways. They are known as stirring rods although there are no precise indications that they were indeed used for stirring liquids. The lengths of objects of this form are estimated at c. 12-18cm.

None of the fragments from Thessaloniki have preserved ends and thus their exact shape remains obscure. For that reason, all types of rods known are presented below.

Rods in general can be divided into the following subgroups:

A) Rods bent at one end to form a closed ring, probably used as a handle. They have a cylindrical body and the second, straight, end is delineated by a small disk. Occasionally a relief decoration was added to this disc: either a simpler bead-shaped motif, or more elaborate creations in the shape, i.e., of a bird, juglet, or dolphin.[2252] They are quite sturdy implements which could indeed be used as stirring rods for mixing e.g. water with wine. A different use for these rods can be assumed on the basis of two 3rd-century Phrygian funerary stelae:[2253] a distaff with a whorl is depicted on one of them, which is connected by the thread with a shaft (another distaff or probably a reel) that ends at one end in a ring, and at other end in a disk with a bead on it. The second stele depicts, placed in a kalathos, a distaff with thread and next to it a shaft with thread reeled around it, the only visible end of which is a bead on which is based a bird-shaped element. There is no indication that the illustrated shafts represented glass objects. However the similarity of the shape with the glass rods of this subgroup is so obvious that we may assume that the glass rods were imitating utilitarian, wooden distaffs or reels, and that they were placed in female burials as symbols of this feminine activity.[2254]
B) Rods placed in the hole of glass conical whorls, which probably functioned as sealings for the bottle in which the rod was placed. In this case the rod was used as a shaft for withdrawing some of the contents of the vessel, and known in the sources as siphōn (σίφων), although siphōn is hollow and actually a tube. Already in the iconography of Classical Greek vases they are depicted on representations of perfumers. In each of the perfumer's vases another rod was placed so that potential buyers could 'test' different fragrances without mixing them up. In addition, there are scenes where the rods are dabbed onto the wrist of the buyer or they are kept in front of the nose of another one.[2255] Some of the finds of this type have been identified as real distaffs, although the fragile nature of glass quite probably rules out such a use.
C) Another group represents the small, thin stirring rods used for the extraction of a liquid cosmetic or medicine from close-shaped vessels.

---

[2243] They were found with coins reg. no. MΘ 3667-3679.
[2244] Found with a glass vessel of the form 49, Cat. no. 205, dated to the 2nd or 3rd century.
[2245] Found with glass vessels Cat. nos. 199, 270, 698, which are dated to the late 3rd and 4th century, and with 4 unidentifiable copper coins.
[2246] Found with 3 bulbous unguentaria, which according to Nalpantis (2003, 123, pl. 39), are ascribed to his form 1.α and are dated to the first half of the 4th century. Also, a single-handled cup was found with them and which has parallels dated to the late 3rd or into the early 4th century, according to Nalpantis 2003, 125, pl. 40.
[2247] Quite probably from the *cubiculum* dated to the second half of the 3rd century, according to the archaeological context in which it was found, as cited by Marki 2006, 211.
[2248] The grave is dated to the second half of the 3rd century on the basis of pottery finds, and also on the style of the wall-paintings that decorated the grave.
[2249] From a grave precisely dated by coin to the reign of Maximian (286-305).
[2250] Makropoulou 2007, 69, from a destruction layer in which coins of the second half of the 4th century were found.
[2251] Makropoulou 2007, 340, pl. 27, from a grave dated to the 3rd or 4th century on typological grounds.
[2252] Spaer 2001, no. 631; Stern 2001, no. 228; Whitehouse 2003a, no. 972; Arveiller-Dulong, Nenna 2011, 306-307, 330-331, with further bibliography; Fremersdorf and Polonyi-Fremersdorf 1984, 111, no. 249.
[2253] Campbell 1997, 21-23, Vlizos 2003, 25-35.
[2254] Tzachili 1997, 136-137.
[2255] Black-figured *pelike* with relevant illustrations of *siphones* used by *myropoles* include, *inter alia*, Florence, Museo Archeologico 72732: Scheibler 1983, 19, fig. 7. Tarquinia, Museo Archeologico RC 1063: Tronchetti 1983, 113, pl. XLVa-b. Agrigento Museo Archeologico Regionale 34: *ARV2*, 185, 30. Pernice 1893, 182-183. Fundamental works on fragrances in antiquity are Brunn 1999 and Brunn 2000. For a review on fragrances and unguentaria in Archaic and Classical Greece, see Aktseli and Manakidou 1997, especially on stirring rods, see page 90, fig. 11, and page 103, figs. 23-26 with further bibliography; more recently on the same topic, see Verbanck and Massar 2013, 273-300 (esp. page 276) where stirring rods are discussed.

D) A fourth group represents pointed needles, smooth or twisted.

Two subgroups are discerned:

a) Those without a hole, for which has been proposed the ancient Greek term kalamis (καλαμίς) or the Latin discerniculum, i.e. a shaft used by women to divide the locks of their hair to arrange a complicated hairstyle.[2256]
b) This subgroup comprises long glass needles with a hole at one end. They were probably decorative rather than utilitarian, as they would be unable to withstand the pressure of penetrating a thick textile (corresponding to their thickness).

All needles, with an eye or not, could have been used for extracting contents (i.e. cosmetics or medicines) from a vessel and applying to the face or body.[2257] The use here would be similar probably to that of a dropper/pipette, possibly identified as a siphōn (σίφων),[2258] a term known from ancient Greek sources, a form of hollow cane, tube, or large needle. These are extremely rare finds, with published examples from Thessaloniki,[2259] Dion,[2260] and two from the Serbian site of Orašije (Margum).[2261]

E) Fragments with unpreserved ends could be parts of rectilinear bands for intarsia decoration, either on the walls of a building or on wooden objects or furniture.[2262]

Generally speaking most of the rods are dated to the Roman period, although they appear from the Hellenistic era;[2263] they are more common in the West than the East.[2264] In Greece published examples have been reported from Athens,[2265] Sifnos,[2266] Naxos,[2267] Delos,[2268] Kastelli Kissamou,[2269] Meniis,[2270] Aghialos,[2271] and Edessa.[2272]

Among the finds from Thessaloniki that can be independently dated are Cat. no. 746[2273] and Cat. nos. 744-745,[2274] which are dated to the 1st century.

### 149. Appliqués/Medallions, with relief masks

(Cat. nos. 753-754)
(Plates 45, XXIV)

This group comprises discoid medallions decorated with relief theatrical masks. All three examples in our catalogue feature comical masks with a wide, smiling, open mouth.[2275] The hair is rendered as rows of parallel ribs that start from the outline of the face; in one

2256 Oxford Latin Dictionary, s.v. 'Discerniculum', 'the bodkin in a woman's head-dress, which parted the hair, a hair-bodkin', Lucil. ap. Non. 35, 31; cf. Varr. L. L. 5, § 129. Liddell and Scott, s.v. 'καλαμίς'. For finds of this type, see Davaras 1985, 201; Platz-Horster 1976, 82, no. 166, smooth example dated to the 1st-2nd century according to analogous finds; Gojković 2015, 267-268, two twisted pins with bird-shaped finials from a 2nd- or early 3rd-century grave.
2257 Spaer 2001, 262, 264-265, dated to the 1st-2nd century.
2258 Liddell and Scott, s.v. 'Σίφων', *Hippon.56, PEleph.5.4.*
2259 Tsimpidou-Avloniti 2003, 236, fig. 9, find from a jar grave of a baby dated to the 1st or 2nd century according to the general dating of the excavation.
2260 Pantermalis 2001, 353, fig. 12.
2261 Karović 1999, 93, pl. VII/4.
2262 Grose 1989, 349, 357-358, nos. 657-673.
2263 Spaer 2001, 262, finds from Rhodes, Delos and Jerusalem are cited, dated respectivly to the 3rd, 2nd and 1st centuries BC.
2264 Isings 1957, 94-95, form 79; Spaer 2001, 262, 264-265 with detailed bibliography, finds from Kostolac (Viminacium) and other Serbian sites, dated by coins and their archaeological context mainly to the 1st-2n century, and also to the early 3r century, are published by Ropkić, 2006, 375-380; Ružić 1994, 56, pl. 43/13, find from Ravna (Timacum Minus); Karović 1999, 93, pl. VII/4, finds from Margum; Stamenković, Ivanišević, Pešić 2016, p. 28-30, ill. 16:1, from a cremation at Mala Kopašica cemetery, dated between the 2nd and the late 3rd century; Buljević 002αα, 297-312, on 44 examples from Solin (Salona), among which appear all variants of stirring rods so far known, and they are dated by archaeological context to the 1st and 2nd centuries.
2265 Alexandri 1973-74c, 128-131, pl. 102, two examples, one smooth and one twisted, from a Roman cemetery.
2266 Mackworth-Young 1949, 80-81, 90, grave 20 pl. 36. 1, a 1st-century, smooth example, found with coins of Vespasian, minted in AD 70-71, and Titus, minted in AD 80.
2267 Zafeiropoulou 1999, pl. 86α, dated according to analogous finds.
2268 Nenna 1998, 151-152, pl. 55, detailing 9 fragmentarily preserved examples.
2269 Tzedakis 1978, 370-372, pl. 191, twisted examples from a 2nd-century grave.
2270 Chrysostomou 2000, 455-467, a twisted example from a 2nd- to 3rd-century burial complex.
2271 Kourkoutidou 1970, 231, pl. 232:b.
2272 Chrysostomou A. 2000, 284, pl. 321, two twisted examples with a ring-shaped finial from a grave dated on the basis of its content to the second half of the 1st century. On the dating of the grave see op. cit., 279.
2273 Find from a grave with 1st-century bird-shaped and globular unguentaria, Forms 120 and 116.
2274 Finds from the baths in the ancient agora of Thessaloniki, which was destroyed in the last 20 years of the 1st century, see Adam-Veleni *et al.* 1998, 85-102; Mavromichali 2001, 141.
2275 The third find was unearthed in an unpublished excavation of the *fossa*, the external ditch of the western city wall of Thessaloniki, from a site that provided insufficient dating evidence.

of the examples they are presented vertically and almost surround the face like rays, while in the second example they bend and follow the outline of the head.

They are made by mould-pressing and they are parts of vessels which have not survived. It is quite probable that they were intended for decorating jugs, although cups and bowl-shaped lamps are known with this type of decoration also.

The well-known Classical Greek and Hellenistic[2276] decoration of metal vessels, whereby a medallion bearing a face, or a theatrical mask, was applied below the base of the handle, was a feature of glassworking adopted already in the 1st century AD; the earliest known examples are preserved on glass cameo vessels.[2277]

In the 1st century, stamped medallions were mostly used for the decoration of free-blown vessels, mostly jugs of Isings Forms 52 and 55, and they were placed below the base of the handle.[2278] Preserved examples are known from Colchester (Camulodunum),[2279] the Oppenländer Collection in Germany,[2280] Bonn (Bonna),[2281] Cologne,[2282] Stein in Holland,[2283] Windisch (Vindonissa),[2284] Budapest,[2285] Strasbourg (Augusta Trebecorum, Argentoratum),[2286] Besançon (Vesontio),[2287] and Corsica.[2288]

One medallion of this form is held in the Newark Museum;[2289] there are also examples made of opaque glass, e.g. in Valladas, Saint-Paul-Trois-Châteaux (Drôme),[2290] and Italian sites.[2291]

Almost none of the medallions bears an inscription or other symbol identifying the glassmaker or workshop. The only known exception is one Amaranthus, a glassmaker (or rather owner of a glass-workshop) who encircled with his name the figure of Medusa on the medallions he used to embellish his jugs. He was professionally active in the second half of the 1st century in Burgundy, SE France, and his products were distributed up to Normandy and the Rhineland.[2292]

Medallions in the shape of lion heads[2293] or horse busts[2294] appear below the bases of handles in the 4th century. In the same period vessels reappear with a medallion depicting a female face placed at the base of the handle. Examples are found in Romania, at Piatra Frecătei and at Capidava.[2295] A jug from Israel, ascribed on stylistic grounds to 4th-century Egyptian production, is also decorated with a mask.[2296]

Already from the end of the 2nd century some vessels appear on which relief masks were arranged freely around the body, regardless of the position of the handle, or even if there was a handle at all.[2297] Two identical bowl-shaped

2276 Themelis 1997, 175-178, figs. 46-49, a plethora of medallions is depicted decorated with various motifs.

2277 Harden *et al.* 1987, 63, Portland vase; Koltes 1982, 31-37, no. 83, pl. 46, a vessel from Besançon.

2278 Isings 1957, 69-71, 72-74.

2279 Cool, Price 1995, 118-120, medallions with a wreathed figure of Dionysus. Some of them reveal that after the vessels were broken the body parts were diligently removed and the medallion reused, probably as a pendant.

2280 Saldern *et al.* 1974, 193-194, nos. 529, 531, medallions dated to the 1st century according to analogous metal and glass finds, and are ascribed to the production of the Eastern Mediterranean.

2281 Kisa 1908, 413 fig. 201; Follmann-Schulz 1992, 21-22, no. 10.

2282 Fremersdorf 1958b, 24-25, pl. 13; Fremersdorf 1961, pls. 73 and 101; Doppelfeld 1966, pl. 21.

2283 Isings 1971, 36, no. 115, pl. 8, bluish-green medallion with a mask dated to the end of the 1st and early 2nd century according to analogous finds. It was found in a sarcophagus with no other traces of the vessel, and thus was considered to be proof that the medallions were used as pendants long after the vessels were broken.

2284 Berger 1960, 42, nos. 90-93, pl. 6, presenting 3 medallions (light-bluish, greenish and amber) with female faces and another (amber) that probably depicts Silenus.

2285 Barkóczi 1996, 105, cat. no. 336, pls. XXXIII, LXX, olive-green comical mask in high relief. It is not dated from the excavation results, but rather by the analogous finds, to the second half of the 1st century.

2286 Arveiller-Dulong, Arveiller 1985, 53-54, no. 69 (photo on page 231), blue-green medallion with relief female figure. It is not dated by excavation results, but nevertheless is presented as a 1st-century form.

2287 Koltes 1982, 31, no. 82, pl. 9, blue-green medallion with relief female figure. It is not dated according to excavation results, but by analogous finds to the second half of the 1st century.

2288 Foy, Nenna 2001, 88, no. 102, from the site of Mariana, near Lucciana. A blue-green handle with a female figure, dated to the second half of the 1st century.

2289 Auth 1976, 107, no. 122, dark-blue medallion with comical mask, which is assumed to be from Egypt. It is mentioned as 'Early Roman', but due to the high relief and the simplified way of rendering the figure, it might, probably, be dated to the later production of the 4th century.

2290 Foy, Nenna 2001, 125, no. 161, opaque white mask on a dark-purple vessel.

2291 Harden *et al.* 1968, 60, cat. no. 75; Tatton-Brown 1991, 62, figs. 83, 84, two-handled vessel in the British Museum from Santelipidio in Campania. The figures are well shaped, although they are rendered in high relief.

2292 Sennequier 1986, 11-18.

2293 See comments on Form 85.

2294 Ružić 2001, 413-422 (esp. page 42). Cylindrical beaker found in Komini kod Pljevalja (Municipium 'S...'), decorated with a rhomboid web. At the centre of each lozenge is a medallion bearing a relief equine bust, dated to the second half of the 3rd or to the early 4th century.

2295 Bucovală 1984, 62-63, fig. 8.

2296 Israeli 2003b, 190, no. 224, from an olive-green vessel with turquoise base and mask identical to Cat. no. 754.

2297 Isings 1964, 63, note 23, find from Bexhill (S England); Saldern 1968, no. 55, a single-handled vessel with a medallion of a comical mask placed

glass lamps dated to the 4th century are decorated around the body with medallions bearing female faces. They were found in Horrem and in Cologne[2298] respectively. Theatrical masks arranged around the vessel in three rows are preserved on a large, 4th-century cup from Cuijk (Ceuclum)[2299] in Holland. A 'sprinkler' held at the Newark Museum, datable to the 3rd-4th century, is decorated around its body with four (touching) comical theatrical masks.[2300] A further stamped theatrical mask is preserved on a medallion-shaped pendant from Solin (Salona),[2301] which is dated to the end of the 3rd or to the 4th century.

Observing diachronically the appearance of the medallions, it becomes clear how they developed and changed. Early on, cameo examples were made in very low relief, almost flat, probably because of the peculiarities of the production technique of the vessel. In addition, the medallions which decorated the free-blown 1st-century vessels were also quite low in profile compared to the almost hemispherical medallions of the 4th century. Furthermore, it is quite obvious how superior in artistic quality the 1st-century examples were, in terms of design and skillfully executed facial characteristics (e.g. hairstyles), compared to the cruder medallions of the 4th century. The two medallions from Thessaloniki cannot be dated precisely by their own archaeological contexts, but on stylistic grounds they seem to belong to production of the 4th century.

The figures on the medallions quite probably also have a symbolic character connected to the apotropaic attributes of Medusa, and associations with the Maenads, Silenoi, and the Dionysiac cult in general, including the contents of the vessel, i.e. the wine. The same motifs also appear on medallions on bronze vessels of the same time, which were apparently the prototypes for the glass and clay vessels.[2302] Bronze medallions often feature the heads of Medusa, Silenus, Satyrs, and several objects connected to the cult of Dionysus (e.g. thyrsus and Pan pipes), as well as lion heads and unidentifiable female figures.[2303]

on the body opposite the handle; it is held in Boston and was dated to the 3rd century.

2298 Isings 1964, 62, note 17; Harden *et al.* 1987, 204-205, no. 113, decorated with six identical medallions constituting a female figure with a diadem.

2299 Isings 1964, 59, photo 1.

2300 Auth 1976, 75, no. 76.

2301 Cambi 1976, 142, 144, 149, fig. II. 4.

2302 Isings 1964, 63; Cermanović-Kuzmanović 198b, 21, with relevant bibliography.

2303 Sennequier 1986, 17-18; Themelis 1997, 175-178, figs. 46-49.

# 5. Decoration of the Studied Material

A. Engraving

The technique of engraving or wheel-cutting, which was already known in glassware from Classical Greek and Hellenistic times,[1] involves the removal of areas of the vessel's surface using a bow-drill or cutting-wheel, with the use of a powdered abrasive, and/or by the use of pointed shafts and tools.[2] With this technique, glass vessels, known in ancient sources as *toreumata*,[3] were connected with the production and decoration of silver and bone vessels and objects;[4] it was quite possible to render on them composite figural representations and simpler geometrical compositions, just like on their more valuable crystal and metal prototypes.

***a. Figural motifs***

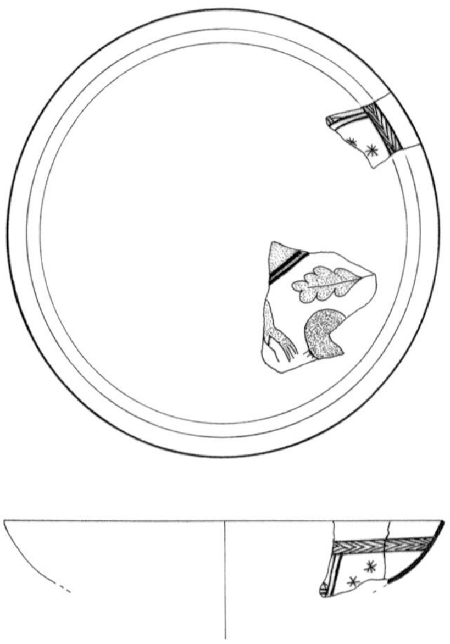

Composite inscribed motifs appear from Thessaloniki amost exclusively on shallow bowls of Form 16, i.e. the following fragments:

Cat. no. 75: On the bowl a standing figure is preserved, pointing with the right hand to a plant that looks like a poppy (?) and showing what would appear to be the seed pod.[5] Perhaps also from the same vessel is a herringbone band preserved around the rim, and parts of what once were eight-pointed stars.

On a bowl with the depiction of Christ and a rooster,[6] which symbolizes Peter's act of denial, the hand of Christ has the same posture and is rendered technically in a quite similar way to the find from Thessaloniki. On the same vessel the background is filled with eight-rayed stars of two types, and the simpler ones are identical to the ones from our vessel.

Identical stars appear also on a vessel from Serbia;[7] they fill the field between two male figures standing in front of an unidentifiable motif, probably a building, with which possibly can be identified the only slightly-preserved vertical motif on the fragment from Thessaloniki.

An analogous fragment with herringbone band, just a little simpler, comes from Luni (Luna);[8] it is generally dated to the early 4th century.

The vessel from Thessaloniki, which consists of two fragments, which do not adjoin and could even possibly be parts of two different vessels of the same form, lacks any specific stratigraphic clues that would allow its more precise dating, but on the basis of all the aforementioned parallels it can be dated to the 4th century, and even probably be ascribed to the production of a workshop in Italy.

Cat. no. 78: From another bowl a male, beardless, head is preserved in profile, over which the letters 'ZH' can be read. The figure appears with a disproportionately large triangular eye and upright hair, which is rendered by short, straight strokes. These features are typical for a trend known on other examples from Italy,[9] and these vessels are considered to be products of a workshop operating in Rome in the second half of the 4th century.[10]

[1] Ignatiadou 2013, 52-53.
[2] This technique was examined in depth by Paolucci 1997, 17-20 and passim.
[3] Martial, *Epigr.* XI.11.1, XII.74. 5; Clement of Alexandria, *Paedagogus*, II.3.35; Trowbridge 1930, 109, 166.
[4] Paolucci 1997, 63-80.
[5] Fremersdorf 1975, 117, nos. 1047-1048, 1061-1062, pl. 82, where similarly rendered plants in rock crystal are illustrated.
[6] Paolucci 1997; Desenzano, 150-151, second half of the 4th or early 5th century.
[7] Ružić 1994, 57, no. 1195, pl. XLIII/14, unknown provenance, kept in Bor and dated to the 4th century.
[8] Paolucci 1997, 164-165.
[9] Paolucci 1997: Desenzano (pages 150-151), Aquileia (pages 149-150, fig. 129), Ravenna (pages 152-153).
[10] On this workshop and its stylistic elements, see Paolucci 1997, 57-62. A very detailed study of the workshops operating in the region of Rome in the late 4th and early 5th century has been conducted by Lucia Saguì, see Saguì 1996, passim.

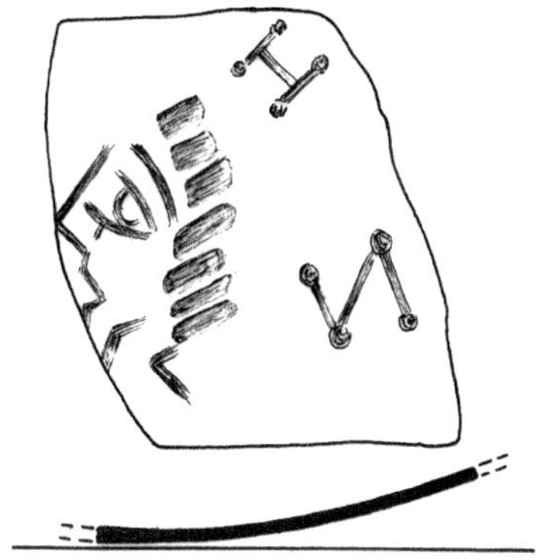

The preserved letters 'ZH' are part of an inscription, quite probably part of the word ZHSHS that often appears on vessels of this form as part of a salutation for the owner of the vessel. Inscriptions with the wish in the form of *pie zeses* (πίε ζήσης)[11] were incised on glass vessels made in various regions, such as Syria, Palestine, Egypt, Rome and Cologne. It seems that Greek letters and language were used more often for the rendering of this greeting,[12] although Latin transcriptions and translations of it are not rare.[13] They appear incised on simpler utilitarian vessels, such as funnel-mouth tableware flasks,[14] bowls,[15] and conical[16] and cylindrical[17] beakers, as well as lavishly decorated vessels such as *fondi d'oro*[18] and *diatreta.*[19] Judging by the preserved examples, this greeting was used equally for pagan, Jewish, and Christian scenes.[20] It seems that it could either be a simple toast or it might signify a remarkable event in its owner's life, such as marriage, etc. It might also have been used in burial rituals, as a grave good. Generally, the inscriptions of this type appear on glass vessels from the middle of the 3rd and up to the end of the 4th century.[21]

From the inscription in Greek it seems probable that the vessel was intended for the Greek-speaking areas of the Empire, or that it was an imitation of a Roman prototype from the Greek-speaking East.

Cat. no. 79: A small fragment of an open-shaped vessel, probably a shallow bowl. The letter 'H' and an indiscernible motif, probably part of the base and lower part of an item of furniture, possibly the leg of a throne, or cathedra, are visible on it.

The preserved letter is quite probably part of the word ΖΗΣΗΣ, which often appears on vessels of this form as part of a greeting for the vessel's owner.[22]

Cat. no. 74: The best-preserved sample of a scene on the base of a shallow bowl represents a young, beardless, standing figure with a nimbus, holding a long staff with his right hand, which ends in a chrism. Around the nimbus can be seen the letter 'M', and possibly a second letter which is barely visible today. The figure is flanked by two *tabulae ansatae* (tablets with handles), possibly standards, placed perpendicularly on each side.

The way the hair is rendered (three short rows of short incisions), as well as the rest of the facial characteristics, are quite characteristic.[23] In addition, the simple depiction of the nimbus and the head are identical to the decoration on the well-known 'Christ and the Rooster' bowl.[24] A typical feature of this technique is the deeper cutting of

[11] Amongst others, it appears in the following forms: ΠΙΕ ΖΗΣΗΝ, ΠΙΕ ΖΗΣΗΣ ΑΕΙ, ΠΙΕ ΖΗΣΗΣ ΚΑΛΩΣ ΑΕΙ, ΠΙΕ ΖΗΣΗΣ ΔΙΑ ΠΑΝΤΩΝ ΑΕΙ, ΡΟΥΦΙΕ ΠΙΕ ΖΗΣΗΣ ΜΕΤΑ ΣΩΝ ΠΑΝ[ΤΩΝ] ΒΟΙΥ (*sic*, ΒΙΟΥ is meant), see Auth 1996 for the relevant bibliography.

[12] Auth 1996, 103-110, where most of the bibliography on the matter is listed

[13] Šaranović-Svetek 1986, 18, presents a conical beaker (Isings 106b) from Kruševlje near Sviloš, on which appears the inscription ACCIPE CALICE PIE ZES, found in a grave with a coin of Constantius II (337-361); Ružić 1994, 57, no. 1196, pl. XLIII/15: a similar conical beaker from Karataš (Diana) preserves part of an inscription '. . VIC...', dated to the second half of the 4th century; Ružić 1994, 57, no. 1197, pl. XLIII/16: an identical beaker from Belgrade, dated to the 4th century, with the the inscription, under a frieze of fish and running lions: '. . . BI. . . M. LT. . . ', and which may be completed as BIBE MULTIS ANNIS.

[14] These vessels, which are often decorated with bands of incised decoration, e.g. geometrical patterns, garlands and grape bunches, are considered to be Syrian products. On one similar vessel, which is also decorated with bunches of grapes, is written only the word ΕΥΤΥΧΙ, Auth 1996, 103-105.

[15] Centrally underneath, the bowls often bear a geometrical or Christian decoration surrounded by an inscription. Vessels of this form are considered to be products from Palestine and Cologne, Auth 1996, 107-108.

[16] Vessels of Isings form 106b; Šaranović-Svetek 1986, 18; Ružić 1994, 57, nos. 1196-97, pl. XLIII/15, 16.

[17] The cylindrical beakers, which feature inscriptions written in gold, are considered to be products of Alexandria because of the character of their decoration, which is related to Egyptian religion, Auth 1996, 105-106.

[18] Engemann 1969, 18, 19 pl. 5a, 6a; Cermanović-Kuzmanović 1987, 8, fig. 29.

[19] Auth 1996, 108-109.

[20] Cabrol, Leclerq, *DACL* 14 (1939), s.v. 'Pie zeses', cols. 1023-1031; Auth 1996, 109.

[21] Šaranović-Svetek 1986, 18; Ružić 1994, 57, nos. 1196-97; Auth 1996, 110.

[22] See discussion of fragment Cat. no. 78.

[23] Paolucci 2002, 21-28, figs. 27-34.

[24] Paolucci 1997, 150-151, Desenzano, dated to the second half of the 4th or to the early 5th century.

the main outlines, while the entire motif itself is more shallow. The details, such as the folds of the tunic, have been incised with a pointed tool held free-hand. This plate is very likely the product of a workshop operating in Rome at the end of the 4th to the early 5th centuries, which specialized in Christian iconographical themes similar to the one on the Thessalonean find.[25]

The motif of the standing figure with nimbus, holding a long staff that ends in a cross, also appears on North African pottery, but from a later period.[26]

Identifying the figure is not an easy task as it lacks an inscription or some other indisputable feature of identification; accordingly, only a few suggestions on this are offered. This form of staff in Early Christian art can be seen in the hands of Christ or St. Peter, and less often held by St. Paul or St. John the Forerunner.[27] Additionally, some saints highly venerated locally are given this *insignium*, such as St. Laurent in the mausoleum of Galla Placidia in Ravenna, and on a glass bowl from Ostia.[28]

If indeed the letter 'M' above his head is an indication of his name, although it is quite clear that no other letters were ever inscribed on the plate, then possibly the figure might be identified with the Archangel Michael or some other saint with an 'M' in his name. *Tabulae ansatae*, or possibly standards, which are presented around the figure, have also been used in association with other archangels and suggest an identification with the (admittedly wingless) figure of the Archangel Michael.

Judging by the active period of the workshop, the vessel was made in the late 4th or in the 5th century.

Handleless globular vessels with unworked rim (Form 145) represent the second form of vessels often decorated with complex figural motifs. Those preserved from the Thessaloniki region include:

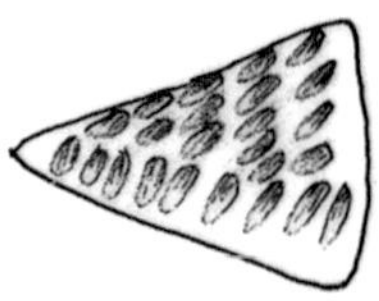

Cat. no. 81: A slightly curved fragment covered by densely arranged short strokes which look like foliage or scales. The mild curving suggests that it was quite probably part of the base of an open-shape bowl. However, because of its remarkabl similarity to a missing fragment from a vessel of Form 145, its final attribution remains unclear.[29] It seems that the decoration represents part of a motif similar to the one used for rendering the foliage of the tree in the hunting scene preserved on Cat. no. 695.

Cat. no. 695: The most significant example of engraved decoration in the region of Thessaloniki is preserved on a close-shaped vessel of Form 145. On the vessel is depicted a hunting scene. A standing female figure of an archer is presented with a quiver on the right shoulder, aiming the bow in front of her. She wears a tunic, and the end of her himation blows out behind her; her feet are protected by boots (*emvades*). In front of her a dog chases two deer that cross before a tree in full leaf, and continue their course towards a horizontally placed net that stretches behind

[25] Paolucci 2002, 27-28. A very detailed study of the workshops operating in the region of Rome in the late 4th and early 5th century has been conducted by Lucia Saguì, see Saguì 1996, passim.
[26] Hayes 1972, 266, form 104, no. 239b, c, dated to the second or third quarter of the 6th century.
[27] Schäfer 1936, 67-104; Kitzinger 1960, 33.
[28] Paolucci 1997, 52, fig. 18.
[29] The foliage of the tree is rendered in an identical way on Cat. no. 695, with a hunting scene, a large part of which is not preserved. The curvature of the fragment might coincide with that of the close-shaped vessel at the part close to the shoulders. It should also be noted that these two objects were found in the same year in practically neighbouring excavations at the eastern necropolis, on the site of the Administration Building and Theological Department of the Aristotle University, respectively.

the hunter and encloses the scene. All around and between the animals, the rich vegetation is emphasized by small, schematized bushes.

Above the scene a partly preserved inscription reads: 'ΥΣΥΧΙ ΠΙΕ. . ΗΣ ΜΕΤΑ. . . . /Ν ΠΙΕ. . . Σ'. The inscription was presented by Petsas,[30] published by Pallas,[31] and later by Feisel,[32] with the lacunae filled with words of a Christian nature. This reconstruction is quite, if not highly, inappropriate, given the pagan scene covering the body of the vessel. It seems very probable that the depicted hunter is the goddess Diana, and therefore that the text, with its salutation to a certain Hesychi[o]s, the owner of the bottle, could be completed as ΤΩΝ ΣΩΝ, or ΤΩΝ ΟΙΚΕΙΩΝ, or ΤΩΝ ΣΩΝ ΟΙΚΕΙΩΝ ΠΑΝΤΩΝ, or, given the size of the vessel's missing part, even with the names of these ΟΙΚΕΙΟΙ – a pattern well known from other similar finds.[33] Therefore, the inscription might be reconstructed as: 'Υσύχι, πίε ζής μετὰ τῶν σῶν οἰκείων πάντων. Πίε ζήσης. ('Hesychios drink; may you live with all your intimates. Drink; may you live!!')

During the 4th century hunting scenes were especially popular on engraved glass vessels, particularly shallow bowls.

The standing figure of the archer-hunter is in direct contrast with the hunters represented on glass vessels, which as a whole are depicted mounted, holding a spear or a sling.[34] The only other figure found hunting in the same way, and wearing the same clothes, has been identified with Diana.[35]

This peculiar way of rendering vegetation, the foliage of trees and, especially, bushes, appears also on other vessels of this period,[36] representing probably a mannerism used by engravers of the time for a generalized schematic rendering of flora.

Hunting scenes in several cases are supplemented by inscriptions – acclamations for good luck or good fortune, either in Greek[37] or Latin,[38] in which the owner of the vessel is mentioned by name, as are relatives.

### *b. Geometrical Motifs*

#### *1. Cross*

Cat. no. 82-83: A Greek cross inscribed in a medallion formed by two concentric circles decorates the bottom of the bowl. Identical plates with the incised central motif of a Greek cross have been found in Israel,[39] and the closest

---

[30] Petsas 1967a, 391, pls. 296ε and 298, drawings 12ε.

[31] Pallas 1977, 75 as Ε]ΥΣΥΧΙΠΙΕ [-----]ΗΣΜΕΤΑ [---]/[---] ΩΝΠΙΕΖ[---]Σ and he reads it (Footnote 163 on the same page) as: Ε]ὐψύχει. Πίε [ζήσ]ης μετὰ/ [τῶν/ ἁγί]ων. Πίε ζ[ήσης].

[32] Feissel 1983, no. 127, as 'Υσύχι πίε [ζήσ]ης μετὰ [τῶν]/ [ἁγί]ων πίε ζ[ήση]ς, and gives the 4th century as a probable date.

[33] See discussion of fragment, Cat. no. 78.

[34] Harden 1960, 44-81, bowls of various origins with representations of hunting hares, deer, and boars; these were produced until the second third of the 4th century. Cologne: Fremersdorf 1967, 189, pls. 214-216, pl. 271, decorated with representations of hare hunting with dogs and nets; Bonn (Bonna): Follmann-Schulz 1988, 119-120, no. 459, pl. 51, find dated to the 4th century with a scene of hare hunting with a mounted hunter with dogs and net; Mainz (Moguntiacum): Harter 1999, 50-51, form A17, subform 3, no. 213, pl. 9, find dated to the 4th century which is decorated with a scene of an unmounted male figure, holding a spear, hunting a boar; Syracuse: Fremersdorf 1967, 190-191, pls. 272-273, decorated with a representation of a boar, bear, and two deer being trapped by nets during a hunt.

[35] Whitehouse 1997, 268-269, no. 457, a vessel already ascribed by Smith (1957, 177-178, no. 358), to the Eastern Mediterranean production of the 4th century. A globular jar, with the inscription VITA BONA PRUAMUR FELICES and the representation of an archer taking aim at some stags. The bushes are rendered in the same characteristic schematized manner. A similar representation of the goddess is found on a 4th-century incised shallow glass bowl, but in a different scene however, in Cologne, see Loeschcke, Willers 1911, pl. 28 and Fremersdorf 1967, 165-66, pl. 218. Also, very similar is the representation, overall and particularly the figure of the goddess, to the hunting scene shown on a faience *kalathos*/situla dated to the 2nd century BC, which was found in a grave in Neapolis, Thessaloniki, where there is no doubt as to the identification of the hunting figure, see Daffa-Nikonanou 1987, 269-275, drawing 3, pl. 56. The iconographical prototype of the representation derives from the well-known type hunting scene featuring Artemis of the mid 2nd-century BC, see *LIMC* II, 648, no. 296, s.v. 'Artemis' (L. Kahil).

[36] Paolucci 1997, 150-151, bowl from Desenzano with a presentation of Christ and rooster, dated to the second half of the 4th or to the 5th century. Fremersdorf 1967, 190-191, pls. 272-273, close-shaped, handless vessel form 145 from Syracuse, with a representation of hunting and an inscription; the bushes are rendered with the same characteristic motif, as is the rich foliage of the trees, with short, densely-spaced strokes. Fremersdorf 1967, 189, pl. 271, close-shaped, handleless vessel form 145 with an inscription, from Cologne with a representation of a hunting scene, with a depiction of identical bushes.

[37] Fremersdorf 1967, 190-191, pls. 272-273, close-shaped, handleless vessel form 145 from Syracuse with the inscription: ΦΟΡΤΟΥΝΑΤΙΩΝ ΠΙΕ ΖΗΣΑΙΣ, and showing a boar, bear, and two deer being trapped by nets during a hunt. Fremersdorf 1967, 189, pl. 271, close-shaped, handless vessel form 145 (Isings Form 103) from Cologne with a hunting scene and the inscription: ΕΡΙΘΑΙΑΝΗ ΠΙΕ ΖΗΣΑΙΣ.

[38] Fremersdorf 1967, pl. 214, bowl showing a hare being hunted with dogs and nets; there is also the inscription: PIE VIVAS CUM TUIS.

[39] Barag 1970c, pl. 52/5.

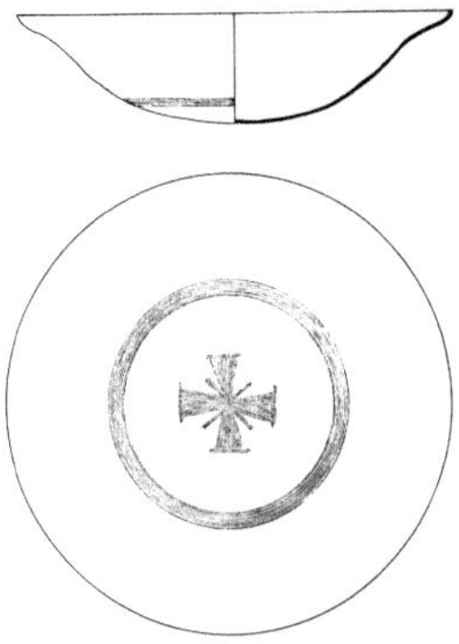

parallel was found at El-Bassa.[40] The decorative motif of a Greek cross inscribed in a medallion appears also in North African pottery;[41] it has been proposed[42] that it shows the only motif so far known, which first appeared on pottery and glassware later, that seems unlike the known general tendency of potters to imitate glassware; however it should be borne in mind that the dating of the clay vessels might not be accurate. Silver vessels with similar motifs appear only quite later and thus it is not possible to verify if the phenomenon of *skevomorfism*[43] in this case is true, and therefore if the artisans were imitating on cheaper materials those motifs and shapes originally used on more precious materials. The vessel was found in the ruins of a 6th-century church that was most probably destroyed in the 7th century. The production date of the vessel, however, according to all the aforementioned parallels, is quite a little earlier, i.e. the 4th century.

*2. Ovular strokes*

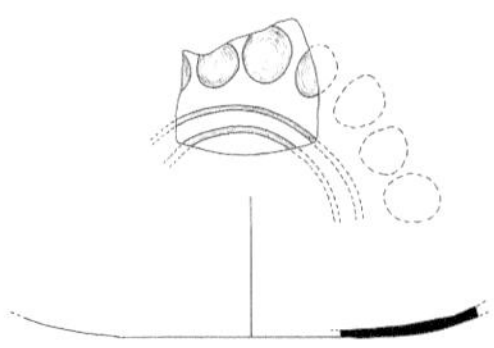

The large and quite deep, ovular, occasionally almost circular strokes, arranged on the surface of the vessels forming parallel bands of concentric circles, present another discernible decorative motif.

Cat. no. 77: Shallow bowl. The underneath of the vessel is decorated with concentric circles and it is surrounded by relatively deep ovular strokes. Analogous finds are reported from several excavations and collections, e.g. Augst,[44] Mainz (Moguntiacum),[45] Heddernheim (Nida),[46] Trier (Treviri), Straubing (Sorviodurum),[47] Szőny (Brigetio),[48] as well as the Corning Museum of Glass;[49] the production of the find is dated at around the 3rd or 4th century. According to the excavation data the vessel was buried after the second half of the 4th, and most probably into the 5th centuse-shapedVertical strokes

*3. Rise-shaped strokes*

Among the engraved vessels is a group characterized by the use of bands of elongated ellipsoid or vertical strokes.

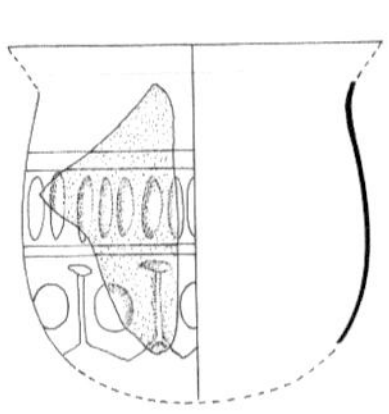

Cat. no. 62: A fragmentarily preserved hemispherical bowl (Form 12) preserves incised decoration that includes ellipsoid strokes. Two bands of unequal height, which are delineated by a double line, decorate the bowl. The higher and narrower band is covered by a dense row of vertical oval strokes. The lower and wider band, whose lower end is not preserved, is filled with a continuous row of touching pentagons, opening upwards. The vertical sides of the pentagons are fmed by ana horizontal ellipsoid stroke and one large, almost circular stroke appears in the centre of each pentagon.

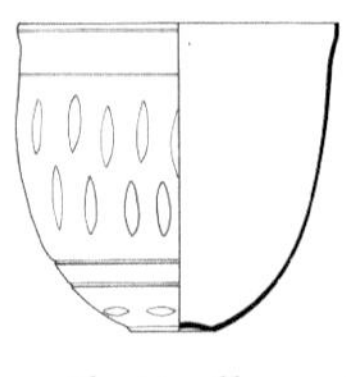

Cat. no. 63-64: Hemispherical bowls from Thessaloniki (Form 12c) preserve decoration comprising of two successive rows of ellipsoid strokes which are delineated either from both sides or towards the bottom by an horizontal, shallow band.[50] Only one of these two vessels (Cat. no. 63) can be dated by its archaeological context to, approximately, the 4th century.

The vessels with decoration consisting of bands of ellipsoid or rising, horizontal strokes present an aesthetically and chronologically discernible group. The decoration shows an evolution of 1st-century vessels with

[40] Iliffe 1933, 81-91, from a grave dated to the last decade of the 4th century, illustrated in Paolucci 1997, 72-73, fig. 36.

[41] The motif is known on clay vessels dated between AD 360-440, see Hayes 1972, 277-278, form 81, fig. 56.

[42] Paolucci 1997, 72-73.

[43] Vickers 1992, 53-72.

[44] Rütti 1991, 65, form AR 58, no. 1263, pl. 55/1263, a find dated from excavation results to the second quarter of the 2nd to the middle of the 3rd century. The vessel appears to be identical with the example from Thessaloniki.

[45] Harter 1999, 50-51, 187, form A17, subform 1, no. 212, pl. 8, find of unknown provenance dated to the 3rd century according to analogous finds.

[46] Welker 1985, 27, form 27b, pl. 19, a find dated by its archaeological context between AD 140-170. The decoration on that is denser than on the Thessalonian find, witnessing that it is older than the latter.

[47] Paolucci 1997, 103-104, a find dated by its archaeological context to the first half of the 3rd century.

[48] Barkóczi 1988, 60, no. 27, pl. III, a find dated by its archaeological context to the 3rd century.

[49] Whitehouse 1997, no. 442, a find, dated on the basis of similar finds to the 4th century that has a different shape, i.e. the bottom is not flat and the body is almost hemispherical.

[50] Kolonas 2002, 118, no. 24, similar to an example from Patras, dated on the basis of analogous finds to the 3rd century.

multifaceted decoration. Compared to that original technique, the arrangement of the strokes in bands required less effort and permitted the creation of more complex decorative motifs. This type met with great popularity and it is believed that examples were produced in at least four different centres: Tanais, which ceased to operate in AD 240 as a result of invasions by the Goths;[51] Syria, possibly Dura-Europos, which ceased to operate after the destruction of the city in AD 256;[52] Pannonia, probably in the area of modern Hungary, which seems to have ceased operating around AD 260;[53] Cologne, which continued into the 4th century, during which it started to produce more elaborate geometrical motifs on vessels of different shapes and forms.[54] Finds from Tanais have proved, at least as far as Black Sea region production is concerned, that decoration was initially impressed on the vessel by mould-blowing and only the final finishing was carried out by the cutter with wheel-cuts.[55] Identical decoration is known from several other sites and collections and it is dated between the 3rd and 5th centuries.[56]

*4. Lozenge-web/rhomboid fields*

A further incised decoration motif appearing in Thessaloniki takes the form of a lozenge-web or rhomboid field.

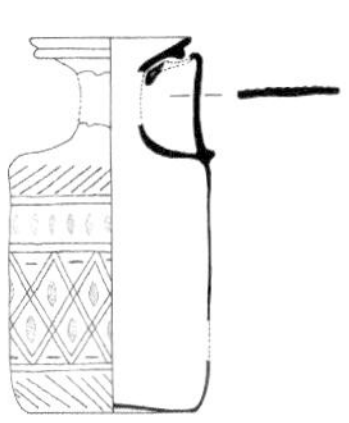

Cat. no. 427: Among the finds from Thessaloniki, the most complex geometrical decoration of this type is preserved on a cylindrical jug (Form 95) and covers the entire body. The body is divided into two unequal bands, the lower is twice the size of the upper one, and both are delineated by a double line. The upper band is filled with vertical ellipsoid strokes. The lower is filled by a double row of lozenges, each one formed by a double line. At the centre of each lozenge there is a vertical ellipsoid stroke, while in the triangular fields, which are formed between the ends of the lozenges at the edge of the band, is a thin, horizontal incised line. The rest of the body, above and below these bands, is covered by rows of parallel oblique incisions forming an oblique hatching. This decorative motif is well known and popular mainly during the 4th century, the same period as the grave in which it was found.[57]

The same motif, but somewhat simpler, is preserved on an unpublished bowl from the eastern necropolis. On this the motif takes the form of a single row of rhomboid fields and at the centre of each field is an ellipsoid stroke.[58]

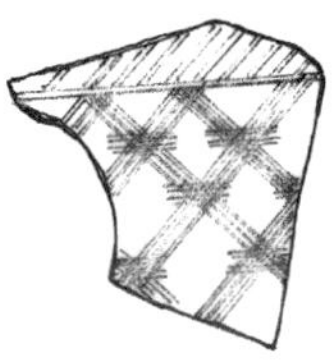

Cat. no. 429: Incised decoration of the same type is preserved on a small fragment of a probably cylindrical vessel. Still visible is the upper zone with oblique hatching, and, lower down, at least four rows of touching lozenges, partly preserved, each one formed by a triple line. The corners of the lozenges are crossed out by a short, triple horizontal line which defines the horizontal axis of the composition. This fragment was found in the ruins of a 6th-century church which was destroyed in the 7th.[59] This late date probably indicates the destruction date of the vessel and not the production date, which is probably 4th-century.

The web of rhomboid fields, quite widespread in the West and East, as the numbers and geographical distribution of preserved examples indicate, creates a wicker pattern and replicates the wicker casings used around glass vessels to minimize the danger of breakage. These casings later were probably occasionally replaced by more precious ones, e.g. in metal and set with gems, which were replicated faithfully by the 4th-century cutters.[60]

[51] Sorokina 1977b, 111-122.
[52] Clairmont 1963, 65-68, forms B2c-d, ascribed to production of the 2nd and 3rd centuries.
[53] Barkóczi 1986, 188.
[54] Stern 2001, 137, with a synthesis and synopsis of all known data on the matter.
[55] Sorokina 1977b, 119.
[56] See analogous finds mentioned in the discussion of Form 12.ii.a.2.
[57] See Chapter 9 (Find-Spots), 'Kallitheas Avenue, Elementary School of Ampelokipoi, 1997'. On the decoration, see Klein 1995, 151-162, with examples from Germany, dated by excavation evidence to the 4th century; Paolucci 1997, 139-140, almost identical example from Cloz, dated to the early 4th century; Stern 2001, nos. 59-60, with a great number of analogous finds from the East and West dated mainly to the 4th century.
[58] For a published analogous find see Whitehouse 1997, 264-275, no. 452, dated according to analogous finds to the 4th century.
[59] See Chapter 9 (Find-Spots), 'Agiou Dimitriou and Agias Sofias Street, at their junction, 1969'.
[60] Stern 2001, 162-163, nos. 59-60, with a great number of well-dated, analogous finds. For examples of glass vessels which retain their wicker-work covering, see Whitehouse 1997, 183, no. 323, a single-handled, cubical vessel dated according to analogous finds to the 3rd century; Foy, Nenna 2001, 115, examples dated to the 1st century and to the era after the advent of Islam.

*5. Concentric circles*

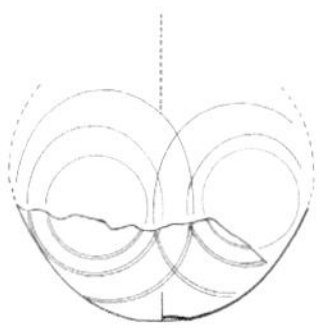

Cat. nos 208, 21: On vessels of Form 50 quite often geometrical decoration appears consisting of tangential or intersecting concentric circles.[61] Two examples preserved in a very fragmentary condition in Thessaloniki depict this motif.

According to the archaeological context, Cat. no. 214 was buried after the second half of the 4th, and more probably into the 5th century.[62]

*6. Deep, isolated grooves*

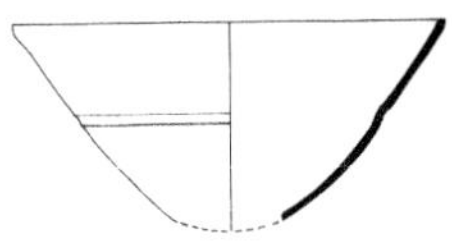

Cat. nos. 49, 51-52, 58-61, 76, 98-104: In the 1st and 2nd centuries some groups of beakers (Forms 24 and 26) were decorated under the rim and n the body with an horizontal groove that divides them into bands.[63] This decoration was also used on bowls, dishes and beakers in the 4th and 5th centuries, quite rarely however, as in this later period shallow incised bands generally prevail.[64]

*7. Shallow, incised grooves*

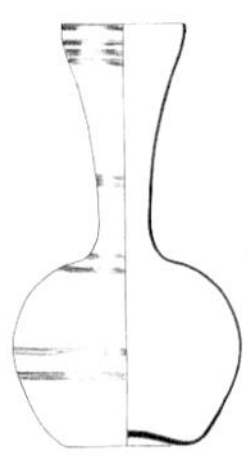

Bowls: Cat. nos. 50, 54-56.
Beakers: Cat. nos. 110, 111-112, 114, 116-117.
Funnel-mouth: Cat. nos. 318, 337, 339-344, 367-369, 374.
Jugs: Cat. no. 417.
Unguentaria: Cat. nos. 690-691, 693.

The simplest and most widespread form of incised decoration in the 3rd and 4th centuries consists of bands of fine, shallow horizontal incisions which form grooves. These grooves embellished the (almost always unworked) rim of the vessels and divided their body into often equal bands. It was used both in close-saped vesselform 50 and forms Forms 62 and 63, as well as in open-shaped drinking vessels (Form 31) and it is quite probable that local Thessalonean workshops preferred it, as some of the vessels bearing this decoration are most probably local products.[65]

B. Application of relief elements

Application of plastic elements on the surface of glass vessels, usually with some regularity creating some form of geometrical motif, was a quite common decorative technique, easy and not at all costly.

***a. Relief medallions***

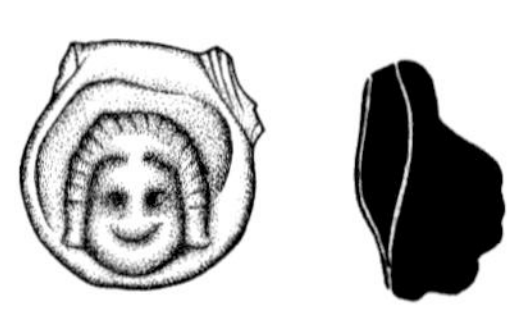

Cat. nos. 412, 753-754.
Glassworkers, following the long-standing practice of silversmiths in the 1st, and again in the 4th century, recognized that it was easy to stamp a figure on a hot lump of glass. Thus they shaped small medallions with relief decoration and applied these on the still-hot vessel, decorating it either singly or in combination with other decorative elements. Medallions have been found dispatched from vessels in Thessaloniki, and only in one case in the wider region is preserved the entire jug with a medallion below the base of its handle. Theatrical masks and a lion head appear on these medallions;[66] they are ascribed

[61] Among others, published examples come from: Dura-Europos: Clairmont 1963, 112, no. 540, dated from excavation results to the end of the 2nd-3rd century; The Black Sea region: Sorokina 1967, 72, no. 18, dated from excavation results to the 3rd century; Bonn (Bonna): Follmann-Schulz 1988, 38, no. 88, pl. 7, dated from excavation results to the end of the 3rd century; Ljubljana: Petru 1972, 62, 127, nos. 6, 17, from burials dated to the 3rd century; Ptuj (Poetovio): Lazar 2004a, 74, no. 81, from a burial dated to the first half of the 3rd century; Cologne: Fremersdorf 1967, 109, 110, nos. 110, 112, dated to the 3rd century; Turin (Augusta Iulia Taurinorum) and Bologna: Paolucci 1997, 114-115 (and with exhaustive bibliography of published examples), finds dated to the 3rd century.

[62] See Chapter 9 (Find-Spots),'Sokratous Street, 1968 (today Moirarchou Koufitsa St.)'.

[63] Isings 1957, 27-30, 48-49, forms 12 and 34, dated to the 1st and to the early 2nd century.

[64] See analogous finds mentioned in this present work, in particular the discussion of forms 12.ii.a.1. and 31.

[65] For the dating, see Chapter 4 (Typology), Form 63.

[66] See the discussion on applied medallions, Form 149, where analogous finds are listed.

to the late 4th century and most probably adorned the base of a jug handle, although open-shaped vessels with similar decoration are also known.[67]

### *b. Blobs*[68]

Bowls: Cat. nos. 65, 66.
Beakers: Cat. nos. 118-119, 145.

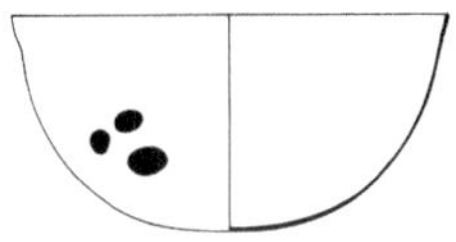

This type of decoration imitates *potoria gemata*, i.e. the expensive metal vessels decorated with enamel or with inset semiprecious stones and colourful glass gems.[69]

The modelling technique of the decoration involves the following: while the vessel was attached on the blowpipe and hot enough, it was marvered on a smooth surface where tiny lumps of glass were arranged to create the desired outcome. Once the glass lumps were attached, the vessel was reheated and marvered again to polish the surface and give the blobs their final smoothness. Another probable method for applying the blobs is for the heated tip of a rod of glass to come into brief contact with the vessel and then swiftly remove it, occasionally leaving the surface of the blob somewhat anomalous. After repeating this procedure as many times as necessary to finish the application of the desired number of blobs for motif, the vessel was reheated and the blobs smoothened.[70] For this decoration dark-blue blobs of glass were usually used arranged around the body in a single or double row (Cat. no. 119), or in triangular groups (Cat. nos. 65, 66), which sometimes alternated with larger, single blobs[71] (Cat. no. 145). The decoration with polychrome blobs is not totally unknown in this area, although it is considered as a Germanic product[72] and occasionally such vessels appear in the region as well (Cat. no. 118).

The vessels as a whole are dated between the end of the 3rd and the middle of the 5th century, with higher concentrations in the 4th and in the first half of the 5th century.[73]

### *c. Applied threads*

Bowls: Cat. no. 67.
Jugs: Cat. nos. 375, 411-412, 417, 421, 425, 427, 428.[74]
Funnel-mouth flasks: Cat. nos. 324, 331, 335, 350, 354.[75]
Unguentaria: Cat. nos. 540-541.[76]

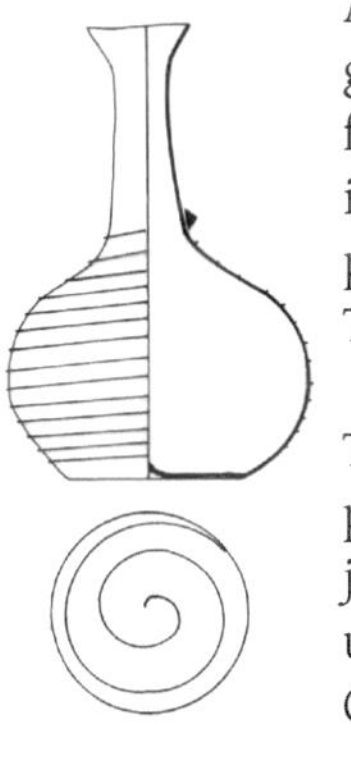

Already from the very early stages, free-blown vessels, continuing a millennia-long tradition in glassware, were decorated with applied fine threads of glass. The thread was of a different colour from the body and was spirally wound around it from the lip to the centre of its base. Sometimes the thread was left in relief or it was flattened and practically incorporated into the thickness of the body. Among the finds from Thessaloniki are preserved two such 1st-century examples (Cat. nos. 540-541).

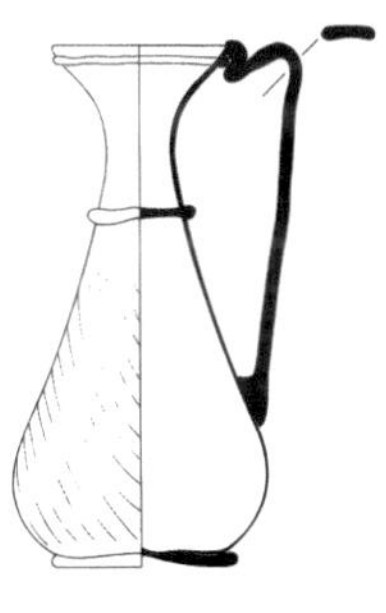

This decorative technique met with much wider popularity in the Late Roman period, especially during the 4th century, a period in which vessels, mainly jugs and funnel-mouth flasks, were very often decorated around the neck and under the rim with a thick trail of glass of the same colour as the body (Jugs: Cat. nos. 375, 411-412, 417, 421, 425, 427, 428; Funnel-mouth flasks: Cat. nos.

[67] See the discussion of Forms 85 and 149.
[68] It is a subgroup of Isings Form 96b2, see Isings 1957, 133. For illustrated, intact examples from Trier (Treviri), see Goethert-Polaschek 1977, 50-62, form 49, nos. 188, 210, 233-234, pl. 39, dated from excavation results to the 4th century.
[69] Fremersdorf 1962, 11.
[70] I thank E. Marianne Stern for enlightening me on this method.
[71] For a colour illustration of an intact conical beaker from Pieria, with relevant decoration, dated to the Roman Imperial period, see Adam-Veleni, Poulaki, Tzanavari 2003, 212, no. 195 (E. Poulaki). For a detailed classification of different variants of this decoration, see Sorokina 1970.
[72] Fremersdorf 1962, 9, the polychromy of the glass blobs is considered to be result of the aesthetics of German tribes, which starts to spread around AD 260. For examples in colour see Goethert-Polaschek 1977, colour plate IV, nos. 268-269.
[73] See analogous finds mentioned in the discussion of forms 12.ii.b.1. and 31.
[74] See the discussion of form 85, where analogous finds are listed.
[75] See the discussion of form 66, where analogous finds are listed.
[76] See the discussion of form 122, where analogous finds are listed.

324, 331, 335, 350, 354). The thread in many cases was used as an additional decorative medium and was combined with dip-mould-blowing and/or incising or engraving.

In a very few cases a thread of glass different in colour from the one used for the body, mostly dark-blue, singly or in twisted pairs, is wound around the vessel under the rim or at the point of the maximum diameter of the body. The fragment from Thessaloniki (Cat. no. 67), which is probably part of a hemispherical bowl, could also belong to a concal vessel of our Form 31,[77] both of which are dated to the end of the 4th to the early 5th century.[78]

The thread was attached to the vessel, very swiftly, while attached to the blowpipe and rotating, so as to avoid the danger of fracture resulting from any thermic shock caused by the difference in the temperature between the body and the thread. This technical feature explains the simplicity, and even apparent casualness, that is quite often visible with this decoration.

C. Indenting

Bowls: Cat. nos. 71-73
Beakers: Cat. nos. 105, 108-109.
Bottle: Cat. no. 205.
Funnel-mouth: Cat. no. 352.
Unguentaria/flasks: Cat. nos. 442, 626-627, 668-681.

Glassworkers aimed to imitate hammered silver vessels, which were often decorated with smaller or bigger indentations. Already from the 1st century they experimented with and exploited the opportunities offered by the malleability of glass while it was still hot. They used gentle pressure while forming the vessel to create oblong indentations that changed the shape of the body, making it more elaborate. Among the finds from Thessaloniki, indentations appear on certain cylindrical beakers (Cat. nos. 105, 108-109).[79]

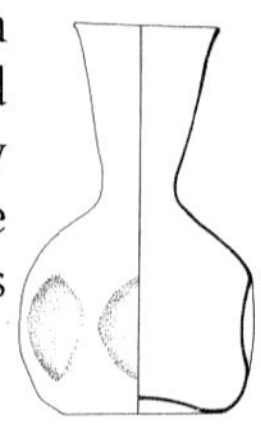

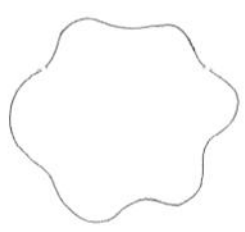

During the 2nd and 3rd centuries this technique was also used on close-shaped vessels and decorated tableware vessels (Cat. no. 205),[80] while mainly during the 4th century it became very popular mostly on small unguentaria whose bodies were rendered practically square by indentations (Cat. nos. 442, 668-681).[81] This form appears in a very rare variant with pointed base (Cat. nos. 626-627).[82] The method is not unknown on tableware vessels, e.g. hemispherical bowls (Cat. nos. 71-73)[83] and the funnel-mouth vessels (Cat. no. 352).[84]

D. Pinching

Bowls: Cat. no. 85.
Beakers: Ct. nos. 125, 129.
Ffunnel-mouth flasks: Cat. no. 353.
Jars: Cat. nos. 447, 451, 453.

The high viscosity of the glass vessels during modelling inspired another decorative method, probably as a result of observation and experimentation after mishaps and unsuccessful handling, which is also evident among the vessels found in Thessaloniki. It concerns the creation of protuberances and folds around the body using pointed pincers on the vessel surface.

Pointed pincers formed smaller or bigger folds on the vessel surface. These protuberances were mostly flattened and sometimes they were even pierced by the excessive pressure of the tool.[85]

[77] Edgar 1905, pl. III 32. 487, a find from Hawara.
[78] See analogous finds mentioned in the discussion of forms 12.ii.b.2. and 31.
[79] See the discussion of form 27, where analogous finds are listed.
[80] See the discussion of form 49, where analogous finds are listed.
[81] See the discussion of Form 141, where analogous finds are listed.
[82] See the discussion of Form 134, where analogous finds are listed.
[83] See the discussion of Form 15, where analogous finds are listed.
[84] See the discussion of Form 67, where analogous finds are listed.
[85] See Stern 2001, 248, no. 134, with relevant bibliography.

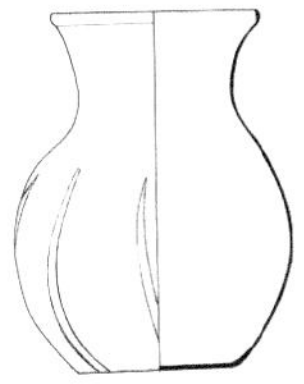

Pinching appears in the Eastern Mediterranean at the end of the 2nd century[86] and was continuusly used until the Eearly Byzantine period[87] (Cat. nos. 85, 125, 129, 453).

By the use of pincers, and before the vessel was expanded to its final dimensions, an even more sophisticated decorative motif was formed. In a simpler variant it consists of vertical ribs decorating the vessel body from the shoulders to the bottom. In Thessaloniki, this variant appears on jars (Cat. no. 447) and globular flasks (Cat. no. 353), which are dated at around the 4th century.

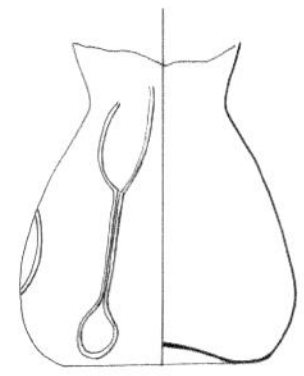

In a more composite variant the ribs are compressed at regular intervals, forming a rhomboid web, usually arranged horizontally and less often vertically, as in the pear-shaped funnel-mouth vessel (Cat. no. 451) from Thessaloniki[88] which is dated around the 4th century. There are also known examples – none from Thessaloniki however – with identical decoration formed by the application of threads on the vessel and further marvering to flatten them.[89] This decoration appears already from the 1st and 2nd centuries.[90] It becomes more popular from the 3rd century onwards. There are published examples of this period from Colchester (Camulodunum),[91] Bonn (Bonna),[92] Cologne,[93] Trier (Treviri),[94] Hungary,[95] Kerch (Panticapaeum),[96] and Cagliari (Caralis).[97] Examples were found in excavations of eastern sites, e.g. Sardis,[98] Samaria-Sebaste,[99] Dura-Europos,[100] Hanita,[101] Bet She'an,[102] Jalame[103] and Kom Aushim (Karanis).[104] In addition, vessels provenanced from the Eastern Mediterranean are held in Western museums and private collections, e.g. the Oppenländer Collection[105] and Cohn,[106] the Corning Museum of Glass,[107] the Antikemuseum, Berlin[108] and the Metropolitan Museum, New York.[109] The decoration survived in 6th-century Western Europe as part of the artistic repertoire of Merovingian, Frankish and Longobard workshops,[110] and, of course, it is also well known in the Arab Caliphates.[111]

E. Mould-blowing

Bowls: Cat. nos. 30-36.

[86] See Price, Cottam 1998a, 32-33; Stern 2001, 249-251, nos. 135-137.

[87] See Dussart 1998, page 128, 158, 161, BX.111a, pl. 32/1, BX 83, pl. 46/21, BXII.1, pl. 49/1. Stern 2001, 354, no. 201.

[88] This decoration is known by the more generic German term as *fadendekor* or *bifurcated*, see Weinberg 1988, 3, 80-81, cat. nos. 351-353, drawing on pl. 4-39, photo on pl. 4-15/351, 352, where an older bibliography is cited along with technical details of its construction.

[89] Doppelfeld 1966, 58-59, pls. 130-131, a pair of funnel-mouthed vessels from Cologne dated to the second half of the 4th century.

[90] Haevernick 1978, 328-30, fig. 3a, from the Field Museum of Natural History, Chicago, no. 26146, the vessel comes from Palestrina and the general dating of the form gives a chronological time-frame for the decoration as well, see Isings 1957, form 37.

[91] Harden *et al.* 1968, 81, cat. no. 111.

[92] Follmann-Schulz 1992, 52, 127, cat. nos. 29, 493, pl. 54, vessels dated to the early and late 3rd century, respectively.

[93] Fremersdorf 1959, 73-74, pl. 114, beaker dated to the 4th century; Doppelfeld 1966, 59, pls. 132-133, beakers which are considered to be local products and dated to the 3rd-4th century.

[94] Goethert-Polaschek 1977, 230, form 136, photo in pl. 73, no. 1381; aryballos dated to the 4th century; beaker pages 76, 304-305, form 58a, cat. no. 323, drawing in pl. 16/176b, photo in the pl. 44/323; rhyton, pages 260, 305-306, form 165, cat. no. 1542, drawing in pl. 17/180m, photo in pl. 81/1542, finds dated to the first half of the 4th century.

[95] Barkóczi 1988, 91, form 57a, cat. nos. 127-128, pls. XII/127, 128, LXXV/127, 128, finds from Szombathely (Savaria) and Egerszeg of the 4th century; page 192, form 173, cat. no. 474, pl. XLVIII/474, find from Dunaújváros (Intercisa) dated to the second half of the 4th century.

[96] Sorokina 1971, 94, drawing in the pl. 5/1; Sorokina 1962, photo 7.4, where the caption dates it to the late 4th century and on page 234 the author mentions as original publication *IAK* 25 (1907), 54. These two vessels are identical to the find from Thessaloniki.

[97] Stiaffini, Borghetti 1994, 135, no. 385, pl. 91, 2nd-4th century; Weinberg 1988, 80-81, on an example from Jalame with vertical decoration, as it is on the example from Thessaloniki.

[98] Saldern 1980a, 20-21, nos. 92-93, two fragments, one was found with a coin of Constantius II (337-361).

[99] Crowfoot 1957, 404, 410, fig. 94.12, vessel Form 145, dated to the 3rd century.

[100] Clairmont 1963, 48-50. The fragments are divided into four groups according to their decoration. It is considered that they appear in the 2nd century and locally continued up to AD 256.

[101] Barag 1978, 26-27, no. 50, a 3rd- or 4th-century vessel.

[102] Bagatti 1967, 229, fig. 3/78, funnel-mouthed vessel, probably 4th-century.

[103] Weinberg 1988, 3, 80-81, cat. nos. 351-353, drawings on pl. 4-39, photo on pl. 4-15/351, 352, fragments from late 3rd- or 4th-century layers (AD 275-350 and AD 351-383).

[104] Harden 1936, 208, cat. no. 593, pls. VIII, XVIII, a vessel dated to the 4th century.

[105] Saldern *et al.* 1974, 234, no. 686, dated to the 3rd-4th century.

[106] Saldern 1980b, 90-94, cat. nos. 85-87, juglets and funnel-mouth flasks, decorated with a rhomboid thread 'fadendekor', which are ascribed to the production of the Syro-Palestinian region of the 3rd-4th century.

[107] Harden *et al.* 1987, 148, no. 76, dated to the 4th century.

[108] Platz-Horster 1976, 58, no. 104, 3rd- or 4th-century bowl, probably from Anatolia.

[109] Vessberg 1952, 142, 144, pls. VI/28, XVII/3, amforisk from the Cesnola Collection, Cyprus.

[110] Ružić 2001, 417, with further relevant bibliography.

[111] Saldern 1980b, 152-153, cat. no. 146.

Bowls: Cat. nos. 69-70.
Beakers: Cat. nos. 86-88, 91-95.
Kantharoi: Cat. nos. 188-189.
Bottles: Cat. nos. 254-273, 297-300.
Lamp-fillers/'Baby-feeders': Cat. no. 307.
Flasks: Cat. nos. 319-320.
Amphorisks: Cat. nos. 375-377, 381.
Prismatic bottles: Cat. nos. 385-396, 427-428, 431-434.
Jugs: Cat. nos. 405, 406, 410-413, 419, 424, 430-431.
Unguentaria: Cat. nos. 439-441, 464, 494-497, 700-712.
Jars: Cat. no. 452.

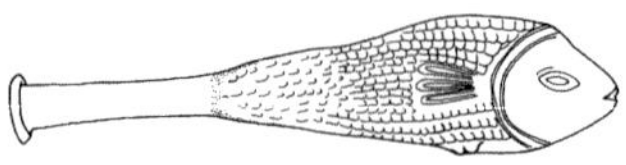

Vessels modelled by mould-blowing always represented a small percentage of the total production of glass vessels;[112] in Thessaloniki especially they account for merely 11% of the total number of preserved examples.

In the wider region of Thessaloniki, vessels of this type are preserved dated to the early production of the 1st century, a period when glass products were still fairly expensive items and were decorated with sophisticated motifs. These were figural vessels rendering fruits, such as dates (Cat. nos. 494-495) and grapes (Cat. no. 496); and animals, e.g. fish (Cat. no. 497). Furthermore, six-sided unguentaria/*pyxides* with relief Dionysiac themes (Cat. no. 464), conical 'mythological' beakers with representations of gods (Cat. no. 95), and probably Hercules' club (Cat. nos. 91-94), calyx-shaped bowls with vertical ribs (Cat. nos. 30-36) and inscribed beakers (Cat. nos. 86-88).

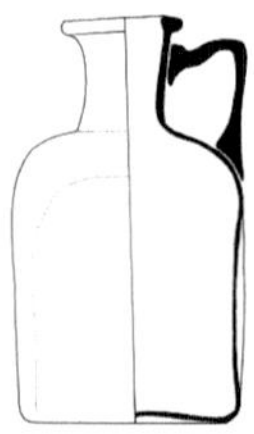

A simpler, cheaper version of the same technique is found on prismatic transportation vessels which occasionally have relief motifs on their bases, mould-blown base marks that usually consist of a simple geometrical pattern, or, less often, a more sophisticated motif that occasionally also includes the name of the glassworker/producer.[113] In excavations in Thessaloniki the following motifs have been found: 1) A Greek cross that divides the bottom into four equal quadrants, each with a dot in the centre, and 6 or 7 concentric semicircles with their centres at the corner; 2) An indiscernible symbol identified as the Greek letter 'Π', but it seems that it might be part of a very fragmentarily preserved ellipsoidal motif, e.g. a rosette petal; 3) A *pelte* shield and the Latin letter 'L' in the centre of a base delineated all around with large dots; 4) A square base of a vessel organized in concentric rings; part of the exterior ring is preserved, decorated with a vine scroll and *pelte* shields; partly preserved geometrical motifs can be seen on the corners. These examples are all dated to the last twenty years of the 1st century (Cat. nos. 385-396).[114]

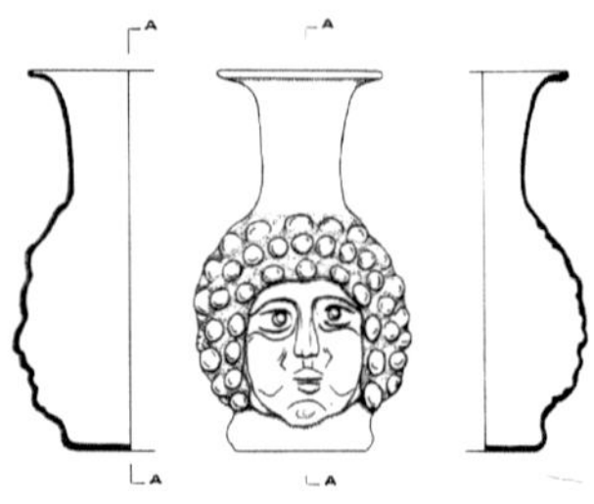

The only form of fully mould-blown vessels that continued to be produced into the 4th century were the vessels, flasks and mostly unguentaria, which formed a human head (Cat. nos. 430, 700-712).[115] A form of cubed unguentaria that appears among the finds from Thessaloniki (Cat. nos. 439-441) is also ascribed to the same period.

A variant of this technique, was widely spread in the Late Roman period and especially during the 4th century, probably because it could be easily manufactured in a type of almost mass production that was typical for this era. This, of course, was dip-mould-blowing, i.e. the initial blowing of the paraison in a small, ribbed mould and then the expansion of the vessel to its final dimensions by free-blowing.[116] This way vertical ribs were imprinted on the body, which by further expansion and rotation of the vessel became faint and often oblique. This decoration appears on globular bottles (Cat. nos. 254-273), pear-shaped bottles (Cat. nos. 297-300), jugs (Cat. nos. 405, 406, 410-413, 419), pear-shaped unguentaria (Cat. nos. 319-320) jars (Cat. no. 452), and lamp-fillers/'baby-feeders' (Cat. no. 307).

The only vessels bearing complex dip-mould-blown decorations are honeycomb bowls and kantharoi (Cat. nos. 69, 188-189). The relief decoration on the vessels usually consists of vertical ribbing, starting right from the slightly flaring rim, and covering the entire circumference. Lower down, the body and base are covered with a web/network

[112] Fleming 1999, 2. For forms of vessels present in the region of Thessaloniki, see Antonaras 2010g, 241-252.
[113] For relevant examples of clay and stone moulds, see: France, Foy, Nenna 2001, 51, 82-83, 118-20; Switzerland, Rütti 1991, 150-152, fig. 103; Germany, Sternini 1995b, 90-91, figs. 129-130, 134; Romania, Baluta 1981, 111-14, figs. 1-2; Turkey, Akat, Firatli, Kocabaş 1984, 55, pl. 16, fig. 45.
[114] Adam-Veleni *et al.* 1998, 85-102; Mavromichali 2001, 141.
[115] Stern 1995, 201-246.
[116] Price, Cottam 1998a, 13.

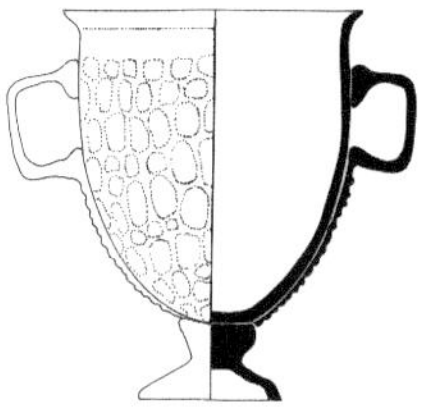

of ovular and hexagonal cells that represent the honeycomb. In some cases the web is supplemented by concentric circles,[117] applied blobs,[118] or with a rosette located centrally underneath.[119] Because of the modelling technique, the honeycomb motif is usually distorted and the cells closer to the bottom are clearly larger than those near the rim. Their production site has not been located, although several relevant assumptions have been published.[120]

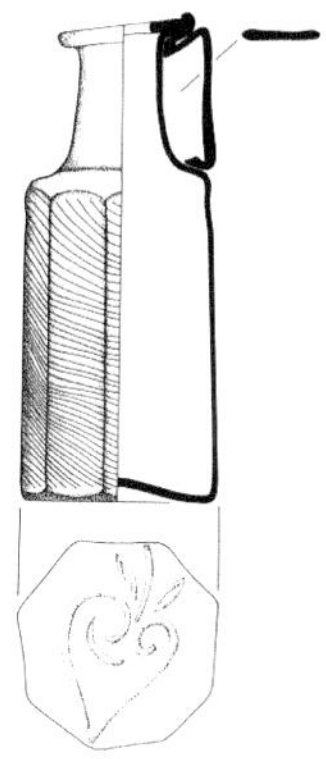

Relatively large 4th-century vessels are also known to be fully mould-blown. It has been assumed that they were blown in a mould because this technique allowed for the relatively precise production of vessels with predefined capacities. Finds in the region of Thessaloniki include eight-sided jugs (Cat. no. 424), amphorisks (Cat. nos. 375-377, 381); and cylindrical (Cat. nos. 427-428) and eight-sided (Cat. no. 431-434) single-handled vessels. The latter featured underneath an ivy leaf or ring-inscribed, four-petaled rosette, probably representing the trademarks of a local workshop.[121]

Finally, a local form of free-blown stemmed beaker produced in the 6th century was also decorated, or rather supplemented, with a combination of stamped decoration, stamped inscription - either a plain numeral or a numeric ligature - placed on the upper side of the base.[122]

[117] Istria (Istros): Minchev 1988, 48, 57, form XXIII, pl. VIII/39, dated from excavation results between the first and third quarters of the 4th century; Museum of Fine Arts, Boston: Saldern 1968, no. 41, dated on the basis of analogous finds to the 4th century.

[118] Hayes 1975, 147, 154, no. 643 (photo 643 on page 226), dated on the basis of analogous finds to the early or mid-4th century.

[119] Doppelfeld 1966, 46, and photos 54-55.

[120] See discussion of Form 14 (Isings 107a), where analogous finds are listed.

[121] See discussion of Form 97.

[122] Antonaras 2011a, 126-128; Antonaras 2011b, 239-242.

# 6. Uses of Glass Vessels

A. Transport and storage vessels

Glass, the fragile medium *par excellence*, was never widely used for the production of transport vessels. Transportation of goods was carried out mostly in clay, wood and leather receptacles because of their resilience to shocks and great pressures; these allowed for stacking and saving valuable cargo space in ships and transport wagons over great distances. Glass transport vessels, on the contrary, most probably served local needs, e.g. within an urban centre, or mainly for storing smaller quantities of liquids to meet household needs.

Around the middle of the 1st century the first rectangular and prismatic single-handled bottles appeared, more or less specifically for the transportation of goods (Form 75).[1] They are made of transparent, bluish or greenish glass and the earlier examples are small, while later on they appein also in larger sizes. The most common are the rectangular forms, but polygonal vessels (three to thirteen sides) are also common. Some of the rectangular vessels are actually free-blown and the walls have been flattened by free-hand tooling. Also quite common are the mould-blown cylindrical bottles. The majority of the finds, and especially the larger examples, appear in the western part of the Empire, although they do occasionally turn up at Eastern sites too.[2]

They are purely utilitarian objects meant for the transportation and storage of liquids. Their main drawback was that they were fragile and that they were far more expensive than their clay counterparts. The fact, however, that they weot impermeable, and thus no odours were transmitted to their contents and that it was easy to check on the contents without breaking the sealing, led to their wide use, at least by the more prosperous sectors of society, and most probably always for specific uses. Very often these vessels bear stamped base marks, rendering geometrical or floral motifs, as well as letters and inscriptions probably referring to their maker. The shape of these vessels was mainly determined by practical considerations, such as the easy handling of a large number of vessels in wooden crates, or in cylindrical leather or wicker cases, in order to secure them during long-distance travel. The body of the vessel is usually left undecorated and when they do carry some decoration this usually consists of a few shallow, incised grooves, which were probably originally also of utilitarian nature, i.e. they might prevent the vessel from slipping out of the hands of the user. Generally these vessels remained in use in the West at the latest until the late 3rd century, except for the cylindrical ones which probably ceased to be produced around AD 110-120. These later examples are smaller and mainly shorter than 1st-century items. In the Eastern Mediterranean single-handled prismatic bottles circulate even into the 4th century.

Additionally, it should be noted that these same vessels were also used for burial uses as grave goods, containing liquids that would sustain the deceased at the banquets on the Elysian Fields. They are also found in cemeteries outside the graves, witnessing to the memorial meals served there by both pagans and Christians in memory of the dead.

B. Tableware

One of the main and probably most common uses for the glass vessels was to adorn the tables of both middle-class and wealthy Romans. At their dinners, polychrome glass vessels made a cheaper alternative to vessels featuring semi-precious stones, while colourless glass could substitute for crystal and silver prototypes, which were used only by elites, military dignitaries and officials.[3] Quite soon the reduction in their prices led to a much wider distribution and made their transparency more important, permitting banqueters to marvel at the sophisticated meals and enticing colours of wines served to them.

The prices of plates, dishes (Forms 1-3) and glass trays (Form 4)[4] steadily fell, but this was matched by a respective decline in standards of modelling and decorative technique. While in the 1st century they were decorated, obviously imitating semi-precious stone and silver vessels, they were later to become ordinary, utilitarian objects made in

[1] Isings 1957, 63-69, 108, forms 50, 51, 90.
[2] Harden *et al.* 1987, 97.
[3] Pliny, *Nat. Hist.*, 36.192, where he mentions that transparent and decolourized glass vessels were more appreciated in his time, because they resembled rock crystal. Grose 1989, 254-256, a group of cast vessels which appeared and were in use only during the first half of the 1st century, and most of the examples were found at sites in the western provinces of the Empire. On that matter, Vickers 1996, passim, is very enlightening.
[4] Isings 1957, 37, 116-118, 148, forms 19, 97, 118.

the simplest, quickest and cheapest possible way; it mattered little if they were off-centre or if their bases showed distinct traces of tool marks and other signs of their production.

From the 2nd century onwards, glass amphorisks (Form 74) began to appear, both pointed and flat-bottomed varieties, which were used for serving liquids, especially wine, at table.[5] Jugs (Forms 76-97)[6] and handleless funnel-mouth flasks (Forms 54-73),[7] which were used for serving wine, were extremely popular vessels and they represented one of the most commonly preserved groups of tableware vessels of this period. The tendency to decorate them is obvious, and even the mere shape of the body or the shape of the handle often has such elegant proportions and curves that they immediately stand out. In some cases, apart from the oblique ribbing and the applied threads of glass, medallions with stamped lion heads or theatrical masks were applied on the base of the handle or even around the body in order to decorate them.[8] In addition, mainly in the 3rd and 4th centuries, twin- or triple-vessels were produced, known as 'dilekythoi' (Form 78).[9] These are vessels produced as a single and united vases internally divided into two or three compartments by the appropriate tooling skills of the glassworker.

Many, probably the majority, of the preserved tableware vessels were plain handleless bottles, with globular or pear-shaped body (Forms 44-52).[10] These vessels are often decorated, especially at the end of the 3rd and during the 4th century, with dip-mould-blown oblique ribbing. The vessels with purely spherical body, an intense constriction at the bottom of the neck and unworked vertical rim, constitute another contemporaneous group (Form 50).[11]

Glass drinking vessels are present in a great variety of size and shape. Apart from kantharoi (Forms 42-43), which are very rare, almost all of them are handleless vessels (Forms 5-37). The group of mould-blown vessels includes low, cylindrical forms bearing Greek relief inscriptions, usually toasts to the health and well-being of the drinker;[12] these commonly appear in the first half of the 1st century (Form 19). Two more forms, known across the entire Mediterranean, were made by mould-blowing during the 1st and into the early 2nd century. These are tall, conical beakers with flat base, and sometimes featuring mythological scenes (Form 22); on occasion they were adorned with almond-shaped protuberances, believed to represent either lotus buds or the knots on Hercules' club (Form 21).[13]

In addition, tall conical free-blown beakers, indented or smooth, represent a very common find and were particularly popular in the 1st and 2nd centuries,[14] although the shape continues into the 3rd and 4th centuries, but in a slightly altered style (Forms 24-32).[15] Stemmed beakers, with a tall stem and conical base (Forms 35, 37) and simpler conical beakers (Forms 31-32) are used from the 4th and particularly in the 5th century as tall drinking vessels and they actually supersede any other earlier form of drinking vessel.

The group of short drinking vessels in the 1st and 2nd century mostly cover the strongly-coloured (in various shades), ribbed bowls (Forms 7-8),[16] which continue the long tradition of their rotary pressed (Form 5-6)[17] prototypes, and the hemispherical bowls (Forms 9-11).[18] From the 3rd century onwards hemispherical bowls, often bearing incised or applied decoration (Forms 12, 14),[19] and shallow, smooth or indented bowls (Forms 15-16)[20] meet the needs for such vessels.

Glass ladles (Form 39) and *trullae* (Form 40), which are quite rare finds, were used for decanting wine or/and water, from larger vessels, e.g. kraters, and they reflect the sumptuousness of the banquets for which they were made.

---

[5] Isings 1957, 157-159, forms 127-129.
[6] Isings 1957, 30-31, 72-74, 104-106, 118, 149-156, forms 13, 55, 88, 98, 120-125.
[7] Isings 1957, 110, 122-124, 161-162, forms 92, 104, 133.
[8] Harden *et al.* 1987, 118-119.
[9] Cool, Price 1995, 203-204.
[10] Isings 1957, 119, form 101.
[11] Isings 1957, 121-122, form 103.
[12] Stern 1995, 97.
[13] Harden *et al.* 1987, 162-163.
[14] Isings 1957, 44-50, forms 29-35.
[15] Isings 1957, 126-131, form 106.
[16] Isings 1957, 35-36, form 17.
[17] Isings 1957, 17-21, form 3.
[18] Isings 1957, 27-30, 97, 101-102, 104, forms 12, 81, 85, 87,
[19] Isings 1957, 133-134, form 107.
[20] Isings 1957, 144-148, form 117.

C. Unguentaria

Since earliest times men and women have searched for substances to protect and cure them. Probably from the same time they started collecting and using fragrances for apotropaic-healing or/and religious reasons. These aromatic and therapeutic substances were precious because they were rare in nature or they were the outcome of complex chemical procedures. Miniature flasks made of different media were used for containing them. Glass, as a non-porous material, is characterized by the fact that vessels made of it prevent of their contents and do not retain odours from previous contents that might eventually contaminate their current contents. These properties made glass flasks almost the exclusive medium for keeping different types of myrrhs, fragrances and medicaments.

Globular aryballoi (Form 111) represent one of the earliest free-blown forms that were widely used as unguentaria. The majority of these vessels were plain, small vials, handleless, encased in wicker, leather, or some other material, to protect them from breaking. They were produced in all geometrical shapes, globular, fusiform, tear-shaped, cylindrical, conical, rectangular and ovular (Forms 121-136, 139-142, 145).[21]

Bands of parallel incisions, or indentations on the body, which transform the originally globular body into a four-sided vessel (Form 141) are simple forms of decoration commonly found on unguentaria. There are more elaborately decorated vessels, such as those bearing painted motifs,[22] either simpler, geometrical ones or occasionally more complex styles. In addition, vessels with wheel-cut decoration[23] are also known, some with entire scenes of myths or everyday life, e.g. hunting. Figural vessels appear from the 1st century and evolve from simpler solutions towards more elaborate and complex creations that were achieved with the use of moulds.

Fusiform unguentaria (Form 147),[24] dated between the late 3rd and late 5th centuries, are among the largest examples of this group. These vessels, as is evident by their totally unstable shape, were designed to be kept upright, placed in wicker or leather basket-shaped holders. Unguentaria forms, as well as the simpler handleless vessels, included miniatures of tableware vessels, such as kraterisks (Form 137) and juglets (Form 143).[25] Amphorisks,[26] either in the shape of large transport vessels (Forms 112-113, 138,) or, occasionally, more original shapes (Forms 114, 144), represented a quite widely distributed form of unguentaria.

A special group are the 1st-century bird-shaped (Form 120)[27] vessels and the fine, egg-shell spheres (Form 116),[28] which were made in north Italian workshops, and probably also locally in Thessaloniki; they represent the first known vessels meant for a single use. They were made and sealed in such a way that one had to break off the tip to access the contents. These vessels are the only ones among unguentaria that clearly contained solids and not liquid contents, i.e. some sort of coloured cosmetics.

Relatively common finds are mould-blown unguentaria made in the shape of fruits or animals.[29] From Thessaloniki are known vessels in the shape of dates (Form 117), grapes (Form 118), and fish (Form 119), which appear with small variants up to the 2nd century. They were Eastern inventions that were equally widely spread in the West. Vessels in the shape of a human head are among the very few mould-blown flasks that survive after the 2nd century and they continued to exist into the 4th century. Those among these vessels with their body rendered in the shape of two heads (Form 146), are known as 'Janiform', the two back-to-back faces recalling the Roman god Janus.[30] In the same, mould-blown technique are also produced small, polygonal pyxides and other vessels, usually hexagonal, which are decorated with floral motifs or with other themes from the Dionysus cult (Form 115).[31] They are made either of dark transparent glass in different shades, or of opaque, blue or white glass. Research and typological classification of the decorative motifs has shown that they were products of Eastern workshops.[32]

[21] De Tommaso 1990, passim.
[22] Rütti 1991, 122-136.
[23] Fremersdorf 1967, passim.
[24] Isings 1957, 126, form 105.
[25] Weinberg, McClellan 1992, 118-119.
[26] Weinberg, McClellan 1992, 118-119.
[27] Isings 1957, 27, form 11.
[28] Isings 1957, 25-26, form 10.
[29] Isings 1957, 93-94, form 78c-d; Stern 1995, passim.
[30] Isings 1957, 93-94, form 78a-b.
[31] Weinberg, McClellan 1992, 71, 123-125.
[32] Stern 1995, 113-148, 161-166.

The different types of unguents were either liquids or creams and powders; these required either quite wide-mouthed vessels to contain them, so that it was easier to extract the contents with some implement, or even a finger. For narrow necks a pointed implement, such as a *spatula*, was needed, and these are common find with glass unguentaria.

Jars and jar-shaped unguentaria (Forms 98-110) are included in the group of wide-mouthed vessels.[33] Most probably they were used for solid cosmetics or creams. Their typically wide and almost horizontal rim was useful for returning any excess contents from the user's hand back to the vessel.

For stirring, removing and applying fragrances, it seems that, *inter alia*, fine glass rods (Form 148) were used that could be easily dipped into often long-necked flasks and unguent vessels.

D. Lamps and oil-fillers

The use of glass for the production of lamps seems to have begun only in Late Roman times. However open-shaped drinking vessels were used as lamps already from the 2nd-3rd century. Of primary significance for the use of the vessel as a lamp must have been that glass would help radiate more light, not a feature of regular clay and metal lamps. Moreover, glass lamps burnt oil floating on a layer of water, thus transforming the convex body of the glass lamp into a kind of large magnifying lens. Such lamps must have made attractive products, despite the relatively higher cost and risk of breakage.

Lamps, at least lamp fragments, are very common finds and they can be classified into the following groups:[34]

Replicas f mMould-made clay lamps (although they do appear in glass fom,[35], it is not represented among the finds from Macedonia or Thessaloniki).[36]

Conical, pointed lamps,[37] dated to the second half of the 4th and to the 5th century. Quite common in Egypt and the Syro-Palestinian region, as well as at sites such as Cologne and in central Europe, here they are quite rare. Their body is made of colourless, yellowish or greenish glass and they are often decorated around their body with blue blobs. Although they are generally ascribed to the group of lamps, vessels of the same form were equally used it seems as drinking vessels, as was, of course, their original use.

An early example of glass lamp, indeed of a particularly large size, represents the extremely rare three-handled calyx-shaped form (Form 41).

Another form of a Late Imperial lamp that appears quite often in Thessaloniki is the simple hemispherical or conical bowl, mainly without handles (Form 12),[38] which are occasionally supplemented with three handles arranged around their rim (Form 13); both variants are dated at around the 4th century.[39]

It appears that stemmed lamps, either with inherent or with an applied stem, become more and more common (Form 38)[40] from the 5th century onwards. They were suspended either singly from metal rings and chains, or more often in groups from metal *polykandela*.[41]

Lamps in the shape of stemmed conical beakers with pushed-in base, supplemented with three small handles around the rim, appear after the 5th century (Form 36). Plain, handleless examples are also identified as lamps and they are very common finds in secular and religious buildings of the Early Christian period (Form 37).[42] The form

[33] Isings 1957, 88-89, 111, forms 68, 94.

[34] The first classification of glass lamps was attempted in Crowfoot, Harden 1931, 196-208. Among the more recent studies the most thorough is Uboldi 1995, 93-145, and a systematization of the Dalmatian finds has been made by Chevalier 1997-98, 161-199. For a short presentation of the finds in the region of Thessaloniki, see Antonaras 2008b, 23-30.

[35] Goethert-Polaschek 1977, 259, form 163, pl. 81; Lazar 2004a, 64, no. 49.

[36] Lazar 2004a, 67, no. 59.

[37] Uboldi 1995, 113-115; Isings 1957, 126-131, form 106.

[38] On their use as lamps, in metal perforated cases see Boyd 1988, passim.

[39] Uboldi 1995, 104-108.

[40] Uboldi 1995, 120-124; Papanikola-Bakirtzi 2002 (ed.), 285, no. 298, (A. Dina).

[41] On similar examples, see Marijanović-Vujović 1973, 13-24; Papanikola-Bakirtzi 2002 (ed.), 283-284, no. 296; Chrzanovski 2003, 30-31, 95, cat. no. 124; Wamser (ed.) 2004, 99-103, nos. 131-139.

[42] Isings 1957, 139-140, form 111.

with handles combines the plain beaker with the shorter, almost hemispherical three-handled lamp; these are generally rare finds.[43]

It worth noting that in the wider region of Thessaloniki no lamps featuring a low, truncated conical body have been uncovered;[44] these are quite usual finds in other regions.[45] In addition, no small, tubular fragments or parts of wick holders attached to the bottom of lamps bowls are traced;[46] such finds are very widely distributed in the East, and certain regions of southern Greece.[47]

The so-called 'baby-feeders' (Form 53)[48] corresponding to the clay vessels of similar shape to which there are references in ancient sources suggesting that they were indeed used for the feeding of babies, represent a special group. However it is not clear whether the glass examples of this form were intended for such use. On the contrary, it is very probable that they were used for filling clay, metal, and, perhaps, glass lamps with oil; such lamps had small filling holes and the filler vessels had low and conical spouts well suited to this purpose.

E. Burial uses

Although no form or use of vessel was invented originally to be used in burials and with burial customs, yet almost all forms of vessels have indeed been used in burial and/or commemorative feasts and memorials (i.e. *mnēmosyna*). It must also be noted that almost every better preserved glass vessel was found during excavations of burials.

Spouted vessels, pseudo-baby-feeders / lamp-fillers, as well as examples of the entire repertoire of glass lamps have been found in large quantities in the cemeteries of Thessaloniki. These finds witness the well-attested custom of letting a lamp burn over the grave for the first forty days after the burial and additionally on special occasions and fixed dates during the year.[49] Furthermore, it is well known that several graves were equipped with a special rectangular niche (*lucernarium*) where the lamp was placed.[50]

The presence of unguentaria and partly of larger, close-shaped tableware vessels (e.g. jugs, flasks, and bottles) is connected with the equally pagan and Christian custom of anointing the dead – to the extent allowed by the Church.[51] The vessels must also have contained oil used in the rites, specifically the symbolic act performed by the priest, pouring it in the shape of a cross over the corpse at the funeral to indicate that the deceased lived and died in accordance with his or her faith.[52] This custom was also connected with the desire of relatives to spoil the often sumptuous shrouds and garments[53] to discourage the desecration of the grave.[54]

Tableware fragments, found in and around the graves, or scattered around the cemeteries, are connected with the memorial ceremonies and the meals (*makaries*) offered by Christians above the graves.[55] All the aforementioned vessels, used during the burial or in the cemetery, were not supposed to be brought back to the deceased's house,

---

[43] Papanikola-Bakirtzi 2002 (ed), 285, no. 299 (A. Antonaras).

[44] Isings 1957, 162, form 134.

[45] Uboldi 1995, 104-108, form I. 1, fig. 2. /I. 1, with further bibliography.

[46] Stern 2001, 273, where detailed presentations of relevant finds date their invention around the 5th century.

[47] Finds of this form were located in excavations at: Gortyna, see Sternini 1997, 232-233, 241, no. 7, fig. 32:o; in the Athenian Agora, see Weinberg, Stern 2009, 153-156, nos 372-381; on the site of the new Acropolis Museum, see Eleftheratou (ed.) 2006, 93; and in Piraeus, see D. Grigoropoulos, unpublished Master's thesis, University of Durham, on the excavation and finds from the site of the Piraeus courthouse.

[48] Isings 1957, 118, handleless variant of form 99.

[49] Koukoules 1948-52, Vol. IV, 208-214. For an early, 4th-century reference to a constant lighting of a lamp on a grave, see Wilkinson 1971, 123-124, § 47.3.

[50] Marki 2006, 115-116, 208 where finds from Thessaloniki are cited. On presentations of lighting lamps and candelabra, see Marki 2006, 167-168, drawings 113-115, pl. 15β, 171, drawings 123, pl. 17β.

[51] Koukoules 1940, 3-80 (esp. pages 8-9, 13); Kyriakakis 1974, 37-72.

[52] K. Kallinikos 1969, 560-561; Koukoules 1940, 43-44; Loukatos 1940, 47.

[53] Antonaras 2001, 45-47; Antonaras 2002a, 18-19.

[54] Koukoules 1940, 47.

[55] Koukoules 1948-52, Vol. IV, 208-214. To the needs of this custom are ascribed the flat surfaces, occasionally supplemented with built benches, which have been traced atop barrel-vaulted graves, see Laskaris 2000, 268, also Marki 2006, 114, 209. For an independent building dated to the late 6th-7th century, with such a use in the cemetery of Carpathos, see Geraskli 2004, 389-402. For depictions of funeral suppers in graves in Thessaloniki, see Marki 2006, 140-141, drawings 74-75, pl. 5β-γ. On the meaning of these common meals, the *agape*, in the life of the Christian community, see Cabrol, Leclerq, s.v. 'Agape', *DACL* 1, (1939), cols. 848-849. Since the deceased continued to represent part of the community, funeral suppers were for the Christians absolutely normal and acceptable practices, see Ascough 2004, 509-530. In addition, funeral suppers had a long tradition in the Greco-Roman world, already known from Homer (Ψ.29, γ.328), while in Roman Macedonia they were connected to *parentalia* and *rosalia*, which also included meals at cemeteries and which were imported by the Italian colonists, see Ascough, 2004, 513-514, where the relevant bibliography is listed.

as they were considered unclean in some way, and they had to be broken and left on site.[56] This belief might partly explain why vessels found in the graves are usually new, bearing no traces of use, even if they are sometimes found broken; it seems they were bought to be placed in the graves. To some extent, this practice explains their often lower quality, or slight imperfections, such as body deformation or poor balance, that might have excluded them from the normal market.

[56] Loukatos 1940, 75-79. On similar finds from Thessaloniki's cemeteries, see Makropoulou 2001, 266-267.

# 7. Chronological Overview of the Finds – Conclusions[1]

During the Hellenistic and Republican periods Thessaloniki was favoured by the Roman administration, and grew gradually to be the capital of the entire province of Macedonia. This political and, supposedly, economical progress, which would allow for the import and wide distributions of the, in those days, precious glass vessels, does not reflect on the glass finds, which, anyway, are rarely present in the strata of the 2nd and 1st centuries BC.

Very few vessels can be dated to the late 1st century BC, which was a period of political fluctuations and turmoil in Macedonia and in Thessaloniki. During this period, and because of the extended Roman conquests further north of the Balkans, Thessaloniki obtained the status of free city in 42 BC, so its habitants were released from taxation and the city's economy flourished, attracting traders from Italy and other regions at the same time, a fact that gradually gave to the city the cosmopolitan character of a large Mediterranean port.

According to the preserved material, the situation changed radically during the Imperial period, when Thessaloniki, as a port and a commercial hub, benefited considerably from the new economic conditions created by the *pax romana* delivered by Octavian Augustus with his victory at the battle of Actium in 31 BC.

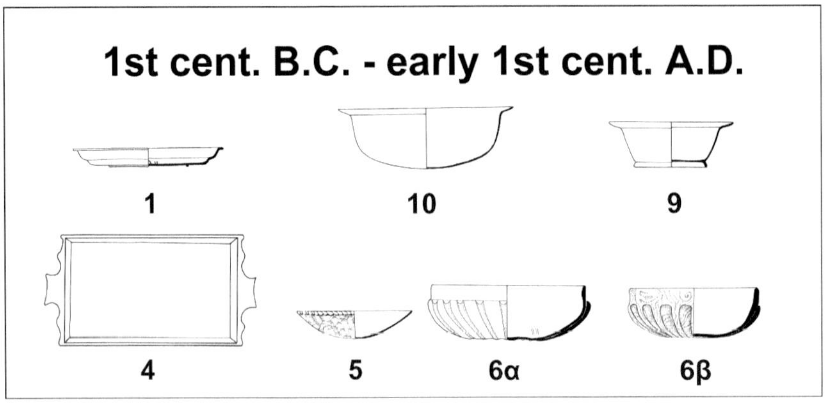


Glass finds appear for the first time in considerable quantities at the end of the 1st century BC. It is known that already in the Augustan era an Italian community of *negotiatores*, i.e. traders, settled in Thessaloniki. Judging by the fact that large numbers of glass vessels from this stratum are of Italian origin, these traders must have imported the glass vessels as trade objects *an sich*, or as glass containers for cosmetics, etc. Since the time of the Apostles, a Jewish community is also known to have been active in trading with the Syro-Palestinian region, probably also in glass.

The most noticeable vessels used in Thessaloniki during the late 1st century BC to the early 1st century AD included decolourized, cast vessels, dishes[2](Form 2), hemispherical[3] (Form 10) and conical bowls[4] (Form 9), a rectangular tray[5] (Form 4), and mosaic bowls, either plain[6] (Form 5) or ribbed[7] (Form 6b). Almost all of these were found around the city's agora and most probably originated from Italy. Bluish, ribbed bowls[8] (Form 6a), popular both in the East and the West, appear in a significantly smaller number, compared to the quantities of finds from other, far less important, sites.

A few finds came from the settlement area, like those from the destruction layer of the bath complex dating to the second half of the 1st century AD. The public baths, destroyed in the Flavian period, were unearthed next to the agora. In its destruction layers large numbers of glass sherds were found. Judging by the published material, it

---

[1] A shorter version of this chapter appeared in Antonaras 2009, 74-79. Furthermore, an overview of all types of glass products (vessels, window panes and jewellery), present in Late Roman and Early Christian Thessaloniki, has been published in Antonaras 2010c, 301-334. The overall production and uses of glass in Byzantine Thessaloniki was published in Antonaras 2014b, 189-198.

[2] Isings 1957, 61-62, form 46a, c.

[3] Grose 1991, 15-18.

[4] Grose 1991, 12-18; Rütti 1991, 40, form AR 21, pl. 39; Massabò 2001, 48-49.

[5] Isings 1957, 37, form 19; Stern, Schlick-Nolte 1994, 326-327, no. 98.

[6] Grose 1989, family II, 249 ff.

[7] Isings 1957, 17-21, form 3.

[8] Isings 1957, 17-21, form 3.

seems, that tableware prevails among the glass finds, i.e. vessels for drinking and serving liquids, a factor that can be anticipated by the character of the building.

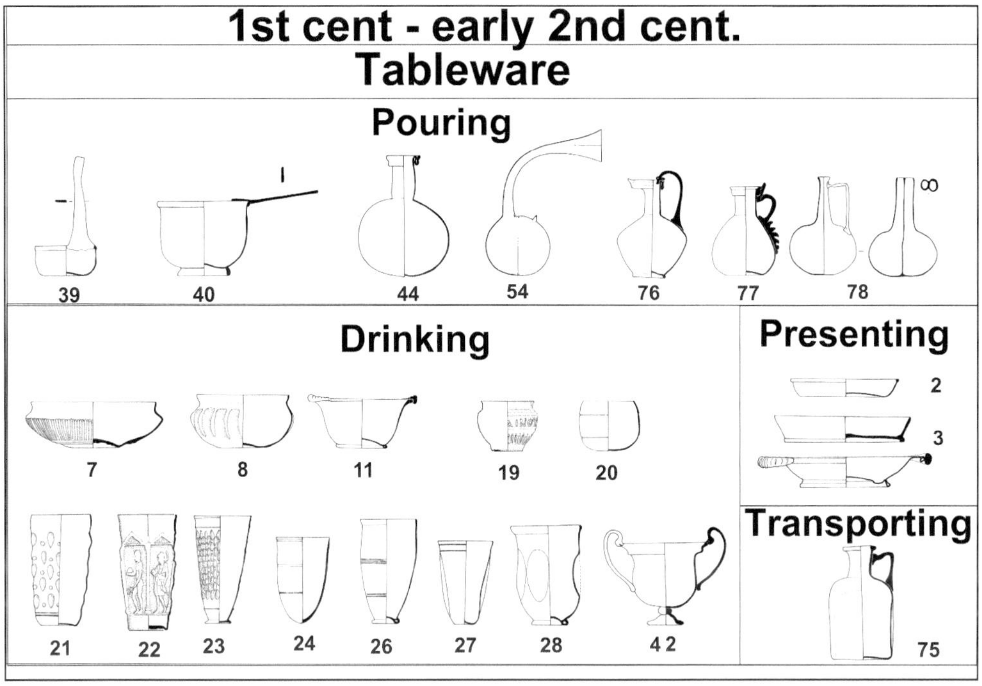


The recognized jugs can be subdivided as jugs with trefoil-mouth and spherical body[9] (Form 76), with conical body and undulating handle[10] (Form 77), and *dilekithoi*[11] (Form 78). Three shapes are known from both the West and East. The flasks with collar shaped rim[12] (Form 44), on the other hand, are to be seen rather as Aegean or Anatolian products. Spherical flasks, characteristic of Italy, are also found, as well as the type with a long, curved neck and two tiny spouts on the shoulders[13] (Form 54).

The few fragmentary ladles[14] (Form 39), *trullae*[15] (Form 40) and the relatively numerous stirring rods[16] (Form 148) are also of Italian origin.

Drinking vessels appear in a variety of shapes. From Italy there are *kantharoi*[17] (Form 42), but also shallow, ribbed bowls, mould-blown[18] (Form 7) and free-blown varieties[19] (Form 8). Shapes from the Eastern Mediterranean in general, but mainly assumed to have been produced in Sidon, include hemispherical bowls, short mould-blown beakers with inscriptions[20] (Form 19), hemispherical beakers, with undulating handles[21] (Form 11), or without handles, cylindrical free-blown beakers decorated with horizontal engraved bands[22] (Form 20).

Tall conical beakers – plain or decorated with impressions[23] (Forms 27, 28) or engraved[24] (Form 23), with ring base[25] (Form 26) or without a foot[26] (Form 27) – and pointed, engraved beakers[27] (Form 24).

[9] Isings 1957, 75-76, form 56b.
[10] Stern 2001, 50, 95, no. 32.
[11] Cool, Price 1995, 203-204.
[12] Isings 1957, 90, form 70; Barag 1970c, Vol. 2, pl. 41, form 15. 3.
[13] La Baume, Salomonson 1976, 44, no. 104, pl. 13/2.
[14] De Franciscis 1963, 138, fig. 1.
[15] Scatozza-Höricht 1986, 37-38, forms 15-16, pl. XXVII.
[16] Isings 1957, 94-95, form 79; Spaer 2001, 262, 264-265.
[17] Isings 1957, 53-54, form 38a; Van Lith 1978/79, 66.
[18] Stern 1995, 111-113, nos. 13-14.
[19] Isings 1957, 35-36, form 17; Haevernick, Pfeffer 1958; Haevernick 1967.
[20] Harden 1935, forms Gli, ii.
[21] Isings 1957, 59, form 43.
[22] Isings 1957, 27-29, form 12.
[23] Isings 1957, 46-47, form 32; Isings 1985, 49-50, form 35.
[24] Oliver 1984, 38-39.
[25] Isings 1957, 48-49, form 34.
[26] Isings 1957, 46-47, form 32; Harden 1936, 391-401, no. 376-82.
[27] Isings 1957, 44, form 29.

Fragments of several of the exquisite mould-blown *lotus-bud*[28] (Form 21), and of one *mythological*[29] (Form 22) beaker have also been found. These two types were produced in both parts of the Empire. The bowls with undulating handles and the shallow handless ones in decolourized glass are possibly of Italian origin.

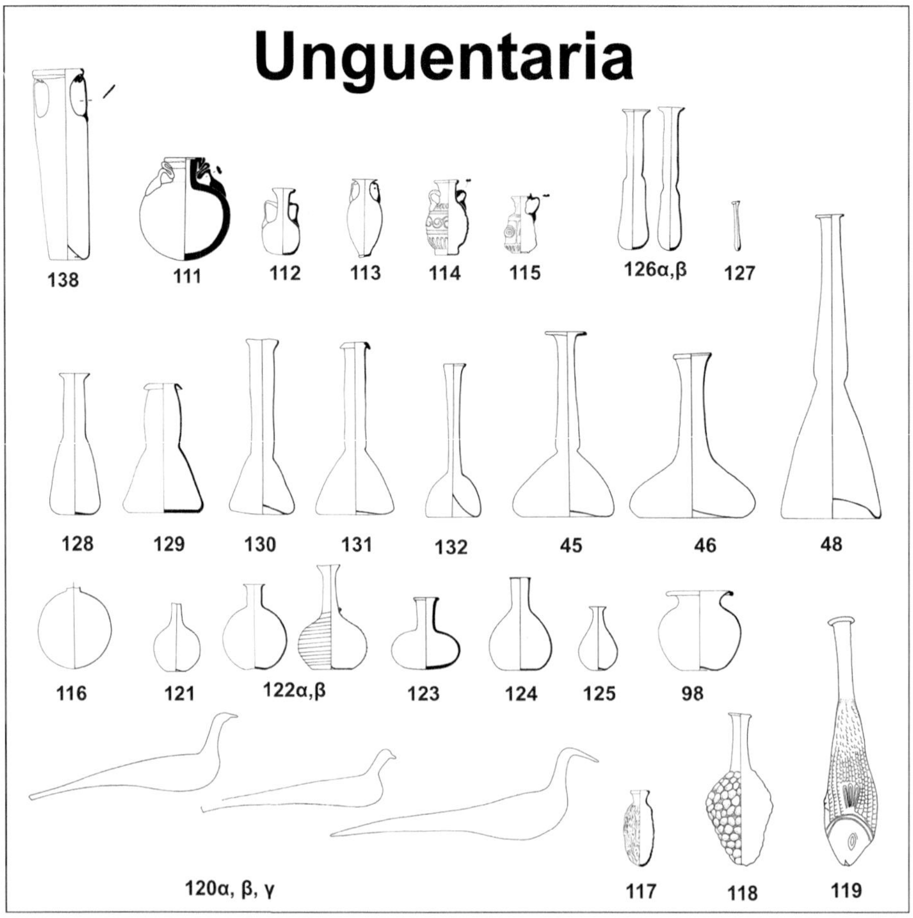


Shallow conical dishes[30] (Forms 2, 3) for presenting food are present as well and occur in different variations: with straight or curved walls, with ring base or without foot, occasionally with horizontal undulating handles but mainly without handles.

Toilet-ware is also catalogued – mainly simple unguentaria without handles[31] (Forms 45-46, 48, 128-132), but also tear-shaped varieties[32] (Forms 126-127) and bulbous[33] and ovular unguentaria[34] (Forms 124-125), some aryballoi[35] (Form 111), and amphoriskoi, and occasionally mould-blown single or two-handled vials[36] (Form 115), all of them known in both East and West.

Another shape recorded from Thessaloniki, but known from both Western and Eastern parts of the Empire, is the mould-blown square bottle (Form 75), occasionally bearing a stamped mark on the base.[37]

The contemporaneous graves of this period revealed chiefly unguentaria in many different forms. The majority are without handles: conical with straight or curved walls[38] (Forms 45-46, 48, 128-132), spherical (Forms 116,[39] 121[40]-122[41]), squat spherical[42] (Form 123), bulbous[43] (Form 124), ovular[44] (Form 125) and tear-shaped[45] (Forms 126-127),

[28] Isings 1957, 45-46, form 31.
[29] Davidson-Weinberg 1972, 29-30.
[30] Isings 1957, 61-63, forms 46a, c, 47 and 49; Whitehouse 1997, 69 no. 79.
[31] Isings 1957, 42-43, 97-99, forms 82A2, 82B2, 28a, 28b.
[32] Isings 1957, 24, 41, forms 8 and 27.
[33] Isings 1957, 34-35, form 16; De Tommaso 1990, 52, form 21.
[34] De Tommaso 1990, 52, form 21.
[35] Isings 1957, 78-80, form 61; Cool, Price 1995, 156-159.
[36] Weinberg, McClellan 1992, 71, 123-125.
[37] Isings 1957, 63-69, 108, forms 50, 51, 90.
[38] Isings 1957, 42-43, 97-99, forms 82A2, 82B2, 28a.
[39] Isings 1957, 25-26, form 10.
[40] Israeli 2003b, 113, nos. 94-95.
[41] Isings 1957, 22-23, form 6; Stern 1977, 47 ff, no. 11 pl. 2; De Tommaso 1990, 39-40, 42-43, forms 4, 5, 7.
[42] Scatozza-Höricht 1986, 63, 127, pls. XX, XXXV; De Tommaso 1990, 38, form 2.
[43] Isings 1957, 34-35, form 16.
[44] De Tommaso 1990, 52, form 21.
[45] Isings 1957, 24, 41, forms 8, 27.

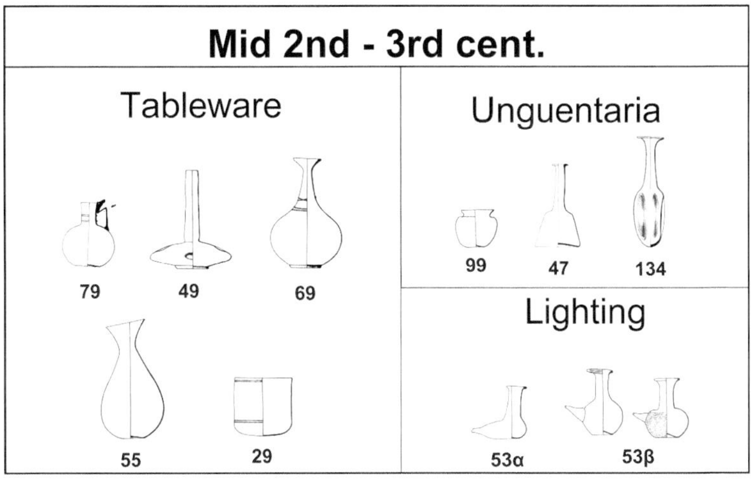


usually with the neck far longer than the body. Amphoriskoi are rather rare, spherical[46] (Form 112) and ovular[47] (Form 113), and occasionally mould-blown[48] (Form 114). Also, during this period bird-shaped unguentaria[49] (Form 120) appear in large numbers. The majority of the unguentaria are imported from northern Italy, although there are Syro-Palestinian products as well, like the occasionally seen mould-blown unguentaria: date-shape[50] (Form 117), grape-shaped[51] (Form 118), and fish-shaped[52] (Form 119).

Thessaloniki flourished from the middle of the 2nd and during the 3rd century. The population increased considerably and a large forum complex was constructed. But, as far as glass finds are concerned, the number of preserved examples diminished considerably, as did the number of forms in use.

On the whole it was the tableware that was preserved, i.e. aryballoid jugs (Form 79), biconical bottles with indentations[53] (Form 49) and pear-shaped flasks[54] (Form 69). Large jar-like unguentaria[55] (Form 99) are also found, possibly of Western origin. Finally, during this period there are finds of vessels with characteristic inbuilt spouts, known as baby-feeders, but most probably used as lamp-fillers[56] (Form 53). These vessels were produced both in the West and East, although the handless variant with applied spout, which appears in Thessaloniki, seems to be a late 3rd to 4th century Eastern product, although they continue to be used and produced in the 4th century as well. Around the end of the 2nd and during the 3rd century new forms appear: bulbous flasks[57] (Form 55), cylindrical beakers with simple engraving[58] (Form 29), as as well as indented ovoid unguentaria[59] (Form 134) and conical ones with horizontal shoulders[60] (Form 47), which were Western products.

From the second half of the 3rd century a new period of development started in Thessaloniki. For the city this is again a prosperous period, apparent from the newly erected monumental buildings in the centre. Its strategic position on the Via Egnatia, important in terms of the military events of the period, along with the prominence of its sea-port, were the main assets of Thessaloniki, which became the Imperial capital under Gaius Galerius.

After the middle of the 3rd century the use of glass evolved, becoming a more common material and now mostly exploited in the production of larger vessels. The forms from Thessaloniki and its wider area are characteristic for the middle of the 3rd century to the beginning of the 5th, and they are equally widespread in the East and West.

Large tableware forms appear in great numbers: jugs in new shapes known both from the East and West, with a spherical body (Forms 80,[61] 82), an ovular body[62] (Form 83), an inverted conical variety (Forms 84,[63] 89) and a

[46] Mackworth-Young 1949, 87, grave 6, vessel no. 4, pl. 29:1.6(4).
[47] Calvi 1968, 24, pls. 1. 2 - 1. 4.
[48] Stern 1995, 152, no. 55.
[49] Isings 1957, 27, form 11; Antonaras 2005, 241-248; Antonaras 2010a, 27-33.
[50] Isings 1957, 94, form 78d.
[51] Isings 1957, 94, form 78e.
[52] Kunina 1997, 281, colour photo on page 37, no. 152 and photo 152.
[53] Whitehouse 2001, 181, no. 319.
[54] Goethert-Polaschek 1977, 149, form 91, pls. 14, 157d. 59.
[55] Isings 1957, 111, form 94.
[56] Isings 1957, 118, handleless variant of form 99.
[57] Cermanović *et al.* 1975, 153, form I/1A.
[58] Sorokina 1965, 205, with further bibliography in footnote 9; Kunina 1997, 319, no. 327.
[59] Goethert-Polaschek 1977, 122, form 76a, nos. 667-668, pl. 53.
[60] De Tommaso 2001, 160-161.
[61] Barkóczi 1988, 168, no. 395, pls. XXXVI, XCIV.
[62] Isings 1957, 149-150, form 120a.
[63] Isings 1957, 150-151, form 120b.

cylindrical one with incised decoration[64] (Form 95); pointed and flat-bottomed amphoriskoi[65] (Form 74); spherical bottles with cracked-off rim and a constriction at the base of their neck[66] (Form 50); spherical and pear-shaped bottles with out-splayed rim or funnel-mouth, ribbed or plain types[67] (Form 5;, 52). These forms were widely used and produced both in the West and East, but it is quite probable that the majority was produced locally. Flasks also arrive in many forms: pear-shaped, squat or slender, spherical with short or long neck (Forms 56,[68] 61,[69] 62,[70] 63,[71] 64,[72] 108,[73]) ovular[74] (Form 71) and cylindrical varieties (Forms 65,[75] 72[76]). Although they are known both in the East and West, there is evidence that some of them were locally produced.

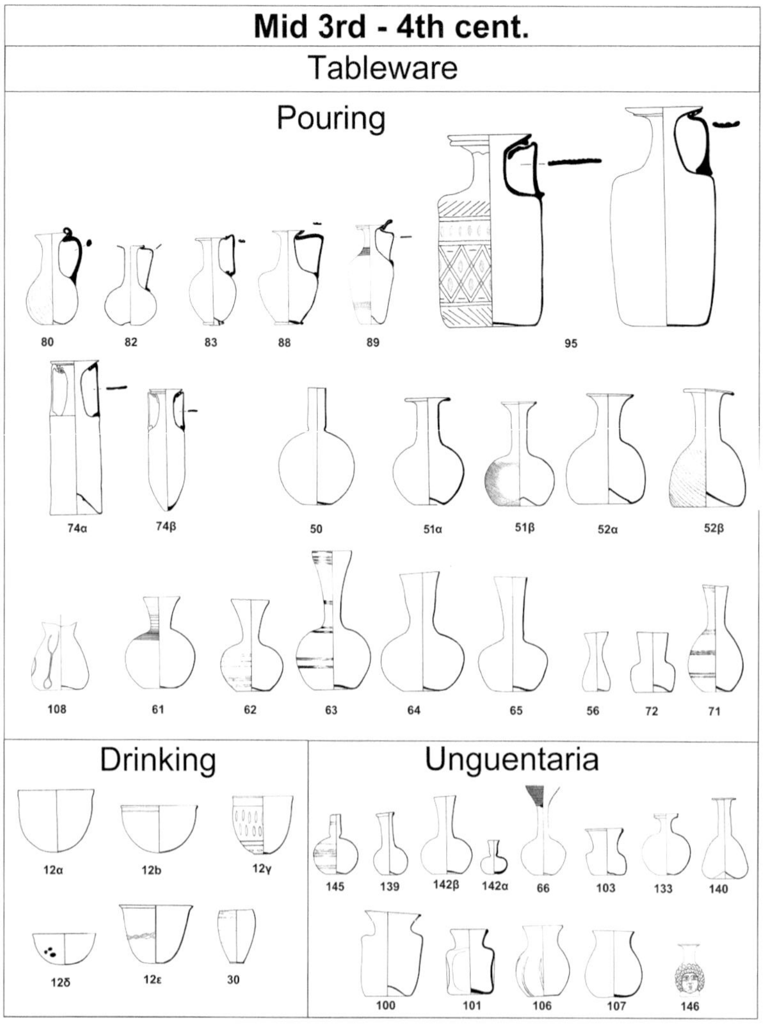


The hemispherical bowl is the prevailing shape of all drinking vessels[77] (Form 12), while ovular beakers[78] (Form 30) appear only sporadically. They often bear an engraved or applied decoration.

The unguentaria appear as miniature versions of tableware vessels, e.g. cylindrical amphorae[79] (Form 138), spherical with cracked-off rim[80] (Form 145), out-splayed rim or funnel-mouthed ones with ring-shaped rim[81] (Form 139) or polished rim (Form 142), spherical and pear-shaped with out-splayed rim (Forms 133, 141[82]), wide cylindrical ones (Forms 100,[83] 103), bulbous ones[84] (Form 107), and jar-like ones with vertical ribs[85] (Form 106). At the same time appear squat[86] (Form 101) and

[64] Isings 1957, 156-157, form 126.
[65] Isings 1957, 157-159, form 127-129.
[66] Isings 1957, 121-122, form 103.
[67] Isings 1957, 119-120, form 101.
[68] Whitehouse 1997, 278, no. 161.
[69] Isings 1957, 123-125, form 104b.
[70] Isings 1957, 123-125, form 104b.
[71] Cf. Isings 1957, 123-125, form 104b.
[72] Barag 1970c, pl. 43, forms 24, 28.
[73] Sorokina 1971, 94, drawings on pl. 5/1; Sorokina 1962, photo 7. 4.
[74] Sokolovska 1986, 127, pl. 45/1.
[75] Barkóczi 1988, 142, 143, no. 310, pls. XXVI, LXXXVI.
[76] Barag 1970c, 43, pl. 5:15, 7. 4e, 6, forms XV:25, 27-1.
[77] Isings 1957, 113-114, 131-133, form 96
[78] Ružić 1994, 50, no. 967.
[79] Isings 1957, 157-158, form 127, small-sized examples.
[80] Isings 1957, 122-123, form 103, small-sized examples.
[81] Isings 1957, 120, form 101, larger examples.
[82] Burger 1966, 137; Stern 1977, 100-101, no. 28; Barkóczi 1988, 151, form 131; Gluščević 1998, 184-185.
[83] Isings 1957, 159-160, form 130b.
[84] Isings 1957, 109, form 94.
[85] Dussart 1998, 92, form BVII. 2432a1.
[86] Barkóczi 1988, 155-156, no. 349, pls. XXXII, XCI; Antonaras 2006a, 413.

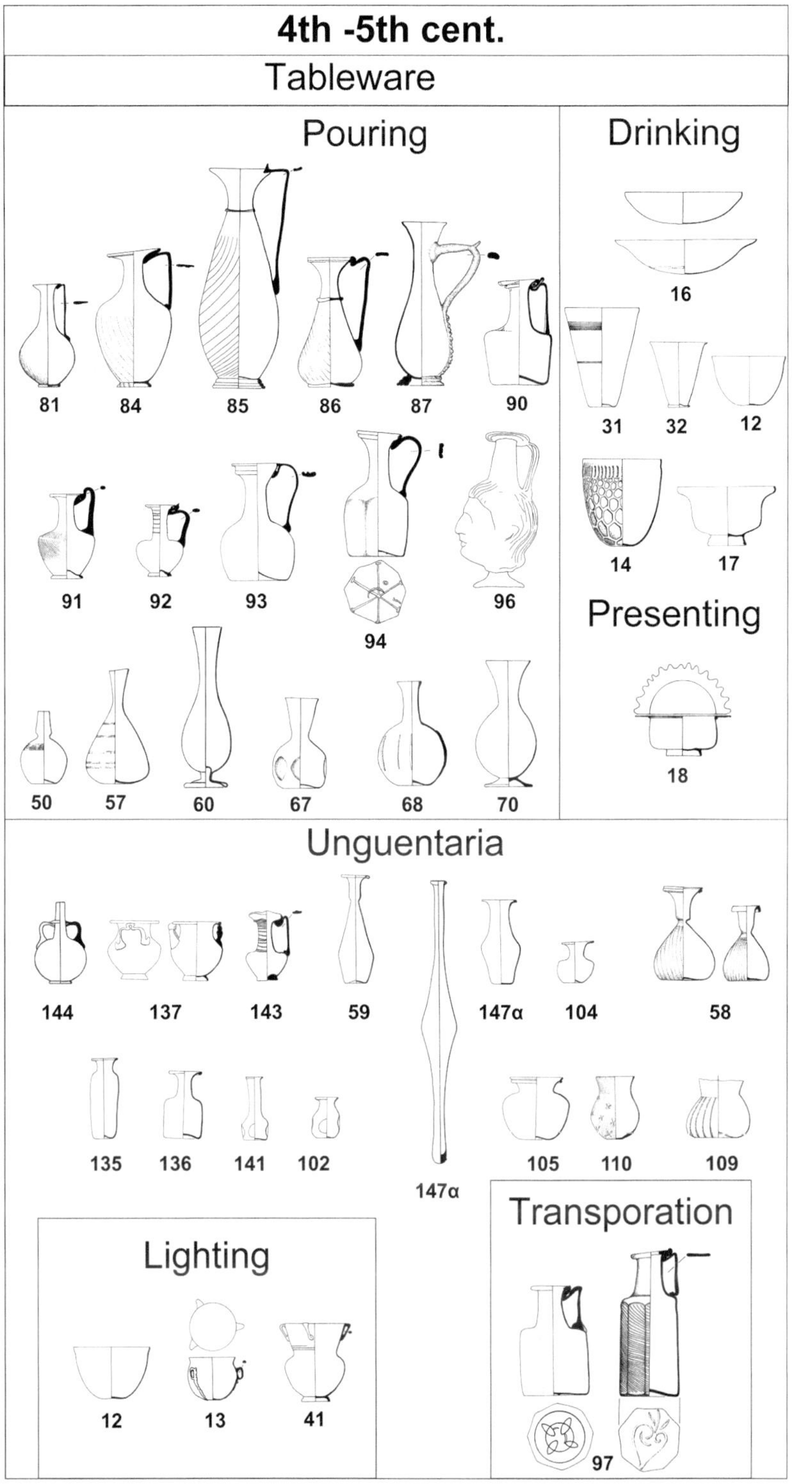


double-faced shapes, i.e. *Janiform*[87] (Form 146) mould-blown unguentaria. All these shapes survive throughout the 4th century and appear to originate in the West or even in the Balkans, while only few give the impression of being of Oriental origin.

A wide use of glass vessels continues in the 4th century AD, and these vessels appear to be mainly Syro-Palestinian shapes, or imitations of them.

Bottles with cracked-off, or ring-shaped rims[88] continue to appear in great numbers[89] (Forms 50-52). Vessels for serving liquids appear in a great variety of forms. There are jugs in the following shapes: spherical[90] (Form 81), ovular[91] (Form 84), fusiform[92] (Form 85), pear-shaped[93] (Form 86), inverted conical (Forms 90,[94] 91, 92[95]), hemispherical[96] (Form 93), eight-sided[97] (Form 94), and in the shape of a male head[98] (Form 96). Also there are flasks in the following shapes: pear-shaped with plain base[99] (Form 57) and slender with a base[100] (Form 60), spherical with indentations[101] (Form 67) or with vertical ribs[102] (Form 68) and ovular with applied coil base (Form 70).

Drinking vessels are usually bowls, shallow[103] (Form 16) or deeper ones (Form 31,[104] 32[105]) and tall conical beakers with plain base, which were probably used also as lamps. The

[87] Isings 1957, 93-94, form 78a-b.
[88] Isings 1957, 121-122, form 103.
[89] Isings 1957, 119-120, form 101.
[90] Barkóczi 1988, 189, no. 464, pl. XLVI.
[91] Isings 1957, 150-151, form 120b.
[92] Kunina 1997, 208, fig. 103, 175a, 6.
[93] Isings 1957, 152, form 121b.
[94] Isings 1957, 156-157, form 126, a short version of the form.
[95] Isings 1957, 152, form 121a.
[96] Dussart 1998, 178, form BXIV. 221, pl. 74, photo 60.9.
[97] Israeli 2003b, 256, no. 330; Antonaras 2006a, 413.
[98] Stern 1995, 214, fig. 99, note 75.
[99] Dussart 1998, 132, pl. 34/1, no. BX 1122.
[100] Stern 2001, 262, 268-269, 297.
[101] Barag 1970c, pl. 40, form XV 12-2.
[102] Cf. Harden 1936, 200, pear-shaped examples, nos. 542 and 544, pl. XVII.
[103] Isings 1957, 144-145, form 116.
[104] Isings 1957, 129, form 106a, b, c.
[105] Isings 1957, 119-120, form 109c.

hemispherical bowls are free-blown[106] (Form 12) or mould-blown with honeycomb pattern[107] (Form 14). Calyx-shaped bowls[108] (Form 17) and short cylindrical ones with scalloped rim[109] (Form 18) used for presenting food also occur.

Unguentaria turn up in a great diversity of sizes and forms, e.g. spherical amphoriskoi[110] (Form 144), craterisks[111] (Form 137), juglets (Form 143), handleless biconical varieties[112] (Form 59), fusiforms[113] (Form 147), squat spherical (Form 104), pear-shaped[114] (Form 58), cylindrical with short and tall neck (Forms 135[115] and 136[116]), with square body and tall or short neck[117] (Forms 141, 102); there are jar-like types, spherical, bulbous and pear-shaped (Forms 105,[118] 109, 110).

Glass was also used for the production of lamps, e.g. the three-handled hemispherical[119] (Form 13) and calyx-shaped (Form 41) bowls.

A special group consists of eight-sided jugs, plain and with oblique ribbing[120] (Form 97) that seem to be of local production during the 4th and 5th centuries.

At the end of the 4th century and during the 5th, tall conical beakers with ring base[121] (Form 32), indented bowls[122] (Form 15) and tall cylindrical flasks[123] (Form 73) arrive, possibly from the north-western provinces.

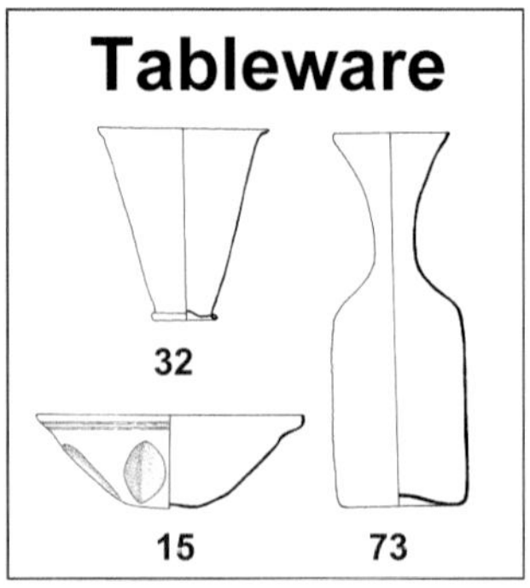


Quite possibly several shapes of 4th-century tableware survive into the 5th and 6th centuries, i.e. jugs, bottles, and drinking vessels, the latter also used as lamps. The well-known form of stemmed beakers[124] (Form 35) prevails. The other forms of glass lamps known from the Thessaloniki area are equally well documented, as they are all over the Mediterranean region, i.e. conical beakers[125] (Form 31), three-handled stemmed beakers[126] (Form 36), hemispherical or calyx-shaped stemmed lamps[127] (Form 38), and pointed cylindrical vessels with a small knop[128] (Form 33) or with a massive knop[129] (Form 34).

Examples of vessels found in settlement areas are very few, dating in the second half of the 4th and 5th centuries. Vessels for presenting liquids, e.g. jugs[130] (Form 31), or for drinking, e.g. beakers[131] (Form 35), deep bowls[132] (Form 12) and shallow bowls[133] (Form 16) predominate. Some bear engraved decoration, suggesting they belonged to local elites. These engraved vessels, which are presumably

[106] Isings 1957, 113-114, 131-133, form 96.
[107] Isings 1957, 133, form 107a.
[108] Cf. Whitehouse 1997, 79-80, no. 104.
[109] Isings 1957, 58, form 42d.
[110] Isings 1957, 159, form 129.
[111] Follmann-Schulz 1988, 76, no. 254, pl. 32; Follmann-Schulz 1992, 89-90, no. 48.
[112] Cf. Fremersdorf, Polónyi-Fremersdorf 1984, nos. 115-116.
[113] Isings 1957, 126, form 105.
[114] Edgar 1905, 42, no. 32588, pl. VI, finds from Fayum (Arsinoe).
[115] Stern 1977, 79, form 21.
[116] Isings 1957, 120-121, no. 102a.
[117] Burger 1966, 137; Stern 1977, 100-101, no. 28; Barkóczi 1988, 151, form 131; Gluščević 1998, 184-185.
[118] Cf. Dussart 1998, 92, form BVII. 2431, pl. 19. Hayes 1975, 94-95, nos. 325, 326, pl. 22.
[119] Whitehouse 2001, 227-228, no. 799.
[120] Antonaras 2006a, 413-414.
[121] Isings 1957, 137-138, form 109c.
[122] Isings 1957, 147-149, form 117.
[123] Isings 1957, 160-161, form 132.
[124] Isings 1957, 139-140, form 111.
[125] Isings 1957, 129, form 106a, b, c.
[126] Uboldi 1995, 124-125, form V.
[127] Uboldi 1995, 120-124.
[128] Uboldi 1995, 119-120, form III. 2, fig. 4. 22-27.
[129] Edgar 1905, 19, nos. 32483-32485.
[130] Isings 1957, 129, form 106a, b, c.
[131] Isings 1957, 139-140, form 111.
[132] Isings 1957, 113-114, 131-133, form 96.
[133] Isings 1957, 144-145, form 116.

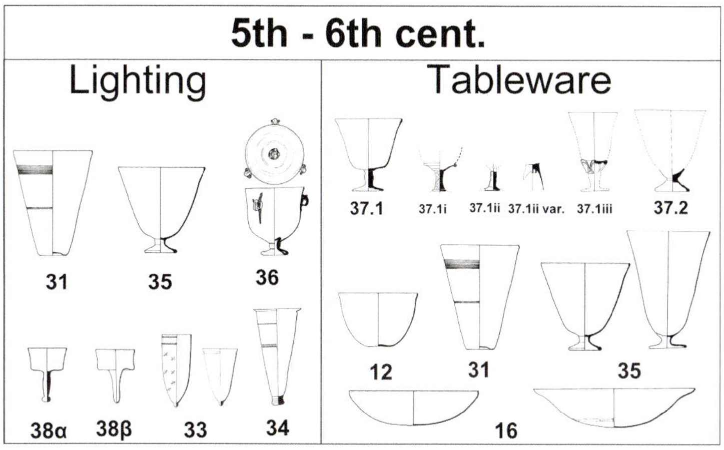


for the most part of Italian origin, also say something about Thessaloniki's relationships with Rome and the Italian peninsula in general.

Summing up, and before presenting the arithmetical data, it should be underlined that the percentages of the samples under study can easily be questioned. This is possible because the results are based on an open sum that reflects only the amount of glass collected from the excavations that simply happened to be on-going during this period of time; the data will definitely alter over time, as excavations are still being conducted in the same areas and new finds keep coming to light.

During the late 1st century BC and up into the early 2nd century AD, glass material is very diverse, in terms of both quantity and shape, covering respectively 38.5% and 39.5% of the total number of vessels examined. The most common groups are unguentaria and drinking vessels, of which a considerable number appear to be purely imports from the West, mainly from Italy, while the remaining shapes can equally be of Western and Eastern origin.

During the 2nd century, and into the very early 3rd century, the total number of shapes and quantities of vessels diminishes significantly, covering 7.5% of the total number of shapes and only 3% of the total number of vessels. Tablewares remain frequent, while the number of vessels for serving liquids increases and the number of unguentaria drastically diminishes. A new shape is the spouted jug. Most of the vessels whose origin can be traced appear to be of Western origin.

Between the middle of the 3rd and early 5th century there was an increase in the use of glass vessels. Again the unguentaria appear in large numbers, however, tableware prevails, in particular for presenting liquids. Few transport vessels and some lamps are also present. Vessels of this period cover 50% of the total number of shapes and 50% of the total number of vessels discussed here. It seems that the imported glass originates from the East, especially from Syro-Palestine, while local production rises at the same time.

Lamps dominate the 5th and 6th centuries glass series, as they do throughout the whole Byzantine era, however, the number of finds in Thessaloniki and its surroundings is quite limited, covering 3% of the total number of shapes and 2.7% of the total number of studied vessels.

7a. Charts Referring to Typological or Chronological Distribution

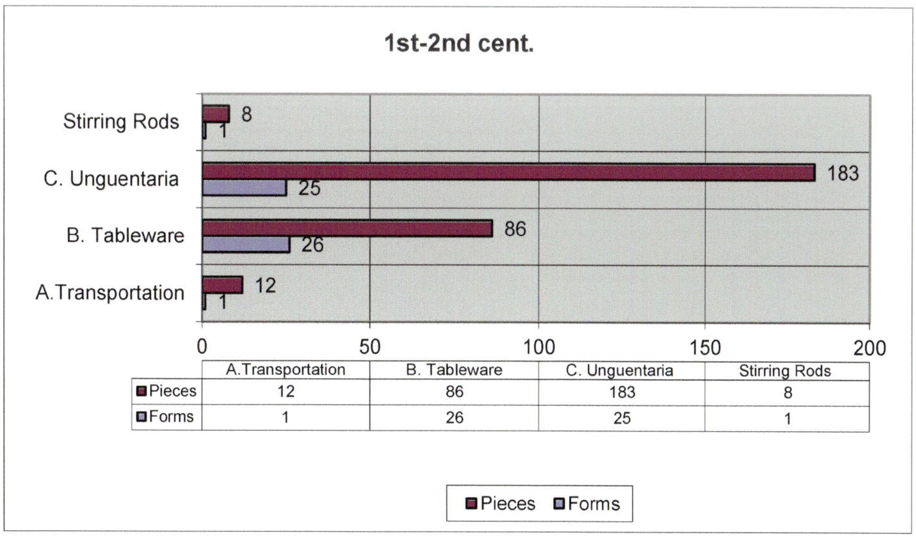


| | A.Transportation | B. Tableware | C. Unguentaria | Stirring Rods |
|---|---|---|---|---|
| Pieces | 12 | 86 | 183 | 8 |
| Forms | 1 | 26 | 25 | 1 |

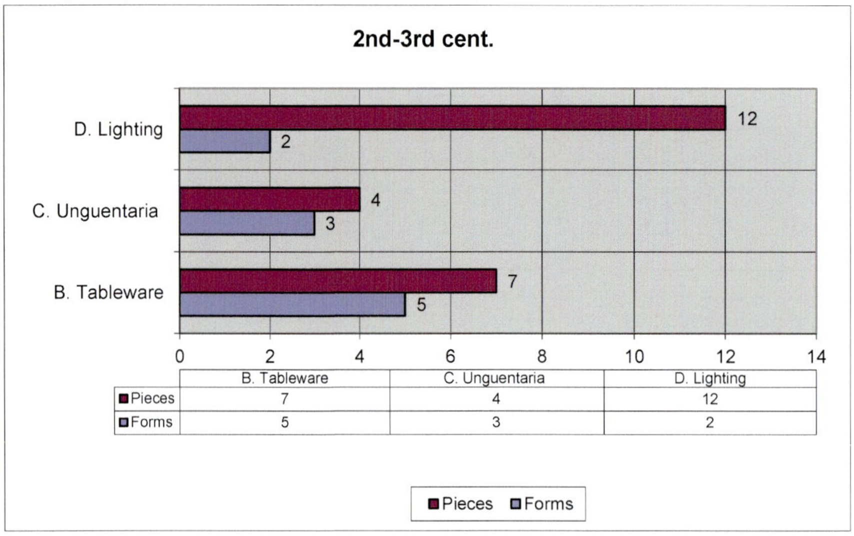
2nd-3rd cent.
D. Lighting
C. Unguentaria
B. Tableware
12
2
4
3
7
5
0 2 4 6 8 10 12 14
B. Tableware | C. Unguentaria | D. Lighting
Pieces 7 4 12
Forms 5 3 2
Pieces Forms

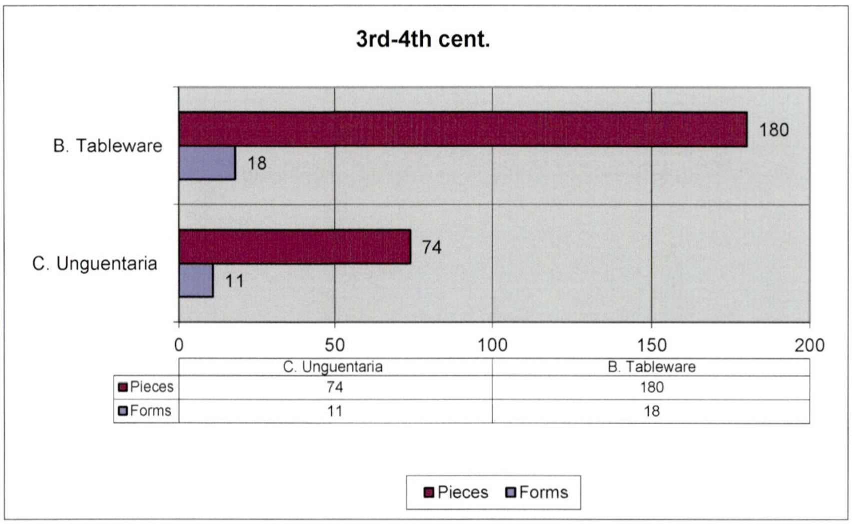
3rd-4th cent.
B. Tableware
C. Unguentaria
180
18
74
11
0 50 100 150 200
C. Unguentaria | B. Tableware
Pieces 74 180
Forms 11 18
Pieces Forms

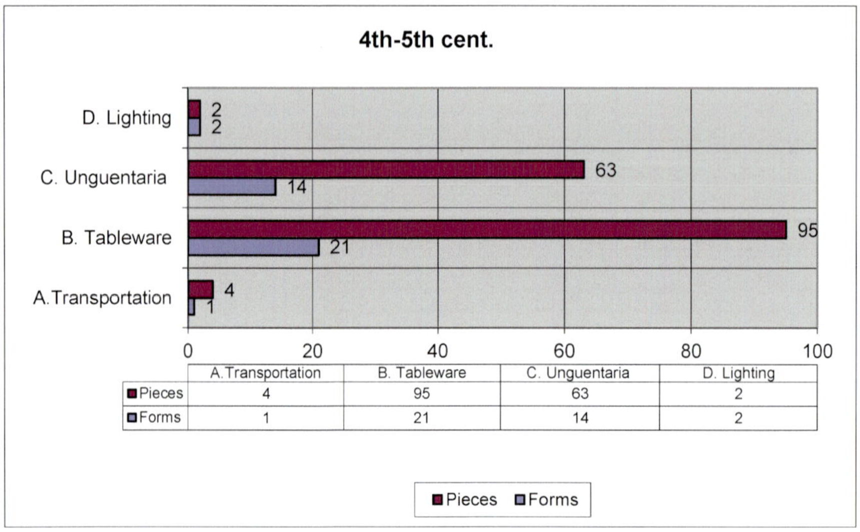
4th-5th cent.
D. Lighting
C. Unguentaria
B. Tableware
A.Transportation
2
2
63
14
95
21
4
1
0 20 40 60 80 100
A.Transportation | B. Tableware | C. Unguentaria | D. Lighting
Pieces 4 95 63 2
Forms 1 21 14 2
Pieces Forms

# 8. Catalogue of the Glass Vessels

Abbreviations used in the Catalogue:
H.= Height
Diam.= Diameter
Th.= Thickness
Max.= Maximum
Pres.= Preserved
Est.= Estimated
AUTH= Aristotle University of Thessaloniki (ΑΠΘ)
TIF= Thessaloniki International Fair (ΔΕΘ)
MBC= Museum of Byzantine Culture (ΜΒΠ)
IST' EPKA= 16th Ephorate of Prehistoric and Classical Antiquities (ΙΣΤ' ΕΠΚΑ)
AMTH= Archaeological Museum of Thessaloniki (ΑΜΘ)
ΠΑΜ= Old Archaeological Museum collection, today part of the Archaeological Museum of Thessaloniki collection
BAA= Ancient Agora Bathhouse finds. Collection of the Ancient Agora Museum, Ephorate of Antiquities of Thessaloniki

1. (ΠΑΜ 59) // AUTH Campus. Theological Department Building. 11-1965 // Fragment // Green and white. // Dimensions: 4.2X2.4, Th.: 0.18 // Unpublished // 1st c.
2. (BAA 83) // Ancient Agora Bathhouse. Rectangular room, north of the Bathhouse. Destruction Layer. 1998 // Fragment // Non vidi. // Mavromichali 2001, draw. 8 β. From a bath complex destroyed in the last twenty years of the 1st c., see Adam-Veleni (ed.) 2001, 86. // 1st c.
3. (BAA 85) // Ancient Agora Bathhouse. Rectangular room, north of the Bathhouse. Destruction Layer. 1998 // Fragment // Non vidi // // Mavromichali 2001, draw. 8 δ. From a bath complex destroyed in the last twenty years of the 1st c., see Adam-Veleni (ed.) 2001, 86. // 1st c.
4. (BAA 82) // Ancient Agora Bathhouse. Rectangular room, north of the Bathhouse. Destruction Layer. 1998 // Fragment // Non vidi. // Mavromichali 2001, draw. 8 α. From a bath complex destroyed in the last twenty years of the 1st c., see Adam-Veleni (ed.) 2001, 86. // 1st c.
5. (BAA 84) // Ancient Agora Bathhouse. Rectangular room, north of the Bathhouse. Destruction Layer. 1998 // Fragment // Non vidi. // Mavromichali 2001, draw. 8 γ. From a bath complex destroyed in the last twenty years of the 1st c., see Adam-Veleni (ed.) 2001, 86. // 1st c.
6. (BAA 86) // Ancient Agora Bathhouse. Rectangular room, north of the Bathhouse. Destruction Layer. 1998 // Fragment // Non vidi. // Mavromichali 2001, draw. 8 ε. From a bath complex destroyed in the last twenty years of the 1st c., see Adam-Veleni (ed.) 2001, 86. // 1st c.
7. (BY 325/9) // Ancient Agora. ΦΦ3, Dig B, Northern side, depth 3.9 μ. 1965; // Fragment // Decolourized. No impurities or pinprick bubbles // H.: 4, Pres. L.: 8.5 // Unpublished // 1st c.
8. (BY 325/3) // Ancient Agora. On the envelope was faintly visible 'S1/1067'. 1967 // Fragment weathered on the exterior // Dark blue, yellow, purple and green. No impurities or pinprick bubbles // H.: 2, Diam. rim: 9, Th.: 0.2 // Unpublished // 1st c.
9. (BAA 25) // Ancient Agora Bathhouse. Rectangular room, north of the Bathhouse. Destruction Layer. 1998 // Fragment // Non vidi. // Mavromichali 2001, fig. 8. From a bath complex destroyed in the last twenty years of the 1st c., see Adam-Veleni (ed.) 2001, 86. // 1st c.
10. (BAA 26) // Ancient Agora Bathhouse. Rectangular room, north of the Bathhouse. Destruction Layer. 1998 // Fragment // Non vidi. // Mavromichali 2001, fig. 8. From a bath complex destroyed in the last twenty years of the 1st c., see Adam-Veleni (ed.) 2001, 86. // 1st c.
11. (BAA 27) // Ancient Agora Bathhouse. Rectangular room, north of the Bathhouse. Destruction Layer. 1998 // Fragment // Non vidi. // Mavromichali 2001, fig. 8. From a bath complex destroyed in the last twenty years of the 1st c., see Adam-Veleni (ed.) 2001, 86. // 1st c.
12. (BAA 28) // Ancient Agora Bathhouse. Rectangular room, north of the Bathhouse. Destruction Layer. 1998 // Fragment // Non vidi. // Mavromichali 2001, fig. 8. From a bath complex destroyed in the last twenty years of the 1st c., see Adam-Veleni (ed.) 2001, 86. // 1st c.
13. (BAA 29) // Ancient Agora Bathhouse. Rectangular room, north of the Bathhouse. Destruction Layer. 1998 // Fragment // Non vidi. // Mavromichali 2001, fig. 8. From a bath complex destroyed in the last twenty years of the 1st c., see Adam-Veleni (ed.) 2001, 86. // 1st c.
14. (BAA 30) // Ancient Agora Bathhouse. Rectangular room, north of the Bathhouse. Destruction Layer. 1998 // Fragment // Non vidi. // Mavromichali 2001, fig. 8. From a bath complex destroyed in the last twenty years of the 1st c., see Adam-Veleni (ed.) 2001, 86. // 1st c.
15. (BAA 31) // Ancient Agora Bathhouse. Rectangular room, north of the Bathhouse. Destruction Layer. 1998 // Fragment // Non vidi. // Mavromichali 2001, fig. 8. From a bath complex destroyed in the last twenty years of the 1st c., see Adam-Veleni (ed.) 2001, 86. // 1st c.
16. (BAA 32) // Ancient Agora Bathhouse. Rectangular room, north of the Bathhouse. Destruction Layer. 1998 // Fragment // Non vidi. // Mavromichali 2001, fig. 8. From a bath complex destroyed in the last twenty years of the 1st c., see Adam-Veleni (ed.) 2001, 86. // 1st c.
17. (BAA 33) // Ancient Agora Bathhouse. Rectangular room, north of the Bathhouse. Destruction Layer. 1998 // Fragment // Non vidi. // Mavromichali 2001, fig. 8. From a bath complex destroyed in the last twenty years of the 1st c., see Adam-Veleni (ed.) 2001, 86. // 1st c.
18. (BY 268/1) // Monastiriou Street, 5 and Odysseos Street. Clearing of the surface of the N. Wall of the plot. 21-6-95 // Fragment // Translucent bluish. No impurities or pinprick bubbles // Dimensions: 5.5X4, Th.: 0.3 // Unpublished // 1st c.
19. (BY 325/6) // Ancient Agora. Ancient Agora./ Dikastirion Square. dig B', depth 0.85. 20-6-62 // Fragment // Translucent amber. No impurities or pinprick bubbles // Diam. rim: 11, Pres. H.: 3, Th.: 0.3 // Unpublished // 1st c.
20. (BY 328/2) // Unknown exact provenance. From the finds of old excavations in Thessaloniki given to MBC from IST' EPKA. // Fragment // Translucent bluish. // Unpublished // 1st c.
21. (ΜΘ 21456 α) // Farmhouse A. 2001 // Large parts missing, mended fragments // Translucent bluish. // H.: 5.9, Diam. rim: 13.7 // Tzanavari, Filis 2000, 162, fig. 16. *Glass Cosmos* 2010, 339 no. 351 // 1st c.
22. (ΠΑΜ 92) // AUTH Campus. Theological Department Building. Probably stray find. 1965 // Fragment // Translucent amber. No impurities or pinprick bubbles // Dimensions: 3X2, Th.: 0.25 // Unpublished // 1st c.
23. (ΠΑΜ 94) // AUTH Campus. Theological Department Building. Probably a stray find. 1965 // Fragment // Translucent bluish. No impurities or pinprick bubbles // Dimensions: 3X2, Th.: 0.3 // Unpublished // 1st c.
24. (ΠΑΜ 95) // AUTH Campus. Theological Department Building. Probably a stray find. 1965 // Fragment // Translucent greenish. No impurities or pinprick bubbles // Dimensions: 2X1.5, Th.: 0.3 // Unpublished // 1st c.
25. (BAA 34) // Ancient Agora Bathhouse. Rectangular room, north of the Bathhouse. Destruction Layer. 1998 // Fragment // Non vidi. // Mavromichali 2001, fig. 9. From a bath complex destroyed in the last twenty years of the 1st c., see Adam-Veleni (ed.) 2001, 86. // 1st c.
26. (BAA 35) // Ancient Agora Bathhouse. Rectangular room, north of the Bathhouse. Destruction Layer. 1998 // Fragment // Non vidi. // Mavromichali 2001, fig. 9. From a bath complex destroyed in the last twenty years of the 1st c., see Adam-Veleni (ed.) 2001, 86. // 1st c.
27. (BAA 36) // Ancient Agora Bathhouse. Rectangular room, north of the Bathhouse. Destruction Layer. 1998 // Fragment // Non vidi. // Mavromichali 2001, fig. 9. From a bath complex destroyed in the last twenty years of the 1st c., see Adam-Veleni (ed.) 2001, 86. // 1st c.

28. (RA 623) // Ancient Agora Bathhouse. Rectangular room, north of the Bathhouse. Destruction Layer. 1998 // Fragment // Non vidi. // Mavromichali 2001, fig. 9. From a bath complex destroyed in the last twenty years of the 1st c., see Adam-Veleni (ed.) 2001, 86. *Glass Cosmos* 2010, 265, no. 185 // 1st c.
29. (ΜΘ ΑΚΑΤΑΓΡΑΦΟ 4) // Unknown exact provenance. // Fragment // Translucent dark-blue and opaque white. // Dimensions: 5X8, Th.: 0.2 // Unpublished // 1st c.
30. (BAA 18) // Ancient Agora Bathhouse. Rectangular room, north of the Bat house. Destruction Layer. 1998 // Fragment // Non vidi. // Mavromichali 2001, fig. 7. From a bath complex destroyed in the last twenty years of the 1st c., see Adam-Veleni (ed.) 2001, 86. // 1st c.
31. (RA 605) // Ancient Agora Bathhouse. Rectangular room, north of the Bathhouse. Destruction Layer. 1998 // Fragment // Non vidi. // Mavromichali 2001, fig. 7. From a bath complex destroyed in the last twenty years of the 1st c., see Adam-Veleni (ed.) 2001, 86. *Glass Cosmos* 2010, 264 no. 184 // 1st c.
32. (BAA 20) // Ancient Agora Bathhouse. Rectangular room, north of the Bathhouse. Destruction Layer. 1998 // Fragment // Non vidi. // Mavromichali 2001, fig. 7. From a bath complex destroyed in the last twenty years of the 1st c., see Adam-Veleni (ed.) 2001, 86. // 1st c.
33. (BAA 21) // Ancient Agora Bathhouse. Rectangular room, north of the Bathhouse. Destruction Layer. 1998 // Fragment // Non vidi. // Mavromichali 2001, fig. 7. From a bath complex destroyed in the last twenty years of the 1st c., see Adam-Veleni (ed.) 2001, 86. // 1st c.
34. (BAA 22) // Ancient Agora Bathhouse. Rectangular room, north of the Bathhouse. Destruction Layer. 1998 // Fragment // Non vidi. // Mavromichali 2001, fig. 7. From a bath complex destroyed in the last twenty years of the 1st c., see Adam-Veleni (ed.) 2001, 86. // 1st c.
35. (BAA 23) // Ancient Agora Bathhouse. Rectangular room, north of the Bathhouse. Destruction Layer. 1998 // Fragment // Non vidi. // Mavromichali 2001, fig. 7. From a bath complex destroyed in the last twenty years of the 1st c., see Adam-Veleni (ed.) 2001, 86. // 1st c.
36. (BAA 24) // Ancient Agora Bathhouse. Rectangular room, north of the Bathhouse. Destruction Layer. 1998 // Fragment // Non vidi. // Mavromichali 2001, fig. 7. From a bath complex destroyed in the last twenty years of the 1st c., see Adam-Veleni (ed.) 2001, 86. // 1st c.
37. (BY 325/5) // Ancient Agora. Ancient Agora./ Dikastirion Square. dig A', Well β', depth 5.20. 26-6-62 // Fragment // Translucent amber. No impurities or pinprick bubbles // H.: 4, Diam. rim: 9, Th.: 0.1 // Unpublished // 1st c.
38. (ΠΑΜ 60) // AUTH Campus. Theological Department Building. 11-1965 // Shoulder fragment // Translucent olive-green. No impurities or pinprick bubbles // Dimensions: 5.5X3.5, Th.: 0.13 // Unpublished // 1st c.
39. (BAA 54) // Ancient Agora Bathhouse. Rectangular room, north of the Bathhouse. Destruction Layer. 1998 // Mended and filled // Non vidi. // Mavromichali 2001, fig. 16. From a bath complex destroyed in the last twenty years of the 1st c., see Adam-Veleni (ed.) 2001, 86. // 1st c.
40. (BY 325/10) // Ancient Agora. ΦΦ3, Dig B, Northern side, depth 3.9 μ. 1965; // Fragment // Decolourized. No impurities or pinprick bubbles // Diam. rim: 19, Pres. H.: 4 // Unpublished // 1st c.
41. (BY 325/11) // Ancient Agora. ΦΦ3, Dig B, Northern side, depth 3.9 μ. 1965; // Fragment // Decolourized. No impurities or pinprick bubbles // Diam. rim: 9, Pres. H.: 3 // Unpublished // 1st c.
42. (BAA 55) // Ancient Agora Bathhouse. Rectangular room, north of the Bathhouse. Destruction Layer. 1998 // Mended // Non vidi. // Mavromichali 2001, fig. 17. From a bath complex destroyed in the last twenty years of the 1st c., see Adam-Veleni (ed.) 2001, 86. // 1st c.
43. (BY 69) // Junction of Agiou Dimitriou Street and Sofokleous. // Part of the rim is preserved // Translucent greenish. No impurities few pinprick bubbles // Diam. rim: 10, Est. H.: 6.2, Th.: 0.12 // Unpublished // 3rd end - early 5th c.
44. (BY 214/2) // Northern part of Evangelistrias Cemetery, area of former Lyssiatreion Clinic. Grave Δ. 25-11-1991 // Large part of the body is missing // Colourless slightly greenish. // H.: 7.5, Diam. rim: 12, Diam. Base: 4.5, Th.: 0.1 // Pelekanidou 1993, 376, photo 6, από τάφο με νόμισμα του 2ου-3ου αι. // 3rd end - early 5th c.
45. (BY 227/53) // Northern part of Evangelistrias Cemetery, area of former Lyssiatreion Clinic. South cluster Γ. Grave 2 (pit grave). 19-4-1995 // Fragments // Translucent greenish. No impurities few pinprick bubbles // H.: 7.5, Diam. rim: 10, Th.: 0.02, Weight: 29.3 // Unpublished // 3rd end - early 5th c.
46. (BY 325/1) // Ancient Agora. // Mended and filled // Translucent greenish. // // Unpublished // 3rd end - early 5th c.
47. (BY 325/4) // Ancient Agora. On the envelope was faintly visible 'S1/1067'. 1967 // Fragment // Translucent greenish with reddish striations. Many pinprick bubbles and few larger one. Some impurities // H.: 7, Diam. rim: 8 // Unpublished // 3rd end - early 5th c.
48. (ΜΘ 21392) // Sindos, Industrial Zone of Thessaloniki. O.T. Goodyear, plot of ΑΘΕΛΚΟ plant, Grave 2. 23-6-2000 // Intact. Iridescent // Translucent greenish. // H.: 7, Diam. rim: 8.7, Th.: 0.11, Weight: 76.1 // Unpublished // 3rd end - early 5th c.
49. (BY 56) // Junction of Agiou Dimitriou Street and Sofokleous. // Part of the rim is preserved // Translucent greenish. No impurities few pinprick bubbles // Diam. rim: 8.7, Diam. Base: 3, Est. H.: 4.5, Th.: 0.12 // Unpublished // 3rd end - early 5th c.
50. (BY 227/54) // Northern part of Evangelistrias Cemetery, area of former Lyssiatreion Clinic. South cluster Γ. Grave 2 (pit grave). 19-4-1995 // Fragments // Translucent greenish. No impurities few pinprick bubbles // H.: 5, Diam. rim: 11, Th.: 0.02, Weight: 4.8 // Unpublished // 3rd end - early 5th c.
51. (ΜΘ 510) // Vardari Square. // Intact // Translucent greenish. No impurities or pinprick bubbles // H.: 7, Diam. rim: 12.2, Diam. Neck: 1, Diam. Base: 3, Th.: 0.2, Weight: 139 // Unpublished // 3rd end - early 5th c.
52. (ΜΘ 511) // Vardari Square. // Intact. Internally iridescent // Translucent greenish. No impurities or pinprick bubbles // H.: 8.2, Diam. Max.: 11.7, Diam. rim: 11.7, Diam. Neck: 1, Diam. Base: 3.2, Th.: 0.3 // E. Trakosopoulou-Salakidou, Thessaloniki apo ta proïstorika mechri ta christianika chronia 1986, 125. *Glass Cosmos* 2010, 282, no. 229 // 3rd end - early 5th c.
53. (ΜΘ 21961) // Unknown exact provenance. // Fragment // Translucent bluish. Very few impurities and pinprick bubbles // H.: 5, Diam. rim: 14, Diam. Base: 2, Th.: 0.12 // Unpublished // 3rd end - early 5th c.
54. (ΜΘ 21965) // Sokratous Street, 47-49. Plot property of Velikas (today Moirarchou Koufitsa Street, 19). 1968 // Fragment // Translucent greenish. No impurities very few pinprick bubbles // H.: 7, Diam. Max.: 14.5, Diam. rim: 14.5, Th.: 0.2 // Unpublished // 3rd end - early 5th c.
55. (ΜΘ 21967) // Sokratous Street, 47-49. Plot property of Velikas (today Moirarchou Koufitsa Street, 19). 1968 // Fragment // Translucent greenish. No impurities or pinprick bubbles // H.: 4, Diam. rim: 14, Th.: 0.12 // Unpublished // 3rd end - early 5th c.
56. (ΜΘ 21968) // Sokratous Street, 47-49. Plot property of Velikas (today Moirarchou Koufitsa Street, 19). 1968 // Fragment // Translucent olive-green. No impurities few pinprick bubbles // H.: 5, Diam. rim: 9, Th.: 0.15 // Unpublished // 3rd end - early 5th c.
57. (ΜΘ 21971) // Sokratous Street, 47-49. Plot property of Velikas (today Moirarchou Koufitsa Street, 19). 1968 // Fragment // Translucent greenish. No impurities few pinprick bubbles // H.: 5.2, Diam. rim: 13, Th.: 0.18 // Unpublished // 3rd end - early 5th c.
58. (ΜΘ 21972) // Sokratous Street, 47-49. Plot property of Velikas (today Moirarchou Koufitsa Street, 19). 1968 // Fragment // Translucent greenish. No impurities few pinprick bubbles // H.: 2.5, Diam. rim: 10, Th.: 0.15 // Unpublished // 3rd end - early 5th c.
59. (ΜΘ 21977) // Sokratous Street, 47-49. Plot property of Velikas (today Moirarchou Koufitsa Street, 19). (4-6)-(Δ-E) ca. at the depth of the floor. 22-6-1968 // Fragment // Translucent greenish. No impurities few pinprick bubbles // H.: 5.5, Diam. rim: 14 // Unpublished // 3rd end - early 5th c.
60. (ΠΑΜ 90) // Junction of Agiou Dimitriou Street and Agias Sophias Street, Byzantine building. 3rd triangular space. 11-6-1969 // Fragment // Translucent greenish. No impurities few pinprick bubbles // H.: 5, Diam. rim: 11, Th.: 0.25 // Unpublished // 3rd end - early 5th c.
61. (ΠΑΜ 91) // Junction of Agiou Dimitriou Street and Agias Sophias Street, Byzantine building. 3rd triangular space. 11-6-1969 // Fragment // Translucent greenish. No impurities few pinprick bubbles // H.: 5, Diam. rim: 10.4, Th.: 0.25 // Unpublished // 3rd end - early 5th c.

62. (BY 328/1) // Unknown exact provenance. From the finds of old excavations in Thessaloniki given to MBC from IST' EPKA. // Fragment // Translucent greenish. No impurities or pinprick bubbles // // Unpublished // 3rd end - early 5th c.

63. (MΘ 5474) // Vardari Square. Found in a marble sarcophagus. 12-2-1920 // Less than half is preserved // Translucent olive-green. // H.: 9.4, Diam. Max.: 9.7, Diam. rim: 9.7, Diam. Neck: 1, Diam. Base: 2.3, Th.: 0.18, Weight: 49.59 // Unpublished // 3rd end - early 5th c.

64. (ΠΑΜ 21) // Ramona Area;. 1934; // Large part of the body is missing // Translucent olive-green. No impurities few pinprick bubbles // H.: 9, Diam. rim: 10, Diam. Base: 3.4, Th.: 0.2 // Unpublished // 3rd end - early 5th c.

65. (RGM Koln 62/58) // Unknown exact provenance. From a grave dated by coins (neither the grave nor the coins were identified). // Intact // Translucent olive-greenish and blue. // H.: 8.2, Diam. rim: 9.1 // La Baume 1973, D 89, pl. 37.4. // 3rd end - early 5th c.

66. (BY 227/18) // Northern part of Evangelistrias Cemetery, area of former Lyssiatreion Clinic. // Part of the body and the rim are missing // Translucent greenish and dark-blue. // H.: 5, Diam. rim: 10 // Pelekanidou 1993, 376, photo 6 // 3rd end - early 5th c.

67. (ΠΑΜ 22) // Ramona Area;. 1934; // Large part of the body is missing // Translucent olive-green body, blue thread. No impurities few pinprick bubbles // Diam. rim: 8, Pres. H.: 5 Th.: 0.2 // Unpublished // 3rd end - early 5th c.

68. (BY 320/1) // Rentina Castle. E2, NE cornet, at the layer of the N. threshold. Φ. 105 Γ́ . 13-10-1979 // Complete, mended // Translucent olive-green. Few pinprick bubbles and impurities // H.: 7.2, Diam. rim: 8.2, Diam. Base: 4, Th.: 0.2 // Unpublished // 4th - 6th c.

69. (BY 320/6) // Rentina Castle. Sector E. Western side of the complex. Φ. 382. 23-9-1981 // Fragment // Dark blue. No impurities or pinprick bubbles // Dimensions: 2.5X6 Th.: 0.12 // Unpublished // 4th c.

70. (ΠΑΜ 57) // AUTH Campus. Theological Department Building. 11-1965 // Fragment // Translucent greenish. Few impurities and pinprick bubbles // Dimensions: 6X2.5, Th.: 0.12 // Unpublished // 4th c.

71. (BY 36/1) // AUTH Campus. Sorority Building (Φ.Ε.Θ.). West of Grave ΙΔ. 1968 // Part of the body and the bottom is preserved // Translucent greenish. No impurities or pinprick bubbles // // Unpublished // 4th end - 5th c.

72. (BY 70) // Junction of Agiou Dimitriou Street and Sofokleous. // Part of the rim and the body are preserved // Translucent greenish. No impurities few pinprick bubbles // Diam. rim: 17, Est. H.: 7.7, Th.: 0.12 // Unpublished // 4th end - 5th c.

73. (ΠΑΜ 96) // AUTH Campus. Theological Department Building. Probably stray find. 1965 // Small part of the body is preserved // Colourless Decolourized. No impurities or pinprick bubbles // Dimensions: 3X6, Th.: 0.1 // Unpublished // 4th end - 5th c.

74. (BY 327/1) // Zakynthou Street, plot 3b. Sector 3 (area between walls 1-3-4). 22-7-2002 // Mended, parts of the body are missing // Translucent bluish. No impurities or pinprick bubbles // H.: 3.7, Diam. Max.: 13.8, Diam. rim: 13.8, Th.: 0.2 // Mentioned at Marki, Akrivopoulou 2003, 293. Find from a building dated to the 5th-6th c. // 4th - 5th c.

75. (MΘ 21960) // Christopoulou Street, 12. Plot property of Mr. Proferis, surface layer. 29-7-1975 // Two non-mending fragments are preserved // Translucent greenish. No impurities or pinprick bubbles // H.: 3, Diam. rim: 14, Th.: 0.12 // Unpublished // 4th - 5th c.

76. (MΘ 21975) // Sokratous Street, 47-49. Plot property of Velikas (today Moirarchou Koufitsa Street, 19). (4-6)-(Δ-Ε) ca. at the depth of the floor. 22-6-1968 // Fragment // Translucent greenish. No impurities, very few pinprick bubbles // H.: 2.6, Diam. rim: 16, Th.: 0.2 // Unpublished // 4th - 5th c.

77. (MΘ 21976) // Sokratous Street, 47-49. Plot property of Velikas (today Moirarchou Koufitsa Street, 19). (4-6)-(Δ-Ε) ca. at the depth of the floor. 22-6-1968 // Fragment // Translucent greenish. No impurities or pinprick bubbles // Dimensions: Pres. diam: 8 // Unpublished // 4th - 5th c.

78. (MΘ 21978) // Sokratous Street, 47-49. Plot property of Velikas (today Moirarchou Koufitsa Street, 19). 1968 // Fragment // Translucent bluish. No impurities very few pinprick bubbles // Dimensions: 4X3, Th.: 0.12 // Unpublished // 4th - 5th c.

79. (ΠΑΜ 36) // AUTH Campus. Administration Building. 2nd Sarcophagus. 8-9-1965 // Body fragment // Translucent slightly greenish. No impurities or pinprick bubbles // Dimensions: 3X2 // Unpublished // 4th - 5th c.

80. (ΠΑΜ 58) // AUTH Campus. Theological Department Building. Near grave T. 11 depth 0.40. 5-8-1965 // Fragment // Translucent greenish. No impurities, several pinprick bubbles // Dimensions: 3X3, Th.: 0.1 // Unpublished // 4th - 5th c.

81. (ΠΑΜ 72) // AUTH Campus. Theological Department Building. Sector 18, depth 1 m. 2-6-1965 // Fragment // Translucent Decolourized. No impurities or pinprick bubbles // Dimensions: 2X1.5 cm // Unpublished // 4th - 5th c.

82. (ΠΑΜ 79) // Junction of Agiou Dimitriou Street and Agias Sophias Street, Byzantine building. 3rd triangular space. 11-6-1969 // Fragment // Translucent greenish. Few impurities and pinprick bubbles // Diam. rim: 13, Pres. H.: 3 // Unpublished // 4th - 5th c.

83. (ΠΑΜ 81) // Junction of Agiou Dimitriou Street and Agias Sophias Street, Byzantine building. 3rd triangular space. 11-6-1969 // Fragment // Translucent bluish. Few impurities and pinprick bubbles // Dimensions: 6.5X3.5 // Unpublished // 4th - 5th c.

84. (MΘ 21962) // Kalochori, plot of D. Gerakoudis. Grave 2. 27-3-1986 // Intact // Translucent dark green. No impurities, several pinprick bubbles // H.: 7.5, Diam. rim: 12, Diam. Base: 4.7, Th.: 0.15 // *Glass Cosmos* 2010, 307, no. 285 // 4th - 6th c.

85. (MΘ ΑΚΑΤΑΓΡΑΦΟ 3) // Mitsaion. 27-4-1984 // Preserved in fragments // Translucent slightly greenish. No impurities few pinprick bubbles // H.: 5.5, Diam. Max.: 14, Diam. rim: 14, Diam. Base: 5, Th.: 0.1 // Unpublished // 4th - 5th c.

86. (PA 934) // Ancient Agora Bathhouse. Rectangular room, north of the Bathhouse. Destruction Layer. 1998 // Fragment // Translucent greenish. // Mavromichali 2001, fig. 2α. From a bath complex destroyed in the last twenty years of the 1st c., see Adam-Veleni (ed.) 2001, 86. *Glass Cosmos* 2010, 261, no. 177 // 1st c.

87. (PA 935) // Ancient Agora Bathhouse. Rectangular room, north of the Bathhouse. Destruction Layer. 1998 // Fragment // Translucent greenish. // Mavromichali 2001, fig. 2β. From a bath complex destroyed in the last twenty years of the 1st c., see Adam-Veleni (ed.) 2001, 86. *Glass Cosmos* 2010, 261, no. 178 // 1st c.

88. (BY 237/1) // Lapithon Street, 7. Layer of the bedrock at the SW of the dig. 13-2-1990 // Fragment // Translucent bluish. No impurities or pinprick bubbles // Unpublished // 1st c.

89. (BAA 12) // Ancient Agora Bathhouse. Rectangular room, north of the Bathhouse. Destruction Layer. 1998 // Base and body fragments // Non vidi. // Mavromichali 2001, fig. 6. From a bath complex destroyed in the last twenty years of the 1st c., see Adam-Veleni (ed.) 2001, 86. // 1st c.

90. (BAA 13) // Ancient Agora Bathhouse. Rectangular room, north of the Bathhouse. Destruction Layer. 1998 // Base and body fragments // Non vidi. // Mavromichali 2001, fig. 6. From a bath complex destroyed in the last twenty years of the 1st c., see Adam-Veleni (ed.) 2001, 86. // 1st c.

91. (BAA 4) // Ancient Agora Bathhouse. Rectangular room, north of the Bathhouse. Destruction Layer. 1998 // Base and body fragments // Translucent greenish. // Mavromichali 2001, fig. 3. From a bath complex destroyed in the last twenty years of the 1st c., see Adam-Veleni (ed.) 2001, 86. // 1st c.

92. (BAA 5) // Ancient Agora Bathhouse. Rectangular room, north of the Bathhouse. Destruction Layer. 1998 // Base and body fragments // Translucent greenish. // Mavromichali 2001, fig. 3. From a bath complex destroyed in the last twenty years of the 1st c., see Adam-Veleni (ed.) 2001, 86. // 1st c.

93. (RA 702) // Ancient Agora Bathhouse. Rectangular room, north of the Bathhouse. Destruction Layer. 1998 // Base and body fragments // Translucent greenish. // Mavromichali 2001, fig. 3. From a bath complex destroyed in the last twenty years of the 1st c., see Adam-Veleni (ed.) 2001, 86. *Glass Cosmos* 2010, 262, no. 179 // 1st c.

94. (BY 325/7) // Ancient Agora. 1962 // Fragment // Translucent greenish. No impurities few pinprick bubbles // Diam. rim: 6X4 // Unpublished // 1st c.

95. (BAA 1) // Ancient Agora Bathhouse. Rectangular room, north of the Bathhouse. Destruction Layer. 1998 // Base fragments, body is missing // Translucent greenish. // Mavromichali 2001, fig. 1. From a bath complex destroyed in the last twenty years of the 1st c., see Adam-Veleni (ed.) 2001, 86. // 1st c.

96. (BY 325/2) // Ancient Agora. With a tag: '1967/11/32'. 29-5-1967 // Fragment // Translucent. No impurities or pinprick bubbles // Dimensions: 3.5X2 // Unpublished // 1st c.

97. (BAA 6 a) // Ancient Agora Bathhouse. Rectangular room, north of the Bathhouse. Destruction Layer. 1998 // Base and body fragments // Translucent greenish. // Mavromichali 2001, fig. 3. From a bath complex destroyed in the last twenty years of the 1st c., see Adam-Veleni (ed.) 2001, 86. // 1st c.

98. (MΘ 20912) // Unknown exact provenance. With a tag: 'Unregistered, from the labs to storeroom at 2-7-1977'. // Intact // Translucent greenish. No impurities few pinprick bubbles // H.: 10.2, Diam. Max.: 7.4, Diam. rim: 7.4, Diam. Base: 1.8, Th.: 0.3, Weight: 94.9 // *Glass Cosmos* 2010, 255, no. 161 // 1st c.

99. (BAA 14) // Ancient Agora Bathhouse. Rectangular room, north of the Bathhouse. Destruction Layer. 1998 // Base and body fragments // Non vidi. // Mavromichali 2001, fig. 6. From a bath complex destroyed in the last twenty years of the 1st c., see Adam-Veleni (ed.) 2001, 86. // 1st c.

100. (BAA 16) // Ancient Agora Bathhouse. Rectangular room, north of the Bathhouse. Destruction Layer. 1998 // Base and body fragments // Non vidi. // Mavromichali 2001, fig. 6. From a bath complex destroyed in the last twenty years of the 1st c., see Adam-Veleni (ed.) 2001, 86. // 1st c.

101. (BAA 10) // Ancient Agora Bathhouse. Rectangular room, north of the Bathhouse. Destruction Layer. 1998 // Base and body fragments // Non vidi. // Mavromichali 2001, fig. 6. From a bath complex destroyed in the last twenty years of the 1st c., see Adam-Veleni (ed.) 2001, 86. // 1st c.

102. (BAA 11) // Ancient Agora Bathhouse. Rectangular room, north of the Bathhouse. Destruction Layer. 1998 // Base and body fragments // Non vidi. // Mavromichali 2001, fig. 6. From a bath complex destroyed in the last twenty years of the 1st c., see Adam-Veleni (ed.) 2001, 86. // 1st c.

103. (RA 710) // Ancient Agora Bathhouse. Rectangular room, north of the Bathhouse. Destruction Layer. 1998 // Base and body fragments // Non vidi. // Mavromichali 2001, fig. 6. From a bath complex destroyed in the last twenty years of the 1st c., see Adam-Veleni (ed.) 2001, 86. *Glass Cosmos* 2010, 263, no. 182 // 1st c.

104. (BAA 17) // Ancient Agora Bathhouse. Rectangular room, north of the Bathhouse. Destruction Layer. 1998 // Base and body fragments // Non vidi. // Mavromichali 2001, fig. 6. From a bath complex destroyed in the last twenty years of the 1st c., see Adam-Veleni (ed.) 2001, 86. // 1st c.

105. (MΘ 5951) // AUTH Campus. Theological Department Building. 1965 // Complete, filled // Translucent olive-green. No impurities or pinprick bubbles // H.: 10, Diam. rim: 7.1, Diam. Base: 2.7, Th.: 0.2 // Petsas 1966, 334-339, pl. 351 ε. E. Trakosopoulou-Salakidou, Thessaloniki apo ta proïstorika mechri ta christianika chronia, 1986, 125. // 1st end - end 2oυ c.

106. (BAA 7) // Ancient Agora Bathhouse. Rectangular room, north of the Bathhouse. Destruction Layer. 1998 // Base and rim fragments // Non vidi. // Mavromichali 2001, fig. 4. From a bath complex destroyed in the last twenty years of the 1st c., see Adam-Veleni (ed.) 2001, 86. // 1st c.

107. (BAA 8) // Ancient Agora Bathhouse. Rectangular room, north of the Bathhouse. Destruction Layer. 1998 // Base and rim fragments // Non vidi. // Mavromichali 2001, fig. 4. From a bath complex destroyed in the last twenty years of the 1st c., see Adam-Veleni (ed.) 2001, 86. // 1st c.

108. (BAA 9) // Ancient Agora Bathhouse. Rectangular room, north of the Bathhouse. Destruction Layer. 1998 // Base and body fragments // Non vidi. // Mavromichali 2001, fig. 5. From a bath complex destroyed in the last twenty years of the 1st c., see Adam-Veleni (ed.) 2001, 86. // 1st c.

109. (ΠΑΜ 65) // AUTH Campus. Theological Department Building. Sector 4, Φ. 20. 22-6-1965 // Fragments. Large part of the body is missing // Translucent slightly greenish. No impurities few pinprick bubbles // H.: 12, Diam. rim: 5, Diam. Base: 4, Th.: 0.1 // Unpublished // 1st - 2nd c.

110. (BY 227/52) // Northern part of Evangelistrias Cemetery, area of former Lyssiatreion Clinic. Stray find. 27-4-1994 // Fragments // Colourless. No impurities, several pinprick bubbles // H.: 3.5, Diam. rim: 6, Diam. Base: 4 // Unpublished // 2nd end - early 3rd c.

111. (MΘ 3763) // Unknown exact provenance. // Intact // Translucent smoky. No impurities few pinprick bubbles // H.: 8, Diam. Max.: 8, Diam. rim: 8, Diam. Neck: 1, Diam. Base: 8, Th.: 0.12, Weight: 92.5 // *Glass Cosmos* 2010, 272 no. 201 // 2nd end - early 3rd c.

112. (MΘ 5952) // AUTH Campus. Theological Department Building. Grave (13) built, rectangular grave, Sector 8. 1965 // Intact? // Translucent greenish. Very few impurities (2 pieces) // H.: 7.2, Diam. rim: 5.1, Diam. Neck: 1, Diam. Base: 2, Th.: 0.1, Weight: 32.5 // Petsas 1966, 334-339, pl. 351 ζ. // 3rd end - early 5th c.

113. (BY 165) // AUTH Campus. Theological Department Building. Expansion of the School, Grave II. 1977 // Intact // Translucent greenish. No impurities few pinprick bubbles // H.: 9, Diam. rim: 10, Diam. Base: 3.8 // *Glass Cosmos* 2010, 277 no. 216 // 4th - early 5th c.

114. (MΘ 513) // Vardari Square. // Intact // Translucent greenish. No impurities or pinprick bubbles // H.: 14, Diam. rim: 10.6, Diam. Base: 3.7 // E. Trakosopoulou-Salakidou, Thessaloniki apo ta proïstorika mechri ta christianika chronia 1986, 125. *Glass Cosmos* 2010, 282, no. 230 // 4th - early 5th c.

115. (MΘ 7167) // AUTH Campus. Administration Building. 2nd Sarcophagus. 8-9-1965 // Half of the vessel is missing, filled // Translucent greenish. // H.: 8.8, Diam. Max.: 8, Diam. rim: 8, Diam. Neck: 1, Diam. Base: 2.2, Th.: 0.15, Weight: 46.35 // Unpublished // 4th - early 5th c.

116. (MΘ 21393) // Sindos, Industrial Zone of Thessaloniki. 51B, Τετραγ. B. T. 20. 1-5-1999 // Intact // Translucent greenish. No impurities few pinprick bubbles // H.: 8, Diam. rim: 7.8, Th.: 0.1, Weight: 40.2 // Unpublished // 4th - early 5th c.

117. (MΘ 21958) // Unknown street, probably Vardari Square area. In a box signed 'unregistered, probably from Vardaris Cemetery' wrapped in 1919 newspaper. Envelop 'Φ', designating probably the grave number. 1920; // Mended, small parts of the body are missing // Translucent greenish. Indiscernible due to weathering // H.: 12, Diam. rim: 9, Diam. Base: 3.5, Th.: 0.1 // *Glass Cosmos* 2010, 294, no. 262 // 4th - early 5th c.

118. (ΠΑΜ 70) // Sokratous Street. Eastern plot. Northern wall, left next to the well. 15-16/11-1968 // Fragment // Colourless Decolourized, dark-blue and brown. No impurities or pinprick bubbles // Diam. rim: 9, Pres. H.: 4 // Unpublished // 4th - early 5th c.

119. (ΠΑΜ 93) // AUTH Campus. Theological Department Building. Probably stray find. 1965 // Fragment // Decolourized and dark-blue. No impurities or pinprick bubbles // Dimensions: 3X3 Th.: 0.12 // Unpublished // 4th - early 5th c.

120. (BY 317/8) // Kallitheas Avenue, Elementary School of Ampelokipi. Grave T. 93, burial layer. 26-11-1997 // Mended // Translucent greenish. Few pinprick bubbles no impurities // H.: 9.3, Diam. rim: 8.5, Diam. Base: 3.4, Th.: 0.01 // Makropoulou 2007, 323, pl. 27, 69, from a grave dated to the first half of the 4th c. by the pottery finds // 4th c.

121. (MΘ 21007) // Plot property of Tsirakoglou. 1940 // Parts missing, mended // Translucent bluish. No impurities or pinprick bubbles // H.: 10.5, Diam. rim: 7.7, Diam. Base: 4, Est. H.: 10.5, Th.: 0.06 // Walter 1940, fig. 35. Kotzias 1953-54, 167-175. // 4th early - early 5th c.

122. (MΘ 21957) // Unknown street, probably Vardari Square area. In a box signed 'unregistered, probably from Vardaris Cemetery' wrapped in 1919 newspaper. Envelop 'T', designating probably the grave number. 1920; // Mended, small parts of the body are missing // Translucent greenish. Indiscernible due to weathering // H.: 10, Diam. rim: 4.1, Diam. Base: 2, Th.: 0.08 // Unpublished // 4th early - early 5th c.

123. (BY 38) // Unknown exact provenance. // Part of the base is preserved // Translucent greenish. No impurities few pinprick bubbles // Diam. Max.: 4, Pres. H.: 4.5, Th.: 0.08 // Unpublished // 5th - 6th c.

124. (MΘ ΑΚΑΤΑΓΡΑΦΟ 5) // Unknown exact provenance. // Only the lower part of the body is preserved // Decolourized, slightly greenish. // Dimensions: Pres. H.: 4, Th.: 0.08 // Unpublished // 5th - 6th c.

125. (MΘ 21959) // Unknown exact provenance. // Only the lower part of the body is preserved // Decolourized, slightly greenish. // Diam. Max.: 7, Pres. H.: 4, Th.: 0.08 // Unpublished // 5th - 6th c.

126. (ΠΑΜ 68) // Junction of Agiou Dimitriou Street and Agias Sophias Street, Byzantine building. X4 on the bedrock. 16-6-1969 // Fragment // Translucent olive-greenish and blue. No impurities few pinprick bubbles // H.: 8.5, Diam. rim: 5.2, Diam. Base: 0.3, Th.: 0.2 // Unpublished // 5th - 7th c.

127. (ΠΑΜ 85) // Junction of Agiou Dimitriou Street and Agias Sophias Street, Byzantine building. 3rd triangular space. 11-6-1969 // Fragment // Translucent bluish. No impurities few pinprick bubbles // Dimensions: 5X4, Th.: 0.12 // Unpublished // 5th - 7th c.

128. (ΠΑΜ 86) // Junction of Agiou Dimitriou Street and Agias Sophias Street, Byzantine building. 3rd triangular space. 11-6-1969 // Fragment // Translucent bluish. No impurities few pinprick bubbles // Dimensions: 3X1, Th.: 0.12 // Unpublished // 5th - 7th c.
129. (ΠΑΜ 87) // Junction of Agiou Dimitriou Street and Agias Sophias Street, Byzantine building. 3rd triangular space. 11-6-1969 // Fragment // Translucent bluish. No impurities few pinprick bubbles // Dimensions: 4X3, Th.: 0.12 // Unpublished // 5th - 7th c.
130. (ΜΘ 21970) // Sokratous Street, 47-49. Plot property of Velikas (today Moirarchou Koufitsa Street, 19). 1968 // Fragment // Translucent olive-green. No impurities very few pinprick bubbles // H.: 4.5, Diam. Base: 2 // Unpublished // 5th - 6th c.
131. (ΠΑΜ 64) // AUTH Campus. Theological Department Building. Grave T11, depth 0.80. Σεπτέμβρης 1965 // Fragment // Translucent olive-greenish. Few pinprick bubbles no impurities // Diam. Max.: 4.2, Diam. Base: 2.3 // Unpublished // 5th - 6th c.
132. (ΠΑΜ 69) // Junction of Agiou Dimitriou Street and Agias Sophias Street, Byzantine building. X4 on the bedrock. 16-6- 1969 // Fragment // Translucent olive-greenish. No impurities few pinprick bubbles // H.: 4, Diam. Max.: 3, Diam. Base: 2, Th.: 0.2 // Unpublished // 5th - 7th c.
133a-b. (BY 48) // Basilica in Panorama. Exterior of the apse. 1971 // Base fragment // Translucent greenish. No impurities or pinprick bubbles // Dimensions: α) Diam. 3.8, H. 1.1 B) Diam. 3.5, H. 2.5 // Unpublished // 5th - 6th c.
134a-b. (BY 49) // Basilica in Panorama. Room III. 1971 // Base fragment // Translucent greenish. No impurities or pinprick bubbles // Dimensions: α) Diam. 4.6, H. 1.5 β) Diam. 4.5, H. 2 // Unpublished // 5th - 6th c.
135a-b. (BY 50) // Basilica in Panorama. 1971 // Base fragment // Translucent greenish. No impurities or pinprick bubbles // α) Diam.4.5, H. 2 β) Diam. 4, H. 2 γ) Diam.3.8, H. 1 δ) Diam. 4.5, H. 2.5 // Unpublished // 5th - 6th c.
136. (BY 227/5) // Northern part of Evangelistrias Cemetery, area of former Lyssiatreion Clinic. Stray find. // Fragment // Translucent greenish. // W. 4.8, H.: 1.7 // Unpublished // 5th - 6th c.
137. (BY 227/6) // Northern part of Evangelistrias Cemetery, area of former Lyssiatreion Clinic. Stray find. // Fragment // Translucent greenish. // W. 4.8, Pres. H.: 2 // Unpublished // 5th - 6th c.
138. (BY 227/7) // Northern part of Evangelistrias Cemetery, area of former Lyssiatreion Clinic. Stray find. // Fragment // Translucent greenish. // W. 4.9, H.: 1.8 // Unpublished // 5th - 6th c.
139. (BY 227/8) // Northern part of Evangelistrias Cemetery, area of former Lyssiatreion Clinic. Stray find. // Fragment // Translucent greenish. // W. 4.2, Pres. H.: 1.5 // Unpublished // 5th - 6th c.
140. (BY 227/9) // Northern part of Evangelistrias Cemetery, area of former Lyssiatreion Clinic. Stray find. // Fragment // Translucent greenish. // W. 3.8, Pres. H.: 3 // Unpublished // 5th - 6th c.
141. (BY 227/10) // Northern part of Evangelistrias Cemetery, area of former Lyssiatreion Clinic. Stray find. // Fragment // Translucent greenish. // W. 4.5, Pres. H.: 4.5 // Unpublished // 5th - 6th c.
142a-b. (BY 227/48) // Northern part of Evangelistrias Cemetery, area of former Lyssiatreion Clinic. Stray find. 28-4-1995 // Fragment // Translucent greenish. Few impurities, several pinprick bubbles // // Unpublished // 5th - 6th c.
143. (BY 245/7) // Lagkada Street, Northern plot of Axios Complex. Room Z', case No 1. 24-5-1991 // Fragment // Translucent greenish. // 5 Dimensions: Pres. H.: 1 // From a grave firstly used in the first half of the 4th c. and again in the 5th-6th c. The fragment probably belongs to the 491-565 period. // 5th c., end - 6th c.
144. (BY 302/2) // Chapel south of Trigonion Tower. Grave IV, on the floor, left of the entrance. 1972 // Fragment // Translucent bluish. No impurities few pinprick bubbles // Unpublished // 5th - 6th c.
145. (BY 320/9) // Rentina Castle. Sector E. EΔ9- 4o layer. Φ. 1276.26-10-1991 // Large parts of the body are missing // Translucent olive-green and dark-blue. Several pinprick bubbles, no impurities // H.: 8.5, Diam. rim: 6, Diam. Base: 4, Th.: 0.12 // Unpublished // 5th - 6th c.
146. (BY 325/8) // Ancient Agora. // Body fragment // Translucent greenish. // // Unpublished // 5th - 6th c.
147. (ΜΘ 3061) // Ramona Area. Square. Grave Cluster II, grave (1). 8-1963 // Part of the foot and the base is preserved. Iridescent // Translucent greenish. No impurities few pinprick bubbles // H.: 2.5, Diam. Base: 3.5 // Unpublished // 5th - 6th c.
148. (ΜΘ 3202) // Ramona Area. Square. 8-1963 // Part of the foot and the base is preserved. Iridescent // Translucent greenish. // H.: 1.2, Diam. Max.: 2 // Unpublished // 5th - 6th c.
149. (ΜΘ 3685) // Lagkada Street, 67. In the soil stirred up during the transportation of a sarcophagus from Ramona Area. 26-6-1963 // Iridescent // Translucent greenish. // H.: 2.2, Diam. Base: 4 // Unpublished // 5th - 6th c.
150. (ΜΘ 3686) // Lagkada Street, 67. In the soil stirred up during the transportation of a sarcophagus from Ramona Area. 26-6-1963 // Iridescent // Translucent greenish. // H.: 2, Diam. Base: 4 // Unpublished // 5th - 6th c.
151. (ΜΘ 20939) // Unknown street, probably Vardari Square area. In a box signed 'unregistered, probably from Vardaris Cemetery'. // Missing part of the body // Translucent greenish. No impurities or pinprick bubbles // H.: 11, Diam. Max.: 11.4, Diam. rim: 11.4, Diam. Base: 4.8, Th.: 0.06, Weight: 45 // Unpublished // 5th - 6th c.
152. (ΜΘ 20940) // Unknown street, probably Vardari Square area. In a box signed 'unregistered, probably from Vardaris Cemetery'. // Large parts of the body are missing // Translucent yellowish. No impurities or pinprick bubbles // Diam. Max.: 11.5, Diam. rim: 11.5, Diam. Neck: 1, Diam. Base: 4.9, Th.: 0.03, Weight: 51.2 // Unpublished // 5th - 6th c.
153. (ΜΘ 20941) // Unknown street, probably Vardari Square area. In a box signed 'unregistered, probably from Vardaris Cemetery'. // Base and foot are preserved // Translucent greenish. // H.: 2.9, Diam. Base: 5.5, Th.: 0.2, Weight: 16.8 // Unpublished // 5th - 6th c.
154. (ΜΘ 20942) // Unknown street, probably Vardari Square area. In a box signed 'unregistered, probably from Vardaris Cemetery'. // Base and foot are preserved // Translucent greenish. // H.: 1.8, Diam. Base: 4.2, Weight: 10.7 // Unpublished // 5th - 6th c.
155. (ΜΘ 20943) // Unknown street, probably Vardari Square area. In a box signed 'unregistered, probably from Vardaris Cemetery'. // Base and foot are preserved // Translucent greenish. // H.: 1.6, Diam. Base: 4.3, Weight: 12.1 // Unpublished // 5th - 6th c.
156. (ΜΘ 20944) // Unknown street, probably Vardari Square area. In a box signed 'unregistered, probably from Vardaris Cemetery'. // Base and foot are preserved // Translucent greenish. // H.: 1.5, Diam. Base: 5.6, Weight: 24.1 // Unpublished // 5th - 6th c.
157. (ΜΘ 20945) // Unknown street, probably Vardari Square area. In a box signed 'unregistered, probably from Vardaris Cemetery'. // Base and foot are preserved // Translucent greenish. // H.: 1.9, Diam. Base: 4.6, Weight: 12.7 // Unpublished // 5th - 6th c.
158. (ΜΘ 20946) // Unknown street, probably Vardari Square area. In a box signed 'unregistered, probably from Vardaris Cemetery'. // Base and foot are preserved // Translucent greenish. // H.: 2, Diam. Base: 4.1, Weight: 14.1 // Unpublished // 5th - 6th c.
159. (ΜΘ 20947) // Unknown street, probably Vardari Square area. In a box signed 'unregistered, probably from Vardaris Cemetery'. // Base and foot are preserved // Translucent greenish. // H.: 2.4, Diam. Base: 4.65 , Weight: 6.3 // Unpublished // 5th - 6th c.
160. (ΜΘ 20948) // Unknown street, probably Vardari Square area. In a box signed 'unregistered, probably from Vardaris Cemetery'. // Base and foot are preserved // Translucent greenish. // H.: 2.6, Diam. Base: 4.7, Weight: 14.9 // Unpublished // 5th - 6th c.
161. (ΜΘ 20949) // Unknown street, probably Vardari Square area. In a box signed 'unregistered, probably from Vardaris Cemetery'. // Base and foot are preserved // Translucent greenish. // H.: 2.3, Diam. Base: 3.8, Weight: 7.9 // Unpublished // 5th - 6th c.
162. (ΠΑΜ 51) // Vardari Square. 1970 // Only the base is preserved // Translucent greenish. No impurities few pinprick bubbles // W. 4, Th.: 0.2 // Unpublished // 5th - 6th c.
163. (ΠΑΜ 52) // Vardari Square. 1970 // Only the base is preserved // Translucent greenish. No impurities few pinprick bubbles // W. 3, Th.: 0.2 // Unpublished // 5th - 6th c.
164. (ΠΑΜ 53) // Vardari Square. 1970 // Only the base is preserved // Translucent greenish. No impurities few pinprick bubbles // W. 4.5, Th.: 0.2 // Unpublished // 5th - 6th c.

164a. (BY 298/39) // Agapis 3. 1996 // Only the base is preserved // Translucent greenish. No impurities few pinprick bubbles // W. 4.8, Th.: 0.2 // Unpublished // 6th -7th c.

164b. (uncatalogued) // Vasileos Irakeliou 45. 2012 // Only the base is preserved // Translucent greenish. No impurities few pinprick bubbles // W. 4.4, Th.: 0.2 // Unpublished // 6th c.

164c. (BY 298/41) // Agapis 3. 1996 // Only the base is preserved // Translucent greenish. No impurities few pinprick bubbles // W. 4.5, Th.: 0.2 // Unpublished // 6th -7th c.

164d. (BY 298/35) // Agapis 3. 1996 // Only the base is preserved // Translucent greenish. No impurities few pinprick bubbles // W. 4, Th.: 0.2 // Unpublished // 6th -7th c.

165. (BY 182) // Kolchida, Kilkis. Hanging from a nail on the eastern wall above the head of the deceased. Found fallen on the couch. // Complete, mended from small fragments // Translucent slightly greenish. No impurities or pinprick bubbles // H.: 10, Diam. rim: 8.7, Diam. Base: 4, Th.: 0.1, Weight: 73.2 // Kissas 1988, 209, fig. 4. Papanikola-Bakirtzi (ed.) 2002, no. 299, 285 (A. Antonaras), dated to the 5th c. // 5th c.

166. (BY 245/3) // Lagkada Street, Northern plot of Axios Complex. Room Z', case No 1. 24-5-1991 // Fragment // Translucent greenish. // Dimensions: 3X1.5 // From a grave firstly used in the first half of the 4th c. and again in the 5th-6th c. The fragment probably belongs to the 491-565 period. // 5th c., end - 6th c.

167. (ΔΕΘ 5) // Area of the Thessaloniki International Fair , Southern Plot. Grave 5. 1993 // One handle is preserved // Translucent greenish. // Dimensions: 2X1.5 // Unpublished // 5th c.

168. (ΜΘ 21973) // Sokratous Street, 47-49. Plot property of Velikas (today Moirarchou Koufitsa Street, 19). 1968 // Fragment // Translucent greenish. No impurities few pinprick bubbles // H.: 5.5, Diam. rim: 10, Th.: 0.1 // Unpublished // 5th c.

169. (ΜΘ 21974) // Sokratous Street, 47-49. Plot property of Velikas (today Moirarchou Koufitsa Street, 19). 1968 // Fragment // Translucent greenish. No impurities few pinprick bubbles // H.: 3, Diam. rim: 8, Th.: 0.1 // Unpublished // 5th c.

170. (ΠΑΜ 39) // AUTH Campus. Administration Building. Grave 25, depth 2m. 10-11-1972 // Fragments // Translucent greenish. No impurities or pinprick bubbles // W. 4, Th.: 0.12 // Unpublished // 5th - 6th c.

171. (ΠΑΜ 40) // AUTH Campus. Administration Building. Grave 25, depth 2m. 10-11-1972 // One handle is preserved // Translucent greenish. No impurities or pinprick bubbles // Dimensions: 4X1.5 Th.: 0.13 // Unpublished // 5th - 6th c.

172. (ΠΑΜ 41) // AUTH Campus. Administration Building. Grave 25, depth 2m. 10-11-1972 // Fragment // Translucent greenish. No impurities or pinprick bubbles // Dimensions: 3X2, Th.: 0.12 // Unpublished // 5th - 6th c.

173. (ΠΑΜ 66) // Junction of Agiou Dimitriou Street and Agias Sophias Street, Byzantine building. Under the pavement, between walls B′ and Δ′. 10-6-1969 // Fragment // Translucent olive-greenish. Few impurities and several pinprick bubbles // 1.5, Pres. H.: 6.5 // Unpublished // 5th - 7th c.

174. (ΠΑΜ 88) // Junction of Agiou Dimitriou Street and Agias Sophias Street, Byzantine building. 3rd triangular space. 11-6-1969 // Fragment // Translucent greenish. No impurities few pinprick bubbles // Pres. H.: 2.2, Th.: 0.15 // Unpublished // 5th - 7th c.

175. (ΠΑΜ 89) // Junction of Agiou Dimitriou Street and Agias Sophias Street, Byzantine building. 3rd triangular space. 11-6-1969 // Fragment // Translucent bluish. No impurities few pinprick bubbles // Diam. Max.: 3.5, Diam. Base: 1.5, Pres. H.: 2.5 // Unpublished // 5th - 7th c.

175a. (BY 25/1) //1 Pringipos Nikolaou Street. From the small square room at the centre of the apse. 1961 // Fragment // Translucent greenish. No impurities few pinprick bubbles // Diam. Base: 2, Pres. H.: 6.5 // Antonaras 2016b, 120, fig. 2 // 5th - 7th c.

175b. (BY 25/2) //1 Pringipos Nikolaou Street. From the small square room at the centre of the apse. 1961 // Fragment // Translucent greenish. No impurities few pinprick bubbles // Diam. Base: 2, Pres. H.: 3.8 // Antonaras 2016b, 120, fig. 2 // 5th - 7th c.

175c. (BY 25/3) //1 Pringipos Nikolaou Street. From the small square room at the centre of the apse. 1961 // Fragment // Translucent greenish. No impurities few pinprick bubbles // Diam. Base: 2, Pres. H.: 5 // Antonaras 2016b, 120, fig. 2 // 5th - 7th c.

176. (BAA 56) // Ancient Agora Bathhouse. Rectangular room, north of the Bathhouse. Destruction Layer. 1998 // Fragment // Translucent greenish. // Mavromichali 2001, draw. 10. From a bath complex destroyed in the last twenty years of the 1st c., see Adam-Veleni (ed.) 2001, 86. // 1st c.

177. (BAA 87) // Ancient Agora Bathhouse. Rectangular room, north of the Bathhouse. Destruction Layer. 1998 // Fragment // Translucent bluish. // Mavromichali 2001, draw. 10. From a bath complex destroyed in the last twenty years of the 1st c., see Adam- Veleni (ed.) 2001, 86. // 1st c.

178. (RA 779) // Ancient Agora Bathhouse. Rectangular room, north of the Bathhouse. Destruction Layer. 1998 // Fragment // Non vidi. // Mavromichali 2001, fig. 20. From a bath complex destroyed in the last twenty years of the 1st c., see Adam-Veleni (ed.) 2001, 86. *Glass Cosmos* 2010, 269, no. 194 // 1st c.

179. (BAA 58) // Ancient Agora Bathhouse. Rectangular room, north of the Bathhouse. Destruction Layer. 1998 // Fragment // Non vidi. // Mavromichali 2001, fig. 20. From a bath complex destroyed in the last twenty years of the 1st c., see Adam-Veleni (ed.) 2001, 86. // 1st c.

180. (BAA 59) // Ancient Agora Bathhouse. Rectangular room, north of the Bathhouse. Destruction Layer. 1998 // Fragment // Non vidi. // Mavromichali 2001, fig. 20. From a bath complex destroyed in the last twenty years of the 1st c., see Adam-Veleni (ed.) 2001, 86. // 1st c.

181. (BAA 60) // Ancient Agora Bathhouse. Rectangular room, north of the Bathhouse. Destruction Layer. 1998 // Fragment // Non vidi. // Mavromichali 2001, fig. 20. From a bath complex destroyed in the last twenty years of the 1st c., see Adam-Veleni (ed.) 2001, 86. // 1st c.

182. (BAA 61) // Ancient Agora Bathhouse. Rectangular room, north of the Bathhouse. Destruction Layer. 1998 // Fragment // Non vidi. // Mavromichali 2001, fig. 20. From a bath complex destroyed in the last twenty years of the 1st c., see Adam-Veleni (ed.) 2001, 86. // 1st c.

183. (BAA 62) // Ancient Agora Bathhouse. Rectangular room, north of the Bathhouse. Destruction Layer. 1998 // Fragment // Non vidi. // Mavromichali 2001, fig. 20. From a bath complex destroyed in the last twenty years of the 1st c., see Adam-Veleni (ed.) 2001, 86. // 1st c.

184. (BAA 63) // Ancient Agora Bathhouse. Rectangular room, north of the Bathhouse. Destruction Layer. 1998 // Fragment // Non vidi. // Mavromichali 2001, fig. 20. From a bath complex destroyed in the last twenty years of the 1st c., see Adam-Veleni (ed.) 2001, 86. // 1st c.

185. (ΜΘ 3759) // Vasilissis Sofias Street, 44. 3rd Sarcophagus. 25-4-1960 // Complete, mended // Translucent greenish. Very few impurities, many pinprick bubbles, few larger ones // H.: 14, Diam. Max.: 11.2, Diam. rim: 11.3, Diam. Neck: 8.5, Diam. Base: 5.6, Th.: 0.3 // E. Trakosopoulou-Salakidou, Thessaloniki apo ta proïstorika mechri ta christianika chronia 1986, 125, fig. 123. *Glass Cosmos* 2010, 272-273. no. 203 // 4th c.

186. (BAA 69) // Ancient Agora Bathhouse. Rectangular room, north of the Bathhouse. Destruction Layer. 1998 // Fragment // Non vidi. // Mavromichali 2001, fig. 22. From a bath complex destroyed in the last twenty years of the 1st c., see Adam-Veleni (ed.) 2001, 86. // 1st c.

187. (BAA 70) // Ancient Agora Bathhouse. Rectangular room, north of the Bathhouse. Destruction Layer. 1998 // Fragment // Non vidi. // Mavromichali 2001, fig. 22. From a bath complex destroyed in the last twenty years of the 1st c., see Adam-Veleni (ed.) 2001, 86. // 1st c.

188. (BY 320/7) // Rentina Castle. Sector E. ΕΔ1, dig IV. Φ. 251. 18-9-1980 // Fragment // Translucent greenish. No impurities or pinprick bubbles // Diam. rim: 12, WXL:4X2.5, Th.: 0.12 // Unpublished // 4th c.

189. (BY 320/8) // Rentina Castle. Sector E. Western side of the complex. Fill west of the south staircase. Φ. 380. 2-2-1981 // Fragment // Dark blue. No impurities or pinprick bubbles // Diam. rim: 12, WXL:5X7, Th.: 0.12 // Unpublished // 4th c.

190. (BAA 41) // Ancient Agora Bathhouse. Rectangular room, north of the Bathhouse. Destruction Layer. 1998 // Fragment // Non vidi. // Mavromichali 2001, fig. 10. From a bath complex destroyed in the last twenty years of the 1st c., see Adam-Veleni (ed.) 2001, 86. // 1st c.

191. (BAA 42) // Ancient Agora Bathhouse. Rectangular room, north of the Bathhouse. Destruction Layer. 1998 // Fragment // Non vidi. // Mavromichali 2001, fig. 10. From a bath complex destroyed in the last twenty years of the 1st c., see Adam-Veleni (ed.) 2001, 86. // 1st c.

192. (BAA 43) // Ancient Agora Bathhouse. Rectangular room, north of the Bathhouse. Destruction Layer. 1998 // Fragment // Non vidi. // Mavromichali 2001, fig. 10. From a bath complex destroyed in the last twenty years of the 1st c., see Adam-Veleni (ed.) 2001, 86. // 1st c.

193. (BY 321/4) // Agiou Nestoros Street, 26-28. Water main works. Grave T. 39, near the right arm. 6-10-2000 // Intact // Translucent greenish. Few pinprick bubbles and impurities // H.: 17.3, Diam. Max.: 10, Diam. rim: 4.1, Diam. Neck: 1.8, Diam. Base: 9.4, Th.: 0.1 // Papanikola-Bakirtzi 2002 (ed.), no. 262 225, dated στον 3ο αι. // 1st early - early 3rd c.

194. (ΜΘ 541) // Vardari Square. // Intact. Iridescent // Translucent greenish. Few impurities, several pinprick bubbles // H.: 20, Diam. Max.: 10.2, Diam. rim: 4.2, Diam. Neck: 2.1, Diam. Base: 10.2 // E. Trakosopoulou-Salakidou, Thessaloniki apo ta proïstorika mechri ta christianika chronia 1986, 125. // 1st early - early 3rd c.

195. (ΜΘ 5945) // AUTH Campus. Theological Department Building. Grave (3κ) Tile grave, Sector 13. *Glass Cosmos* 2010, 272, no. 200 // Complete, filled // Translucent greenish. // H.: 11, Diam. rim: 1, Diam. Neck: 2.7, Diam. Base: 9, Th.: 0.1, Weight: 111.75 // Unpublished // 1st early - early 3rd c.

196. (ΜΘ 7168) // AUTH Campus. Administration Building. 2nd Sarcophagus. 8-9-1965 // Filled in the base and the body // Translucent greenish. // H.: 15, Diam. rim: 3.1, Diam. Neck: 2.5, Diam. Base: 9, Th.: 0.1, Weight: 64.51 // Unpublished // 1st early - early 3rd c.

197. (BY 227/20) // Northern part of Evangelistrias Cemetery, area of former Lyssiatreion Clinic. Grave 122. 25-6-1993 // In fragments // Translucent bluish. No impurities few pinprick bubbles // H.: 16.7, Diam. Max.: 10.5, Diam. rim: 5.2, Diam. Neck: 3.4, Diam. Base: 9.5, Th.: 0.1 // Unpublished. From a grave with a clay lamp dated to the second half or to the end of the 2nd c. // 1st early - early 3rd c.

198. (ΜΘ 5478) // Unknown provenance. // Large part of the body is missing // Translucent greenish. Few impurities, few pinprick bubbles // H.: 13, Diam. rim: 3.7, Diam. Neck: 2.5, Diam. Base: 11, Th.: 0.08, Weight: 49.92 // Unpublished // 1st early - early 3rd c.

199. (ΜΘ 23963) // Ramona Area. 1971 // Intact // Translucent greenish. No impurities few pinprick bubbles // H.: 16.2, Diam. Max.: 9.5, Diam. rim: 3.5, Diam. Neck: 2, Diam. Base: 9, Th.: 0.1 // *Glass Cosmos* 2010, 287, no. 240 // 2nd end - 3rd c.

200. (BY 255/4) // Margaropoulou Street, 20. Grave 103. 11-5-1995 // Body is missing // Translucent greenish. // Diam. rim: 3.4, Diam. Neck: 2, Pres. H.: 11 // Makropoulou 2007, 234, pl. 27. From a 3rd-century grave // 3rd c.

201. (ΜΘ 540) // Vardari Square. // Intact // Translucent greenish. Few impurities, few pinprick bubbles // H.: 16, Diam. Max.: 5.1, Diam. rim: 2.6, Diam. Neck: 1.8, Diam. Base: 5.1 // *Glass Cosmos* 2010, 271, no. 198 // 1st early - early 3rd c.

202. (ΜΘ 5946) // AUTH Campus. Theological Department Building. 3-1965 // Complete, mended // Translucent greenish. // H.: 19.5, Diam. rim: 2.3, Diam. Neck: 1.4, Diam. Base: 5.7, Th.: 0.07, Weight: 43.79 // Petsas 1966, 334-339, pl. 351 δ. // 1st early - early 3rd c.

203. (ΜΘ 11167) // Ramona Area. Building 4, marble covered grave. // Intact // Translucent greenish. Few impurities, few pinprick bubbles // H.: 23, Diam. rim: 3.5, Diam. Neck: 2, Diam. Base: 7, Th.: 0.08, Weight: 68.01 // *Glass Cosmos* 2010, 270, no. 196 // 1st early - early 3rd c.

204. (ΜΘ 11168) // Ramona Area. Building 4, marble covered grave. // Complete, mended // Translucent greenish. No impurities few pinprick bubbles // H.: 24, Diam. rim: 3, Diam. Neck: 2.4, Diam. Base: 7.5, Th.: 0.08, Weight: 94.04 // *Glass Cosmos* 2010, 271 no. 197 // 1st early - early 3rd c.

205. (ΜΘ 3757) // Angelaki Street, plot property of Zervas. In a sarcophagus. 28-4-1960 // Intact // Translucent greenish. No impurities or pinprick bubbles // H.: 15, Diam. Max.: 12.9, Diam. rim: 2.2, Diam. Neck: 2.2, Diam. Base: 5.6 // E. Trakosopoulou-Salakidou, Thessaloniki apo ta proïstorika mechri ta christianika chronia 1986 fig. 125. *Glass Cosmos* 2010, 272, no. 202 // 2nd - 3rd c.

206. (BY 244/7) // Stratou Avenue 2, plot of the Museum of Byzantine Culture. Grave 89. 13-10-1992 // Complete, mended // Translucent greenish. No impurities few pinprick bubbles // H.: 15.4, Diam. rim: 2, Diam. Base: 3.5, Th.: 0.1 // Nalpantis 2003, 80, 129-130, fig. 45, pl. 43, from a grave with coin of Probus (276-282), which probably dates the time of the interment of the vessel, and late 3rd - early 4th c. hoards. // 3rd end -early 4th c.

207. (BY 255/6) // Margaropoulou Street, 20. Grave 111. 3-7-1995 // Intact // Translucent greenish. // H.: 16.5, Diam. Max.: 10.5, Diam. rim: 2.5, Diam. Neck: 2.5, Diam. Base: 4.5, Th.: 0.1 // Makropoulou 2007, 214, pls. 26, 68. From a late 3rd-century burial, dated by a coin of Probus, struck between 276 and 282. // 3rd end c.

208. (BY 301/3) // Syntagmatarchi Fryzi Street, 31. Clearing of enclosure wall A. // Body fragments mended // Colourless Decolourized. No impurities very few pinprick bubbles // H.: 4 // Makropoulou 2007, 284, pl. 25., from a cubiculum which was in use from the end of the 3rd to the late 4th c. // 3rd end - early 5th c.

209. (ΜΘ 521) // Vardari Square. // Large parts of the body are missing. Mended // Translucent greenish. Very few impurities, several pinprick bubbles, few large ones // H.: 10.4, Diam. Max.: 7.1, Diam. rim: 1.7, Diam. Neck: 2.3, Diam. Base: 2.8, Th.: 0.05, Weight: 26.32 // Unpublished // 3rd end - early 5th c.

210. (ΜΘ 3683) // Lagkada Street, 67. In the soil stirred up during the transportation of a sarcophagus from Ramona Area. 26-6-1963 // Parts missing // Translucent greenish. Few impurities, several pinprick bubbles // H.: 8, Diam. Max.: 10, Diam. rim: 2, Diam. Neck: 2.8, Th.: 0.08, Weight: 35.71 // Unpublished // 3rd end - early 5th c.

211. (ΜΘ 3684) // Lagkada Street, 67. In the soil stirred up during the transportation of a sarcophagus from Ramona Area. 26-6-1963 // Part of the base and the body are preserved // Translucent greenish. Few impurities, several pinprick bubbles // Diam. Max.: 12, Diam. Base: 5, Pres. H.: 4.4, Th.: 0.1 // Unpublished // 3rd end - early 5th c.

212. (ΜΘ 9906) // Area of the Thessaloniki International Fair , Italian Pavilion. Grave T. 80. // Upper part of the neck is missing. Iridescent on the body // Translucent greenish. Few pinprick bubbles and impurities // H.: 12, Diam. Max.: 8.6, Diam. rim: 2.3, Diam. Neck: 2.3, Diam. Base: 4, Th.: 0.12 // Unpublished // 3rd end - early 5th c.

213. (ΜΘ 20997) // Plot property of Tsirakoglou. 1940 // Part of the rim, neck and body is missing. Covered with a dark crust // Translucent greenish. // H.: 14, Diam. Max.: 9.2, Diam. rim: 2, Diam. Neck: 2, Diam. Base: 3.3, Th.: 0.13, Weight: 48 // Walter 1940, fig. 35. Kotzias 1953-54, 167-175. // 3rd end - early 5th c.

214. (ΜΘ 21966) // Sokratous Street, 47-49. Plot property of Velikas (today Moirarchou Koufitsa Street, 19). 1968 // Fragment // Translucent slightly greenish. No impurities or pinprick bubbles // Dimensions: 2.5X2.5, Th.: 0.1 // Unpublished // 3rd end - early 5th c.

215. (BY 1) // AUTH Campus. Law School Building. // Fragments // Translucent greenish. // // Unpublished. Possibly the same fragment mentioned in the notebook of the Cubibulum, dated to the second half of the 3rd c. by coins of Antonius Pius (136-161) and Galienus (253-268). See Marki 2006, 211 // 3rd end - early 5th c.

216. (BY 36/2) // AUTH Campus. Sorority Building (Φ.Ε.Θ.). West of Grave ΙΔ. 1968 // Only part of the body is preserved // Translucent greenish. No impurities many pinprick bubbles // Unpublished // 3rd end - early 5th c.

217. (BY 207/1) // Lagkada Street, Ivanof Park. Grave XIX. 1991 // Large part of the body is missing // Translucent greenish. // H.: 15.5, Diam. rim: 3.8, Diam. Base: 5 // Makropoulou 1991, 257-270, photo 21. Burial dated by pottery to the 4th c. Makropoulou 2007, 50, pl. 68. // 4th c.

218. (BY 207/5) // Lagkada Street, Ivanof Park. Grave 37. 1991 // Mended, small parts of the body and rim are missing // Translucent greenish. No impurities few pinprick bubbles // H.: 10.2, Diam. Max.: 9.4, Diam. Neck: 2.7, Diam. Base: 5.2 // Makropoulou 2007, 45, pl. 24. Burial dated by pottery to the first half of the 4th c. // 4th c.

219. (BY 214/3) // Northern part of Evangelistrias Cemetery, area of former Lyssiatreion Clinic. Grave Δ. 25-11-1991 // Mended, small parts of the body and neck are missing // Translucent greenish. No impurities or pinprick bubbles // H.: 13, Diam. Max.: 10, Diam. rim: 5, Diam. Neck: 2, Th.: 0.1 // Unpublished. From a grave with a 2nd-3rd- century coin // 3rd end - early 5th c.

220. (BY 224) // Doxatou Street, 8. Grave 12. 23-2-1993 // Almost complete, mended // Translucent greenish. // Diam. Max.: 8, Diam. rim: 3.7, Diam. Neck: 2, Est. H.: 17, Th.: 0.1 // Simple note of the vessel at Makroupoulou 1994, 508, where is dated to the 4th c. Makropoulou 2007, 272, dated to the early 4th c. according to the type of the grave // 4th c.

221. (BY 227/19) // Northern part of Evangelistrias Cemetery, area of former Lyssiatreion Clinic. // Large part of the body preserved, mended // Translucent greenish. // Unpublished // 3rd end - early 5th c.

222. (BY 227/43) // Northern part of Evangelistrias Cemetery, area of former Lyssiatreion Clinic. In the NW corner a cist grave at the interior of the Eastern wall of the Stoa, depth 2.5 m. // Fragments. Upper part of the neck and rim are missing // Translucent greenish. // H.: 12, Diam. Max.: 9, Diam. Neck: 3, Diam. Base: 5, Th.: 0.1, Weight: 43.5 // Unpublished // 3rd end - early 5th c.

223. (BY 227/46 α,β) // Northern part of Evangelistrias Cemetery, area of former Lyssiatreion Clinic. Grave 231. 16-5-1995 // Almost complete, in large pieces // Translucent greenish. // Diam. Neck: 2, Diam. Base: 5 // Unpublished // 3rd end - early 5th c.

224. (BY 244/10) // Stratou Avenue 2, plot of the Museum of Byzantine Culture. Grave 27. 12-5-1989 // Fragments // Translucent greenish. No impurities few pinprick bubbles // Pres. H.: 4, Th.: 0.1 // Nalpantis 2003, 130, pl. 43, from a grave with coins of 334-348. // 4th c.

225. (BY 244/13) // Stratou Avenue 2, plot of the Museum of Byzantine Culture. Grave 16, SE corner. 13-6-1989 // Fragments // Translucent greenish. No impurities few pinprick bubbles // Diam. Neck: 2.5, Pres. H.: 5.3, Th.: 0.1 // Nalpantis 2003, 34, 130, fig. 8, from a grave with coins of 317-348. // 4th c.

226. (BY 253/1) // Junction of Efxeinou and Panagopoulou Streets. In a pit burial (No. 8). Left of the head of the deceased. // Intact // Translucent greenish. // H.: 15.5, Diam. rim: 5, Diam. Base: 5.5 // Karydas 1994, 501, pl. 156β, from a 4th-century cemetery // 4th c.

227. (BY 253/2) // Junction of Efxeinou and Panagopoulou Streets. Barrel-vaulted grave (No. 3) at the feet of the deceased at the N-NW corner. // Intact // Translucent greenish. // H.: 17.5, Diam. rim: 6.5, Diam. Base: 5.5 // Karydas 1994, 501, pl. 156γ, from a 4th-century cemetery // 4th c.

228. (BY 253/4) // Junction of Efxeinou and Panagopoulou Streets. Tile grave 6, next to the left temple bone. // Fragments // Translucent greenish. // H.: 14, Diam. Max.: 10, Diam. Neck: 4 // Karydas 1994, 501, from a 4th-century cemetery // 4th c.

229. (BY 257/3) // Municipal Hospital, 'Agios Dimitrios', court. Grave 4. 9-2-1987 // Mended, parts of the body are missing // Translucent greenish. No impurities or pinprick bubbles // H.: 13, Diam. Max.: 11, Diam. rim: 5, Diam. Neck: 2.5, Diam. Base: 5, Th.: 0.1, Weight: 83.2 // Makropoulou 1990a, 133, where a simple note to the finds is made // 3rd end - early 5th c.

230. (BY 317/1) // Kallitheas Avenue, Elementary School of Ampelokipi. Grave T. 42, burial layer. 3-9-1997 // Part of the rim missing // Translucent greenish. Few pinprick bubbles and impurities // H.: 15, Diam. Max.: 10.5, Diam. rim: 5, Diam. Neck: 2.5, Diam. Base: 6.5, Th.: 0.2 // Makropoulou 2007, 300, pl. 24., From a 3rd-century grave // 3rd - 4th c.

231. (BY 317/10) // Kallitheas Avenue, Elementary School of Ampelokipi. Grave T. 58, burial layer. 8-10-1997 // Mended // Translucent greenish. Few pinprick bubbles no impurities // H.: 18, Diam. Max.: 10, Diam. rim: 5, Diam. Neck: 2.5, Diam. Base: 4.5, Th.: 0.01 // Makropoulou 2007, 312, pl. 24, From a probably 4th-century grave // 4th c.

232. (BY 319/1) // Lagkada Street, middle plot of Axios Complex. Grave T. 9, burial layer. 14-1-1998 // Intact // Translucent olive-green. Very few pinprick bubbles no impurities // H.: 14, Diam. Max.: 9, Diam. rim: 6, Diam. Neck: 2.2, Diam. Base: 5.5, Th.: 0.13 // Makropoulou 2007, 138, pl. 24, from a 4th-century grave, on the basis of the grave's construction style. // 4th c.

233. (BY 321/3) // Agiou Nestoros Street, 32. Water main works. Grave T. 31 (N.0.40-E.0.50). -10-2000 // Intact // Translucent greenish. Few impurities and few pinprick bubbles some of which are larger // H.: 13.5, Diam. Max.: 10, Diam. rim: 5, Diam. Neck: 2.5, Diam. Base: 5, Th.: 0.15 // Unpublished // 3rd end - early 5th c.

234. (ΔΕΘ 44) // Area of the Thessaloniki International Fair , Southern Plot. Grave 44. 1993 // Complete, mendable // Translucent greenish. No impurities few pinprick bubbles // H.: 15.5, Diam. Max.: 6.8, Diam. rim: 4, Diam. Neck: 2.6, Diam. Base: 3.5, Th.: 0.1 // Unpublished // 3rd end - early 5th c.

235. (Θ 1/01-Υ3 ΑΓΚ 7212) // Thiseos Street, 2. Grave 14 (on the floor of the grave, depth -4.45). 22-3-2001 // Almost complete, broken. Small part of the body is missing // Translucent greenish. No impurities few pinprick bubbles // H.: 16, Diam. Max.: 10.3, Diam. rim: 5, Diam. Neck: 2.1, Diam. Base: 6, Th.: 0.1 // Unpublished // 3rd end - early 5th c.

236. (ΜΘ 526) // Vardari Square. // Complete, mended // Translucent greenish. No impurities few pinprick bubbles // H.: 18, Diam. Max.: 12, Diam. rim: 5.2, Diam. Neck: 2.5, Diam. Base: 7, Th.: 0.08, Weight: 119.41 // *Glass Cosmos* 2010, 291 no. 252 // 3rd end - early 5th c.

237. (ΜΘ 534) // Vardari Square. // Fragments. Large part of the body is missing // Translucent greenish. // H.: 14, Diam. Max.: 10, Diam. rim: 6.3, Diam. Neck: 2.5, Diam. Base: 5, Th.: 0.08, Weight: 112.33 // Unpublished // 3rd end - early 5th c.

238. (ΜΘ 3234) // Ramona Area. Grave 20. 27-9-1963 // Fragments. Large part of the body is missing // Translucent bluish. // Diam. Max.: 8, Diam. Neck: 2, Diam. Base: 4.5 // Unpublished // 3rd end - early 5th c.

239. (ΜΘ 3754) // Vasilissis Sofias Street, 44. 3rd Sarcophagus. 25-4-1960 // Small part of the rim is missing. // Translucent greenish. No impurities few pinprick bubbles // H.: 16, Diam. Max.: 11, Diam. rim: 4.7, Diam. Neck: 2.5, Diam. Base: 5.5, Th.: 0.08, Weight: 97.04 // *Glass Cosmos* 2010, 288-289, no. 245 // 3rd end - early 5th c.

240. (ΜΘ 5467) // Nea Apollonia. // Almost complete, mended. Parts of the body are missing // Translucent greenish. No impurities few large bubbles // H.: 11, Diam. Max.: 15, Diam. rim: 5, Diam. Neck: 3, Diam. Base: 7, Th.: 0.08, Weight: 92.95 // Unpublished // 3rd end - early 5th c.

241. (ΜΘ 5489) // Unknown exact provenance. // Only the lower part of the body is preserved // Translucent olive-green. Several impurities and pinprick bubbles // H.: 8, Diam. rim: 3, Diam. Neck: 2.2, Th.: 0.08, Weight: 25.42 // Unpublished // 3rd end - early 5th c.

242. (ΜΘ 5932) // AUTH Campus. Theological Department Building. Grave (26) built, rectangular grave, Sector 13. // Complete, filled // Translucent greenish. No impurities, pinprick bubbles on the body, larger ones on the neck // H.: 17, Diam. Max.: 12, Diam. rim: 3.6, Diam. Neck: 2.4, Diam. Base: 4.8, Th.: 0.08, Weight: 82 // Unpublished // 3rd end - early 5th c.

243. (ΜΘ 5933) // AUTH Campus. Theological Department Building. Grave (ψ) barrel-vaulted, Sector 3. // Complete, filled // Translucent greenish. No impurities few pinprick bubbles // H.: 16, Diam. rim: 4, Diam. Neck: 2.5, Diam. Base: 5 , Weight: 67 // Unpublished // 3rd end - early 5th c.

244. (ΜΘ 5934) // AUTH Campus. Theological Department Building. Grave (1η) Tile grave, Sector 6. // Complete, filled // Translucent greenish. // H.: 16, Diam. rim: 5.5, Diam. Neck: 3, Diam. Base: 6.5, Th.: 0.1, Weight: 94.09 // *Glass Cosmos* 2010, 292, no. 256 // 3rd end - early 5th c.

245. (ΜΘ 5935) // AUTH Campus. Theological Department Building. Grave (1λ) Tile grave, Sector 8. 5-1965 // Small part of the body is missing // Translucent greenish. Few impurities, few pinprick bubbles // H.: 16, Diam. Max.: 12, Diam. rim: 6.5, Diam. Neck: 2.5, Diam. Base: 5.6, Th.: 0.1, Weight: 99.37 // *Glass Cosmos* 2010, 292, no. 257 // 3rd end - early 5th c.

246. (ΜΘ 5936) // AUTH Campus. Theological Department Building. Grave (1λ) Tile grave, Sector 8. 1965 // Complete, filled // Translucent yellow greenish. No impurities, several large bubbles // H.: 14, Diam. Max.: 11, Diam. rim: 4.2, Diam. Neck: 2.3, Diam. Base: 6.5, Th.: 0.1, Weight: 127.96 // *Glass Cosmos* 2010, 293, no. 258 // 3rd end - early 5th c.

247. (ΜΘ 10921) // Ramona Area. // Complete with fractures // Translucent greenish. Few pinprick bubbles and impurities // H.: 12.5, Diam. rim: 4.7, Diam. Neck: 2.4, Diam. Base: 5.5, Th.: 0.06, Weight: 80.12 // Petsas 1974, 329, pl. 24 β. *Glass Cosmos* 2010, 291, no. 253 // 3rd end - early 5th c.

248. (ΜΘ 10923) // Ramona Area. // Complete, mended // Translucent greenish. Few impurities, few pinprick bubbles // H.: 16, Diam. rim: 5, Diam. Neck: 3, Diam. Base: 5, Th.: 0.06, Weight: 61.89 // Petsas 1974, 329, pl. 24 γ. *Glass Cosmos* 2010, 289, no. 247 // 3rd end - early 5th c.

249. (ΜΘ 20917) // Ramona Area. 2-3-1972 // Almost complete, mended, parts of the body are missing // Translucent greenish. // H.: 22.8, Diam. Max.: 12, Diam. rim: 5.6, Diam. Neck: 2.7, Diam. Base: 6.35, Th.: 0.08, Weight: 168 // Unpublished // 3rd end - early 5th c.

250. (ΜΘ 20995) // Plot property of Tsirakoglou. 1940 // Rim and part of the neck are missing. Covered by a hart crust // Translucent greenish. // H.: 11.4, Diam. Max.: 8.8, Diam. Neck: 2.2, Diam. Base: 4.1, Th.: 0.15, Weight: 58.3 // Walter 1940, fig. 35. Kotzias 1953-54, 167-175. // 3rd end - early 5th c.

251. (ΜΘ 20996) // Plot property of Tsirakoglou. // Rim and part of the neck are missing. Covered by a hart crust // Translucent greenish. // H.: 12, Diam. Max.: 7.8, Diam. Neck: 1.75, Diam. Base: 3.5, Th.: 0.15, Weight: 51.2 // Walter 1940, fig. 35. Kotzias 1953-54, 167-175. // 3rd end - early 5th c.
252. (ΜΘ 21000) // Unknown street, probably Vardari Square area. In a box signed 'unregistered, probably from Vardaris Cemetery'. // Parts missing, mended // Translucent greenish. No impurities few pinprick bubbles // H.: 12.3, Diam. Max.: 8.3, Diam. rim: 3.9, Diam. Neck: 2.3, Diam. Base: 3.8, Th.: 0.05, Weight: 51.7 // Unpublished // 3rd end - early 5th c.
253. (ΜΘ 21390) // Junction of Margaropoulou Street and Fryzis Street 2. Angelidis plot. Grave 11, K1. 16-10-98 // Intact // Translucent greenish. No impurities few pinprick bubbles // H.: 16, Diam. Max.: 11, Diam. rim: 4.4, Diam. Neck: 2.4, Diam. Base: 6, Th.: 0.08, Weight: 113.6 // Unpublished. From a grave with a coin of Constantine I' // 4th c.
254. (BY 114) // Unknown exact provenance. // Mended. Parts of the body are missing // Translucent greenish. No impurities, several pinprick bubbles // H.: 13, Weight: 79.1 // E. Kourkoutidou-Nikolaidou, Thessaloniki Istoria kai techni 1986, 58 // 3rd end - early 5th c.
255. (BY 210/1) // Junction of Aristatou, Panagopoulou and Patroklou Streets. Cist grave II. 20-5-1991 // Complete, mended // Translucent greenish. // H.: 15.5, Diam. Max.: 9.2, Diam. rim: 5.5, Diam. Neck: 2.5, Diam. Base: 5.5, Th.: 0.1 // Makropoulou 1992b, 426-428, fig. 15, noted but not illustrated. From the fill of the burial complex dated to the second half of the 4th c. // 3rd end - early 5th c.
256. (BY 227/29) // Northern part of Evangelistrias Cemetery, area of former Lyssiatreion Clinic. Grave 194. 12-4-1994 // Part of the rim is missing // Translucent greenish. // H.: 18, Diam. rim: 5, Diam. Base: 5.5 // Unpublished // 3rd end - early 5th c.
257. (BY 238/3) // Louloudies, Kitros; // // Translucent greenish. // H.: 19.5, Diam. Max.: 10, Diam. rim: 5.2, Diam. Neck: 2.5, Diam. Base: 7, Th.: 0.15 // Unpublished // 3rd end - early 5th c.
258. (BY 317/5) // Kallitheas Avenue, Elementary School of Ampelokipi. Grave T. 52, burial layer. 30-9-1997 // Mended, parts of the body are missing // Translucent greenish. Few pinprick bubbles no impurities // H.: 15.2, Diam. Max.: 9, Diam. rim: 4.7, Diam. Neck: 3, Diam. Base: 6, Th.: 0.1, Weight: 74.8 // Makropoulou 2007, 310, pl. 24, from a grave dated to the first half of the 4th c. by the pottery finds // 4th c.
259. (BY 321/2) // Agiou Nestoros Street, 26-28. Water main works. Grave T. 40, near the right thigh. 6-10-2000 // Intact // Translucent bluish. Few impurities several pinprick bubbles // H.: 16.3, Diam. Max.: 12, Diam. rim: 5.6, Diam. Neck: 2, Diam. Base: 5.5, Th.: 0.1 // Papanikola-Bakirtzi 2002 (ed.), no. 764, 556, dated to the 4th c. // 4th c.
260. (ΔΕΘ 57) // Area of the Thessaloniki International Fair , Southern Plot. Grave 57. 1993 // Part of the neck and the rim are missing // Translucent greenish. No impurities few pinprick bubbles // Diam. Max.: 9.2, Diam. rim: 6, Diam. Neck: 2.3, Diam. Base: 5, Pres. H.: 10.5. Th.: 0.1 // Unpublished // 3rd end - early 5th c.
261. (ΔΕΘ 92) // Area of the Thessaloniki International Fair, Southern Plot. Draining pipe 92. 1993 // Part of the rim missing // Translucent greenish. No impurities few pinprick bubbles // H.: 15, Diam. Max.: 9.2, Diam. rim: 5.5, Diam. Neck: 2.5, Diam. Base: 5, Th.: 0.1 // Unpublished // 3rd end - early 5th c.
262. (ΜΘ 3225) // Ramona Area. Grave 19. 24-9-1963 // Neck, base and small fragments of the body are preserved // Translucent greenish. // H.: 16, Diam. Max.: 12, Diam. rim: 5.2, Diam. Neck: 2.7, Diam. Base: 6.5, Th.: 0.08, Weight: 95.2 // Unpublished // 3rd end - early 5th c.
263. (ΜΘ 3755) // Unknown exact provenance. // Part of the body is missing // Translucent olive-green. Few impurities // H.: 15, Diam. rim: 3.2, Diam. Neck: 2.5, Diam. Base: 4.5, Th.: 0.08, Weight: 44.08 // *Glass Cosmos* 2010, 289, no. 246 // 3rd end - early 5th c.
264. (ΜΘ 5937) // AUTH Campus. Theological Department Building. Grave (VIII) barrel-vaulted. Phase α'. 1965 // Intact // Translucent greenish. // H.: 13, Diam. Max.: 10, Diam. rim: 4.1, Diam. Neck: 2.5, Diam. Base: 5, Th.: 0.08, Weight: 82.38 // Unpublished // 3rd end - early 5th c.
265. (ΜΘ 10802) // Ramona Area, Axios complex. Grave 7. 18-12-1968 // In large fragments. Large parts of the body are missing // Translucent greenish. No impurities, several pinprick bubbles // H.: 14, Diam. Max.: 10, Diam. rim: 5.6, Diam. Neck: 2.5, Diam. Base: 6, Th.: 0.08 // Unpublished // 3rd end - early 5th c.
266. (ΜΘ 10922) // Ramona Area. // Almost complete, mended. Parts of the body are missing // Translucent greenish. No impurities few pinprick bubbles // H.: 17, Diam. rim: 5.7, Diam. Neck: 3.5, Diam. Base: 6, Th.: 0.08, Weight: 187.71 // Petsas 1974, 329, pl. 24 γ. // 3rd end - early 5th c.
267. (ΜΘ 20915) // Ramona Area. Built grave. 14-10-1969 // Mendable fragments // Translucent greenish. No impurities few pinprick bubbles // H.: 15.2, Diam. Max.: 12, Diam. rim: 5.5, Diam. Neck: 2.7, Diam. Base: 5.5, Th.: 0.06, Weight: 95.8 // Unpublished // 3rd end - early 5th c.
268. (ΜΘ 20920) // Ramona Area. With a tag: '2', probably number of a grave. 2-3-1972 // Intact // Translucent olive-green. No impurities few pinprick and larger bubbles // H.: 14.4, Diam. Max.: 10, Diam. rim: 5.8, Diam. Neck: 2.9, Diam. Base: 5.5, Th.: 0.06, Weight: 107 // Unpublished // 3rd end - early 5th c.
269. (ΜΘ 20921) // Ramona Area. With a tag '1', probably number of the grave. 2-3-1972 // Almost complete, mended. Small part of the body is missing // Translucent greenish. // H.: 14.3, Diam. Max.: 10.1, Diam. rim: 4.6, Diam. Neck: 2.2, Diam. Base: 4.8, Th.: 0.06, Weight: 87.6 // *Glass Cosmos* 2010, 289-290, no. 248 // 3rd end - early 5th c.
270. (ΠΑΜ 9) // Ramona Area. 1971 // Fragments // Translucent greenish. No impurities, several pinprick bubbles // H.: 17, Diam. Max.: 10, Diam. rim: 5, Diam. Neck: 2.2, Diam. Base: 5, Th.: 0.12 // Unpublished // 3rd end - early 5th c.
271. (ΠΑΜ 48) // Ramona Area. 1971 // Large fragments // Translucent greenish. No impurities few pinprick bubbles // H.: 15.2, Diam. Max.: 9, Diam. rim: 5.5, Diam. Neck: 2.7, Diam. Base: 6, Th.: 0.1 // Unpublished // 3rd end - early 5th c.
272. (ΠΑΜ 55) // Vardari Square. 1970 // Shoulder fragment // Translucent greenish. No impurities few pinprick bubbles // Dimensions: 7X4, Th.: 0.1 // Unpublished // 3rd end - early 5th c.
273. (ΠΑΜ 78) // Giantison Street. Clinics of Veterinary Faculty of AUTH. // Fragments // Translucent greenish. Few impurities, several pinprick bubbles // H.: 15.5, Diam. Max.: 10, Diam. rim: 5.2, Diam. Neck: 2.5, Diam. Base: 4, Th.: 0.15 // Unpublished // 3rd end - early 5th c.
274. (BY 207/2) // Lagkada Street, Ivanof Park. Grave 11. 1991 // Intact // Translucent greenish. // H.: 17, Diam. rim: 5.1, Diam. Base: 11 // Makropoulou 1991, 257-270, photo 19, grave dated to the end of the 3rd- early 4th c. by its context // 3rd end - early 5th c.
275. (BY 207/7) // Lagkada Street, Ivanof Park. Grave XXIII. 1991 // Mended // Translucent greenish. // H.: 17, Diam. Max.: 9.5, Diam. rim: 6.7, Diam. Neck: 3, Diam. Base: 8, Th.: 0.1, Weight: 75 // Makropoulou 2007, 54, pl. 25. The grave, which was used as ossuary, is dated by pottery to the first half of the 4th c. // 4th c.
276. (BY 227/30) // Northern part of Evangelistrias Cemetery, area of former Lyssiatreion Clinic. 1994 // Intact // Translucent greenish. // H.: 17.3, Diam. rim: 5, Diam. Base: 7.5 // Pelekanidou 1993, 376, photo 5 // 3rd end - early 5th c.
277. (BY 227/50) // Northern part of Evangelistrias Cemetery, area of former Lyssiatreion Clinic. Grave 215. 6-12-1994 // Mended // Translucent greenish. // H.: 17, Diam. rim: 5, Diam. Neck: 3, Diam. Base: 9, Th.: 0.1 // Unpublished. From a grave with coins of Constantine I' // 4th c.
278. (BY 244/6) // Stratou Avenue 2, plot of the Museum of Byzantine Culture. Grave 72. 28-6-1989 // Parts of the body are missing // Translucent greenish. No impurities few pinprick bubbles // H.: 15, Diam. rim: 4.8, Diam. Base: 8, Th.: 0.1 // Nalpantis 2003, 67, 128, drawing 36, pl. 43, from a grave with coins of Diocletian, minted around 293-95 and Licinius, minted around 312-17, dating the grave probably at the second decade of the 4th c. or even later // 4th c.
279. (BY 244/8) // Stratou Avenue 2, plot of the Museum of Byzantine Culture. Stray find. 29-5-1989 // Fragments // Translucent greenish. No impurities few pinprick bubbles // Diam. rim: 2.4, Pres. H.: 6.5 Th.: 0.1 // Nalpantis 2003, 128, drawing, possibly dated to the 3rd-4th

c. according to an analogous find from the plot at the junction of Koloniaris, Galanaki and Papathanasiou Streets which was found with coin minted around 335-337, βλ. Nalpantis 2003, 129 note. 267. // 4th c.

280. (BY 244/9) // Stratou Avenue 2, plot of the Museum of Byzantine Culture. 29-5-1989 // Fragments. Large part of the body is missing // Translucent greenish. No impurities, few pinprick bubbles // Diam. rim: 5, Pres. H.: 7.3 Th.: 0.1 // Nalpantis 2003, 56, 128, draw. 24, pl. 43, probably dated to the 3rd-4th c. // 3rd end - early 5th c.

281. (BY 247/1) // Regkoukou Street, 16. Grave 16. 2-3-1994 // Intact // Translucent greenish. No impurities, few pinprick bubbles // H.: 25.7, Diam. Max.: 17, Diam. rim: 7, Diam. Base: 14, Th.: 0.13 // Makropoulou 2007, 241, pl. 23, 68, grave and burial are dated to the first half of the 4th c. by the pottery // 4th c.

282. (BY 310/1) // Junction of Kallitheas and Skeparni Street. Grave T. 18. 18-12-1996 // Mended // Translucent greenish. // H.: 16, Diam. Max.: 9.5, Diam. rim: 4.2, Diam. Neck: 2.5, Diam. Base: 8, Th.: 0.12 // Makropoulou 2007, 303, pl. 23, from a grave dated to the first half of the 4th century // 4th c.

283. (MΘ 530) // Vardari Square. // In large fragments // Translucent greenish. Few impurities, few pinprick bubbles // H.: 17, Diam. Max.: 10, Diam. rim: 5, Diam. Neck: 3, Diam. Base: 9, Th.: 0.08, Weight: 116 // Unpublished // 3rd end - early 5th c.

284. (MΘ 533) // Vardari Square. // Broken in neck and rim // Translucent greenish. // H.: 14, Diam. Max.: 10, Diam. rim: 4.6, Diam. Neck: 2.4, Diam. Base: 8, Th.: 0.09, Weight: 78.59 // Unpublished // 3rd end - early 5th c.

285. (MΘ 535) // Vardari Square. // Part of the neck and the mouth are missing // Translucent greenish. Few impurities several pinprick bubbles, some up to 1X0.8 // H.: 13, Diam. Max.: 10, Diam. rim: 5, Diam. Neck: 2.3, Diam. Base: 7, Th.: 0.1, Weight: 91.22 // Unpublished // 3rd end - early 5th c.

286. (MΘ 537) // Vardari Square. // Complete, mended // Translucent greenish. No impurities, few pinprick bubbles // H.: 20, Diam. Max.: 12, Diam. rim: 5.6, Diam. Neck: 4, Diam. Base: 9.5, Th.: 0.1, Weight: 206 // *Glass Cosmos* 2010, 292, no. 254 // 3rd end - early 5th c.

287. (MΘ 5469) // Vardari Square. Found in a marble sarcophagus. 12-2-1920 // It has many fractures in the body // Brownish green. No impurities, several pinprick bubbles // H.: 17.5, Diam. Max.: 14.5, Diam. rim: 6.5, Diam. Neck: 3.3, Diam. Base: 11, Th.: 0.12, Weight: 233 // Unpublished // 3rd end - early 5th c.

288. (MΘ 5473) // Vardari Square. Found in a marble sarcophagus. 12-2-1920 // Half of the body is missing // Translucent greenish. Few impurities, few pinprick bubbles // H.: 20, Diam. Max.: 8, Diam. rim: 4.8, Diam. Neck: 2.4, Diam. Base: 7.2, Th.: 0.08, Weight: 103.19 // Unpublished // 3rd end - early 5th c.

289. (MΘ 21001) // Unknown street, probably Vardari Square area. In a box signed 'unregistered, probably from Vardaris Cemetery'. // Parts missing, mended // Translucent greenish. Few impurities, few pinprick bubbles // H.: 15.5, Diam. Max.: 9.5, Diam. rim: 5, Diam. Neck: 2.7, Diam. Base: 6.6, Th.: 0.05, Weight: 74 // Unpublished // 3rd end - early 5th c.

290. (MΘ 21002) // Unknown street, probably Vardari Square area. In a box signed 'unregistered, probably from Vardaris Cemetery'. // Parts missing, mended // Translucent greenish. Few impurities, several pinprick bubbles // H.: 14.1, Diam. Max.: 9.5, Diam. rim: 5.2, Diam. Neck: 2.5, Diam. Base: 6.3, Th.: 0.08, Weight: 92 // Unpublished // 3rd end - early 5th c.

291. (MΘ 21003) // Unknown street, probably Vardari Square area. In a box signed 'unregistered, probably from Vardaris Cemetery'. // Parts missing, mended // Translucent greenish. Few impurities, few pinprick bubbles // H.: 10.5, Diam. Max.: 8, Diam. rim: 4.8, Diam. Neck: 2.4, Weight: 38 // Unpublished // 3rd end - early 5th c.

292. (MΘ 21004) // Unknown street, probably Vardari Square area. In a box signed 'unregistered, probably from Vardaris Cemetery'. // Rim and parts of the neck and body are missing. Mended // Translucent greenish. No impurities, several pinprick bubbles and few larger ones // H.: 12, Diam. Max.: 9.4, Diam. Neck: 2.7, Diam. Base: 8.5, Th.: 0.05, Weight: 36.8 // Unpublished // 3rd end - early 5th c.

293. (MΘ 21005) // Unknown street, probably Vardari Square area. In a box signed 'unregistered, probably from Vardaris Cemetery'. // Rim and parts of the neck and body are missing. Mended // Translucent greenish. Several impurities and pinprick bubbles, some larger ones // H.: 13, Diam. Max.: 9.7, Diam. Neck: 2.6, Diam. Base: 7.7, Th.: 0.08, Weight: 112.7 // Unpublished // 3rd end - early 5th c.

294. (MΘ 21391) // Junction of Margaropoulou Street and Fryzis Street 2. Angelidis plot. Grave 26, K4. 23-10-98 // Intact // Translucent greenish. // H.: 17, Diam. Max.: 10, Diam. rim: 4.5, Diam. Neck: 2.2, Diam. Base: 7, Th.: 0.08, Weight: 129.6 // Unpublished. From a grave with a coin of Constantine I. *Glass Cosmos* 2010, 281, no. 227 // 4th c.

295. (ΠΑΜ 4) // Ramona Area, Lagkada Street, tall, twelve-storey tower. Built grave. 9-9-1971 // Intact // Translucent greenish. Few impurities, few pinprick bubbles // H.: 14.7, Diam. Max.: 9.5, Diam. rim: 5.5, Diam. Neck: 3.2, Diam. Base: 7.5, Th.: 0.1 // Unpublished // 3rd end - early 5th c.

296. (ΠΑΜ 5) // Ramona Area. 1971 // Fragmentary in large pieces // Translucent greenish. Few impurities, few pinprick bubbles // H.: 17, Diam. Max.: 10, Diam. rim: 5, Diam. Neck: 2.5, Diam. Base: 8, Th.: 0.1 // Unpublished // 3rd end - early 5th c.

297. (BY 244/2) // Stratou Avenue 2, plot of the Museum of Byzantine Culture. Grave 35. 14-6-1989 // Intact // Translucent greenish. No impurities, several pinprick bubbles // H.: 14.5, Diam. rim: 5.5, Diam. Base: 5, Th.: 0.1 // Nalpantis 2003, 48, 128, fig. 17, pl. 43, from a grave with coins dated to 330-340, dating the grave to the second quarter of the 4th c. // 4th c.

298. (BY 244/5) // Stratou Avenue 2, plot of the Museum of Byzantine Culture. Grave 35. 13-6-1989 // The neck and the rim are missing // Translucent greenish. No impurities, few pinprick bubbles // W. 5.1, Pres. H.: 9.1, Th.: 0.1 // Nalpantis 2003, 48, 128, fig. 18, pl. 43 from a grave with coins date to 330-340, dating the grave to the second quarter of the 4th c. // 4th c.

299. (MΘ 532) // Vardari Square. // Part of the body is missing // Translucent greenish. // H.: 10, Diam. Max.: 10, Diam. rim: 5.4, Diam. Neck: 2.7, Diam. Base: 8, Th.: 0.06, Weight: 87.6 // E. Trakosopoulou-Salakidou, Thessaloniki apo ta proïstorika mechri ta christianika chronia 1986, 125. *Glass Cosmos* 2010, 291, no. 254 // 3rd end - early 5th c.

300. (MΘ 20999) // Unknown street, probably Vardari Square area. In a box signed 'unregistered, probably from Vardaris Cemetery'. // Parts missing, mended // Translucent greenish. No impurities or pinprick bubbles // H.: 13.3, Diam. Max.: 8.9, Diam. rim: 4.8, Diam. Neck: 2.7, Diam. Base: 7, Th.: 0.06, Weight: 66.2 // Unpublished // 3rd end - early 5th c.

301. (MΘ 520) // Vardari Square. // Intact // Translucent greenish. No impurities, few pinprick bubbles // H.: 7.5, Diam. Max.: 5.3, Diam. rim: 3, Diam. Neck: 2.2, Diam. Base: 3.1, Th.: 0.18 // E. Trakosopoulou-Salakidou, Thessaloniki apo ta proïstorika mechri ta christianika chronia 1986, 125. *Glass Cosmos* 2010, 276, no. 213 // 2nd - end 4th c.

302. (MΘ 12320) // Sindos, Good Year plant. In the fill of the tile grave T9. // Iridescent // Translucent greenish. // H.: 11, Diam. rim: 4 // Unpublished // 2nd - end 4th c.

303. (MΘ 19266) // Junction of Melenikou and Chatziandreou Streets. Grave 101. 10-6-1994 // Intact // Translucent bluish. Several impurities and few pinprick bubbles // H.: 8, Diam. rim: 3.3, Diam. Neck: 2, Th.: 0.1 // Tsimpidou-Auloniti, Lioutas 1994, 436, pl. 138β // 2nd - end 4th c.

304. (BY 217/1) // Junction of Koloniari, Galanaki and Papathanasiou Streets. Stray find. 2-7-1987 // Intact // Translucent greenish. No impurities, few pinprick bubbles // H.: 8.6, Diam. Max.: 6.5, Diam. rim: 4.6, Diam. Neck: 2.5, Diam. Base: 4, Th.: 0.12, Weight: 47.7 // Papanikola-Bakirtzi (ed.) 2002, no. 809 574, 3rd-4th c., from a cemetery used in the 4th c., and some of the built-graves again in the 5th c. // 3rd -4th c.

305. (Θ 1/01-Υ1 ΑΓΚ 7216) // Thiseos Street, 2. Grave 4, burial 1 and reburial of bones 1 (depth -4.45. 1.37 from the SE corner T. 4 and 1.40 from NE corner T. 4). // Small part of the body is missing // Translucent greenish. Very few pinprick bubbles, no impurities // H.: 11.4, Diam. Max.: 7.3, Diam. rim: 5.5, Diam. Neck: 2.2, Diam. Base: 5, Th.: 0.12 // Unpublished // 2nd - end 4th c.

306. (Θ 18/01 Y1) // N. Kapatou street, 4. Grave 25 (Vessel Y1, depth -1.40/-2.10. Undisturbed vessel). 24-8-01 // Large part of it is preserved in small fragments // Translucent greenish. No impurities, few pinprick bubbles // H.: 13.2, Diam. Max.: 8, Diam. rim: 5.2, Diam. Neck: 2.1, Diam. Base: 5, Th.: 0.1 // Unpublished // 2nd - end 4th c.

307. (ΜΘ 517) // Vardari Square. // Intact // Translucent greenish and dark-blue. Few impurities, few pinprick bubbles // H.: 10.2, Diam. Max.: 7.5, Diam. rim: 4.6, Diam. Neck: 2.5, Diam. Base: 4.3, Th.: 0.3 // E. Trakosopoulou-Salakidou, Thessaloniki apo ta proïstorika mechri ta christianika chronia 1986, 125, fig. 123. *Glass Cosmos* 2010, 275, no. 210 // 2nd - end 4th c.

308. (ΜΘ 518) // Vardari Square. // Intact // Translucent greenish. No impurities or pinprick bubbles // H.: 10, Diam. Max.: 6.6, Diam. rim: 4.4, Diam. Neck: 2, Diam. Base: 3.5, Th.: 0.28 // E. Trakosopoulou-Salakidou, Thessaloniki apo ta proïstorika mechri ta christianika chronia 1986, 125. *Glass Cosmos* 2010, 275, no. 211 // 2nd - end 4th c.

309. (ΜΘ 519) // Vardari Square. // Intact // Translucent greenish. Impurities only on the spout. Few pinprick bubbles on the body // H.: 11, Diam. Max.: 6, Diam. rim: 4, Diam. Neck: 2, Diam. Base: 4.6, Th.: 0.1, Weight: 43.03 // *Glass Cosmos* 2010, 276, no. 212 // 2nd - end 4th c.

310. (ΜΘ 5481) // Unknown provenance. // Rim and neck are missing // Translucent olive-green. // H.: 6, Diam. Max.: 7, Diam. Neck: 2.5, Diam. Base: 5, Th.: 0.08, Weight: 37.67 // Unpublished // 2nd - end 4th c.

311. (ΜΘ 20871) // Area of the Thessaloniki International Fair. Grave 1, next to the disintegrated baby bones. 4-4-1979 // Upper part of the neck and the rim are missing // Translucent yellowish and blue. // H.: 7.8, Diam. Max.: 6.4, Diam. Neck: 2.2, Diam. Base: 3.3 , Weight: 24.78 // Unpublished // 2nd - end 4th c.

312. (ΠΑΜ 49) // Ramona Area. 1971 // Only the spout is preserved // Translucent yellowish. No impurities or pinprick bubbles // Dimensions: 6.5X3 // Unpublished // 2nd - end 4th c.

313. (ΜΘ 21456 β) // Farmhouse A. 2000 // Fragment // Translucent greenish. ; // // Adam-Veleni, Poulaki, Tzanavari 2003, 212 no. 193 (K. Tzanavari). *Glass Cosmos* 2010, 339, no. 352 // 1st end - early 2nd c.

314. (ΒΥ 4) // AUTH Campus. Law School Building. // Part of the neck and the rim are missing // Translucent greenish. // H.: 14, Weight: 56.6 // E. Kourkoutidou-Nikolaidou, Thessaloniki Istoria kai techni 1986, 59 // 2nd - 3rd c.

315. (ΜΘ 509) // Vardari Square. // Parts of the body are missing // Translucent greenish. No impurities, few pinprick bubbles // H.: 17.2, Diam. Max.: 9.5, Diam. rim: 5.7, Diam. Neck: 2.7, Diam. Base: 4.2, Th.: 0.02, Weight: 42.58 // Unpublished // 2nd - 3rd c.

316. (ΒΥ 207/10) // Lagkada Street, Ivanof Park. Grave IV. 16-4-1991 // Mended, small parts of the body and rim are missing // Translucent greenish. // Diam. Max.: 8.5, Diam. Neck: 4, Diam. Base: 6, Pres. H.: 15, Th.: 0.1 // Makropoulou 2007, 46, pl. 25. Burial dated by pottery to the first half of the 4th c. // 4th c.

317. (ΜΘ 5482) // Unknown provenance. // Complete, filled // Translucent greenish. No impurities, few pinprick bubbles // H.: 9, Diam. Max.: 5, Diam. rim: 3.3, Diam. Neck: 2, Diam. Base: 3.3, Th.: 0.08, Weight: 23.19 // Unpublished // 3rd end - 5th c.

318. (ΜΘ 20918) // Unknown exact provenance. // Almost complete, mended. Small part of the body is missing // Translucent slightly greenish. No impurities, few pinprick bubbles // H.: 13.9, Diam. Max.: 8.2, Diam. rim: 2.4, Diam. Neck: 1.1, Diam. Base: 4, Th.: 0.06, Weight: 46.9 // Unpublished // 4th c.

319. (Θ 18/01 Y2) // N. Kapatou street, 4. Grave 37, (vessel Y 2, inner fill of the open burial. depth -2.25). 29-8-01 // Broken in large pieces // Translucent olive-greenish. No impurities or pinprick bubbles // H.: 13, Diam. Max.: 8.6, Diam. rim: 5.5, Diam. Neck: 2.2, Diam. Base: 6.5, Th.: 0.12 // Unpublished // 4th-5th c.

320. (ΜΘ 3074) // Ramona Area. Burial cluster I. 8-1963 // Fragments. Large part of the body is missing. Iridescent // Translucent olive-green. No impurities, few pinprick bubbles // H.: 11.1, Diam. Max.: 7, Diam. rim: 4.1, Diam. Neck: 2.3, Diam. Base: 5.8, Th.: 0.08 // *Glass Cosmos* 2010, 281-282, no. 228 // 4th-5th c.

321. (ΒΥ 217/5) // Junction of Koloniari, Galanaki and Papathanasiou Streets. Grave 1, eastern side of the grave. 4-6-1987 // Intact // Translucent greenish. No impurities, few pinprick bubbles // H.: 14.4, Diam. rim: 3.8, Diam. Base: 3.5 // Unpublished. Unpublished. From a cemetery used in the 4th c. Some of the built-graves were re-used in the 5th c. In the same grave were found 3 flasks Is. 101min // 4th-5th c.

322. (ΜΘ 5479) // Unknown provenance. // Rim and parts of the neck are missing // Translucent olive-green yellowish. // H.: 12, Diam. Max.: 4, Diam. Neck: 1.5, Diam. Base: 3, Th.: 0.08, Weight: 18.21 // Unpublished // 4th-5th c.

323. (ΔΕΘ 115) // Area of the Thessaloniki International Fair, Southern Plot. Grave 115. 1993 // Rim is missing // Translucent greenish. No impurities, few pinprick bubbles // Diam. Max.: 6.2, Diam. Neck: 2.4, Diam. Base: 4.2, Pres. H.: 14, Th.: 0.1 // Unpublished // 4th c.

324. (Av-Pi /1) // Municipal Hospital, 'Agios Dimitrios', plot of the Water Company. // Intact // Silver grey. Non vidi // H.: 6.5, Diam. Max.: 4 // Avezou, Picard 1912, 337-361 // 3rd end - early 5th c.

325. (ΒΥ 213/2) // Municipal Hospital, 'Agios Dimitrios'. // Large part of the body is missing // Translucent greenish. // H.: 9.5, Diam. Max.: 7.5, Diam. rim: 4.2, Diam. Neck: 1.6, Diam. Base: 4.2, Th.: 0.1 // Unpublished // 3rd end - early 5th c.

326. (ΒΥ 244/3) // Stratou Avenue 2, plot of the Museum of Byzantine Culture. Grave 93. 19-10-1992 // Complete, mended // Translucent greenish. No impurities, few pinprick bubbles // H.: 12, Diam. Max.: 9.5, Diam. rim: 4, Diam. Base: 3, Th.: 0.1 // Nalpantis 2003, 86, 129, drawing 50, pl. 43, from a grave with coins dated between 267 and 282, which date the grave to the last quarter of the 3rd c. // end of the 3rd c.

327. (ΒΥ 255/7) // Margaropoulou Street, 20. Grave 111. 3-7-1995 // Complete, mended // Translucent greenish. // H.: 13.5, Diam. Max.: 9, Diam. rim: 5.5, Diam. Base: 4, Th.: 0.1, Weight: 49.6 // Makropoulou 2007, 214, pl. 25. From a late 3rd-century burial, dated by a coin of Probus, struck between 276 and 282. // end 3rd c.

328. (ΜΘ 5470) // Vardari Square. Found in a marble sarcophagus. 12-2-1920 // Missing parts of the body and the neck // Translucent greenish. // H.: 13, Diam. Max.: 10.5, Diam. rim: 5, Diam. Neck: 3, Diam. Base: 4, Th.: 0.06, Weight: 43.11 // Unpublished // 3rd end - early 5th c.

329. (ΜΘ 5475) // Vardari Square. Found in a marble sarcophagus. 12-2-1920 // Complete, mended and filled // Translucent greenish. No impurities or pinprick bubbles // H.: 11, Diam. Max.: 8, Diam. rim: 3.9, Diam. Neck: 2, Diam. Base: 2.7, Th.: 0.1, Weight: 94.98 // Unpublished // 3rd end - early 5th c.

330. (ΜΘ 5477) // Unknown provenance. // Broken rim // Translucent greenish. // H.: 8, Diam. Max.: 5, Diam. Neck: 2.2, Diam. Base: 2.5, Th.: 0.08, Weight: 20.03 // Unpublished // 3rd end - early 5th c.

331. (ΜΘ 5485) // Unknown provenance. // Body and neck fragment // Translucent greenish. // H.: 4, Diam. Max.: 6, Diam. rim: 6.4, Diam. Neck: 3.5, Th.: 0.17, Weight: 23.15 // Unpublished // 3rd end - early 5th c.

332. (ΜΘ 5486) // Unknown provenance. // Upper part of the vessel // Translucent olive-green. // H.: 4, Diam. rim: 6.4, Diam. Neck: 4.5, Weight: 28.4 // Unpublished // 3rd end - early 5th c.

333. (ΜΘ 5487) // Unknown provenance. // Lower body part is missing // Translucent greenish. // H.: 6, Diam. Max.: 6.6, Diam. rim: 4, Diam. Neck: 1.8, Th.: 0.06, Weight: 18.75 // Unpublished // 3rd end - early 5th c.

334. (ΜΘ 5488) // Unknown provenance. // Body and neck fragment // Translucent greenish. // H.: 4, Diam. Max.: 7.2, Diam. Neck: 2.5, Th.: 0.06, Weight: 8.74 // Unpublished // 3rd end - early 5th c.

335. (ΜΘ 11141) // Ramona Area. Grave 80ς. 8-4-1969 // Broken rim // Translucent greenish. // H.: 12, Diam. rim: 4.8, Diam. Neck: 2.2, Diam. Base: 4, Th.: 0.08, Weight: 48.73 // Petsas 1974, 329, pl. 25 ε. *Glass Cosmos* 2010, 288, no. 244 // 3rd end - early 5th c.

336. (ΠΑΜ 45) // Ramona Area. 27-3-1969 // Parts of the body are missing // Translucent greenish. No impurities or pinprick bubbles // Diam. Max.: 10.5, Diam. rim: 5.4, Diam. Neck: 2.1, Diam. Base: 6, Th.: 0.1 // Unpublished // 3rd end - early 5th c.

337. (ΜΘ 5483) // Unknown provenance. // Mended, large parts of the body are missing // Translucent greenish. // H.: 13, Diam. rim: 5.6, Diam. Neck: 2.6, Diam. Base: 5.5, Th.: 0.06, Weight: 29.7 // Unpublished // 3rd end - early 5th c.

338. (ΒΥ 91 α) // Agios Pavlos Aquaduct. Room B, grave A,B. December 1966 // Intact // Translucent greenish. // H.: 22 // E. Kourkoutidou-Nikolaidou, Thessaloniki Istoria kai techni 1986, 58 // 3rd-4th c.

339. (ΒΥ 174) // AUTH Campus. Theological Department Building. // Intact // Translucent greenish. Several pinprick bubbles, few impurities // H.: 20, Diam. Max.: 10.8, Diam. rim: 5.8, Diam. Neck: 2.2, Diam. Base: 6, Th.: 0.18, Weight: 134.6 // E. Kourkoutidou-Nikolaidou, Thessaloniki Istoria kai techni 1986, 59. Papanikola-Bakirtzi (ed.) 2002, no. 349 318 (A. Antonaras), dated to the 4th c. // 4th c.

340. (ΜΘ 5942) // AUTH Campus. Theological Department Building. 1965 // Complete, mended // Translucent greenish. No impurities, few pinprick bubbles // H.: 17, Diam. rim: 5.7, Diam. Base: 4.5, Th.: 0.1, Weight: 115.29 // Petsas 1966, 334-339, pl. 351 γ. *Glass Cosmos* 2010, 274, no. 208 // 3rd-4th c.

341. (ΜΘ 10924) // Ramona Area. // Part of the rim and neck are missing // Translucent greenish. No impurities or pinprick bubbles // H.: 15.5, Diam. rim: 5, Diam. Neck: 2, Diam. Base: 4, Th.: 0.15, Weight: 69.29 // Petsas 1974, 329, pl. 24 β. // 3rd-4th c.

342. (ΜΘ 11267) // Ramona Area. Building 35, tile grave. // Mouth and part of the neck are missing // Translucent greenish. // H.: 12, Diam. Max.: 9.5, Diam. Neck: 2.5, Diam. Base: 5, Th.: 0.1, Weight: 78.4 // Unpublished // 3rd-4th c.

343. (ΜΘ 20872) // Area of the Thessaloniki International Fair. Grave 1. 4-4-1979 // Rim and the upper part of the neck are missing // Translucent bluish. // H.: 12.6, Diam. Max.: 8.2, Diam. Neck: 4, Diam. Base: 4.2, Th.: 0.18, Weight: 80.16 // Unpublished // 3rd-4th c.

344. (ΜΘ 20916) // Ramona Area, Lagkada Street, tall, twelve-storey tower. Vaulted grave A'. 2-3-1972 // Rim and parts of the body are missing // Translucent bluish. No impurities, few pinprick bubbles // H.: 14, Diam. Max.: 8.2, Diam. Neck: 2.2, Diam. Base: 3.1, Th.: 0.08, Weight: 63.2 // Unpublished // 3rd-4th c.

345. (ΒΥ 319/5) // Lagkada Street, middle plot of Axios Complex. Grave 65, burial layer. 13-2-1998 // Mended // Translucent greenish. No impurities, few pinprick bubbles // H.: 17, Diam. Max.: 9.5, Diam. rim: 6, Diam. Neck: 3, Diam. Base: 5.5, Th.: 0.1 // Makropoulou 2007, 162, from a barrel-vaulted grave with a worn out, 4th-century coin // 4th c.

346. (ΜΘ 3753) // Unknown exact provenance. // Intact // Translucent bluish. No impurities, few pinprick bubbles // H.: 16.2, Diam. rim: 5.5, Diam. Neck: 3.8, Diam. Base: 5.5, Th.: 0.12, Weight: 153.37 // *Glass Cosmos* 2010, 288, no. 243 // 3rd end - early 5th c.

347. (ΜΘ 5941) // AUTH Campus. Theological Department Building. Grave (2ω) Tile grave, Sector 11. 6-2-1965 // Complete, mended // Translucent yellowish. // H.: 15.3, Diam. rim: 6.4, Diam. Base: 5.5, Th.: 0.08, Weight: 97.1 // Petsas 1966, 334-339, pl. 351 γ. // 3rd end - early 5th c.

348. (ΠΑΜ 42) // AUTH Campus. Swimming pool. In an envelope with single marking a 'T'. Probably a grave number. // Broken, probably complete // Translucent bluish. No impurities, few pinprick bubbles // H.: 13, Diam. Max.: 6, Diam. rim: 6, Diam. Neck: 2.8, Diam. Base: 4.5, Th.: 0.1 // Tsigaridas 1972, 563, where a simple note to the vessel is given // 3rd end - early 5th c.

349. (Θ 18/01-Υ4 ΑΓΚ 7253) // N. Kapatou street, 4. Grave 1 cist grave (Vessel Y4, depth -2.68. On the floor 0.30 m. from NE corner of the cist grave). 29-8-01 // Intact // Translucent greenish. Several pinprick bubbles, no impurities // H.: 15.5, Diam. Max.: 9.5, Diam. rim: 4.5, Diam. Neck: 2.8, Diam. Base: 5.5, Th.: 0.14 // Unpublished // 3rd end - early 5th c.

350. (ΒΥ 319/4) // Lagkada Street, middle plot of Axios Complex. Grave 65, burial layer. 13-2-1998 // Mended // Translucent greenish and bluish. Few pinprick bubbles no impurities // H.: 9, Diam. Max.: 7.3, Diam. rim: 5.5, Diam. Base: 3.2, Th.: 0.15 // Makropoulou 2007, 162, pl. 25, from a barrel-vaulted grave with a worn out, 4th century coin // 4th c.

351. (ΠΑΜ 43) // Junction of Apolloniados 31 and Michalitsiou Streets. Grave 5. 24-10-1970 // Large parts of the body are missing // Translucent greenish. No impurities or pinprick bubbles // Diam. rim: 10, Diam. Neck: 3.2, Diam. Base: 6, Th.: 0.12 // Unpublished // 3rd-4th c.

352. (ΜΘ 539) // Vardari Square. // Part of the neck is missing // Translucent greenish. No impurities, few pinprick bubbles. Clear // H.: 12, Diam. Max.: 8, Diam. rim: 4.1, Diam. Neck: 2.5, Diam. Base: 5.5, Th.: 0.12, Weight: 64.32 // *Glass Cosmos* 2010, 283-284, no. 233 // 4th c.

353. (ΒΥ 317/2) // Kallitheas Avenue, Elementary School of Ampelokipi. Grave T. 46, burial layer. 5-9-1997 // Mended // Translucent greenish. Few pinprick bubbles and impurities // H.: 13, Diam. Max.: 9, Diam. rim: 3.3, Diam. Base: 4.4, Th.: 0.1, Weight: 36.6 // Makropoulou 2007, 308, from a 4th-century grave // 4th c.

354. (ΒΥ 227/1) // Northern part of Evangelistrias Cemetery, area of former Lyssiatreion Clinic. // Intact // Translucent greenish. // H.: 14.5, Diam. rim: 3.7, Diam. Base: 4, Th.: 0.08 // Pelekanidou 1993, 376, photo 5, 6. // 2nd end - 3rd c.

355. (ΒΥ 27) // Agiou Dimitriou Street, cemetery found at the extension of the street. // Large part of the base and the neck are missing // Translucent olive-green. // H.: 18, Weight: 97.6 // E. Kourkoutidou-Nikolaidou, Thessaloniki Istoria kai techni 1986, 59 // 4th c.

356. (ΒΥ 229/1) // Lagkada Street. Grave 23. 5-2-1990 // Almost complete, mended // Translucent olive-green. // H.: 20, Diam. Max.: 9, Diam. rim: 6, Diam. Neck: 2.8, Diam. Base: 5, Th.: 0.2 // Makropoulou 2007, 57-58, the burial is dated by a worn coin to the end of the 3rd c., by the pottery to the early 4th, and by the glass find to the middle of the 4th c. // 4th c.

357. (ΒΥ 247/2) // Regkoukou Street, 16. Grave 18. 14-1-1994 // Intact // Translucent olive-green. // H.: 22, Diam. Max.: 9, Diam. rim: 6.8, Diam. Neck: 2.7, Diam. Base: 6.3, Th.: 0.2 // Makropoulou 1994, 509-511, wall-painted grave dated probably to the second half of the 4th c. see Makropoulou 2007, 242, pls. 26, 68. // 4th c.

358. (ΒΥ 273/1) // Lagkada Street, Southern plot of Axios Complex. Grave 1. 20-9-1995 // Intact // Translucent olive-green. // H.: 20.5, Diam. Max.: 8, Diam. rim: 6.9, Diam. Base: 5.5, Th.: 0.2 // Makropoulou 2007, 108, pls. 26, 68, from a group of graves dated after the end of the 4th c. // 4th-5th c.

359. (ΒΥ 309/1) // Peleiadou Street, 5. Grave 9, tile grave. // The base is missing // Translucent olive-greenish. No impurities or pinprick bubbles // H.: 20, Diam. Max.: 7.8, Diam. rim: 5.8, Diam. Neck: 2.5, Diam. Base: 2.6, Th.: 0.20 // Makropoulou 2007, 253, pl. 26, from a grave generally dated to the end of the 3rd or to the 4th c. // 4th c.

360. (ΔΕΘ 97) // Area of the Thessaloniki International Fair, Southern Plot. Grave 97. 1993 // Complete, mendable // Translucent greenish. No impurities, few pinprick bubbles // H.: 18, Diam. Max.: 7, Diam. rim: 5.5, Diam. Neck: 2.7, Diam. Base: 5.7, Th.: 0.2 // Unpublished // 4th c.

361. (ΜΘ 512) // Vardari Square. // Intact // Translucent olive-green. No impurities or pinprick bubbles // H.: 17.5, Diam. Max.: 8, Diam. rim: 6.4, Diam. Neck: 2.9, Diam. Base: 6.2, Th.: 0.3 // E. Trakosopoulou-Salakidou, Thessaloniki apo ta proïstorika mechri ta christianika chronia 1986, 125, fig. 124. // 4th c.

362. (ΜΘ 3760) // Agiou Dimitriou Street, cemetery found at the extension of the street. // Intact // Translucent greenish. Very few pinprick bubbles only on the neck. The base has many impurities and pinprick bubbles // H.: 20, Diam. rim: 6, Diam. Neck: 2.5, Diam. Base: 5.5, Th.: 0.15, Weight: 131.28 // Unpublished // 4th c.

363. (ΜΘ 7169) // AUTH Campus. Administration Building. 2nd Sarcophagus. 8-9-1965 // Complete, reconstructed // Translucent olive-green. // H.: 2, Diam. rim: 6.2, Diam. Neck: 3, Diam. Base: 6.5, Th.: 0.08, Weight: 155.1 // *Glass Cosmos* 2010, 285, no. 237 // 4th c.

364. (ΠΑΜ 3) // Ramona Area, Lagkada Street, tall, twelve-storey tower. Built grave. 9-9-1971 // Intact // Translucent olive-greenish. No impurities or pinprick bubbles // H.: 17.7, Diam. Max.: 7.7, Diam. rim: 5.5, Diam. Neck: 2.2, Diam. Base: 5, Th.: 0.15 // *Glass Cosmos* 2010, 287, no. 241 // 4th c.

365. (ΠΑΜ 23) // Ramona Area;. 1934 // Only the base is preserved // Translucent olive-green. No impurities, few pinprick bubbles // Diam. Max.: 5, Pres. H.: 2, Th.: 0.2 // Unpublished // 4th c.
366. (BY 321/1) // Agiou Nestoros Street, 8. Water main works. Grave 66 (burial Δ') depth 3.90. 10-11-2000 // Intact // Translucent bluish. No impurities very few pinprick bubbles // H.: 13.6, Diam. Max.: 8, Diam. rim: 3.2, Diam. Neck: 1.8, Diam. Base: 3.9, Th.: 0.1, Weight: 31.4 // Papanikola-Bakirtzi 2002 (ed.), no. 643 470, dated to the 4th c. // 3rd end - 4th c.
367. (ΜΘ 5943) // AUTH Campus. Theological Department Building. Grave (4γ) Tile grave, Sector 9. 1965 // Complete, filled // Translucent greenish. // H.: 18, Diam. rim: 4, Diam. Neck: 2.5, Diam. Base: 4.8, Th.: 0.08, Weight: 84.81 // Petsas 1966, 334-339, pl. 351 γ. // 3rd end - 4th c.
368. (ΜΘ 5944) // AUTH Campus. Theological Department Building. Grave (XV) barrel-vaulted, Sector 5. 1965 // The neck and the shoulder are missing part of the body is broken // Translucent light greenish. // H.: 12, Diam. Neck: 2.5, Diam. Base: 4.5, Th.: 0.1, Weight: 64 // Unpublished // 3rd end - 4th c.
369. (ΠΑΜ 1) // Ramona Area. From a group of finds from several tile-covered graves. 20-8-1971 // Upper part of the neck is missing // Translucent greenish. No impurities, few pinprick bubbles // H.: 13, Diam. Max.: 7, Diam. Neck: 2.5, Diam. Base: 2.7, Th.: 0.12 // Unpublished // 3rd end - 4th c.
370. (ΠΑΜ 56) // Vardari Square. 1970 // Shoulder fragment // Translucent greenish. // Diam. rim: 5X3 Th.: 0.1 // Unpublished // 3rd end - 4th c.
371. (ΠΑΜ 82) // Junction of Agiou Dimitriou Street and Agias Sophias Street, Byzantine building. 3rd triangular space. 11-6- 1969 // Fragments, base is missing // Translucent bluish. // H.: 11.5, Diam. Max.: 6.5, Diam. rim: 3.8, Diam. Neck: 2, Th.: 1 // Unpublished // 3rd end - 4th c.
372. (ΠΑΜ 83) // Junction of Agiou Dimitriou Street and Agias Sophias Street, Byzantine building. 3rd triangular space. 11-6- 1969 // Fragment // Translucent greenish. // Pres. H.: 9, Th.: 0.1 // Unpublished // 3rd end - 4th c.
373. (BY 213/1) // Municipal Hospital, 'Agios Dimitrios'. // Intact // Colourless decolourized. No impurities, few pinprick bubbles // H.: 9.5, Diam. Max.: 7, Diam. rim: 5, Diam. Neck: 4, Diam. Base: 5.8, Th.: 0.1 // Unpublished // 3rd end - 6th c.
374. (BY 217/8) // Junction of Koloniari, Galanaki and Papathanasiou Streets. Tile grave 9. 16-6-1987 // Rim is missing // Translucent greenish. No impurities, few pinprick bubbles // H.: 17, Diam. Base: 6.6 // Unpublished. From a cemetery used in the 4th c. Some of the built-graves were re-used in the 5th c. // 4th end - 5th c.
375. (BY 255/3) // Margaropoulou Street, 20. Grave 23, burial layer. 14-12-1994 // Mended, small parts of the body are missing // Translucent bluish. No impurities, few pinprick bubbles // H.: 20, Diam. Max.: 7, Diam. rim: 8.4, Diam. Neck: 2.4, Diam. Base: 6, Th.: 0.1 // Makropoulou 2007, 197, pl. 28, from a grave with a worn out 4th-century coin // 3rd middle - early 5th c.
376. (BY 297/1) // Peleiadou Street, 6. Grave 24. // Lower body part is missing // Translucent bluish and translucent greenish. No impurities, few pinprick bubbles // Diam. Max.: 7.5, Diam. rim: 7, Diam. Neck: 2.5, Pres. H.: 14 Th.: 0.1, Weight: 137 // Makropoulou 2007, 269, pls. 28, 69, from a 3rd-century grave // 3rd middle - early 5th c.
377. (ΜΘ 507) // Vardari Square. // Broken and missing part of the neck and the rim // Translucent greenish. No impurities, several pinprick bubbles // H.: 25, Diam. Max.: 8.5, Diam. rim: 8.2, Diam. Neck: 3.9, Diam. Base: 8.5, Th.: 0.3 // E. Trakosopoulou-Salakidou, Thessaloniki apo ta proïstorika mechri ta christianika chronia 1986, 125. *Glass Cosmos* 2010, 290, no. 249 // 3rd middle - early 5th c.
378. (ΜΘ 810) // Vardari Square. // Large part of the body and one of the handles is missing // Translucent greenish. Few impurities, and pinprick bubbles // H.: 16, Diam. rim: 4.7, Diam. Neck: 1.7, Diam. Base: 5.5, Th.: 0.1, Weight: 77.5 // Unpublished // 3rd middle - early 5th c.
379. (ΜΘ 5465) // Unknown exact provenance. // Broken in the lip // Translucent greenish. // H.: 24, Diam. Base: 9 // Unpublished // 3rd middle - early 5th c.
380. (ΜΘ 5466) // Palaia Pella. // Mended at the base // Translucent greenish. // H.: 20, Diam. rim: 6.4, Diam. Neck: 2.5, Diam. Base: 6.6, Th.: 0.08, Weight: 130.04 // *Glass Cosmos* 2010, 212, no. 71 // 3rd middle - early 5th c.
381. (ΜΘ 9973) // Area of the Thessaloniki International Fair, Italian Pavilion. Grave 31A, from N. side. 0.79m on the western side. // Intact // Translucent greenish. No impurities or pinprick bubbles // H.: 16, Diam. Max.: 4.6, Diam. rim: 4.2, Diam. Neck: 1.8, Diam. Base: 4, Th.: 0.1 0 // Unpublished // 3rd middle - early 5th c.
382. (BY 309/5) // Peleiadou Street, 5. Grave T. 27, burial layer. 29-9-1997 // Parts of the body are missing // Translucent greenish. Few impurities and pinprick bubbles // H.: 18.5, Diam. Max.: 5.5, Diam. rim: 5, Diam. Neck: 2.3, Diam. Base: 0.7, Th.: 0.1, Weight: 77.6 // Makropoulou 2007, 265, pl. 28, from a grave with a clay vessel dated to the first half of the 4th c. // 4th c.
383. (ΜΘ 21985) // Unknown street, probably Vardari Square area. // Base fragment // Translucent greenish. No impurities, few pinprick bubbles // Diam. Max.: 7.2, Pres. H.: 5.7 // Unpublished // 3rd middle - early 5th c.
384. (ΜΘ 23957) // Ramona Area. 1971 // Broken in the neck, one handle is missing // Translucent greenish. No impurities or pinprick bubbles // H.: 27.5, Diam. Max.: 8.8, Diam. rim: 8, Diam. Neck: 3.2, Diam. Base: 1.8, Th.: 0.1 // *Glass Cosmos* 2010, 294, no. 261 // 3rd middle - early 5th c.
385. (RA 481) // Ancient Agora Bathhouse. Rectangular room, north of the Bathhouse. Destruction Layer. 1998 // Fragment // Non vidi. // Mavromichali 2001, fig. 10. From a bath complex destroyed in the last twenty years of the 1st c., see Adam-Veleni (ed.) 2001, 86. *Glass Cosmos* 2010, 267 no. 190 // 1st c.
386. (RA 475) // Ancient Agora Bathhouse. Rectangular room, north of the Bathhouse. Destruction Layer. 1998 // Fragment // Non vidi. // Mavromichali 2001, fig. 10. From a bath complex destroyed in the last twenty years of the 1st c., see Adam-Veleni (ed.) 2001, 86. *Glass Cosmos* 2010, 268, no. 191 // 1st c.
387. (BAA 40) // Ancient Agora Bathhouse. Rectangular room, north of the Bathhouse. Destruction Layer. 1998 // Fragment // Non vidi. // Mavromichali 2001, fig. 10. From a bath complex destroyed in the last twenty years of the 1st c., see Adam-Veleni (ed.) 2001, 86. // 1st c.
388. (BAA 48) // Ancient Agora Bathhouse. Rectangular room, north of the Bathhouse. Destruction Layer. 1998 // Fragment // Non vidi. // Mavromichali 2001, fig. 10. From a bath complex destroyed in the last twenty years of the 1st c., see Adam-Veleni (ed.) 2001, 86. // 1st c.
389. (BAA 49) // Ancient Agora Bathhouse. Rectangular room, north of the Bathhouse. Destruction Layer. 1998 // Fragment // Non vidi. // Mavromichali 2001, fig. 10. From a bath complex destroyed in the last twenty years of the 1st c., see Adam-Veleni (ed.) 2001, 86. // 1st c.
390. (BAA 50) // Ancient Agora Bathhouse. Rectangular room, north of the Bathhouse. Destruction Layer. 1998 // Fragment // Non vidi. // Mavromichali 2001, fig. 10. From a bath complex destroyed in the last twenty years of the 1st c., see Adam-Veleni (ed.) 2001, 86. // 1st c.
391. (BAA 51) // Ancient Agora Bathhouse. Rectangular room, north of the Bathhouse. Destruction Layer. 1998 // Fragment // Non vidi. // Mavromichali 2001, fig. 11. From a bath complex destroyed in the last twenty years of the 1st c., see Adam-Veleni (ed.) 2001, 86. // 1st c.
392. (BAA 52) // Ancient Agora Bathhouse. Rectangular room, north of the Bathhouse. Destruction Layer. 1998 // Fragment // Non vidi. // Mavromichali 2001, fig. 11. From a bath complex destroyed in the last twenty years of the 1st c., see Adam-Veleni (ed.) 2001, 86. // 1st c.
393. (BAA 53) // Ancient Agora Bathhouse. Rectangular room, north of the Bathhouse. Destruction Layer. 1998 // Fragment // Non vidi. // Mavromichali 2001, fig. 11. From a bath complex destroyed in the last twenty years of the 1st c., see Adam-Veleni (ed.) 2001, 86. // 1st c.
394. (ΜΘ ΣιντριβάνιΤ9Α2a) // Junction of Egnatia Street and K. Melenikou Street. Grave 9 A2. 1969 // Fragment // Translucent greenish. // Th.: 0.25 // Unpublished // 1st - 2nd c.
395. (ΜΘ 5490) // Unknown provenance. // Fragment // Translucent bluish. // H.: 5, Diam. Max.: 5.5, Diam. rim: 5, Diam. Neck: 3, Th.: 0.25, Weight: 73.64 // Unpublished // 1st - 2nd c.
396. (ΠΑΜ 97) // AUTH Campus. Administration Building. Stray find. 4-10-1965 // Small part of the base is preserved // Translucent greenish. // Dimensions: 4X4, Th.: 0.2 // Antonaras 2006, 413, 416. Dated on the basis of analogous finds, to the early Imperial period // 1st - 2nd c.

397. (RA 931 α-γ) // Ancient Agora Bathhouse. Rectangular room, north of the Bathhouse. Destruction Layer. 1998 // Fragments. Large part of the body is missing // Translucent greenish. // Mavromichali 2001, fig. 23. From a bath complex destroyed in the last twenty years of the 1st c., see Adam-Veleni (ed.) 2001, 86. *Glass Cosmos* 2010, 268 no. 192 // 1st c.

398. (BAA 64) // Ancient Agora Bathhouse. Rectangular room, north of the Bathhouse. Destruction Layer. 1998 // Fragment // Non vidi. // Mavromichali 2001, fig. 21. From a bath complex destroyed in the last twenty years of the 1st c., see Adam-Veleni (ed.) 2001, 86. // 1st c.

399. (BAA 65) // Ancient Agora Bathhouse. Rectangular room, north of the Bathhouse. Destruction Layer. 1998 // Fragment // Non vidi. // Mavromichali 2001, fig. 21. From a bath complex destroyed in the last twenty years of the 1st c., see Adam-Veleni (ed.) 2001, 86. // 1st c.

400. (BAA 66) // Ancient Agora Bathhouse. Rectangular room, north of the Bathhouse. Destruction Layer. 1998 // Fragment // Non vidi. // Mavromichali 2001, fig. 21. From a bath complex destroyed in the last twenty years of the 1st c., see Adam-Veleni (ed.) 2001, 86. // 1st c.

401. (BAA 67) // Ancient Agora Bathhouse. Rectangular room, north of the Bathhouse. Destruction Layer. 1998 // Fragment // Non vidi. // Mavromichali 2001, fig. 21. From a bath complex destroyed in the last twenty years of the 1st c., see Adam-Veleni (ed.) 2001, 86. // 1st c.

402. (BAA 68) // Ancient Agora Bathhouse. Rectangular room, north of the Bathhouse. Destruction Layer. 1998 // Fragment // Non vidi. // Mavromichali 2001, fig. 21. From a bath complex destroyed in the last twenty years of the 1st c., see Adam-Veleni (ed.) 2001, 86. // 1st c.

403. (ΠΑΜ 71) // AUTH Campus. Theological Department Building. Sector 18, depth 1 m. 2-6-1965 // Fragment // Translucent bluish. No impurities or pinprick bubbles // // Unpublished // 1st - 2nd c.

404. (ΠΑΜ 30) // Stratou Avenue, Building of *ΥΕΝΕΔ/ΕΡΤ3* TV Station. Grave 1. 15-3-1969 // Mended. Large part of the body are missing // Translucent greenish. No impurities or pinprick bubbles // Diam. Max.: 10, Diam. rim: 3.4, Diam. Neck: 3, Diam. Base: 5, Weight: 195 // Unpublished // 2nd - 3rd c.

405. (ΜΘ 3761) // Agiou Dimitriou Street, cemetery found at the extension of the street. // Iridescent // Translucent greenish. // H.: 17.5, Diam. rim: 6.5, Diam. Neck: 3, Diam. Base: 5.5, Th.: 0.1, Weight: 195 // Unpublished // 3rd end - 4th c.

406. (ΜΘ 514) // Vardari Square. // Complete. Mended at the rim, neck and body // Translucent greenish. Few pinprick bubbles. Impurities only at the handle // H.: 14.6, Diam. Max.: 9.3, Diam. rim: 5.4, Diam. Neck: 2.4, Diam. Base: 4, Th.: 0.2 // E. Trakosopoulou-Salakidou, Thessaloniki apo ta proïstorika mechri ta christianika chronia 1986, 125, fig. 124. // 4th c.

407. (BY 310/2) // Junction of Kallitheas and Skeparni Street. Grave T. 25. 30-12-1996 // Mended // Translucent greenish. No impurities, few pinprick bubbles // H.: 15, Diam. Max.: 9, Diam. rim: 4.2, Diam. Neck: 2.5, Diam. Base: 4.5, Th.: 0.12 // Makropoulou 2007, 304, pl. 69., from a grave dated by pottery between the end of the 3rd and the end of the 4th c. // 3rd end - end 4th c.

408. (BY 317/3) // Kallitheas Avenue, Elementary School of Ampelokipi. Grave T. 31, burial layer. 29-9-1997 // Intact // Translucent greenish. Very few pinprick bubbles no impurities // H.: 14.5, Diam. Max.: 9, Diam. rim: 5.5, Diam. Neck: 3, Diam. Base: 4.5, Th.: 0.1 // Makropoulou 2007, 306, pl. 29., From a probably 4th century grave // 3rd-4th c.

409. (BY 317/9) // Kallitheas Avenue, Elementary School of Ampelokipi. Grave T. 83-B, burial layer. 2-12-1997 // Mended // Translucent greenish. Few pinprick bubbles no impurities // H.: 14, Diam. Max.: 7.5, Diam. rim: 5.5, Diam. Base: 3.5, Th.: 0.01 // Makropoulou 2007, 320, pls. 29, 68, from a grave dated to the first half of the 4th c. by the pottery finds // 4th c.

410. (ΜΘ 542) // Vardari Square. // Mouth, body and the handle and small part is preserved // Translucent greenish and dark-blue. No impurities, few pinprick bubbles // Diam. rim: 8, Diam. Neck: 5.2, Diam. Base: 5.7, Th.: 0.1, Weight: 82 // Unpublished // 4th c.

411. (ΜΘ 20923) // Ramona Area. 10-2-1971 // Almost complete, mended. Small part of the shoulder and the base are missing // Translucent bluish. No impurities, few pinprick bubbles // H.: 19, Diam. Max.: 9.5, Diam. rim: 6.7, Diam. Neck: 2.3, Diam. Base: 5.1, Th.: 0.06 // *Glass Cosmos* 2010, 280-281, no. 225 // 4th c.

412. (ΜΘ 23964) // Plot property of Tsirakoglou. 1940 // Broken, mendable // Translucent olive-greenish. No impurities, several pinprick bubbles // H.: 27.5, Diam. Max.: 10, Diam. rim: 8, Diam. Neck: 3.5, Diam. Base: 6.8, Th.: 0.1 // Walter 1940, col. 173 fig. 35, 36. Kotzias 1953-54, 167-175. *Glass Cosmos* 2010, 205, no. 58 // 4th-5th c.

413. (BY 253/3) // Junction of Efxeinou and Panagopoulou Streets. At the eastern corner of the SE side of the barrel-vaulted grave (No 2). // Parts missing, mended fragments // Translucent olive-greenish. // H.: 20, Diam. rim: 6, Diam. Base: 7 // Karydas 1994, 501, pl. 156α, from a 4th-century cemetery // 4th c.

414. (BY 319/3) // Lagkada Street, middle plot of Axios Complex. // Mended. Small part of the body and the larger part of the handle are missing // Dark blue. // H.: 21.1, Diam. Max.: 8, Diam. rim: 5.2, Diam. Base: 7, Th.: 0.1 // Makropoulou 2007, 162, pl. 29, from a 4th-century grave, dated by pottery // 4th c.

415. (BY 244/1) // Stratou Avenue 2, plot of the Museum of Byzantine Culture. Grave 63. 13-6-1989 // Intact // Translucent greenish. No impurities, few pinprick bubbles // H.: 15.9, Diam. Max.: 10.2, Diam. rim: 5.3, Diam. Neck: 2.4, Diam. Base: 4.8, Th.: 0.1, Weight: 136.3 // Papanikola-Bakirtzi 2002 (ed.), no. 763 556, dated to the 3rd-4th c. Nalpantis 2003, 61-62, 127-128, drawing 30, pl. 43, probably dated to the 4th c. // 3rd-4th c.

416. (ΜΘ 7229) // Gefyra, Thessaloniki. 1965 // Parts of the body are missing // Translucent greenish. // H.: 17, Diam. rim: 4, Diam. Base: 3, Th.: 0.1, Weight: 99.1 // *Glass Cosmos* 2010, 307, no. 284 // 3rd-4th c.

417. (ΜΘ 11531) // Ramona Area, Lagkada Street, tall, twelve-storey tower. 27-1-1972 // Complete, mended // Translucent greenish. // H.: 18, Diam. Max.: 7.7, Diam. rim: 5, Diam. Neck: 2.1, Diam. Base: 4.2 // E. Trakosopoulou- Salakidou, Thessaloniki apo ta proïstorika mechri ta christianika chronia 1986, 125, fig. 124. *Glass Cosmos* 2010, 286-87, no. 239 // 3rd-4th c.

418. (ΜΘ 5471) // Vardari Square. Found in a marble sarcophagus. 12-2-1920 // Parts of the body and neck are missing // Translucent greenish. // H.: 15, Diam. Max.: 10.3, Diam. rim: 6, Diam. Neck: 3.3, Diam. Base: 7, Th.: 0.08, Weight: 86.31 // Unpublished // 4th c.

419. (BY 309/7) // Peleiadou Street, 5. Grave T. 52 burial layer. 26-11-1997 // Complete, mended // Translucent greenish. Few pinprick bubbles no impurities // H.: 15.5, Diam. Max.: 10.6, Diam. rim: 6, Diam. Neck: 3, Diam. Base: 6, Th.: 0.12 // Makropoulou 2007, 264, pl. 29, from a grave generally dated to the 4th c. // 4th c.

420. (BY 309/4) // Peleiadou Street, 5. Grave T. 20, burial layer. 21-10-1997 // Mended, parts of the rim and neck are missing // Translucent greenish. // H.: 12.5, Diam. Max.: 8, Diam. rim: 5, Diam. Neck: 2.4, Diam. Base: 4, Th.: 0.15 // Makropoulou 2007, 249, pl. 65., from a grave dated by pottery and coins of Constantine I and Constans II (329-333) to the first half of the 4th c. // 4th c.

421. (BY 109) // Unknown exact provenance. // Intact // Translucent olive-green. No impurities, many pinprick bubbles // H.: 14.5, Diam. Max.: 9.7, Diam. rim: 6.2, Diam. Neck: 2.2, Diam. Base: 7.5, Th.: 0.15 // E. Kourkoutidou-Nikolaidou, Thessaloniki Istoria kai techni 1986, 55, 58. // 4th c.

422. (ΜΘ 1492) // Agios Pavlos area. 3-1931 // Intact // Translucent olive-greenish. No impurities or pinprick bubbles // H.: 16, Diam. Max.: 10, Diam. rim: 6.7, Diam. Neck: 2.8, Diam. Base: 7.5, Th.: 0.3 // E. Trakosopoulou-Salakidou, Thessaloniki apo ta proïstorika mechri ta christianika chronia 1986, 125, fig. 124. *Glass Cosmos* 2010, 280, no. 223 // 4th c.

423. (ΜΘ 20910) // Eleftherias Street, 13, Ampelokipi. Grave 3. 8-12-1977 // Almost complete, mended. Parts of the mouth and the neck are missing // Translucent greenish. Few pinprick bubbles. Impurities only at the handle // H.: 15.3, Diam. Max.: 9.2, Diam. rim: 6.6, Diam. Neck: 2.5, Diam. Base: 6.6, Th.: 0.12, Weight: 184.5 // Unpublished // 4th c.

424. (ΜΘ 1491) // Agios Pavlos area. With a tag: 'From Thessaloniki, across the Eptapyrgion. Found in a grave at Sykies Site.' March 1931 // Fragmentary, missing in lip and body // Translucent greenish. No impurities or pinprick bubbles // H.: 16, Diam. rim: 5, Diam. Neck: 2.5, Diam. Base: 8, Th.: 1.2, Weight: 104 // Antonaras 2006, 414, 416, dated on the basis of analogous finds, to the 4th c. // 4th c.

425. (ΜΘ 5494) // Unknown provenance. // Handle, and part of rim and mouth are preserved // Translucent greenish. // H.: 8, Diam. Max.: 2, Diam. Neck: 1, Weight: 45 // Unpublished // 4th c.

426. (BY 276/1) // Junction of Kallitheas and Agathonos Street. Fill of Grave 1. 13-9-1995 // Fragments mended. Large part of the body is missing // Translucent olive-greenish and bluish. Few pinprick bubbles and impurities // H.: 19.6, Diam. Max.: 9.6, Diam. rim: 7.5, Diam. Base: 8, Th.: 0.1 // Makropoulou 2007, 329, pl. 69, from a grave dated probably to the 4th century. // 3rd end - early 5th c.
427. (BY 317/7) // Kallitheas Avenue, Elementary School of Ampelokipi. Grave T. 87, burial layer. // Fragments mended, missing large part of the body // Translucent olive-greenish. Few pinprick bubbles and impurities // H.: 24, Diam. Max.: 13.2, Diam. rim: 10.4, Diam. Neck: 4.1, Diam. Base: 11.2, Th.: 0.5 // Makropoulou 2007, 340, pl. 29, from a 3rd-4th century grave // 3rd- 4th c.
428. (ΜΘ 5493) // Unknown provenance. // The largest part of the body and the base are missing // Translucent greenish. Few pinprick bubbles. Impurities only at the handle and on the ring // H.: 7.5, Diam. Max.: 10, Diam. rim: 7, Diam. Neck: 2.5, Th.: 0.12, Weight: 144 // Unpublished // 3rd end - early 5th c.
429. (ΠΑΜ 80) // Junction of Agiou Dimitriou Street and Agias Sophias Street, Byzantine building. 3rd triangular space. 11-6- 1969 // Fragment // Colourless decolourized. Indiscernible due to weathering // Dimensions: 3X3 // Unpublished // 3rd end - early 5th c.
430. (BY 51) // Basilica in Panorama. 1971 // Fragments mended // Translucent greenish. No impurities very few pinprick bubbles // H.: 4 // Unpublished // 4th - early 5th c.
431. (BY 173) // AUTH Campus. Theological Department Building. // Intact // Translucent greenish. Several pinprick bubbles, few impurities // H.: 16.5, Diam. Max.: 4.5, Diam. rim: 5, Diam. Neck: 3.3, Diam. Base: 7.5, Th.: 0.2, Weight: 225 // E. Kourkoutidou-Nikolaidou, Thessaloniki Istoria kai techni 1986, 53. Papanikola-Bakirtzi(ed) 2002, no. 348, 317-318 (A. Antonaras), dated to the 4th c. Antonaras 2006, 414, 416. // 4th-5th c.
432. (BY 323/1) // Peleiadou Street, 3. Square Γ2-Γ3. Grave 11. (SW corner). 5-12-2000 // Intact // Translucent greenish. Very few impurities, several pinprick bubbles // H.: 18, Diam. Max.: 8, Diam. rim: 5.3, Diam. Neck: 3.5, Diam. Base: 7, Th.: 0.12, Weight: 280.3 // Papanikola-Bakirtzi (ed.) 2002, no. 350 318 (E. Marki), dated to the 4th-5th c. Antonaras 2006, 414, 416 // 4th-5th c.
433. (ΜΘ 9925) // Regkoukou Street, 14. Grave 9. 1979; // Intact // Translucent greenish. No impurities, several pinprick bubbles // H.: 13.6, Diam. Max.: 8.5, Diam. rim: 5, Diam. Neck: 3.1, Diam. Base: 8.3, Th.: 0.1 // Vokotopoulou 1980, 365-7, fig. 3, pl. 215 γ-δ. E. Trakosopoulou-Salakidou, Thessaloniki apo ta proïstorika mechri ta christianika chronia 1986, 13. Marki 2006, 218, Grave of the first half of the 4th c. Antonaras 2006, 414, 416 // 4th
434. (ΜΘ 23961) // Ramona Area. From a group of finds from several tile-covered graves. 20-8-1971 // Intact // Translucent greenish. No impurities, few pinprick bubbles // H.: 12, Diam. Max.: 8.5, Diam. rim: 5.4, Diam. Neck: 3.5, Diam. Base: 8.5, Th.: 0.18 // Antonaras 2006, 414, 416, dated on the basis of analogous finds, to the 4th c. *Glass Cosmos* 2010, 294-295, no. 263 // 4th c.
435. (BY 227/17) // Northern part of Evangelistrias Cemetery, area of former Lyssiatreion Clinic. 1993 // Intact // Translucent bluish. No impurities, few pinprick bubbles // H.: 7.5, Diam. rim: 6, Diam. Base: 3.5 // Pelekanidou 1993, 376, photo 5, 6 // 1st - early 2nd c.
436. (ΜΘ 3155) // Ramona Area. Bone cist (β). 8-1963 // Intact. Iridescent // Translucent bluish. No impurities, few pinprick bubbles // H.: 4.8, Diam. Max.: 5.6, Diam. rim: 4.5, Diam. Neck: 3.6, Diam. Base: 3.2, Th.: 0.2 // E. Trakosopoulou- Salakidou, Thessaloniki apo ta proïstorika mechri ta christianika chronia 1986, 125. *Glass Cosmos* 2010, 274, no. 206 // 2nd - 4th c.
437. (ΜΘ 544) // Vardari Square. // Small parts of the body are missing // Translucent yellowish. No impurities, few pinprick bubbles // H.: 12, Diam. Max.: 7.7, Diam. rim: 7.5, Diam. Neck: 5, Diam. Base: 5.5, Th.: 0.07, Weight: 97.68 // *Glass Cosmos* 2010, 276, no. 214 // 3rd-4th c.
438. (ΜΘ 5484) // Unknown provenance. // Complete, mended // Translucent greenish. // // *Glass Cosmos* 2010, 276 no. 214 // 3rd-4th c.
439. (BY 207/3) // Lagkada Street, Ivanof Park. Grave 35. 1991 // Part of the body is missing // Translucent greenish. Few pinprick bubbles and impurities // H.: 8.5, Diam. rim: 5, Diam. Neck: 3, Th.: 0.2, Weight: 95.8 // Makropoulou 1991, 257-270, photo 22. Papanikola-Bakirtzi (ed.) 2002, no. 810, 574 (D Makropoulou), dated to the end of the 3rd- early 4th c., a bronze belt buckle found in the grave dates the burial more probably to the early 4th c. // 3rd-4th c.
440. (BY 309/3) // Peleiadou Street, 5. Grave T. 38 (barrel-vaulted, burial layer). // Intact // Translucent greenish. Few impurities and pinprick bubbles // H.: 8.7, Diam. Max.: 6.5, Diam. rim: 5.3, Diam. Neck: 4, Diam. Base: 6, Th.: 0.2 // Makropoulou 2007, 262, pls. 28, 68, from a grave with a wall-painted Chrism and a mid 4th century coin // 4th c.
441. (ΠΑΜ 84) // Junction of Agiou Dimitriou Street and Agias Sophias Street, Byzantine building. 3rd triangular space. 11-6- 1969 // Fragments. Part of the body is missing // Translucent greenish. Few impurities and pinprick bubbles // Diam. Max.: 4, Diam. rim: 3.5, Diam. Neck: 2.5, Diam. Base: 3.7, Pres. H.: 4.5, Th.: 0.2 // Antonaras 2006, 414, 416 // 3rd-4th c.
442. (ΜΘ 10800) // Ramona Area, Axios complex. Grave 4. // Complete, mended // Translucent greenish. No impurities, several pinprick bubbles // H.: 6, Diam. rim: 3, Diam. Neck: 2.3, Diam. Base: 2.5, Th.: 0.08, Weight: 15.3 // Petsas 1969a, 300, pl. 309 α. *Glass Cosmos* 2010, 273 no. 204 // 4th c.
443. (ΜΘ 5472) // Vardari Square. Found in a marble sarcophagus. 12-2-1920 // Missing parts of the body and neck // Translucent greenish. // H.: 7, Diam. Max.: 6, Diam. rim: 5.5, Diam. Neck: 4, Diam. Base: 4.5, Th.: 0.08, Weight: 23.9 // With a paper note in it: OL XIII (60, 46). *Glass Cosmos* 2010, 276-277 no. 215 // 3rd-4th c.
444. (ΜΘ 516) // Vardari Square. // Intact // Translucent greenish. // H.: 5, Diam. Max.: 6 // Unpublished // 4th c.
445. (ΜΘ 3777) // Agiou Dimitriou Street, cemetery found at the extension of the street. 1-2-1960 // Part of the body missing // Purple. Few impurities, few pinprick bubbles // H.: 5, Diam. Max.: 4.7, Diam. rim: 4.3, Diam. Neck: 2.3, Diam. Base: 3, Th.: 0.6 // *Glass Cosmos* 2010, 274, no. 207 // 4th c.
446. (BY 255/2) // Margaropoulou Street, 20. Stray find during the μηχανική/mechanical dig. 1994 // Intact // Translucent greenish. // H.: 9, Diam. Max.: 9.7, Diam. rim: 8, Diam. Base: 9, Th.: 0.2 // Makropoulou 2007, 235, pl. 25. Dated to the 3rd-4th c. by the rest of the context // 4th c.
447. (ΜΘ 508) // Vardari Square. // Part of the body is missing // Translucent greenish. Few impurities, few pinprick bubbles // H.: 10, Diam. Max.: 8.3, Diam. rim: 6.4, Diam. Neck: 4.9, Diam. Base: 5.5, Th.: 0.3 // E. Trakosopoulou-Salakidou, Thessaloniki apo ta proïstorika mechri ta christianika chronia 1986, 125. // 3rd end - early 5th c.
448. (BY 214/1) // Northern part of Evangelistrias Cemetery, area of former Lyssiatreion Clinic. Grave Δ. 25-11-1991 // Intact // Translucent greenish. // H.: 10.2, Diam. rim: 6.8, Diam. Base: 6.5 // Pelekanidou 1993, 376, photo 5, from a grave with 2nd-3rd c. coins. // 3rd-4th c.
449. (BY 321/5) // Agiou Nestoros Street, 26. Water main works. Grave T. 48. Near the skull of burial A'. 13-10-2000 // Part of the head and the body are missing // Translucent greenish. Few impurities several pinprick bubbles // H.: 9, Diam. Max.: 8, Diam. rim: 5, Diam. Base: 4, Th.: 0.12 // 9th EBA 2003, 3, where it is illustrated. // 3rd-4th c.
450. (BY 93) // Agiou Nestoros Street, 8. Larnax α'. 5-4-1977 // Parts of the body are missing // Translucent greenish. No impurities, few pinprick bubbles // Diam. Max.: 10, Diam. Neck: 4, Diam. Base: 8, Pres. H.: 10.5, Th.: 0.2 // Unpublished, mentioned in Pazaras 1981, 372-387. Grave dated in the second half of the 3rd c. by the pottery finds and the wall-paintings // end of 3rd c.
451. (ΜΘ 3078) // Ramona Area. Burial cluster I. 8-1963 // Non-joining base and body fragments are preserved. // Translucent greenish. // H.: 7, Diam. Max.: 6 // Unpublished // 3rd end - 4th c.
452. (ΜΘ 20924) // Unknown exact provenance. From a tile grave. 19-3-1971 // Complete, mended // Translucent olive-green. No impurities, few pinprick bubbles // H.: 5.7, Diam. Max.: 8.2, Diam. rim: 5.3, Diam. Neck: 4.8, Diam. Base: 5.6, Th.: 0.12 // *Glass Cosmos* 2010, 281, no. 226 // 4th-5th c.
453. (BY 227/35) // Northern part of Evangelistrias Cemetery, area of former Lyssiatreion Clinic. South cluster Γ. Grave 2 (pit grave). 19-4-1995 // Fragments // Translucent bluish. No impurities or pinprick bubbles // H.: 8.5, Diam. Max.: 7, Diam. rim: 5.5, Diam. Base: 2, Weight: 55 // Unpublished. From a grave with a 2nd-3rd-century coin // 4th-5th c.

454. (BAA 75) // Ancient Agora Bathhouse. Rectangular room, north of the Bathhouse. Destruction Layer. 1998 // Body and neck fragment // Non vidi. // Mavromichali 2001, fig. 27. From a bath complex destroyed in the last twenty years of the 1st c., see Adam-Veleni (ed.) 2001, 86. // 1st c.

455. (BAA 76) // Ancient Agora Bathhouse. Rectangular room, north of the Bathhouse. Destruction Layer. 1998 // Body and neck fragment // Non vidi. // Mavromichali 2001, fig. 27. From a bath complex destroyed in the last twenty years of the 1st c., see Adam-Veleni (ed.) 2001, 86. // 1st c.

456. (BAA 77) // Ancient Agora Bathhouse. Rectangular room, north of the Bathhouse. Destruction Layer. 1998 // Body and neck fragment // Non vidi. // Mavromichali 2001, fig. 27. From a bath complex destroyed in the last twenty years of the 1st c., see Adam-Veleni (ed.) 2001, 86. // 1st c.

457. (BAA 78) // Ancient Agora Bathhouse. Rectangular room, north of the Bathhouse. Destruction Layer. 1998 // Body and neck fragment // Non vidi. // Mavromichali 2001, fig. 27. From a bath complex destroyed in the last twenty years of the 1st c., see Adam-Veleni (ed.) 2001, 86. // 1st c.

458. (BAA 79) // Ancient Agora Bathhouse. Rectangular room, north of the Bathhouse. Destruction Layer. 1998 // Body and neck fragment // Non vidi. // Mavromichali 2001, fig. 27. From a bath complex destroyed in the last twenty years of the 1st c., see Adam-Veleni (ed.) 2001, 86. // 1st c.

459. (BY 192) // AUTH Campus. Theological Department Building. // Intact // Purple body, translucent greenish handles. // H.: 6.5 // E. Kourkoutidou-Nikolaidou, Thessaloniki Istoria kai techni 1986, 52 // 1st c.

460. (IST 1077) // Area of the Thessaloniki International Fair, Macedonian Museum of Contemporary Art. Grave 160, vessel no. 5. 2001 // Intact // Opaque dark-blue and white. // H.: 5.2, Diam. Max.: 4, Diam. rim: 2.2, Diam. Neck: 1, Diam. Base: 2.2 // *Glass Cosmos* 2010, 251, no. 146 // 1st c.

461. (IST 1081) // Area of the Thessaloniki International Fair, Macedonian Museum of Contemporary Art. Grave 160, vessel no. 1. 2001 // Intact // Translucent slightly greenish. No impurities, few pinprick bubbles // H.: 7, Diam. Max.: 3.5, Diam. rim: 2.4, Diam. Neck: 1.2, Diam. Base: 1.5, Th.: 0.15 // *Glass Cosmos* 2010, 252, no. 150 // 1st c.

462. (IST 1073) // Area of the Thessaloniki International Fair, Macedonian Museum of Contemporary Art. Grave 160, vessel no. 10. 2001 // Intact // Colourless decolourized. No impurities, few pinprick bubbles // H.: 6.7, Diam. Max.: 3.5, Diam. rim: 2.3, Diam. Neck: 1, Diam. Base: 1.7, Th.: 0.1 // *Glass Cosmos* 2010, 250, no. 142 // 1st c.

463. (ΜΘ 11258 β) // Sintrivani Square. Excavation of the Walls. In built, cist Grave T2. 6-3-1970 // The neck and the rim are missing // Purple. // H.: 6, Weight: 7.38 // Vavritsas 1971, 378, pl. 371B.; E.M. Stern, Roman mould-blown glass, 152, no. 55, note 6f. *Glass Cosmos* 2010, 254, no. 158 // 1st c.

464. (BAA 74) // Ancient Agora Bathhouse. Rectangular room, north of the Bathhouse. Destruction Layer. 1998 // Fragment // Non vidi // Dimensions: 3X4 // Mavromichali 2001, fig. 26. From a bath complex destroyed in the last twenty years of the 1st c., see Adam-Veleni (ed.) 2001, 86. // 1st c.

465. (ΔΕΘ 28/1) // Area of the Thessaloniki International Fair, Southern Plot. Dig 28, layer 001. 1993 // Fragments // Translucent greenish. Few pinprick bubbles no impurities // Est. H.: 8, Th.: 0.1 // Unpublished // 1st c.

466. (ΔΕΘ 28/2) // Area of the Thessaloniki International Fair, Southern Plot. Dig 28, layer 001. 1993 // Fragments // Translucent bluish. Few pinprick bubbles no impurities // Est. H.: 8, Th.: 0.1 // Unpublished // 1st c.

467. (ΔΕΘ 28/3) // Area of the Thessaloniki International Fair, Southern Plot. Dig 28, layer 001. 1993 // Fragments // Translucent greenish. Few pinprick bubbles no impurities // Est. H.: 8, Th.: 0.1 // Unpublished // 1st c.

468. (ΔΕΘ 158/1) // Area of the Thessaloniki International Fair, Southern Plot. Grave 158, vessel 1. 1993 // Fragments. Large part of the body is missing // Translucent bluish. // H.: 8, Diam. Max.: 7, Diam. rim: 2.1, Diam. Neck: 2.1, Th.: 0.08 // Unpublished // 1st c.

469. (ΔΕΘ 158/2) // Area of the Thessaloniki International Fair, Southern Plot. Grave 158, vessel 2. 1993 // Fragments. Large part of the body is missing // Translucent bluish. // H.: 7, Diam. Max.: 6, Diam. rim: 1.7, Diam. Neck: 1.7, Th.: 0.08 // Unpublished // 1st c.

470. (ΔΕΘ 158/3) // Area of the Thessaloniki International Fair, Southern Plot. Grave 158, vessel 3. 1993 // Fragments. Large part of the body is missing // Translucent bluish. // H.: 8, Diam. Max.: 7, Diam. rim: 1.7, Diam. Neck: 1.7, Th.: 0.08 // Unpublished // 1st c.

471. (ΔΕΘ 187/2) // Area of the Thessaloniki International Fair, Southern Plot. Grave 187, vessel 2. 1993 // Fragments // Translucent greenish. // H.: 7, Diam. Max.: 6, Diam. rim: 1.7, Diam. Neck: 1.7, Th.: 0.08 // Unpublished // 1st c.

472. (ΔΕΘ 216/3) // Area of the Thessaloniki International Fair, Southern Plot. Grave 216, vessel 3. 1993 // Fragments // Translucent bluish. Few pinprick bubbles no impurities // Diam. Max.: 7, Diam. rim: 1.3, Diam. Neck: 1.3, Est. H.: 8.3, Th.: 0.1 // Unpublished // 1st c.

473. (ΔΕΘ 216/4 α) // Area of the Thessaloniki International Fair, Southern Plot. Grave 216, vessel 4. 1993 // Fragments // Translucent bluish. Few pinprick bubbles no impurities // Diam. Max.: 7, Diam. rim: 1.3, Diam. Neck: 1.3, Est. H.: 8.3, Th.: 0.1 // Unpublished // 1st c.

474. (ΔΕΘ 216/4 β) // Area of the Thessaloniki International Fair, Southern Plot. Grave 216, vessel 4. 1993 // Fragments // Translucent bluish. Few pinprick bubbles no impurities // Diam. Max.: 7, Diam. rim: 1.3, Diam. Neck: 1.3, Est. H.: 8.3 Th.: 0.1 // Unpublished // 1st c.

475. (ΔΕΘ 216/6) // Area of the Thessaloniki International Fair, Southern Plot. Grave 216, vessel 6. 1993 // Fragments. Large part of the body is missing // Translucent greenish. // H.: 8, Diam. rim: 1.8, Diam. Neck: 1.8, Th.: 0.08 // Unpublished // 1st c.

476. (ΜΘ 10295) // Filikis Etaireias. Excavation of the Hippodrome. Plot property of Nastos. Near the Eastern Wall. 11-9- 1967 // Large part of the body preserved, mended // Purple. No impurities, several pinprick bubbles // H.: 9, Diam. Max.: 8.5, Diam. rim: 1.2, Diam. Neck: 1.2, Th.: 0.1 // Petsas 1968b, 332-334, pl. 273β. Petsas 1974, 329. Simple note to the reg. no. and reference to Petsas 1966/67, No. 54, pl. 53α. *Glass Cosmos* 2010, 248, no. 137 // 1st c.

477. (ΜΘ 10296) // Filikis Etaireias. Excavation of the Hippodrome. Plot property of Nastos. Near the Eastern Wall. 11-9- 1967 // Fragments // Translucent greenish. No impurities, several pinprick bubbles // Est. H.: ca 8, Th.: 0.1 // Petsas 1968b, 332-334, pl. 273β. Petsas 1974, 329. Simple note to the reg. no. and reference to Petsas 1966/67, no. 54, pl. 53α. // 1st c.

478. (ΜΘ 11575) // Unknown exact provenance. // Parts of the body are missing // Translucent greenish. No impurities, few pinprick bubbles // H.: 7, Diam. Max.: 7, Diam. Neck: 1.5, Th.: 0.03, Weight: 11.05 // E. Trakosopoulou-Salakidou, Thessaloniki apo ta proïstorika mechri ta christianika chronia 1986, 134. *Glass Cosmos* 2010, 248, no. 138 // 1st c.

479. (ΜΘ 11576) // Filikis Etaireias Street, 39, plot property of D. Margaritis. Grave II. // Fragments. Large part of the body is missing // Translucent slightly greenish. No impurities or pinprick bubbles // H.: 3.5, Diam. Max.: 7.3, Diam. Neck: 1.7, Th.: 0.04, Weight: 4.12 // *Glass Cosmos* 2010, 249, no. 139 // 1st c.

480. (ΜΜΣΤ 2/4) // Area of the Thessaloniki International Fair, Macedonian Museum of Contemporary Art. Grave 2. 18-9-2000 // Preserved in fragments // Translucent greenish. No impurities, few pinprick bubbles // H.: 7, Diam. Max.: 7, Diam. rim: 1.3, Diam. Neck: 1.3, Th.: 0.07 // Unpublished // 1st c.

481. (ΜΜΣΤ 2/17) // Area of the Thessaloniki International Fair, Macedonian Museum of Contemporary Art. Grave 2. // Preserved in fragments // Translucent bluish. No impurities, few pinprick bubbles // H.: 6, Diam. Max.: 5, Diam. rim: 1, Diam. Neck: 1 // Unpublished // 1st c.

482. (ΜΜΣΤ 2/21) // Area of the Thessaloniki International Fair, Macedonian Museum of Contemporary Art. Grave 2. // In small fragments // Translucent bluish. No impurities, few pinprick bubbles // Max. pres. frag.: L. 3, Th.: 0.1 // Unpublished // 1st c.

483. (ΜΜΣΤ 2/22 α) // Area of the Thessaloniki International Fair, Macedonian Museum of Contemporary Art. Grave 2. // Preserved in fragments // Translucent bluish. No impurities, few pinprick bubbles // H.: 6, Diam. Max.: 5.5, Diam. rim: 0.9, Th.: 0.07 // Unpublished // 1st c.

484. (ΜΜΣΤ 2/22 β) // Area of the Thessaloniki International Fair, Macedonian Museum of Contemporary Art. Grave 2. // Preserved in fragments // Translucent bluish with purplish mixture. No impurities, few pinprick bubbles // H.: 6, Diam. Max.: 5.5, Diam. rim: 1.3, Th.: 0.07 // Unpublished // 1st c.

485. (ΜΜΣΤ 53/3) // Area of the Thessaloniki International Fair, Macedonian Museum of Contemporary Art. Grave 53 vessel no. 3. // Fragments. Large part of the body is missing // Translucent greenish. No impurities, few pinprick bubbles // H.: 8.5, Diam. Max.: 8, Diam. rim: 1.3, Diam. Neck: 1.3, Th.: 0.06 // Unpublished // 1st c.

486. (ΜΜΣΤ 53/4) // Area of the Thessaloniki International Fair, Macedonian Museum of Contemporary Art. Grave 53 vessel no. 4. // Fragments. Large part of the body is missing // Translucent greenish. No impurities, few pinprick bubbles // H.: 8.5, Diam. Max.: 8, Diam. rim: 1.6, Diam. Neck: 1.6, Th.: 0.06 // Unpublished // 1st c.

487. (ΠΑΜ 12) // Junction of Egnatia Street and K. Melenikou Street. Cenotaph. 24-4-1969 // Fragments. Part of the body is missing // Translucent bluish. No impurities, few pinprick bubbles // ca 5- 6 // Unpublished // 1st c.

488. (ΠΑΜ 13) // Junction of Egnatia Street and K. Melenikou Street. Cenotaph. 24-4-1969 // Fragments. Part of the body is missing // Translucent bluish. No impurities, few pinprick bubbles // ca 5- 6 // Unpublished // 1st c.

489. (ΠΑΜ 14) // Junction of Egnatia Street and K. Melenikou Street. Cenotaph. 24-4-1969 // Fragments. Part of the body is missing // Translucent bluish. No impurities, few pinprick bubbles // ca 5- 6 // Unpublished // 1st c.

490. (ΠΑΜ 15) // Junction of Egnatia Street and K. Melenikou Street. Cenotaph. 24-4-1969 // Fragments. Part of the body is missing // Translucent bluish. No impurities, few pinprick bubbles // ca 5- 6 // Unpublished // 1st c.

491. (ΠΑΜ 16) // Junction of Egnatia Street and K. Melenikou Street. Cenotaph. 24-4-1969 // Fragments. Part of the body is missing // Translucent bluish. No impurities, few pinprick bubbles // ca 5- 6 // Unpublished // 1st c.

492. (ΠΑΜ 29) // Filikis Etaireias. Excavation of the Hippodrome. Plot property of Nastos. Near the western face of the Eastern wall. 11-9-1967 // Shoulder and neck fragment // Translucent bluish. No impurities, few pinprick bubbles // Diam. Max.: 5.5, Weight: 11.05 // Unpublished // 1st c.

493. (ΠΑΜ 75) // Giantison Street. Clinics of Veterinary Faculty of AUTH. Among finds from several graves. 23-1-1970 // Fragment // Purple. // Diam. rim: 1.2, Est. H.: 7-8, Th.: 0.1 // Unpublished // 1st c.

494. (ΜΘ 20938) // Unknown street, probably Vardari Square area. In a box signed 'unregistered, probably from Vardaris Cemetery'. // Parts missing, Rim, neck and body partly preserved // Translucent amber. // H.: 9, Diam. Max.: 3, Diam. rim: 1.4, Diam. Neck: 1.2, Th.: 0.08, Weight: 5 // Unpublished // 1st c.

495. (ΠΑΜ 20) // Ramona Area;. 1934; // Only the base is preserved // Translucent amber. // Diam. Base: 2 // Unpublished // 1st c.

496. (ΜΘ 23805) // Confiscated in hands of a villager (2) in Katerini. Probably from the region of Lofos in Pieria. 1973 // Complete, in two pieces. Deformed in fire // Opaque light blue. No impurities, several pinprick bubbles // // *Glass Cosmos* 2010, 230, no. 99 // 1st c.

497. (ΜΘ 21963) // Unknown exact provenance. In AMTH was kept in a box with post-Byzantine pottery from old excavations in Thessaloniki. // Body fragment // Translucent greenish. // Max Dim. 6X7, Th.: 0.13 // *Glass Cosmos* 2010, 254, no. 159 // 1st - early 2nd c.

498. (ΜΘ 11105 β) // Giantison Street. Clinics of Veterinary Faculty of AUTH. // Head and neck are preserved // Translucent bluish. // H.: 6 // Petsas 1974, 329 // 1st c.

499. (ΜΘ 11197) // Giantison Street. Clinics of Veterinary Faculty of AUTH. Altar-like construction. Tile grave. // Parts missing // Translucent bluish. // H.: 6, L. 8 // *Glass Cosmos* 2010, 245, no. 129 // 1st c.

500. (ΜΘ 11199) // Giantison Street. Clinics of Veterinary Faculty of AUTH. Altar-like construction. Tile grave. // Head and neck fragment // Translucent greenish. Few impurities, several pinprick bubbles // // *Glass Cosmos* 2010, 246, no. 131 // 1st c.

501. (ΜΘ 11200) // Giantison Street. Clinics of Veterinary Faculty of AUTH. Altar-like construction. Tile grave. // Part of the body and the tail are missing // Translucent greenish. Few impurities, several pinprick bubbles // // E. Trakosopoulou-Salakidou, Thessaloniki apo ta proïstorika mechri ta christianika chronia 1986, 125, fig. 120. *Glass Cosmos* 2010, 246, no. 132 // 1st c.

502. (ΜΘ 11195) // Giantison Street. Clinics of Veterinary Faculty of AUTH. Altar-like construction. Tile grave. // Part of the head and the body are missing // Translucent bluish. // H.: 13, Diam. Max.: 7 // E. Trakosopoulou-Salakidou, Thessaloniki apo ta proïstorika mechri ta christianika chronia 1986 125. fig. 120. *Glass Cosmos* 2010, 245, no. 127 // 1st c.

503. (ΜΘ 11196) // Giantison Street. Clinics of Veterinary Faculty of AUTH. Altar-like construction. Tile grave. // Large part of the body is missing // Translucent bluish. // H.: 13, Diam. Max.: 7 / // *Glass Cosmos* 2010, 245, no. 128 // 1st c.

504. (ΜΘ 11551) // Filikis Etaireias Street, 39, plot property of D. Margaritis. // Three pieces mended. Large part of the tail is missing // Translucent bluish. No impurities, many pinprick bubbles // H.: 17 // E. Trakosopoulou-Salakidou, Thessaloniki apo ta proïstorika mechri ta christianika chronia 1986, 134. *Glass Cosmos* 2010, 247, no. 133 // 1st c.

505. (ΜΘ 11552) // Filikis Etaireias Street, 39, plot property of D. Margaritis. // Mended, parts of the tail are missing // Translucent greenish. // H.: 19 // E. Trakosopoulou-Salakidou, Thessaloniki apo ta proïstorika mechri ta christianika chronia 1986, 134. *Glass Cosmos* 2010, 247, no. 134 // 1st c.

506. (ΜΘ 20929) // Unknown exact provenance. From an excavation conducted before 1970. // Broken, parts of the head and the tail are missing // Translucent. No impurities or pinprick bubbles // H.: 5, Th.: 0.06, Weight: 7.4 // Unpublished // 1st c.

507. (ΠΑΜ 28) // Filikis Etaireias. Excavation of the Hippodrome. Plot property of Nastos. Near the western face of the Eastern wall. 11-9-1967 // Small fragment // Translucent bluish. No impurities, many pinprick bubbles // Diam. rim: 3X2, Th.: 0.1 // Unpublished // 1st c.

508. (ΒΥ 227/40 b) // Northern part of Evangelistrias Cemetery, area of former Lyssiatreion Clinic. Grave 185. 19-4-1994 // Head fragment // Translucent bluish. No impurities or pinprick bubbles // Dimensions: 3X1.2 // Unpublished // 1st c.

509. (ΔΕΘ 158/4) // Area of the Thessaloniki International Fair, Southern Plot. Grave 158, vessel 4. 1993 // Head and neck fragment // Translucent bluish. // Pres. L.: 4 // Unpublished // 1st c.

510. (ΔΕΘ 158/5) // Area of the Thessaloniki International Fair, Southern Plot. Grave 158, vessel 5. 1993 // Head and neck fragment // Translucent bluish. // Pres. L.: 4 // Unpublished // 1st c.

511. (ΔΕΘ 187/1) // Area of the Thessaloniki International Fair, Southern Plot. Grave 187, vessel 1. 1993 // Head and neck fragment // Translucent bluish. // Pres. L.: 4 // Unpublished // 1st c.

512. (ΜΘ 10297) // Filikis Etaireias. Excavation of the Hippodrome. Plot property of Nastos. Near the Eastern Wall. 11-9- 1967 // Iridescent // Colourless. // H.: 3.8 // Petsas 1968b, 332-334, pl. 273β. Petsas 1974, 329. Simple note to the reg. no. and reference to Petsas 1966/67, No. 54, pl. 53α. // 1st c.

513. (ΜΘ 10298) // Filikis Etaireias. Excavation of the Hippodrome. Plot property of Nastos. Near the Eastern Wall. 11-9-1967 // Filled in the shoulder. Iridescent // Dark coloured. // H.: 4, Diam. Neck: 1.3 // Petsas 1968b, 332-334, pl. 273β. Petsas 1974, 329. Simple note to the reg. no. and reference to Petsas 1966/67, no. 54, pl. 53α. // 1st c.

514. (ΜΘ 11574) // Filikis Etaireias Street, 39, plot property of D. Margaritis. Grave II. // Mended, large parts of the body are missing // Translucent greenish. No impurities, few pinprick bubbles // H.: 18 // E. Trakosopoulou-Salakidou, Thessaloniki apo ta proïstorika mechri ta christianika chronia 1986, 134. *Glass Cosmos* 2010, 248, no. 136 // 1st c.

515. (ΠΑΜ 76) // Giantison Street. Clinics of Veterinary Faculty of AUTH. Among finds from several graves. 23-1-1970 // Fragment // Colourless decolourized. No impurities or pinprick bubbles // Pres. L.: 5, Th.: 0.1 // Unpublished // 1st c.

516. (Av-Pi /3) // Municipal Hospital, 'Agios Dimitrios', plot of the Water Company. // Intact // Silver grey. Non vidi // Diam. Max.: 4, L. 26, Max. body diam. 4 // Avezou, Picard 1912, 337-361 // 1st c.

517. (BY 316/1) // Tritis Septemvriou Street, Cemeterial Basilica. Sector III. From a wide grave near the ring. // Head, neck and body fragments // Translucent greenish. // H.: 4 // Unpublished // 1st c.

518. (ΔΕΘ 216/1) // Area of the Thessaloniki International Fair, Southern Plot. Grave 216, vessel 1. 1993 // Head and neck fragment // Translucent bluish. // Pres. L.: 9 // Unpublished // 1st c.

519. (ΔΕΘ 216/2) // Area of the Thessaloniki International Fair, Southern Plot. Grave 216, vessel 2. 1993 // Head, neck and body fragments // Translucent bluish. // Pres. L.: 6 // Kousoulakou 1993, 311, 317 fig. 2β. // 1st c.

520. (ΔΕΘ 216/5) // Area of the Thessaloniki International Fair, Southern Plot. Grave 216, vessel 5. 1993 // Head and neck fragment // Translucent bluish. // Pres. L.: 8 // Kousoulakou 1993, 311, 317 fig. 2β. // 1st c.

521. (ΔΕΘ 216/7) // Area of the Thessaloniki International Fair, Southern Plot. Grave 216, vessel 7. 1993 // Head, neck and body fragments // Translucent bluish. // Pres. L.: 16 // Unpublished // 1st c.

522. (ΔΕΘ 216/8) // Area of the Thessaloniki International Fair, Southern Plot. Grave 216, vessel 8. 1993 // Head and neck fragment // Translucent bluish. // Pres. L.: 7 // Unpublished // 1st c.

523. (ΜΘ 11198) // Giantison Street. Clinics of Veterinary Faculty of AUTH. Altar-like construction. Tile grave. // The back and part of the body are missing // Translucent greenish. // H.: 6, L.12 // *Glass Cosmos* 2010, 246, no. 130 // 1st c.

524. (ΜΘ 11573) // Filikis Etaireias Street, 39, plot property of D. Margaritis. Grave III. // Large part of the tail is missing // Translucent greenish. No impurities, few pinprick bubbles // H.: 12 // E. Trakosopoulou-Salakidou, Thessaloniki apo ta proïstorika mechri ta christianika chronia 1986, 134. *Glass Cosmos* 2010, 247, no. 135 // 1st c.

525. (ΜΘ 20928) // Unknown exact provenance. From an excavation conducted before 1970. // Front half of the vessel is preserved // Translucent. No impurities or pinprick bubbles // H.: 8.6, Diam. Max.: 4.2, Th.: 0.06, Weight: 11.6 // Unpublished // 1st c.

526. (ΜΘ 20930) // Unknown exact provenance. From an excavation conducted before 1970. // Head and neck fragment // Translucent. No impurities or pinprick bubbles // H.: 6.5, Th.: 0.06, Weight: 1.9 // Unpublished // 1st c.

527. (ΜΜΣΤ 2/16) // Area of the Thessaloniki International Fair, Macedonian Museum of Contemporary Art. Grave 2. // Portion of the upper part are preserved // Translucent bluish. No impurities, few pinprick bubbles // H.: 3, Pres. L.: 6 // Unpublished // 1st c.

528. (ΜΜΣΤ 2/28) // Area of the Thessaloniki International Fair, Macedonian Museum of Contemporary Art. Grave 2. // In small fragments // Translucent bluish. Few impurities, several pinprick bubbles // Max. pres. frag.: L. 3, Th.: 0.1 // Unpublished // 1st c.

529. (ΜΜΣΤ 53/1) // Area of the Thessaloniki International Fair, Macedonian Museum of Contemporary Art. Grave 53 vessel no. 1. // Large parts of the body are missing // Translucent bluish. // H.: 5, Max. Pres. L.: 8 // Unpublished // 1st c.

530. (ΜΜΣΤ 53/2) // Area of the Thessaloniki International Fair, Macedonian Museum of Contemporary Art. Grave 53 vessel no. 2. // Large parts of the body are missing // Translucent bluish. // H.: 6, Pres. L.: 12 // Unpublished // 1st c.

531. (ΜΜΣΤ 123/1) // Area of the Thessaloniki International Fair, Macedonian Museum of Contemporary Art. Grave 123, vessel no. 1. // Large parts of the body are missing // Translucent bluish. Few impurities, several pinprick bubbles // Max. pres. frag.: L. 3, Th.: 0.1 // Unpublished // 1st c.

532. (ΜΜΣΤ 123/2) // Area of the Thessaloniki International Fair, Macedonian Museum of Contemporary Art. Grave 123, vessel no. 2. // Large parts of the body are missing // Translucent bluish. Few impurities, several pinprick bubbles // Max. pres. frag.: L. 3, Th.: 0.1 // Unpublished // 1st c.

533. (ΜΜΣΤ 123/3) // Area of the Thessaloniki International Fair, Macedonian Museum of Contemporary Art. Grave 123, vessel no. 3. // Large parts of the body are missing // Translucent bluish. Few impurities, several pinprick bubbles // Max. pres. frag.: L. 3, Th.: 0.1 // Unpublished // 1st c.

534. (ΜΜΣΤ 123/4) // Area of the Thessaloniki International Fair, Macedonian Museum of Contemporary Art. Grave 123, vessel no. 4. // Large parts of the body are missing // Translucent bluish. Few impurities, several pinprick bubbles // Max. pres. frag.: L. 3, Th.: 0.1 // Unpublished // 1st c.

535. (ΜΜΣΤ 123/5) // Area of the Thessaloniki International Fair, Macedonian Museum of Contemporary Art. Grave 123, vessel no. 5. // Large parts of the body are missing // Translucent bluish. Few impurities, several pinprick bubbles // Max. pres. frag.: L. 3, Th.: 0.1 // Unpublished // 1st c.

536. (ΠΑΜ 17) // Junction of Egnatia Street and K. Melenikou Street. Cenotaph. 24-4-1969 // Fragments. Part of the body is missing // Translucent bluish. Few impurities, several pinprick bubbles // Pres. L.: 7, Th.: 0.1 // Unpublished // 1st c.

537. (ΠΑΜ 18) // Junction of Egnatia Street and K. Melenikou Street. Cenotaph. 24-4-1969 // Fragments. Part of the body is missing // Translucent bluish. Few impurities, several pinprick bubbles // Pres. L.: 5, Th.: 0.1 // Unpublished // 1st c.

538. (ΜΘ 4906) // Lagkada Street. 17-6-1961 // Intact // Opaque milky-white. // H.: 6, Diam. Max.: 4, Diam. rim: 1, Diam. Neck: 1.4, Diam. Base: 1.8, Th.: 0.08, Weight: 12 // *Glass Cosmos* 2010, 253, no. 153 // 1st c.

539. (Av-Pi /2) // Municipal Hospital, 'Agios Dimitrios', plot of the Water Company. // Intact // grey-blue. Non vidi // Non vidi // Avezou, Picard 1912, 337-361 // 1st - early 2nd c.

540. (ΜΘ 808) // Confiscated in hands of a villager (4). Possibly from Olynthos. // Filled very carefully. Professionally // Dark blue and white. // H.: 9, Diam. rim: 2.2, Diam. Neck: 1, Diam. Base: 3.7, Th.: 0.09, Weight: 24.91 // *Glass Cosmos* 2010, 360, no. 393 // 1st - early 2nd c.

541. (ΜΘ 809) // Confiscated in hands of a villager (4). Possibly from Olynthos. // Missing parts of the body // Translucent greenish and dark-blue thread. // H.: 9, Diam. Max.: 6.5, Diam. rim: 5, Diam. Neck: 2.4, Diam. Base: 3, Th.: 0.08 // Unpublished // 1st - early 2nd c.

542. (BY 227/33) // Northern part of Evangelistrias Cemetery, area of former Lyssiatreion Clinic. Burial chamber II, cist grave. 19-5-1995 // Almost complete, deformed in fire // milky-white. // H.: 6, Diam. rim: 2, Diam. Neck: 1.8, Diam. Base: 3 // Unpublished // 1st - early 2nd c.

543. (BY 227/41) // Northern part of Evangelistrias Cemetery, area of former Lyssiatreion Clinic. Burial chamber II, cist grave. 19-5-1995 // Intact // Translucent bluish. No impurities or pinprick bubbles // H.: 7.5, Diam. Max.: 5.5, Diam. rim: 2, Diam. Neck: 2.5, Diam. Base: 3 // Pelekanidou 1993, 376, photo 5, 6 // 1st - early 2nd c.

544. (ΜΘ ΣιντριβάνιΤ9Β5) // Junction of Egnatia Street and K. Melenikou Street. Grave 9 B5. 1969 // Deformed in fire // Translucent greenish. // H.: 8, Diam. rim: 1.5, Diam. Neck: 1, Th.: 0.15 // Unpublished // 1st - early 2nd c.

545. (ΜΘ 11172) // Ramona Area. Tile grave. // Mended, large parts of the body are missing // Translucent amber. // H.: 9, Diam. Max.: 5.5, Diam. rim: 2.8, Diam. Neck: 1, Diam. Base: 3.5, Th.: 0.05, Weight: 16 // Unpublished // 1st - early 2nd c.

546. (IST 1042) // Area of the Thessaloniki International Fair, Macedonian Museum of Contemporary Art. Grave 2. // Intact // Translucent bluish. Indiscernible due to weathering // H.: 7.5, Diam. Max.: 5.4, Diam. rim: 2.2, Diam. Neck: 1.5, Diam. Base: 2.2, Th.: 0.1 // *Glass Cosmos* 2010, 249, no. 140 // 1st - early 2nd c.

547. (ΜΜΣΤ 2/20) // Area of the Thessaloniki International Fair, Macedonian Museum of Contemporary Art. Grave 2. // Preserved in fragments, the base is missing // Translucent yellowish. Indiscernible due to weathering // H.: 5.5, Diam. Max.: 4, Diam. rim: 16, Diam. Neck: 1.1, Th.: 0.1 // Unpublished // 1st - early 2nd c.

548. (ΠΑΜ 73) // Giantison Street. Clinics of Veterinary Faculty of AUTH. Among finds from several graves. 23-1-1970 // Complete, deformed // Translucent greenish. // H.: 5.7, Diam. Max.: 3.5, Diam. rim: 2, Diam. Neck: 1.5, Diam. Base: 2, Th.: 0.2 // *Glass Cosmos* 2010, 253, no. 154 // 1st - early 2nd c.
549. (ΠΑΜ 74) // Giantison Street. Clinics of Veterinary Faculty of AUTH. Among finds from several graves. 23-1-1970 // Almost complete, deformed // Opaque dark-blue. // H.: 5.2, Diam. Max.: 3, Diam. rim: 1.7, Diam. Neck: 1.1, Diam. Base: 1, Th.: 0.2 // *Glass Cosmos* 2010, 253, no. 154 // 1st - early 2nd c.
550. (ΠΑΜ 77) // Giantison Street. Clinics of Veterinary Faculty of AUTH. Grave XI. 29-5-1970 // Almost complete, deformed // Opaque red. // H.: 7, Diam. Max.: 3 // *Glass Cosmos* 2010, 253, no. 156 // 1st - early 2nd c.
551. (BY 227/42) // Northern part of Evangelistrias Cemetery, area of former Lyssiatreion Clinic. Burial chamber II, cist grave. 19-5-1995 // Intact // Turquoise. No impurities or pinprick bubbles // H.: 6.8, Diam. Max.: 6, Diam. rim: 2.5, Diam. Neck: 1.8, Diam. Base: 3.5 // Pelekanidou 1993, 376, photo 5, 6 // 1st c.
552. (IST 1043) // Area of the Thessaloniki International Fair, Macedonian Museum of Contemporary Art. Grave 2. // Intact // Purple. // H.: 3.5, Diam. rim: 1.3, Diam. Neck: 1, Diam. Base: 2.8, Th.: 0.1 // *Glass Cosmos* 2010, 249, no. 141 // 1st c.
553. (BY 206) // Unknown exact provenance. 17-12-1949 // Mended, small parts of the body are missing // Translucent bluish. // H.: 8.5 // Unpublished // 1st end - early 2nd c.
554. (BY 227/39 β) // Northern part of Evangelistrias Cemetery, area of former Lyssiatreion Clinic. 16-8-1993 // Fragments // Translucent greenish. // H.: 3.2, Diam. Max.: 4, Diam. Base: 2.5 // Unpublished // 1st c.
555. (BY 321/7) // Agiou Nestoros Street, 26-28. Water main works. Grave T.72, near the skull (0.80WX0.18S). 17-11-2000 // Intact // Dark blue. No impurities, few pinprick bubbles // H.: 6.9, Diam. Max.: 4.5, Diam. rim: 2.1, Diam. Neck: 1.8, Diam. Base: 1.4, Th.: 0.2, Weight: 41.7 // Papanikola-Bakirtzi 2002 (ed.), 470, no. 644, dated to the 3rd c. // 1st c.
556. (BY 321/8) // Agiou Nestoros Street, 26-28. Water main works. Grave 72, (0.55ΔX0.23N), north of the right thigh. 17-11- 2000 // Intact // Translucent amber. No impurities, few pinprick bubbles // H.: 5, Diam. Max.: 2.5, Diam. rim: 1.4, Diam. Neck: 0.7, Diam. Base: 0.7, Th.: 0.1, Weight: 4 // Papanikola-Bakirtzi 2002 (ed.), 470, no. 645, dated to the 3rd c. // 1st c.
557. (ΜΘ 11539 β) // Ramona Area. Tile grave. // Part of the neck is missing. Covered by grey crust // Translucent (?) greenish. // H.: 5, Diam. Max.: 2.8, Diam. rim: 2.2, Diam. Neck: 1.1, Diam. Base: 1.8 // Thessaloniki apo ta proïstorika mechri ta christianika chronia 1986, 134 (E. Trakosopoulou-Salakidou) // 1st c.
558. (ΜΘ 20927) // Unknown street, probably Vardari Square area. In a box signed 'unregistered, probably from Vardaris Cemetery'. // Part of the body is missing // Translucent (?) bluish. No impurities, several pinprick bubbles // H.: 4.8, Diam. Max.: 2.9, Diam. rim: 1.5, Diam. Neck: 0.9, Diam. Base: 1.4, Th.: 0.06, Weight: 4.1 // *Glass Cosmos* 2010, 255, no. 160 // 1st c.
559. (BY 251/3 A-B) // Apolloniados street, 13. Grave 13. Disturbed burial, it was found on the floor relatively close to the pottery vessels. // Complete, mended // Translucent greenish. // H.: 10.5 // Nalpantis 1992, 316-317, photo 18. Grave dated to the second period of use of the cemetery in 5th - 6th c. // 1st c.
560. (BY 259/2) // AUTH Campus. North of the Central Library. Cremation 13 and between the closing slabs of grave 9. 26-7- 94 και 18-8-94 // Parts missing, mended // Translucent amber. No impurities or pinprick bubbles // H.: 9, Diam. Max.: 2, Diam. rim: 1.5, Diam. Base: 2, Weight: 8 // Unpublished // 1st c.
561. (ΜΘ 7225) // Unknown exact provenance. 1976 // Intact // Light blue. Few impurities // H.: 13, Diam. Max.: 2.7, Diam. rim: 2.7, Diam. Neck: 1.5, Diam. Base: 1.5, Th.: 0.1, Weight: 30 // *Glass Cosmos* 2010, 260, no. 175 // 1st c.
562. (ΜΘ 10792) // Ramona Area, Axios complex. Grave 4. // Intact // Translucent olive-green. Few pinprick bubbles and impurities // H.: 12, Diam. rim: 2.6, Diam. Neck: 1.4, Diam. Base: 1.5, Th.: 0.1, Weight: 23.37 // Petsas 1969a, 300, pl. 309 γ. *Glass Cosmos* 2010, 258 no. 169 // 1st c.
563. (ΜΘ 19348) // Junction of Melenikou and Chatziandreou Streets. Grave ?. -1994 // Small parts of the body and the neck are missing // Translucent dark-blue. No impurities, several pinprick bubbles // H.: 7.5, Diam. Max.: 1.4, Diam. Neck: 0.7, Diam. Base: 0.6, Th.: 0.1 // Unpublished // 1st c.
564. (ΜΘ 20933) // Unknown exact provenance. // Body fragment // Colourless. No impurities or pinprick bubbles // Diam. rim: 3.5X3.2, Th.: 0.12, Weight: 0 // Unpublished // 1st c.
565. (ΜΘ 20934) // Unknown exact provenance. // Part of the body is missing // Translucent bluish. No impurities, few pinprick bubbles // H.: 5.9, Diam. Max.: 1.4, Diam. rim: 1.6, Diam. Neck: 0.9, Th.: 0.08, Weight: 5.4 // Unpublished // 1st c.
566. (ΜΘ 20936) // Unknown exact provenance. // Body fragment // Translucent bluish. No impurities or pinprick bubbles // Dimensions: 3X2, Th.: 0.12 // Unpublished // 1st c.
567. (ΜΜΣΤ 61/7) // Area of the Thessaloniki International Fair, Macedonian Museum of Contemporary Art. Grave 61 vessel no. 7. // Rim and parts of the neck and body are missing. Traces of cremation are visible // Translucent greenish. No impurities, few pinprick bubbles // H.: 3.5, Diam. Max.: 2, Diam. Base: 1.4, Th.: 0.1 // Unpublished // 1st c.
568. (IST 1083) // Area of the Thessaloniki International Fair, Macedonian Museum of Contemporary Art. Grave 160, vessel no. 8. 2001 // Intact // Translucent greenish. // H.: 3, Diam. Max.: 1.2, Diam. rim: 0.7, Diam. Neck: 0.4, Diam. Base: 0.5, Th.: 0.1 // *Glass Cosmos* 2010, 252, no. 152 // 1st c.
569. (IST 1080) // Area of the Thessaloniki International Fair, Macedonian Museum of Contemporary Art. Grave 160, vessel no. 11. 2001 // Intact // Translucent bluish. No impurities, few pinprick bubbles // H.: 12.5, Diam. Max.: 2, Diam. rim: 2.2, Diam. Neck: 1, Diam. Base: 1.3 // *Glass Cosmos* 2010, 252, no. 149 // 1st c.
570. (ΜΜΣΤ 161) // Area of the Thessaloniki International Fair, Macedonian Museum of Contemporary Art. Grave 161.// Fragments // Translucent greenish. Few impurities, several pinprick bubbles // Max. pres. frag.: L. 2 // Unpublished // 1st c.
571. (ΠΑΜ 62) // AUTH Campus. Theological Department Building. Grave 1κ, Sector 7. 7-7-1965 // Rim is missing // Translucent greenish. Few pinprick bubbles no impurities // H.: 4.8, Diam. Max.: 1.2, Diam. Neck: 0.5 // Unpublished // 1st c.
572. (BY 227/2) // Northern part of Evangelistrias Cemetery, area of former Lyssiatreion Clinic. // Intact // Translucent amber. // H.: 6.5, Diam. rim: 1.8, Diam. Base: 1.6 // Pelekanidou 1993, 376, photo 5, 6. // 1st c.
573. (BY 227/32) // Northern part of Evangelistrias Cemetery, area of former Lyssiatreion Clinic. Grave 229. 29-6-1994 // Intact // Translucent bluish. // H.: 10, Diam. rim: 1.3, Diam. Base: 1.3 // Pelekanidou 1993, 376, photo 5, 6. From a grave with multiple burials. // 1st c.
574. (BY 227/49) // Northern part of Evangelistrias Cemetery, area of former Lyssiatreion Clinic. 27-4-1994 // Intact // Translucent greenish. // H.: 3, Diam. Max.: 1.8, Diam. rim: 1.2 // Pelekanidou 1993, 376, photo 5, 6 // 1st c.
575. (ΜΘ 83) // Amphipolis. 14.05.1913 // Intact // Translucent bluish. // H.: 11, Diam. Max.: 3.5, Diam. rim: 2, Diam. Neck: 1.4, Diam. Base: 1.5, Th.: 0.1, Weight: 17 // Unpublished // 1st c.
576. (ΜΘ 16944) // Area of the Thessaloniki International Fair, Italian Pavilion. // Almost complete // Translucent greenish. // H.: 9, Diam. Max.: 2, Diam. rim: 1.5, Th.: 0.13 // Unpublished // 1st c.
577. (IST 1076) // Area of the Thessaloniki International Fair, Macedonian Museum of Contemporary Art. Grave 160, vessel no. 4. 2001 // Intact // Translucent bluish. No impurities, few pinprick bubbles // H.: 13, Diam. Max.: 2.5, Diam. rim: 2.5, Diam. Neck: 1.7, Diam. Base: 0.7, Th.: 0.07 // *Glass Cosmos* 2010, 251, no. 145 // 1st c.

578. (IST 1082) // Area of the Thessaloniki International Fair, Macedonian Museum of Contemporary Art. Grave 160, vessel no. 9. 2001 // Intact // Translucent bluish. No impurities, few pinprick bubbles // H.: 8.2, Diam. Max.: 1.7, Diam. rim: 2, Diam. Neck: 1, Diam. Base: 1.5 // *Glass Cosmos* 2010, 252, no. 151 // 1st c.

579. (BY 319/2) // Lagkada Street, middle plot of Axios Complex. Fill atop of grave 97. 20-2-1998 // Intact // Translucent greenish. No impurities, many pinprick bubbles // H.: 6, Diam. Max.: 8, Diam. rim: 1, Diam. Base: 0.8, Th.: 0.08, Weight: 1.4 // Makropoulou 2007, 152, pl. 68, without dating // 1st - 4th c.

580. (BY 320/4) // Rentina Castle. Basilica Complex, K39- layer α' ( probably a reburial of bones). Φ. 1328. 14-10-1991 // Fragment // Translucent greenish. Few pinprick bubbles no impurities // H.: 4.5, Diam. rim: 1.2, Diam. Neck: 1, Diam. Base: 0.8, Th.: 0.1 // Unpublished // 1st - 4th c.

581. (BAA 72) // Ancient Agora Bathhouse. Rectangular room, north of the Bathhouse. Destruction Layer. 1998 // Complete // Non vidi. // Mavromichali 2001, fig. 24. From a bath complex destroyed in the last twenty years of the 1st c., see Adam-Veleni (ed.) 2001, 86. // 1st c.

582. (BAA 73) // Ancient Agora Bathhouse. Rectangular room, north of the Bathhouse. Destruction Layer. 1998 // Complete // Non vidi. // Mavromichali 2001, fig. 25. From a bath complex destroyed in the last twenty years of the 1st c., see Adam-Veleni (ed.) 2001, 86. // 1st c.

583. (BY 227/40 a) // Northern part of Evangelistrias Cemetery, area of former Lyssiatreion Clinic. Grave 185. 19-4-1994 // Large parts missing, mended fragments // Opaque turquoise. Very few impurities and pinprick bubbles // H.: 5, Diam. Max.: 3.2, Diam. rim: 1.2, Diam. Neck: 1, Diam. Base: 2.3 // Unpublished // 1st - early 2nd c.

584. (BY 257/2) // Municipal Hospital, 'Agios Dimitrios', court. Grave 12. 23-4-87 // Intact // Translucent bluish. // H.: 13, Diam. Max.: 4.8, Diam. rim: 2.8, Diam. Neck: 2, Diam. Base: 3.5, Th.: 0.15 // Makropoulou 1990, 133, pl. 1β // 1st - early 2nd c.

585. (ΜΘ 11250 στ) // Sintrivani Square. Excavation of the Walls. In tile grave T1, (Diary no. T1, 2). 6-3-1970 // Tiny bit of the rim is missing // Translucent greenish. // H.: 4, Diam. rim: 2.1, Diam. Neck: 1.3, Diam. Base: 2, Th.: 0.03, Weight: 6 // Vavritsas 1971, 377. *Glass Cosmos* 2010, 254, no. 157 // 1st - early 2nd c.

586. (Av-Pi /4) // Municipal Hospital, 'Agios Dimitrios', plot of the Water Company. // Intact // grey-blue // H.: 7, Diam. Max.: 4, Diam. rim: 1.5, Non vidi // Avezou, Picard 1912, 337-361 // 1st - early 2nd c.

587. (Av-Pi /5) // Municipal Hospital, 'Agios Dimitrios', plot of the Water Company. // Intact // grey-blue. Non vidi // H.: 7.5, Diam. Max.: 4, Diam. rim: 1.5, Non vidi // Avezou, Picard 1912, 337-361 // 1st - early 2nd c.

588. (BY 227/13) // Northern part of Evangelistrias Cemetery, area of former Lyssiatreion Clinic. Grave 122. 25-6-1993 // Intact // Translucent bluish. // H.: 12, Diam. rim: 3.5, Diam. Base: 7.3 // Unpublished. From a grave with a clay lamp dated to the second half or to the end of the 2nd c. // 1st - early 2nd c.

589. (BY 227/14) // Northern part of Evangelistrias Cemetery, area of former Lyssiatreion Clinic. Grave 122. 25-6-1993 // Complete, mended // Translucent bluish. // H.: 11, Diam. rim: 3.2, Diam. Base: 7 // Pelekanidou 1993, 376, photo 5. From a grave with a clay lamp dated to the second half of the 2nd c. or the late 2nd c. // 1st - early 2nd c.

590. (BY 227/15) // Northern part of Evangelistrias Cemetery, area of former Lyssiatreion Clinic. Grave 122. 25-6-1993 // Fragmentary in pieces // Translucent bluish. No impurities, few pinprick bubbles // H.: 10, Diam. Max.: 7.8, Diam. rim: 3.5, Diam. Neck: 3.2, Diam. Base: 7.5, Th.: 0.15 // Unpublished. From a grave with a clay lamp dated to the second half or to the end of the 2nd c. // 1st - early 2nd c.

591. (BY 227/16) // Northern part of Evangelistrias Cemetery, area of former Lyssiatreion Clinic. Grave 122. 25-6-1993 // Fragmentary // milky-white. // // Unpublished. From a grave with a clay lamp dated to the second half or to the end of the 2nd c. // 1st - early 2nd c.

592. (BY 257/1) // Municipal Hospital, 'Agios Dimitrios', court. Grave 12. 23-4-87 // Intact // Translucent greenish. // H.: 13, Diam. Max.: 6.5, Diam. rim: 3, Diam. Neck: 2.5, Diam. Base: 5, Th.: 0.2 // Makropoulou 1990, 133, pl. 1β // 1st - early 2nd c.

593. (ΔΕΘ 159/2) // Area of the Thessaloniki International Fair, Southern Plot. Grave 159, vessel 2. 1993 // Complete, mendable // Translucent greenish. No impurities, few pinprick bubbles // H.: 11, Diam. Max.: 5.4, Diam. rim: 3.4, Diam. Neck: 2.3, Diam. Base: 4.5, Th.: 0.18 // Unpublished // 1st - early 2nd c.

594. (ΜΘ 21955) // Junction of Melenikou and Chatziandreou Streets. Grave 85. 22-4-1994 // Mended, small parts of the neck are missing // Probably translucent greenish. Indiscernible due to weathering // H.: 15.3, Diam. Max.: 7.5, Diam. rim: 3.7, Diam. Neck: 33, Diam. Base: 5, Th.: 0.15 // Tsimpidou-Auloniti, Lioutas 1994, 436, pl. 138β // 1st - early 2nd c.

595. (IST 1075) // Area of the Thessaloniki International Fair, Macedonian Museum of Contemporary Art. Grave 160, vessel no. 3. 2001 // Intact // Colourless decolourized. Few pinprick bubbles and impurities // H.: 7, Diam. Max.: 3.5, Diam. rim: 1.8, Diam. Neck: 1.3, Diam. Base: 1.2, Th.: 0.15 // *Glass Cosmos* 2010, 251, no. 144 // 1st - early 2nd c.

596. (ΠΑΜ 63) // AUTH Campus. Theological Department Building. Grave γX1. Σεπτέμβρης 1965 // Fragments // Translucent greenish. Few pinprick bubbles no impurities // Diam. rim: 2, Diam. Neck: 3, Diam. Base: 7, Est. H.: ca. 12-14 // Unpublished // 1st - early 2nd c.

597. (BY 227/34) // Northern part of Evangelistrias Cemetery, area of former Lyssiatreion Clinic. Burial chamber II, cist grave. 19-5-1995 // Almost complete, deformed in fire // Translucent greenish. // Unpublished // 1st - early 2nd c.

598. (BY 227/39 α) // Northern part of Evangelistrias Cemetery, area of former Lyssiatreion Clinic. 16-8-1993 // Fragment // Translucent bluish. // H.: 6.5, Diam. rim: 2.6, Diam. Neck: 1.7 // Unpublished // 1st - early 2nd c.

599. (ΔΕΘ 159/1) // Area of the Thessaloniki International Fair, Southern Plot. Grave 159, vessel 1. 1993 // Intact // Translucent greenish. Indiscernible due to weathering // H.: 8.7, Diam. Max.: 2.8, Diam. rim: 2.2, Diam. Neck: 1.2, Diam. Base: 2.5, Th.: 0.2 // Unpublished // 1st - early 2nd c.

600. (ΜΘ ΣιντριβάνιΤ2) // Junction of Egnatia Street and K. Melenikou Street. Grave 2. 1969 // Lower body part is missing // Translucent greenish. // H.: 8, Diam. rim: 1.3, Diam. Neck: 1, Diam. Base: 3, Th.: 0.1 // Unpublished // 1st - early 2nd c.

601. (ΜΘ ΣιντριβάνιΤ9Β8) // Junction of Egnatia Street and K. Melenikou Street. Grave 9 B8. 1969 // Deformed in fire // Translucent greenish. // H.: 7, Diam. rim: 1.5, Diam. Neck: 1, Th.: 0.15 // Unpublished // 1st - early 2nd c.

602. (ΜΘ ΣιντριβάνιΤ9Α2b) // Junction of Egnatia Street and K. Melenikou Street. Grave 9 A2. 1969 // Deformed in fire // Translucent greenish. // Diam. rim: 1.5, Diam. Neck: 1, Pres. H.: 5 Th.: 0.1 // Unpublished // 1st - early 2nd c.

603. (ΜΘ 6308) // Junction of Egnatia Street and K. Melenikou Street. Outside of grave 2. -4-1969 // Intact // Translucent bluish. No impurities, few pinprick bubbles // H.: 16.2, Diam. Max.: 6.6, Diam. rim: 3.2, Diam. Neck: 2.2, Diam. Base: 6, Th.: 0.2, Weight: 97.82 // Unpublished // 1st - early 2nd c.

604. (ΜΘ 11093) // Giantison Street. Clinics of Veterinary Faculty of AUTH. Grave Γ'. // Half of the rim is missing // Translucent greenish. // H.: 4, Diam. Neck: 1.1, Diam. Base: 1.5, Th.: 0.1, Weight: 3.62 // Petsas 1974, 329. *Glass Cosmos* 2010, 259, no. 172 // 1st - early 2nd c.

605. (ΜΘ 20935) // Unknown exact provenance. // In fragments // Translucent bluish. No impurities, few pinprick bubbles // H.: 3.6, Diam. Max.: 1.5, Diam. rim: 1.4, Diam. Neck: 0.8, Diam. Base: 1.3, Th.: 0.05 // Unpublished // 1st - early 2nd c.

606. (IST 1074) // Area of the Thessaloniki International Fair, Macedonian Museum of Contemporary Art. Grave 160, vessel no. 2. 2001 // Intact // Translucent bluish. No impurities or pinprick bubbles // H.: 3.1, Diam. Max.: 1.3, Diam. rim: 1.3, Diam. Neck: 0.4, Diam. Base: 0.8 // *Glass Cosmos* 2010, 250, no. 143 // 1st - early 2nd c.

607. (IST 1078) // Area of the Thessaloniki International Fair, Macedonian Museum of Contemporary Art. Grave 160, vessel no. 6. 2001 // Part of the neck and the rim are missing // Translucent slightly greenish. // H.: 6.6, Diam. Max.: 3, Diam. Neck: 1.2, Diam. Base: 3 Dimensions: Pres. H.: 6.6 Th.: 0.12 // *Glass Cosmos* 2010, 251, no. 147 // 1st - early 2nd c.

608. (ΜΘ 3000) // Agiou Dimitriou Street, 232, next to the enclosure wall of 'Kentrikon' Hospital. Chryse Gate, Sarcophagus of T. Claudius Lycus. 7-3-1962 // Intact // Translucent greenish. // H.: 3/ // Themelis 1963, 447, pl. Ivβ. *Glass Cosmos* 2010, 258, no. 170 // 1st - early 2nd c.

609. (ΜΘ 5480) // Unknown provenance. // Rim and neck are missing // Translucent olive-green. // H.: 5.5, Diam. Neck: 2, Diam. Base: 6.2, Th.: 0.13, Weight: 40.9 // Unpublished // 1st - early 2nd c.

610. (ΜΘ 5947) // AUTH Campus. Theological Department Building. Grave (2), Phase α'. 1965 // Large parts of the body and the neck are missing // Dark blue greenish. // H.: 15, Diam. rim: 3, Diam. Neck: 1.5, Diam. Base: 4.5, Th.: 0.1, Weight: 46.15 // Petsas 1966, 334-339, pl. 354 δ. // 1st - early 2nd c.

611. (ΜΘ 5955) // AUTH Campus. Theological Department Building. Grave under grave (3β), Sector 11. 7-1965 // Intact // Translucent bluish. // H.: 10, Diam. rim: 2, Diam. Neck: 1.2, Diam. Base: 2.7, Th.: 0.08, Weight: 15.37 // Unpublished // 1st - early 2nd c.

612. (ΜΘ 11174) // Ramona Area. Building 1, pit grave // Parts of the body and neck are missing. Blue greenish iridescent // Opaque dark-blue. // H.: 6.5, Diam. Max.: 2.8, Diam. rim: 3, Diam. Neck: 1.4, Diam. Base: 2.8, Th.: 0.1 // Unpublished // 1st - early 2nd c.

613. (ΜΘ 11241) // Sintrivani Square. Excavation of the Walls. West of grave T12 at the eastern edge of K. Melenikou Street. // Intact. Iridescent // Translucent bluish green. No impurities or pinprick bubbles // H.: 14.8, Diam. Max.: 7.25, Diam. rim: 3.4, Diam. Neck: 2.5, Diam. Base: 7.25, Th.: 0.35 // Vavritsas 1971, 377, pl. 370ζ. Thessaloniki apo ta proïstorika mechri ta christianika chronia 1986, 125 (E. Trakosopoulou- Salakidou). *Glass Cosmos* 2010, 271, no. 199 // 1st - early 2nd c.

614. (ΜΘ 11539 α) // Ramona Area. Tile grave. // Mended, tiny part of the body is missing. Covered with white patina // Translucent yellowish. // H.: 5, Diam. Max.: 2.7, Diam. rim: 1.3, Diam. Neck: 1, Diam. Base: 2.4, Th.: 0.12 // E. Trakosopoulou-Salakidou, Thessaloniki apo ta proïstorika mechri ta christianika chronia 1986, 134. *Glass Cosmos* 2010, 259, no. 173 // 1st - early 2nd c.

615. (ΜΘ 11539 γ) // Ramona Area. Tile grave. // Covered by dark crust // Translucent greenish. // H.: 2.8, Diam. Max.: 1.6, Diam. rim: 1.2, Diam. Neck: 0.9, Diam. Base: 1 // E. Trakosopoulou-Salakidou, Thessaloniki apo ta proïstorika mechri ta christianika chronia 1986, 134. *Glass Cosmos* 2010, 260, no. 174 // 1st - early 2nd c.

616. (ΜΘ 19258) // Junction of Melenikou and Chatziandreou Streets. Grave 46. 13-4-1994 // Mended // Translucent greenish. No impurities, few pinprick bubbles // H.: 15.5, Diam. Max.: 6, Diam. rim: 3.3, Diam. Neck: 1.7, Diam. Base: 6, Th.: 0.15 // Tsimpidou-Auloniti, Lioutas 1994, 436, pl. 138β // 1st - early 2nd c.

617. (ΜΘ 19264) // Junction of Melenikou and Chatziandreou Streets. Grave 91. 23-5-1994 // Intact // Translucent bluish. No impurities, several pinprick bubbles // H.: 15.5, Diam. Max.: 6, Diam. rim: 3.3, Diam. Neck: 1.7, Diam. Base: 6, Th.: 0.15 // Tsimpidou-Auloniti, Lioutas 1994, 436, pl. 138β // 1st - early 2nd c.

618. (ΜΘ 19265) // Junction of Melenikou and Chatziandreou Streets. Grave 91. 23-5-1994 // Mended, small parts of the neck are missing // Translucent bluish. No impurities, several pinprick bubbles // H.: 14, Diam. Max.: 5.2, Diam. rim: 3.2, Diam. Neck: 1.7, Diam. Base: 5.2, Th.: 0.15 // Tsimpidou-Auloniti, Lioutas 1994, 436, pl. 138β // 1st - early 2nd c.

619. (ΜΘ 20931) // Unknown exact provenance. // Several parts of the body are missing // Translucent greenish. No impurities, several pinprick bubbles // H.: 14, Diam. Max.: 7, Diam. rim: 3.6, Diam. Neck: 2.8, Diam. Base: 4.7, Th.: 0.18, Weight: 71.6 // Unpublished // 1st - early 2nd c.

620. (ΜΘ 20932) // Unknown exact provenance. // Several parts of the body are missing // Translucent greenish. No impurities, several pinprick bubbles // H.: 14.8, Diam. Max.: 6.7, Diam. rim: 3.6, Diam. Neck: 2.8, Diam. Base: 4.2, Th.: 0.18, Weight: 68 // Unpublished // 1st - early 2nd c.

621. (ΒΥ 255/5) // Margaropoulou Street, 20. Grave 103. 11-5-1995 // Body is missing // Translucent greenish. // Diam. rim: 2.5, Diam. Neck: 1.5, Pres. H.: 16 // Makropoulou 2007, 235, pl. 27. From a 3rd-century grave // 3rd c.

622. (ΜΘ ΣιντριβάνιΤ22) // Junction of Egnatia Street and K. Melenikou Street. Grave 22. 1969 // Broken, almost complete // Translucent greenish. // H.: 17, Diam. rim: 2, Diam. Neck: 1.5, Diam. Base: 4, Th.: 0.12 // Unpublished // 1st early - early 3rd c.

623. (ΜΘ 20998) // Plot property of Tsirakoglou. 1940 // Part of the body is missing. Covered by dark crust // Translucent greenish. // H.: 14.5, Diam. Max.: 4.9, Diam. rim: 2.15, Diam. Neck: 1.3, Diam. Base: 4.6, Th.: 0.05, Weight: 37 // Walter 1940, fig. 35. Kotzias 1953-54, 167-175. // 1st early - early 3rd c.

624. (ΜΜΣΤ 2/19) // Area of the Thessaloniki International Fair, Macedonian Museum of Contemporary Art. Grave 2. // Rim is missing. The rest of the vessel in fragments // Purple. No impurities, few pinprick bubbles // H.: 7, Diam. Max.: 4.5, Diam. Neck: 1.2, Diam. Base: 4 // Unpublished // 1st early - early 3rd c.

625. (ΒΥ 209/2) // Lagkada Street, Southern plot of Axios Complex. Grave 3. 12-9-1994 // Intact // Translucent greenish. No impurities, few pinprick bubbles // H.: 9.8, Diam. Max.: 8, Diam. rim: 4, Diam. Base: 3, Th.: 0.12 // Makropoulou 2007, 102, pl. 25, 68. grave dated to the 3rd or early 4th c, on the basis of the grave's construction style. // 3rd end - early 4th c.

626. (ΒΥ 5) // AUTH Campus. Law School Building. // Intact // Translucent greenish. No impurities or pinprick bubbles // H.: 16.5, Diam. Max.: 4.5, Diam. rim: 4.2, Diam. Neck: 1.5, Diam. Base: 2, Th.: 0.15, Weight: 49.5 // E. Kourkoutidou-Nikolaidou, Thessaloniki Istoria kai techni 1986, 53. Papanikola-Bakirtzi (ed.) 2002, 224, no. 260, (A. Antonaras), dated to the 4th c. // 2nd end - early 4th c.

627. (ΒΥ 98) // Agiou Nestoros Street, 8. Larnax α'. 5-4-1977 // Parts of the body are preserved // Translucent greenish. No impurities, few pinprick bubbles // Diam. Max.: 4, Diam. Base: 1.4 Dimensions: Pres. H.: 13 Th.: 0.12 // Unpublished, mentioned at Pazaras 1981, 372-387. Grave dated in the second half of the 3rd c. by the pottery finds and the wall-paintings // 3rd c., end

628. (ΜΘ 543) // Vardari Square. // Part of the rim missing // Translucent greenish and reddish. Few impurities // H.: 12.5, Diam. Max.: 4.5, Diam. rim: 4.4, Diam. Neck: 2.3, Diam. Base: 3.2, Th.: 0.08, Weight: 51.32 // E. Trakosopoulou-Salakidou, Thessaloniki apo ta proïstorika mechri ta christianika chronia 1986, 125. *Glass Cosmos* 2010, 284, no. 234 // 4th c.

629. (ΜΘ 545) // Vardari Square. // Intact // Purple. No impurities or pinprick bubbles // H.: 13, Diam. Max.: 4, Diam. rim: 4.6, Diam. Neck: 2.5, Diam. Base: 3.7, Th.: 0.2 // E. Trakosopoulou-Salakidou, Thessaloniki apo ta proïstorika mechri ta christianika chronia 1986, 125, fig. 123. *Glass Cosmos* 2010, 284-285, no. 235 // 4th c.

630. (ΒΥ 207/4) // Lagkada Street, Ivanof Park. Grave IV. 1991 // Complete, small bit of lip missing // Translucent greenish. Several pinprick bubbles and impurities // H.: 9, Diam. rim: 4.5, Diam. Base: 5, Th.: 0.13 // Makropoulou 1991, 257-270, photo 24. Burial dated by pottery to the 4th c. Makropoulou 2007, 46, pl. 27, 68. // 4th c.

631. (ΜΘ 11528) // Ramona Area. Vaulted grave. 3-4-1971 // Intact // Green and blue. Few pinprick bubbles // H.: 8, Diam. rim: 7, Diam. Base: 4 // E. Trakosopoulou-Salakidou, Thessaloniki apo ta proïstorika mechri ta christianika chronia 1986, 125, fig. 123. *Glass Cosmos* 2010, 275, no. 209 // 4th c.

632. (ΒΥ 10) // AUTH Campus. Law School Building. // Rim, neck and a handle are preserved // Translucent greenish and bluish. No impurities or pinprick bubbles // Diam. Max.: 4.2, Diam. rim: 3.4, Diam. Neck: 1.3, Pres. H.: 4.5 Th.: 0.1 // Unpublished // 3rd middle - early 5th c.

633. (ΜΘ 5491) // Unknown provenance. // Rim, handles and shoulders are preserved // Translucent greenish and bluish. // H.: 4, Diam. rim: 3, Diam. Neck: 1.5, Th.: 0.1, Weight: 13.05 // Unpublished // 3rd middle - early 5th c.

634. (ΜΘ 5492) // Unknown provenance. // Upper part of the amphora. Preserved are neck, shoulder, part of the body and the handles // Colourless, handles blue. No impurities very few pinprick bubbles // H.: 10, Diam. Max.: 7.5, Diam. rim: 5, Th.: 0.06, Weight: 70.9 // Unpublished // 3rd middle - early 5th c.

635. (MΘ 20922) // Unknown exact provenance. // Almost complete, mended. Small part of the body is missing // Translucent greenish. Few pinprick bubbles. Impurities only at the handles // H.: 17, Diam. Max.: 4.8, Diam. rim: 5.2, Diam. Neck: 2.4, Diam. Base: 4.4, Th.: 0.06, Weight: 112.9 // *Glass Cosmos* 2010, 290-291, no. 251 // 3rd middle - early 5th c.

636. (BY 95) // Apolloniados Street 18. Clearing grave 5. 1976 // Part of the body is missing // Translucent greenish. Few impurities, few pinprick bubbles // H.: 12, Weight: 59.5 // E. Kourkoutidou-Nikolaidou, Thessaloniki Istoria kai techni 1986, 53 // 3rd end - early 5th c.

637. (BY 217/6) // Junction of Koloniari, Galanaki and Papathanasiou Streets. Grave 1, eastern side of the grave. 4-6-1987 // Intact // Translucent greenish. No impurities, few pinprick bubbles // H.: 7.2, Diam. rim: 3.9 // Unpublished. From a cemetery used in the 4th c. Some of the built-graves were re-used in the 5th c. // 3rd end - early 5th c.

638. (BY 244/4) // Stratou Avenue 2, plot of the Museum of Byzantine Culture. Grave 49. 8-6-1989 // Mended // Translucent greenish. No impurities, few pinprick bubbles // H.: 14.5, Diam. rim: 2.3, Diam. Base: 4, Th.: 0.1 // Nalpantis 2003, 53, 129, draw. 22, pl. 43, from a grave with a coin dated between 218 and 222, and probably dated in the first quarter of the 3rd c. // 3rd c.

639. (BY 251/1) // Apolloniados Street 13. Grave 9. Undisturbed grave with multiple burials. The vessel was found in the NE corner with the bones from prior inhumation. // Mended, small parts of the body are missing // Translucent greenish. // H.: 14, Diam. rim: 5.2, Diam. Base: 4 // Nalpantis 1992, 316-317, photo 17, grave from the second phase of the cemetery, dated to the 5th-6th c. // 5th-6th c.

640. (BY 255/8) // Margaropoulou Street, 20. Grave 122. 11-7-1995 // Mended, small parts of the body are missing // Translucent greenish. No impurities or pinprick bubbles // H.: 10.1, Diam. Max.: 8, Diam. rim: 4.2, Diam. Neck: 2.5, Diam. Base: 3.6, Th.: 0.2 // Makropoulou 2007, 235, pl. 24, χ without dating // 3rd end - early 5th c.

641. (BY 259/1) // AUTH Campus. North of the Central Library. Grave 52. 6-9-1994 // Almost complete // Translucent greenish. Several impurities and pinprick bubbles // H.: 8.5, Diam. Max.: 4, Diam. rim: 1.5, Diam. Neck: 5, Diam. Base: 3, Th.: 0.1, Weight: 17.2 // Unpublished // 3rd end - early 5th c.

642. (BY 323/2) // Peleiadou Street, 3. Grave 14 0, 30 from the NE corner 0.70 SE corner. 5-12-2000 // Small part of the neck and the rim are missing // Translucent greenish. Several pinprick bubbles no impurities // H.: 15.5, Diam. rim: 5, Diam. Neck: 3.5, Diam. Base: 6.5, Th.: 0.12 // Unpublished // 3rd end - early 5th c.

643. (ΔΕΘ 6 α) // Area of the Thessaloniki International Fair, Southern Plot. Grave 6α. 1993 // Intact // Translucent greenish. Few impurities, few pinprick bubbles // H.: 11.5, Diam. Max.: 5.6, Diam. rim: 3.6, Diam. Neck: 1.4, Diam. Base: 4, Th.: 0.12 // Unpublished // 3rd end - early 5th c.

644. (ΔΕΘ 30) // Area of the Thessaloniki International Fair, Southern Plot. Grave 30. 1993 // Complete, mendable // Translucent greenish. No impurities, few pinprick bubbles // H.: 11, Diam. Max.: 7, Diam. rim: 3, Diam. Neck: 2.5, Diam. Base: 4.5, Th.: 0.1 // Unpublished // 3rd end - early 5th c.

645. (ΔΕΘ 94/1) // Area of the Thessaloniki International Fair, Southern Plot. Grave 94 vessel 1. 1993 // Rim is missing // Translucent greenish. No impurities, few pinprick bubbles // Diam. Max.: 7, Diam. Neck: 1.6, Diam. Base: 3, Pres. H.: 9.6, Th.: 0.1 // Unpublished // 3rd end - early 5th c.

646. (ΔΕΘ 94/2) // Area of the Thessaloniki International Fair, Southern Plot. Grave 94 vessel 2. 1993 // Intact // Translucent greenish. No impurities, few pinprick bubbles // H.: 13.6, Diam. Max.: 6.2, Diam. rim: 3.2, Diam. Neck: 1.6, Diam. Base: 2.7, Th.: 0.1 // Unpublished // 3rd end - early 5th c.

647. (Θ 14/99-Υ2 ΑΓΚ 7114) // Agion Panton Street, 24. Grave 21 (in the grave). Depth 5.57. Vessel 2. 3-2-2000 // Intact // Translucent greenish. No impurities, few pinprick bubbles // H.: 10.2, Diam. Max.: 6.5, Diam. rim: 3.7, Diam. Neck: 1.7, Diam. Base: 4, Th.: 0.1 // 9th EBA 2001, 6 // 3rd end - early 5th c.

648. (MΘ 523) // Vardari Square. // The upper part of the cylindrical neck is broken // Translucent greenish. No impurities, few pinprick bubbles // H.: 6, Diam. Max.: 6, Diam. Neck: 1.5, Diam. Base: 3.5, Th.: 0.12, Weight: 20.63 // Unpublished // 3rd end - early 5th c.

649. (MΘ 525) // Vardari Square. // Intact // Translucent greenish. // H.: 12, Diam. Max.: 8, Diam. rim: 4.7, Diam. Neck: 2.5, Diam. Base: 4, Th.: 0.1, Weight: 69 // Unpublished // 3rd end - early 5th c.

650. (MΘ 527) // Vardari Square. // Intact // Translucent greenish. No impurities, several pinprick bubbles // H.: 9, Diam. Max.: 5, Diam. rim: 2.5, Diam. Neck: 1.8, Diam. Base: 3.5, Th.: 0.15, Weight: 29 // Unpublished // 3rd end - early 5th c.

651. (MΘ 528) // Vardari Square. // Part of the body missing // Translucent greenish. No impurities, few pinprick bubbles // H.: 10, Diam. Max.: 5, Diam. rim: 2.7, Diam. Neck: 1.8, Diam. Base: 2.5, Th.: 0.08, Weight: 31.31 // Unpublished // 3rd end - early 5th c.

652. (MΘ 529) // Vardari Square. // Fair part of the body is missing // Translucent greenish. // H.: 14, Diam. Max.: 7, Diam. rim: 4.4, Diam. Neck: 2.2, Diam. Base: 4.5, Th.: 0.06, Weight: 60.39 // Unpublished // 3rd end - early 5th c.

653. (MΘ 5468) // Nea Apollonia. // Missing parts of the body, neck and rim // Translucent greenish. // H.: 12, Diam. Max.: 7, Diam. rim: 3.8, Diam. Neck: 2, Diam. Base: 3.6, Th.: 0.1, Weight: 52 // Unpublished // 3rd end - early 5th c.

654. (MΘ 5939) // AUTH Campus. Theological Department Building. Grave (III) barrel-vaulted, Phase α'. // Small part of the rim is missing // Translucent greenish. No impurities, few pinprick bubbles // H.: 13, Diam. Max.: 5, Diam. rim: 3, Diam. Neck: 1.5, Diam. Base: 3.5, Th.: 0.06, Weight: 49.92 // Unpublished // 3rd end - early 5th c.

655. (MΘ 5940) // AUTH Campus. Theological Department Building. Grave (XXVII), space 4. 10-1965 // Complete with fractures // Translucent olive-green. No impurities, few pinprick bubbles // H.: 11.5, Diam. Max.: 7, Diam. rim: 4.1, Diam. Neck: 2, Diam. Base: 4, Th.: 0.1, Weight: 53.18 // Unpublished // 3rd end - early 5th c.

656. (MΘ 7052) // AUTH Campus. Administration Building. Grave 19. 15-9-1965 // Complete, mended // Translucent greenish. Very few impurities, few pinprick bubbles // H.: 11, Diam. rim: 3.8, Diam. Neck: 2, Diam. Base: 3, Th.: 0.06, Weight: 40 // *Glass Cosmos* 2010, 293, no. 259 // 3rd end - early 5th c.

657. (MΘ 11094) // Giantison Street. Clinics of Veterinary Faculty of AUTH. Grave K4. // Rim and part of the neck and the body are missing // Translucent greenish. // H.: 8, Diam. Neck: 2, Diam. Base: 3, Th.: 0.08, Weight: 15.14 // Petsas 1974, 329 // 3rd end - early 5th c.

658. (MΘ 11144) // Ramona Area. Built grave. // Intact? // Translucent greenish. // H.: 10, Diam. Max.: 6.5, Diam. rim: 2.7, Diam. Neck: 2.2, Diam. Base: 4, Th.: 0.1, Weight: 40 // *Glass Cosmos* 2010, 293, no. 260 // 3rd end - early 5th c.

659. (MΘ 20911) // Eleftherias Street.13, Ampelokipi. Grave 3. 8-12-1977 // Intact // Translucent greenish. Few impurities, few pinprick bubbles // H.: 9, Diam. Max.: 6, Diam. rim: 3.2, Diam. Neck: 2.2, Diam. Base: 4, Th.: 0.12, Weight: 43.4 // Unpublished // 3rd end - early 5th c.

660. (MΘ 20914) // AUTH Campus. Theological Department Building. Grave Θ'. 23-5-1979 // Complete // Translucent greenish. // H.: 11.9, Diam. Max.: 6.7, Diam. rim: 2.8, Diam. Neck: 0.8, Diam. Base: 7.6, Th.: 0.06, Weight: 64.8 // *Glass Cosmos* 2010, 280, no. 224 // 3rd end - early 5th c.

661. (MΘ 20919) // Ramona Area. Stored with finds from Ramona, 'Grave 1st with brooch and coin' (Unidentified grave). 18-8-1972 // Intact // Translucent greenish. Very few impurities and pinprick bubbles // H.: 11.2, Diam. Max.: 5.7, Diam. rim: 2.5, Diam. Neck: 1.7, Diam. Base: 3.4, Weight: 31.1 // *Glass Cosmos* 2010, 287-288, no. 242 // 3rd end - early 5th c.

662. (MΘ 20993) // Plot property of Tsirakoglou. 1940 // Intact. Covered by hard crust // Translucent greenish. Probably no impurities and pinprick bubbles // H.: 9.25, Diam. Max.: 6.5, Diam. rim: 3, Diam. Neck: 1.6, Diam. Base: 3.6, Th.: 0.1, Weight: 27.4 // Walter 1940, fig. 35. Kotzias 1953-54, 167-175. // 3rd end - early 5th c.

663. (MΘ 20994) // Plot property of Tsirakoglou. 1940 // Intact. Covered by hard crust // Translucent greenish. Probably no impurities and pinprick bubbles // H.: 7.56, Diam. Max.: 4.8, Diam. rim: 3.6, Diam. Neck: 2, Diam. Base: 3, Th.: 0.06, Weight: 24.3 // Walter 1940, fig. 35. Kotzias 1953-54, 167-175. // 3rd end - early 5th c.

664. (ΠAM 19) // Area of the Thessaloniki International Fair. 28-4-1972 // Rim and base are missing // Translucent yellowish. No impurities, few pinprick bubbles // H.: 13.5, Diam. Max.: 8, Diam. Neck: 2.7 // Unpublished // 3rd end - early 5th c.

665. (ΠAM 32) // Stratou Avenue, Building of *YENEΔ/EPT3* TV Station. Grave 1. 15-3-1969 // Large part of the body is missing // Translucent greenish. Probably no impurities and pinprick bubbles // Diam. rim: 3, Diam. Neck: 1.2, Weight: 27.4 // Unpublished // 3rd end - early 5th c.

666. (ΠAM 44) // Junction of Apolloniados 31 and Michalitsiou Streets. Grave 5. 24-10-1970 // Intact // Translucent greenish. No impurities few elongated pinprick bubbles // H.: 13, Diam. Max.: 8, Diam. rim: 4.6, Diam. Neck: 2, Diam. Base: 4.5, Th.: 0.12 // Unpublished // 3rd end - early 5th c.

667. (BY 309/6) // Peleiadou Street, 5. Grave T. 44 burial layer. 31-10-1997 // Intact // Colourless decolourized. Few pinprick bubbles, no impurities // H.: 13.5, Diam. Max.: 7.7, Diam. rim: 4, Diam. Neck: 2.5, Diam. Base: 7, Th.: 0.12 // Makropoulou 2007, 251, pl. 23, from a grave with coins of Constantine I and Constans II (325-330) // 4th c.

668. (BY 2) // AUTH Campus. Law School Building. // Small part of the neck is missing // Translucent greenish. // H.: 7.3, Diam. Base: 4, Weight: 18.1 // E. Kourkoutidou-Nikolaidou, Thessaloniki Istoria kai techni 1986, 56 // 4th c.

669. (BY 34) // AUTH Campus. Sorority Building (Φ.Ε.Θ.). Grave A. 1968 // Intact // Translucent greenish. // H.: 11 // E. Kourkoutidou-Nikolaidou, Thessaloniki Istoria kai techni 1986, 52 // 4th c.

670. (BY 217/3) // Junction of Koloniari, Galanaki and Papathanasiou Streets. Stray find. 9-7-1987 // Mended // Translucent greenish. No impurities, few pinprick bubbles // H.: 7, Diam. Max.: 4, Diam. rim: 3, Diam. Neck: 2, Diam. Base: 3.5, Th.: 0.1 // Unpublished. From a cemetery used in the 4th c. Some of the built-graves were re-used in the 5th c. // 4th c.

671. (BY 255/1) // Margaropoulou Street, 20. Grave 8. 31-10-1994 // Intact // Translucent greenish. // H.: 11, Diam. rim: 2.2, Diam. Neck: 1.5, Diam. Base: 3.3, Th.: 0.12 // Makropoulou 2007, 224, pl. 28. From a grave probably dated to the 4th c. // 4th c.

672. (BY 317/4) // Kallitheas Avenue, Elementary School of Ampelokipi. Grave T. 48, burial layer. 24-9-1997 // Intact // Translucent greenish. // H.: 7, Diam. rim: 2.5, Diam. Base: 3.5 // Makropoulou 2007, 309, pl. 28, From a probably 3rd century grave // 4th c.

673. (BY 321/9) // Agiou Nestoros Street, 26-28. Water main works. Grave 41. At the height of the right arm of burial A' (0.75WX0.10S). 10-11-2000 // Intact // Translucent greenish. No impurities, several pinprick bubbles // H.: 6, Diam. rim: 2.9, Diam. Neck: 1.5, Diam. Base: 3.8, Th.: 0.12, Weight: 21.4 // Papanikola-Bakirtzi 2002 (ed.), no. 646, 470, dated to the 4th c. // 4th c.

674. (BY 321/10) // Agiou Nestoros Street, 26-28. Water main works. Grave 49. SE corner of the corner, south of the right foot of burial A'. 17-10-2000 // Intact // Translucent bluish. Few impurities, several pinprick bubbles // H.: 7.5, Diam. rim: 3.5, Diam. Neck: 1.8, Diam. Base: 2.6, Th.: 0.12, Weight: 27.1 // Papanikola-Bakirtzi 2002 (ed.), no. 261, 225, dated to the 4th c. // 4th c.

675. (MΘ 538) // Vardari Square. // Complete // Translucent greenish. No impurities very few pinprick bubbles // H.: 10, Diam. Max.: 4, Diam. rim: 3, Diam. Neck: 1.5, Diam. Base: 3.8, Th.: 0.05, Weight: 26.35 // Unpublished // 4th c.

676. (MΘ 5948) // AUTH Campus. Theological Department Building. Grave (VIII) barrel-vaulted. Phase α'. 1965 // Complete, mended // Translucent greenish. No impurities, few pinprick bubbles // H.: 9, Diam. rim: 2.5, Diam. Neck: 1.8, Diam. Base: 2.8, Th.: 0.08, Weight: 21.1 // Unpublished // 4th c.

677. (MΘ 5949) // AUTH Campus. Theological Department Building. Grave (2σ) built, rectangular grave, Sector 9. 1965 // Intact // Translucent slightly greenish. No impurities, few pinprick bubbles // H.: 8, Diam. Max.: 5.5, Diam. rim: 3, Diam. Neck: 2.5, Diam. Base: 3.8, Th.: 0.1, Weight: 28.03 // Petsas 1966, 334-339, pl. 351 στ. // 4th c.

678. (MΘ 5950) // AUTH Campus. Theological Department Building. Grave (VII) Phase α'. 1965 // Complete, filled // Translucent greenish. // H.: 7.7, Diam. Max.: 4, Diam. rim: 2.5, Diam. Neck: 1.8, Diam. Base: 3.2, Th.: 0.1, Weight: 17.9 // *Glass Cosmos* 2010, 285, no. 236 // 4th c.

679. (MΘ 10798) // Ramona Area, Axios complex. Grave 3. // Complete, mended // Translucent olive-green. No impurities very few pinprick bubbles // H.: 11, Diam. rim: 4.7, Diam. Neck: 1.7, Diam. Base: 4, Th.: 0.1, Weight: 51.01 // Petsas 1969a, 300, pl. 309 α. *Glass Cosmos* 2010, 278 no. 218 // 4th c.

680. (MΘ 19349) // Junction of Melenikou and Chatziandreou Streets. Grave. -1994 // Mended, small parts of the body and rim are missing // Translucent dark-blue. No impurities, several pinprick bubbles // H.: 3.5, Diam. Max.: 1.8, Diam. Neck: 0.9, Diam. Base: 1.2, Th.: 0.1 // Unpublished // 4th c.

681. (MΘ 20925) // Ramona Area, Lagkada Street, tall, twelve-storey tower. Tile grave. // Rim and part of the neck are missing // Translucent greenish. No impurities, few pinprick bubbles // H.: 6.5, Diam. Max.: 4.2, Diam. Neck: 1.8, Diam. Base: 2.2, Th.: 0.1 // Unpublished // 4th c.

682. (MΘ 21006) // Unknown street, probably Vardari Square area. In a box signed 'unregistered, probably from Vardaris Cemetery'. // Only the lower part of the body is preserved // Translucent greenish. // H.: 7.1, Diam. Max.: 7.5, Th.: 0.1, Weight: 36.8 // Unpublished // 4th c.

683. (BY 227/31) // Northern part of Evangelistrias Cemetery, area of former Lyssiatreion Clinic. Burial chamber I. 1994 // Intact // Translucent greenish. // H.: 5.5, Diam. rim: 1.9, Diam. Base: 2.4 // Pelekanidou 1993, 376, photo 5, 6. From a grave with a 4th-century cross-bow fibula. // 4th c.

684. (BY 317/16) // Kallitheas Avenue, Elementary School of Ampelokipi. Grave T. 61, fill of the grave. // Almost complete, part of the body is missing // Translucent greenish. Several pinprick bubbles no impurities // H.: 7.8, Diam. Max.: 5, Diam. rim: 2.6, Diam. Base: 3, Th.: 0.02 // Makropoulou 2007, 335, pl. 69, from a grave dated to the end of the 3rd or of the 4th c. // 3rd-4th c.

685. (BY 35/1) // AUTH Campus. Sorority Building (Φ.Ε.Θ.). Grave Z (1968). // Intact // Translucent olive-green. // H.: 13, Diam. Base: 5 // E. Kourkoutidou-Nikolaidou, Thessaloniki Istoria kai techni 1986, 56 // 3rd-4th c.

686. (BY 301/1) // Syntagmatarchi Fryzi Street, 31. Grave 3. 24-10-1996 // Intact // Colourless decolourized. Few pinprick bubbles and impurities // H.: 9.8, Diam. Max.: 6.5, Diam. rim: 3.5, Diam. Neck: 2, Diam. Base: 5, Th.: 0.2 // Makropoulou 2007, 286, pl. 24., from a cist grave used, according to the pottery finds, in the first half of the 4th c. // 4th c.

687. (BY 309/8) // Peleiadou Street, 5. Grave T. 35, burial layer. 21-10-1997 // Mended // Translucent greenish. Few pinprick bubbles, no impurities // H.: 11, Diam. rim: 3.5, Diam. Base: 4.5, Th.: 0.1 // Makropoulou 2007, 258, from a grave with coins of Constantine I (336-337) // 4th c.

688. (BY 217/32) // Junction of Koloniari, Galanaki and Papathanasiou Streets. Grave 5. 18-6-1987 // Mended, parts of the rim are missing // Purple and translucent greenish. No pinprick bubbles, several large inclusions // H.: 9.5, Diam. Max.: 5.7, Diam. rim: 3.6, Diam. Neck: 1.8, Diam. Base: 3.2, Th.: 0.18, Weight: 64.8 // Papanikola-Bakirtzi 2002 (ed.), no. 259, 224, dated to the 4th c. From grave dated by coins of Constantine I, minted around 335-336, and of Constans, minted around 337-40. // 4th c.

689. (BY 217/40) // Junction of Koloniari, Galanaki and Papathanasiou Streets. Tile grave 247. 16-7-1987 // The tip of the neck is missing // Translucent greenish. No impurities, few pinprick bubbles // Diam. Max.: 5.6, Diam. Neck: 1.8, Diam. Base: 2.8, Pres. H.: 10, Th.: 0.1 // Unpublished. From a grave with a bulbous clay vessel (ΜΒΠ, ΒΚ 4518/39), a form dated to the first half of the 4th c. (Nalpantis 2003, 123) // 4th c.

690. (BY 227/37) // Northern part of Evangelistrias Cemetery, area of former Lyssiatreion Clinic. Stray find. 27-4-1994 // Fragments mended // Colourless. No impurities very few pinprick bubbles // H.: 7, Diam. Max.: 6, Diam. Neck: 1.4, Diam. Base: 4, Th.: 0.1 // Unpublished // 3rd-4th c.

691. (BY 227/51) // Northern part of Evangelistrias Cemetery, area of former Lyssiatreion Clinic. Stray find. 27-4-1994 // Fragments mended // Colourless. // H.: 6.6, Diam. Max.: 6, Diam. rim: 1.5, Diam. Neck: 1.6, Diam. Base: 4, Th.: 0.1 // Unpublished // 3rd-4th c.

692. (BY 324/1) // Borou Street, 4. Grave 53, vessel 1. 30-6-2000 // Intact // Translucent greenish. No impurities, few pinprick bubbles // H.: 3.7, Diam. Max.: 3, Diam. rim: 1, Diam. Neck: 0.9, Diam. Base: 1, Th.: 0.18, Weight: 8 // Papanikola-Bakirtzi 2002 (ed.), 470, no. 647 , dated 3rd c. // 3rd-4th c.

693. (ΜΘ 522) // Vardari Square. // Almost complete. Small part of the neck is missing // Translucent greenish. // H.: 10, Diam. Max.: 7, Diam. rim: 1.9, Diam. Neck: 1.9, Diam. Base: 3.2, Th.: 0.1, Weight: 42.8 // Unpublished // 3rd-4th c.

694. (ΜΘ 5938) // AUTH Campus. Theological Department Building. Grave (29) built, rectangular grave, Sector 1-2. // Rim and part of the neck are completed with resin // Translucent greenish. No impurities or pinprick bubbles // H.: 12, Diam. Max.: 8.3, Diam. Neck: 1.8, Diam. Base: 3.5, Th.: 0.3 // Unpublished // 3rd-4th c.

695. (ΜΘ 7224) // AUTH Campus. Administration Building. 1965 // Rim is missing // Translucent greenish. // H.: 14.5, Diam. Max.: 9, Diam. rim: 2.3, Diam. Neck: 2.3, Diam. Base: 5, Th.: 0.25 // Petsas 1967, 391, pl. 296E and 298, fig. 12. Thessaloniki apo ta proïstorika mechri ta christianika chronia, 1986, 125. *Glass Cosmos* 2010, 285-286, no. 238 // 3rd-4th c.

696. (ΜΘ 11047) // Ramona Area. Building 5, marble grave. // Intact // Translucent greenish. No impurities, few pinprick bubbles // H.: 8, Diam. rim: 1.5, Diam. Neck: 1.5, Diam. Base: 2.5, Th.: 0.1, Weight: 26.47 // Petsas 1974, 329, pl. 25 δ. // 3rd-4th c.

697. (ΜΜΣΤ 99/2) // Area of the Thessaloniki International Fair, Macedonian Museum of Contemporary Art. Grave 99, vessel 2. // Parts of the body are missing // Colourless decolourized. // H.: 7, Diam. Max.: 5, Diam. rim: 1.2, Diam. Neck: 1.2, Diam. Base: 3, Th.: 0.1 // Unpublished // 3rd-4th c.

698. (ΠΑΜ 11) // Ramona Area. 1971 // Lower body part is missing // Translucent greenish. No impurities, several pinprick bubbles // H.: 9, Diam. Max.: 6.8, Diam. rim: 1.6, Diam. Neck: 1.6, Th.: 0.12 // Unpublished // 3rd-4th c.

699. (ΠΑΜ 31) // Stratou Avenue, Building of *ΥΕΝΕΔ/ΕΡΤ3* TV Station. Grave 1. 15-3-1969 // Large part of the body is missing // Translucent greenish. No impurities, few pinprick bubbles // Diam. Max.: 5, Diam. rim: 1.7, Diam. Neck: 1.4 // Unpublished // 3rd-4th c.

700. (BY 21) // Agiou Dimitriou Street, cemetery found at the extension of the street. 14-3-1960 // Fragments of body and neck are preserved // Translucent greyish. // Th.: 0.05 // Unpublished // 3rd-4th c.

701. (BY 180) // Kallitheas Avenue, 5. Grave 10. 1983 // Intact // Translucent greenish. Few impurities, several pinprick bubbles // H.: 8, Diam. Max.: 4.7, Diam. rim: 3.5, Diam. Neck: 1.8, Diam. Base: 3, Th.: 0.1, Weight: 23.3 // Papanikola-Bakirtzi 2002 (ed.), 224, no. 258, dated to the 4th c. // 3rd-4th c.

702. (BY 209) // Lagkada Street, Southern plot of Axios Complex. Grave XXVII. 1991 // Intact // Translucent greenish. No impurities, few pinprick bubbles // H.: 9, Diam. Max.: 4.7, Diam. rim: 4.2, Diam. Neck: 2, Diam. Base: 3.5, Th.: 0.12 // Makropoulou 1991, 257-270, photo 29. Makropoulou 2007, 131, pl. 27, 68. grave dated to the 3rd-4th c. and before the end of the 4th c // 3rd-4th c.

703. (BY 309/2) // Peleiadou Street, 5. Grave 10, burial layer. 24-12-1996 // Intact // Translucent olive-green. // H.: 8.5, Diam. Max.: 5, Diam. rim: 3.5, Diam. Neck: 2.8, Diam. Base: 3, Th.: 0.12 // Makropoulou 2007, 261, pl. 27, from a grave dated for stylistic reasons to the 4th c. // 3rd-4th c.

704. (Θ 1/01-Υ2 ΑΓΚ 7215) // Thiseos Street, 2. Grave 4, burial 1 and reburial of bones 1 (Revealing of burial 1 and reburial of bones 1. depth -4.40/-4.60. 1.46 from the SE corner and 1.43 from the NE corner of T. 4). // Intact // Translucent greenish. Few pinprick bubbles and impurities // H.: 8.5, Diam. Max.: 4.8, Diam. rim: 3.5, Diam. Neck: 2.2, Diam. Base: 3.5, Th.: 0.1 // Unpublished // 3rd-4th c.

705. (Θ 18/01 Υ3) // N. Kapatou street, 4. Grave 1 cist grave (Vessel Υ3, depth -2.68, (second layer of burials, disordered group)). 29-8-01 // Partly preserved in fragments // Translucent greenish. Few pinprick bubbles, no impurities // Diam. rim: 3.2, Diam. Neck: 1.5, Diam. Base: 3.3, Pres. H.: 4.5, Th.: 0.1 // Unpublished // 3rd-4th c.

706. (ΜΘ ΑΚΑΤΑΓΡΑΦΟ 2) // Unknown street, probably Vardari Square area. In a box signed 'unregistered, probably from Vardaris Cemetery' wrapped in 1919 newspaper. Envelop 'Σ', designating probably grave number. 1920 // The lower part of the body is preserved in fragments // Translucent greenish. Few impurities, few pinprick bubbles // H.: 3.6, Diam. Max.: 4.7, Diam. Base: 3.5, Th.: 0.08 // Unpublished // 3rd-4th c.

707. (ΜΘ 5953) // AUTH Campus. Theological Department Building. Grave (13) built, rectangular grave, Sector 8. // Intact // Translucent olive-green. No impurities or pinprick bubbles // H.: 9, Diam. Max.: 4.9, Diam. rim: 3.9, Diam. Neck: 1.8, Diam. Base: 3.6 // Petsas 1966, 334-339, pl. 351 α. E. Trakosopoulou-Salakidou, Thessaloniki apo ta proïstorika mechri ta christianika chronia, 1986, 125. // 3rd-4th c.

708. (ΜΘ 5954) // AUTH Campus. Theological Department Building. Grave (13), Sector 8. 7-1965 // Filled in the rim and near the base // Translucent olive-green. Few impurities // H.: 8, Diam. rim: 3.7, Diam. Neck: 1.7, Diam. Base: 3.4, Th.: 0.1, Weight: 19 // Petsas 1966, 334-339, pl. 351 β. E. Trakosopoulou-Salakidou, Thessaloniki apo ta proïstorika mechri ta christianika chronia, 1986, fig. 123. *Glass Cosmos* 2010, 277-278, no. 217 // 3rd-4th c.

709. (ΜΘ 10878) // AUTH Campus. Administration Building. During the opening of a ditch, in a built cist grave. // Missing parts of the body and rim // Translucent olive-green. // H.: 9, Diam. rim: 3.6, Diam. Neck: 1.7, Diam. Base: 3, Th.: 0.1, Weight: 22 // Petsas 1974, 329. *Glass Cosmos* 2010, 273, no. 205 // 3rd-4th c.

710. (ΜΘ 11053) // Ramona Area. Built grave. 19-3-1969 // Neck is missing. Iridescent // Purple. // H.: 5.7, Diam. Max.: 4.2, Diam. rim: 2.9, Diam. Neck: 1.6, Diam. Base: 3, Th.: 0.12 // Petsas 1974, 329, pl. 25 γ. *Glass Cosmos* 2010, 279, no. 221 // 3rd-4th c.

711. (ΜΘ 11527) // Ramona Area. Grave 4. 2-3-1972 // Complete, mended // Translucent olive-green. No impurities, several pinprick bubbles // H.: 9.2, Diam. Max.: 5.3, Diam. rim: 3.2, Diam. Neck: 1.8, Diam. Base: 4 // E. Trakosopoulou-Salakidou, Thessaloniki apo ta proïstorika mechri ta christianika chronia 1986, 125. // 3rd-4th c.

712. (ΜΘ 21956) // Agiou Dimitriou Street, cemetery found at the extension of the street. 14-3-1960 // The lower part of the body is preserved // Translucent greenish. Few impurities, few pinprick bubbles // H.: 3.6, Diam. Max.: 4.7, Diam. Base: 3.5, Th.: 0.08 // Unpublished // 3rd-4th c.

713. (BY 15) // AUTH Campus. Law School Building. Quite probably from the cubiculum, according to Marki 2000, 191-192. // Base and body fragments are preserved // Translucent greenish. No impurities a lot of pinprick bubbles // H.: 46, Diam. Max.: 2.8, Diam. rim: 2.5, Diam. Neck: 1.6, Est. H.: 37 // E. Kourkoutidou-Nikolaidou, Thessaloniki Istoria kai techni 1986, 56. If it is indeed from the cubiculum, then it is dated to the second half of the 3rd c. // 4th-5th c.

714. (BY 94) // Agiou Nestoros Street, 8. Larnax α'. 5-4-1977 // Parts of the rim, neck and body are preserved // Translucent greenish. // // Unpublished, mentioned in Pazaras 1981, 372-387. Grave dated in the second half of the 3rd c. by the pottery finds and the wall-paintings // 3rd c., end

715. (BY 227/3) // Northern part of Evangelistrias Cemetery, area of former Lyssiatreion Clinic. Grave 32. 10-11-1993 // Intact // Translucent greenish. No impurities, many pinprick bubbles // H.: 46, Diam. rim: 3 // Pelekanidou 1993, 376, photo 5. From a grave with a coin of Alexander Severus (222-235). // 3rd c.

716. (BY 227/4) // Northern part of Evangelistrias Cemetery, area of former Lyssiatreion Clinic. Grave 11. 9-11-1992 // Fragmentary // Translucent greenish. // Pres. H.: 30 // Unpublished // 4th-5th c.

717. (BY 227/11) // Northern part of Evangelistrias Cemetery, area of former Lyssiatreion Clinic. Stray find. // Fragment // Translucent greenish. // Pres. H.: 4.8 // Unpublished // 4th-5th c.

718. (BY 227/21 α) // Northern part of Evangelistrias Cemetery, area of former Lyssiatreion Clinic. Grave 147. 1-3-1993 // Fragments mended. Large part of the body is missing // Translucent greenish. Probably several impurities and pinprick bubbles. Difficult to say due to

the weathering // H.: 50, Diam. Max.: 8.4, Diam. Neck: 2.6, Diam. Base: 2.7, Weight: 59 // Unpublished. From a grave with a coin of Maximianus (286-305). // 4th c.

719. (BY 227/21 β) // Northern part of Evangelistrias Cemetery, area of former Lyssiatreion Clinic. Grave 147. 1-3-1993 // Fragments mended. Large part of the body is missing // Translucent greenish. Probably several impurities and pinprick bubbles. Difficult to say due to the weathering // H.: 56, Diam. Max.: 10, Diam. Neck: 2.8, Diam. Base: 2.4, Weight: 51.5 // Unpublished. From a grave with a coin of Maximianus (286-305). // 4th c.

720. (BY 244/12) // Stratou Avenue 2, plot of the Museum of Byzantine Culture. Grave 16, SE corner. 17-4-1989 // Fragment // Translucent greenish. No impurities, few pinprick bubbles // Pres. H.: 3.5, Th.: 2.5 // Nalpantis 2003, 34, 130, fig. 8, from tomb with coins dated 317-348. // 4th c.

721. (BY 245/1) // Lagkada Street, Northern plot of Axios Complex. Northern half of the antechamber, layer B. 12-11-1989 // Fragment // Translucent greenish. // Pres. H.: 3 // Makropoulou 2007, 69, It is considered an intrusion from the surrounding area, after the destruction of the building. // 4th-5th c.

722. (BY 317/6) // Kallitheas Avenue, Elementary School of Ampelokipi. Grave T. 87, burial layer. 4-11-1997 // Mended, parts of the body are missing // Translucent greenish. Many pinprick bubbles, few impurities // H.: 54.5, Diam. Max.: 5, Diam. rim: 2.2, Diam. Neck: 1.8, Diam. Base: 2.3, Th.: 0.1 // Makropoulou 2007, 340, pl. 27, from a 3rd-4th century grave // 3rd-4th c.

723. (BY 321/6) // Agiou Nestoros Street, 26-28. Water main works. Grave T- 50. Along the right arm of burial β'. 18-10-2000 // Intact // Translucent greenish. Few impurities, several pinprick bubbles // H.: 47.5, Diam. Max.: 3, Diam. rim: 2.9, Diam. Neck: 2, Diam. Base: 2, Th.: 0.1 // Papanikola-Bakirtzi 2002 (ed.), 556, no. 765, dated to the 3rd-4th c. // 4th-5th c.

724. (ΜΘ 515) // Vardari Square. // Iridescent // Translucent greenish. Very few impurities, several pinprick bubbles // H.: 40.5, Diam. Max.: 3.8, Diam. rim: 2.9, Diam. Neck: 1.1, Diam. Base: 2.2 // E. Trakosopoulou-Salakidou, Thessaloniki apo ta proïstorika mechri ta christianika chronia 1986, 125. *Glass Cosmos* 2010, 282-283, no. 231 // 4th-5th c.

725. (ΜΘ 3682) // Lagkada Street, 67. In the soil stirred up during the transportation of a sarcophagus from Ramona Area. 26- 6-1963 // Front half of the vessel is preserved // Translucent greenish. No impurities, many pinprick bubbles // // Unpublished // 4th-5th c.

726. (ΜΘ 3697) // Ramona Area. 8-1963 // Lower part of the body is preserved // Translucent greenish. No impurities, many pinprick bubbles // // Unpublished // 4th-5th c.

727. (ΜΘ 3756) // Angelaki Street, plot property of Zervas. In a sarcophagus. 28-4-1960 // Rim is missing // Translucent greenish. Several pinprick bubbles and few impurities. // H.: 47.8, Diam. Max.: 3.4, Diam. rim: 3.3, Diam. Neck: 1.9, Diam. Base: 2.7 // E. Trakosopoulou-Salakidou, Thessaloniki apo ta proïstorika mechri ta christianika chronia 1986, 125. // 4th-5th c.

728. (ΜΘ 10797) // Ramona Area, Axios complex. Grave 2. 12-1968 // Complete, mended at the neck // Translucent greenish. Few impurities, several pinprick bubbles // H.: 44, Diam. Max.: 3.5, Diam. rim: 3, Diam. Neck: 1.5, Diam. Base: 2, Weight: 75 // Petsas 1969a, 300, pl. 309 δ. *Glass Cosmos* 2010, 283, no. 232 // 4th-5th c.

729. (ΜΘ 10925) // Ramona Area. // Complete, mended // Translucent greenish. Few pinprick bubbles and impurities // H.: 40, Diam. Max.: 6.1, Diam. rim: 2.7, Diam. Neck: 2, Diam. Base: 2.2 // Petsas 1974, 329, pl. 24 α. *Glass Cosmos* 2010, 278-79, no. 219 // 4th-5th c.

730. (ΜΘ 11054) // Ramona Area. Tile grave. 19.3.1969 // Body is missing // Translucent greenish. No impurities, several pinprick bubbles // H.: 37, Diam. Max.: 3.5, Diam. rim: 2.5, Diam. Neck: 1.5, Diam. Base: 2.2, Th.: 0.08, Weight: 62.8 // Unpublished // 4th-5th c.

731. (ΜΘ 11165) // Ramona Area. Tile grave. // Iridescent // Colourless. // H.: 40, Diam. Max.: 6, Diam. rim: 2.7, Diam. Neck: 1.7, Diam. Base: 1.8, Th.: 0.12, Weight: 78.31 // *Glass Cosmos* 2010, 279, no. 220 // 4th-5th c.

732. (ΜΘ 11166) // Ramona Area. Tile grave. // Broken, iridescent // Translucent greenish. No impurities, several pinprick bubbles // H.: 35, Diam. Max.: 5.5, Diam. Neck: 1.7, Diam. Base: 1.5, Pres. H.: 35 Th.: 0.06, Weight: 58 // *Glass Cosmos* 2010, 279-280, no. 222 // 4th-5th c.

733. (ΜΘ 20870) // AUTH Campus. Law School Building. In front of the School, Grave 12. 3-1966 // Large parts missing, very weathered // Translucent greenish. // H.: 10, Diam. Max.: 5.5, Diam. Base: 2.1, Weight: 20 // Unpublished // 4th-5th c.

734. (ΜΘ 20926) // Ramona Area, Lagkada Street, tall, twelve-storey tower. 27-1-1972 // Mended, small parts of the body are missing // Translucent greenish. No impurities, several pinprick bubbles // H.: 43.5, Diam. Max.: 3.5, Diam. rim: 2.5, Diam. Neck: 1.5, Diam. Base: 2.2, Th.: 0.08 // Unpublished // 4th-5th c.

735. (ΜΘ 20937) // Unknown street, probably Vardari Square area. In a box signed 'unregistered, probably from Vardaris Cemetery'. // Half of the body, the neck and the rim are missing // Translucent greenish. No impurities, few pinprick bubbles // H.: 25.4, Diam. Max.: 7, Diam. Base: 2.7, Th.: 0.06, Weight: 102.8 // Unpublished // 4th-5th c.

736. (ΠΑΜ 7) // Ramona Area. 1971 // Rim and parts of the body are missing // Translucent greenish. Few impurities, several pinprick bubbles // Diam. Max.: 8, Est. H.: 45-50, Th.: 0.12 // Unpublished // 4th-5th c.

737. (ΠΑΜ 8) // Ramona Area. 1971 // Rim and parts of the body are missing // Translucent greenish. Few impurities, several pinprick bubbles // Diam. Max.: 6, Est. H.: 45-50, Th.: 0.12 // Unpublished // 4th-5th c.

738. (ΠΑΜ 33) // AUTH Campus. Administration Building. Grave V. 9-9-1965 // Fragments // Translucent greenish. Several impurities and few pinprick bubbles // Est. H.: 40-45, Th.: 0.12 // Unpublished // 4th-5th c.

739. (ΠΑΜ 34) // AUTH Campus. Administration Building. Grave V. 9-9-1965 // Fragments // Translucent greenish. Several impurities and few pinprick bubbles // Est. H.: 40-45, Th.: 0.12 // Unpublished // 4th-5th c.

740. (ΠΑΜ 50) // Ramona Area. Tile grave. 22-3-1971 // Parts of the body are missing // Translucent greenish. No impurities, several pinprick bubbles // Diam. rim: 2.6, Diam. Neck: 1.5, Diam. Base: 2 // Unpublished // 4th-5th c.

741. (ΠΑΜ 61) // AUTH Campus. Theological Department Building. Grave 3 Tile grave T'. 22-6-1965 // Fragment // Translucent greenish. Several pinprick bubbles no impurities // H.: 4.5, Diam. Base: 3 // Unpublished // 4th-5th c.

742. (ΜΘ 20992) // Plot property of Tsirakoglou. 1940 // Intact. Covered by hard crust // Translucent greenish. // H.: 22.1, Diam. Max.: 4.15, Diam. rim: 2.3, Diam. Neck: 1.9, Diam. Base: 1.3, Weight: 31.2 // Walter 1940, fig. 35. Kotzias 1953- 54, 167-175. *Glass Cosmos* 2010, 206, no. 59 // 4th-5th c.

743. (ΜΜΣΤ 148/2) // Area of the Thessaloniki International Fair, Macedonian Museum of Contemporary Art. Grave 148. // Complete, small bit of lip missing // Translucent greenish. // H.: 11.5, Diam. Max.: 5.6, Diam. rim: 5, Diam. Neck: 2.7, Diam. Base: 3.3, Th.: 0.1 // Unpublished // 4th-5th c.

744. (BAA 80) // Ancient Agora Bathhouse. Rectangular room, north of the Bathhouse. Destruction Layer. 1998 // Fragment // Non vidi. // Mavromichali 2001, fig. 28. From a bath complex destroyed in the last twenty years of the 1st c., see Adam-Veleni (ed.) 2001, 86. // 1st c.

745. (BAA 81) // Ancient Agora Bathhouse. Rectangular room, north of the Bathhouse. Destruction Layer. 1998 // Fragment // Non vidi. // Mavromichali 2001, fig. 28. From a bath complex destroyed in the last twenty years of the 1st c., see Adam-Veleni (ed.) 2001, 86. // 1st c.

746. (ΜΘ 10299) // Filikis Etaireias. Excavation of the Hippodrome. Plot property of Nastos. Near the Eastern Wall. 11-9- 1967 // Mended // Translucent yellowish with white thread. // H.: 18, Diam. Max.: 0.5 / // Petsas 1968b, 332-334, plate 273β. Petsas 1974, 329.
Simple note to the reg. no. and reference to Petsas 1966/67, No. 54, pl. 53α. *Glass Cosmos* 2010, pp. 296-297, no. 270 // 1st c.

747. (ΜΘ 21980) // Unknown street, probably Vardari Square area. // Fragment // Translucent purple. No impurities, few pinprick bubbles // Dimensions: 12X0.5-0.35 // *Glass Cosmos* 2010, 295, no. 265 // 1st c.

748. (ΜΘ 21981) // Unknown street, probably Vardari Square area. // Fragment // Colourless and light blue. No impurities, few pinprick bubbles // Dimensions: 9X0.6-0.7 // *Glass Cosmos* 2010, pp. 295-296, no. 266 // 1st c.

749. VACAT

750. (MΘ 21982) // Unknown street, probably Vardari Square area. // Fragment // Translucent yellowish. No impurities, few pinprick bubbles // Dimensions: 4X0.6-0.4 // *Glass Cosmos* 2010, 296, no. 267 // 1st c.

751. (MΘ 21983) // Unknown street, probably Vardari Square area. // Fragment // Translucent greenish. No impurities, few pinprick bubbles // Pres. L.: 4, Th.: 0.5 // *Glass Cosmos* 2010, 296, no. 268 // 1st c.

752. (MΘ 21984) // Unknown street, probably Vardari Square area. // Fragment // Translucent purple. No impurities, few pinprick bubbles // Pres. L.: 2, Th.: 0.4 // *Glass Cosmos* 2010, 296, no. 269 // 1st c.

753. (BY 228/11) // Dioikitiriou Square. 22-7-1993 // Fragment // Translucent greenish. No impurities, and pinprick bubbles // Diam. rim: 4X3, Th.: 3 // Unpublished // 4th c.

754. (BY 282/1) // Thessaloniki. Unknown exact provenance. Find from the excavations of Prof. Xyngopoulos. // Fragment // Colourless, slightly greenish. No impurities and pinprick bubbles // Dimensions: 3X3, Th.: 3 // Unpublished // 4th c.

# 9. Catalogue and Brief History of the Excavations and Find-Spots of the Glass Vessels

## Thessaloniki Prefecture
## Thessaloniki, extra muros
## Eastern Necropolis

1. '*Agios Dimitrios*' Municipal Hospital and Water Company Site (*Εταιρεία Υδάτων*), 1911

Parts of the ancient necropolis of Thessaloniki were unearthed during the works that were carried out in the plot of the Municipal Hospital and in the adjacent property of the Water Company (*Εταιρεία Υδάτων*) in various periods before 1911. It is noted that most of the graves were rock-hewn and that some of them were covered with large roof tiles. Some of the movable finds were collected selectively e.g. inscriptions, clay vessels and figurines as well as glass vessels.[1]

| Cat. No. | Form | Other excavation indications | Find Date |
|---|---|---|---|
| 324 | 061 | | |
| 516 | 120c | | |
| 539 | 122a | | |
| 586-587 | 129 | | |

2-3. '*Agios Dimitrios*', Municipal Hospital, 1978, 1986-1988

During the works for the expansion of the hospital in 1978, in the hospital yard, 15 graves were revealed: cist-graves and barrel-vaulted graves. They are dated to the 3rd and 4th centuries.[2]

During the works for the arrangement of the hospital yard in 1986, 12 graves were revealed: probably 5 cist-graves, 2 tile-graves, 3 built cist-graves, and 2 more which are not described in the reports. They are dated to the Late Roman period.[3]

In the area of the new wing of the hospital, excavations conducted in 1987 revealed 31 graves arranged in two clusters: 15 barrel-vaulted graves, 9 cist-graves and 7 rock-hewn. On the surface, on top of two graves, two constructions in the shape of an altar were detected. Some of the graves featured wall-paintings and some were paved with crude mosaics and *opus sectile*. In the graves were found bronze coins; gold and bronze jewellery; and glass and clay vessels. These are dated between the 1st and 4th centuries.[4]

During the widening of the street north and east of the hospital, south of the area of 'Kipoi tou Pasa' (*Pasha's Gardens)*, 152 graves were revealed: 129 barrel-vaulted graves, 15 cist-graves, 1 tile-grave and 7 unidentified due to their bad state of preservation. They are dated to the Late Roman and mainly to the Early Christian period.[5]

2. Unknown find-spots

| Cat. No. | Form | Other excavation indications | Find Date |
|---|---|---|---|
| 325 | 061 | | |
| 373 | 072 | | |

3. Hospital courtyard

| Cat. No. | Form | Other excavation indications | Find Date |
|---|---|---|---|
| 229 | 051a | Grave 49 | 2-1987 |
| 584 | 128 | Grave 12 | 23-4-1987 |
| 592 | 129 | Grave 12 | 23-4-1987 |

4. Agios Pavlos (*Sikies* Site, *Heptapyrgion*), 1931

The exact find-spot or the conditions of discovery are not known. What is known however is that in this area extends the eastern necropolis of the city, and that Late Roman and Early Christian graves have occasionally been unearthed.

| Cat. No. | Form | Other excavation indications | Find Date |
|---|---|---|---|
| 422 | 093 | | March 1931 |

[1] Avezou, Picard 1912, 337-361.
[2] Kourkoutidou-Nikolaidou, Mavropoulou-Tsioumi 1978, 240, pl. 108 α-γ.
[3] Eleftheriadou 1987, 407, pl. 228β.
[4] Makropoulou 1987, 396-397, pl. 225α, β; Makropoulou 1990a, 190-208.
[5] Eleftheriadou 1989a, 271-282; Eleftheriadou 1988, 393-394; Eleftheriadou 1989b, 336-337.

| 424 | 094 | 'From Thessaloniki, across the *Heptapyrgion*. | March 1931 |
|---|---|---|---|

Found in a grave at the site Sikies'

5. Aqueduct, Agios Pavlos, 1966

Northeast of the Evangelistria cemetery, close to the reservoir of the Water Supply Organization (*Οργανισμός 'Υδρευσης*), 12 barrel-vaulted, Early Christian graves were unearthed.[6] They are quite probably geographically connected to the graves which were unearthed in the early 20th century, in a nearby plot that belonged to the Water Company (*Εταιρεία Υδάτων*).[7]

| Cat. No. | Form | Other excavation indications | Find Date |
|---|---|---|---|
| 338 | 063 | Room B, Grave A, B | December 1966 |

6. Agkelaki Street, 'Zervas' Plot, 1960

The exact site of the excavation is not known.[8] It probably coincides with the excavation conducted in 1959, during which a 3rd-century sarcophagus bearing inscriptions and relief decorations, and a single Roman built-grave were unearthed.[9]

| Cat. No. | Form | Other excavation indications | Find Date |
|---|---|---|---|
| 205 | 049 | Within a sarcophagus | 28-4-1960 |
| 727 | 147a | Within a sarcophagus | 28-4-1960 |

7. Agiou Dimitriou Street, East Extension, 1960

During works to widen of Agiou Dimitriou Street, north of the Physico-mathematical and Geoponics and Forestry Departments of the Aristotle University of Thessaloniki, many graves of all sorts were unearthed. In the graves were found clay and glass vessels, jewellery and coins. The graves are dated to the Late Roman and Early Christian periods.[10]

| Cat. No. | Form | Other excavation indications | Find Date |
|---|---|---|---|
| 355, 362 | 070 | | |
| 405 | 080 | | |
| 445 | 104 | | 1-2-1960 |
| 700, 712 | 146 | | 14-3-1960 |

8. No. 232 Agiou Dimitriou Street (Near Wall of the ***'Kentrikon'*** Hospital), 1962

A sarcophagus was unearthed at this spot, which is dated to the last quarter of the 2nd century. Judging by the coins found inside it was used 4-5 times; the coins are dated up to the reign of Licinius (308-324). The glass unguentarium belongs to the latest burial, i.e. from the era of Licinius.[11]

| Cat. No. | Form | Other excavation indications | Find Date |
|---|---|---|---|
| 608 | 131 | Golden Gate, Sarcophagus of T. Claudius Lycus | 7-3-1962 |

9. Aristotle University Campus, North of Central Library, 1994

In the area north of the University's Central Library, during works to expand the Library, 56 graves and 2 burial complexes were revealed. 39 are dated from the 3rd to the 6th century: 1 barrel-vaulted, 1 tile-grave, 20 cist-graves, 10 pit-graves, 2 jar-burials and 5 cremations. 17 pit-graves are dated between the 16th and the 19th centuries.[12]

| Cat. No. | Form | Other excavation indications | Find Date |
|---|---|---|---|
| 560 | 126a | Cremation 13 and between the cover slabs of grave 92 | 6-7-94 and 18-8-94 |
| 641 | 139 | Grave 52 | 6-9-1994 |

10. Aristotle University Campus, Swimming Pool, 1970

In the area where the University swimming pool was to be built in 1966, 3 cist-graves, 1 barrel-vaulted grave and 2 other graves (with unidentified upper parts due to their bad state of preservation) were unearthed.[13] In addition, in 1970, in the same area 43

[6] Michailidis 1967, 438, pl. 325.
[7] Avezou, Picard 1912, 338.
[8] The Archive of the 16th Ephorate of Prehistoric and Classical Antiquities in Thessaloniki does not contain any information on the excavation of the plot.
[9] Petsas 1966/67, 277-368 (esp. page 298).
[10] Pelekanidis 1960, 223-224, pls. 194β, 195a, 196β; Touratsoglou, Loverdou-Tsigarida 1976, 25-26, 29-31; Marki 2006, 267-271, 274-275.
[11] Themelis 1963, 447, pl. Ινβ.
[12] Marki, Karydas 1994, 523.
[13] Michailidis 1967, 437-438, pl. 324.

barrel-vaulted graves and 1 cist-grave were revealed. They are dated between the 4th century and the reign of Justinian I, but they had been re-used in the middle Byzantine era,[14] and certainly again in modern times.[15] No other offerings were found in them other than a few pottery fragments and one glass vessel.[16] In the same area was also revealed a building with mosaic and *opus sectile* pavements, dated to the middle, or in the second half of the 5th century,[17] the largest part of which was destroyed by the construction of the later graves.[18]

| Cat. No. | Form | Other excavation indications | Find Date |
|---|---|---|---|
| 348 | 064a | 'T' is written on the envelope that contained it; probably the number of the grave | |

11-12. Aristotle University Campus, Administration Building, 1965, 1972

During the works for the new administration building in 1965, several graves were unearthed.[19] 8 barrel-vaulted graves, one with wall-paintings, 23 cist-graves ('rectangularly built'), 1 tile-grave, 7 pit-graves, 8 jar-burials and 3 cremations; at least 2 sarcophagi were also found.[20] The graves are dated from the Augustan period up to the reign of Justinian, with the highest density/concentration in the time of the Tetrarchy.[21]

On the site of the administration building 'by the street leading to the Ano Toumba', 82 graves were uncovered in 1972: 60 pit-graves, with one or more burials and secondary burials, most of them with no offerings, 9 tile-graves, each containing one burial and a very few offerings, 6 built cist-graves, containing one or more burials, 5 jar-burials, 1 cenotaph and 1 barrel-vaulted grave.[22]

11. Aristotle University Campus, Administration Building, 1965

| Cat. No. | Form | Other excavation indications | Find Date |
|---|---|---|---|
| 79 | 016 | Sarcophagus no 2 | 8-9-1965 |
| 115 | 031 | Sarcophagus no 2 | 8-9-1965 |
| 196 | 045 | Sarcophagus no 2 | 8-9-1965 |
| 363 | 070 | Sarcophagus no 2 | 8-9-1965 |
| 396 | 075 | Stray find | 4-10-1965 |
| 656 | 139 | Grave 19 | 15-9-1965 |
| 695 | 145 | | 1965 |
| 709 | 146 | During the opening of the ditch, within a built cist-grave | |
| 738-739 | 147a | Grave V | 9-9-1965 |

12. Aristotle University Campus, Administration Building, 1972

| | | | |
|---|---|---|---|
| 170-17 | 2036 | Grave 25, depth 2 μ. | 10-11-1972 |

13-15. Aristotle University Campus, Theological Department, 1965, 1977, 1979

During the works for the foundation of the new Theological Department in 1965, ancient graves were revealed. These works were only partly overseen by the Archaeological Service. Approximately 28 barrel-vaulted graves were found, one of them with wall-paintings. There were 31 cist-graves, described as 'rectangularly built', and more than 100 tile-graves. There were no pit-graves, according to the excavator 'probably because they were more ancient and were destroyed during the opening of the newer graves, which were very dense on that site, and only scattered offerings from them were found'. There were 29 jar-burials and one burial complex.[23] The graves are dated from the Augustan period to the reign of Justinian I, with higher density, probably, in the era of the Tetrarchy.[24]

A building was revealed during the excavations in 1965, among and above the graves, which was identified by the excavator as a warehouse. The building showed consequent phases and it was dated to the early Byzantine period, at least in part, since its walls were founded on graves dated to Late Antiquity and up to Justinian's time.[25] Clay moulds for clay lamps were found walled up the building, a fact that, along with the nature of the rest of the clay vessels and clay pipes found in it, led to the assumption that it was a building with a practical use, and near which a clay-lamp workshop must had existed. From the same excavations an amount of glassworking waste was also collected.

---

[14] Marki 2006, 75-78, 227-228.

[15] Marki 2006, 225, Russian coin struck in 1871. Marki dates one of the barrel-vaulted graves to the 6th century on the basis of stylistic and typological indications.

[16] Tsigaridas 1972, 559-564, pls. 509α, 510γ; Archive of the 16th Ephorate of Prehistoric and Classical Antiquities, Thessaloniki.

[17] Asimakopoulou-Atzaka 1998, 151, 264-65, pls. 190-191.

[18] Michailidis 1967, 437.

[19] Petsas 1966, 334-339, pls. 345-359.

[20] Only one sarcophagus is mentioned in the report in *Archaeologicon Deltion*. The existence of a second sarcophagus is concluded/conjectured on the basis of the labels which were found on, and in the glass vessels which found in it.

[21] Petsas 1966, 337.

[22] Alexandri 1973-74a, 669-670, pls. 482-483.

[23] Petsas 1966, 334-339, pls. 345-359.

[24] Petsas 1966, 337.

[25] Petsas 1966, 334-339, pls. 345-359 (esp. pages 338-339, pl. 347 β).

During the opening of the ditch for an electricity cable between the Philosophy and Theological Departments, in 1977, 4 Late Roman/Early Christian barrel-vaulted graves and two tile-graves were found.[26]

During the widening of Egnatia street in 1978-1979, in the area between the Law School and the Theological Department, near the administration offices of the International Fair of Thessaloniki (*ΔΕΘ*), and in front of the Children's Asylum (*Άσυλο του Παιδιού*), excavations were conducted which revealed 18 Late Roman and Early Christian graves: 14 barrel-vaulted graves and 4 tile-graves. The burials contained no, or very few, movable finds.[27]

Part of the eastern necropolis was also unearthed during the expansion to the west of the building in 1981. 12 mixed graves were found, some of them decorated with Christian wall-paintings. They are dated to the Late Roman and Early Christian era.[28]

13. Aristotle University Campus, Theological Department, 1965

| Cat. No. | Form | Other excavation indications | Find Date |
|---|---|---|---|
| 1 | 001 | | November 1965 |
| 22-24 | 006a | Probably stray find | 1965 |
| 380 | 08 | | November 1965 |
| 70 | 014 | | November 1965 |
| 73 | 015 | Probably stray find | 1965 |
| 80 | 016 | Near the grave T. 11, depth 0. 40 | 5-8-1965 |
| 81 | 016 | Sector 18, depth 1m | 2-6-1965 |
| 105 | 027 | | 1965 |
| 109 | 028 | Sector 4, Φ. 202 | 2-6-1965 |
| 112 | 030 | Grave (13) built, rectangular, Sector 8 | 1965 |
| 119 | 031 | Probably a stray find | 1965 |
| 131 | 034 | Grave T11, depth 0.80September | 1965 |
| 195 | 045 | Grave (3κ) tile-grave, Sector 13 | |
| 202 | 048 | | March 1965 |
| 242 | 051a | Grave (26) built, rectangular, Sector 13 | |
| 243 | 051a | Grave (ψ) barrel-vaulted grave, Sector 3 | |
| 244-246 | 051a | Grave (1η) tile-grave, Sector 6 | May 1965 |
| 264 | 051b | Grave (VIII) barrel-vaulted grave. Phase α' | 1965 |
| 339 | 063 | | |
| 340 | 063 | | 1965 |
| 347 | 064 | Grave (2ω) tile-grave, Sector 1 | 16-2-1965 |
| 367 | 071 | Grave (4γ) tile-grave, Sector 91965 | |
| 368 | 071 | Grave (XV) barrel-vaulted grave, Sector 5 | 1965 |
| 403 | 078 | Sector 18, depth 1 m. | 2-6-1965 |
| 431 | 097 | | |
| 459 | 112 | | |
| 571 | 126a | Grave 1κ, Sector 7 | 7-7-1965 |
| 596 | 129 | Grave γX1 | September 1965 |
| 610 | 131 | Grave (2), phase α' | 1965 |
| 611 | 131 | Grave under the grave (3β), Sector 1 | 1July 1965 |
| 654 | 139 | Grave (III) barrel-vaulted grave, Phase α' | |
| 655 | 139 | Grave (XXVII), room 4 | October 1965 |
| 677 | 141 | Grave (2σ) built, rectangular, Sector 9 | 1965 |
| 678 | 141 | Grave (VII) Phase α' | 1965 |
| 676 | 141 | Grave (VIII) barrel-vaulted grave. Phase α' | 1965 |
| 694 | 145 | Grave (29) built, rectangular, Sector 1-2 | |
| 707-708 | 146 | Grave (13) built, rectangular, Sector 8 | July 1965 |
| 741 | 147a | Grave 3 tile-grave T' | 22-6-1965 |

[26] Mavropoulou-Tsioumi, Kourkoutidou-Nikolaidou, Pazaras 1977b, 233.

[27] Romiopoulou 1979, 275.

[28] Mavropoulou-Tsioumi, Kourkoutidou-Nikolaidou 1981, 311, pl. 203β.

14. Aristotle University Campus, Theological Department, 1977

| Cat. No. | Form | Other excavation indications | Find Date |
|---|---|---|---|
| 113 | 031 | Expansion of the Department, Grave II | 1977 |

15. Aristotle University Campus, Theological Department, 1979

| Cat. No. | Form | Other excavation indications | Find Date |
|---|---|---|---|
| 660 | 139 | Grave Θ', offering | 23-5-1979 |

16. Aristotle University Campus, Law School, 1965, 1966

A large number of barrel-vaulted and cist-graves was unearthed on the site of the Law School. The graves were either in clusters or isolated and some of them had wall-paintings.[29]

In the excavations of 1965, several graves were revealed, but in the archaeological reports most of them were not included. Only a few exceptional graves are mentioned, i.e. a cross-shaped one and one with marble revetments, inscription and wall-paintings.[30] The graves are dated from the Augustan era to the reign of Justinian I, with a greater density in the time of the Tetrarchy.[31]

In 1966 during the excavation next to the Law School, undertaken during the opening of the ditch for a water main, at least 10 graves were revealed. A glass vial with incised hunting scene and inscription was found in tile-grave no. I.[32] 14 built-graves (probably cist-graves), 6 tile-graves, 10 pit-graves and 1 barrel-vaulted grave were unearthed in the same year, during the opening of a ditch for the installation of a further water main for the Law School.[33]

In front of the Law School, during the works for the final arrangement of the site that took place in 1977, 2 Late Roman cist-graves were revealed.[34]

Excavation was conducted in the space between the buildings of the Law School and the Theological Department, and during the widening of the Egnatia street in 1978-1979, near the administration offices of the Thessaloniki International Fair (*ΔΕΘ*) and in front of the Children's Asylum Institute (*Άσυλο του Παιδιού*), where 18 Late Roman or Early Christian graves were unearthed: 14 barrel-vaulted graves and 4 tile-graves. Very few or no offerings were found in the graves.[35]

| Cat. No. | Form | Other excavation indications | Find Date |
|---|---|---|---|
| 215 | 051a | | |
| 314 | 055 | | |
| 626 | 134 | | |
| 632 | 138 | | |
| 668 | 141 | | |
| 713 | 147a | Quite probably from the *cubiculum* | |
| 733 | 147a | In front of the Department, Grave | 12 March 1966 |

17. Aristotle University Campus, Sorority Building (Φ.Ε.Θ.), 1968

During works for the construction of a water main next to the Sorority Building (*ΦΕΘ*), in 1968, a cluster of Early Christian graves was revealed, one of them wall-painted.[36]

| Cat. No. | Form | Other excavation indications | Find Date |
|---|---|---|---|
| 71 | 015 | West of grave ΙΔ' | 1968 |
| 216 | 051a | West of grave ΙΔ' | 1968 |
| 669 | 141 | Grave A' | 1968 |
| 685 | 142b | Grave Z' | 1968 |

18. Egnatia and Melenikou Streets, 1969

During the salvage excavations conducted near the city wall, 6-8m from it, a brick cist-grave was found contained offerings. There were also 6 tile-graves and 2 standard cist-graves. The graves are dated before the era of the Tetrarchy, sometime in the Roman era.[37]

[29] Pelekanidis 1961/2c, 257, pl. 314α, β; Marki 2006, 213-214, 217, graves with glass vessels dated to the second half of the 3rd and to the first half of the 4th century.

[30] Petsas 1966, 334-339, pls. 345-359.

[31] Petsas 1966, 337.

[32] Petsas 1967a, 392, pls. 296ε, 298, drawing 12.

[33] Petsas 1969c, 154, pls. 55α-β, 56α-β, 57, 58, 66α, drawings 2-7; Petsas 1974, 330.

[34] Romiopoulou 1978, 234.

[35] Romiopoulou 1979, 275.

[36] Michailidis 1969, 361.

[37] Petsas 1970b, 352; Petsas 1974, 376.

| Cat. No. | Form | Other excavation indications | Find Date |
|---|---|---|---|
| 394 | 075 | Grave 9 A2 | 1969 |
| 487-49 | 1116 | Cenotaph | 24-4-1969 |
| 536-537 | 120c | Cenotaph | 24-4-1969 |
| 544 | 122b | Grave 9 B5 | 1969 |
| 600 | 130 | Grave 2 | 1969 |
| 601 | 130 | Grave 9 B8 | 1969 |
| 602 | 130 | Grave 9 A2 | 1969 |
| 603 | 130 | Outside Grave 2 | April 1969 |
| 622 | 132 | Grave 22 | 1969 |

19. Northern Area of 'Evangelistria' Cemetery (***'Lyssiatreion')***, 1993-97

The excavation covered the northern part of the 'Evangelistria' cemetery. The Remains of a mosaic-paved monumental entrance/ access to an unknown monument were revealed. A large, built, brick cistern was also found on the site. 262 graves were also uncovered: 32 cist-graves, 71 tile-graves, 135 pit-graves, and 4 jar-burials. In addition, many 'free' pit-burials were revealed, as well as several barrel-vaulted graves, re-used several times, and at some point their vaults were removed and they were transformed into simple cist-graves. A cross-shaped burial monument and three clusters of barrel-vaulted graves (with 6, 8 and 10 barrel-vaulted graves, respectively) came to light, dated to the 6th century. In the graves were finds of jewellery and coins, and many clay and glass vessels. The graves are dated between the 3rd and 6th centuries.[38]

| Cat. No. | Form | Other excavation indications | Find Date |
|---|---|---|---|
| 440 | 12a | Grave Δ' | 25-11-1991 |
| 450 | 12a | Southern cluster Γ'. Grave 2 (pit-grave) | 19-4-1995 |
| 500 | 12b | Southern cluster Γ'. Grave 2 (pit-grave) | 19-4-199566012d |
| 110 | 029 | Stray find | 27-4-1994 |
| 136-141 | 035 | Stray find | |
| 142 | 035 | Stray find | 28-4-1995 |
| 197 | 046 | Grave 122 | 25-6-1993 |
| 219 | 051a | Grave Δ' | 25-11-1991 |
| 221 | 051a | | |
| 222 | 051a | In the NW corner of a cist-grave, in the trench inside the eastern wall of the stoa, at a depth of 2.5m | |
| 223 | 051a | Grave 231 | 16-5-1995 |
| 256 | 051b | Grave 194 | 12-4-1994 |
| 276 | 052a | | 1994 |
| 277 | 052a | Grave 215 | 6-12-1994 |
| 354 | 069 | | |
| 435 | 098 | | 1993 |
| 448 | 107 | Grave Δ' | 25-11-1991 |
| 453 | 110 | Southern cluster Γ'. Grave 2 (pit-grave) | 19-4-1995 |
| 508 | 120b | Grave 185 | 19-4-1994 |
| 542-543 | 122b | Burial Monument II, cist grave | 19-5-1995 |
| 551 | 123 | Burial Monument II, cist grave | 19-5-1995 |
| 554 | 125 | | 16-8-1993 |
| 572 | 126b | | |
| 573 | 126b | Grave 229 | 29-6-1994 |
| 574 | 126b | | 27-4-1994 |
| 583 | 128 | Grave 185 | 19-4-1994 |
| 588-591 | 129 | Grave 122 | 25-6-1993 |
| 597 | 130 | Burial Monument II, cist grave | 19-5-1995 |
| 598 | 130 | | 16-8-1993 |
| 683 | 142a | Burial Monument I | 1994 |
| 690-691 | 145 | Stray find | 27-4-1994 |
| 717 | 147a | Stray find | |
| 716 | 147a | Grave 1 | 19-11-1992 |
| 715 | 147a | Grave 32 | 10-11-1993 |
| 718-719 | 147a | Grave 147 | 1-3-1993 |

20. No. 2 Leoforos Stratou, Museum of Byzantine Culture, 1988-9

On the site of the Museum of Byzantine Culture, at the junction of 3rd Septemvriou Street and Leoforos Stratou, a salvage excavation was conducted that revealed parts of the ancient necropolis. 88 graves, one building, and the remains of artisanal installations were unearthed. 46 of the graves were tile-graves, 24 cist-graves, 12 were pit-graves, 1 barrel-vaulted grave, which had wall-paintings, 3 'free' burials and 2 built-graves, which, due to the bad state of preservation of their upper parts, were unidentifiable. In the graves were found bronze coins, clay and glass vessels, gold, bronze and glass jewellery, as well as metal tools.[39]

[38] Pelekanidou 1992, 439-440; Pelekanidou 1993, 373-387; Pelekanidou 1994, 526-529, pls. 158, 159; Pelekanidou 1995, 534-535, pl. 165.

[39] Nalpantis 1988, 381; Nalpantis 1989, 346; Nalpantis 2003.

| Cat. No. | Form | Other excavation indications | Find Date |
|---|---|---|---|
| 206 | 050 | Grave 89 | 13-10-1992 |
| 225 | 051a | Grave 16, SE corner | 13-6-1989 |
| 224 | 051a | Grave 271 | 2-5-1989 |
| 278 | 052a | Grave 72 | 28-6-1989 |
| 279 | 052a | Stray find | 29-5-1989 |
| 280 | 052a | | 29-5-1989 |
| 297 | 052b | Grave 35 | 14-6-1989 |
| 298 | 052b | Grave 35 | 13-6-1989 |
| 326 | 061 | Grave 93 | 19-10-1992 |
| 415 | 088 | Grave 63 | 13-6-1989 |
| 638 | 139 | Grave 49 | 8-6-1989 |
| 720 | 147a | Grave 16, SE corner | 17-4-1989 |

21. Leoforos Stratou, 3rd Army Corps Site (Building of the *ΥΕΝΕΔ/ΕΡΤ*3 TV Station), 1969

Several graves were revealed during the excavation for the foundations of the building for the Military television station, at the SE corner of a plot that dated from the Hellenistic period up to the 4th century. The finds included 17 tile-graves, 2 barrel-vaulted graves, 11 cist-graves, and 2 Hellenistic cist-graves. In the graves were found bronze coins, clay and glass vessels, gold, bronze and glass jewellery, as well as metal tools.[40]

| Cat. No. | Form | Other excavation indications | Find Date |
|---|---|---|---|
| 404 | 079 | Grave 1 | 15-3-1969 |
| 665 | 139 | Grave 1 | 15-3-1969 |
| 699 | 145 | Grave 1 | 15-3-1969 |

22. Filikis Etaireias Street. 'Hippodrome', Anast. Nastos Site, 1967

Part of the eastern city wall and part of the long, eastern side of the Hippodrome was revealed in the plot at Filikis Etaireias Street, next to the church of SS Constantinos and Eleni. Among the movable finds of this excavation are included fragments of bird-shaped glass vessels and clay figurines.[41]

| Cat. No. | Form | Other excavation indications | Find Date |
|---|---|---|---|
| 476-477 | 116 | Close to the eastern wall | 11-9-1967 |
| 492 | 116 | Near the western face of the eastern wall | 11-9-1967 |
| 507 | 120a | Near the western face of the eastern wall | 11-9-1967 |
| 512-513 | 120b | Close to the eastern wall | 11-9-1967 |
| 746 | 148 | Close to the eastern wall | 11-9-1967 |

23. No. 39 Filikis Etaireias and D. Margariti Streets, 1973 (?)

During the excavations conducted prior to the construction of the new building, a part of the Galerian Walls was uncovered on the plot at the junction of the two streets. Among the discoveries, 4 graves, which were partly covered by the city wall, were identified: 2 cist-graves and 2 tile-graves.[42]

| Cat. No. | Form | Other excavation indications | Find Date |
|---|---|---|---|
| 479 | 116 | Grave II | |
| 504-505 | 120a | | |
| 5141 | 20b | Grave II | |
| 5241 | 20c | Grave III | |

24-27. Thessaloniki International Fair (*ΔΕΘ*), 1972, 1979, 1988, 1992, 2000

Parts of the eastern necropolis were already known (since 1940) in the area occupied by the site of the Thessaloniki International Fair. In 1972 an excavation was conducted for which no written records have been preserved. The photographic material, however, indicates that 11 cist-graves, 1 sarcophagus, 1 marble larnax and 2 square-built constructions in the shape of an altar had been excavated.[43]

During the widening of Egnatia Street, from November 1978 until June 1979, salvage excavations were conducted during which Late Roman and Early Christian many graves were revealed: in the area in front of the Children's Asylum Institute (*'Ασυλο του Παιδιού*), 7 seven graves (Θ', Ι', ΙΑ', and the cluster of graves ΙΒ'-ΙΕ') came to light; between the buildings of the Theological and Law Departments 4 graves (Ε', ΣΤ', Ζ', Η'); near the offices of the International Fair (*ΔΕΘ*) a further 3 were investigated (1-3). 14 are barrel-vaulted and 4 are tile-graves; the burials themselves contained very few or no movable finds.[44]

---

[40] Petsas 1970a, 347-35 (esp. page 349), pls. 301-303 α-δ; Petsas 1974, 375-376, fig. 21, pls. 78α-β, 79α-γ.

[41] Petsas 1968b, 332-334, pl. 273β, ε, 279α-γ; Petsas 1974, 330.

[42] Alexandri 1973-74b, 673-674, pl. 486 α-β. This information is partly provided from unpublished documents in the archives of the 16th Ephorate of Prehistoric and Classical Antiquities in Thessaloniki.

[43] Photo archive of the 16th Ephorate of Prehistoric and Classical Antiquities in Thessaloniki, nos. 1914, 1917-1918, 1921, 1924, 1928, 1930-1931, date of photography: 5-5-1972.

[44] Romiopoulou 1979, 275.

In the area SE of the tower of the Greek Organization of Telecommunications (*OTE*), in 1988, on the grounds of the amusement park, 6 pit-graves, 1 tile-grave, 1 cist-grave and 1 jar-burial were revealed.[45]

Excavations conducted in 1992 on the southern part of the International Fair (*ΔΕΘ*) site, with further finds of 260 graves. The tile-graves were more often located among the upper, Roman horizon of graves, and were mostly without movable finds. Many Hellenistic graves also appear, which are mostly simple pit-graves, while cremations and jar-burials are quite rare.[46] The graves contained bronze coins, clay and glass vessels, as well as gold, bronze and glass jewellery.

During works for the expansion of the Macedonian Museum of Modern Art, part of the ancient necropolis was revealed: cist-graves, tile-graves and pit burials.[47] The graves contained bronze coins, clay and glass vessels, as well as gold, bronze and glass jewellery.

24. Thessaloniki International Fair, Unknown Find-Spots

| Cat. No. | Form | Other excavation indications | Find Date |
|---|---|---|---|
| 311 | 053b | Grave 1, next to the disintegrated bones of a baby | 4-4-1979 |
| 343 | 063 | Grave | 14-4-1979 |
| 664 | 139 | | 28-4-1972 |

25. Italian Pavilion

| Cat. No. | Form | Other excavation indications | Find Date |
|---|---|---|---|
| 212 | 050 | Grave T. 80 | |
| 381 | 074a | Grave 31A, 0. 79m. from N. side, on the W. side. | |
| 576 | 126b | | |

26. Macedonian Museum of Modern Art

| Cat. No. | Form | Other excavation indications | Find Date |
|---|---|---|---|
| 460 | 112 | Grave 160, vessel no. 52001 | |
| 461-462 | 113 | Grave 160, vessels nos. 1, 102001 | |
| 480-484 | 116 | Grave 218-9-2000 | |
| 485-486 | 116 | Grave 53, vessels nos. 3, 4 | |
| 527-528 | 120c | Grave 2 | |
| 529-530 | 120c | Grave 53, vessels nos. 1, 2 | |
| 531-535 | 120c | Grave 123, vessels nos. 1-5 | |
| 546-547 | 122b | Grave 2 | |
| 552 | 123 | Grave 2 | |
| 5671 | 26a | Grave 61, vessel no. 7 | |
| 568-569 | 126a | Grave 160, vessels nos. 8, 112001 | |
| 570 | 126a | Grave 161 | |
| 577-578 | 126b | Grave 160, vessels nos. 4, 92001 | |
| 595 | 129 | Grave 160, vessel no. 32001 | |
| 606-607 | 130 | Grave 160, vessels nos. 2, 62001 | |
| 624 | 132 | Grave 2 | |
| 697 | 145 | Grave 99, vessel 2 | |
| 743 | 147b | Grave 148 | |

27. Thessaloniki International Fair (*ΔΕΘ*), Southern Plot

| Cat. No. | Form | Other excavation indications | Find Date |
|---|---|---|---|
| 167 | 036 | Grave 5 | 1993 |
| 234 | 051a | Grave 44 | 1993 |
| 260 | 051b | Grave 57 | 1993 |
| 261 | 051b | Sewege pipe 92 | 1993 |
| 323 | 060 | Grave 115 | 1993 |
| 360 | 070 | Grave 97 | 1993 |
| 465-467 | 116 | Trench 28, stratum 001 | 1993 |
| 468-470 | 116 | Grave 158, vessels 1-3 | 1993 |
| 471 | 116 | Grave 187, vessel 2 | 1993 |
| 472-475 | 116 | Grave 216, vessels 3-61993 | |
| 509-510 | 120b | Grave 158, vessels 4-51993 | |
| 511 | 120b | Grave 187, vessel 11993 | |
| 518-522 | 120c | Grave 216, vessels 1-2, 5, 7-81993 | |
| 593 | 129 | Grave 159, vessel 21993 | |
| 599 | 130 | Grave 159, vessel 11993 | |
| 643 | 139 | Grave 6a1993 | |
| 644 | 139 | Grave 301993 | |
| 645-646 | 139 | Grave 94, vessels 1, 21993 | |

[45] Moschonisiotou 1988, 354, pl. 204α. In a built cist grave lined with hydraulic mortar a glass vessel was found that was not located during our research.

[46] Kousoulakou 1993, 311-314, 317-319.

[47] Tsimpidou-Avloniti, Galiniki, Anagnostopoulou 2001, 215-228.

28. Melenikou and Chatziandreou Streets, 1994

During excavations here, 110 graves were unearthed: 13 cist-graves, 2 with an above-ground construction, probably a heroon; 38 tile-graves, half of them covering remains of cremations; a few simple tile-graves; 12 tile-graves with an above-ground altar, some of them with a clay pipe for libations; 3 tile-graves; 6 simple pit-graves, 8 with an above-ground altar; 13 free burials; and 3 jar-burials. Almost half the graves contained no offerings. The 16 coins found in the graves were in a bad state of preservation; the clay vessels inside were dated between the 1st and 3rd centuries, as, approximately, were the glass unguentaria as well.[48]

| Cat. No. | Form | Other excavation indications | Find Date |
|---|---|---|---|
| 303 | 053a | Grave 101 | 10-6-1994 |
| 563 | 126a | Grave | 1994 |
| 594 | 129 | Grave 85 | 22-4-1994 |
| 616 | 131 | Grave 46 | 13-4-1994 |
| 617-618 | 131 | Grave 91 | 23-5-1994 |
| 680 | 141 | Grave | 1994 |

29. ***Trigonion*** Tower, Chapel to the South, 1972

During works to clean and restore the eastern city wall,[49] a cluster of 12 barrel-vaulted graves was revealed, attached to the wall's western internal face. The graves were connected with a vaulted burial chapel, which was founded in a square tower of the wall, and from which Early Christian wall-painting and dedicatory inscriptions are preserved.[50]

| Cat. No. | Form | Other excavation indications | Find Date |
|---|---|---|---|
| 144 | 035 | Grave IV, on the floor, left of the entrance | 1972 |

30. Sintrivani Square, City Wall Excavations, 1970

A section of the Galerian Wall, and more than 30 graves, were unearthed during the excavations conducted between Panepistimiou, Egnatia and K. Melenikou Streets in 1970. The graves found included tile-graves, simple and others with above-ground constructions in the shape of altars; built-graves in the shape of an altar (similar to a larnax); and jar-burials.[51] All graves, with the exception of one, contained remains of cremations and are dated earlier than the first half of the 3rd century.[52]

| Cat. No. | Form | Other excavation indications | Find Date |
|---|---|---|---|
| 463 | 114 | Within built cist-grave T2 | 6-3-1970 |
| 585 | 128 | Within tile-grave T1, (numbered in the diary as T1, 2) | 6-3-1970 |
| 613 | 131 | Found west of grave T12 at the eastern edge of K. Melenikou Street | |

31. Tritis Septemvriou Street, Cemeterial Basilica, 1980

A 5th-century cemeterial basilica[53] and a 4th-century cross-shaped martyrium[54] were unearthed in 1980 at the junction of Tritis Septemvriou Street and Leoforos Stratou. At the site of the excavation 70 barrel-vaulted graves came to light: 11 were found in the foundation of the cross-shaped martyrium, 1 on the axis of the *bema* of the basilica, and 58 in the surrounding space of the the cross-shaped martyrium and the basilica. Most of graves were grouped in clusters and only a few were isolated; they were used in the period Constantine I and up to AD 618.

The site, before its (Christian) consecration was used as a cemetery, at least from the 1st century AD, according to a glass find which found outside a grave in the same plot.

| Cat. No. | Form | Other excavation indications | Find Date |
|---|---|---|---|
| 517 | 120c | Sector III. From the wide grave | |

32. No. 44 Vasilissis Sofias Street (Ethnikis Amynis Street), 1960

The exact whereabouts of these excavations have not been identified.[55] Due to the fact that the glass objects were found in a sarcophagus, all excavations conducted along this street, and the sarcophagi found, are listed:

4 sarcophagi were unearthed in Vasilissis Sofias Street in 1957.[56]
2, at least, inscribed sarcophagi (ΜΘ 10040, ΜΘ 10041) were found in the plot once occupied by the Student Centre (*Φοιτητική Λέσχη*) before 1968.[57]

[48] Tsimpidou-Avloniti, Lioutas 1994, 433-436, pls. 137-139α-β.

[49] Tsigaridas 1973a, 478, 496, drawings 1, pl. 439α-β; Lazaridis 1973-74, 729, pl. 524γ.

[50] Tsigaridas, Loverdou-Tsigarida 1979, 86-89, dated to the 5th century.

[51] Vavritsas 1971, 373-382, pls. 367-379.

[52] Vavritsas 1971, 375.

[53] Makropoulou 1983, 25-46, pl. 1-12.

[54] Marki 1981, 64-67, pl. Α'-ΙΔ'.

[55] The archive of the 16th Ephorate of Prehistoric and Classical Antiquities in Thessaloniki does not contain any information on the excavations of this plot. However, in the depots there are some 3rd- and 4th-century coins, clay vessels, jewellery, a wooden comb and a braid, which were all found in graves on the same plot. Furthermore, in the Museum of Byzantine Culture there is a coin, excavated at exactly the same date (25-4-1960), in a sarcophagus on a plot owned at the time by the Zervas family at 3 Angelaki Street. The movable finds of the excavation are being published by the author (Antonaras in press 1 and 2).

[56] Daux 1958, 760-761, figs. 9-10; document no. 432/21-7-1957 in the archives of the 16th Ephorate of Prehistoric and Classical Antiquities, Thessaloniki.

[57] Petsas 1974, 212-381 (esp. page 334). The vessels examined in this present study were found in sarcophagus No. 3, along with a glass bracelet

18 sarcophagi were unearthed in 1965 from the northern part of the same street, south of the junction with Egnatia Street and up to the junction with Pringipos Nikolaou Street.[58]

| Cat. No. | Form | Other excavation indications | Find Date |
|---|---|---|---|
| 185 | 041 | Sarcophagus 3rd | 25-4-1960 |
| 239 | 051a | Sarcophagus 3rd | 25-4-1960 |

**Western Necropolis**

33. No. 8 Agiou Nestoros Street, 1977

In the area of the western walls, during excavations for the foundations of a new building, an unspecified number of barrel-vaulted graves, and 1 cist-grave and 1 tile-grave, were found.[59] An Early Christian, barrel-vaulted and wall-painted burial building was also discovered; 2 cist-graves were opened below the pavement of this building. The building was dated, according to the paintings and the clay vessels found in the graves, to the second half of the 3rd century.[60] In the same plot a barrel-vaulted grave was also traced, featuring wall-paintings dated to the middle of the 4th century.

| Cat. No. | Form | Other excavation indications | Find Date |
|---|---|---|---|
| 450 | 108 | Larnax a' | 5-4-1977 |
| 627 | 134 | Larnax a' | 5-4-1977 |
| 714 | 147a | Larnax a' | 5-4-1977 |

34-37. Nos. 8, 26, 26-28, 32 Agiou Nestoros Street, Water Main Works, 2000

During excavations to lay a water main in Thessaloniki along Agiou Nestoros Street, part of the eastern necropolis was revealed, with Late Roman and Early Christian graves. In the graves were found clay and glass vessels as well as gold and bronze jewellery.[61]

34. No. 8 Agiou Nestoros St., Water Main Works

| Cat. No. | Form | Other excavation indications | Find Date |
|---|---|---|---|
| 366 | 071 | Grave 66 (Burial Δ') depth 3.90 | 10-11-2000 |

35. No. 26 Agiou Nestoros St., Water Main Works

| Cat. No. | Form | Other excavation indications | Find Date |
|---|---|---|---|
| 449 | 107 | Grave T. 48. At the height level of skull, burial A' | 13-10-2000 |

36. Nos. 26-28 Agiou Nestoros St., Water Main Works

| Cat. No. | Form | Other excavation indications | Find Date |
|---|---|---|---|
| 193 | 045 | Grave T. 39, At the height level of the right arm | 6-10-2000 |
| 259 | 051b | Grave T. 40, At the height level of the right thigh | 6-10-2000 |
| 555 | 125 | Grave T. 72, At the height level of the skull (0.80WX0.18S) | 17-11-2000 |
| 556 | 125 | Grave T. 72, (0.55ΔX0.23N), north of the right thigh | 17-11-2000 |
| 673 | 141 | Grave T. 41. At the height level of the right arm, burial A' | 10-11-2000 |
| 674 | 141 | Grave T. 49. SE corner of the grave, south of the right foot instep, burial A' | 17-10-2000 |
| 723 | 14a | Grave T. 50. Along the right hand, burial B' | 18-10-2000 |

37. No. 32 Agiou Nestoros St., Water Main Works

| Cat. No. | Form | Other excavation indications | Find Date |
|---|---|---|---|
| 233 | 05a | Grave T. 31 (B. 0.40-A. 0.50) | October 2000 |

38. No. 24 Agion Panton Street, 1999-2000

24 graves were located here: 9 barrel-vaulted graves, 14 cist-graves, and 2 tile-graves; part of a cistern was also revealed. In the graves were found clay and glass vessels, iron and bronze jewellery, a bone casket and bronze coins. The finds are dated to the 4th century.[62]

| Cat. No. | Form | Other excavation indications | Find Date |
|---|---|---|---|
| 647 | 139 | Grave 21 (In the grave). Depth 5.57. Vessel | 23-2-2000 |

39. No. 13 Apolloniados Street and Oraiopoulou Street, 1991, 1992

17 Early Christian and 5 post-Byzantine graves were revealed. Only 2 were cist-grave, the rest were barrel-vaulted. In the graves were found clay and glass vessels, jewellery and coins. The cemetery was used in two distinct periods, in the 4th, and then again in the 5th and 6th centuries.[63]

(reg. no. vessels MΘ 11548).

58 According to the registries of the 16th Ephorate of Prehistoric and Classical Antiquities, Thessaloniki.

59 Romiopoulou 1977a, 197, pl. 126α; Mavropoulou-Tsioumi, Kourkoutidou-Nikolaidou, Pazaras 1977a, 233, pl. 143β-γ.

60 Pazaras 1981, 375. The wall-paintings have been dated to the last quarter of the 3rd century (see Marki 2006, 114).

61 9th EBA 2003, 3, where glass vessels are illustrated.

62 9th EBA 2001.

63 Nalpantis 1992, 316-317.

| Cat. No. | Form | Other excavation indicationsFind Date |
|---|---|---|
| 559 | 12a | Grave 13. Distorted burial, found on the grave's pavement relatively close to the clay vessels |
| 639 | 139 | Grave 9. Undisturbed grave with multiple burials. The vessel was found in the NE corner next to the secondary burial |

40. No. 18 Apolloniados Street, 1976.

9 Early Christian graves came to light during excavations on this plot: 4 cist-graves and 5 barrel-vaulted.[64]

| Cat. No. | Form | Other excavation indications | Find Date |
|---|---|---|---|
| 636 | 139 | During the clearing of grave 5 | 1976 |

41. No. 31 Apolloniados Street and Michalitsiou Street, 1970

At least 6 graves were excavated in this plot: 2 barrel-vaulted and 4 cist-graves. A bronze jewel, and clay and glass vessels were found in the burials.[65]

| Cat. No. | Form | Other excavation indications | Find Date |
|---|---|---|---|
| 351 | 066 | Grave 5 | 24-10-1970 |
| 666 | 139 | Grave 5 | 24-10-1970 |

42. Aristatou, Panagopoulou and Patroklou Streets, Ampelokipi, 1991

During the excavation of the plot several graves were uncovered: 2 cist-graves dated to the 3rd century; 8 barrel-vaulted graves dated to the 4th century; and 2 burial complexes, each one comprising three-chambers, surrounded by a fence dated to the second half of the 4th century. The movable finds comprise glass unguentaria, bronze coins, small clay, bone and stone artefacts.[66]

| Cat. No. | Form | Other excavation indications | Find Date |
|---|---|---|---|
| 255 | 051b | Cist-grave No. II | 20-5-1991 |

43. No. 4 Borou Street, 2000

During the excavation conducted on the plot before the erection of a new building, part of Thessaloniki's eastern necropolis was unearthed. Glass vessels are among the movable finds from the site.[67]

| Cat. No. | Form | Other excavation indications | Find Date |
|---|---|---|---|
| 692 | 145 | Grave 53, vessel 1 | 30-6-2000 |

44. No. 8 Doxatou Street, 1993

11 isolated graves were revealed in the plot: 1 tile-grave, 4 cist-graves and 6 barrel-vaulted graves. Furthermore, 2 burial enclosures were notes, the first with 1 barrel-vaulted grave and the second with 2 cist-graves. Movable finds comprise a clay burial vessel, a bronze coin, glass unguentaria and small bronze objects, all dating the graves to the 4th and early 5th centuries.[68]

| Cat. No. | Form | Other excavation indications | Find Date |
|---|---|---|---|
| 220 | 051a | Grave 12 | 23-2-1993 |

45. No. 13 Eleftherias Street, Ampelokipi, 1977

During excavations here finds were made of 4 graves dated to the Late Roman and/or Early Christian period. Several cist-graves and 7 barrel-vaulted graves were noted.[69]

| Cat. No. | Form | Other excavation indications | Find Date |
|---|---|---|---|
| 423 | 093 | In Grave 3 | 8-12-1977 |
| 659 | 139 | In Grave 3 | 8-12-1977 |

46. Efxinou and Panagopoulou Streets, 1994

This plot presented 14 graves of various types: 3 barrel-vaulted, 3 cist-graves, 4 tile-graves, and 4 pit-graves. All graves were oriented NW-SE with heads NW. Movable finds were scarce and the graves are dated to the 4th and up to the early 5th century by construction technique and the movable objects they contained: clay and glass vessels, and some bronze jewellery.[70]

| Cat. No. | Form | Other excavation indications | Find Date |
|---|---|---|---|
| 226 | 051a | In pit-grave (No. 8), at the left of head of the deceased | 6-6-1994 |
| 227 | 051a | In barrel-vaulted grave (No. 3), at the feet of the deceased, in the S-SW corner | 2-6-1994 |
| 228 | 051a | In tile-grave grave (No, 6), next to the left temple of the deceased | 5-6-1994 |
| 413 | 086 | At the east corner of the SE side of barrel-vaulted grave (No. 2) | 1-6-1994 |

[64] Kourkoutidou-Nikolaidou, Mauropoulou-Tsioumi, Pazaras 1976, 268, pl. 210α.

[65] Archive of the 16th Ephorate of Prehistoric and Classical Antiquities, Thessaloniki, document no. 2149/29-10-1970.

[66] Makropoulou 1992b, 426-428, drawing 15.

[67] Marki 2000, 745-747.

[68] Makropoulou 1994, 507-509.

[69] Romiopoulou 1977b, 197, pl. 126β; Romiopoulou 1978, 233.

[70] Karydas 1994, 501, pl. 156α-γ.

47. Giannitson Street, Veterinary Department, 1969

Approximately 10 graves were revealed in the area of the clinics of the Veterinary Department; most were brick-built (probably cist-graves) and a few tile-graves.[71] The graves contained glass and other vessels.[72]

| Cat. No. | Form | Other excavation indications | Find Date |
|---|---|---|---|
| 273 | 051b | | |
| 493 | 116 | From different graves | 23-1-1970 |
| 498 | 120 | | |
| 499-501 | 120 | Altar-like construction. Tile-grave | |
| 502-503 | 120a | Altar-like construction. Tile-grave | |
| 515 | 120b | From different graves | 23-1-1970 |
| 523 | 120γ | Altar-like construction. Tile-grave | |
| 548-549 | 122b | From different graves | 23-1-1970 |
| 550 | 122b | Grave XI | 29-5-1970 |
| 604 | 130 | Grave Γ' | |
| 657 | 139 | Grave K4 | |

48. No. 5 Kallitheas Avenue, Ampelokipi, 1983

Here were found 21 barrel-vaulted graves arranged in clusters. They were looted in antiquity and the sole finds were a glass vessel from one grave and coins of Justinian I in another. According to the morphological characteristics of the graves they are dated by their excavator to the 5th century, and the time period of their use is assumed to extend across the entire 5th and into the 6th century.[73]

| Cat. No. | Form | Other excavation indications | Find Date |
|---|---|---|---|
| 701 | 146 | Grave 10 | 1983 |

49. Kallitheas Avenue and Agathonos Street, Ampelokipi, 1995

50. Kallitheas Avenue and Skeparni Street, Ampelokipi, 1996

51. Kallitheas Avenue, Elementary School, Ampelokipi, 1997

In the same plot of the 2nd and 8th Elementary School of Ampelokipi, which was excavated in three campaigns from 1995 until 1997, one hundred and sixteen graves were found: eighty three tile-graves, nine barrel-vaulted graves, twelve cist-graves, one pit burial, three jar burials and eight pit-graves, which are dated from the second half of the 3rd century up to the first half of the 4th century. Movable finds from the graves include clay burial vessels, glass unguentaria, bronze coins, gold-, bronze-, lead-, stone-, bone- and glass jewellery, an iron dagger and nails.[74]

49. Kallitheas Avenue and Agathonos Street

| Cat. No. | Form | Other excavation indications | Find Date |
|---|---|---|---|
| 426 | 095 | Fill of grave1 | 13-9-1995 |

50. Kallitheas Avenue and Skeparni Street

| Cat. No. | Form | Other excavation indications | Find Date |
|---|---|---|---|
| 282 | 052a | Grave T. 18 | 18-12-1996 |
| 407 | 082 | Grave T. 25 | 30-12-1996 |

51. Kallitheas Avenue, Elementary School, Ampelokipi

| Cat. No. | Form | Other excavation indications | Find Date |
|---|---|---|---|
| 120 | 032 | Grave T. 93, burial layer | 26-11-1997 |
| 230 | 051a | Grave T. 42, burial layer | 3-9-1997 |
| 231 | 051a | Grave T. 58, burial layer | 8-10-1997 |
| 2580 | 51b | Grave T. 52, burial layer | 30-9-1997 |
| 353 | 068 | Grave T. 46, burial layer | 5-9-1997 |
| 408 | 083 | Grave T. 31, burial layer | 29-9-1997 |
| 409 | 083 | Grave T. 83-B, burial layer | 2-12-1997 |
| 427 | 095 | Grave T. 87, burial layer | |
| 672 | 141 | Grave T. 48, burial layer | 24-9-1997 |
| 684 | 142a | Grave T. 61, grave fill | |
| 722 | 147a | Grave T. 87, burial layer | 4-11-1997 |

52. No. 4 Kapatou N. Street, 2001

37 graves were unearthed in the plot behind the church of Agios Nikolaos at Xirokrini: 1 cist-grave, 13 tile-graves, 1 pit-grave, 21 pit-burials and 1 jar-burial. Glass vessels were found in the graves which are dated to the 4th-5th century.[75]

---

[71] Petsas 1969a, 300, pls. 309δ-ε, 311, 312α; Petsas 1974, 377-378, pls. 80-84.

[72] Petsas 1974, 329.

[73] Mavropoulou-Tsioumi, Kourkoutidou-Nikolaidou 1983, 285-287, drawings 1-2.

[74] Makropoulou 1997, 701, pl. 256δ.

[75] 9th EBA 2002, 13.

| Cat. No. | Form | Other excavation indications | Find Date |
|---|---|---|---|
| 349 | 065 | Grave 1, cist-grave (vessel Y4, depth -2.68) On the pavement 0. 30m from the NE corner of the grave | 29-8-01 |
| 306 | 053b | Grave 25 (vessel Y1, depth -1.40/-2.10. Undisturbed grave) | 24-8-01 |
| 319 | 058 | Grave 37 (vessel Y2, internal fill of an open burial,29-8-01 depth -2.25) | |
| 705 | 146 | Grave 1 cist-grave (vessel Y3, depth -2.68) (burial layer B', disturbed find) | 29-8-01 |

53. Junction of Koloniari, Galanaki and Papathanasiou Streets, 1987

This plot was the site of 6 barrel-vaulted graves, 3 cist-graves, 310 tile-graves, a few jar-burials, 1 pottery furnace and traces of a second. In the graves were found glass and clay vessels, bronze jewellery and coins. The graves are dated to Late Roman times,[76] mainly to the 4th century, although the graves had been re-used repeatedly in the 5th century.[77]

| Cat. No. | Form | Other excavation indications | Find Date |
|---|---|---|---|
| 304 | 053b | A stray find | 2-7-1987 |
| 321 | 059 | Grave 1, eastern side of the grave | 4-6-1987 |
| 374 | 073 | Tile-grave No. 9 | 16-6-1987 |
| 637 | 139 | Grave 1, eastern side of the grave | 4-6-1987 |
| 670 | 141 | A stray find | 9-7-1987 |
| 688 | 143 | Grave 5 | 18-6-1987 |
| 689 | 144 | Tile-grave 247 | 16-7-1987 |

54. Lagkada Street, 1961

During works to widen Lagkada Street in 1961, several Christian and Roman graves and a sarcophagus were unearthed. The works were not constantly supervised by the Archaeological Service. Movable finds from the graves were collected, however, comprising inscriptions, bronze coins, clay and glass vessels.[78]

| Cat. No. | Form | Other excavation indications | Find Date |
|---|---|---|---|
| 538 | 121 | | 17-6-1961 |

55. No. 67 Lagkada Street, 1963

Clusters of cist- and barrel-vaulted graves, dated between the Hellenistic and Early Christian periods, were unearthed and excavated at the beginning of Lagkada Street, in the vicinity of Vardaris square, while removing marble sarcophagi from the site. Among the most important movable finds are clay figurines, bronze coins, a mirror, clay lamps, and inscriptions.[79]

| Cat. No. | Form | Other excavation indications | Find Date |
|---|---|---|---|
| 149-150 | 035 | In the soil scattered during the removal of a sarcophagus in the area of Ramona | 26-6-1963 |
| 210-211 | 050 | In the soil scattered during the removal of a sarcophagus in the area of Ramona | 26-6-1963 |
| 725 | 147a | In the soil scattered during the removal of a sarcophagus in the area of Ramona | 26-6-1963 |

56. Lagkada Street, The 'Dodekaorofi' (A 12-Storey Tower in Ramona), 1972

No special clues during this excavation were provided. The only reference, probably connected with this find, concerns a Roman grave with wall-paintings, dug by the 16th Ephorate of Prehistoric and Classical Antiquities. The wall-paintings were detached and sent to the 9th Ephorate of Byzantine Antiquities for conservation.[80]

| Cat. No | Form | Other excavation indications | Find Date |
|---|---|---|---|
| 295 | 052a | Built grave | 9-9-1971 |
| 344 | 063 | Vaulted gave A' | 2-3-1972 |
| 364 | 070 | Built grave | 9-9-1971 |
| 417 | 089 | | 27-1-1972 |
| 681 | 141 | Tile-grave | |
| 734 | 147a | | 27-1-1972 |

57. Lagkada Street, The Axios Residential Complex, 1968-70

During rescue excavations conducted on the plot prior to the construction of the Axios residential complex in the area of Xirokrini, more than 100 graves were revealed; these were arranged in clusters, and partly one on top of the other. Some of them are built, barrel-vaulted and covered with schist slabs, while the majority are simpler tile-graves.[81] Significant amounts of clay and glass vessels were found.[82]

[76] Nalpantis 1987, 404-406, pl. 227.

[77] I thank the excavator of the site, Dimitris Nalpantis, for the information provided, as well as for permission to illustrate vessels from his (not yet published) excavations.

[78] Diary Θ9 (1961-1963), pages 3 and 6, held in the archive of the 16th Ephorate of Prehistoric and Classical Antiquities, Thessaloniki.

[79] Papadopoulou 1964b, 334, drawing 6, pl. 381α-β.

[80] Tsigaridas 1973a, 496. It seems that the wall-paintings are those held in the Museum of Byzantine Culture, registered as BT 83-BT 86.

[81] Petsas 1969a, 300, pl. 309; Petsas 1974, 377-378, pls. 80-84.

[82] Petsas 1974, 329, 377-78, pls. 24-25.

| Cat. No. | Form | Other excavation indications | Find Date |
|---|---|---|---|
| 265 | 051b | Grave 7 | 18-12-1968 |
| 442 | 102 | Grave 4 | |
| 562 | 126a | Grave 4 | |
| 679 | 141 | Grave 3 | |
| 728 | 147a | Grave 2 | December 1968 |

58. Lagkada Street, The Axios Residential Complex (Northern Plot), 1989-1991, 1995

95 graves, dated to the Hellenistic and Roman periods were found within the SE part of the northern plot of the Axios residential complex: 45 cist-graves, 32 pit-graves, 14 tile-graves, 3 barrel-vaulted graves, and 1 jar-burial. Altar-like constructions (of different types) were found on top of the graves. In the graves were found bronze coins, clay and glass vessels, bronze, stone and gold jewellery, and tools. The graves are dated between the 1st century BC and the 4th century AD.[83]

The same plot also contained an exceptionally well-preserved burial complex with 9 chambers, used in the second half of the 4th century, as well as an isolated cist-grave dated to the 3rd century, and a cluster of 22 barrel-vaulted graves dated to the reign of Theodosius I (379-395).[84]

In additional excavations conducted in the same plot in 1995, 26 new graves came to light. 6 barrel-vaulted graves, 8 cist-graves, 6 tile-graves, 1 pit-grave, 4 rock-cut graves, and 1 jar-burial. In the graves were found clay and glass vessels, jewellery, coins and tools. They are dated to the Early Christian period up to the end of the 4th century.[85]

| Cat. No. | Form | Other excavation indications | Find Date |
|---|---|---|---|
| 143 | 035 | Room Z', Cist No. 1 | 24-5-1991 |
| 166 | 036 | Room Z', Cist No. 3 | 24-5-1991 |
| 356 | 070 | Grave 23 | 5-2-1990 |
| 721 | 147a | Northern half of the antechamber, Layer B | 12-11-1989 |

59. Lagkada Street, The Axios Residential Complex (Southern Plot), 1991 and 1994-95

In 1991 many graves were unearthed on the western side of Lagkada Street, across from Ivanof Park, at the southern site of the Axios residential complex (Lagkada, Galanaki and Ivanof Streets). The graves included 8 tile-graves (dated to the 3rd and early 4th century), 3 cist-graves (dated to the 3rd century), 2 rock-hewn graves, one of them with a construction for receiving libations offered to the diseased, and 1 pit-burial. From the same site comes a cluster of 47 barrel-vaulted graves, dated to the end of the 4th century, which were used until the end of the 5th and into the 6th century. A further burial complex revealed 12 barrel-vaulted graves dated to the end of the 5th and into the first half of the 6th century.[86]

14 more graves were found in 1994: barrel-vaulted graves, cist-graves, tile-graves, rock-hewn graves and a single jar-burial. The graves are dated between the 2nd and 4th centuries and contained bronze coins, clay and glass vessels, and gold and bronze jewellery.[87]

| Cat. No. | Form | Other excavation indications | Find Date |
|---|---|---|---|
| 358 | 070 | Grave 1 | 20-9-1995 |
| 625 | 133 | Grave 3 | 12-9-1994 |
| 702 | 146 | Grave XXVII | 1991 |

60. Lagkada Street, The Axios Residential Complex (Central Plot), 1998

During the supplementary excavations conducted at the plot in 1998 to link the 'northern' and 'southern' sectors finds were made of 195. Of them, 3 were pre-Christian and the rest are dated between the 3rd and 9th centuries. Many of them were part of the five clusters dated to the 4th century. Two burial enclosures were unearthed that are dated between the middle of the 2nd and the first half of the 3rd century. These graves consisted of 96 barrel-vaulted graves, 40 rock-hewn, 17 tile-graves, 7 pit-graves, 13 cist-graves, 12 jar-burials and 10 pit burials. The movable finds comprise clay and glass vessels, bronze coins, gold, stone, bronze and iron jewellery, and tools.[88]

| Cat. No. | Form | Other excavation indications | Find Date |
|---|---|---|---|
| 232 | 051a | Grave T. 9, burial layer | 14-1-1998 |
| 345 | 064 | Grave 65, burial layer | 13-2-1998 |
| 350 | 066 | Grave 65, burial layer | 13-2-1998 |
| 414 | 087 | | |
| 579 | 127 | Fill above Grave 97 | 20-2-1998 |

61. Lagkada Street, Ivanof Park, 1991

In Lagkada Street, between Agiou Nestoros and Agiou Dimitriou Streets, in front of the Georgiou Ivanof Park, road widening works were undertaken in 1991. During these works 76 graves were unearthed:[89] 17 tile-graves, 32 cist-graves, and 23 barrel-vaulted graves. In the graves were found numerous objects, including bronze coins, bronze, stone, glass and gold jewellery, bronze and silver tools, and clay and glass vessels. The graves were dated to the 3rd and 4th centuries.

---

83 Adam-Veleni, Konstantoulas 1991, 221-228.

84 Makropoulou 1990b, 335-337; Makropoulou 1991, 261.

85 Makropoulou 1995, 530.

86 Makropoulou 1991, 261-265; Makropoulou 1992a, 425.

87 Makropoulou 1994, 510.

88 Makropoulou 2007, 111-172.

89 Makropoulou 1991, 257-261; Makropoulou 1992a, 425-426.

| Cat. No. | Form | Other excavation indications | Find Date |
|---|---|---|---|
| 217 | 051a | Grave XIX | 1991 |
| 218 | 051a | Grave 37 | 1991 |
| 274 | 052a | Grave 11 | 1991 |
| 275 | 052a | Grave XXIII | 1991 |
| 316 | 056 | Grave IV | 16-4-1991 |
| 439 | 101 | Grave 35 | 1991 |
| 630 | 136 | Grave IV | 1991 |

62. No. 20 Margaropoulou Street, 1995

Part of the ancient necropolis of Thessaloniki was unearthed during excavations in this plot. 152 graves were unearthed: 23 barrel-vaulted graves, dated between the 3rd and 6th centuries; 55 cist-graves, dated to the second half of the 3rd and into the 4th century; 37 pit- and 32 tile-graves, dated to the 3rd and up to the early 4th century; and 5 pit-burials. In the graves were found clay and glass vessels, gold, stone and bronze jewellery, and bronze coins. Most of the graves are dated to the 3rd and 4th centuries; the cemetery was abandoned in the 5th century and then a three-aisled Christian basilica was erected on the site.[90]

| Cat. No. | Form | Other excavation indications | Find Date |
|---|---|---|---|
| 200 | 048 | Grave 103 | 11-5-1995 |
| 207 | 050 | Grave 111 | 3-7-1995 |
| 327 | 061 | Grave 111 | 3-7-1995 |
| 375 | 074a | Grave 23, burial layer | 14-12-1994 |
| 446 | 105 | Chance find | 1994 |
| 621 | 132 | Grave 103 | 11-5-1995 |
| 640 | 139 | Grave 122 | 11-7-1995 |
| 671 | 141 | Grave 8 | 31-10-1994 |

63. Margaropoulou and No. 2 Syntagmatarchou Fryzi Streets, Xirokrini, 1998

During the excavations here prior to the construction of a new building on the site, 26 graves were unearthed (15 tile-, 3 cist- and 7 pit-graves, and 1 jar-burial). The graves contained bronze coins, clay and glass vessels, and items of gold and glass jewellery.[91]

| Cat. No. | Form | Other excavation indications | Find Date |
|---|---|---|---|
| 253 | 051a | Grave 11, K1 | 16-10-98 |
| 294 | 052a | Grave 26, K4 | 23-10-98 |

64. No. 5 Monastiriou and Odysseos Streets, 1995

Sections of a wine or oil storehouse, probably attached to a relevant workshop and dated to the Roman Imperial period, were unearthed during rescue excavations on this site. The storehouse was probably a civic building, based on its proximity to the 'Chrysi Gate' complex. The building was destroyed by fire.[92]

| Cat. No. | Form | Other excavation indications | Find Date |
|---|---|---|---|
| 180 | 06a | Surface cleaning of the plot's northern wall | 21-6-95 |

65. No. 3 Peleiadou Street, 2001

25 graves were unearthed during the excavation conducted prior to the construction of a new building on the site: 7 tile-, 3 pit-, 4 cist-, and 10 barrel-vaulted graves. These graves contained clay and glass vessels, bronze jewellery, and bronze and iron tools. The graves are dated to the 4th and 5th centuries, although they were been used for a long period of time.[93]

| Cat. No. | Form | Other excavation indications | Find Date |
|---|---|---|---|
| 432 | 097 | Square Γ2-Γ3. Grave 11 (SW corner) | 5-12-2000 |
| 642 | 139 | Grave 14 (0.30m from the NE corner and 0.70m from the SE corner) | 5-12-2000 |

66. No. 5 Peleiadou Street, 1997

57 graves were located on this site during excavations prior to the construction of a new building: 30 tile-graves, 15 pit-graves, 9 barrel-vaulted graves, 2 pit-burials, and 1 possible jar-burial. The graves are dated to the 4th century[94] and contained clay and glass vessels, gold, bronze and bone jewellery, as well as bronze coins.

| Cat. No. | Form | Other excavation indications | Find Date |
|---|---|---|---|
| 359 | 070 | Grave 9, tile-grave | |
| 382 | 074b | Grave T. 27, burial layer | 29-9-1997 |
| 419 | 091 | Grave T. 52, burial layer | 26-11-1997 |
| 420 | 092 | Grave T. 20, burial layer | 21-10-1997 |
| 440 | 101 | Grave T. 38, barrel-vaulted grave, burial layer | |
| 667 | 140 | Grave T. 44, burial layer | 31-10-1997 |
| 687 | 142b | Grave T. 35, burial layer | 21-10-1997 |
| 703 | 146 | Grave 10, burial layer | 24-12-1996 |

[90] Makropoulou 1995, 527-530, drawing 10.

[91] I thank the excavator of the site, Eleni Lamprothanasi-Korantzi, for letting me use the diaries of this excavation.

[92] Makropoulou 1995, 527.

[93] 9th EBA 2002, 18.

[94] Makropoulou 1997, 700-701, pl. 256γ.

67. No. 6 Peleiadou Street, 1996

During excavations prior to the construction of a new building on this site, 24 graves were found: 17 tile-graves, 4 pit-graves, 1 barrel-vaulted grave, and 2 pit burials. The burials are dated to the 3rd-4th century. The contents included clay and glass vessels, a bronze jewel, and a bronze coin.[95]

| Cat. No. | Form | Other excavation indications | Find Date |
|---|---|---|---|
| 376 | 074a | Grave 24 | |

68-71. The Ramona Area, 1963, 1968, 1969, 1971, 1972

The area known as 'Ramona', in the 1960s and 1970s, comprised the triangular area of the ancient western necropolis stretching west of Lagkada Street, north of Monastiriou Street, extending north and up to the Axios residential complex, and to the west as far as the new railway station, or even to the church of Agioi Pantes.

In 1963 some sarcophagi were unearthed in Ramona, at the start of Lagkada Street, around Vardaris Square. After the removal of the sarcophagi two clusters of graves, dated from Hellenistic to Early Christian times, were located and excavated. They were cist-graves and few barrel-vaulted graves. The only movable finds that specifically mentioned are two clay figurines, a few clay lamps, a bronze mirror and coins.[96]

Excavations in the same area in 1966-67, revealed several graves, where Byzantine coins, gold jewellery, an inscribed gold sheet, etc. were found.[97]

A wall-painted Roman grave was detected in 1972 during excavations conducted by the 16th Ephorate of Prehistoric and Classical Antiquities. The wall-painting was detached from the grave and sent to the 9th Ephorate of Byzantine Antiquities for conservation.[98]

(For the graves and offerings found in the area of the Axios residential complex, see supra (Nos. 57-60).[99]

68. The Ramona Area, 1963

| Cat. No. | Form | Other excavation indications | Find Date |
|---|---|---|---|
| 147 | 035 | Square. Grave cluster II, grave (1) | August 1963 |
| 148 | 035 | Square | August 1963 |
| 262 | 051b | Grave 19 | 24-9-1963 |
| 238 | 051a | Grave 20 | 27-9-1963 |
| 320 | 058 | Cluster of graves I | August 1963 |
| 436 | 099 | *Larnax*/ossuary (β) | August 1963 |
| 451 | 108 | Grave cluster I | August 1963 |
| 726 | 147a | | August 1963 |

69. The Ramona Area, 1969

| Cat. No. | Form | Other excavation indications | Find Date |
|---|---|---|---|
| 267 | 051b | Built-grave | 14-10-1969 |
| 335 | 061 | Grave 8 | 8-4-1969 |
| 336 | 061 | | 27-3-1969 |
| 710 | 146 | Built-grave | 19-3-1969 |
| 730 | 147a | Tile-grave, within19-3-1969 | |

70. The Ramona Area, 1971

| Cat. No. | Form | Other excavation indications | Find Date |
|---|---|---|---|
| 199 | 047 | | 1971 |
| 270-271 | 051b | | 1971 |
| 296 | 052a | | 1971 |
| 312 | 053b | | 1971 |
| 369 | 071 | Tile-grave | 20-8-1971 |
| 384 | 074b | | 1971 |
| 411 | 084 | | 10-2-1971 |
| 434 | 097 | Tile-grave | 20-8-1971 |
| 631 | 137 | Vaulted-grave | 3-4-1971 |
| 736-737 | 147a | | 1971 |
| 740 | 147a | Tile-grave | 22-3-1971 |
| 698 | 145 | | 1971 |

71. The Ramona area, 1972

| Cat. No. | Form | Other excavation indications | Find Date |
|---|---|---|---|
| 249 | 051a | | 2-3-1972 |
| 268 | 051b | Has a paper tag marked '2', probably the grave number | 2-3-1972 |
| 269 | 051b | Has a paper tag marked '1', probably the grave number | 2-3-1972 |

95 Makropoulou 1996, 504.

96 Papadopoulou 1964b, 334, drawing 6, pl. 381α-β.

97 Petsas 1966/67, 294.

98 Tsigaridas 1973a, 496.

99 Petsas 1969a, 300, pls. 308-310; Petsas 1970b, 351-352, pl. 303 ε, στ; Petsas 1974, 377-378, pls. 80-84.

| | | | |
|---|---|---|---|
| 661 | 139 | Found in the depot held with material from Ramona with the tag 'Grave 1st, with a fibula and coin' (this grave has not been identified) | 18-8-1972 |
| 711 | 146 | Grave 4 | 2-3-1972 |

72. The Ramona Area (No Date)

| Cat. No. | Form | Other excavation indications | Find Date |
|---|---|---|---|
| 203-204 | 048 | Building 4, grave with marble revetment | |
| 247-248 | 051a | | |
| 266 | 051b | | |
| 341 | 063 | | |
| 342 | 063 | Building 35, tile-grave | |
| 545 | 122b | Tile-grave | |
| 557 | 125 | Tile-grave | |
| 612 | 131 | Building 1, pit-grave | |
| 614-615 | 131 | Tile-grave | |
| 658 | 139 | Built-grave | |
| 696 | 145 | Building 5, marble grave | |
| 729 | 147a | | |
| 731-732 | 147a | Tile-grave | |

73. Probably from the Ramona Area

| Cat. No. | Form | Other excavation indications | Find Date |
|---|---|---|---|
| 640 | 12c1 | | 934 |
| 670 | 12e | | 1934 |
| 365 | 070 | | 1934 |
| 495 | 117 | | 1934 |

74. No. 14 Regkoukou Street, Xirokrini, 1979

12 graves were unearthed in the excavation of this plot: 1 cist-grave, 1 tile-grave and 10 barrel-vaulted graves. The movable finds are few and are datable to the Late Roman period; they comprise few plain clay vessels, single-handled or handleless, a glass *lagenos* with eight-sided body, a few jewels and 4 bronze coins, the only legible one among them dating to Constantius (struck AD 341-346).[100]

| Cat. No. | Form | Other excavation indications | Find Date |
|---|---|---|---|
| 433 | 097 | Grave 9 | 1979 |

75. No. 16 Regkoukou Street, Xirokrini, 1994

24 graves (tile-, cist-, and barrel-vaulted) were revealed during the excavation of this plot; they are dated from the end of the 3rd to the early 5th century. In the graves were found clay burial vessels, glass unguentaria, a bronze coin, and bronze, lead and glass jewellery.[101]

| Cat. No. | Form | Other excavation indications | Find Date |
|---|---|---|---|
| 281 | 052a | Grave 16 | 2-3-1994 |
| 357 | 070 | Grave 18 | 14-1-1994 |

76. No. 31 Syntagmatarchou Fryzi Street, 1996

This site produced several finds, including a *cubiculum*, the remains of a building or of an enclosure, and 24 graves: 11 cist-graves, 1 barrel-vaulted grave with a stepped entrance/*dromos*, 9 pit-graves, 2 tile-graves, and 1 jar burial; these date from the 3rd to the 5th centuries. Traces of habitation in theLlate Byzantine or earlneOttoman era also were traced. The movable finds from the graves comprise glass unguentaria, clay burial vessels, metal and stone jewellery, and bronze coins.[102]

| Cat. No. | Form | Other excavation indications | Find Date |
|---|---|---|---|
| 208 | 050 | Clearing of enclosure A' | |
| 686 | 142b | Grave 3 | 24-10-1996 |

77. No. 2 Thiseos Street, 2001

During rescue excavations in this plot 18 graves were documented: 4 cist- and 12 barrel-vaulted graves.[103] The graves are dated to the 4th and 5th centuries and remained in use up to the 7th or 8th century. In the graves were found bronze jewellery, glass and clay vessels.

| Cat. No. | Form | Other excavation indications | Find Date |
|---|---|---|---|
| 235 | 051a | Grave 14 (on the pavement of the grave, depth -4.45) | 22-3-2001 |
| 305 | 053b | Grave 4, burial 1 and secondary burial 1 (depth 4.45m. 1.37m from the SE corner and 1.40m from the NE corner) | |
| 704 | 146 | Grave 4, burial 1 and secondary burial 1 (revealing burial 1 and | |

[100] Vokotopoulou 1980, 365-367, pl. 215γ-δ.

[101] Makropoulou 1994, 509-510.

[102] Makropoulou 1997, 701-702.

[103] 9th EBA 2002, 19.

secondary burial 1. Depth -4.40/-4.60m, 1.46m from the SE corner and 1.43m from the NE corner)

78-81. Vardari

**Monolithic sarcophagus, 12-2-1920**

It was not possible to trace any special clues about the discovery of this sarcophagus. In addition, none of the sarcophagi currently known to the Archaeological Service can be connected with this find. Judging by the glass finds, it must have been used during the 3rd or 4th century.

Many more glass vessels are stored with the note 'probably from the cemetery of Vardari'. This excavation is not recorded, apart from a generic handwritten report of 1920 now in the archives of the Archaeological Museum of Thessaloniki. In this report it is mentioned that in the area of Vardari several pit-graves, tile-graves, and built-graves, some of them bearing marble revetments, were excavated. The movable finds are only mentioned as groups, in particular bronze jewellery and coins, clay vessels and glass vessels of the same shape as clay vessels. A simple reference to these works is given in French in 1921.[104] This mentions that some levelling of the site outside the western city wall was carried out after the fire of 1917, and at the same time a section of the Late Roman necropolis was uncovered, much of which was excavated. The tombs were built-graves, some of which had marble revetments. Movable finds comprised Late Roman vessels, *fine technique* glass, bronze jewellery (belt buckles and brooches/ fibulae, some of them were made of silver or gilded bronze).

78. Vardari, Monolithic Sarcophagus, 12-2-1920

| Cat. No. | Form | Other excavation indications | Find Date |
|---|---|---|---|
| 443 | 103 | Found in a marble sarcophagus | 12-2-1920 |
| 418 | 090 | Found in a marble sarcophagus | 12-2-1920 |
| 328-329 | 061 | Found in a marble sarcophagus | 12-2-1920 |
| 287-288 | 052a | Found in a marble sarcophagus | 12-2-1920 |
| 63 | 012 | Found in a marble sarcophagus | 12-2-1920 |

79. Vardari, 1970

| Cat. No. | Form | Other excavation indications | Find Date |
|---|---|---|---|
| 162-164 | 035 | | 1970 |
| 272 | 051b | | 1970 |
| 370 | 071 | | 1970 |

80. Vardari (No Date)

| Cat. No. | Form | Cat. No. | Form |
|---|---|---|---|
| 51-52 | 012b | 361 | 070 |
| 114 | 031 | 377-378 | 074a |
| 194 | 045 | 406 | 081 |
| 201 | 048 | 410 | 084 |
| 209 | 050 | 437 | 100 |
| 236-237 | 051a | 444 | 104 |
| 283-286 | 052a | 447 | 106 |
| 299 | 052b | 628-629 | 135 |
| 301 | 053a | 648-652 | 139 |
| 307-309 | 053b | 675 | 141 |
| 315 | 055 | 693 | 145 |
| 352 | 067 | 724 | 147a |

81. Vardari, Probably from the cemetery

| Cat. No. | Form | Other excavation indications |
|---|---|---|
| 117 | 031 | In a box with the indication 'unregistered, probably from the1920 Vardari Cemetery', wrapped in newspapers of 1919, in an envelope with a 'Φ' written on it, probably the number of the grave |
| 122 | 032 | In a box with the indication 'unregistered, probably from the1920 Vardari Cemetery', wrapped in newspapers of 1919, in an envelope with a 'T' written on it, probably the number of the grave |
| 151-161 | 035 | In a box with the indication 'unregistered, probably from the Vardari Cemetery' |
| 252 | 051a | In a box with the indication 'unregistered, probably from the Vardari Cemetery' |
| 289-293 | 052a | In a box with the indication 'unregistered, probably from the Vardari Cemetery' |
| **Cat. No.** | **Form** | **Other excavation indications** |
| 300 | 052b | In a box with the indication 'unregistered, probably from the Vardari Cemetery' |
| 383 | 074b | In a box with the indication 'unregistered, probably from the Vardari Cemetery' |
| 494 | 117 | In a box with the indication 'unregistered, probably from the Vardari Cemetery' |
| 558 | 125 | In a box with the indication 'unregistered, probably from the Vardari Cemetery' |

[104] Referenced by the Redaction of BCH as: «Chroniques des fouilles et découvertes archéologiques dans l'Orient hellénique» *Bulletin de correspondance hellénique* 45, 1921, 487-568, esp. page 541.

| | | |
|---|---|---|
| 682 | 141 | In a box with the indication 'unregistered, probably from the Vardari Cemetery' |
| 706 | 146 | In a box with the indication 'unregistered, probably from the1920 Vardari Cemetery', wrapped in newspapers of 1919, in an envelope with a 'Σ' written on it, probably the number of the grave |
| 735 | 147a | In a box with the indication 'unregistered, probably from the Vardari Cemetery' |
| 747-752 | 148 | |

### *Thessaloniki, Intra Muros*

82. No. 3 Agapis Street, 1996

During the rescue excavations on this plot a sumptuous Early Christian house with mosaic floors and a small bath were unearthed. In the mid-Byzantine period at least two houses were built on the same site and abundant numismatic and pottery finds are datable to the end of the 11th century. The site was turned into a cemetery in theLlate Byzantine period.[105]

| Cat. No. | Form | Other excavation indications | Find Date |
|---|---|---|---|
| 164d | 37.2 | | 1996 |
| 164a | 37.1 | | 1996 |
| 164c | 37.1 | | 1996 |

83. Acheiropoietos, East of the Church Apse (S. Georgiadis Plot), 1966

At the site where, in the middle of the 5th century, the great Christian basilica of Acheropoietos was built, there existed a large Roman public bath complex; several Early Christian churches in Thessaloniki shared this feature.[106] This bathhouse was revealed during excavations to the east[107] and west[108] of the church, but also under its northern aisle.[109] Three consequent floors are partly preserved which all belong to the bathhouse[110] and they are dated from the second half of the 2nd century to the first half of the 4th century.[111] The eastern and northern parts of the bathhouse remained in use even after the construction of the basilica.[112]

Further ruins of the bathhouse were also found during the 1966 excavations east of the apse of the basilica.[113] The movable finds of these excavations included glass-working waste, indicating the operation of such a unit in this area. It is probable that it was operating during the 5th century, some time before the erection of the basilica, and taking advantage of a temporarily abandoned part of the bathhouse complex, or even during the time of the erection of the basilica itself, to meet various needs of the work-site.

84. Junction/Corner of Agiou Dimitriou and Agias Sophias Street, Byzantine Building, 1969

During works along Agiou Dimitriou Street in 1969, a building was excavated stretching from Agias Sophias Street in the west to Evripidou Street in the east (the southern edge was not located).[114] This building was identified with a cross-shaped church.[115] On the basis of the numismatic evidence[116] and for purely architectural reasons,[117] there being no special stratigraphic clues from the excavation itself, it was assumed that this church was founded at the time of Justinian I (AD 527-565) and destroyed by an earthquake in the early 7th century.[118] The glass finds, along with mainly Early Christian clay finds and coins, were found in the triangular rooms formed between the apse and the corner compartments.[119] The same building was encountered again during excavations for the Thessaloniki water system.[120]

| Cat. No. | Form | Other excavation indications | Find Date |
|---|---|---|---|
| 60-61 | 012b | 3rd triangular room | 11-6-1969 |
| 82-83 | 016 | 3rd triangular room | 11-6-1969 |
| 126 | 033 | X4 on the sterile soil | 16-6-1969 |
| 127-129 | 033 | 3rd triangular room | 11-6-1969 |
| Cat. No. | Form | Other excavation indications | Find Date |
| 132 | 034 | X4 on the sterile soil | 16-6-1969 |
| 173 | 038b | Under the edging of the pavement between walls B′and Δ′ | 10-6-1969 |
| 174-175 | 038b | 3rd triangular room | 11-6-1969 |
| 371-372 | 071 | 3rd triangular room | 11-6-1969 |

[105] For the excavation of the site at No. 3 Agapis Street, see Kanonidis 1996b, 490-493. For mid-Byzantine clay finds, see Kanonidis, 2003, 71-80. For the finds of the glass-workshop, see Antonaras 2016a, 49, 180-181, figs. 277-279.
[106] Bakirtzis 1983, 310-329 (esp. pages 325-326).
[107] Petsas 1967b, 393-396, pls. 299-300; Petsas 1969c, 160-161, pls. 59-64, 66.
[108] Pelekanidis 1961/62a, 252-253; Bakirtzis 1983, 321.
[109] Xyngopoulos 1941/52, 472-487.
[110] Bakirtzis 1983, 323.
[111] Asimakopoulou-Atzaka 1998, 127, 227-228.
[112] Bakirtzis 1983, 324.
[113] Petsas 1967b, 393-396, pls. 299-300.
[114] Vavritsas 1970, 363-364.
[115] Marki 1985, passim.
[116] Marki 1985, 166.
[117] Marki 1985, 166-171.
[118] Marki 1985, 159-188 (esp. page 171).
[119] Marki 1985, 162.
[120] See the DVD: *Σωστική ανασκαφή αγωγού ύδρευσης Thessaloniki II (2000-2002)*. Thessaloniki: ΥΠΠΟ/Επιτροπή Παρακολούθησης Μεγάλων Έργων/Εφορεία Βυζαντινών Αρχαιοτήτων Thessaloniki, 2006, 169-171 (Trench K39 B-191, Ημερολόγιο 310VI, 8-18/6, 16-19/7/2001).

| | | | |
|---|---|---|---|
| 429 | 095 | 3rd triangular room | 11-6-1969 |
| 441 | 101 | 3rd triangular room | 11-6-1969 |

85. Agiou Dimitriou and Sofokleous Streets (No Date)

This excavation was not identified precisely. There is a report dated 1977[121] on the finding of a monolithic urn (probably a sarcophagus), in secondary use, during the opening of the ditch for a water main. The sole published excavation conducted on this street was at Plot No. 6, where parts of a Late Roman bathhouse were revealed.[122] In addition, it is known that Plot No. 35 contained the walls of a building.[123]

| Cat. No. | Form | Other excavation indications | Find Date |
|---|---|---|---|
| 43 | 012a | | |
| 49 | 012b | | |
| 72 | 015 | | |

86-87. Ancient Agora, 1962, 1966, 1989-1999

The site of the ancient agora was used by artisans for their workshops before the founding of the agora in the 1st or 2nd century, and again after the agora lost its official status. In the area there were pottery workshops operating during the 1st century BC and up to the 1st century AD.[124] The city mint was located on the same site during the second half of the 4th century.[125] During the 5th century, probably already by end of the 4th, the period when the agora partly lost its official status, diverse artisanal activities flourished on the site, e.g. pottery, metalworking, glass-working, textile/purple dyeing[126] and fulling mills, as well as bone curving – activities that continued into the 6th century as well.[127] Judging by the dates of the finds (1962 and 1966 respectively), glass-working waste was revealed at the Odeion and in the area of the plazza.[128]

In the southeastern part of the agora there were finds of a Late Hellenistic bathhouse, dated to the 2nd century BC, which was destroyed by fire in the time of Vespasian (around AD 78). A large number of clay and glass vessels were included among the movable finds of this bathhouse.[129]

86. Ancient Agora, 1962-1966

| Cat. No. | Form | Other excavation indications | Find Date |
|---|---|---|---|
| 7 | 004 | ΦΦ3, Trench B, north side, depth 3.9m | 1965 |
| 8 | 005 | On the envelope (hardly legible) 'S1/1067' | 1967 |
| 19 | 006a | Dikastirion Square. Trench B', depth 0.85m | 20-6-62 |
| 37 | 008 | Dikastirion Square. Trench A', well β', depth 5.20m | 26-6-62 |
| 40-41 | 010 | ΦΦ3, Trench B, north side, depth 3.9. m | 1965 |
| 46 | 012a | | |
| 47 | 012a | On the envelope (hardly legible) 'S1/1067' | 1967 |
| 94 | 021 | | 1962 |
| 96 | 023 | On a paper tag: '1967/11/32' | 29-5-1967 |
| 146 | 035 | | |

87. Bath Complex, Ancient Agora

| Cat. No. | Form | Other excavation indications | Find Date |
|---|---|---|---|
| 2-30 | 02 | Rectangular space north of the bathhouse. Destruction layer | 1998 |
| 4-6 | 003 | Rectangular space north of the bathhouse. Destruction layer | 1998 |
| 9-17 | 006a | Rectangular space north of the bathhouse. Destruction layer | 1998 |
| 25-28 | 006b | Rectangular space north of the bathhouse. Destruction layer | 1998 |
| 30-36 | 007 | Rectangular space north of the bathhouse. Destruction layer | 1998 |
| 39 | 009 | Rectangular space north of the bathhouse. Destruction layer | 1998 |
| 42 | 011 | Rectangular space north of the bathhouse. Destruction layer | 1998 |
| 86-87 | 019 | Rectangular space north of the bathhouse. Destruction layer | 1998 |
| 89-90 | 020 | Rectangular space north of the bathhouse. Destruction layer | 1998 |
| 91-93 | 021 | Rectangular space north of the bathhouse. Destruction layer | 1998 |
| 95 | 022 | Rectangular space north of the bathhouse. Destruction layer | 1998 |
| 97 | 024 | Rectangular space north of the bathhouse. Destruction layer | 1998 |
| 99-100 | 025 | Rectangular space north of the bathhouse. Destruction layer | 1998 |

[121] Archive of the 16th Ephorate of Prehistoric and Classical Antiquities,Thessaloniki, Diary Θ 17, pages 80-81.

[122] Alexandri 1973-74b, 673.

[123] Document reg. no. 1334/21-6-1972, from the Archive of the 16th Ephorate of Prehistoric and Classical Antiquities, Thessaloniki.

[124] Valavanidou 2001, 119, 120-123; Boli, Skiadaresis 2001, 91.

[125] Velenis 1996b, 49-60.

[126] 'The association of purple dyers' (*σύλλογος των πορφυροβάφων*) was probably connected with the site of Thessaloniki's ancient agora, on the evidence of an inscription known as 'one of the eighteenth' (της οκτωκαιδεκάτης), that seems to distinguish it from the other associations of purple dyers in the city, and to stress the location where they undertook their professional activities (IG X. 2, no. 291). The inscription is discussed by Robert 1937, 535, note 3. Vitti 1996, 78-79.

[127] Vitti 1996, 102-103; Valavanidou 2001, 123, 127-129.

[128] Papadopoulou 1963, 196-99; Papadopoulou 1964a, 329-331, drawings 1-2; Petsas 1967c, 383. For more on the bibliography and excavation results, see Adam-Veleni 2001, passim and esp. page 36, note 22.

[129] Adam-Veleni *et al.* 1996, 504; Adam-Veleni 1997, 351-364; Adam-Veleni 2003, 139-142.

| Cat. No. | Form | Other excavation indications | Find Date |
|---|---|---|---|
| 101-104 | 026 | Rectangular space north of the bathhouse. Destruction layer | 1998 |
| 106-108 | 028 | Rectangular space north of the bathhouse. Destruction layer | 1998 |
| 176-177 | 039 | Rectangular space north of the bathhouse. Destruction layer | 1998 |
| 178-184 | 040 | Rectangular space north of the bathhouse. Destruction layer | 1998 |
| 186-187 | 042 | Rectangular space north of the bathhouse. Destruction layer | 1998 |
| 190-192 | 044 | Rectangular space north of the bathhouse. Destruction layer | 1998 |
| 385-393 | 075 | Rectangular space north of the bathhouse. Destruction layer | 1998 |
| 397 | 076 | Rectangular space north of the bathhouse. Destruction layer | 1998 |
| 398-402 | 077 | Rectangular space north of the bathhouse. Destruction layer | 1998 |
| 454-458 | 111 | Rectangular space north of the bathhouse. Destruction layer | 1998 |
| 464 | 115 | Rectangular space north of the bathhouse. Destruction layer | 1998 |
| 581-582 | 128 | Rectangular space north of the bathhouse. Destruction layer | 1998 |
| 744-745 | 148 | Rectangular space north of the bathhouse. Destruction layer | 1998 |

88. No. 12 Christopoulou Street, Proferis Plot, 1975

No documents relevant to this excavation and their movable finds were traced in the archive of the 16th Ephorate of Prehistoric and Classical Antiquities.

| Cat. No. | Form | Other excavation indications | Find Date |
|---|---|---|---|
| 75 | 016 | Proferis plot, surface layer | 29-7-1975 |

89. Dioikitiriou Square, 1990-1998

Extended excavation works were conducted in the area of Dioikitiriou Square for a parking lot on the site. Important urban remains were unearthed which are dated from the founding of the city - from the time of King Kassandros up to the 20th century and the great city fire of 1917.[130]

| Cat. No. | Form | Other excavation indications | Find Date |
|---|---|---|---|
| 753 | 149 | | 22-7-1993 |

90. No. 7 Lapithon Street, 1990

A luxurious urban house – with its wall-paintings and mosaic pavements – dated to the second half of the 5th century, was discovered during excavations here.[131] A glass cup was located here, probably from an earlier use of the site.

| Cat. No. | Form | Other excavation indications | Find Date |
|---|---|---|---|
| 88 | 019 | Layer from the uncovered rock at the SW part of the excavation | 13-2-1990 |

91. Mitsaion Street, 1984

No documents relevant to this excavation were traced in the archive of the 16th Ephorate of Prehistoric and Classical Antiquities.

| Cat. No. | Form | Other excavation indications | Find Date |
|---|---|---|---|
| 85 | 018 | | 27-4-1984 |

92. No. 1 Pringipos Nikolaou Street, 1961

During the excavations conducted east of the apse of Agias Sophias part of the apse of the huge Early Christian basilica was found;[132] this was founded between the end of the 4th and the early 6th century[133] and was destroyed, probably, during the early 7th century. The apse, however, or part of it, possibly remained in use even after the foundation of the new church, as in its fill some mid-Byzantine pottery shards, wall-paintings and mosaics were found. Parts of marble and mosaic decoration, along with a large section of a brass decorative arch, a bronze cross-shaped *polykandelon*, and few stemmed glass lamps were all unearthed in the central space of the apse.[134]

| Cat. No. | Form | Other excavation indications | Find Date |
|---|---|---|---|
| 175a | 38a | From the small square room in the centre of the apse | 1961 |
| 175b | 38a | From the small square room in the centre of the apse | 1961 |
| 175c | 38a | From the small square room in the centre of the apse | 1961 |

93-94. Nos. 47-49 and 58-62 Sokratous Street (Mirarchou Koufitsa St.), 1968

In 1968 two plots were excavated in Sokratous Street. The first was Nos. 47-49, owned by the Velikas family (today 19 Mirarchou Koufitsa St.).[135] The second plot is directly opposite at Nos. 58-62.[136] Both sites contained the remains of buildings, with mosaic

[130] A synopsis of the finds of the excavations, as well as references to the annual archaeological reports in *Archaeologicon Deltion* can be found in: Tasia *et al.* 1996, 545-57; Kanonidis 1996b, 558-70; Kanonidis 2002, 143-153.

[131] Kourkoutidou-Nikolaidou 1990, 323; Eleftheriadou 1990, 332, drawings 1, pl. 151β; Asimakopoulou-Atzaka 1998, 258-259.

[132] Pelekanidis 1961-62b, 253-256; Drosogianni 1963, 235-240.

[133] Late 4th century: Theocharidou 1994, 39 and 150-160; 5th century: Pelekanidis 1961-62b, 253-56; Late 5th - early 6th century: Mentzos 1981, 205, 219.

[134] For the movable bronze finds from the site, see Antonaras 2014c, 21-22; Antonaras 2016b, 117-32.

[135] Petsas 1969b, 303, pl. 313β; Petsas 1974, 302, 305, 338, pls. 86-89; Michaud 1971, 958, fig. 350. Asimakopoulou-Atzaka 1998, 248-249, pls. 151-153, 318-319, pl. 268α, XXXVI, XXXIX.

[136] Petsas 1969b, 304; Petsas 1974, 379, pl. 90α.

floors who seem to belong to each other. There are two phases identified in the use of the mosaics. The first phase is dated to the second half of the 4th century and the newer phase to the 5th century.[137]

93. Nos. 47-49 Sokratous Street

| Cat. No. | Form | Other excavation indications | Find Date |
|---|---|---|---|
| 54-58 | 012b | | 1968 |
| 59 | 012b | (4-6)-(Δ-E) approx. at the depth of the pavement | 22-6-1968 |
| 76-77 | 01 | 6(4-6)-(Δ-E) approx. at the depth of the pavement | 22-6-1968 |
| 78 | 016 | | 1968 |
| 130 | 034 | | 1968 |
| 168-169 | 036 | | 1968 |
| 214 | 050 | | 1968 |

94. 58-62 Sokratous Street

| Cat. No. | Form | Other excavation indications | Find Date |
|---|---|---|---|
| 118 | 031 | North wall, adjacent to the left side of the well | 15-16/11-1968 |

95. No. 45 Vassileos Irakleiou Street, 2010

A relatively well-preserved, late 6th-century glass-workshop was excavated in the corner shop of a nicely built, two-storey *insula* at the junction of two streets. Several phases of glass-furnace activity were preserved in it along with much glass waste. It had two entrances, probably extending its business on the pavement in front of it. In the workshop two circular glass furnaces were found, c. 1m wide. One had five consequent phases of use and repair, and two more, older ones, only partly preserved.

More than 5kg of glass finds have been found discarded, mainly in the northeastern corner of the workshop. These included lumps and deformed masses of glass, testing droplets, and many moils, fragments of many types of vessels, stemmed beakers, flasks, stemmed and conical lamps and hemispherical bowls, and a few fragments of window panes.[138]

96. Thessaloniki, Unknown Find-Spots

There are no clues about these objects to connect them with a particular excavation.

| Cat. No. | Form | Other excavation indications | Find Date |
|---|---|---|---|
| 20 | 006a | From old excavations in Thessaloniki, originally kept by the 16th Ephorate of Prehistoric and Classical Antiquities | |
| 29 | 006b | | |
| 53 | 012b | | |
| 62 | 012c | From old excavations in Thessaloniki, originally kept by the 16th Ephorate of Prehistoric and Classical Antiquities | |
| 65 | 012d | From a grave dated by coins (no other indication is held about the nature of the grave or the coins) | |
| 98 | 024 | With a tag: 'Unidentified, moved to *apothiki* 67 from the laboratory. 2-7-1977' | |
| 111 | 029 | | |
| 123-125 | 033 | | |
| 241 | 051a | | |
| 254, 263 | 051b | | |
| 318 | 057 | | |
| 346 | 064 | | |
| 379 | 074a | | |
| 421 | 093 | | |
| 452 | 09 | From a tile-grave | 19-3-1971 |
| 478116 | | | |
| 497 | 119 | In the Archaeological Museum of Thessaloniki they were kept in a box with post-Byzantine pottery sherds from old excavations conducted in Thessaloniki | |
| 506 | 120a | From an excavation conducted before 1970 | |
| 525-526 | 120γ | From an excavation conducted before 1970 | |
| 553 | 124 | | 17-12-1949 |
| 561 | 126a | | 1976 |
| 564-566 | 126a | | |
| 605 | 130 | | |
| 619-62 | 0131 | | |
| 635 | 138 | | |
| 754 | 149 | From excavations conducted by Prof. Andreas Xygopoulos | |

## *Wider Thessaloniki Prefecture Area*

97. Gefyra, Thessaloniki Prefecture, 1965

No documents relevant to this excavation were traced in the archive of the 16th Ephorate of Prehistoric and Classical Antiquities.

[137] Asimakopoulou-Atzaka 1998, 319-320.

[138] Akrivopoulou 2014, 255-262; Antonaras 2011a, 126-128; Antonaras 2014a, 95-113; Antonaras 2014d, 263-272.

| Cat. No. | Form | Other excavation indications | Find Date |
|---|---|---|---|
| 416 | 089 | | 1965 |

98. Kalochori, Thessaloniki Prefecture, D. Gerakoudis Field, 1986

Two cist-graves built with of local green stone were unearthed during a small-scale excavation during the opening of a trench for a reservoir. The sole find was a glass vessel. It is not clear if the plot belongs to the community of Kalochori, Nea Diavata or Sindos.[139]

| Cat. No. | Form | Other excavation indications | Find Date |
|---|---|---|---|
| 84 | 017 | Grave 2 | 27-3-1986 |

99. Lete, Thessaloniki Prefecture, Farmhouse 'A', 1999-2000

Two buildings from a relatively large *villa rustica* were excavated in the area of ancient Lete. The site was inhabited from much earlier, although the villa was founded in the middle of the 2nd century; it was temporarily abandoned in the middle of the 3rd century during the invasion of the Goths and re-inhabited at the end of the 3rd century; it was finally abandoned during the barbaric invasions at the end of the 4th or early 5th century.[140]

| Cat. No. | Form | Other excavation indications | Find Date |
|---|---|---|---|
| 21 | 006a | | 2001 |
| 313 | 054 | | 2000 |

100. Nea Apollonia, Thessaloniki Prefecture (Undated)

No special information was found on this excavation in the archives of the 16th Ephorate of Prehistoric and Classical Antiquities. Due to the good preservation of the vessels we assume that they are grave offerings.

| Cat. No. | Form | Other excavation indications | Find Date |
|---|---|---|---|
| 240 | 051a | | |
| 653 | 139 | | |

101. Palaiokastro, Thessaloniki Prefecture, 2003

A *villa rustica* with it enclosure, a tower, its rooms arranged around an atrium with a well, a *triclinium*, structures with a wine press and a bathhouse, were all revealed in the area of Palaiokastron, SW of Oraiokastron at the junction of Stanisi and Zakinthou streets. The mosaic pavements, the wall-paintings and the marble architectural elements are witness to the wealth of the villa's owners. The complex was founded in the first half of the 5th century, was destroyed during the invasions of the 7th century, was reconstructed and used again until the 'Dark Ages', when it was finally abandoned. Movable finds from this excavation comprise storage and cooking clay vessels, a few clay- and one bronze lamp, a glass bowl with incised decoration, and 20 coins dated to the 5th and mainly to the 6th century.[141]

| Cat. No. | Form | Other excavation indications | Find Date |
|---|---|---|---|
| 74 | 016 | Sector 3 (Room between 1-3-4) | 22-7-2002 |

102. Panorama, Thessaloniki Prefecture, Basilica, 1973

The basilica is a twin church with two phases. The older phase, to which are connected the mosaic pavements, is dated to the second half of the 5th century[142] and the second phase is dated to the 6th.[143] During the construction of the church a burial structure was destroyed, although parts of it are preserved at the right end of the apse of the north aisle.[144] The glass finds were found east of the basilica.[145]

| Cat. No. | Form | Other excavation indications | Find Date |
|---|---|---|---|
| 133 | 035 | External side of the apse | 1971 |
| 134 | 035 | Room III | 1971 |
| 135 | 035 | | 1971 |
| 430 | 096 | | 1971 |

103. Rentina, Thessaloniki Prefecture, 1979 To Present

On the slopes below the castle of Rentina[146] an Early Christian basilica was unearthed, with a cemetery used in the Early Christian period and again in the mid-Byzantine era, at least judging by the movable grave finds.[147] Among these finds were glass vessels and glass jewellery.

| Cat. No. | Form | Other excavation indications | Find Date |
|---|---|---|---|
| 68 | 013 | E2, NE corner, at the level of the N. threshold. Φ. 105Γ́ | 13-10-1979 |

[139] Thanks are due to my colleague Anastasios Keramaris for his help in locating the relevant information about this day-long excavation (27-3-1986) from the archives of the 16th Ephorate of Prehistoric and Classical Antiquities.

[140] Tzanavari, Filis 2000, 153-168; Tzanavari 2003, 78-83.

[141] Marki, Akrivopoulou 2003, 283-298.

[142] Asimakopoulou-Atzaka 1998, 266.

[143] Personal communication with Prof. E. Tsigaridas on 5-12-2001.

[144] Tsigaridas 1973b, 498-501, drawing 12, pl. 462α.

[145] Tsigaridas 1973b, 498-501, drawings 12, pl. 462α.

[146] Moutsopoulos 1996, 605-36; Moutsopoulos 2000, passim.

[147] *6000 Years* 1997, 171-174, 209, 221. Also unpublished finds exhibited in the permanent exhibition of the Museum of Byzantine Culture.

| | | | |
|---|---|---|---|
| 69 | 014 | Sector E, West side of the complex. Φ. 382 | 23-9-1981 |
| 145 | 035 | Sector E, ΕΔ9- 4th layer. Φ. 1276 | 26-10-1991 |
| 188 | 043 | Sector E, ΕΔ1, trench IV. Φ. 251 | 18-9-1980 |
| 189 | 043 | Sector E, West side of the complex. Fill west of the southern staircase. Φ. 380 | 2-2-1981 |
| 58 | 0127 | Basilica complex, K39- α' layer (secondary burial?). Φ. 1328 | 14-10-1991 |

104. Sindos, Thessaloniki Prefecture, Goodyear Plant, 1981

During the excavation of the sewage pipes on the site of the Goodyear plant, 14 graves (2nd or 3rd century) were detected at a depth of 1.80m. A short report on 2 of them was published.[148] The graves included 7 tile-, 3 cist- and 4 tile-graves. In the graves were found clay and glass vessels, jewellery and bronze coins.[149]

| Cat. No. | Form | Other excavation indications | Find Date |
|---|---|---|---|
| 302 | 053a | In the fill of tile-grave T9 | |

105. Sindos, Thessaloniki Prefecture, Thessaloniki Industrial Zone, 1999-2000

A cemetery dated to the Roman period was uncovered during construction works in the wider area of the Goodyear plant in 1999-2000.[150]

On the property of the Mazmoulian family, block 51β, west of the Goodyear plant, 27 graves were found .

44 graves were uncovered on the property of the ΑΘΕΛΚΟ company, in the Goodyear plant.

The graves are oriented E-W. The majority are tile graves, although there are some built cist-graves, and simple, jar-burials and simple pit-graves. In the graves were found clay and glass vessels, clay lamps, bronze, silver and glass jewellery and bronze coins. All coins, clay and glass vessels, as well as the lamps are dated to the 4th century.[151]

| Cat. No. | Form | Other excavation indications | Find Date |
|---|---|---|---|
| 48 | 012a | Goodyear plant, ΑΘΕΛΚΟ plot, Grave 2 | 23-6-2000 |
| 116 | 0315 | 1B, square. B.T. 201-5-1999 | |

### *Imathia Prefecture*

106. Veroia, Tsirakoglou Plot, 1940

Rescue excavations were conducted at St. Tsirakoglou plot, outside of Veroia, at the junction of the rail station and the road to Naousa. A Roman sarcophagus was found there and 110 Early Christian graves of various types. In the graves were found clay and glass vessels, glass, silver and bronze jewellery. The graves were covered with marble *spolia* from earlier reliefs and inscriptions.[152]

| Cat. No. | Form | Other excavation indications | Find Date |
|---|---|---|---|
| 121 | 032 | | 1940 |
| 213 | 050 | | 1940 |
| 250-251 | 051a | | 1940 |
| 412 | 085 | | 1940 |
| 623 | 132 | | 1940 |
| 662-663 | 139 | | 1940 |
| 742 | 147b | | 1940 |

### *Kilkis Prefecture*

107. Kolchis, Kilkis Prefecture, 1983-88

A walled site next to the River Gallikos, which can probably by identified with the Roman station Gallicum or Callicum, was partly excavated. Five different phases in its history are discerned. Early Christian, barrel-vaulted graves with a built platform/couch in them were excavated, one of which contained a glass lamp. During the 5th century a single-aisled church was founded, which was extended in the Justinian period and transformed into a three-aisled basilica. At the end of the 6th century the complex was supplemented by a baptistery, a wine press and other structures. It has been assumed that the destruction of the site took place in the second half of the 7th century. On the site a smaller church was built in the 9th century, which, it appears, was operating at least until the early 13th century.[153]

| Cat. No. | Form | Other excavation indications |
|---|---|---|
| 165 | 036 | Hanging on a nail at the eastern wall of the grave, above the head of the deceased<br>Found laying on the platform/couch |

[148] Romiopoulou 1981, 302.
[149] Diary no. Θ35, pages 2-19, of the 16th Ephorate of Prehistoric and Classical Antiquities, Thessaloniki.
[150] Keramaris, Papagianni 2000, 145-152.
[151] Keramaris, Papagianni 2000, 147-148.
[152] Walter 1940, cols. 172-175; Kotzias 1953-54, 167-175; Makaronas 1941/52, 626-627.
[153] Kissas 1988, 207-212, fig. 1-12.

## *Pella Prefecture*

108. Palaia Pella (No Date)

No documentation relevant to excavations related to the glass vessels is preserved in the archives of the 16th Ephorate of Prehistoric and Classical Antiquities.

| Cat. No. | Form | Other excavation indications | Find Date |
|---|---|---|---|
| 380 | 074a | | |

## *Pieria Prefecture*

109. Confiscated Items, Katerini, 1973

Among the confiscated objects, according to the archives of the 16th Ephorate of Prehistoric and Classical Antiquities, was 'a dark-blue faience lekythos, broken into two pieces, transformed into masses by fire, of no value'.[154] It appears that it is a mould-blown vessel in the shape of a bunch of grapes. As the looters were inhabitants of the village of Lofos in Pieria, and that the confiscation took place in Katerini, it seems quite probable that the looted objects were from that region as well.

| Cat. No. | Form | Other excavation indications | Find Date |
|---|---|---|---|
| 496 | 118 | Probably from the area of Lofos, in Pieria | 1973 |

110. Louloudies Pierias, 1993-1999

The site of Louloudies is an episcopal complex, founded as a result of the settlement of the Goths at Pydna during the period AD 479-483. The people, with their bishop, moved 8km to the south of Pydna, where the *mutatio Anamon* was located, and erected a *quatriburgium*, i.e. four-towered episcopal complex in a form of a four-towered fort, as found in Roman camps. The episcopal complex included a basilica, an episcopal palace and galleries; it was protected by a wall, a double-towered gate and four towers at the corners. Diverse artisanal and commercial activities have been identified as taking place on the site, which was, quite probably, at least partly, in use until the 8th century. On the site, small circular kilns have been found together with glass residues. The kilns belong to the latest phase of the castle (7th and 8th centuries). On the site there are finds of glass droplets and moils that confirm the local production of vessels. Also found were many pieces of glass slabs for the production of glass *tesserae*.[155]

| Cat. No. | Form | Other excavation indications | Find Date |
|---|---|---|---|
| 257 | 51b? | | |

## *Serres Prefecture*

111. Amphipolis, 1913

No documentation relevant to excavations at that period is preserved in the archives of the 16th Ephorate of Prehistoric and Classical Antiquities.

| Cat. No. | Form | Other excavation indications | Find Date |
|---|---|---|---|
| 575 | 126b | | |

## *Chalkidiki Prefecture*

112. Unknown Find-Spots (No Date)

No documentation relevant to these finds is preserved in the archives of the 16th Ephorate of Prehistoric and Classical Antiquities. Some fragments of this group were actually mended with vessels registered in the Archaeological Museum of Thessaloniki many years prior to this group and therefore ascribing the entire group to one excavation seems quite unsound.

| Cat. No. | Form | Cat. No. | Form | Cat. No. | Form | Cat. No. | Form |
|---|---|---|---|---|---|---|---|
| 198 | 046 | 322 | 059 | 395 | 075 | 438 | 100 |
| 310 | 053b | 330-334 | 061 | 425 | 094 | 609 | 131 |
| 317 | 056 | 337 | 062 | 428 | 095 | 633-634 | 138 |

113. Confiscations (Possibly from Olynthos)

No further information was found in the Archives of the 16th Ephorate of Prehistoric and Classical Antiquities.

| Cat. No. | Form | Other excavation indications | Find Date |
|---|---|---|---|
| 540 | 122a | | |
| 541 | 122a | | |

[154] The author would like to thank Dr Despina Ignatiadou for this information.

[155] On the site, see Marki 1996, 239-243; on artisanal activities on the site, see Marki 2002, 65-66. On the causes of the establishment of the episcopal complex at Louloudies, as a result of the settlement of Theodoric's Goths at Pydna, see Marki 2013, 1-10.

1. '*Agios Dimitrios*' Municipal Hospital and Water Company Site (*Εταιρεία Υδάτων*), 1911
2-3. 'Agios Dimitrios', Municipal Hospital, 1978, 1986-1988
4. Agios Pavlos (*Sikies* Site, *Heptapyrgion*), 1931
5. Aqueduct, Agios Pavlos, 1966
6. Agkelaki Street, 'Zervas' Plot, 1960
7. Agiou Dimitriou Street, East Extension, 1960
8. No. 232 Agiou Dimitriou Street (Near Wall of the '*Kentrikon*' Hospital), 1962
9. Aristotle University Campus, North of Central Library, 1994
10. Aristotle University Campus, Swimming Pool, 1970
11-12. Aristotle University Campus, Administration Building, 1965, 1972
13-15. Aristotle University Campus, Theological Department, 1965, 1977, 1979
16. Aristotle University Campus, Law School, 1965, 1966
17. Aristotle University Campus, Sorority Building (Φ.Ε.Θ.), 1968
18. Egnatia and Melenikou Streets, 1969
19. Northern Area of 'Evangelistria' Cemetery ('*Lyssiatreion*'), 1993-97
20. No. 2 Leoforos Stratou, Museum of Byzantine Culture, 1988-9
21. Leoforos Stratou, 3rd Army Corps Site (Building of the *ΥΕΝΕΔ/ΕΡΤ3* TV Station), 1969
22. Filikis Etaireias Street. 'Hippodrome', Anast. Nastos Site, 1967
23. No. 39 Filikis Etaireias and D. Margariti Streets, 1973 (?)
24-27. Thessaloniki International Fair (*ΔΕΘ*), 1972, 1979, 1988, 1992, 2000
28. Melenikou and Chatziandreou Streets, 1994
29. Trigonion Tower, Chapel to the South, 1972
30. Sintrivani Square, City Wall Excavations, 1970
31. Tritis Septemvriou Street, Cemeterial Basilica, 1980
32. No. 44 Vasilissis Sofias Street (Ethnikis Amynis Street), 1960
33. No. 8 Agiou Nestoros Street, 1977
34-37. Nos. 8, 26, 26-28, 32 Agiou Nestoros Street, Water Main Works, 2000
38. No. 24 Agion Panton Street, 1999-2000
39. No. 13 Apolloniados Street and Oraiopoulou Street, 1991, 1992
40. No. 18 Apolloniados Street, 1976.
41. No. 31 Apolloniados Street and Michalitsiou Street, 1970
42. Aristatou, Panagopoulou and Patroklou Streets, Ampelokipi, 1991
43. No. 4 Borou Street, 2000
44. No. 8 Doxatou Street, 1993
45. No. 13 Eleftherias Street, Ampelokipi, 1977
46. Efxinou and Panagopoulou Streets, 1994
47. Giannitson Street, Veterinary Department, 1969
48. No. 5 Kallitheas Avenue, Ampelokipi, 1983
49. Kallitheas Avenue and Agathonos Street, Ampelokipi, 1995
50. Kallitheas Avenue and Skeparni Street, Ampelokipi, 1996
51. Kallitheas Avenue, Elementary School, Ampelokipi
52. No. 4 Kapatou N. Street, 2001
53. Junction of Koloniari, Galanaki and Papathanasiou Streets, 1987
54. Lagkada Street, 1961
55. No. 67 Lagkada Street, 1963
56. Lagkada Street, The 'Dodekaorofi' (A 12-Storey Tower in Ramona), 1972
57. Lagkada Street, The Axios Residential Complex, 1968-70
58. Lagkada Street, The Axios Residential Complex (Northern Plot), 1989-1991, 1995
59. Lagkada Street, The Axios Residential Complex (Southern Plot), 1991 and 1994-95
60. Lagkada Street, The Axios Residential Complex (Central Plot), 1998
61. Lagkada Street, Ivanof Park, 1991
62. No. 20 Margaropoulou Street, 1995
63. Margaropoulou and No. 2 Syntagmatarchou Fryzi Streets, Xirokrini, 1998
64. No. 5 Monastiriou and Odysseos Streets, 1995
65. No. 3 Peleiadou Street, 2001
66. No. 5 Peleiadou Street, 1997
67. No. 6 Peleiadou Street, 1996
68-73. The Ramona Area, 1963, 1968, 1969, 1971, 1972
74. No. 14 Regkoukou Street, Xirokrini, 1979
75. No. 16 Regkoukou Street, Xirokrini, 1994
76. No. 31 Syntagmatarchou Fryzi Street, 1996
77. No. 2 Thiseos Street, 2001
78-81. Vardari
82. No. 3 Agapis Street, 1996
83. Acheiropoietos, East of the Church Apse (S. Georgiadis Plot), 1966
84. Junction/Corner of Agiou Dimitriou and Agias Sophias Street, Byzantine Building, 1969
85. Agiou Dimitriou and Sofokleous Streets (No Date)
86-87. Ancient Agora, 1962, 1966, 1989-1999
88. No. 12 Christopoulou Street, Proferis Plot, 1975
89. Dioikitiriou Square, 1990-1998
90. No. 7 Lapithon Street, 1990
91. Mitsaion Street, 1984
92. No. 1 Pringipos Nikolaou Street, 1961
93-94. Nos. 47-49 and 58-62 Sokratous Street (Mirarchou Koufitsa St.), 1968
95. No. 45 Vassileos Irakleiou Street, 2010

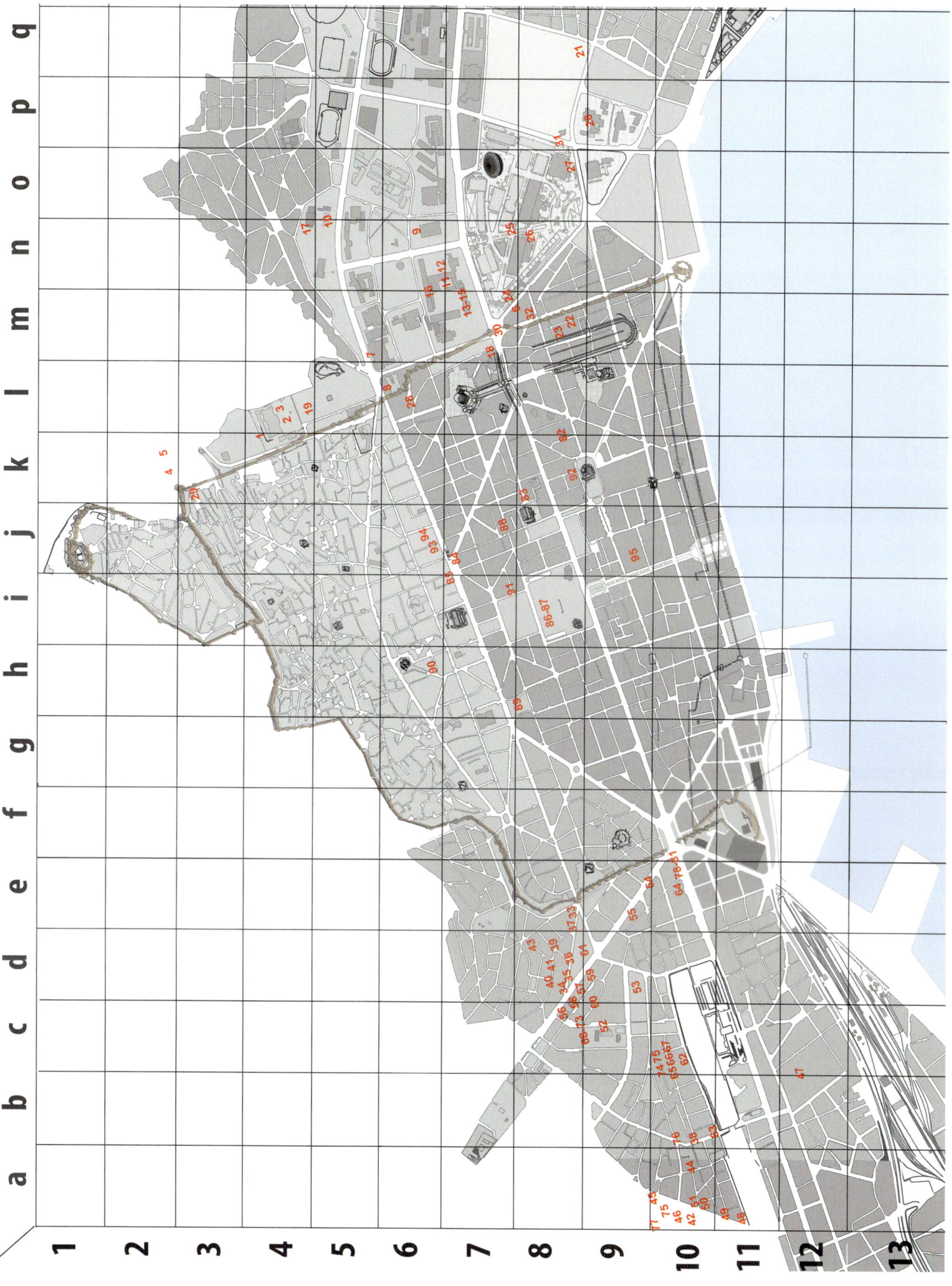
q
p
o
n
m
l
k
j
i
h
g
f
e
d
c
b
a
1
2
3
4
5
6
7
8
9
10
11
12
13

# Glossary of Technological Terms Related to Glassworking[1]

**Abrasion/abrading** — The smoothening of the external surface of glass objects by removing a mass of glass and effacing chance protrusions. Done mechanically, by hand with an **abrasive**, or tool. Constitutes a light form of **grinding**.

**Abrasive/abrasive powder** — The material employed with the **bow-drill** during incision or during the smoothening of the surface. Normally in the form of a fine powder, e.g. emery.

**Annealing** — The process of cooling a glass product. Gradual annealing is necessary in order to release the stresses created during glassworking, as well as due to a difference in contraction rates between surface and interior. Thus fracturing is avoided. Takes place in a chamber of the furnace, as far away as possible from the heat source.

**Annealing oven/ annealing furnace/ lehr/leer** — The place where **annealing**, the gradual cooling of a glass product, takes place. This is a chamber in the **furnace**, as far away as possible from the heat source, normally its top shelf, or in an area off to one side. The temperature at which **annealing** occurs ranges between the temperature in the **furnace** and that in the workshop.

**Band** — Long, narrow strip of glass used for **forming** vessels with **mosaic technique**. The bands are normally monochrome, composite polychrome or gold (with gold leaf encased between two sheets of **transparent** glass).

**Barilla** — A plant belonging to the genus *Salicornia Kali* or *Salsola Kali*. **Plant ash** is used as a source of alkali for glassmaking. Given names for the plant are *glasswort* or *prickly glasswort*

**Batch** — All the raw materials (**silica**, **soda**, **calcium**) that are **fused** in a **crucible** to make glass.

**Blank** — The glass object that has already undergone **annealing** and will be finished off in the next production phase.

**Block** — Wooden, hollowed out, glassworking tool. Used during **blowing** as a **mould** for forming round shapes.

**Blowing/free-blowing** — The technique of **forming** an object by passing air through a mass of hot glass. Blowing is done with the **blowpipe**, on whose lower end the **gob** (molten glass) has already been picked up. The **paraison** is produced by the initial act of blowing, and the final product is then **formed**. A variant of the blowing of the **gob** is the blowing of a glass cylinder, at the closed end of which the **paraison** is formed. Blowing appears in the first half of the 1st c. BC.

► Blowing is differentiated into **free-blowing** and **mould-blowing**.

**Blowpipe** — The hollow metal rod ending in a cone, used for **blowing** glass. Normally around 1.40-1.70m in length and with a diameter ranging between .02 and .025m.

**Bow-drill** — The drill operated with the aid of a bowstring. The string is fastened to the drill and causes it to rotate through the back-and-forth movement of the bow.

**Bow-lathe** — The lathe moved with the aid of a bowstring. The string is fastened to the **lathe**, and causes its rotation through the back-and-forth movement of the bow.

**Bubble** — The air pocket trapped in the glass during the process of **fusion**. The tiny bubbles thus formed are called **pinprick bubbles** or **seeds**.

**Calcium oxide/lime** — The oxide of **calcium** added to the **batch** to increase the stability of the glass .

**Cane/Bar/composite mosaic cane** — The glass **rod** created by the fusion of a bundle of simple **rods** of glass to form a polychrome design, visible in section.

**Carving /cold-cutting** — The mechanical removal (with a tool or the **bow-drill**) of a mass of glass from the body of a glass object.

[1] This glossary refers only to terms used in the present work. It consists of entries from the dictionary of glassworking, which covers Antiquity and Medieval times: Ignatiadou, Antonaras 2007.

**Casing** The process of placing a second layer of glass of a different colour on top of the original **paraison** of a vessel. The technique is used in the production of vessels with outer walls of a colour different from that of the inner ones.

**Casting** The technique of **forming** a glass object through use of a **mould**. The **mould** may be open, bi-, multi-partite, or a **cire-perdue** mould. The term is borrowed from metalworking, but the casting of glass differs from that of metals. In **glassworking**, molten glass is not used directly from the **crucible**, rather, the **mould** is filled with hot, viscous glass or **glass chunks**.

**Chair** The glassworker's long, wide workbench. It has side supports on top of which the **blowpipe** is horizontally rotated. Introduced in the post-Renaissance period.

► The terms 'Chair' and 'Shop' also refer to the team of glassworkers under the supervision of a **gaffer**.

**Chip-casting** The technique of forming a glass object by filling a **mould** with **chips**.

**Chunk/chip** The piece of **raw glass** destined for **secondary production**.

**Cold** A term characterizing every process that occurs after **annealing**, when the glass has already cooled.

**Cold-polishing** **Polishing** which occurs following **annealing**, done by hand with an **abrasive**.

**Collar** The band (round in section) running around the mouth of a vessel. Decorative but probably functional as well, serving to strengthen the mouth and make for safer transport of a vessel.

**Colouring agent** The element that determines the colour of glass. Natural colouring agents include iron, found in the **sand**. The commonest colouring agents added to glass are copper, manganese, and cobalt.

**Colourization** The adding of a **colouring agent** to the **batch**.

**Core forming/core technique** The construction of vessels by **forming** around a **core**.

A **core** of organic and inorganic materials is made at the end of a **metal rod**. The rod is rolled on crushed glass and then inserted in the **kiln** until the glass melts and covers the **core**. The neck and rim of the vessel is formed with tools, and handles and bases are added. The decoration is usually made by winding **threads** of coloured glass. The **threads** are usually **combed** to create **zig-zags** or a **feather pattern**.

**Corrosion/weathering** Alteration to the glass, first on its surface and then below the surface, as a result of the influence of environmental factors. Resulting from the loss of alkaline ions and the creation of a siliceous layer on the surface. It is influenced by the composition of the glass. 'Corrosion' is a general term that includes the following specific occurrences: **discolouration, loss of vitreous nature, incipient crizzling, pitting, milky weathering, crusting, dulling, iridescence, Crazing/crizzling, plowing, cracking/fracturing, sugaring, lamination**.

**Cracked-off rim** The unformed rim of a vessel following the **cracking-off** of the **moil**.

**Cracking-off** The process of separating a glass product from the **blowpipe** or **metal rod/punty** after **forming**. This is done by a light striking of the **blowpipe** after determining the point where the cracking-off will be done by an **incision** around its circumference. The **moil** remains on the **blowpipe**.

**Crucible/melting pot/pot** The container made of **refractory** clay in which the **fusion** of the raw materials takes place, and in which the molten glass is stored within the **furnace**.

**Crystal/rock crystal** Chemically pure **silica**, in the form of a natural mineral. It is **translucent** and colourless.

See also **Lead glass**.

**Cullet** Small pieces of glass destined for recycling. Cullet may be:

1. Broken glass objects collected for adding to a new **batch**.

2. Defective products that are shattered for the same purpose.

3. Remains from a previous **fusion** process.

4. Leftovers from the **forming** of glass products.

**Cut** The object that has undergone **cutting** so as to create an **intaglio** (creation of a design lower than the surface) or **relief** (removal of the original surface in order to allow the design to emerge).

**Decolourizer** The antimony or manganese compound added to the batch in order to neutralize or lessen the natural greenish tint of glass.

**Decolourizing** The neutralizing or lessening of the greenish tint of glass, which is due to the natural impurities in iron. Decolourizing is accomplished by adding **decolourizers** to the **batch**.

**Drawing/pulling** The lengthening, by continuous pulling and simultaneous heating, of a mass of glass so as to form long (many meters in length) **threads** and smaller **rods** of glass. Some **beads** are also **formed** by drawing.

**Drilling** The creation of a **perforation** through the use of the **bow-drill**. Drilling is used to create an interior hollow in some vessels (e.g. possibly alabastra) or the handles of other vessels.

**Engraving/wheel-engraving/wheel- cutting** The technique of decorating glass objects by creating a design with the **bow-drill**. Simultaneous use of some powdered **abrasive** is required. Used chiefly from the Roman period.

**Facet cutting/faceting** The decorative technique in which flat areas (facets) are created on the surface of an object through **abrasion**.

**Fire-polishing** **Polishing** done before **annealing**, by briefly placing the glass object in the **furnace**.

**Firing** Exposure of a material or an object to a temperature below its **melting** point. Firing is required in the case of **enamel** and for forming **game counters**.

**Flute** A vertical **groove** decorating an object, usually a vessel.

**Flux** The alkali (**sodium carbonate**, or **soda** or **natron**, or **nitrum**, or **potassium carbonate**, or **potash**) added to the **batch** during **glassmaking/primary production** in order to lower its **melting** point. For the same reason, flux is also added to **enamel**.

**Forest glass** A type of Medieval glass. Wood **ash** was employed as a source of alkali (**potassium**) for its production

**Former mould** The convex **mould** on which an object is **formed** by the **sagging/slumping** of a **blank**, or by the adding of viscous glass and pressing it down with a **plunger**, during the technique of **rotary pressing**.

**Free-modelling** The **forming** of a glass object with tools, without the use of a mould or **metal rod/pontil/punty/mandrel**. Some parts of **core-formed** vessels (handles, rims, bases) or pendants may have been made using this method in antiquity.

**Frit** The vitreous product of **fritting**, i.e. the extended preparatory heating of the raw materials (in antiquity, **sand** and **natron**). After **annealing**, the frit was crushed and heated in another

**furnace** to **melt** and produce glass. Frit has an appearance similar to that of **faience** and in prehistoric times it was used for the **forming** of small objects.

*Hammonitrum* is the ancient word attested in Pliny the Elder. It accurately renders this material's composition in antiquity.

**Fritting** The extended preparatory heating of the raw materials for **glassmaking** (in antiquity **sand** and **natron**) for **primary production**. The glass-like product resulting from fritting is **frit**, which after **annealing** was crushed and heated in order to **melt** and produce glass.

| | |
|---|---|
| Furnace | The closed construction used for the production and application of heat. The glassworking furnace was employed for fusion, the maintaining of molten glass during **secondary production**, and **fire-polishing**. |
| Fusion | 1. The process of heating a **batch** so as to unite (chemically) its component raw materials. Fusion takes place during **primary production**, and differs from **melting**, which is applied to already-prepared glass.<br><br>2. The process of heating a group of preprepared glass elements, e.g. **bands** or **rosettes**, to join them to another object. |
| Gather/gob | The preheated **chunk** or the mass of molten glass that the glassworker removes from the **crucible** with the end of a **metal rod/pontil/punty** or the **blowpipe**, with which he then proceeds to the **forming** of objects with the techniques of **winding/rod forming**, **drawing/pulling**, or **blowing**. |
| Gilding | The process of decorating glass with gold, which may be in the form of gold leaf, which is attached to the glass, or a gold-powder emulsion which is applied on the glass with a brush. |
| Glass | The inorganic substance, the result of the **fusion** of materials that solidify as an amorphous solid without crystallization having occurred. This 'amorphous' structure classifies glass as a liquid. |
| Glass former | **Silicon** dioxide (**sand** or pebbles) that serves as the basis for making glass. |
| Glass forming | The making of glass objects. |
| Glass making: primary production | **Primary production**, i.e. the **fusion** of raw materials resulting in the creation of glass.<br><br>**Secondary production** follows primary production. |
| Glass working: secondary production | The **forming** of glass objects. It follows the process of **primary production** of the **glass** itself. |
| Glory hole | The opening in the **furnace** through which the **blowpipe** is introduced and the **gather/gob** is removed; also serves for the heating of vessels during the process of their **forming**. |
| Gold band | The long and narrow, flat strip of glass in which gold leaf is encased between two sheets of **transparent** glass. It is employed, together with other **bands** or **rosettes** or special elements, for the **forming** of vessels using the **mosaic technique**. |
| Gold-glass/sandwich gold-glass/*fondi d'oro* | The glass or glass object made of cut-out sheets of gold leaf encased between two layers of **transparent** glass. The technique is employed for the **forming** of vessels, plaques, and **tesserae** in the Hellenistic, Roman, and Byzantine periods. The term is attested in the sources. It is not confirmed whether it refers to the inlaying of gold in glass or to decoration with gold-glass, but in any case it accurately renders the term 'sandwich gold-glass'.<br><br>The modern term *fondi d'oro* specifically describes disk-shaped gold-glass bases for open vessels of the 4th c. AD. The gold leaf cut-outs were complemented by incision and added colour. In a second use these were also employed as decorative wall inlays for funerary monuments in catacombs. |
| Grinding | The smoothening of the surface of glass objects by removing extruding glass. Accomplished mechanically, by hand and with an **abrasive**. |
| Groove | The narrow horizontal channel that adorns an object, normally a vessel. |
| Hot-/fire- | Term characterizing any process that takes place prior to **annealing** while the glass is still hot. |
| Inclusion | A foreign body incorporated into the mass of glass, whether by chance, in the form of an impurity, or as a decorative element. |
| Ingot | The product of **fusion**, i.e. **raw glass** of a specific shape (normally that of the clay **crucible**), which facilitates its transport and trade. |
| Intaglio | Creation of a design lower than the surface. It is done with **cutting** or in a **mould**. |

| | |
|---|---|
| **Iridescence** | The phenomenon of the presence of colours on the **corroded** surface of the glass. A type of **corrosion**. Caused by the refraction of light by the layers of **lamination,** and changes in accordance with the angle from which it is viewed or the angle at which light falls on it. |
| **Jacks** | See **Tool/Jacks/Pucellas**. |
| **Kick** | The depression on the bottom of a vessel. Created by pressure applied by the glassworker in order to strengthen the vessel's construction or decrease the vessel's capacity. |
| **Kiln/oven/calcar/calker/caulker** | The low temperature **furnace** used for converting the **batch** into **frit**, for **secondary production**, and for the **firing** of **enamel**. |
| **Knop/knob** | The round or flattened element, solid or hollow on its interior. Either alone or in groups, used to form the stem of **beakers** or on **lids** and **lamps**. |
| **Lamp-working/flame- working** | The process of **forming** or finishing a glass object with the use of a flame. The flame of a small torch or lamp may be used as a heat source. It is not certain that this process was employed in antiquity. |
| **Lathe-polishing/rotary-polishing** | The **cold-polishing/grinding** that takes place during the rotation of a glass object on a potter's wheel/lathe. Possibly not employed in antiquity. |
| **Lost-wax casting/lost-wax technique** | The technique for **glass forming** based on construction of a wax model of the object. The model is encased in clay or plaster, so as to create a **mould**, and is then heated. The wax melts and runs off via exit channels. The empty space created is then filled with glass **chips**. The **mould** is heated and after **annealing** is removed or destroyed. Decoration of the product can then take place. |
| **Marver** | The flat, level surface on top of which glass objects are rolled in various stages of **secondary production**. The surface is stone, usually marble. |
| **Marvering** | The rolling of glass objects on the **marver**. The purpose of rolling is to achieve the shape's final form, smoothen the surface, or incorporate inlays or attached elements.<br><br>Done only with objects **formed** by **rod- or core-forming**, or **blowing**. |
| **Melting** | The process of heating glass until it reaches a liquid state. It is preceded by **fusion**. |
| **Millefiori** | type of **mosaic glass** consisting of pieces of composite **mosaic** canes, exclusively those having the shape of a flower in section. These pieces are called **rosettes**. Roman period.<br><br>The term was first introduced in the 15th c. by Venetian glassworkers, who used the term to refer to copies of ancient vessels they constructed using the mosaic technique.<br><br>▶ Frequently used incorrectly instead of the general and more accurate term, **mosaic glass**. |
| **Modifier/stabilizer** | The **calcium**, manganese, or barium carbonate added to the **batch** to increase the stability of glass. |
| **Moil/moile** | The upper segment of the **paraison**, i.e. around/near the **blowpipe**, which in the final stage of **forming** is cracked-off as excess. |
| **Mould:** | The plaster, clay, stone, or metal construction within or on top of which glass is pressed, cast, sagged, or blown. It may be: |
| **Open mould/dip-mould** | – A concave, **open mould** (for **pressing/sagging/ blowing**). |
| **Bi-/tri-/multi-partite mould** | – A closed (single, bi-, tri-, multi-partite) mould (for **casting**). |
| **Former mould** | – A convex **former mould** (for **rotary pressing/sagging**). |
| **Mould-pressed/mould-made** | An object **formed** by **pressing** into an open **mould**. |
| **Mould-pressing** | The technique of **forming** an object by the pressing of viscous glass into an open **mould**. |
| **Molten glass** | The liquid glass resulting from the process of **fusion** of the **batch**. |

| | |
|---|---|
| **Mosaic glass** | Glass created by preprepared elements of coloured glass. The elements (**rods, cane sections, rosettes, bands**) are placed in the **mould**, and this is followed by their **fusion**. Roman period. |
| **Mosaic technique** | The construction technique for **mosaic glass**. |
| **Natron/sodium sesquicarbonate/soda** | The mineral formed by **sodium** salts (carbonate, bicarbonate, chloride, and sulfate) that comprises one of the raw materials in glass. Traditionally its sources are Cairo's Wadi Natrun area and ancient Chalastra in Macedonia. |
| **Opacifier** | The component added to the **batch**, rendering glass **opaque** through the formation of micro-crystals after **annealing**. The commonest opacifiers are lead and **calcium** antimonate. |
| **Opaque** | Glass, and any other material impenetrable by light. Opacity is achieved by the addition of **opacifiers**.<br><br>Glass that allows the passage of light is defined as **translucent**, while that which permits visibility as well is defined as **transparent**. |
| **Overblow** | That part of the **paraison** next to the **blowpipe** that overflows from the **mould** during **blowing** and is **cracked-off** during the vessel's final stage of **forming**. |
| **Overflow** | The excess of glass during the filling of the mould in the **mould-pressing** technique. |
| **Oxide** | A compound formed between an element and oxygen. All the components of ancient glass are in the form of oxides. |
| **Paddle/pallet** | The long, wide, flat tool, normally made of wood, with which the flat surfaces of vessels are formed. |
| **Paraison/parison** | The bubble of glass at the tip of the blow pipe, in the method of free-blowing. |
| **Perforation** | The stringing hole through the body of **beads** and the hanging loops of **pendants**. |
| **Pig/yoke** | The support in front of the **glory hole** of the **furnace**, for the placement of **blowpipes** and **metal rods**, to keep them hot and ready for use. |
| **Pincers** | The metal tool shaped like a small **jacks/pucellas**, used for the **drawing/pulling** of glass, or creating **lugs** or folds. |
| **Plant ash** | The ash resulting from the slow burning of particular plants. In antiquity, chiefly from the Roman period and onwards, it was used as an alternative source of sodium, instead of **natron**. Depending on the type of plant, the ash was a source of **sodium**, **potassium**, or both. See also **Barilla**. |
| **Plunger** | The specially shaped tool used during the **mould- pressing** technique to shape the interior of a vessel. |
| **Polishing:**<br>**Cold-polishing**<br>**Fire-polishing** | The **abrading** of the visible/exterior surface of glass objects. When done before annealing it is called **hot**- or **fire-polishing**. When done after **annealing** it is called **cold-polishing**, and distinguished into **grinding** (done by hand) and **lathe-polishing** (done on the lathe). The use of **cold-polishing** was limited in antiquity. |
| **Pontil/punty/metal rod** | The solid metal rod used for:<br><br>1. Picking up the **gather/gob** and its further processing.<br><br>2. Drawing/pulling to create **rods** and **threads**.<br><br>3. The **forming** of a vessel by **winding** around it. The **forming** of **beads** and **pendants** is done with the smaller **mandrel/mandril**. |
| **Potassium carbonate/potash** | The substance employed alternatively as an **alkali** instead of **soda** in glassmaking. Its traditional source was **plant ash**. |
| **Potter's wheel/lathe** | The horizontal, rotating clay disk, comparable to that used for making pottery. On it the **hot-forming** of vessels takes place using the method of **rotary- pressing**, as well as **cold**-processing. |
| **Primary production/glass making** | see **Glassmaking** |
| **Raw glass** | Glass destined for **secondary production**. |

**Refractory** — Material resistant to firing, normally glass mixed with **silica**; **furnaces** and **crucibles** were made of such material.

**Relief** — A design formed higher than its surrounding surface. Distinguished as 'low' or 'high' relief.

**Reverberator furnace** — **Furnace** with a domed roof that reflects heat.

**Reverse painting** — The added painting behind or beneath the face of a glass surface so that the design is visible only from in front or above the object. Encountered from the Hellenistic period.

**Rod** — The delicate cylindrical rod of glass that is joined to others and **fused** in order to form a **composite mosaic cane/bar**.

**Rosette** — A glass **cane section**, part of a **composite mosaic cane/bar** having the form of a flower in section. Used in the **forming** of **millefiori** objects.

**Rotary marks/rotary scratches/rotary striations** — Concentric circles, continuous or discontinuous, visible on the interior or exterior of **mould-pressed** vessels. Due to impurities trapped in the construction material of the **mould**, which leave undesirable traces on a vessel surface. Formerly connected erroneously with the techniques of **casting** and **cold-polishing** of vessels.

**Rotary-pressing/glass pottery** — The technique of **forming** vessels on a **potter's wheel**. The viscous glass is pressed down with a **plunger**, either on the rotating lathe or in a **mould**. The **mould** could be concave, or a convex **former mould**.

**Sagging/slumping** — The method of forming glass vessels by heating a **blank** glass disk in a positive (hollow) **mould**, or on top of a negative (convex) **former mould**. The **blank** gradually assumes the shape of the **mould**, its only impetus being its own weight.

**Sand** — The basic raw material of glass, source of the requisite **silica**. It is taken from shoreline areas or natural deposits. It is prepared by sifting, washing, heating, and grinding.

**Secondary production/glass working** — see **Glassmaking**.

**Seed/pinprick bubble** — The very small **bubble** trapped in the mass of glass during the process of **fusion**.

**Shears** — Tool which was perhaps used for the removal of excess material during the **forming** of glass objects. Similar to the common scissors of the Roman period, i.e. it has the form of **jacks** with flat arms.

**Silica** — **Silicon** dioxide, the basic component of glass, in an amount ranging from 50% to 70%. Its source was **sand**, shells, or **rock crystal**.

**Silicon** — The non-metallic element (Si) encountered in an abundance of combinations in nearly 30% of the earth's crust.

**Slag** — The remains from the fusion of any material, most commonly of metals. Slags are frequently made up of melted, vitreous products.

**Sodium carbonate/soda** — One of the raw materials in glass. Added to the batch as flux. Its source is Natron/Nitrum.

**Tank** — The rectangular container or built construction within which glass is made.

**Test droplet** — The drop-shaped **thread** of glass that results during a check of the fluidity and **colouring** of the **molten glass**.

**Thread/trail** — The delicate filament of glass that forms the decoration on **beads**, **pendants**, and vessels made with the **core** and **free-blowing** techniques. The **thread/trail** is formed by **drawing/pulling**.

**Tool/jacks/pucellas** — A metal tool with two arms joined at the top, used in a variety of ways in **forming** glass objects, particularly the mouth of vessels.

**Translucent** — Glass, and generally any material that allows the passage of light through it, but which does not permit visibility.

**Transparent/clear** — Glass, and generally any material that not only allows the passage of light through it, but also full visibility.

**Waster/glass waste** — 1. A failed or broken product that is rejected during **glassworking**.

2. The remainder from **glassworking**. Glass waste is normally recycled in the form of **cullet**.

# 10. Concordance

1. Registration Number - 2. Catalogue Number

| Registration Number | Catalogue Number |
|---|---|
| Av-Pi /1 | 324 |
| Av-Pi /2 | 539 |
| Av-Pi /3 | 516 |
| Av-Pi /4 | 586 |
| Av-Pi /5 | 587 |
| RGM Koln 62/58 | 65 |
| BAA 1 | 95 |
| BAA 2 | 86 |
| BAA 3 | 87 |
| BAA 4 | 91 |
| BAA 5 | 92 |
| BAA 6 | 93 |
| BAA 6 a | 97 |
| BAA 7 | 106 |
| BAA 8 | 107 |
| BAA 9 | 108 |
| BAA 10 | 101 |
| BAA 11 | 102 |
| BAA 12 | 89 |
| BAA 13 | 90 |
| BAA 14 | 99 |
| BAA 15 | 103 |
| BAA 16 | 100 |
| BAA 17 | 104 |
| BAA 18 | 30 |
| BAA 19 | 31 |
| BAA 20 | 32 |
| BAA 21 | 33 |
| BAA 22 | 34 |
| BAA 23 | 35 |
| BAA 24 | 36 |
| BAA 25 | 9 |
| BAA 26 | 10 |
| BAA 27 | 11 |
| BAA 28 | 12 |
| BAA 29 | 13 |
| BAA 30 | 14 |
| BAA 31 | 15 |
| BAA 32 | 16 |
| BAA 33 | 17 |
| BAA 34 | 25 |
| BAA 35 | 26 |
| BAA 36 | 27 |
| BAA 37 | 28 |
| BAA 38 | 385 |
| BAA 39 | 386 |
| BAA 40 | 387 |
| BAA 41 | 190 |
| BAA 42 | 191 |
| BAA 43 | 192 |
| BAA 48 | 388 |
| BAA 49 | 389 |
| BAA 50 | 390 |
| BAA 51 | 391 |
| BAA 52 | 392 |
| BAA 53 | 393 |
| BAA 54 | 39 |
| BAA 55 | 42 |
| BAA 56 | 176 |
| BAA 57 | 178 |
| BAA 58 | 179 |
| BAA 59 | 180 |
| BAA 60 | 181 |
| BAA 61 | 182 |
| BAA 62 | 183 |
| BAA 63 | 184 |
| BAA 64 | 398 |
| BAA 65 | 399 |
| BAA 66 | 400 |
| BAA 67 | 401 |
| BAA 68 | 402 |
| BAA 69 | 186 |
| BAA 70 | 187 |
| BAA 71 | 397 |
| BAA 72 | 581 |
| BAA 73 | 582 |
| BAA 74 | 464 |
| BAA 75 | 454 |
| BAA 76 | 455 |
| BAA 77 | 456 |
| BAA 78 | 457 |
| BAA 79 | 458 |
| BAA 80 | 744 |
| BAA 81 | 745 |
| BAA 82 | 4 |
| BAA 83 | 2 |
| BAA 84 | 5 |
| BAA 85 | 3 |
| BAA 86 | 6 |
| BAA 87 | 177 |
| BY 1 | 215 |
| BY 2 | 668 |
| BY 4 | 314 |
| BY 5 | 626 |
| BY 10 | 632 |
| BY 15 | 713 |
| BY 21 | 700 |
| BY 25/1 | 175a |
| BY 25/2 | 175b |
| BY 25/3 | 175c |
| BY 27 | 355 |
| BY 34 | 669 |
| BY 35/1 | 685 |
| BY 36/1 | 71 |
| BY 36/2 | 216 |
| BY 38 | 123 |
| BY 48 | 133a-b |
| BY 49 | 134a-b |
| BY 50 | 135a-b |
| BY 51 | 430 |
| BY 56 | 49 |
| BY 69 | 43 |
| BY 70 | 72 |
| BY 91 α | 338 |
| BY 93 | 450 |
| BY 94 | 714 |
| BY 95 | 636 |
| BY 98 | 627 |
| BY 109 | 421 |
| BY 114 | 254 |
| BY 165 | 113 |
| BY 173 | 431 |
| BY 174 | 339 |
| BY 180 | 701 |
| BY 182 | 165 |
| BY 192 | 459 |
| BY 206 | 553 |
| BY 207/1 | 217 |
| BY 207/2 | 274 |
| BY 207/3 | 439 |
| BY 207/4 | 630 |
| BY 207/5 | 218 |
| BY 207/7 | 275 |
| BY 207/10 | 316 |
| BY 209 | 702 |
| BY 209/2 | 625 |
| BY 210/1 | 255 |
| BY 213/1 | 373 |
| BY 213/2 | 325 |
| BY 214/1 | 448 |
| BY 214/2 | 44 |
| BY 214/3 | 219 |
| BY 217/1 | 304 |
| BY 217/3 | 670 |
| BY 217/5 | 321 |
| BY 217/6 | 637 |
| BY 217/8 | 374 |
| BY 217/32 | 688 |
| BY 217/40 | 689 |
| BY 224 | 220 |
| BY 227/1 | 354 |
| BY 227/2 | 572 |
| BY 227/3 | 715 |
| BY 227/4 | 716 |
| BY 227/5 | 136 |
| BY 227/6 | 137 |
| BY 227/7 | 138 |
| BY 227/8 | 139 |
| BY 227/9 | 140 |
| BY 227/10 | 141 |
| BY 227/11 | 717 |
| BY 227/13 | 588 |

| | |
|---|---|
| BY 227/14 | 589 |
| BY 227/15 | 590 |
| BY 227/16 | 591 |
| BY 227/17 | 435 |
| BY 227/18 | 66 |
| BY 227/19 | 221 |
| BY 227/20 | 197 |
| BY 227/21 α | 718 |
| BY 227/21 β | 719 |
| BY 227/29 | 256 |
| BY 227/30 | 276 |
| BY 227/31 | 683 |
| BY 227/32 | 573 |
| BY 227/33 | 542 |
| BY 227/34 | 597 |
| BY 227/35 | 453 |
| BY 227/37 | 690 |
| BY 227/39 α | 598 |
| BY 227/39 β | 554 |
| BY 227/40 α | 583 |
| BY 227/40 β | 508 |
| BY 227/41 | 543 |
| BY 227/42 | 551 |
| BY 227/43 | 222 |
| BY 227/46 α,β | 223 |
| BY 227/48 | 142a-b |
| BY 227/49 | 574 |
| BY 227/50 | 277 |
| BY 227/51 | 691 |
| BY 227/52 | 110 |
| BY 227/53 | 45 |
| BY 227/54 | 50 |
| BY 228/11 | 753 |
| BY 229/1 | 356 |
| BY 237/1 | 88 |
| BY 238/3 | 257 |
| BY 244/1 | 415 |
| BY 244/2 | 297 |
| BY 244/3 | 326 |
| BY 244/4 | 638 |
| BY 244/5 | 298 |
| BY 244/6 | 278 |
| BY 244/7 | 206 |
| BY 244/8 | 279 |
| BY 244/9 | 280 |
| BY 244/10 | 224 |
| BY 244/12 | 720 |
| BY 244/13 | 225 |
| BY 245/1 | 721 |
| BY 245/3 | 166 |
| BY 245/7 | 143 |
| BY 247/1 | 281 |
| BY 247/2 | 357 |
| BY 251/1 | 639 |
| BY 251/3 A-B | 559 |
| BY 253/1 | 226 |
| BY 253/2 | 227 |
| BY 253/3 | 413 |
| BY 253/4 | 228 |
| BY 255/1 | 671 |
| BY 255/2 | 446 |
| BY 255/3 | 375 |
| BY 255/4 | 200 |
| BY 255/5 | 621 |
| BY 255/6 | 207 |
| BY 255/7 | 327 |
| BY 255/8 | 640 |
| BY 257/1 | 592 |
| BY 257/2 | 584 |
| BY 257/3 | 229 |
| BY 259/1 | 641 |
| BY 259/2 | 560 |
| BY 268/1 | 18 |
| BY 273/1 | 358 |
| BY 276/1 | 426 |
| BY 282/1 | 754 |
| BY 297/1 | 376 |
| BY 298/35 | 164d |
| BY 298/39 | 164a |
| BY 298/41 | 164c |
| BY 301/1 | 686 |
| BY 301/3 | 208 |
| BY 302/2 | 144 |
| BY 309/1 | 359 |
| BY 309/2 | 703 |
| BY 309/3 | 440 |
| BY 309/4 | 420 |
| BY 309/5 | 382 |
| BY 309/6 | 667 |
| BY 309/7 | 419 |
| BY 309/8 | 687 |
| BY 310/1 | 282 |
| BY 310/2 | 407 |
| BY 316/1 | 517 |
| BY 317/1 | 230 |
| BY 317/2 | 353 |
| BY 317/3 | 408 |
| BY 317/4 | 672 |
| BY 317/5 | 258 |
| BY 317/6 | 722 |
| BY 317/7 | 427 |
| BY 317/8 | 120 |
| BY 317/9 | 409 |
| BY 317/10 | 231 |
| BY 317/16 | 684 |
| BY 319/1 | 232 |
| BY 319/2 | 579 |
| BY 319/3 | 414 |
| BY 319/4 | 350 |
| BY 319/5 | 345 |
| BY 320/1 | 68 |
| BY 320/4 | 580 |
| BY 320/6 | 69 |
| BY 320/7 | 188 |
| BY 320/8 | 189 |
| BY 320/9 | 145 |
| BY 321/1 | 366 |
| BY 321/2 | 259 |
| BY 321/3 | 233 |
| BY 321/4 | 193 |
| BY 321/5 | 449 |
| BY 321/6 | 723 |
| BY 321/7 | 555 |
| BY 321/8 | 556 |
| BY 321/9 | 673 |
| BY 321/10 | 674 |
| BY 323/1 | 432 |
| BY 323/2 | 642 |
| BY 324/1 | 692 |
| BY 325/1 | 46 |
| BY 325/2 | 96 |
| BY 325/3 | 8 |
| BY 325/4 | 47 |
| BY 325/5 | 37 |
| BY 325/6 | 19 |
| BY 325/7 | 94 |
| BY 325/8 | 146 |
| BY 325/9 | 7 |
| BY 325/10 | 40 |
| BY 325/11 | 41 |
| BY 327/1 | 74 |
| BY 328/1 | 62 |
| BY 328/2 | 20 |
| ΔΕΘ 5 | 167 |
| ΔΕΘ 6 α | 643 |
| ΔΕΘ 28/1 | 465 |
| ΔΕΘ 28/2 | 466 |
| ΔΕΘ 28/3 | 467 |
| ΔΕΘ 30 | 644 |
| ΔΕΘ 44 | 234 |
| ΔΕΘ 57 | 260 |
| ΔΕΘ 92 | 261 |
| ΔΕΘ 94/1 | 645 |
| ΔΕΘ 94/2 | 646 |
| ΔΕΘ 97 | 360 |
| ΔΕΘ 115 | 323 |
| ΔΕΘ 158/1 | 468 |
| ΔΕΘ 158/2 | 469 |
| ΔΕΘ 158/3 | 470 |
| ΔΕΘ 158/4 | 509 |
| ΔΕΘ 158/5 | 510 |
| ΔΕΘ 159/1 | 599 |
| ΔΕΘ 159/2 | 593 |
| ΔΕΘ 187/1 | 511 |
| ΔΕΘ 187/2 | 471 |
| ΔΕΘ 216/1 | 518 |
| ΔΕΘ 216/2 | 519 |
| ΔΕΘ 216/3 | 472 |
| ΔΕΘ 216/4 α | 473 |
| ΔΕΘ 216/4 β | 474 |
| ΔΕΘ 216/5 | 520 |

| | |
|---|---|
| ΔΕΘ 216/6 | 475 |
| ΔΕΘ 216/7 | 521 |
| ΔΕΘ 216/8 | 522 |
| Θ 1/01-Υ1 ΑΓΚ 7216 | 305 |
| Θ 1/01-Υ2 ΑΓΚ 7215 | 704 |
| Θ 1/01-Υ3 ΑΓΚ 7212 | 235 |
| Θ 14/99-Υ2 ΑΓΚ 7114 | 647 |
| Θ 18/01 Υ1 | 306 |
| Θ 18/01 Υ2 | 319 |
| Θ 18/01 Υ3 | 705 |
| Θ 18/01-Υ4 ΑΓΚ 7253 | 349 |
| ΜΘ ΑΚΑΤΑΓΡΑΦΟ 1 | 496 |
| ΜΘ ΑΚΑΤΑΓΡΑΦΟ 2 | 706 |
| ΜΘ ΑΚΑΤΑΓΡΑΦΟ 3 | 85 |
| ΜΘ ΑΚΑΤΑΓΡΑΦΟ 4 | 29 |
| ΜΘ ΑΚΑΤΑΓΡΑΦΟ 5 | 124 |
| ΜΘ ΣιντριβάνιΤ2 | 600 |
| ΜΘ ΣιντριβάνιΤ22 | 622 |
| ΜΘ ΣιντριβάνιΤ9Β5 | 544 |
| ΜΘ ΣιντριβάνιΤ9Β8 | 601 |
| ΜΘ ΣιντριβάνιΤ9Α2a | 394 |
| ΜΘ ΣιντριβάνιΤ9Α2b | 602 |
| ΜΘ 83 | 575 |
| ΜΘ 507 | 377 |
| ΜΘ 508 | 447 |
| ΜΘ 509 | 315 |
| ΜΘ 510 | 51 |
| ΜΘ 511 | 52 |
| ΜΘ 512 | 361 |
| ΜΘ 513 | 114 |
| ΜΘ 514 | 406 |
| ΜΘ 515 | 724 |
| ΜΘ 516 | 444 |
| ΜΘ 517 | 307 |
| ΜΘ 518 | 308 |
| ΜΘ 519 | 309 |
| ΜΘ 520 | 301 |
| ΜΘ 521 | 209 |
| ΜΘ 522 | 693 |
| ΜΘ 523 | 648 |
| ΜΘ 525 | 649 |
| ΜΘ 526 | 236 |
| ΜΘ 527 | 650 |
| ΜΘ 528 | 651 |
| ΜΘ 529 | 652 |
| ΜΘ 530 | 283 |
| ΜΘ 532 | 299 |
| ΜΘ 533 | 284 |
| ΜΘ 534 | 237 |
| ΜΘ 535 | 285 |
| ΜΘ 537 | 286 |
| ΜΘ 538 | 675 |
| ΜΘ 539 | 352 |
| ΜΘ 540 | 201 |
| ΜΘ 541 | 194 |
| ΜΘ 542 | 410 |
| ΜΘ 543 | 628 |
| ΜΘ 544 | 437 |
| ΜΘ 545 | 629 |
| ΜΘ 808 | 540 |
| ΜΘ 809 | 541 |
| ΜΘ 810 | 378 |
| ΜΘ 1491 | 424 |
| ΜΘ 1492 | 422 |
| ΜΘ 3000 | 608 |
| ΜΘ 3061 | 147 |
| ΜΘ 3074 | 320 |
| ΜΘ 3078 | 451 |
| ΜΘ 3155 | 436 |
| ΜΘ 3202 | 148 |
| ΜΘ 3225 | 262 |
| ΜΘ 3234 | 238 |
| ΜΘ 3682 | 725 |
| ΜΘ 3683 | 210 |
| ΜΘ 3684 | 211 |
| ΜΘ 3685 | 149 |
| ΜΘ 3686 | 150 |
| ΜΘ 3697 | 726 |
| ΜΘ 3753 | 346 |
| ΜΘ 3754 | 239 |
| ΜΘ 3755 | 263 |
| ΜΘ 3756 | 727 |
| ΜΘ 3757 | 205 |
| ΜΘ 3759 | 185 |
| ΜΘ 3760 | 362 |
| ΜΘ 3761 | 405 |
| ΜΘ 3763 | 111 |
| ΜΘ 3777 | 445 |
| ΜΘ 4906 | 538 |
| ΜΘ 5465 | 379 |
| ΜΘ 5466 | 380 |
| ΜΘ 5467 | 240 |
| ΜΘ 5468 | 653 |
| ΜΘ 5469 | 287 |
| ΜΘ 5470 | 328 |
| ΜΘ 5471 | 418 |
| ΜΘ 5472 | 443 |
| ΜΘ 5473 | 288 |
| ΜΘ 5474 | 63 |
| ΜΘ 5475 | 329 |
| ΜΘ 5477 | 330 |
| ΜΘ 5478 | 198 |
| ΜΘ 5479 | 322 |
| ΜΘ 5480 | 609 |
| ΜΘ 5481 | 310 |
| ΜΘ 5482 | 317 |
| ΜΘ 5483 | 337 |
| ΜΘ 5484 | 438 |
| ΜΘ 5485 | 331 |
| ΜΘ 5486 | 332 |
| ΜΘ 5487 | 333 |
| ΜΘ 5488 | 334 |
| ΜΘ 5489 | 241 |
| ΜΘ 5490 | 395 |
| ΜΘ 5491 | 633 |
| ΜΘ 5492 | 634 |
| ΜΘ 5493 | 428 |
| ΜΘ 5494 | 425 |
| ΜΘ 5932 | 242 |
| ΜΘ 5933 | 243 |
| ΜΘ 5934 | 244 |
| ΜΘ 5935 | 245 |
| ΜΘ 5936 | 246 |
| ΜΘ 5937 | 264 |
| ΜΘ 5938 | 694 |
| ΜΘ 5939 | 654 |
| ΜΘ 5940 | 655 |
| ΜΘ 5941 | 347 |
| ΜΘ 5942 | 340 |
| ΜΘ 5943 | 367 |
| ΜΘ 5944 | 368 |
| ΜΘ 5945 | 195 |
| ΜΘ 5946 | 202 |
| ΜΘ 5947 | 610 |
| ΜΘ 5948 | 676 |
| ΜΘ 5949 | 677 |
| ΜΘ 5950 | 678 |
| ΜΘ 5951 | 105 |
| ΜΘ 5952 | 112 |
| ΜΘ 5953 | 707 |
| ΜΘ 5954 | 708 |
| ΜΘ 5955 | 611 |
| ΜΘ 6308 | 603 |
| ΜΘ 7052 | 656 |
| ΜΘ 7167 | 115 |
| ΜΘ 7168 | 196 |
| ΜΘ 7169 | 363 |
| ΜΘ 7224 | 695 |
| ΜΘ 7225 | 561 |
| ΜΘ 7229 | 416 |
| ΜΘ 9906 | 212 |
| ΜΘ 9925 | 433 |
| ΜΘ 9973 | 381 |
| ΜΘ 10295 | 476 |
| ΜΘ 10296 | 477 |
| ΜΘ 10297 | 512 |
| ΜΘ 10298 | 513 |
| ΜΘ 10299 | 746 |
| ΜΘ 10792 | 562 |
| ΜΘ 10797 | 728 |
| ΜΘ 10798 | 679 |
| ΜΘ 10800 | 442 |
| ΜΘ 10802 | 265 |
| ΜΘ 10878 | 709 |
| ΜΘ 10921 | 247 |
| ΜΘ 10922 | 266 |
| ΜΘ 10923 | 248 |
| ΜΘ 10924 | 341 |
| ΜΘ 10925 | 729 |
| ΜΘ 11047 | 696 |
| ΜΘ 11053 | 710 |

| | |
|---|---|
| ΜΘ 11054 | 730 |
| ΜΘ 11093 | 604 |
| ΜΘ 11094 | 657 |
| ΜΘ 11105 β | 498 |
| ΜΘ 11141 | 335 |
| ΜΘ 11144 | 658 |
| ΜΘ 11165 | 731 |
| ΜΘ 11166 | 732 |
| ΜΘ 11167 | 203 |
| ΜΘ 11168 | 204 |
| ΜΘ 11172 | 545 |
| ΜΘ 11174 | 612 |
| ΜΘ 11195 | 502 |
| ΜΘ 11196 | 503 |
| ΜΘ 11197 | 499 |
| ΜΘ 11198 | 523 |
| ΜΘ 11199 | 500 |
| ΜΘ 11200 | 501 |
| ΜΘ 11241 | 613 |
| ΜΘ 11250 στ | 585 |
| ΜΘ 11258 β | 463 |
| ΜΘ 11267 | 342 |
| ΜΘ 11527 | 711 |
| ΜΘ 11528 | 631 |
| ΜΘ 11531 | 417 |
| ΜΘ 11539 α | 614 |
| ΜΘ 11539 β | 557 |
| ΜΘ 11539 γ | 615 |
| ΜΘ 11551 | 504 |
| ΜΘ 11552 | 505 |
| ΜΘ 11573 | 524 |
| ΜΘ 11574 | 514 |
| ΜΘ 11575 | 478 |
| ΜΘ 11576 | 479 |
| ΜΘ 12320 | 302 |
| ΜΘ 16944 | 576 |
| ΜΘ 19258 | 616 |
| ΜΘ 19264 | 617 |
| ΜΘ 19265 | 618 |
| ΜΘ 19266 | 303 |
| ΜΘ 19348 | 563 |
| ΜΘ 19349 | 680 |
| ΜΘ 20870 | 733 |
| ΜΘ 20871 | 311 |
| ΜΘ 20872 | 343 |
| ΜΘ 20910 | 423 |
| ΜΘ 20911 | 659 |
| ΜΘ 20912 | 98 |
| ΜΘ 20914 | 660 |
| ΜΘ 20915 | 267 |
| ΜΘ 20916 | 344 |
| ΜΘ 20917 | 249 |
| ΜΘ 20918 | 318 |
| ΜΘ 20919 | 661 |
| ΜΘ 20920 | 268 |
| ΜΘ 20921 | 269 |
| ΜΘ 20922 | 635 |
| ΜΘ 20923 | 411 |
| ΜΘ 20924 | 452 |
| ΜΘ 20925 | 681 |
| ΜΘ 20926 | 734 |
| ΜΘ 20927 | 558 |
| ΜΘ 20928 | 525 |
| ΜΘ 20929 | 506 |
| ΜΘ 20930 | 526 |
| ΜΘ 20931 | 619 |
| ΜΘ 20932 | 620 |
| ΜΘ 20933 | 564 |
| ΜΘ 20934 | 565 |
| ΜΘ 20935 | 605 |
| ΜΘ 20936 | 566 |
| ΜΘ 20937 | 735 |
| ΜΘ 20938 | 494 |
| ΜΘ 20939 | 151 |
| ΜΘ 20940 | 152 |
| ΜΘ 20941 | 153 |
| ΜΘ 20942 | 154 |
| ΜΘ 20943 | 155 |
| ΜΘ 20944 | 156 |
| ΜΘ 20945 | 157 |
| ΜΘ 20946 | 158 |
| ΜΘ 20947 | 159 |
| ΜΘ 20948 | 160 |
| ΜΘ 20949 | 161 |
| ΜΘ 20992 | 742 |
| ΜΘ 20993 | 662 |
| ΜΘ 20994 | 663 |
| ΜΘ 20995 | 250 |
| ΜΘ 20996 | 251 |
| ΜΘ 20997 | 213 |
| ΜΘ 20998 | 623 |
| ΜΘ 20999 | 300 |
| ΜΘ 21000 | 252 |
| ΜΘ 21001 | 289 |
| ΜΘ 21002 | 290 |
| ΜΘ 21003 | 291 |
| ΜΘ 21004 | 292 |
| ΜΘ 21005 | 293 |
| ΜΘ 21006 | 682 |
| ΜΘ 21007 | 121 |
| ΜΘ 21390 | 253 |
| ΜΘ 21391 | 294 |
| ΜΘ 21392 | 48 |
| ΜΘ 21393 | 116 |
| ΜΘ 21456 α | 21 |
| ΜΘ 21456 β | 313 |
| ΜΘ 21955 | 594 |
| ΜΘ 21956 | 712 |
| ΜΘ 21957 | 122 |
| ΜΘ 21958 | 117 |
| ΜΘ 21959 | 125 |
| ΜΘ 21960 | 75 |
| ΜΘ 21961 | 53 |
| ΜΘ 21962 | 84 |
| ΜΘ 21963 | 497 |
| ΜΘ 21965 | 54 |
| ΜΘ 21966 | 214 |
| ΜΘ 21967 | 55 |
| ΜΘ 21968 | 56 |
| ΜΘ 21970 | 130 |
| ΜΘ 21971 | 57 |
| ΜΘ 21972 | 58 |
| ΜΘ 21973 | 168 |
| ΜΘ 21974 | 169 |
| ΜΘ 21975 | 76 |
| ΜΘ 21976 | 77 |
| ΜΘ 21977 | 59 |
| ΜΘ 21978 | 78 |
| ΜΘ 21980 | 747 |
| ΜΘ 21981 | 748 |
| ΜΘ 21982 | 750 |
| ΜΘ 21983 | 751 |
| ΜΘ 21984 | 752 |
| ΜΘ 21985 | 383 |
| ΜΜΣΤ 2/4 | 480 |
| ΜΜΣΤ 2/5 | 546 |
| ΜΜΣΤ 2/8 | 552 |
| ΜΜΣΤ 2/16 | 527 |
| ΜΜΣΤ 2/17 | 481 |
| ΜΜΣΤ 2/19 | 624 |
| ΜΜΣΤ 2/20 | 547 |
| ΜΜΣΤ 2/21 | 482 |
| ΜΜΣΤ 2/22 α | 483 |
| ΜΜΣΤ 2/22 β | 484 |
| ΜΜΣΤ 2/28 | 528 |
| ΜΜΣΤ 53/1 | 529 |
| ΜΜΣΤ 53/2 | 530 |
| ΜΜΣΤ 53/3 | 485 |
| ΜΜΣΤ 53/4 | 486 |
| ΜΜΣΤ 61/7 | 567 |
| ΜΜΣΤ 99/2 | 697 |
| ΜΜΣΤ 123/1 | 531 |
| ΜΜΣΤ 123/2 | 532 |
| ΜΜΣΤ 123/3 | 533 |
| ΜΜΣΤ 123/4 | 534 |
| ΜΜΣΤ 123/5 | 535 |
| ΜΜΣΤ 148/2 | 743 |
| ΜΜΣΤ 160/1 | 461 |
| ΜΜΣΤ 160/2 | 606 |
| ΜΜΣΤ 160/3 | 595 |
| ΜΜΣΤ 160/4 | 577 |
| ΜΜΣΤ 160/5 | 460 |
| ΜΜΣΤ 160/6 | 607 |
| ΜΜΣΤ 160/8 | 568 |
| ΜΜΣΤ 160/9 | 578 |
| ΜΜΣΤ 160/10 | 462 |
| ΜΜΣΤ 160/11 | 569 |
| ΜΜΣΤ 161 | 570 |
| ΠΑΜ 1 | 369 |
| ΠΑΜ 2 | 434 |
| ΠΑΜ 3 | 364 |

| | | | |
|---|---|---|---|
| ПАМ 4 | 295 | ПАМ 55 | 272 |
| ПАМ 5 | 296 | ПАМ 56 | 370 |
| ПАМ 6 | 384 | ПАМ 57 | 70 |
| ПАМ 7 | 736 | ПАМ 58 | 80 |
| ПАМ 8 | 737 | ПАМ 59 | 1 |
| ПАМ 9 | 270 | ПАМ 60 | 38 |
| ПАМ 10 | 199 | ПАМ 61 | 741 |
| ПАМ 11 | 698 | ПАМ 62 | 571 |
| ПАМ 12 | 487 | ПАМ 63 | 596 |
| ПАМ 13 | 488 | ПАМ 64 | 131 |
| ПАМ 14 | 489 | ПАМ 65 | 109 |
| ПАМ 15 | 490 | ПАМ 66 | 173 |
| ПАМ 16 | 491 | ПАМ 68 | 126 |
| ПАМ 17 | 536 | ПАМ 69 | 132 |
| ПАМ 18 | 537 | ПАМ 70 | 118 |
| ПАМ 19 | 664 | ПАМ 71 | 403 |
| ПАМ 20 | 495 | ПАМ 72 | 81 |
| ПАМ 21 | 64 | ПАМ 73 | 548 |
| ПАМ 22 | 67 | ПАМ 74 | 549 |
| ПАМ 23 | 365 | ПАМ 75 | 493 |
| ПАМ 28 | 507 | ПАМ 76 | 515 |
| ПАМ 29 | 492 | ПАМ 77 | 550 |
| ПАМ 30 | 404 | ПАМ 78 | 273 |
| ПАМ 31 | 699 | ПАМ 79 | 82 |
| ПАМ 32 | 665 | ПАМ 80 | 429 |
| ПАМ 33 | 738 | ПАМ 81 | 83 |
| ПАМ 34 | 739 | ПАМ 82 | 371 |
| ПАМ 36 | 79 | ПАМ 83 | 372 |
| ПАМ 39 | 170 | ПАМ 84 | 441 |
| ПАМ 40 | 171 | ПАМ 85 | 127 |
| ПАМ 41 | 172 | ПАМ 86 | 128 |
| ПАМ 42 | 348 | ПАМ 87 | 129 |
| ПАМ 43 | 351 | ПАМ 88 | 174 |
| ПАМ 44 | 666 | ПАМ 89 | 175 |
| ПАМ 45 | 336 | ПАМ 90 | 60 |
| ПАМ 47 | 412 | ПАМ 91 | 61 |
| ПАМ 48 | 271 | ПАМ 92 | 22 |
| ПАМ 49 | 312 | ПАМ 93 | 119 |
| ПАМ 50 | 740 | ПАМ 94 | 23 |
| ПАМ 51 | 162 | ПАМ 95 | 24 |
| ПАМ 52 | 163 | ПАМ 96 | 73 |
| ПАМ 53 | 164 | ПАМ 97 | 396 |

# Typological and Chronological Table

| Type no. | | 1st c. | 2nd c. | 3rd c. | 4th c. | 5th c. | 6th c. | Cat. no. |
|---|---|---|---|---|---|---|---|---|
| **Plates – Dishes** | | | | | | | | |
| 1 | 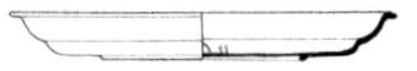 | — | | | | | | 1 |
| 2 | 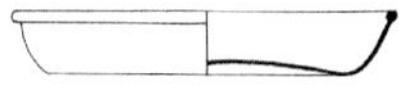 | — | | | | | | 2-3 |
| 3A | 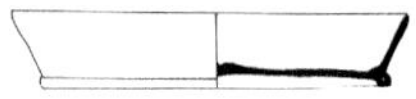 | — | | | | | | 4 |
| 3B | 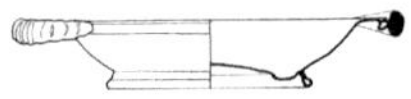 | — | | | | | | 6 |
| 4 |  | — | | | | | | 7 |
| **Bowls** | | | | | | | | |
| 5 |  | — | | | | | | 8 |
| 6a | 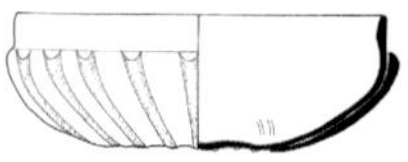 | — | | | | | | 9-24 |
| 6b | 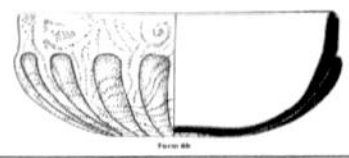 | — | | | | | | 25-29 |
| 7 | 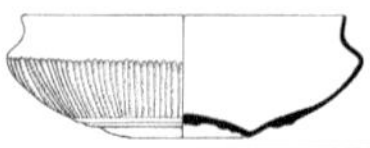 | — | | | | | | 30-36 |
| 8 | 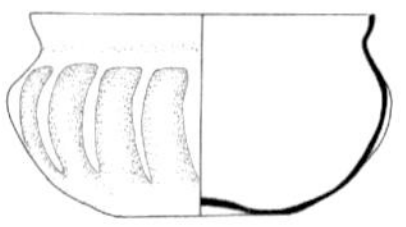 | — | | | | | | 37-38 |
| 9 | 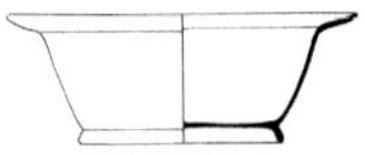 | — | — | | | | | 39 |
| 10 | 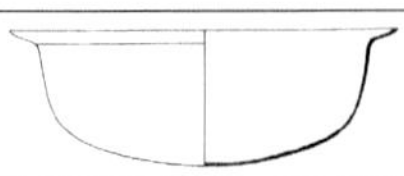 | — | — | | | | | 40-41 |
| 11 | 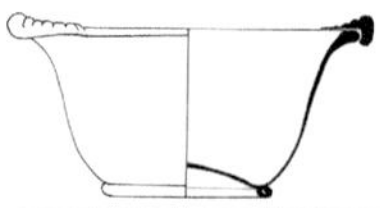 | — | | | | | | 42 |
| 12.1 | 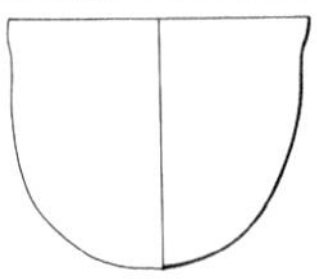 | | | — | — | — | | 43-48 |
| 12.2 Aa | 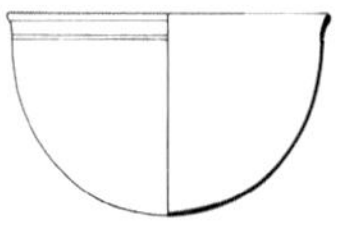 | | | — | — | — | | 49-61 |
| 12 Ab | 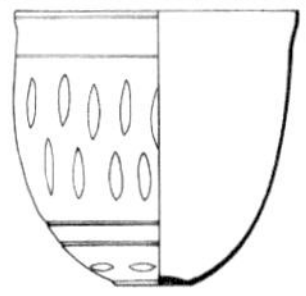 | | | — | — | — | — | 62-64 |
| 12.2 Ba | 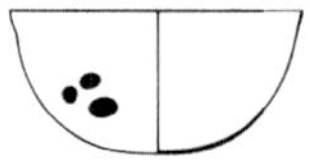 | | | — | — | — | — | 65-66 |
| 12.2 Bb | 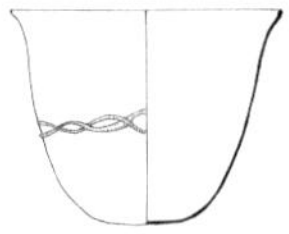 | | | — | — | — | — | 67 |
| 13 | 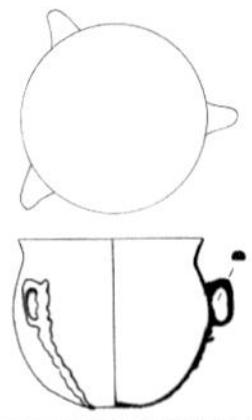 | | | | — | — | — | 68 |
| 14 | 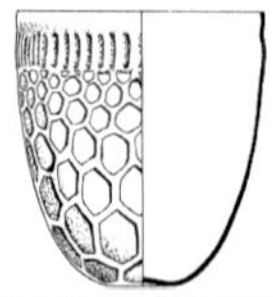 | | | — | — | | | 69-70 |
| 15 | 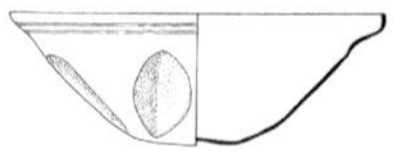 | | | — | — | — | | 71-73 |
| 16 | 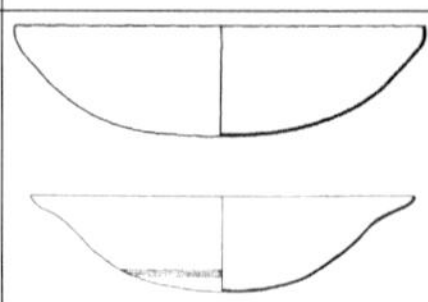 | | | — | — | — | | 74-83 |
| 17 | 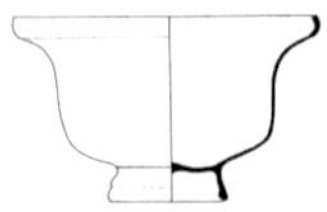 | | | | — | — | — | 84 |
| 18 | 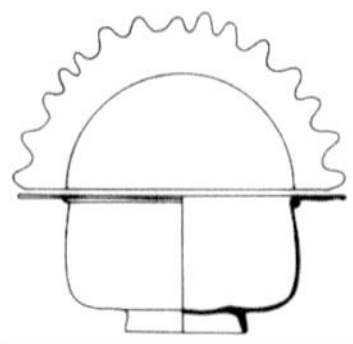 | | | — | — | | | 85 |
| **Beakers** | | | | | | | | |
| 19 |  | — | — | | | | | 86-88 |
| 20 | 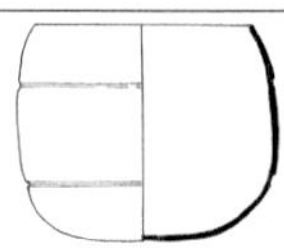 | — | | | | | | 89-90 |

| Type no. | | 1st c. | 2nd c. | 3rd c. | 4th c. | 5th c. | 6th c. | Cat. no. |
|---|---|---|---|---|---|---|---|---|
| 21 | 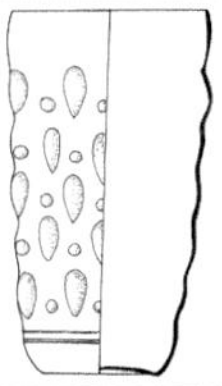 | — | — | | | | | 91-94 |
| 22 |  | — | — | | | | | 95 |
| 23 | 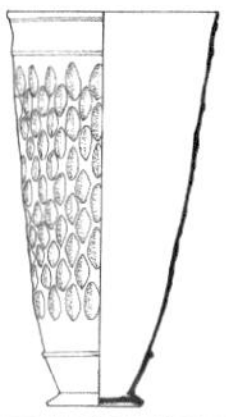 | — | | | | | | 96 |
| 24 | 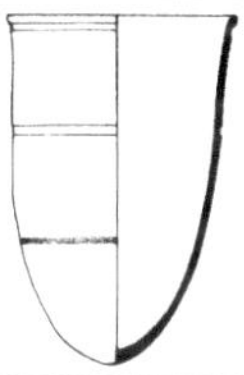 | — | | | | | | 92-98 |
| 25 | 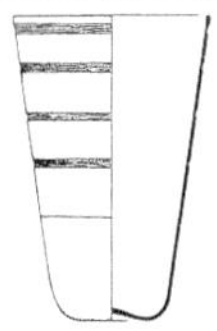 | — | | | | | | 99-100 |
| 26 | 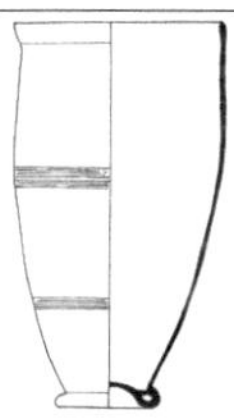 | — | | | | | | 101-104 |
| 27 |  | — | — | | | | | 105 |
| 28 | 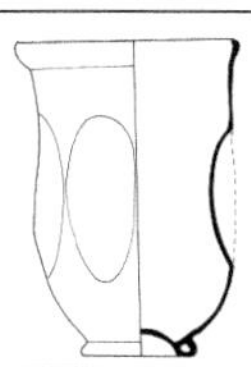 | — | — | | | | | 106-109 |
| 29 | 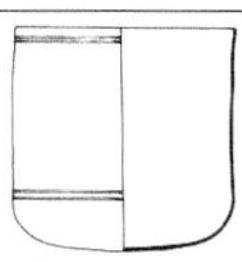 | | — | — | | | | 110-111 |

| Type no. | | 1st c. | 2nd c. | 3rd c. | 4th c. | 5th c. | 6th c. | Cat. no. |
|---|---|---|---|---|---|---|---|---|
| 30 | 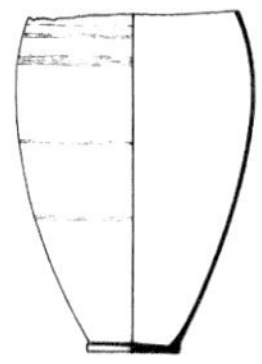 | | | — | — | — | | 112 |
| 31 | 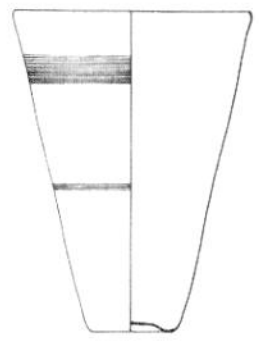 | | | | — | — | | 223-119 |
| 32 | 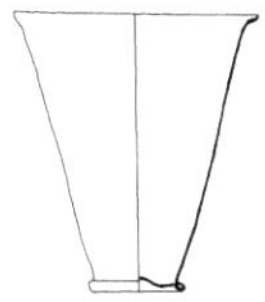 | | | — | — | — | | 120-122 |
| 33 | 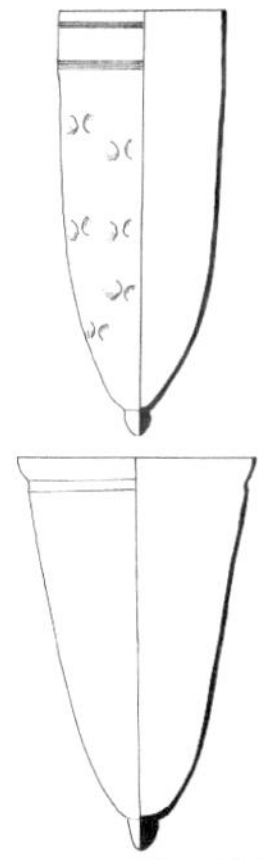 | | | | | — | — | 123-129 |
| 34 | 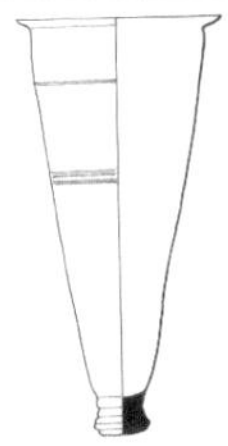 | | | | | — | — | 130-132 |
| 35 | 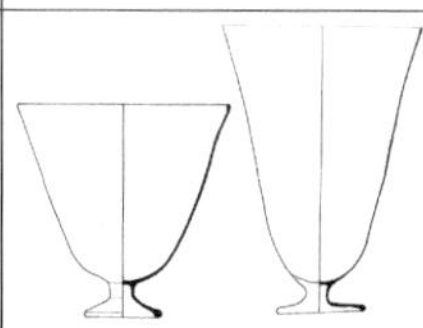 | | | | | | | 133-136<br>138<br>140-142<br>144-145<br>147-152<br>154-167<br>159-164 |
| 36 | 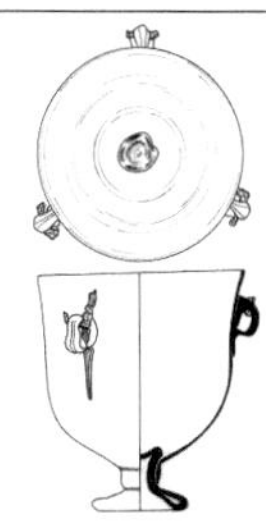 | | | | | — | — | 165-172 |

| Type no. | | 1st c. | 2nd c. | 3rd c. | 4th c. | 5th c. | 6th c. | Cat. no. |
|---|---|---|---|---|---|---|---|---|
| | 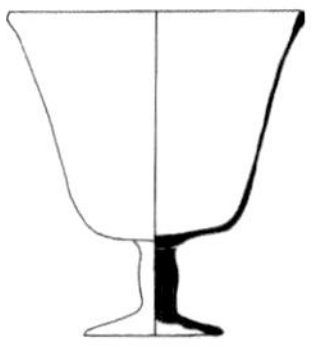 | | | | | — | — | |
| | 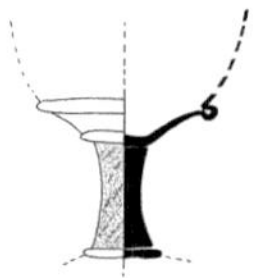 | | | | | | | |
| 37.1 | 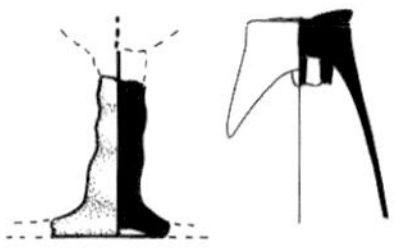 | | | | | | | 146 |
| | 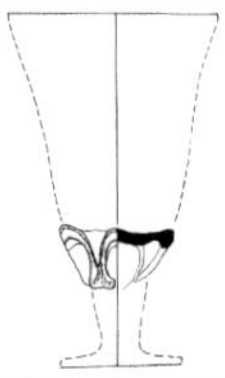 | | | | | | — | |
| 37.2 | 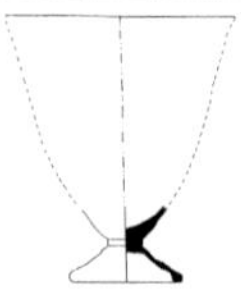 | | | | | | | |
| 38a | 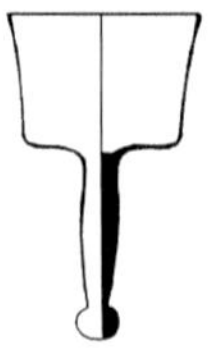 | | | | | | — | 173-175 |
| 38b | 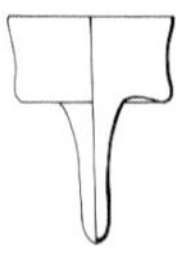 | | | | | | — | |
| **Ladles** | | | | | | | | |
| 39 | 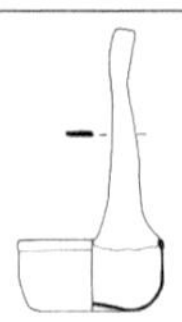 | — | | | | | | 176-177 |
| 40 | 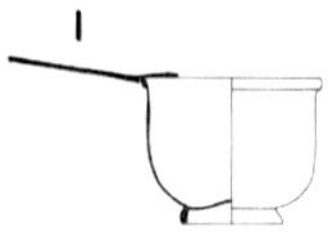 | — | — | | | | | 178-184 |
| **Calix** | | | | | | | | |
| 41 | 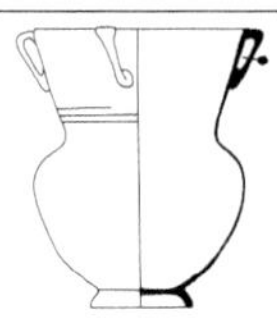 | | | — | — | | | 185 |

| Type no. | | 1st c. | 2nd c. | 3rd c. | 4th c. | 5th c. | 6th c. | Cat. no. |
|---|---|---|---|---|---|---|---|---|
| **Kantharoi** | | | | | | | | |
| 42 | 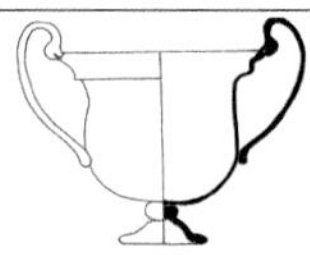 | — | | | | | | 186-187 |
| 43 |  | | | | — | — | — | 188-189 |
| **Flasks-Bottles** | | | | | | | | |
| 44 | 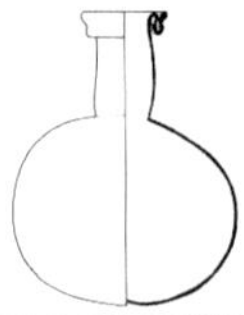 | — | — | | | | | 190-192 |
| 45 | 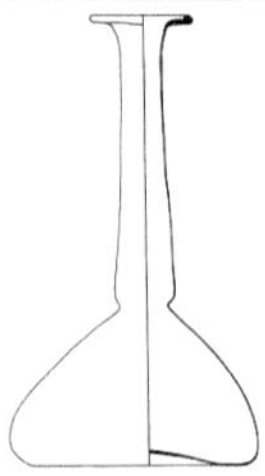 | — | — | — | | | | 193-196 |
| 46 | 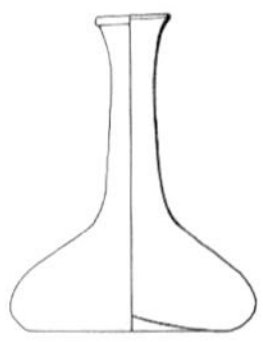 | — | — | — | | | | 197-198 |
| 47 | 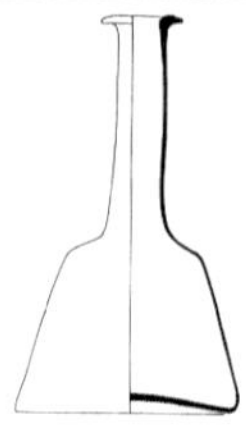 | | — | — | | | | 199 |
| 48 | 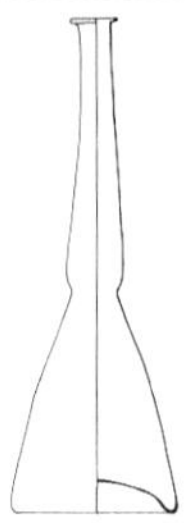 | — | — | — | | | | 200-204 |
| 49 | 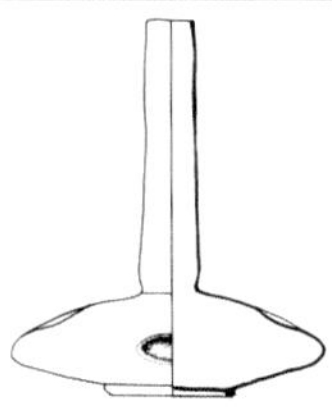 | | — | — | | | | 205 |

| Type no. | | 1st c. | 2nd c. | 3rd c. | 4th c. | 5th c. | 6th c. | Cat. no. |
|---|---|---|---|---|---|---|---|---|
| 50 | 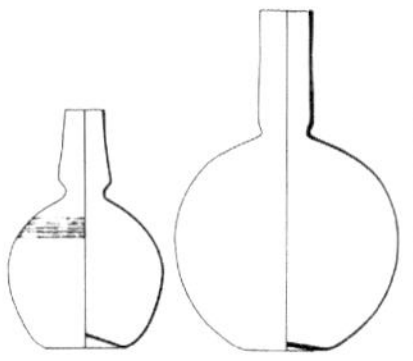 | | | | | | | 206-214 |
| 51a | 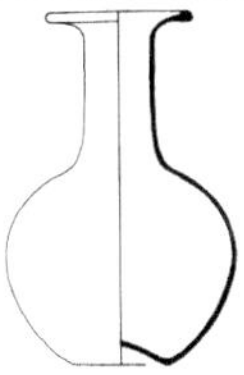 | | | | | | | 215-253 |
| 51b | 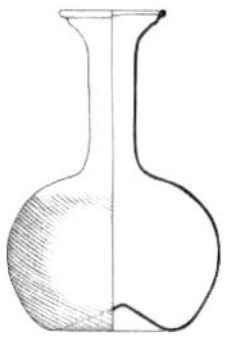 | | | | | | | 254-273 |
| 52a | 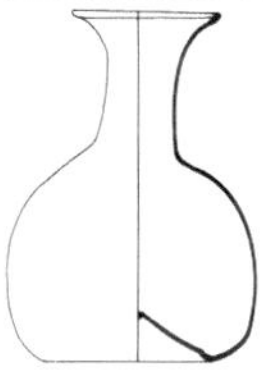 | | | | | | | 274-296 |
| 52b | 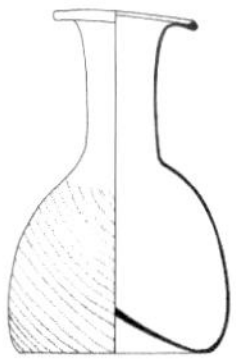 | | | | | | | 197-300 |
| 53a | 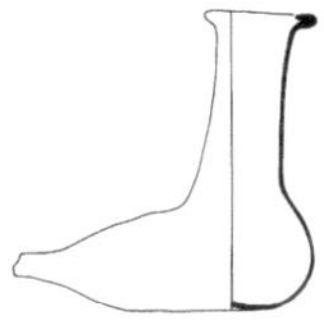 | | | | | | | 301-303 |
| 53b | 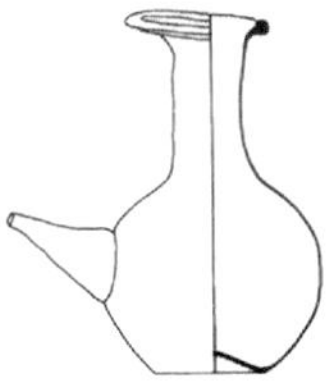 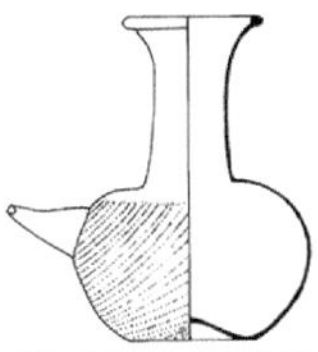 | | | | | | | 304-312 |
| 54 | 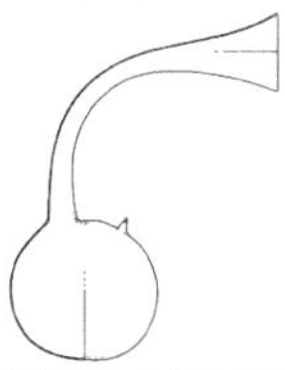 | | | | | | | 313 |
| 55 | 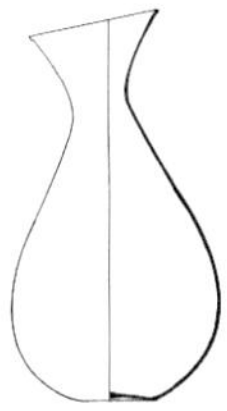 | | | | | | | 314-315 |
| 56 | 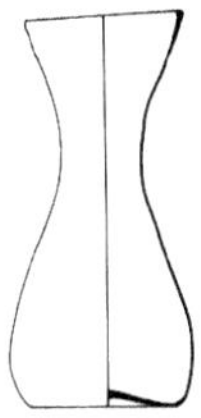 | | | | | | | 316-317 |
| 57 | 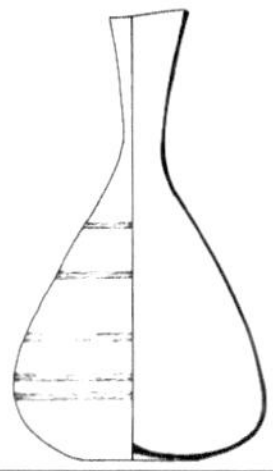 | | | | | | | 318 |
| 58 |  | | | | | | | 319-320 |
| 59 |  | | | | | | | 321-322 |
| 60 |  | | | | | | | 323 |
| 61 | 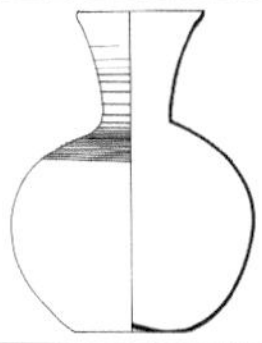 | | | | | | | 324-336 |
| 62 | 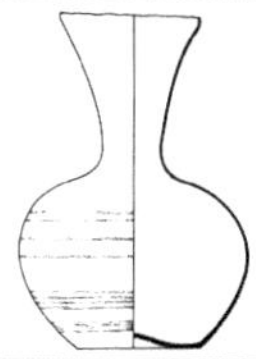 | | | | | | | 337 |

| Type no. | | 1st c. | 2nd c. | 3rd c. | 4th c. | 5th c. | 6th c. | Cat. no. |
|---|---|---|---|---|---|---|---|---|
| 63 | 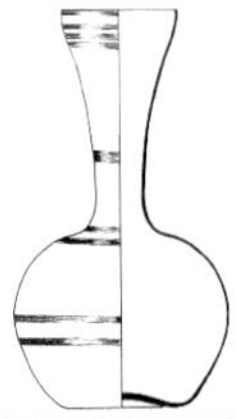 | | | | — | | | 338-344 |
| 64 | 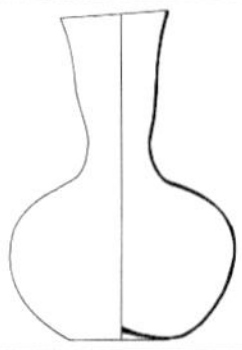 | | | | — | | | 345-348 |
| 65 | 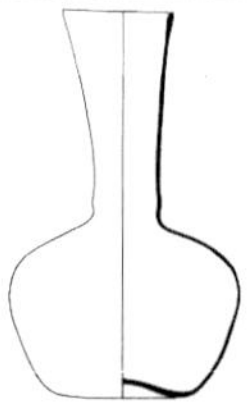 | | | | — | | | 349 |
| 66 | 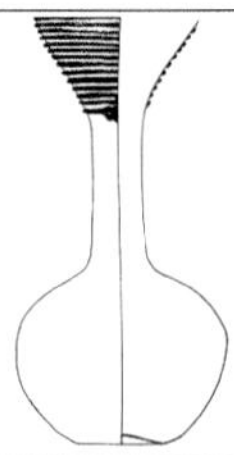 | | | | — | — | | 350-351 |
| 67 | 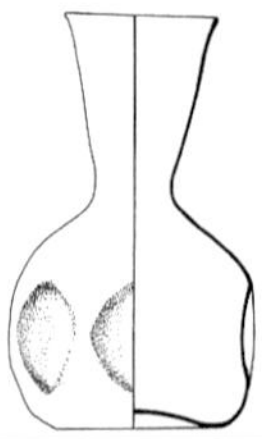 | | | | — | | | 352 |
| 68 | 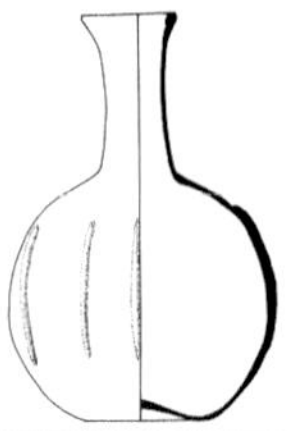 | | | | — | | | 353 |
| 69 | 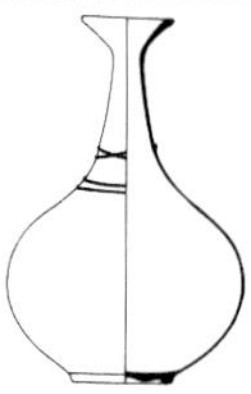 | | — | — | | | | 354 |
| 70 | 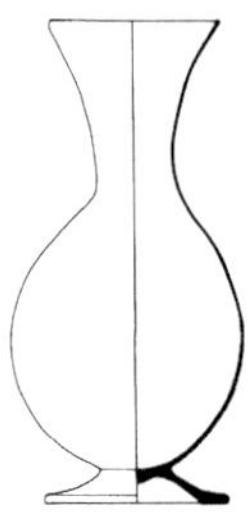 | | | | — | | | 355-365 |

| Type no. | | 1st c. | 2nd c. | 3rd c. | 4th c. | 5th c. | 6th c. | Cat. no. |
|---|---|---|---|---|---|---|---|---|
| 71 | 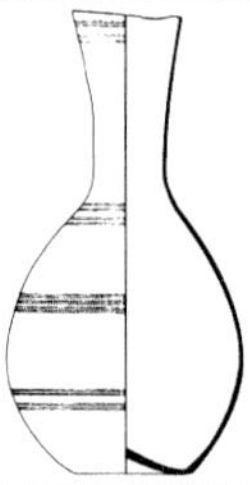 | | | | — | | | 366-372 |
| 72 | 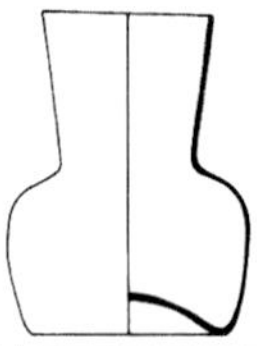 | | | | — | — | — | 373 |
| 73 |  | | | | — | — | | 374 |
| **Amphorisks** | | | | | | | | |
| 74a | 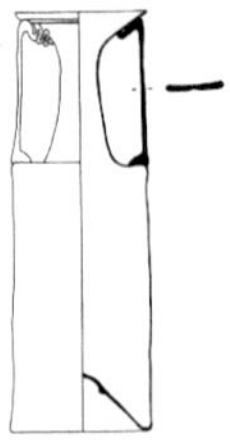 | | | — | — | — | | 375-381 |
| 74b | 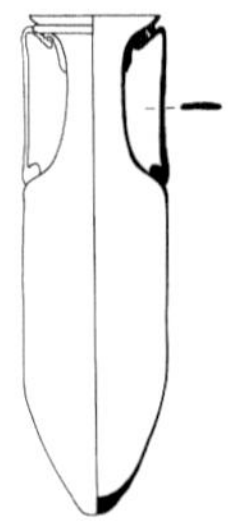 | | | | | | | 382-384 |
| **Jugs** | | | | | | | | |
| 75 | 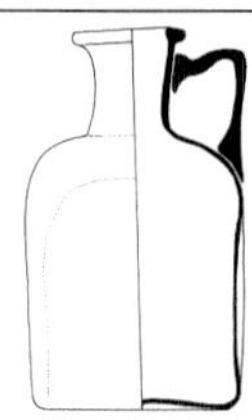 | — | — | | | | | 385-396 |
| 76 | 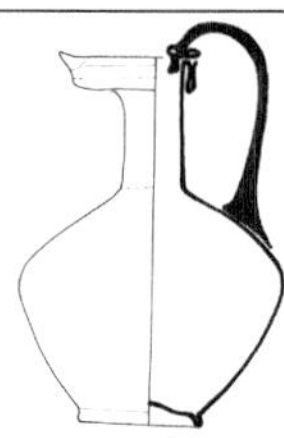 | — | — | | | | | 397 |

| Type no. | | 1st c. | 2nd c. | 3rd c. | 4th c. | 5th c. | 6th c. | Cat. no. |
|---|---|---|---|---|---|---|---|---|
| 77 |  | — | — | | | | | 398-402 |
| 78 | 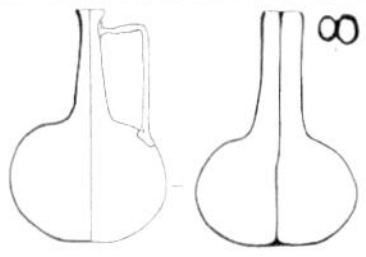 | — | — | — | — | | | 403 |
| 79 | 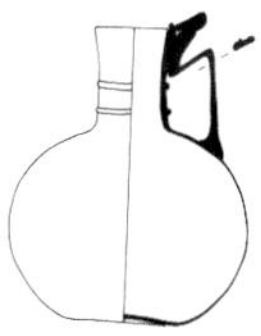 | | — | — | | | | 404 |
| 80 | 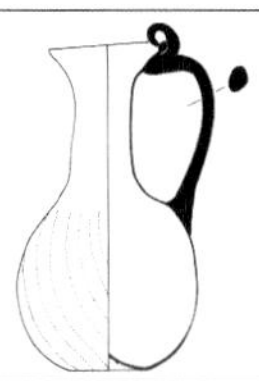 | | | | — | | | 405 |
| 81 | 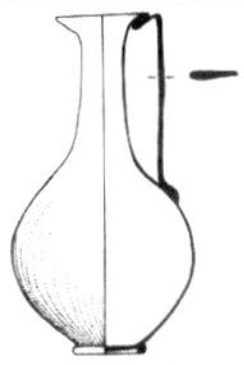 | | | | — | | | 406 |
| 82 | 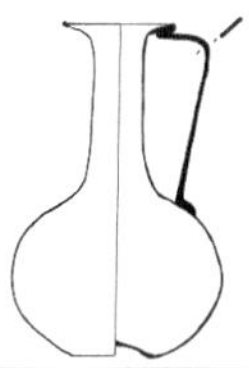 | | | — | — | | | 407 |
| 83 | 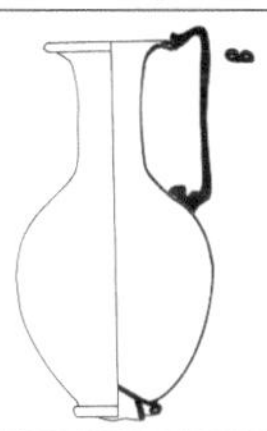 | | | | — | | | 408-409 |
| 84 | 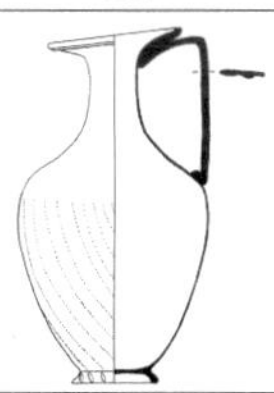 | | | | — | | | 410-411 |
| 85 | 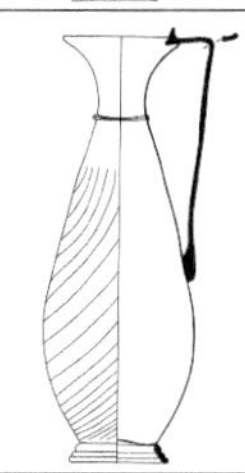 | | | — | — | | | 412 |
| 86 | 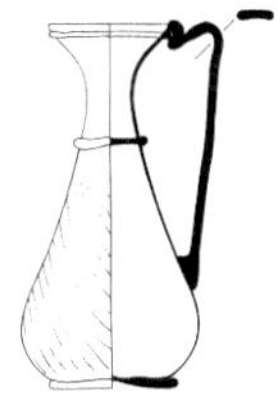 | | | | — | — | | 413 |
| 87 | 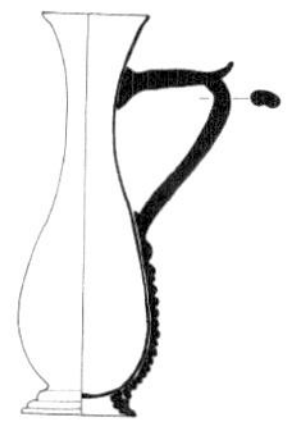 | | | | — | | | 414 |
| 88 | 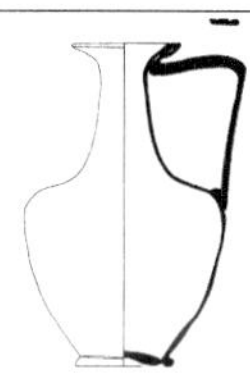 | | | — | — | | | 415 |
| 89 | 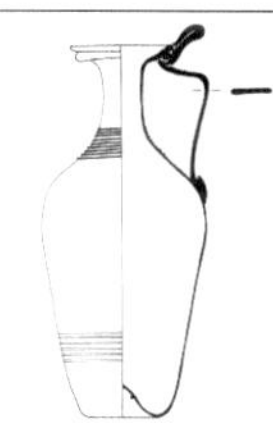 | | | — | — | | | 416-417 |
| 90 | 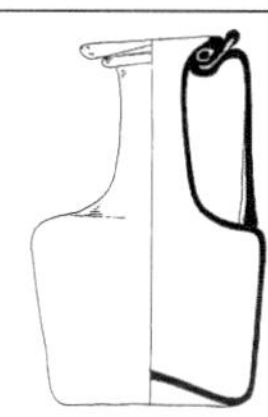 | | | | — | — | | 418 |
| 91 | 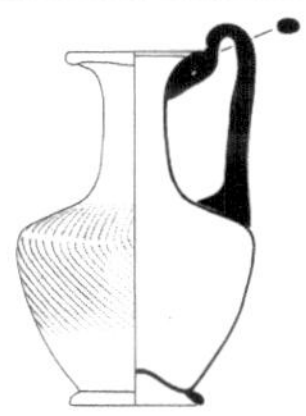 | | | | — | | | 419 |
| 92 | 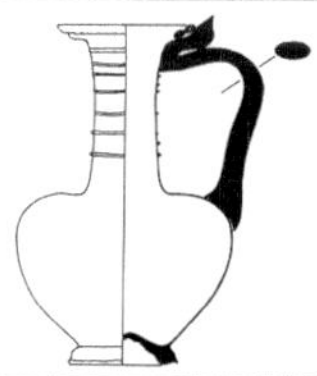 | | | | — | | | 420 |
| 93 | 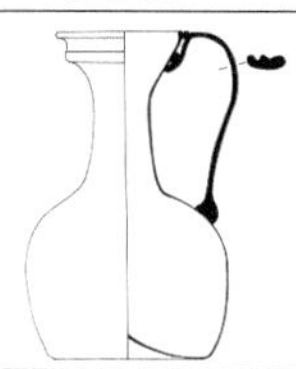 | | | | — | | | 421-423 |

| Type no. | | 1st c. | 2nd c. | 3rd c. | 4th c. | 5th c. | 6th c. | Cat. no. |
|---|---|---|---|---|---|---|---|---|
| 94 | 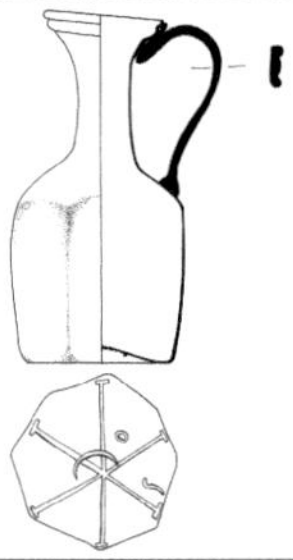 | | | | — | | | 424-425 |
| 95 | 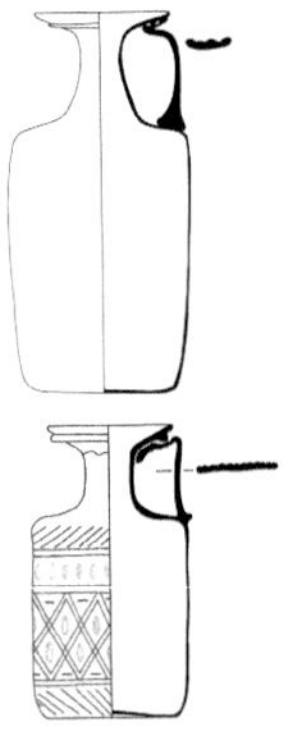 | | — | — | | | | 426-429 |
| 96 |  | | | | | | | 430 |
| 97 |  | | | | — | — | | 431-344 |
| **Jars** | | | | | | | | |
| 98 | 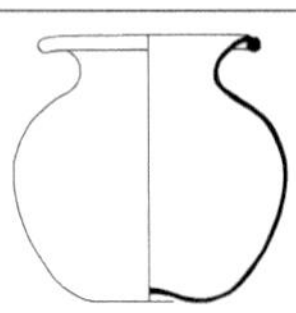 | — | — | | | | | 435 |
| 99 | 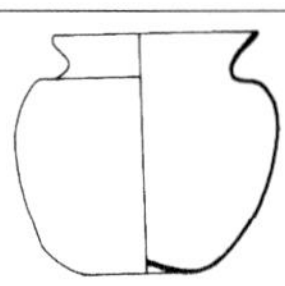 | | | — | — | | | 436 |
| 100 | 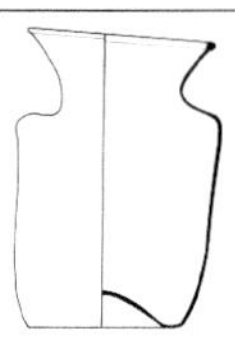 | | | — | — | | | 437-438 |
| 101 | 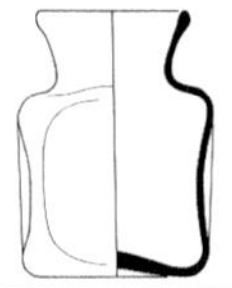 | | | | — | | | 439-441 |
| 102 | 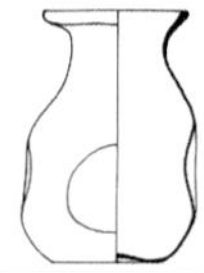 | | | — | — | | | 442 |
| 103 | 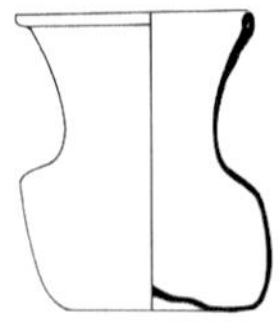 | | | — | — | | | 443 |
| 104 | 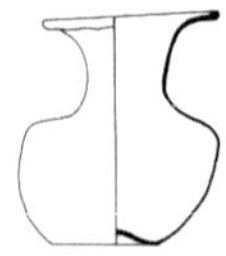 | | | | — | | | 444-445 |
| 105 | 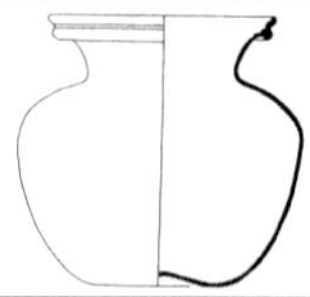 | | | — | — | | | 446 |
| 106 | 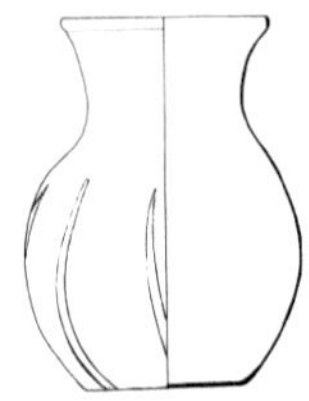 | | | — | — | — | | 447 |
| 107 | 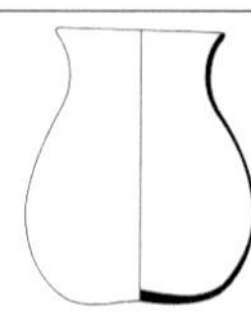 | | | — | — | | | 448-449 |
| 108 |  | | | — | — | | | 450-451 |
| 109 | 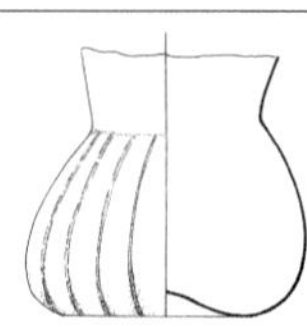 | | | | — | — | | 452 |
| 110 |  | | | | — | — | | 453 |

| Type no. | | 1st c. | 2nd c. | 3rd c. | 4th c. | 5th c. | 6th c. | Cat. no. |
|---|---|---|---|---|---|---|---|---|
| **Unguentaria** | | | | | | | | |
| 111 |  | — | — | | | | | 454-458 |
| 112 |  | — | | | | | | 459-460 |
| 113 |  | — | | | | | | 461-462 |
| 114 | 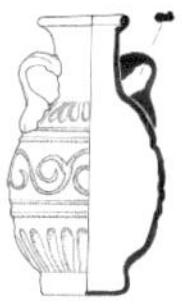 | — | | | | | | 463 |
| 115 | 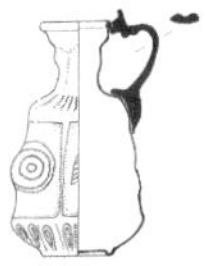 | — | | | | | | 464 |
| 116 |  | — | — | | | | | 465-493 |
| 117 | 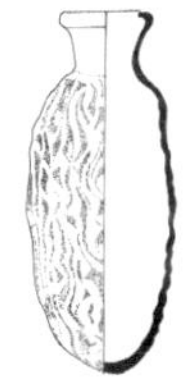 | — | | | | | | 494-495 |
| 118 | 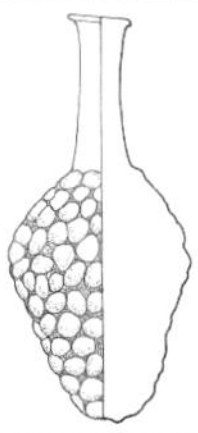 | — | — | | | | | 496 |
| 119 | 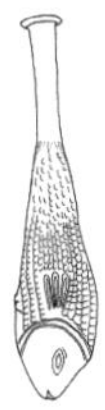 | — | — | | | | | 497 |
| 120 | 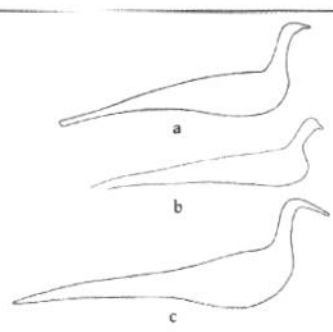 | — | — | | | | | 498-507<br>508-515<br>516-537 |
| 121 | 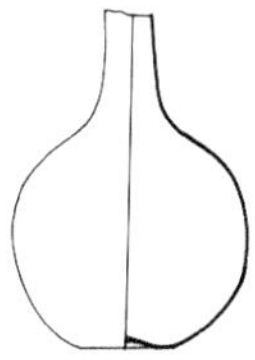 | — | | | | | | 538 |
| 122a | 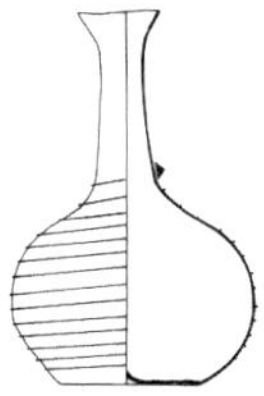 | — | | | | | | 539-541 |
| 122b |  | — | | | | | | 542-550 |
| 123 |  | — | | | | | | 551-552 |
| 124 | 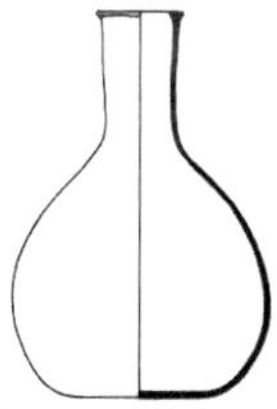 | — | | | | | | 553 |
| 125 | 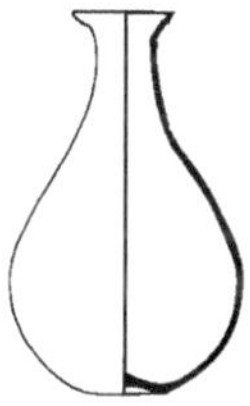 | — | | | | | | 554-558 |
| 126a | 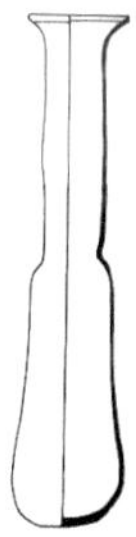 | — | — | | | | | 559-571 |
| 126b |  | — | — | | | | | 572-578 |

| Type no. | | 1st c. | 2nd c. | 3rd c. | 4th c. | 5th c. | 6th c. | Cat. no. |
|---|---|---|---|---|---|---|---|---|
| 127 |  | — | — | — | — | | | 579-580 |
| 128 | 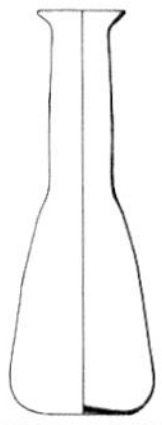 | — | — | | | | | 581-585 |
| 129 | 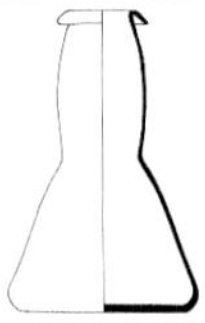 | — | — | | | | | 586-596 |
| 130 | 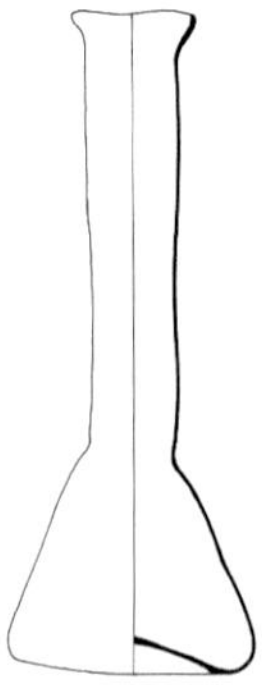 | — | — | | | | | 597-607 |
| 131 | 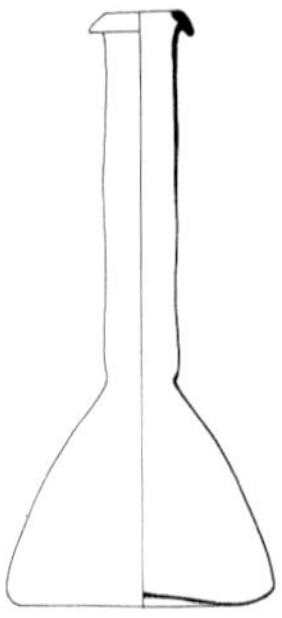 | — | — | | | | | 608-620 |
| 132 | 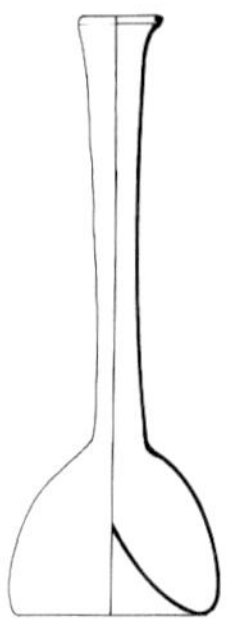 | — | — | — | | | | 621-624 |
| 133 | 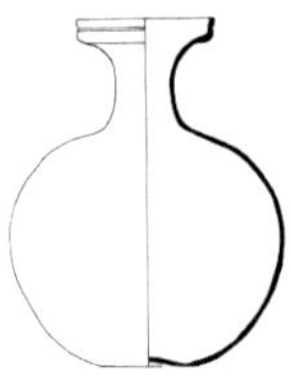 | | | — | — | — | | 625 |

| Type no. | | 1st c. | 2nd c. | 3rd c. | 4th c. | 5th c. | 6th c. | Cat. no. |
|---|---|---|---|---|---|---|---|---|
| 134 | 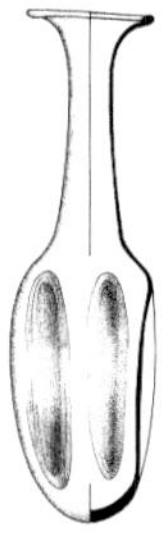 | | — | — | — | | | 626-627 |
| 135 |  | | | | — | | | 628-629 |
| 136 | 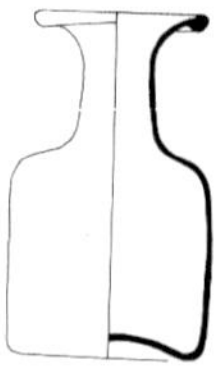 | | | — | — | | | 630 |
| 137 | 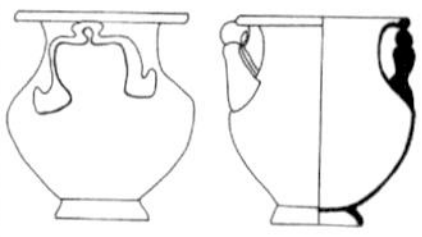 | | | | — | | | 631 |
| 138 | 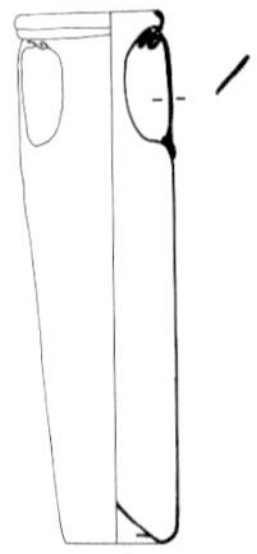 | | | — | — | — | | 632-635 |
| 139 | 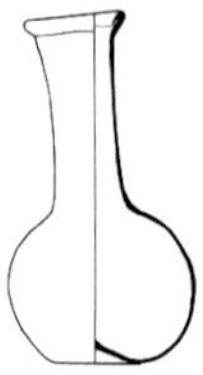 | | | — | — | — | | 636-666 |
| 140 | 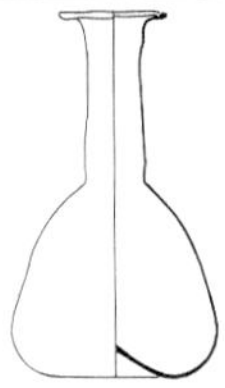 | | | — | — | — | | 667 |
| 141 | 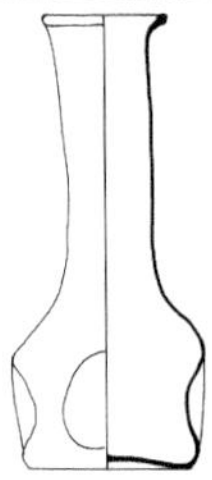 | | | — | — | | | 668-682 |

| Type no. | | 1st c. | 2nd c. | 3rd c. | 4th c. | 5th c. | 6th c. | Cat. no. |
|---|---|---|---|---|---|---|---|---|
| 142a | 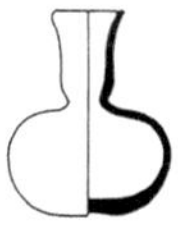 | | | — | — | | | 683-684 |
| 142b | 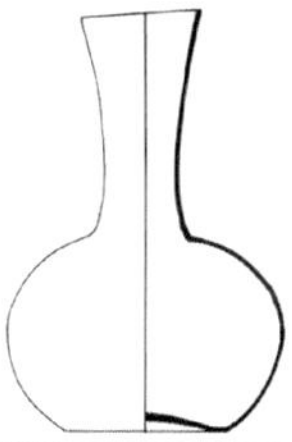 | | | | — | | | 685-687 |
| 143 | 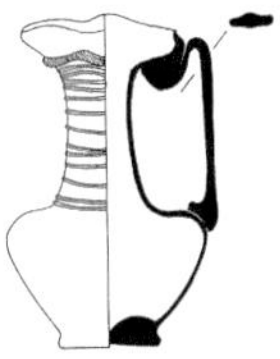 | | | | — | | | 688 |
| 144 |  | | | | — | | | 689 |
| 145 | 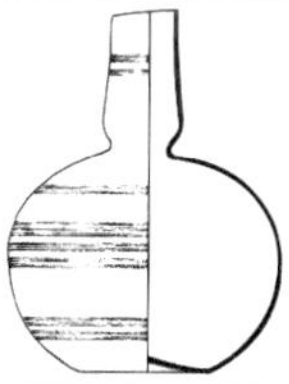 | | | — | — | | | 690-699 |
| 146 | 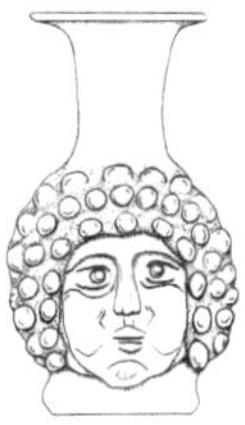 | | | — | — | | | 700-712 |

| Type no. | | 1st c. | 2nd c. | 3rd c. | 4th c. | 5th c. | 6th c. | Cat. no. |
|---|---|---|---|---|---|---|---|---|
| 147a | 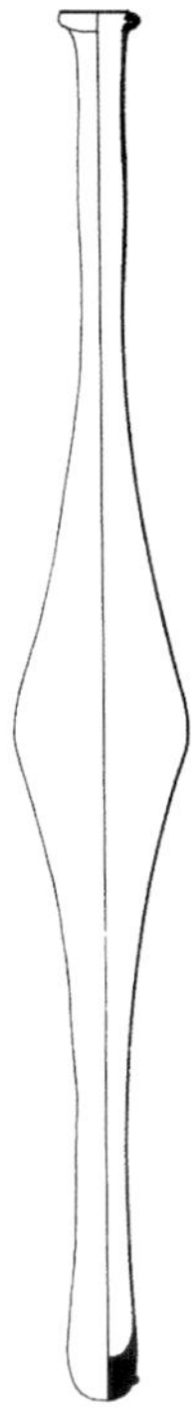 | | | | — | — | | 713-741 |
| 147b | 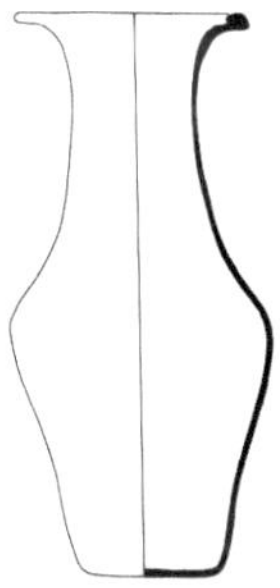 | | | | — | — | | 742-743 |
| 148 |  | — | | | | | | 744-752 |
| 149 |  | | | | — | — | | 753-754 |

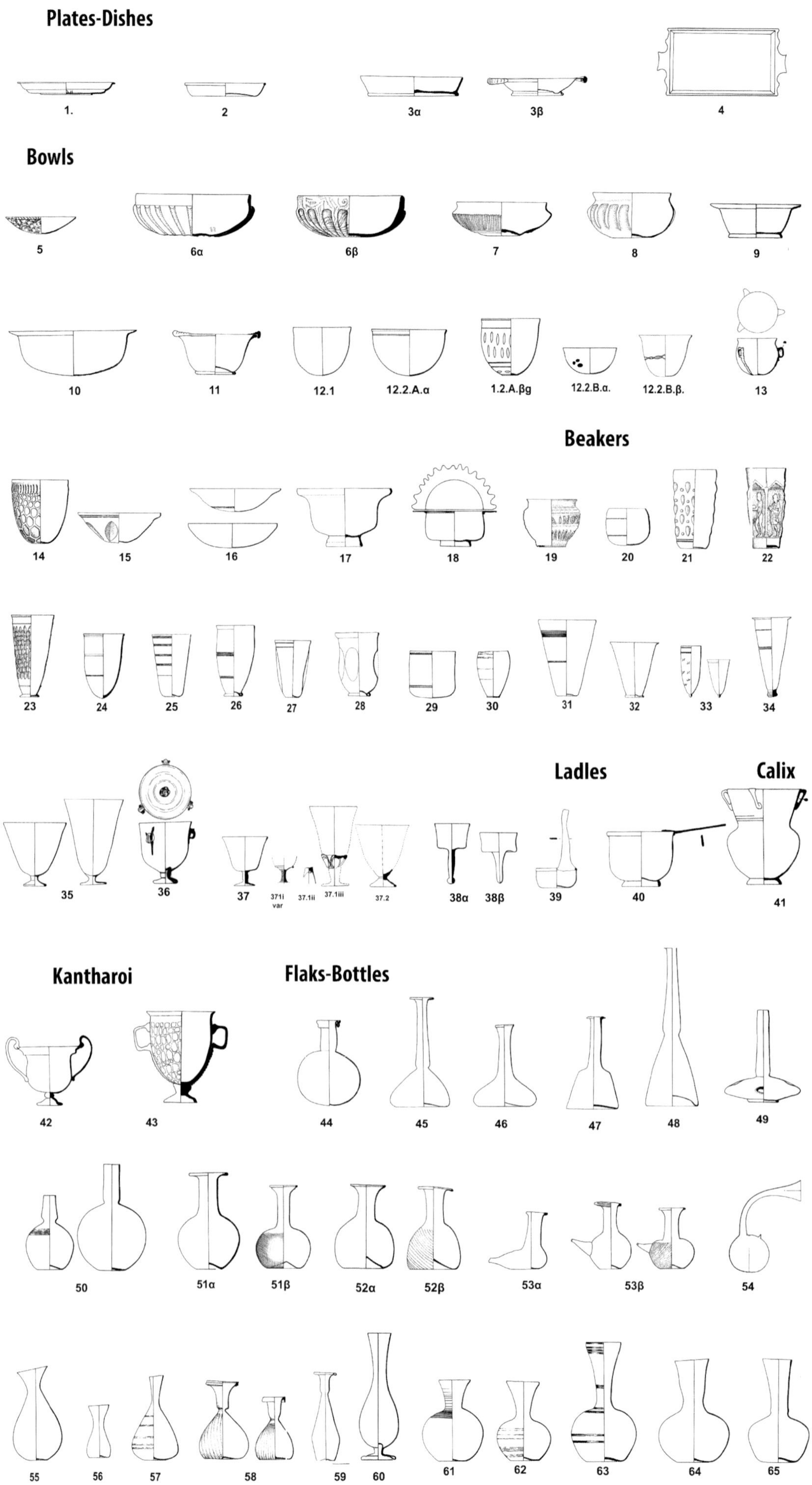
Plates-Dishes
1.
2
3α
3β
4
Bowls
5
6α
6β
7
8
9
10
11
12.1
12.2.A.α
1.2.A.βg
12.2.B.α.
12.2.B.β.
13
Beakers
14
15
16
17
18
19
20
21
22
23
24
25
26
27
28
29
30
31
32
33
34
Ladles
Calix
35
36
37
371i var
37.1ii
37.1iii
37.2
38α
38β
39
40
41
Kantharoi
Flaks-Bottles
42
43
44
45
46
47
48
49
50
51α
51β
52α
52β
53α
53β
54
55
56
57
58
59
60
61
62
63
64
65

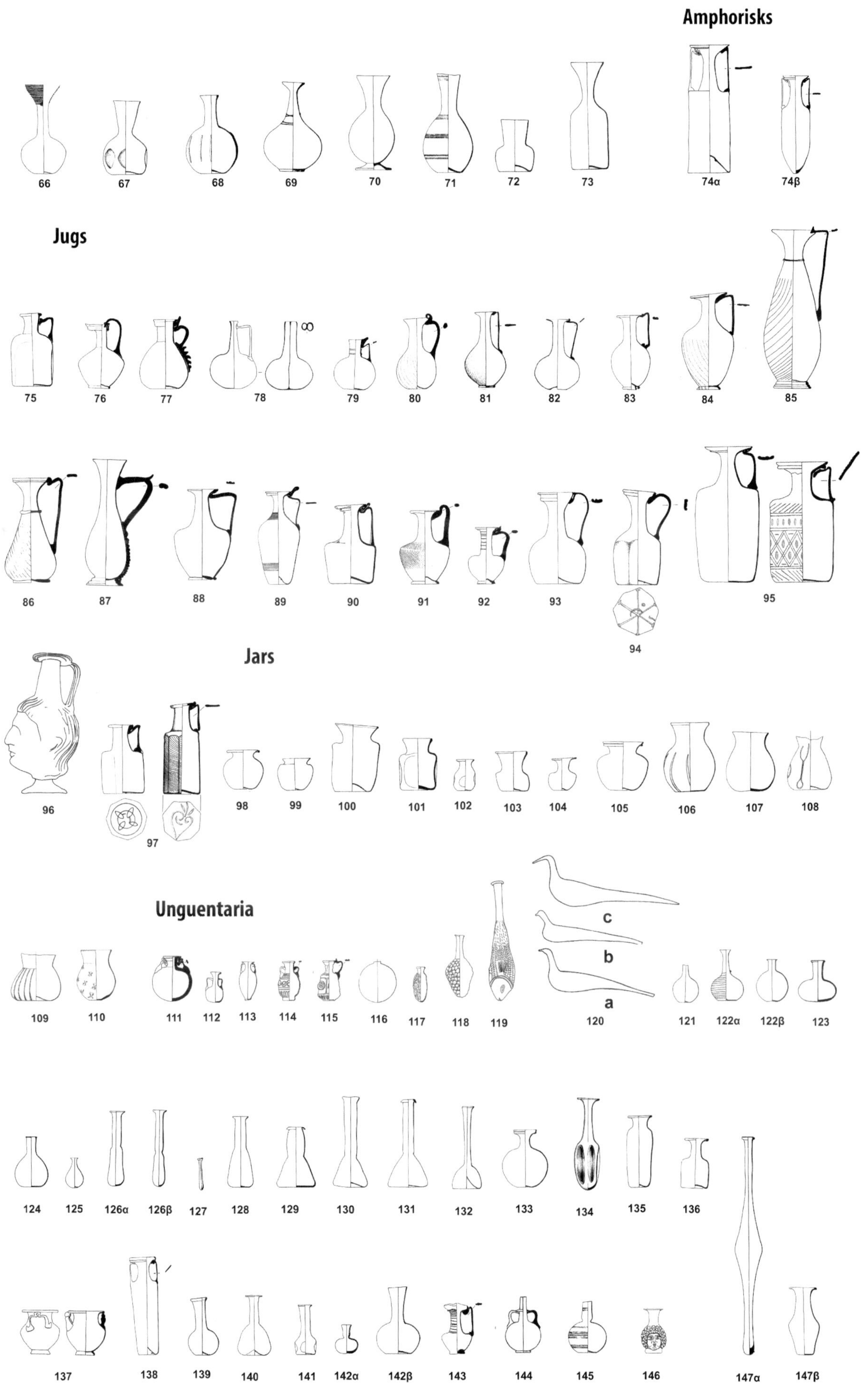
Amphorisks
66 67 68 69 70 71 72 73 74α 74β
Jugs
75 76 77 78 79 80 81 82 83 84 85
86 87 88 89 90 91 92 93 94 95
Jars
96 97 98 99 100 101 102 103 104 105 106 107 108
Unguentaria
c
b
a
109 110 111 112 113 114 115 116 117 118 119 120 121 122α 122β 123
124 125 126α 126β 127 128 129 130 131 132 133 134 135 136
137 138 139 140 141 142α 142β 143 144 145 146 147α 147β

Scale 1:3

Plate 1

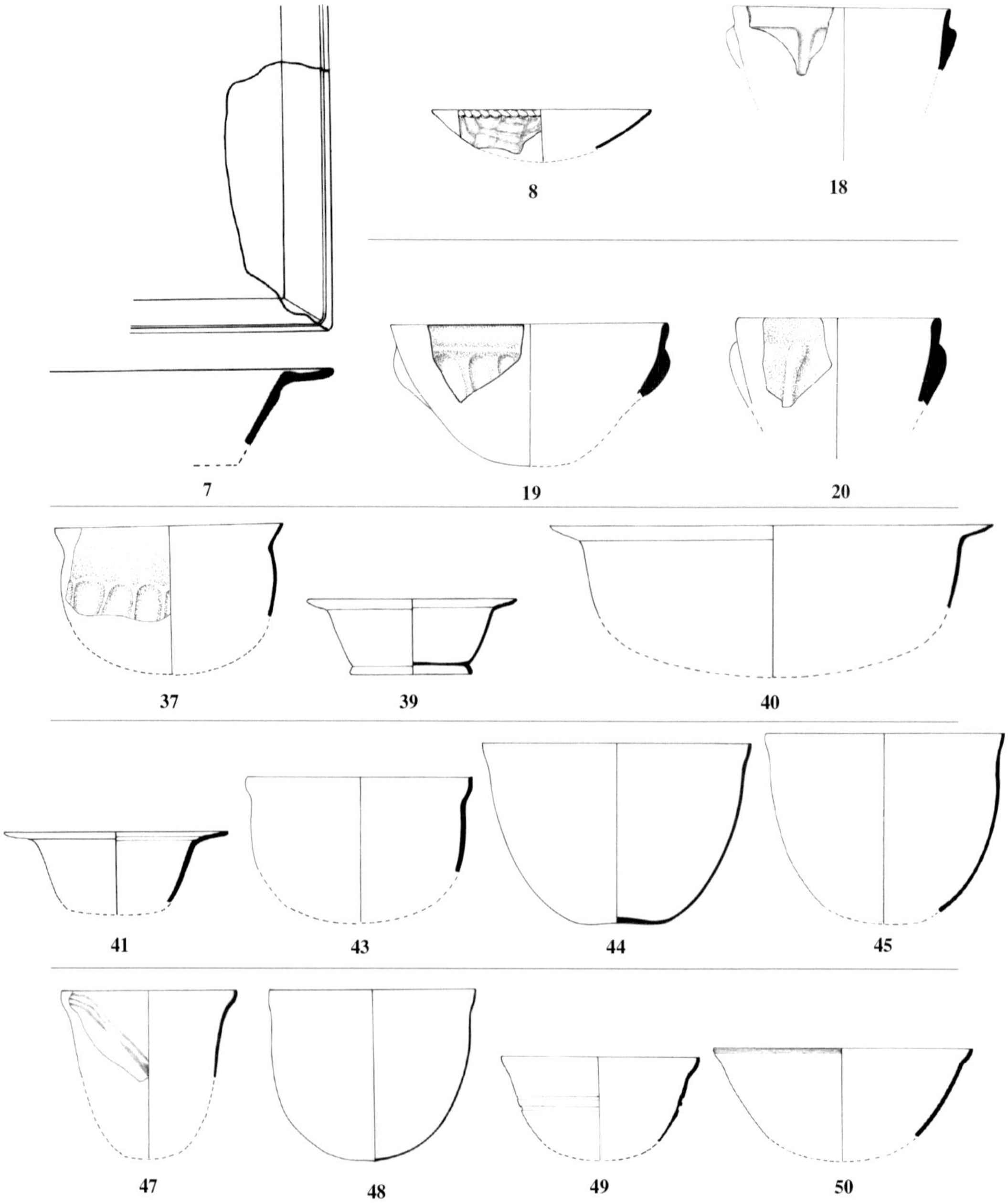

# Plate 2

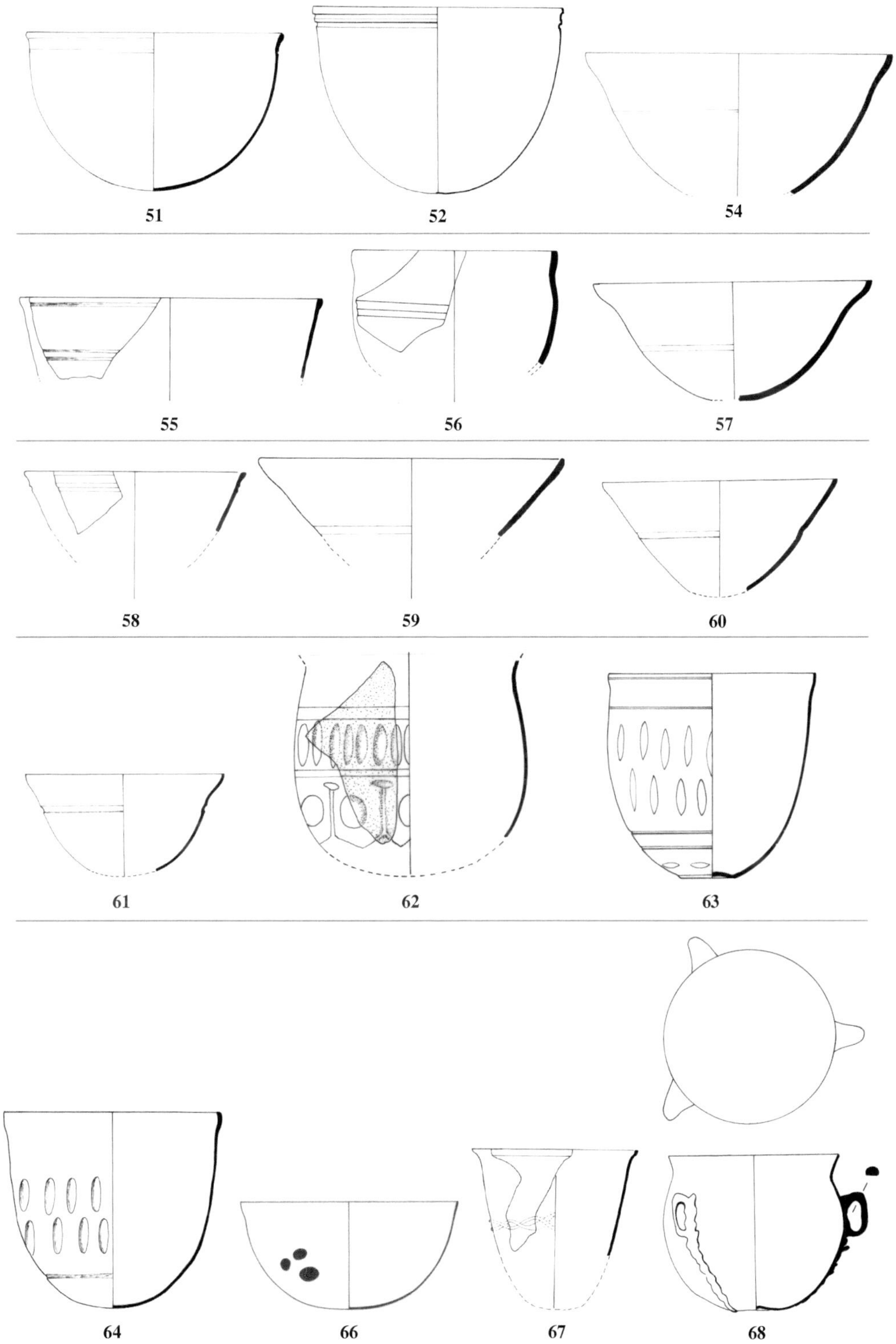

Plate 3

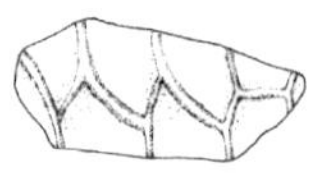

69

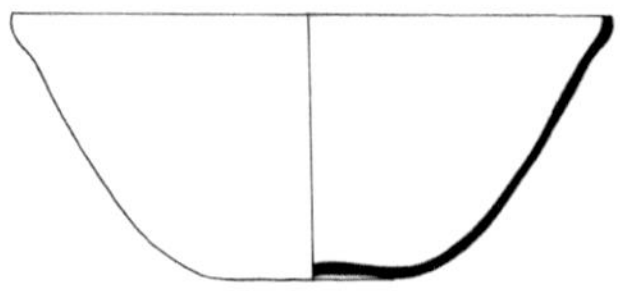

71

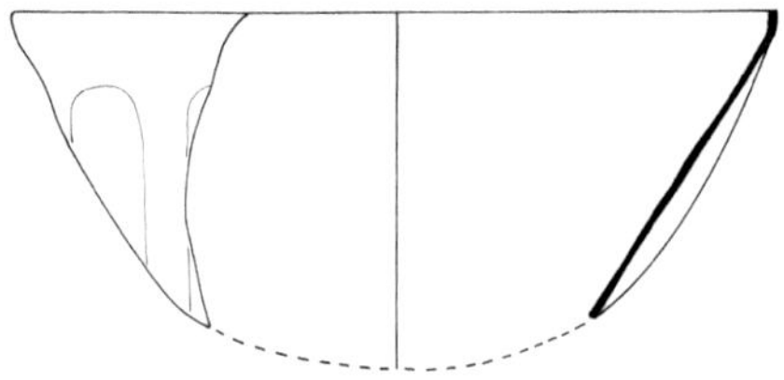

72

74

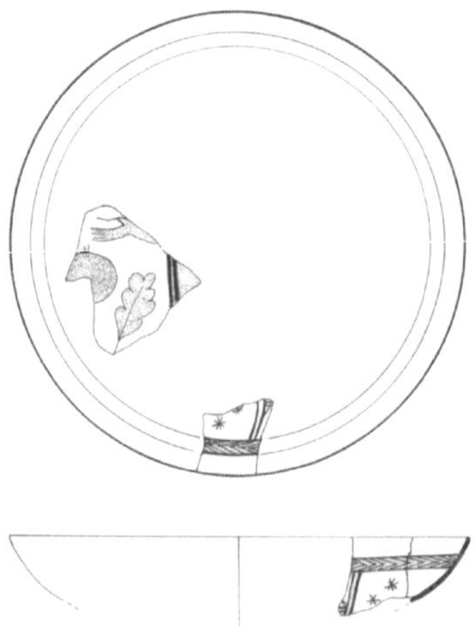

75

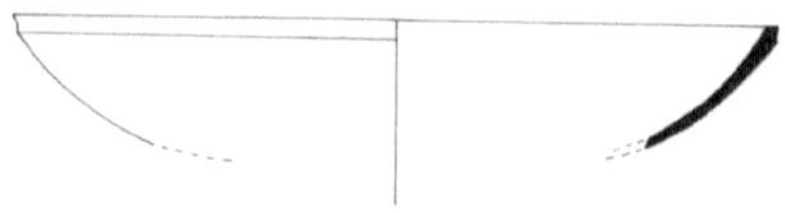

76

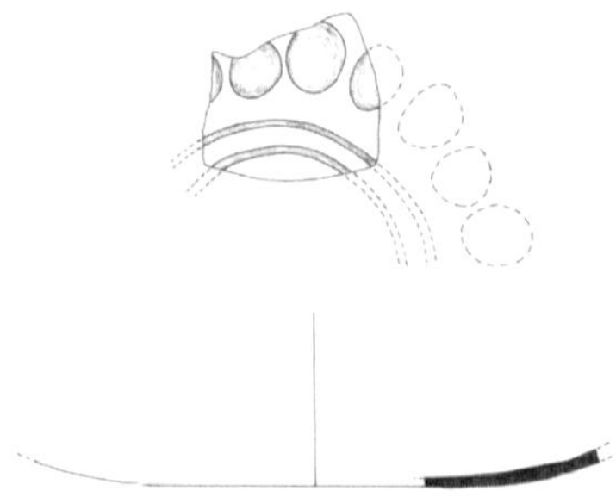

77

78

81

# Plate 4

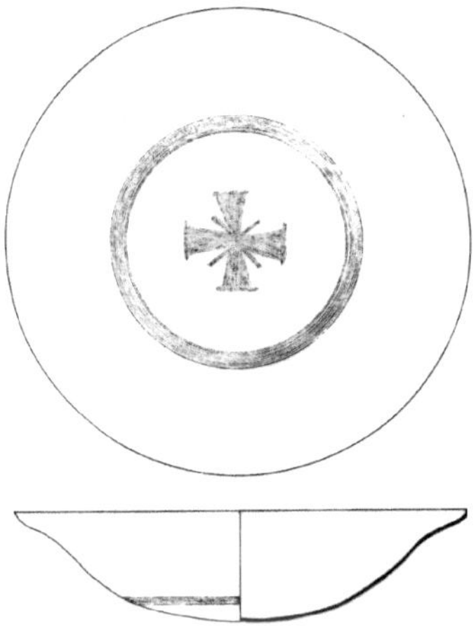

82-83

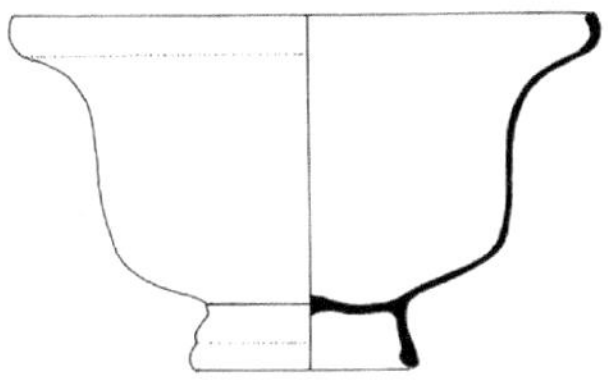

84

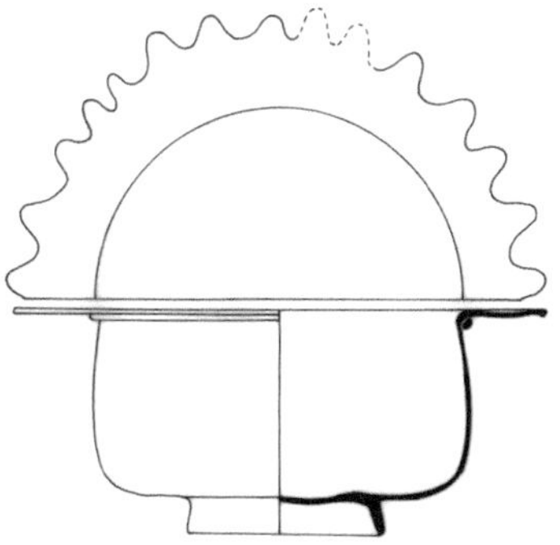

85

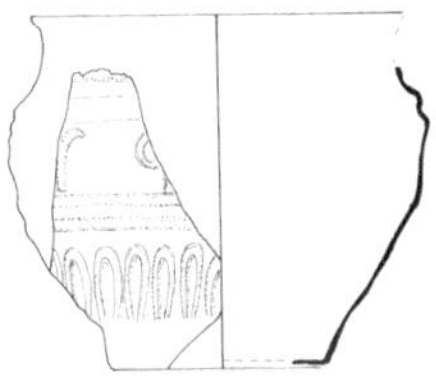

88

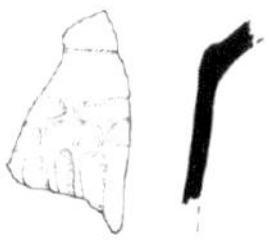

96

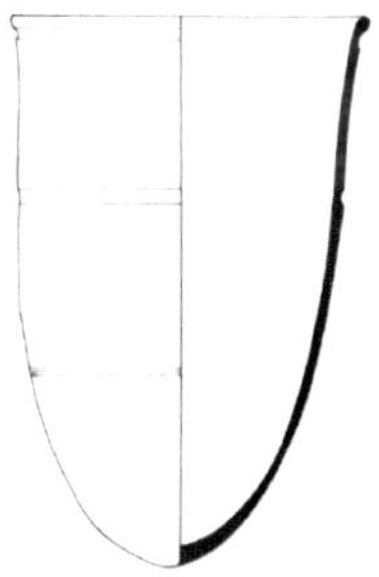

98

105

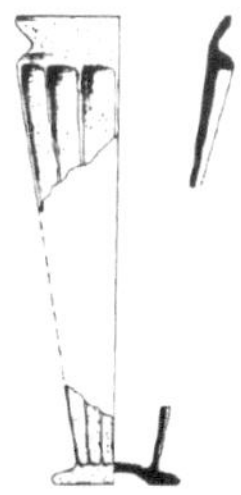

106-107

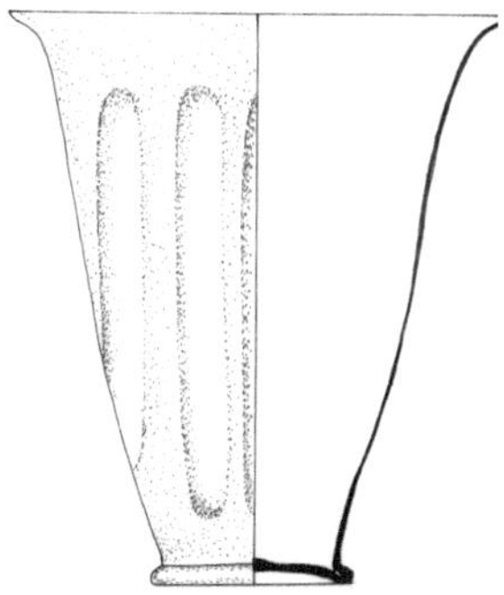

109

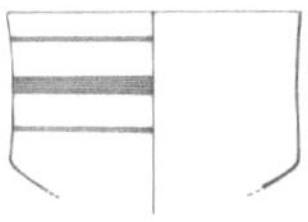

110

111

112

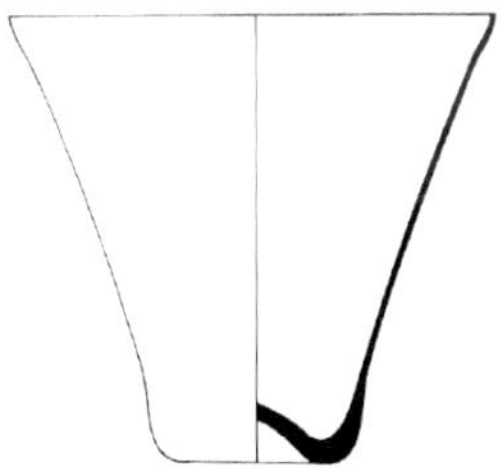

113

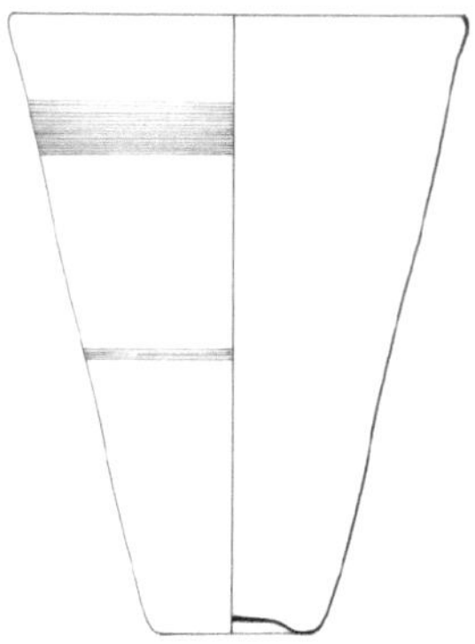

114

Plate 5

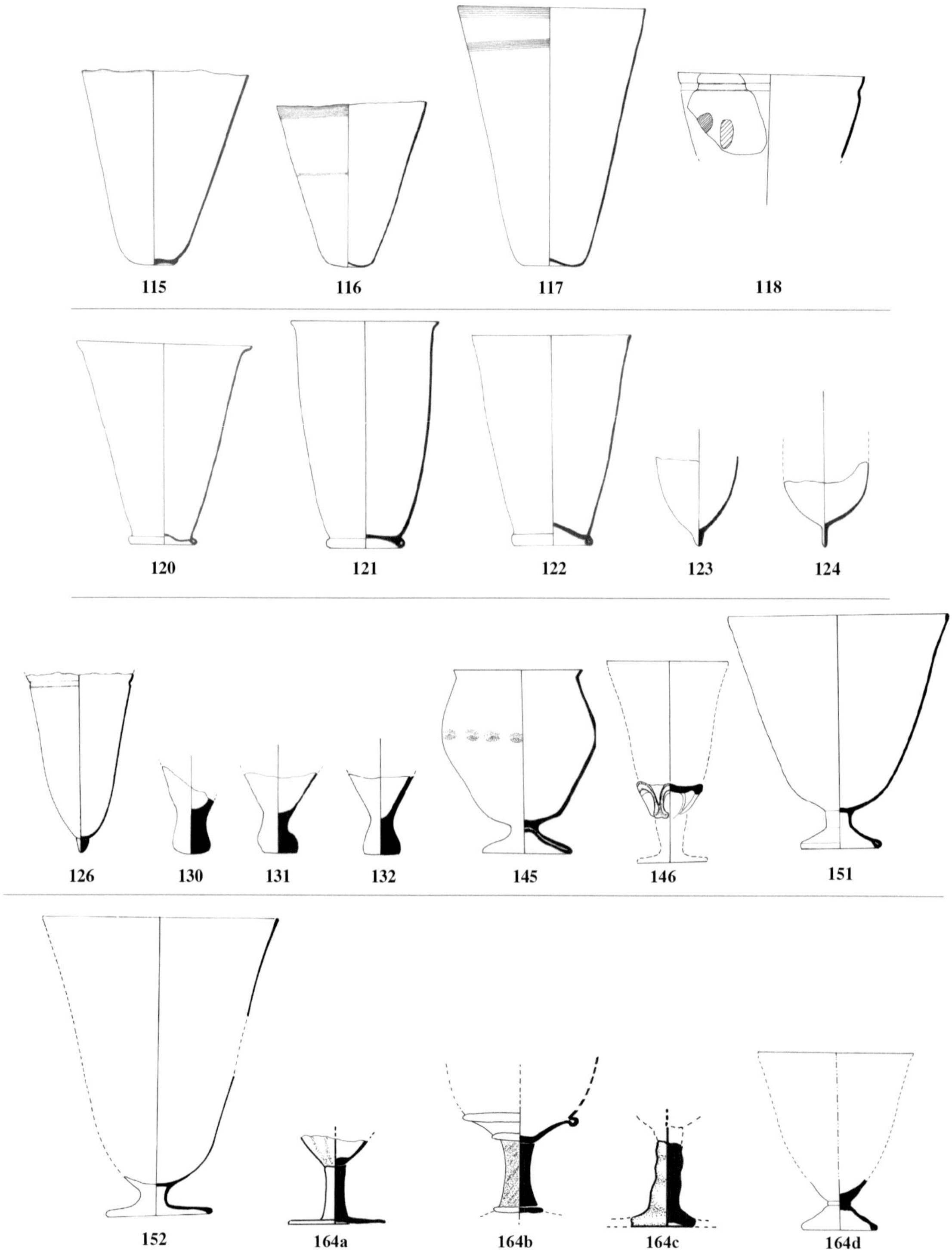
115
116
117
118
120
121
122
123
124
126
130
131
132
145
146
151
152
164a
164b
164c
164d

# Plate 6

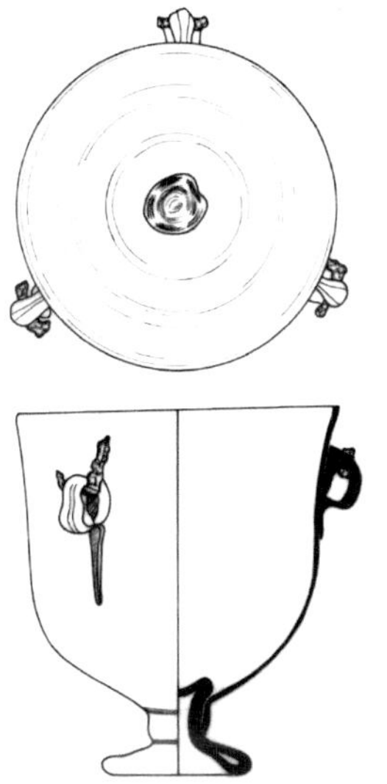

165

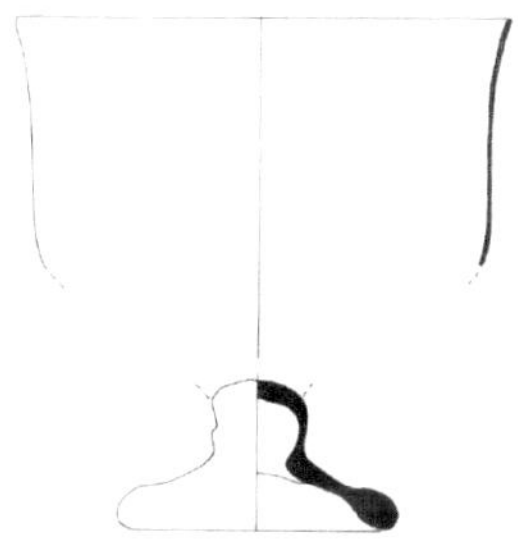

168

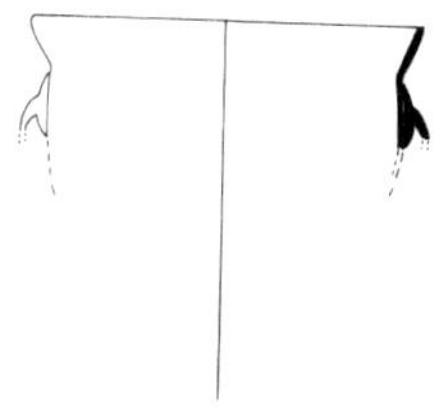

169

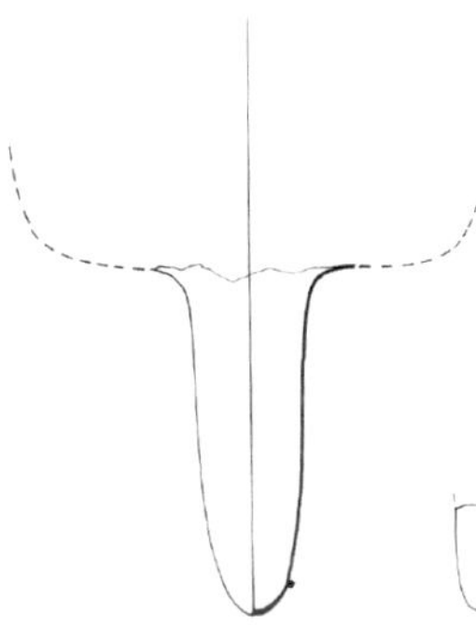

173

174

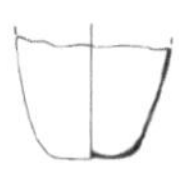

175

175a

175b

175c

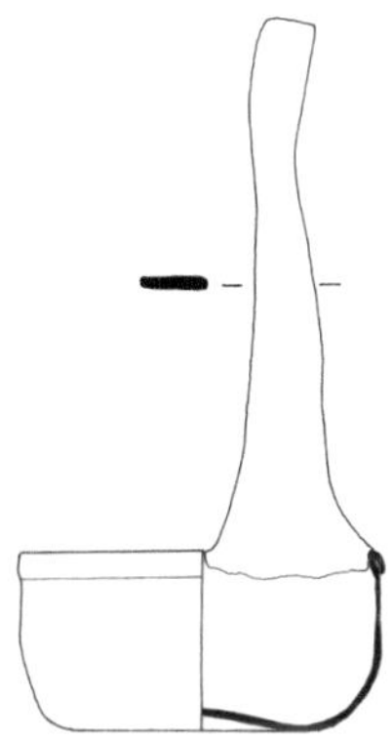

176

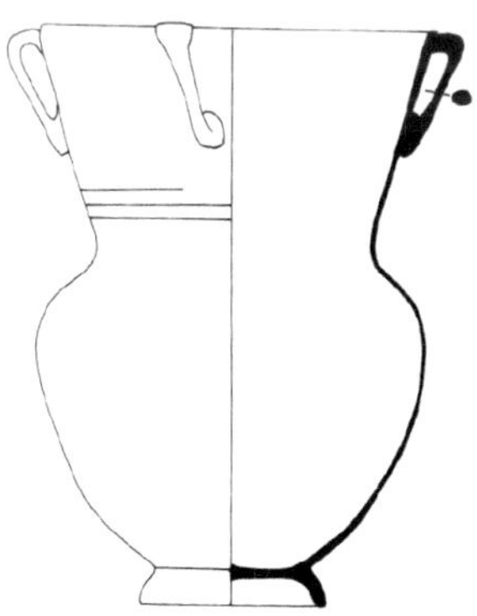

185

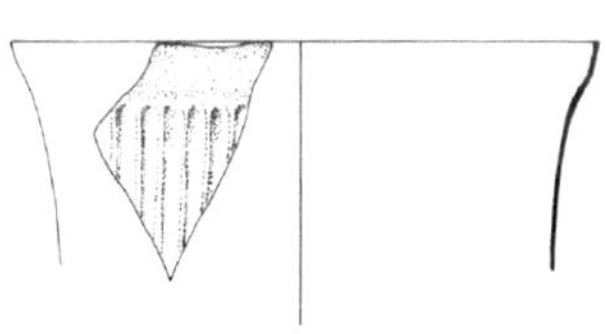

188

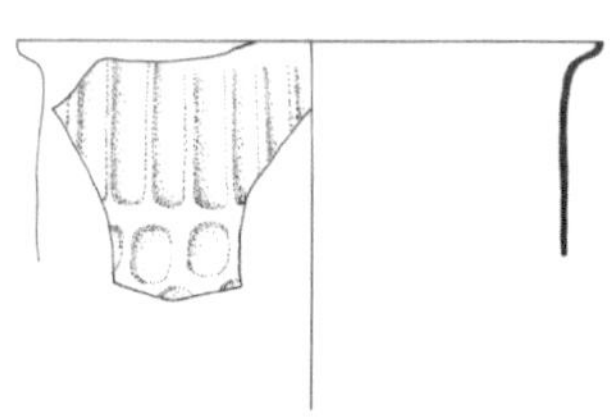

189

# Plate 7

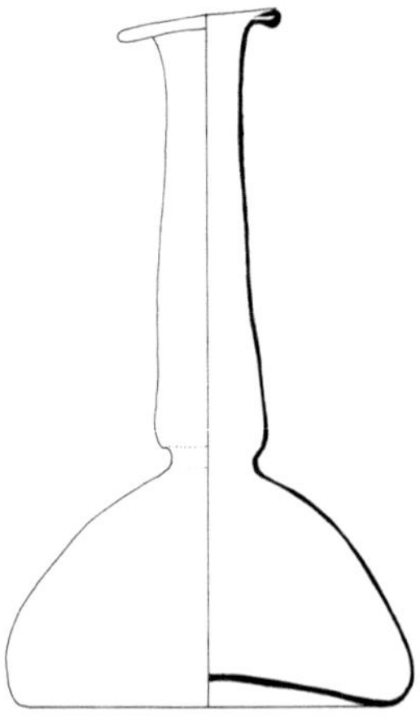

193

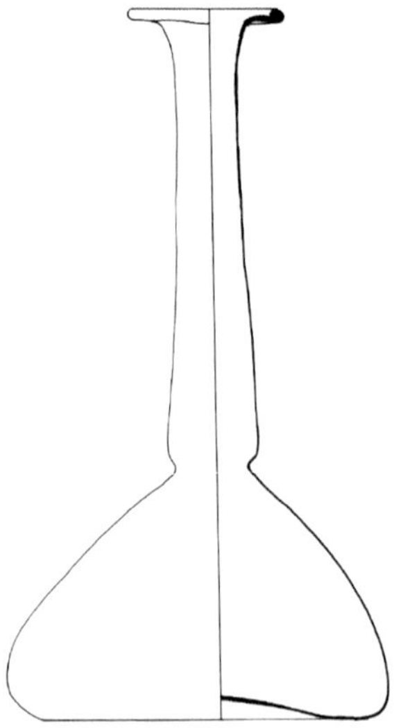

194

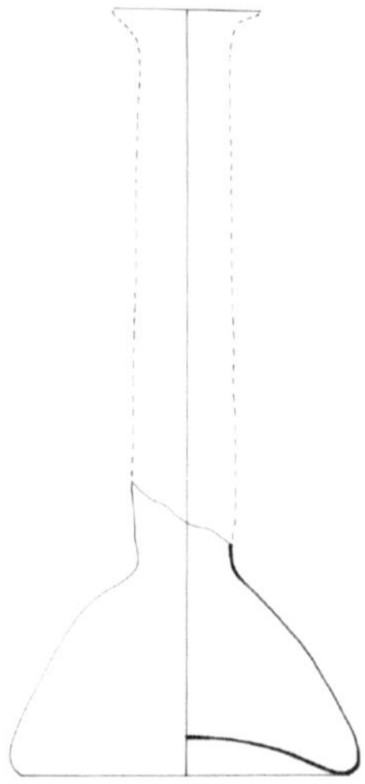

195

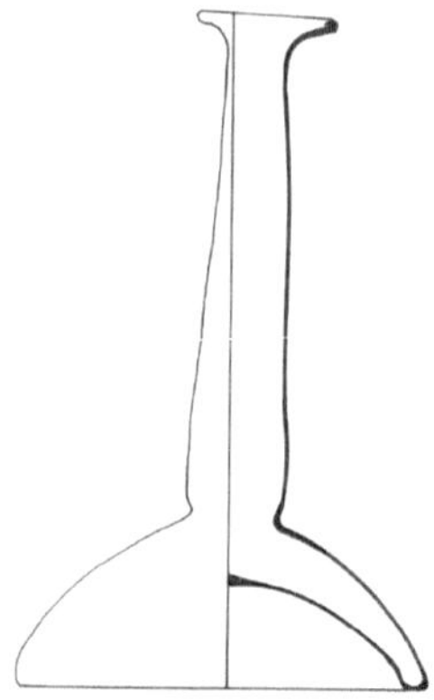

196

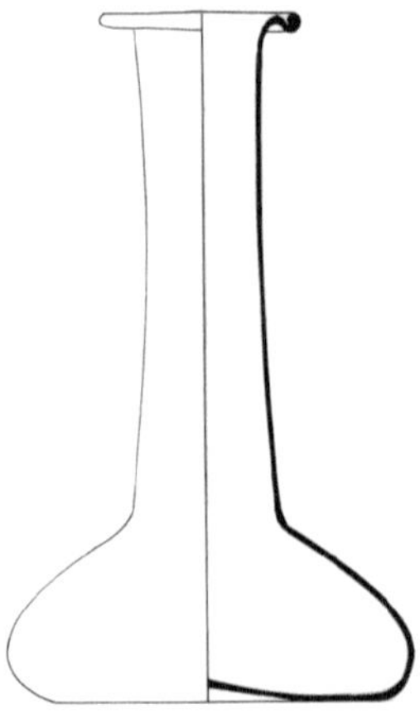

197

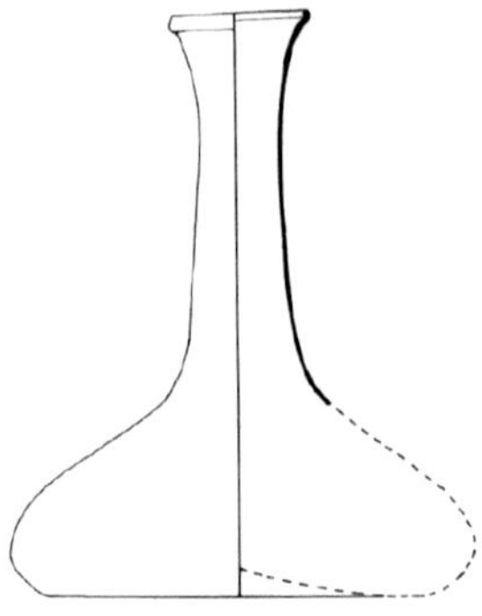

198

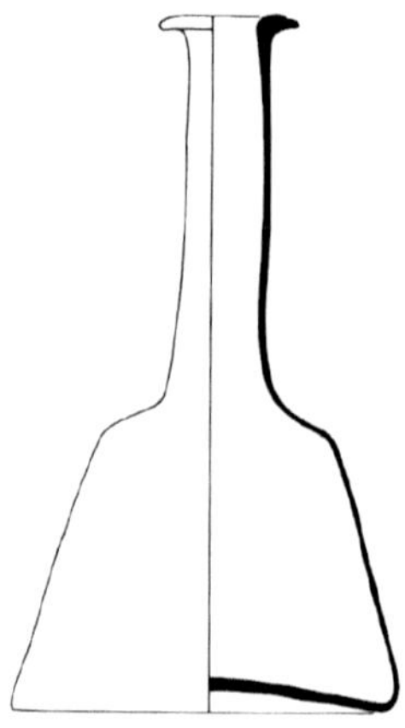

199

Plate 8

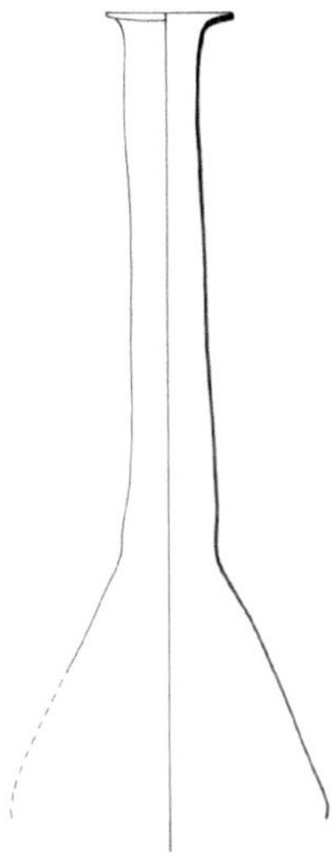

**200**

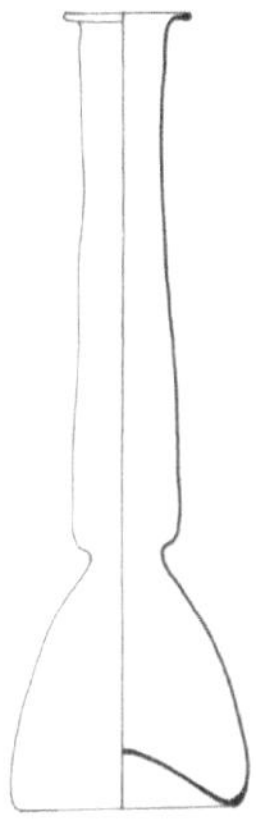

**201**

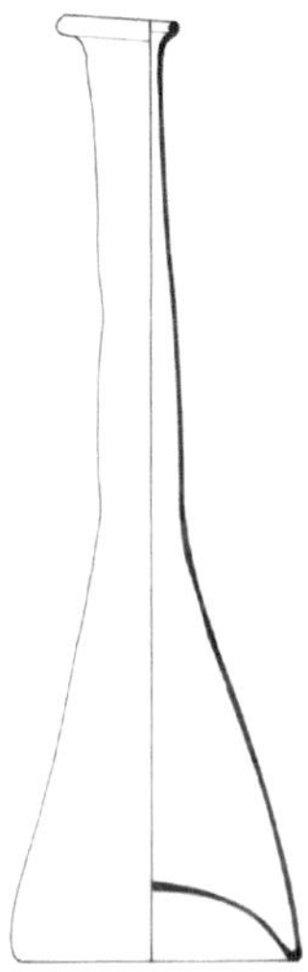

**202**

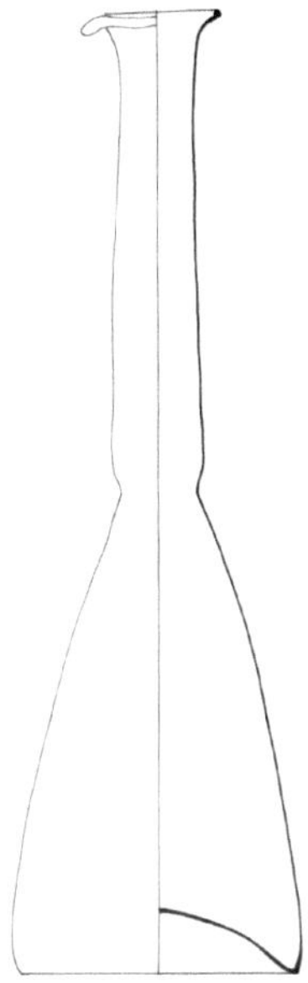

**203**

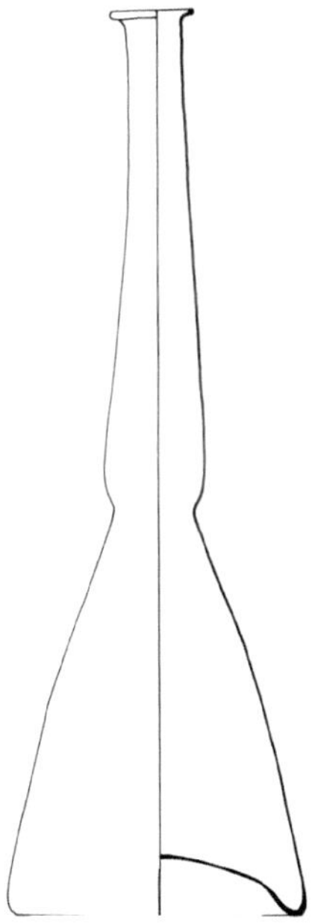

**204**

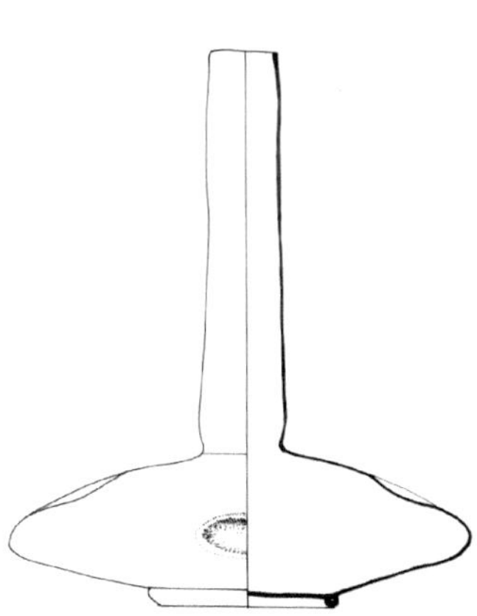

**205**

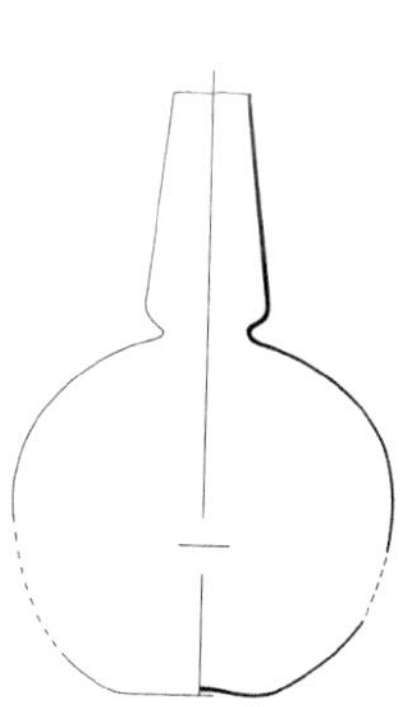

**206**

Plate 9

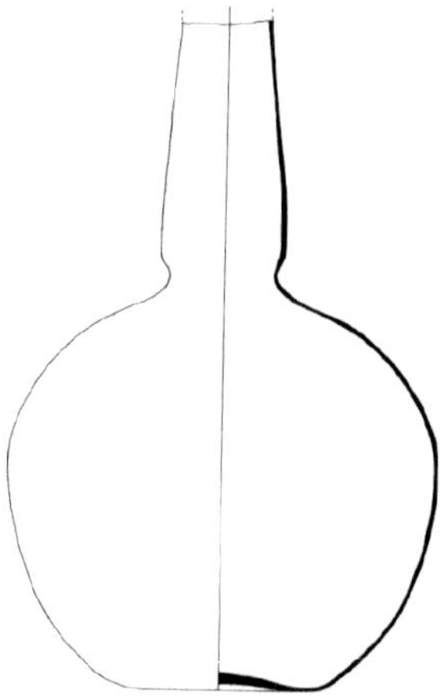
207

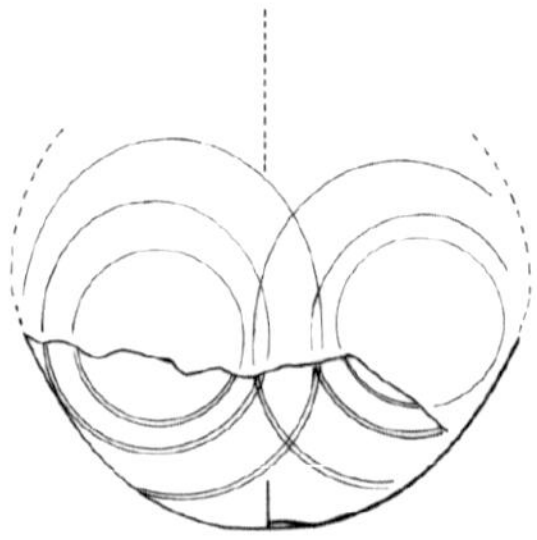
208

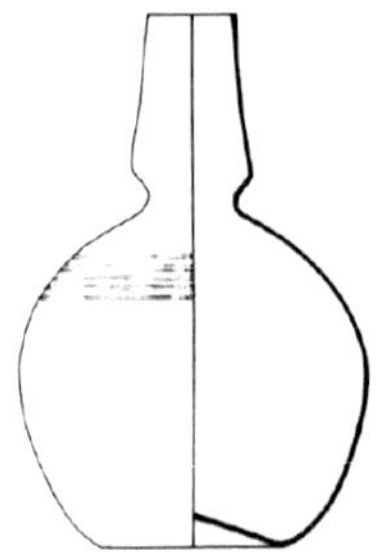
209

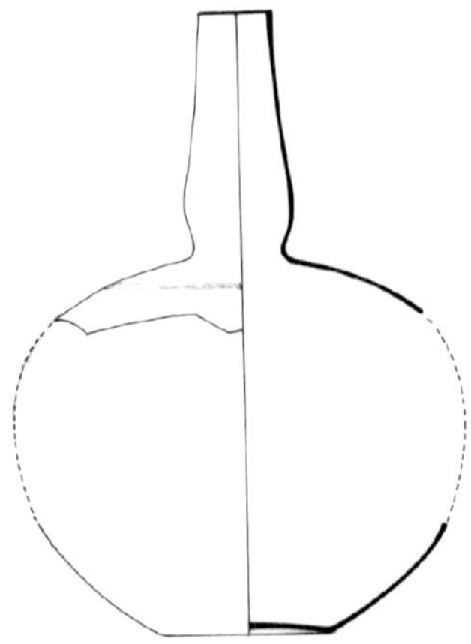
210-211

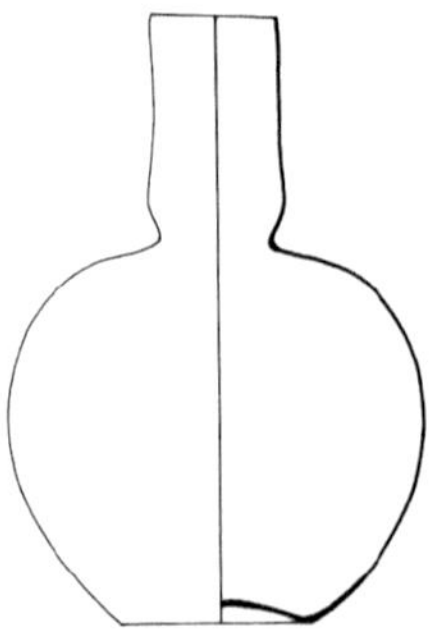
212

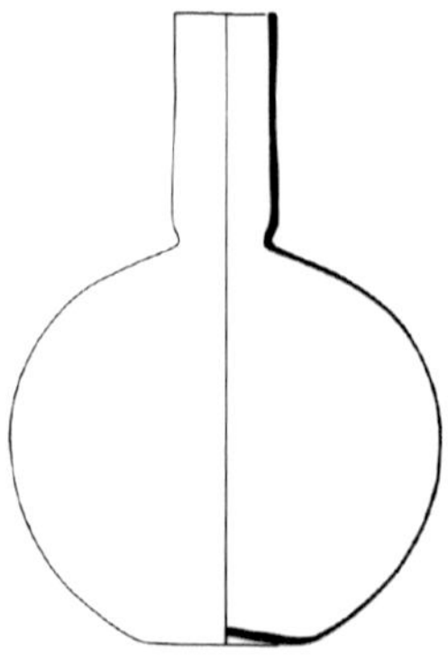
213

214

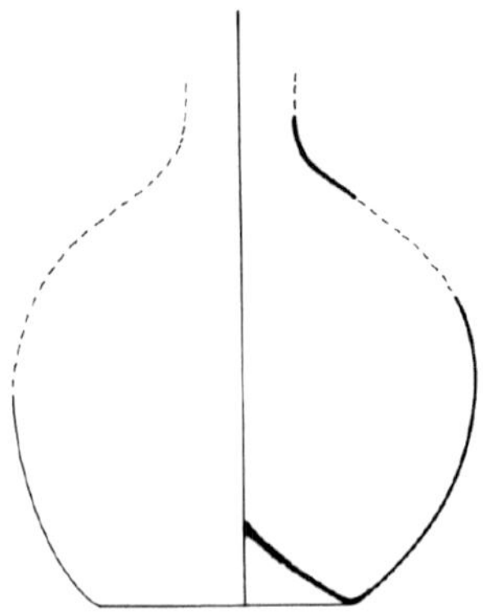
215

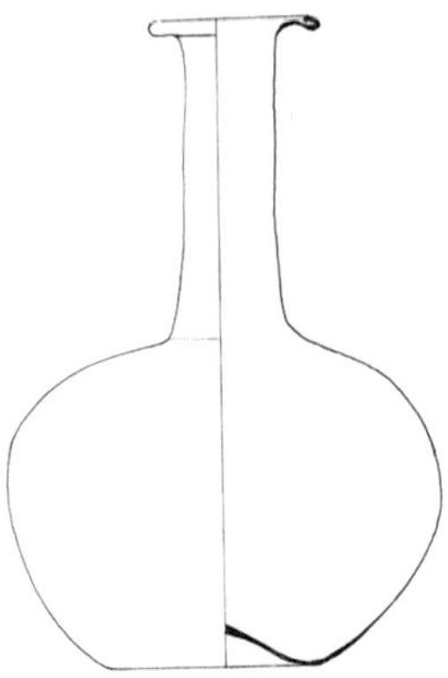
217

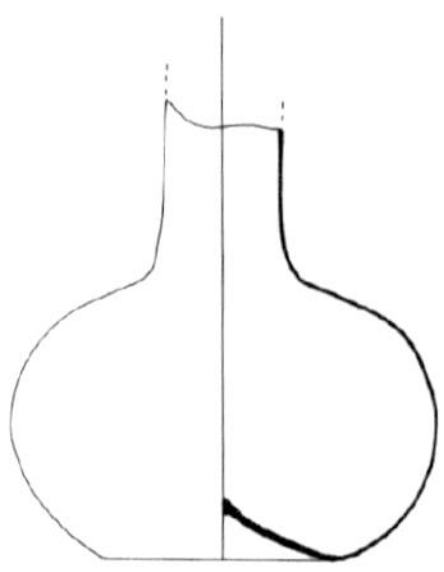
218

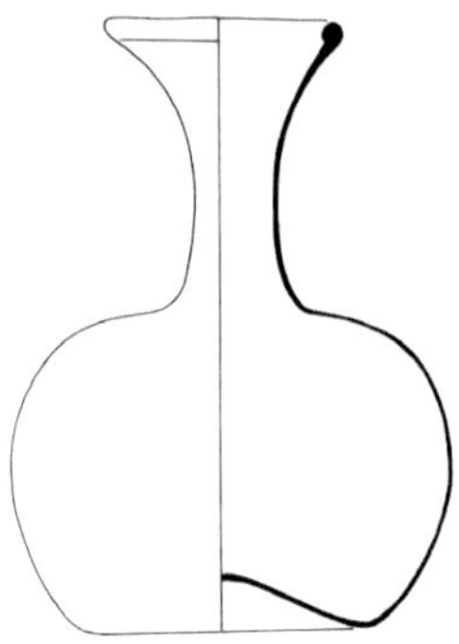
219

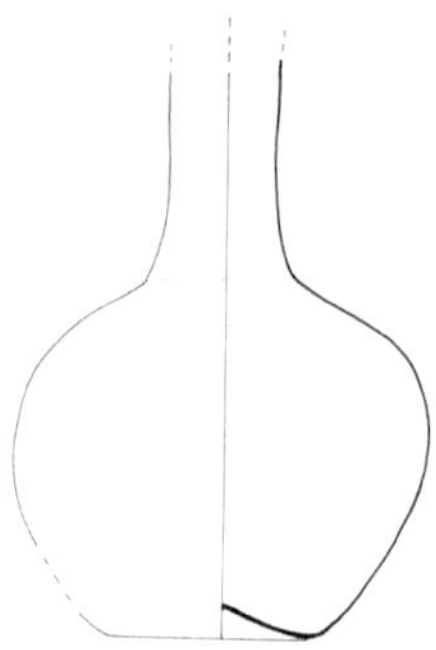
221

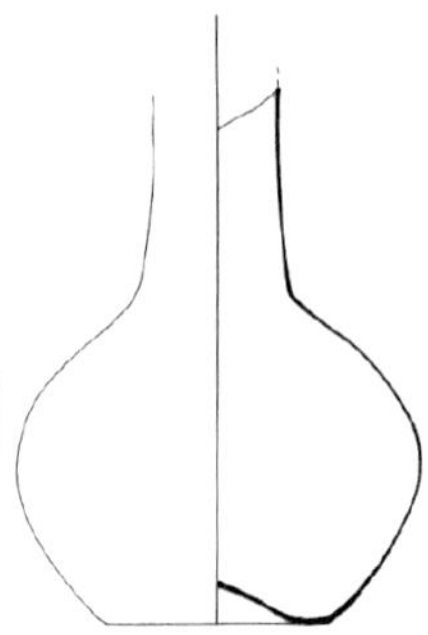
222

# Plate 10

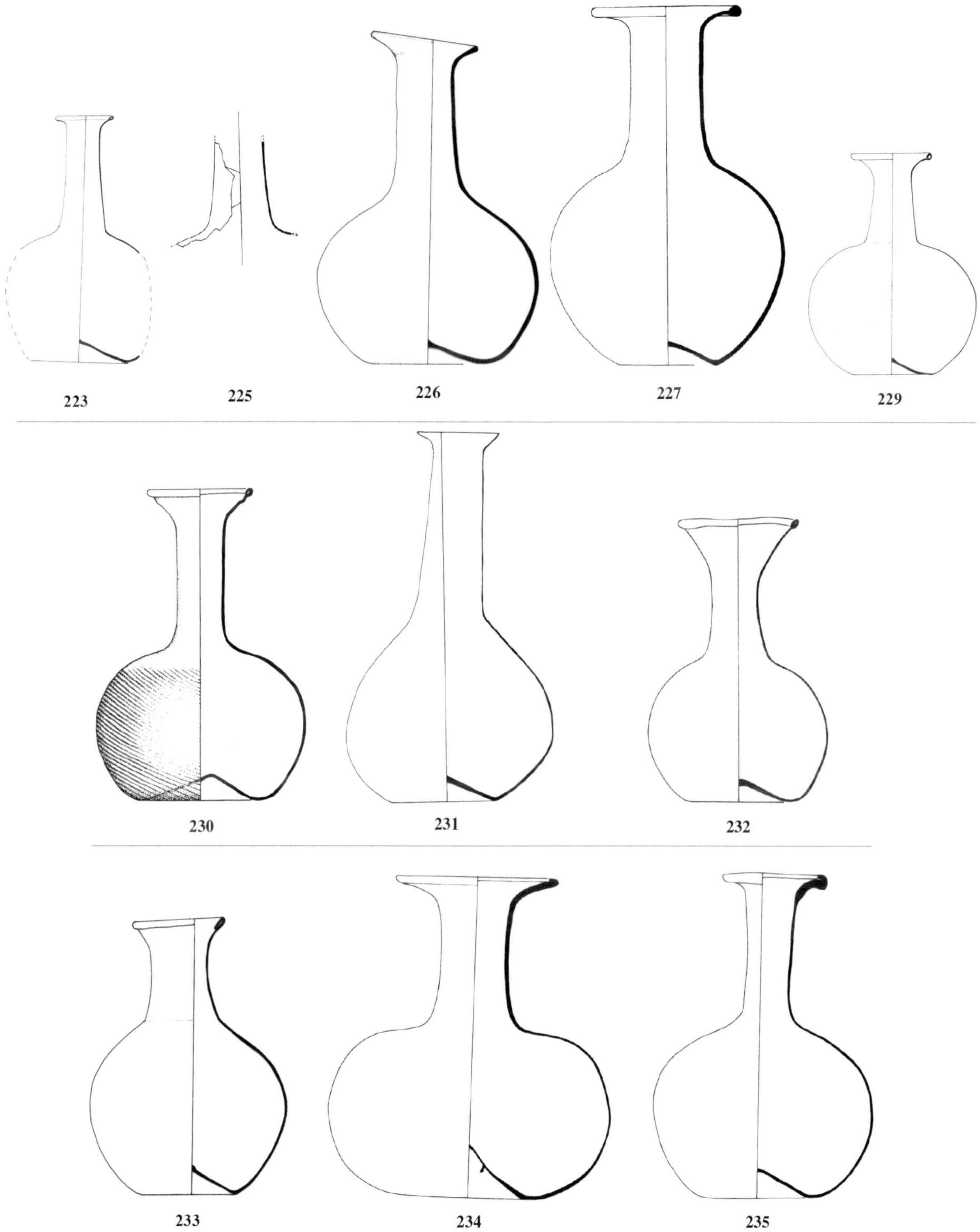

# Plate 11

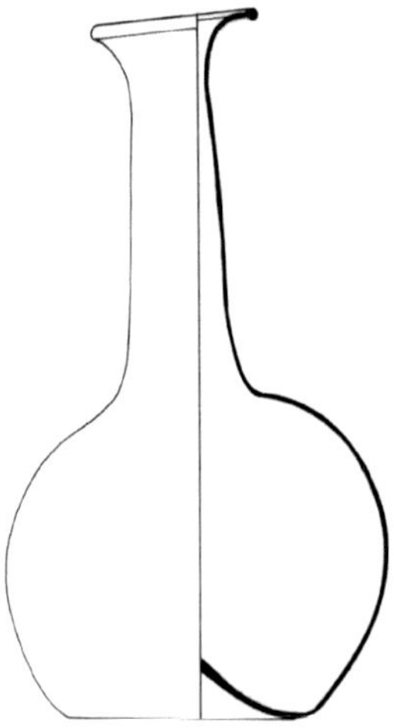

236

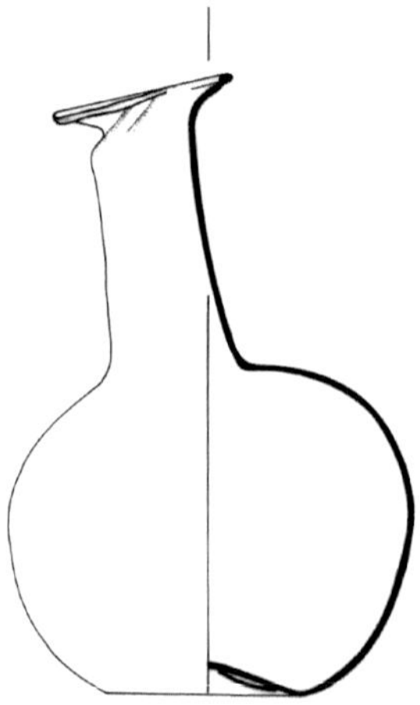

239

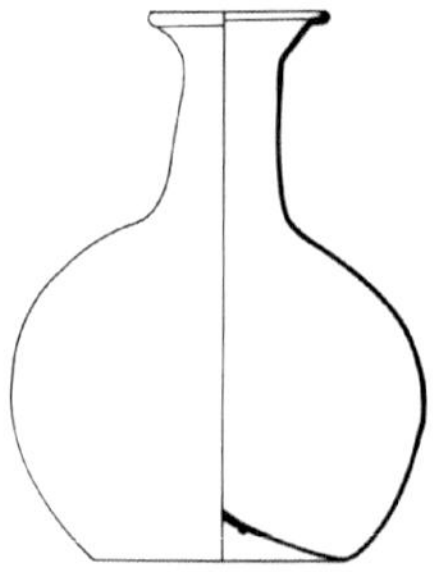

240

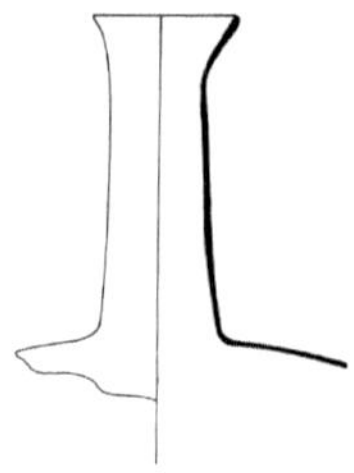

241

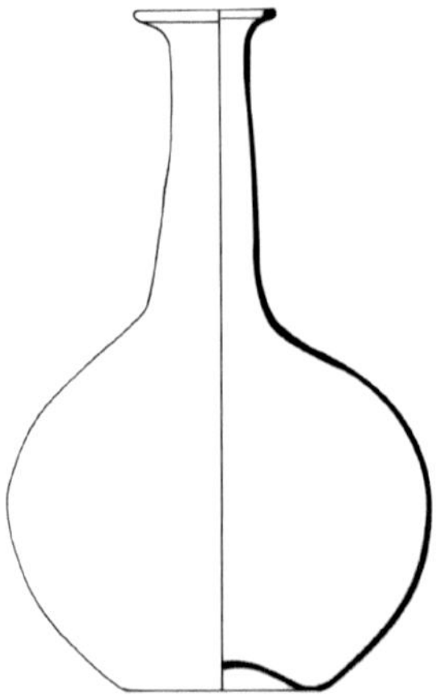

242

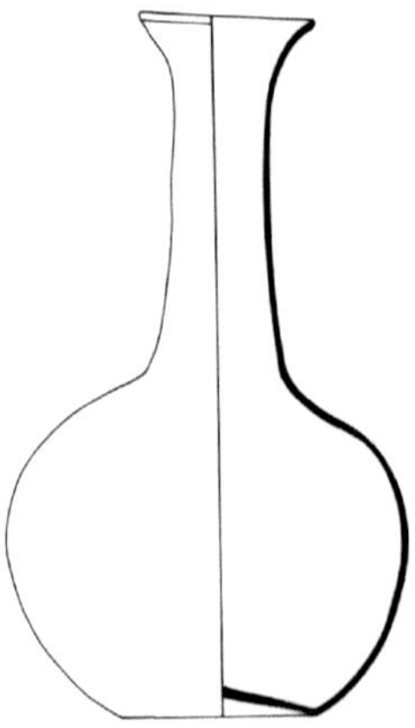

243

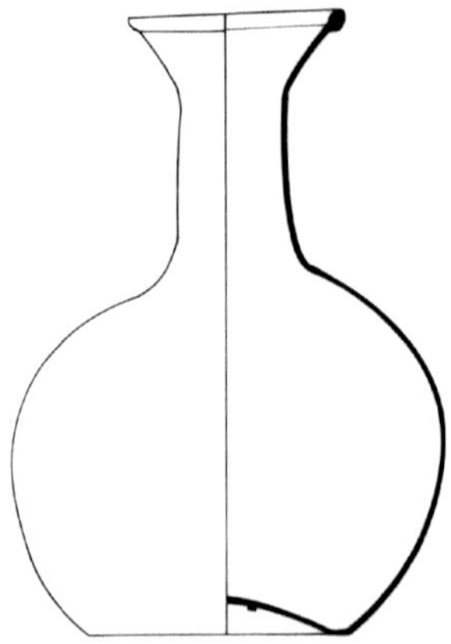

244

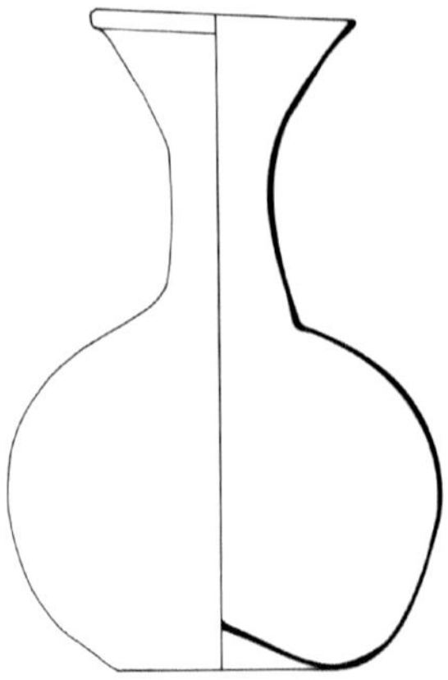

245

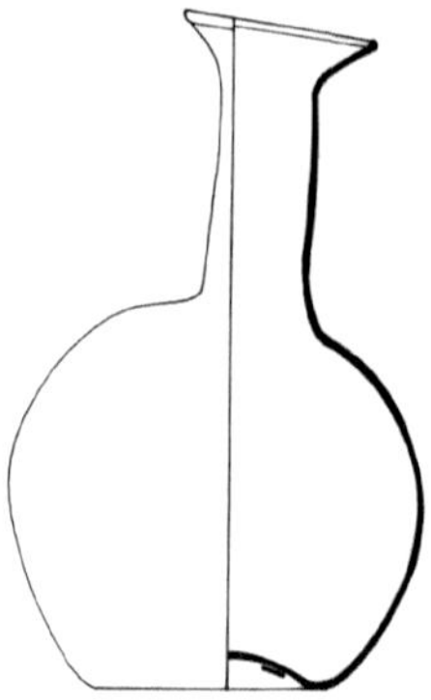

246

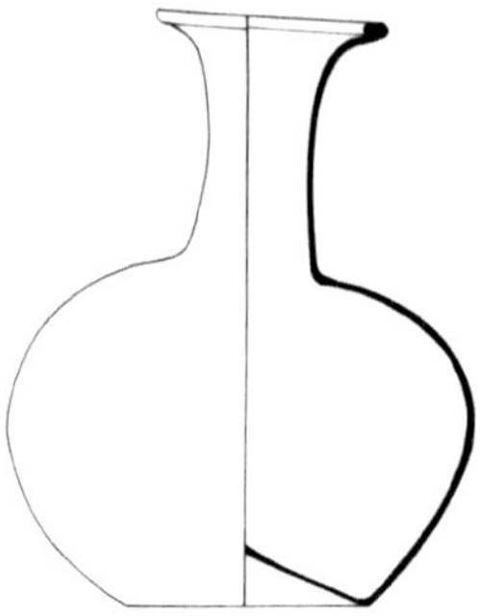

247

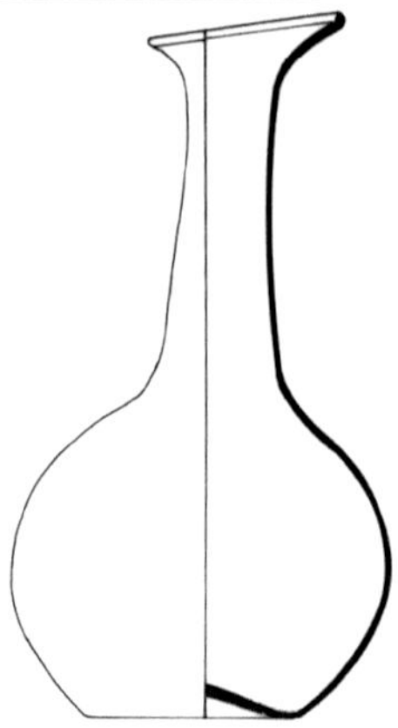

248

# Plate 12

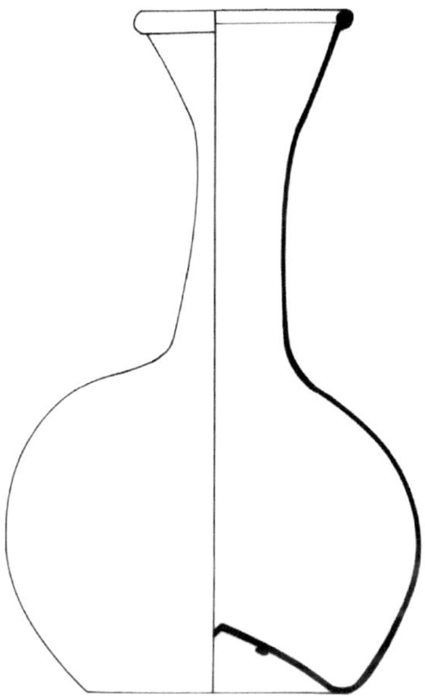

249

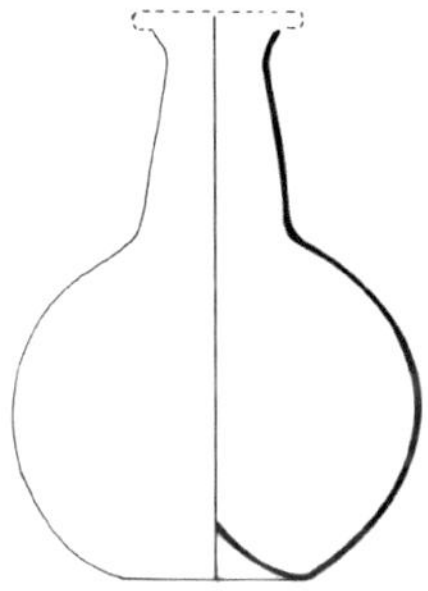

250

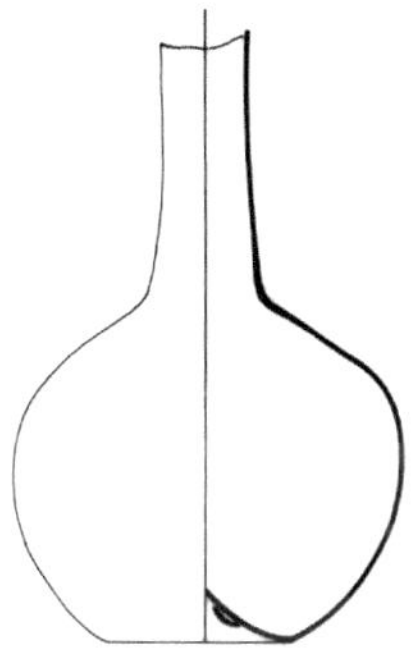

251

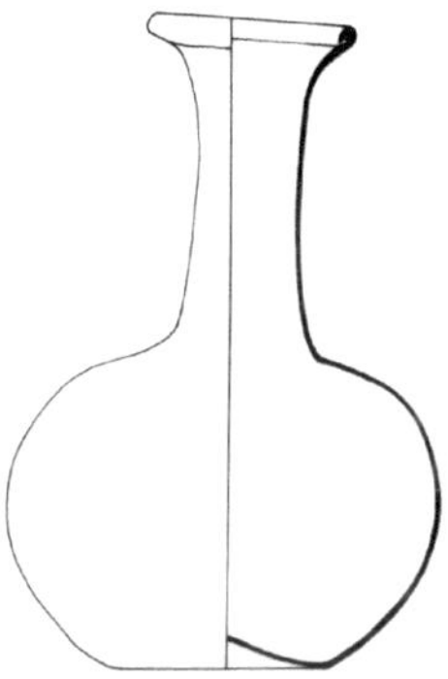

252

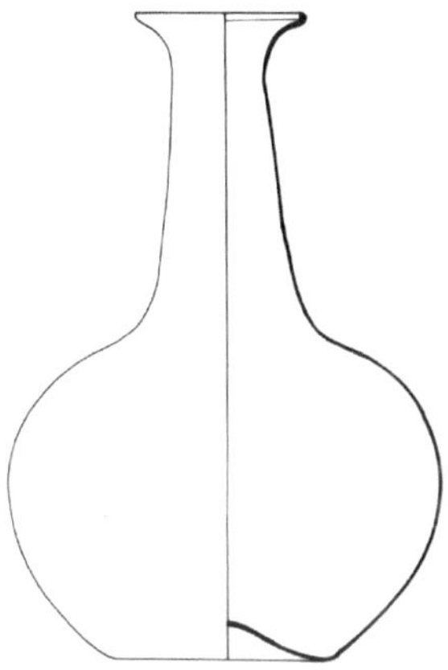

253

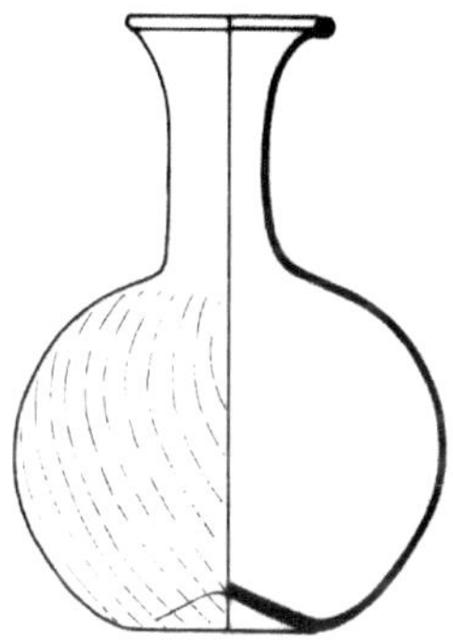

254

# Plate 13

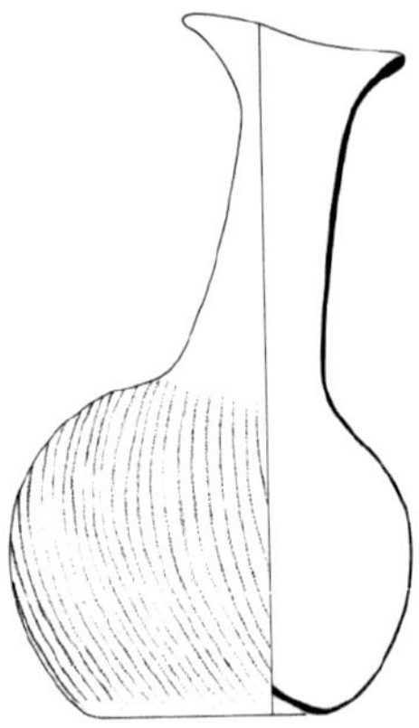
255

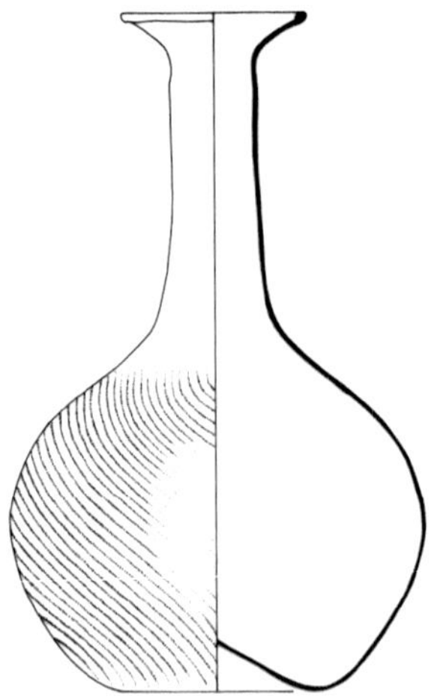
256

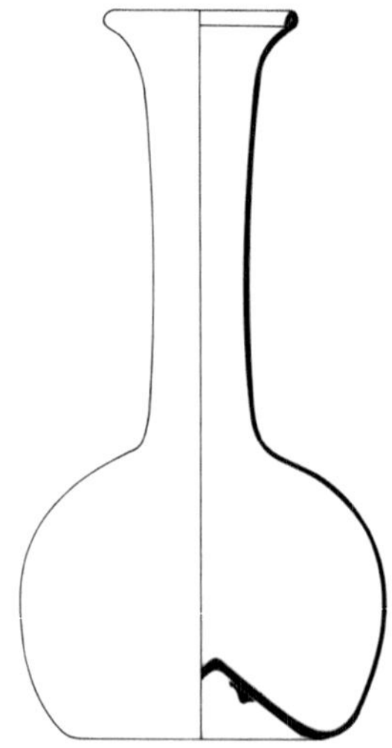
257

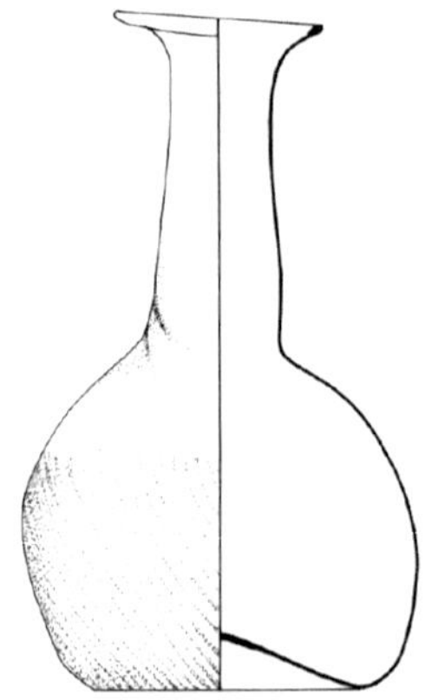
258

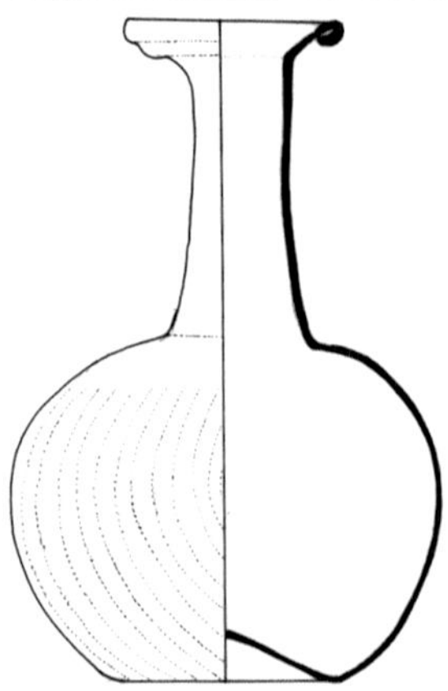
259

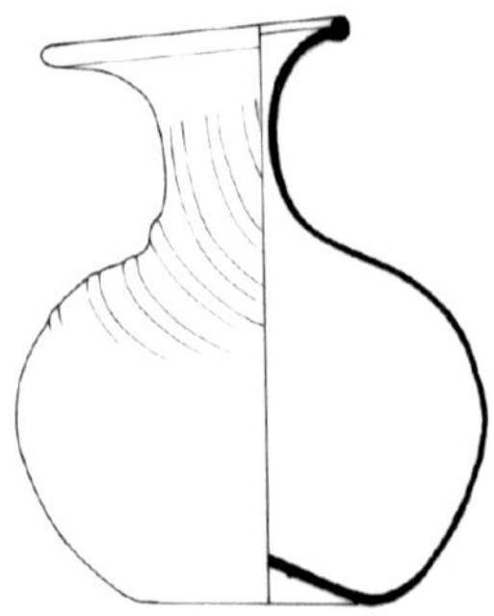
260

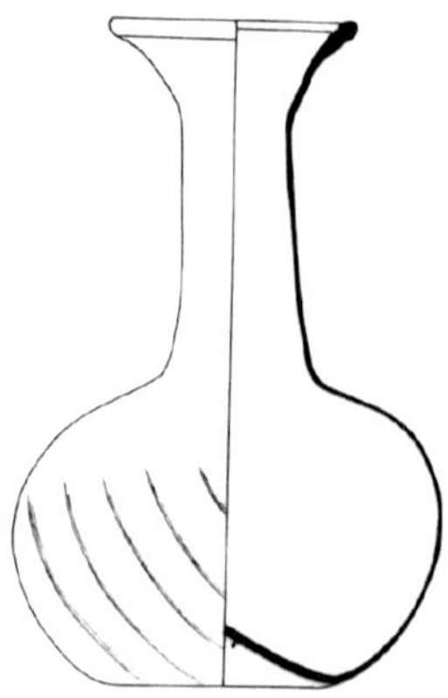
261

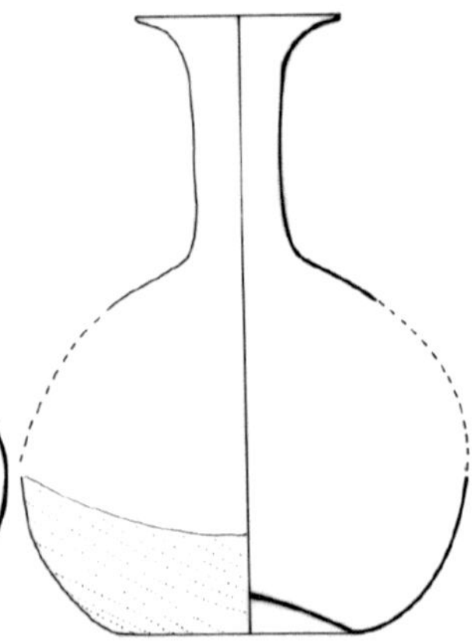
262

263

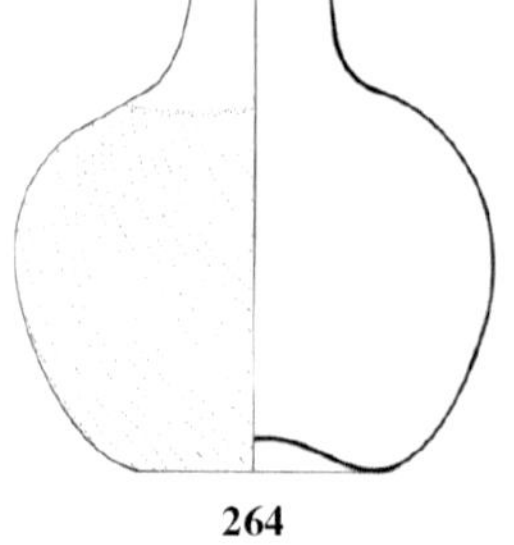
264

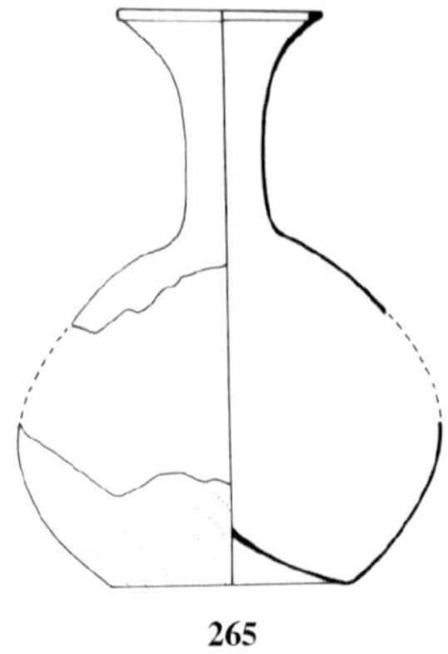
265

# Plate 14

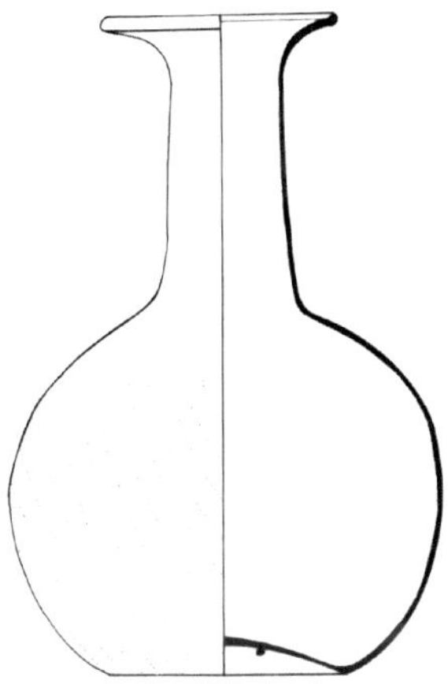

266

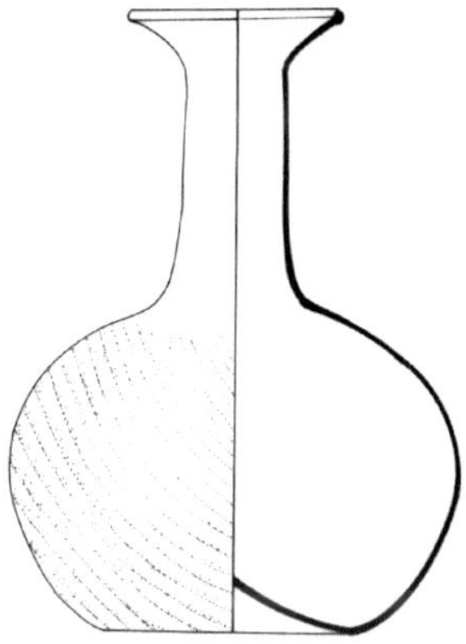

267

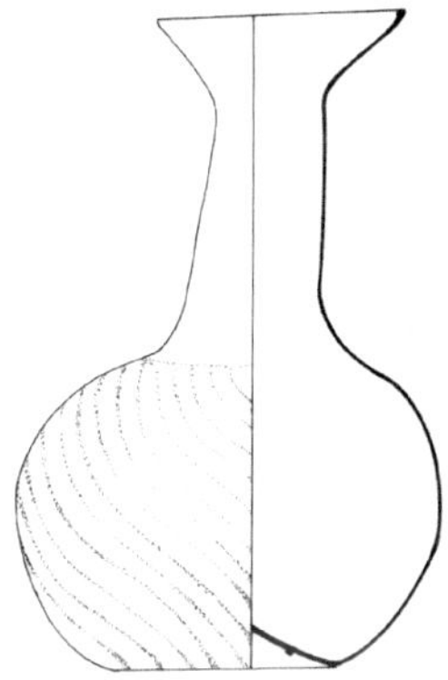

268

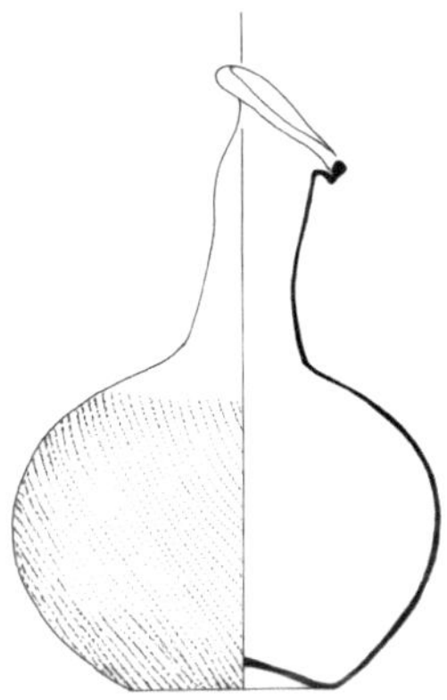

269

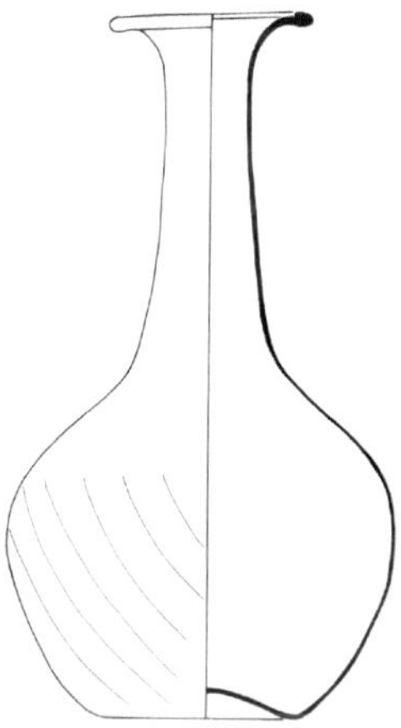

270

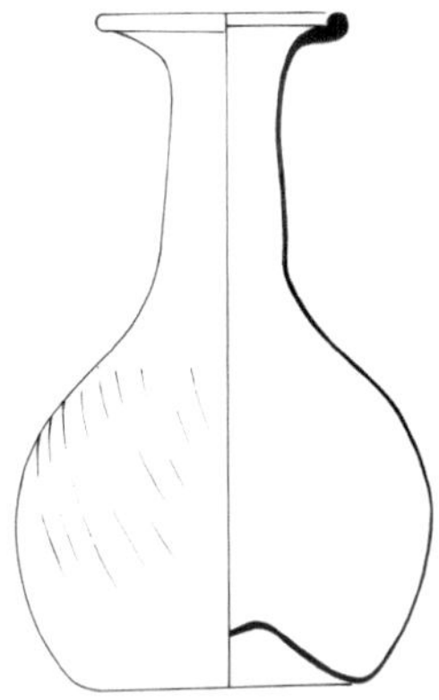

271

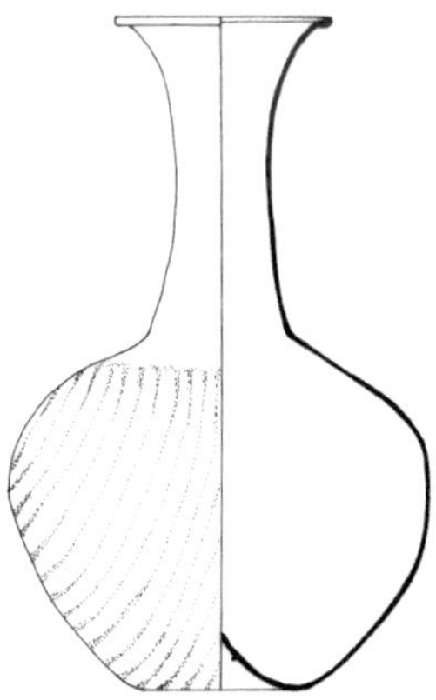

273

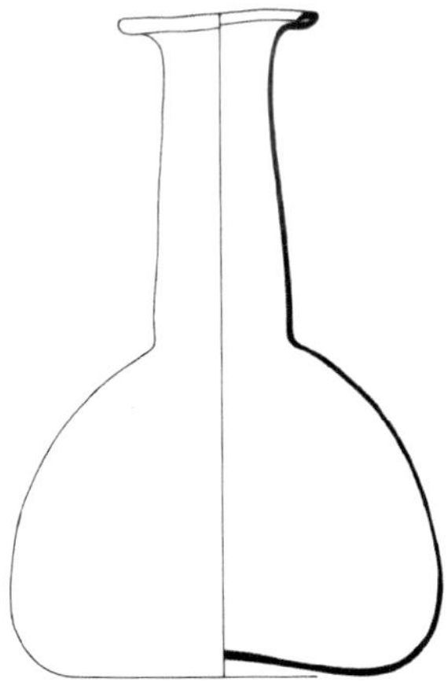

274

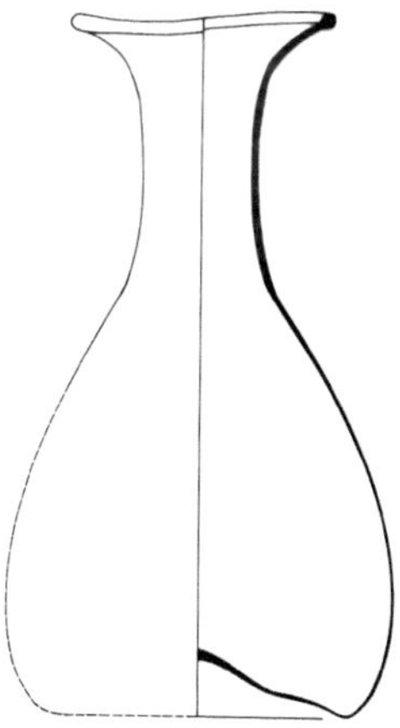

275

Plate 15

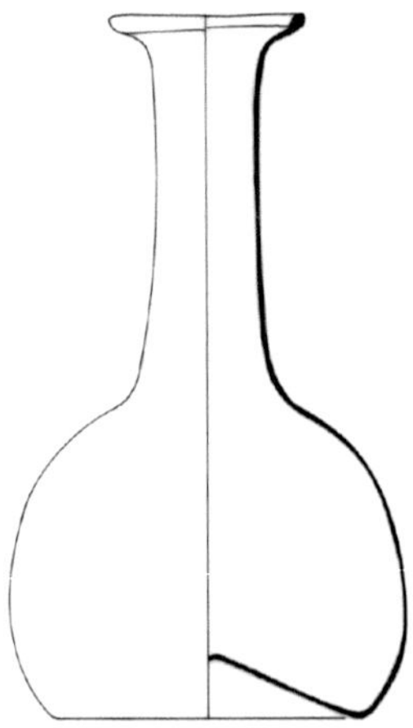
276

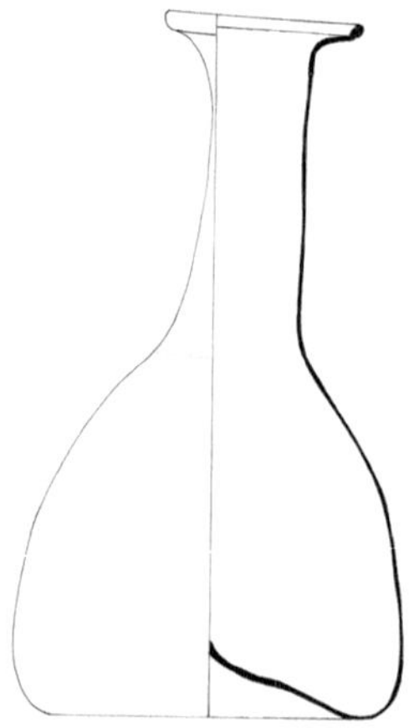
277

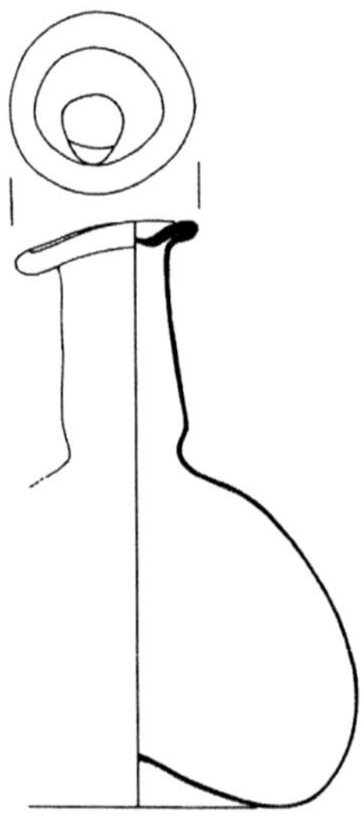
278

279

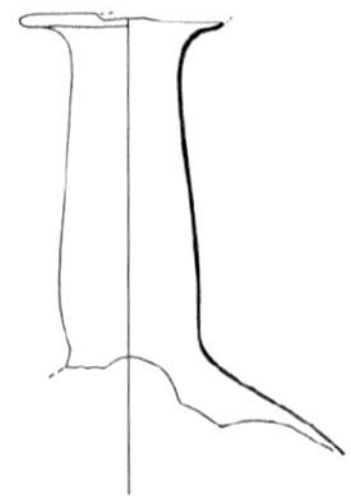
280

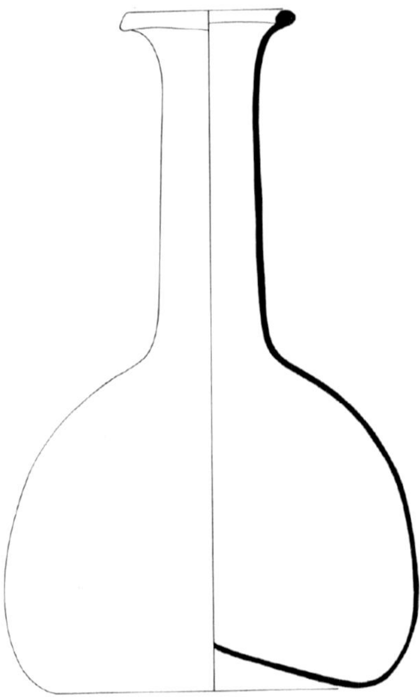
281

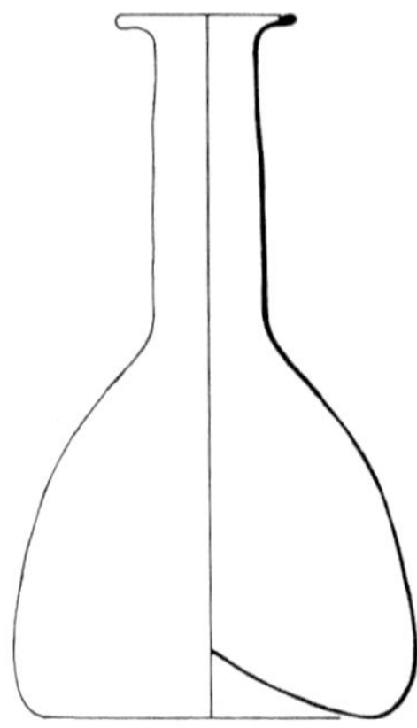
282

283

# Plate 16

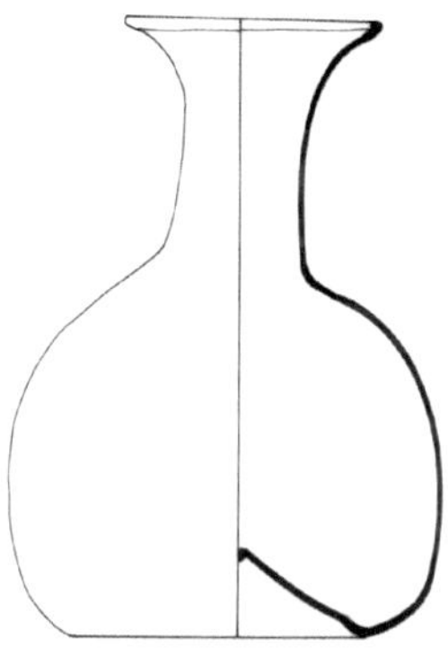

284

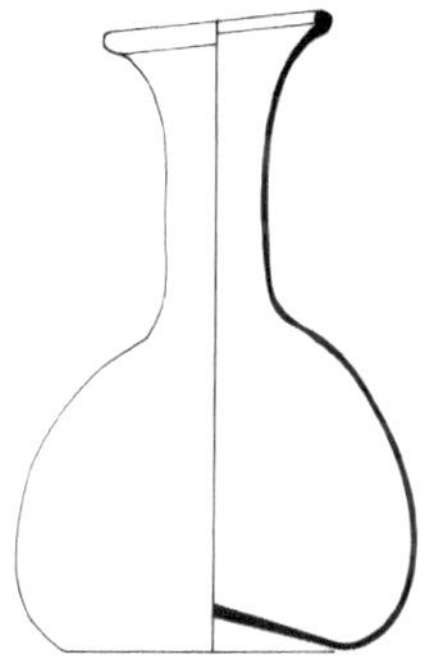

285

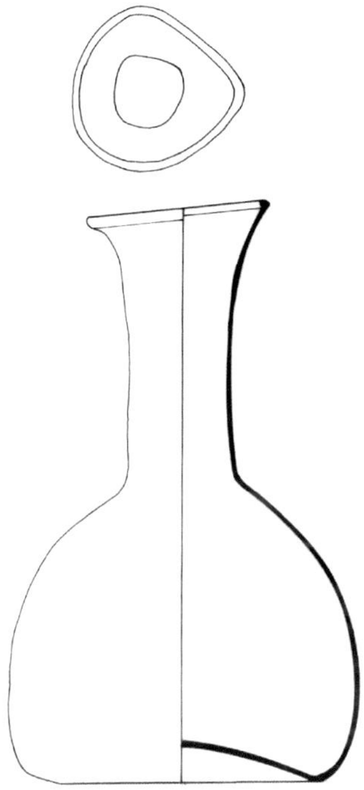

286

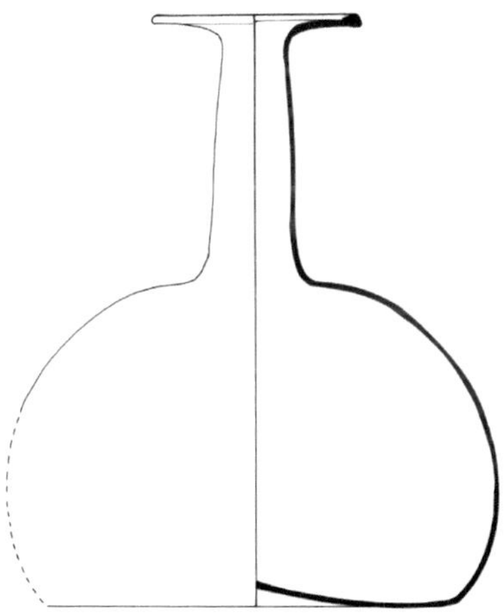

287

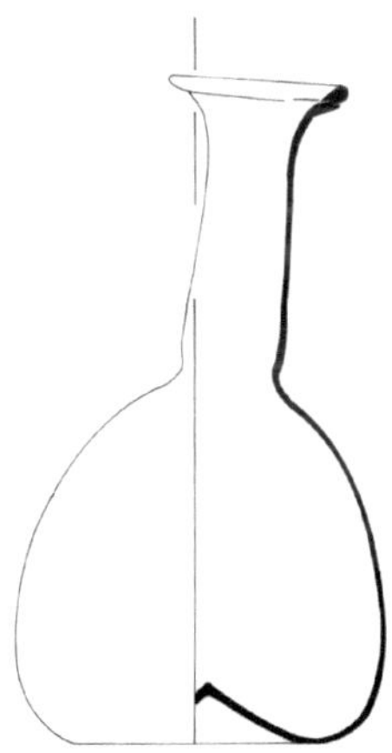

289

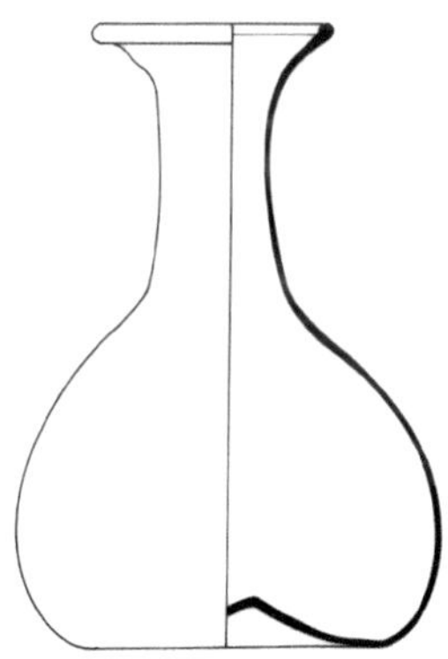

290

Plate 17

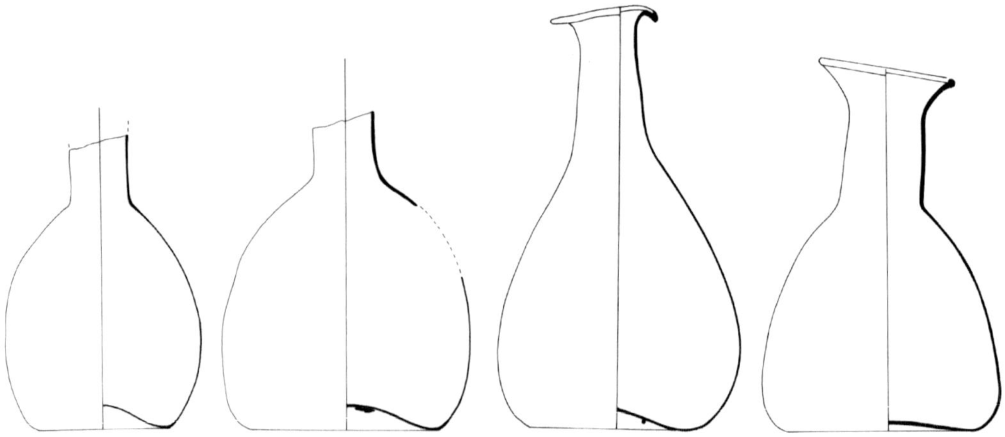

292 293 294 295

296 297 298 299

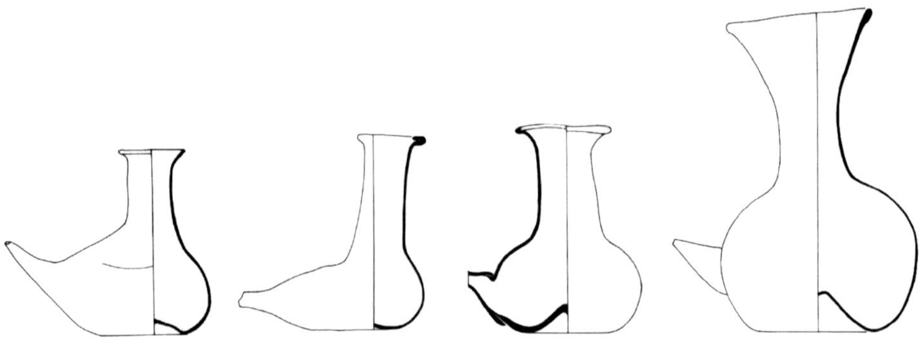

301 303 304 305

Plate 18

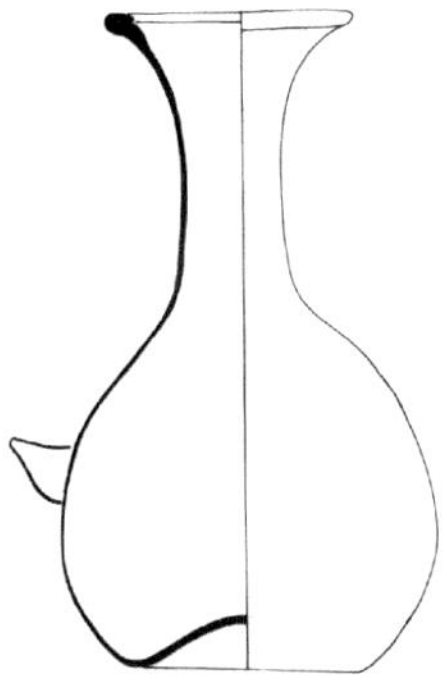
306

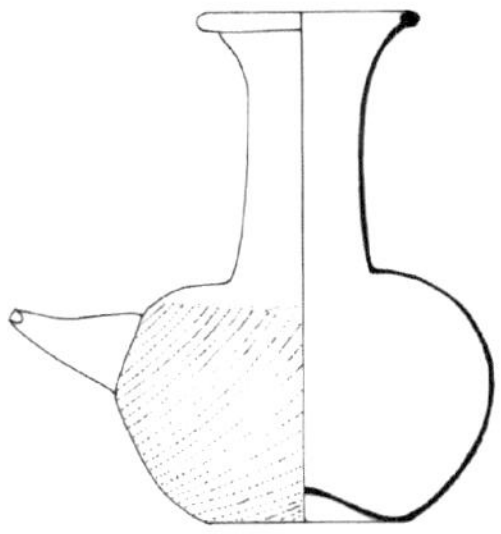
307

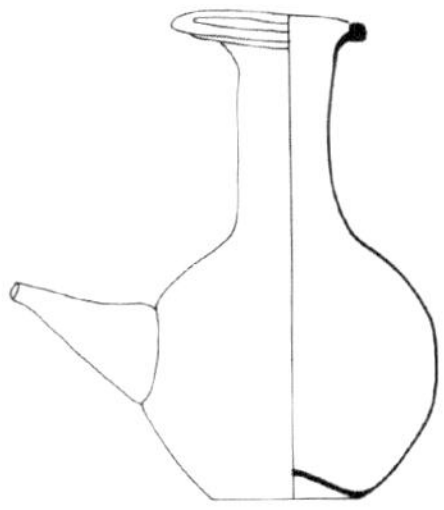
308

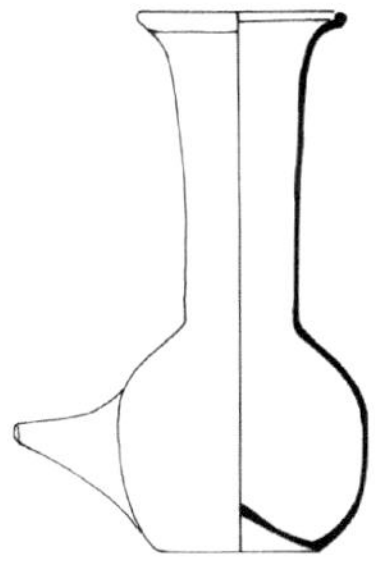
309

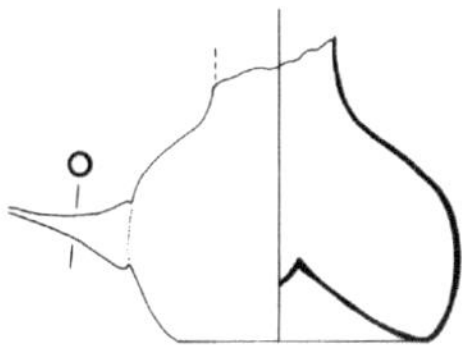
310

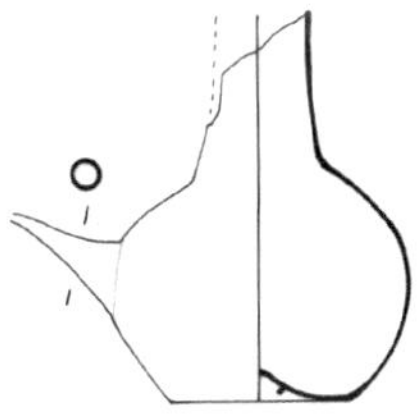
311

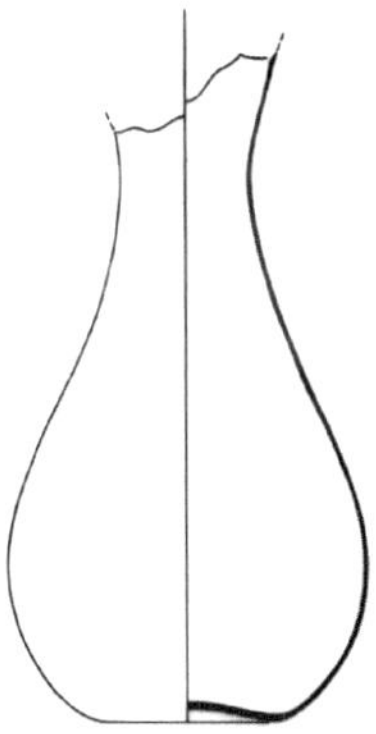
314

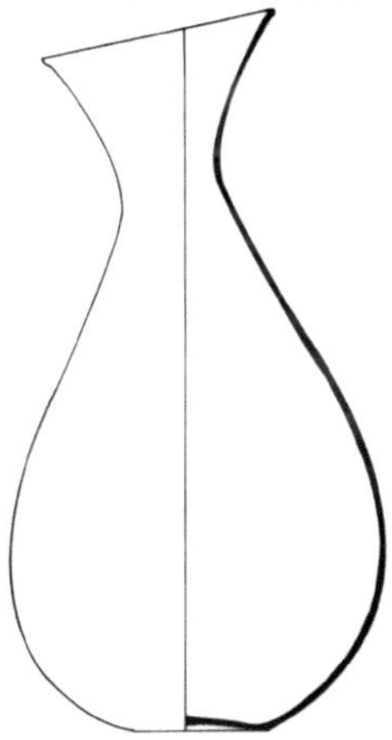
315

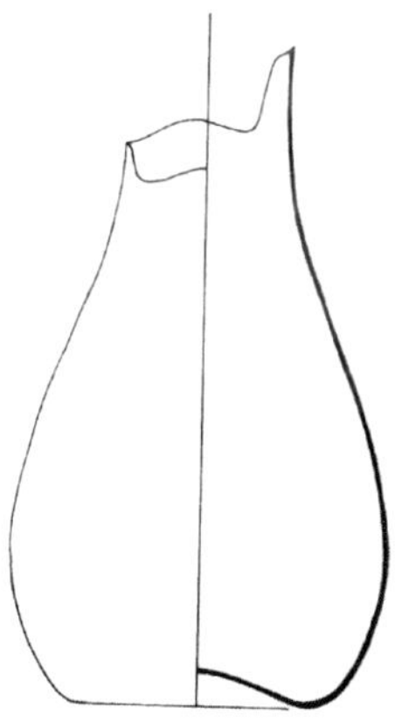
316

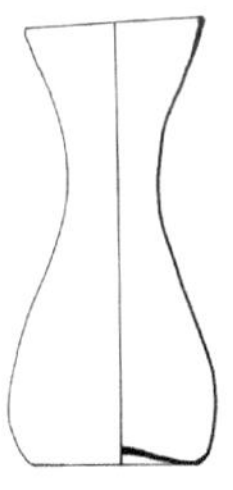
317

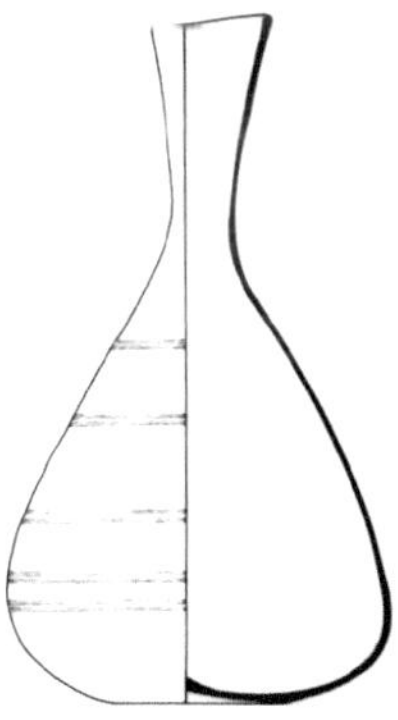
318

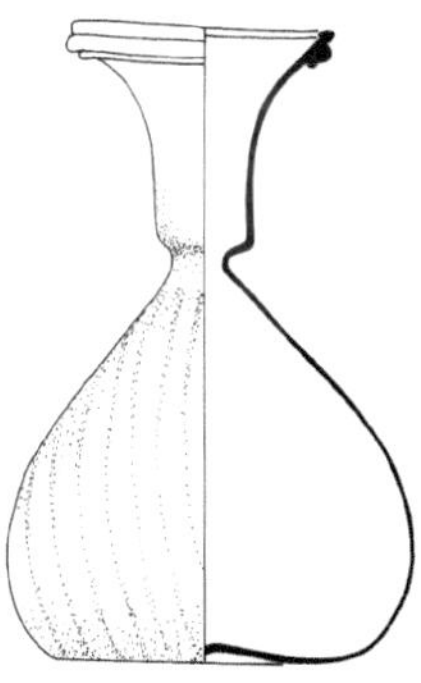
319

# Plate 19

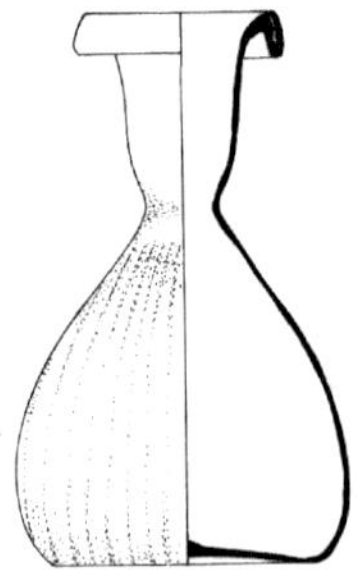
320

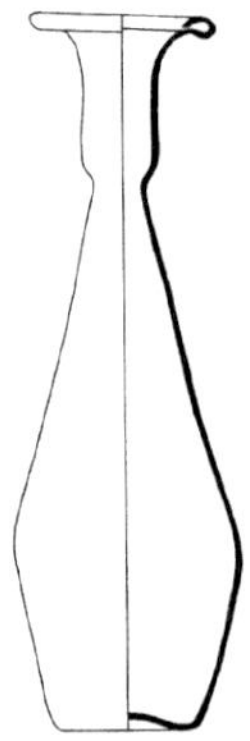
321

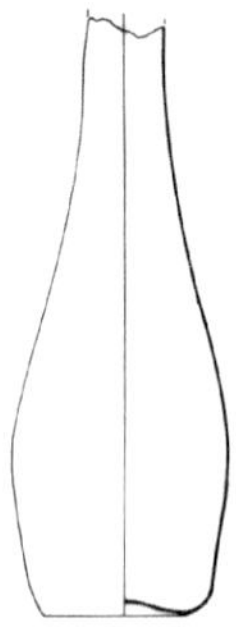
322

323

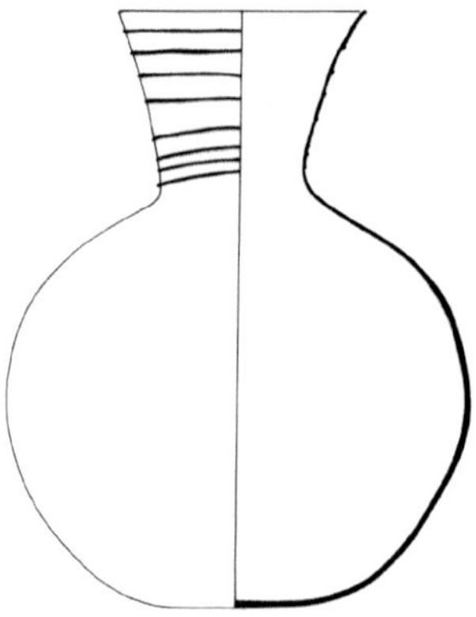
324

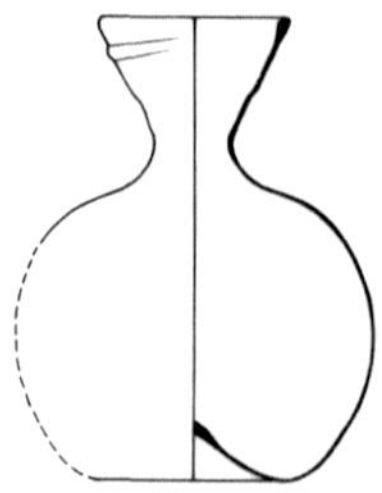
325

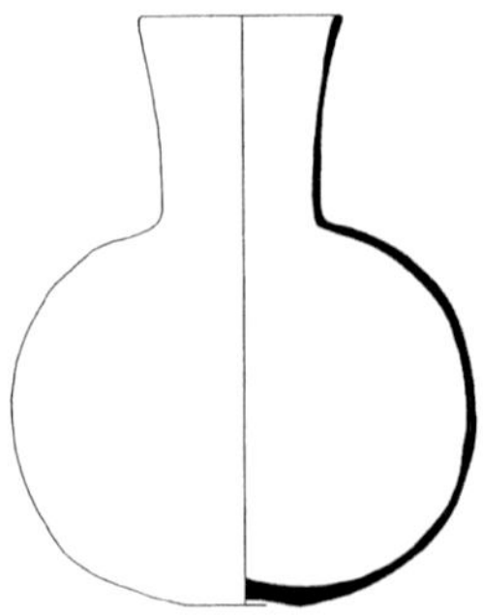
326

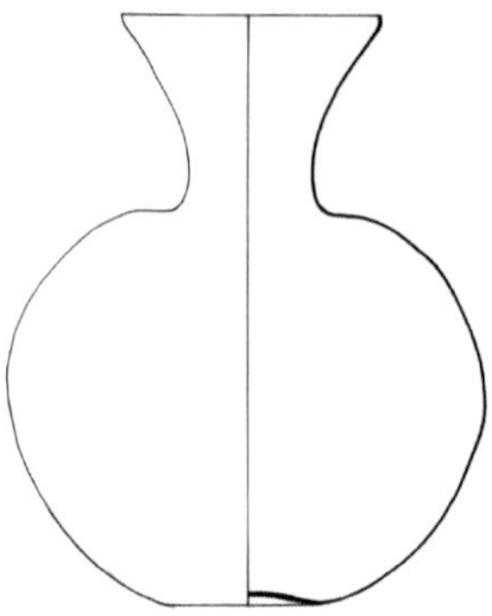
327

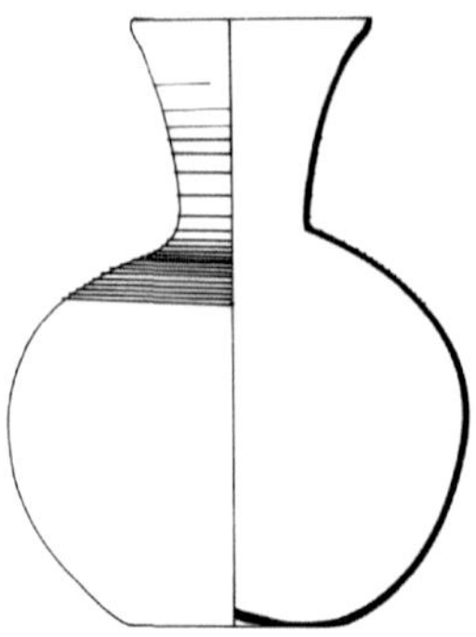
328

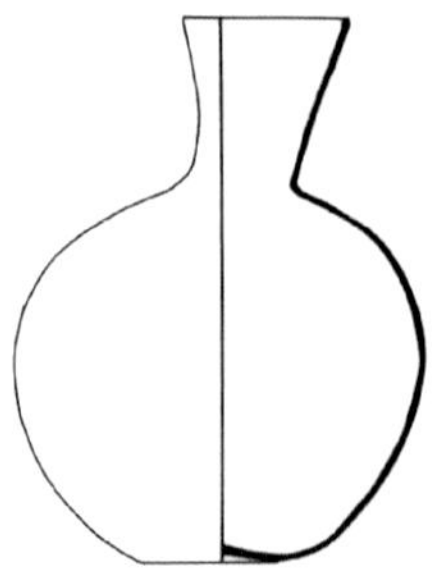
329

# Plate 20

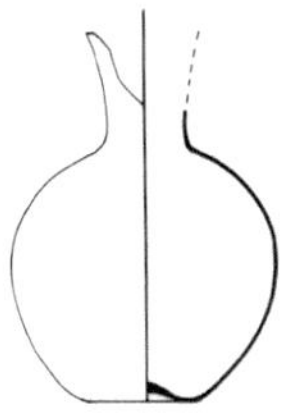

330

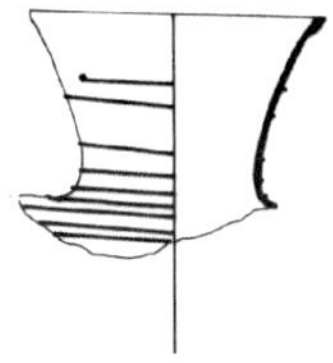

331

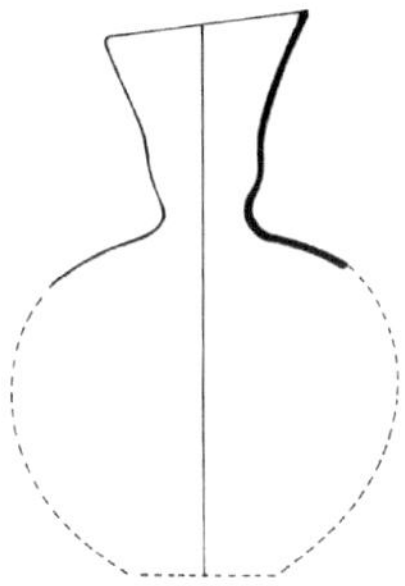

333

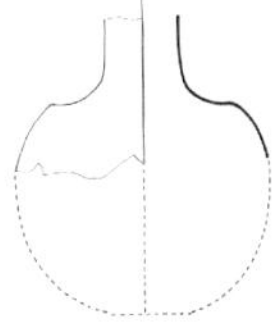

334

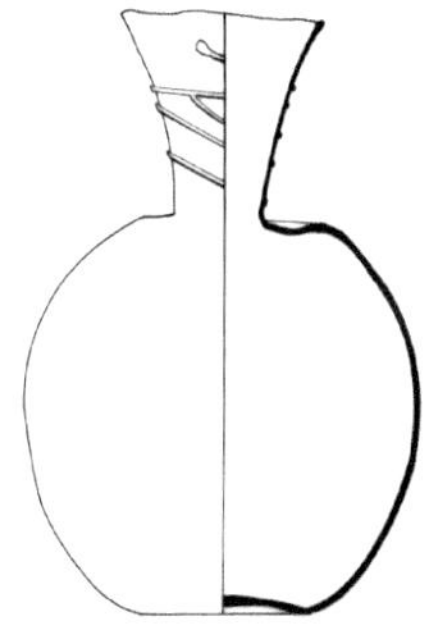

335

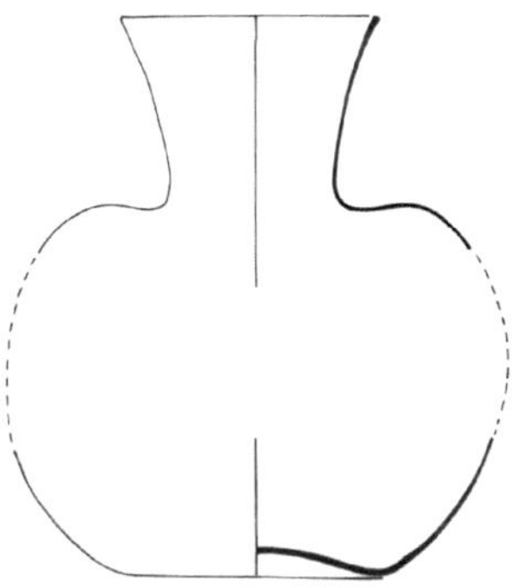

336

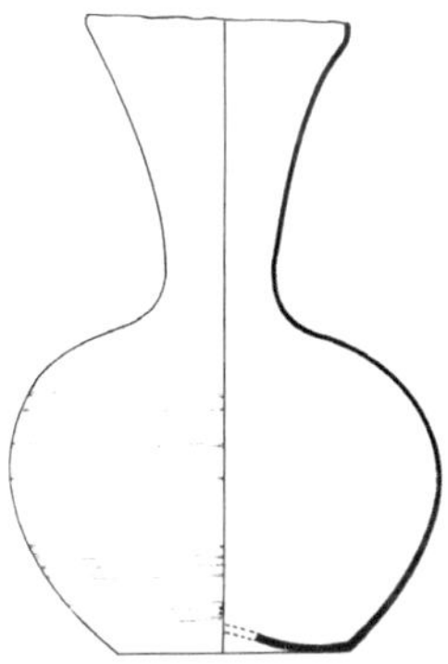

337

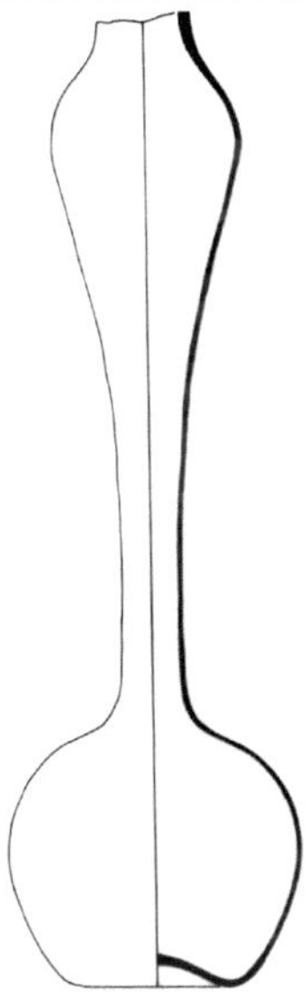

338

339

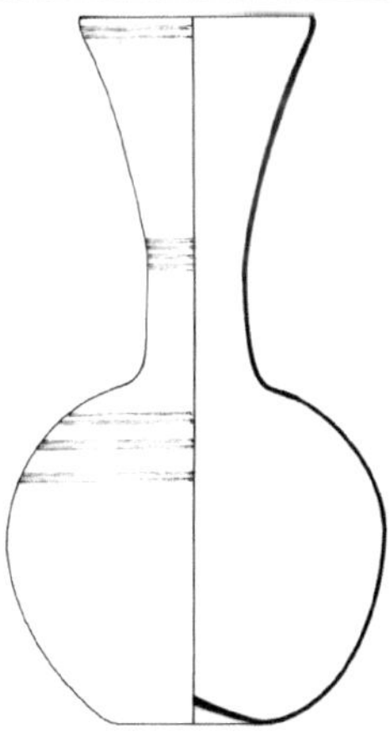

340

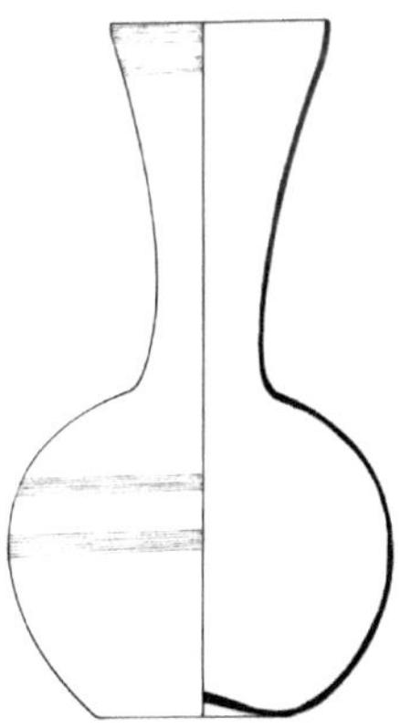

341

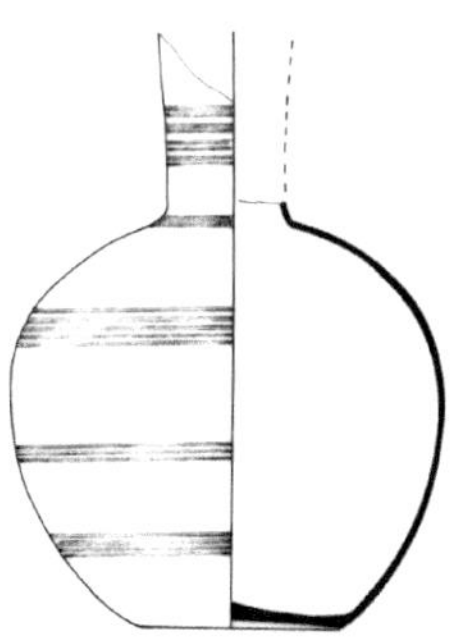

342

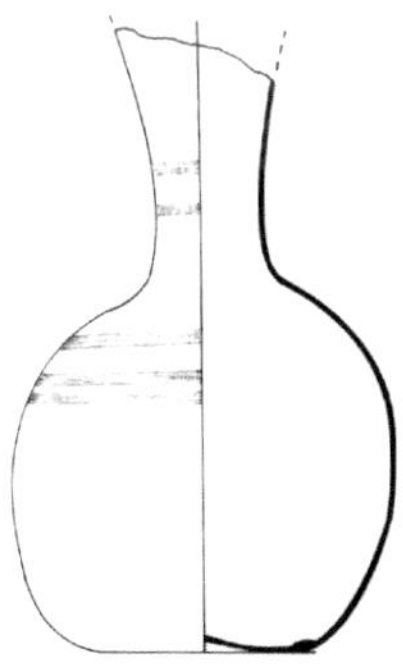

343

# Plate 21

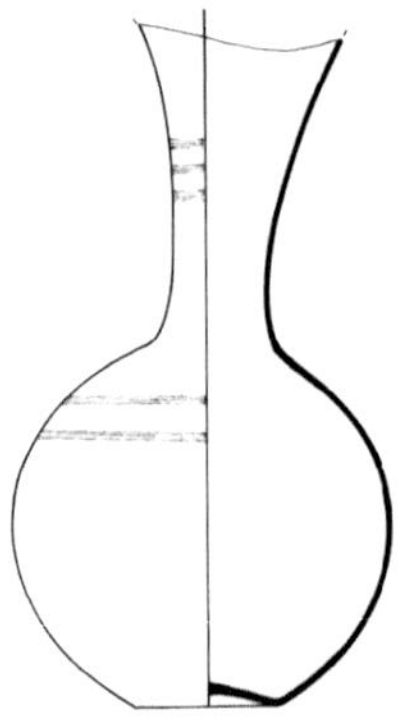

344

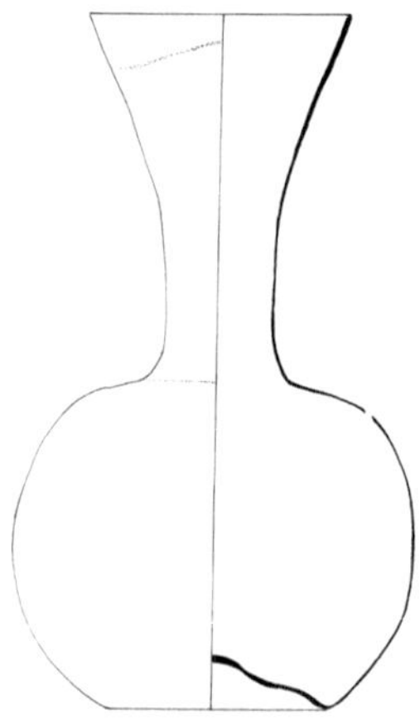

345

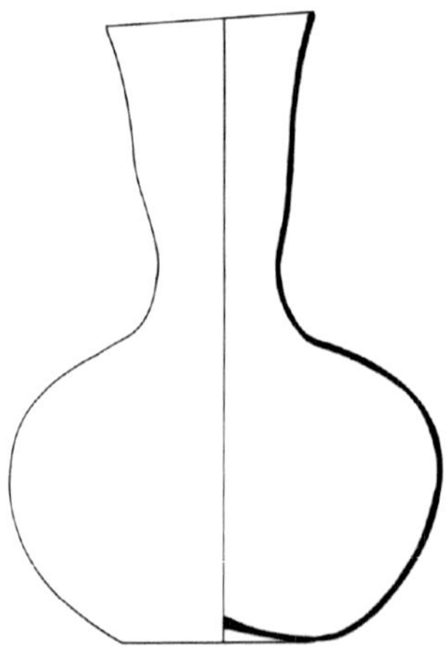

346

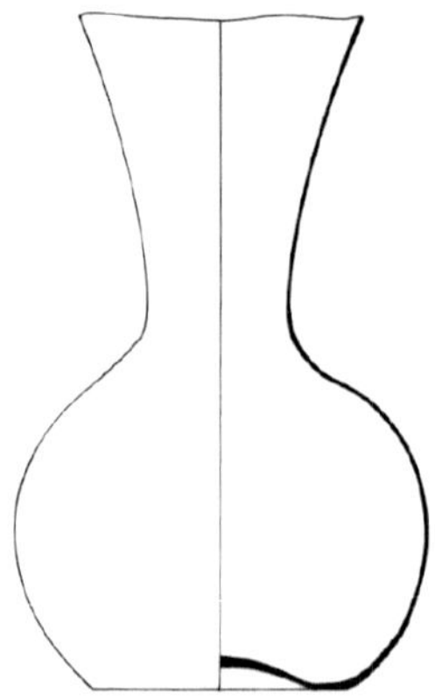

347

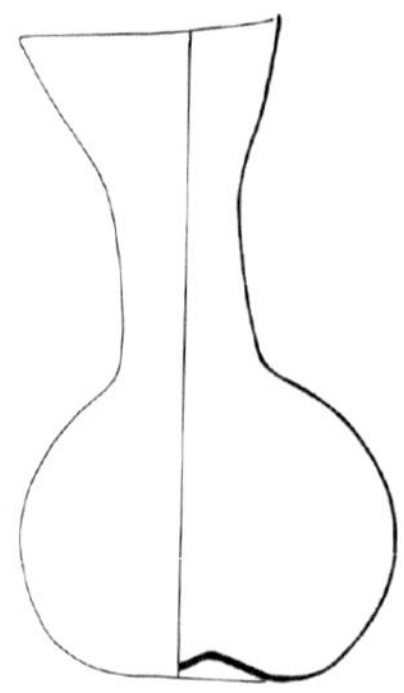

348

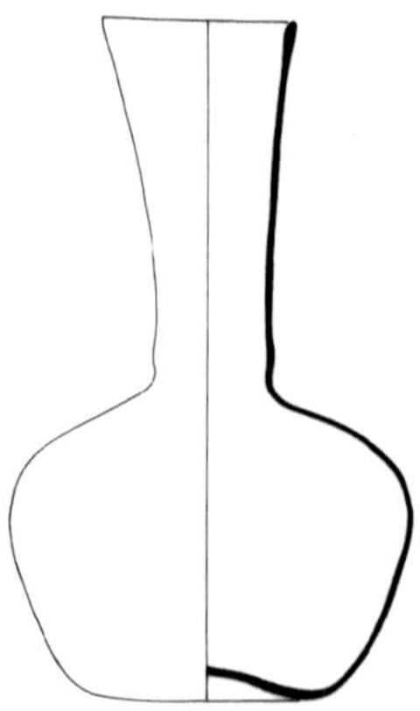

349

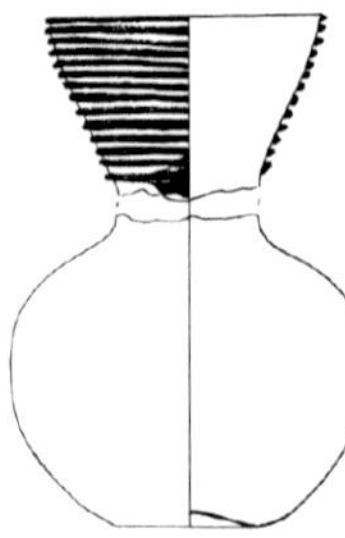

350

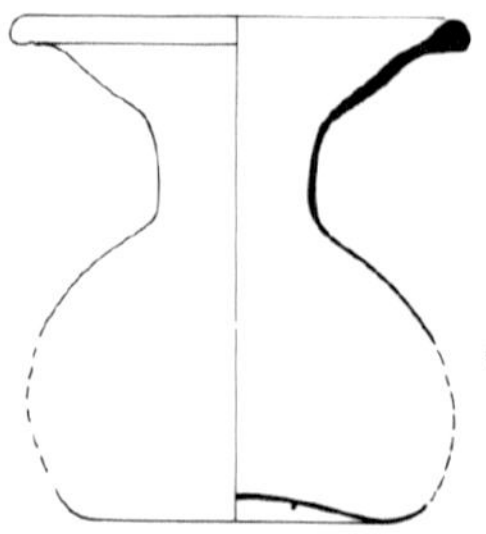

351

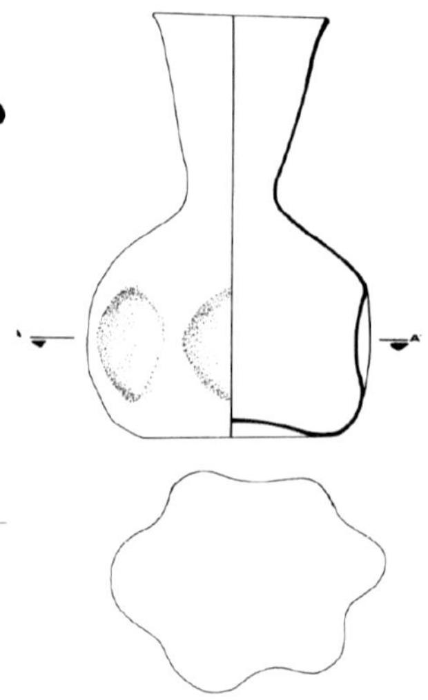

352

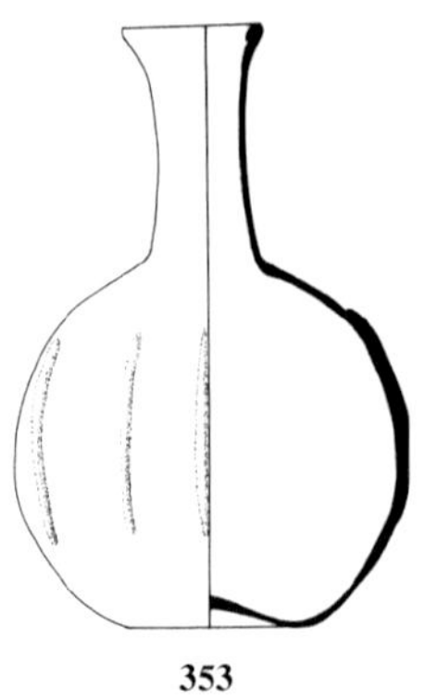

353

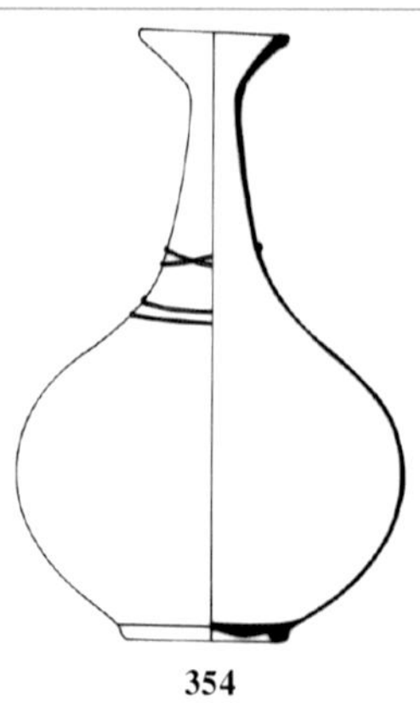

354

# Plate 22

355

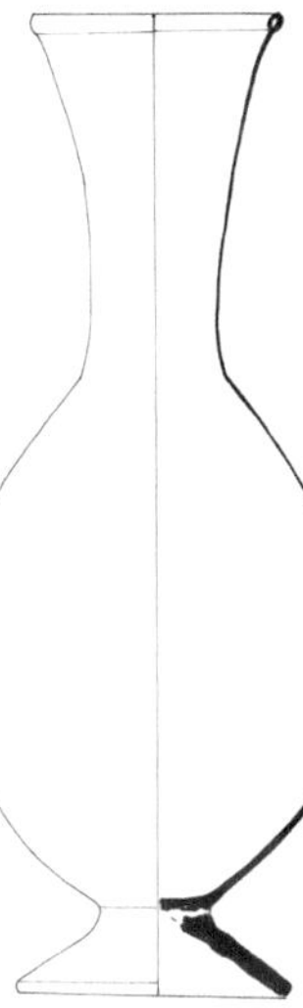
356

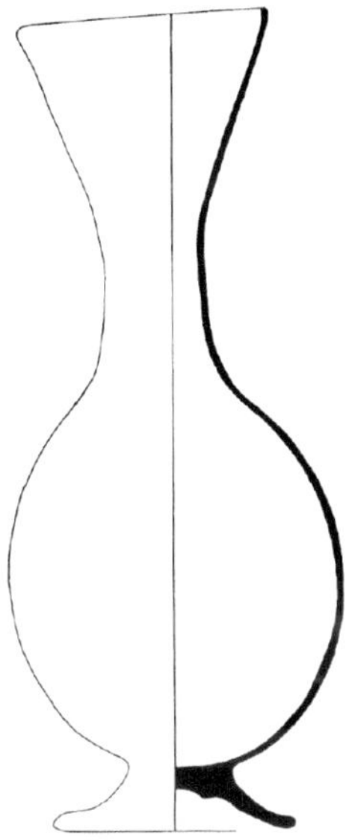
357

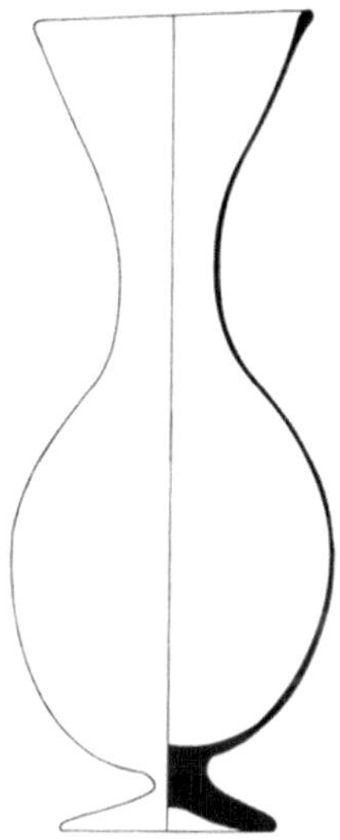
358

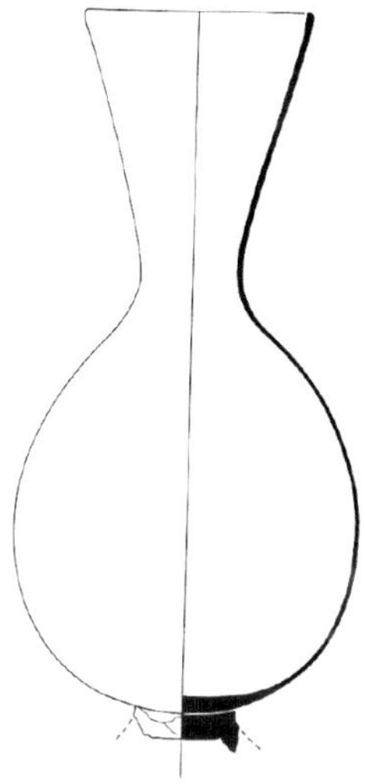
359

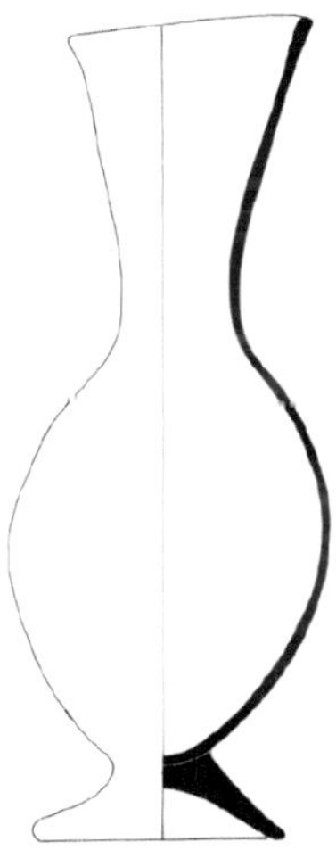
360

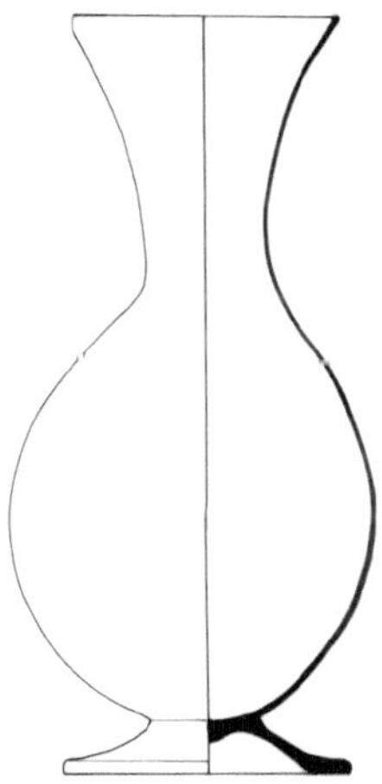
361

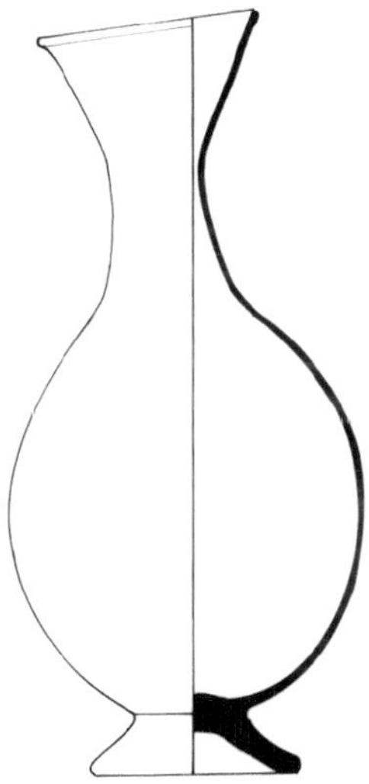
362

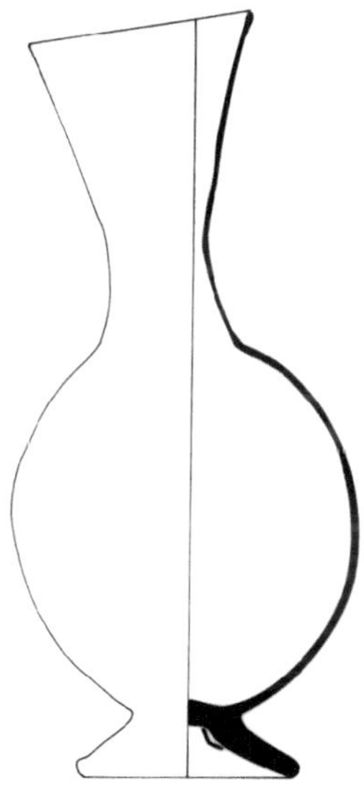
363

## Plate 23

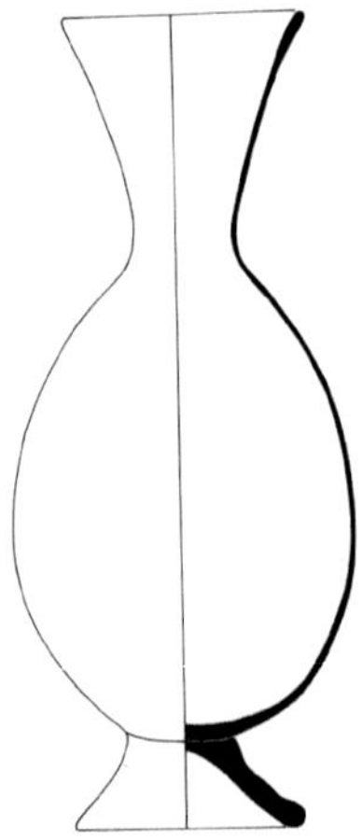

**364**

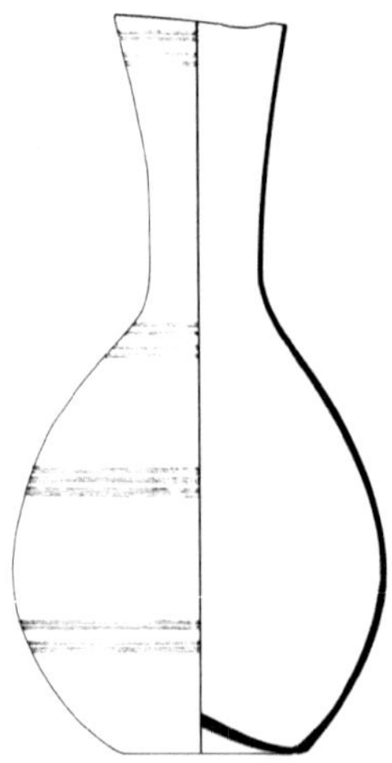

**367**

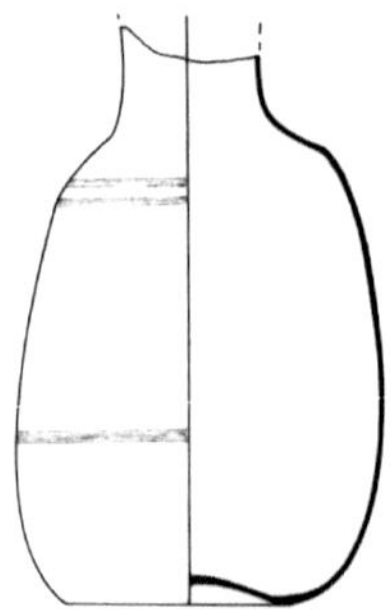

**368**

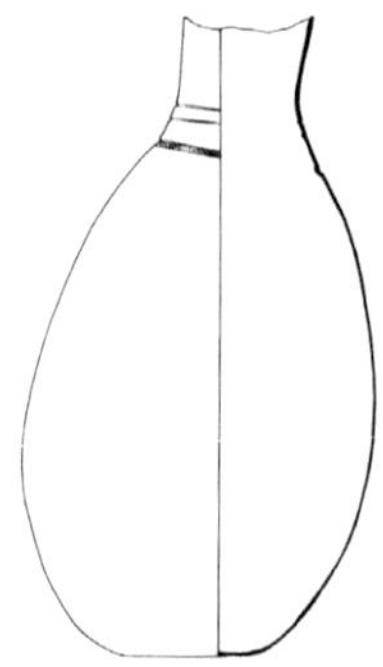

**369**

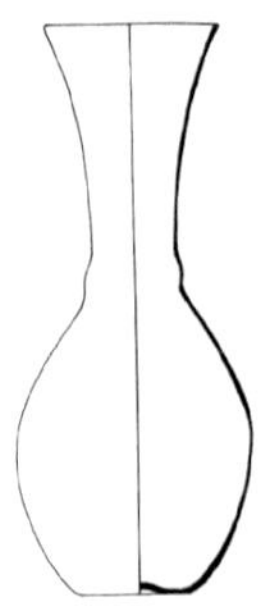

**371**

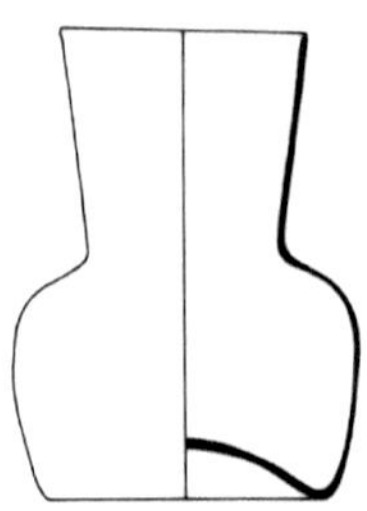

**373**

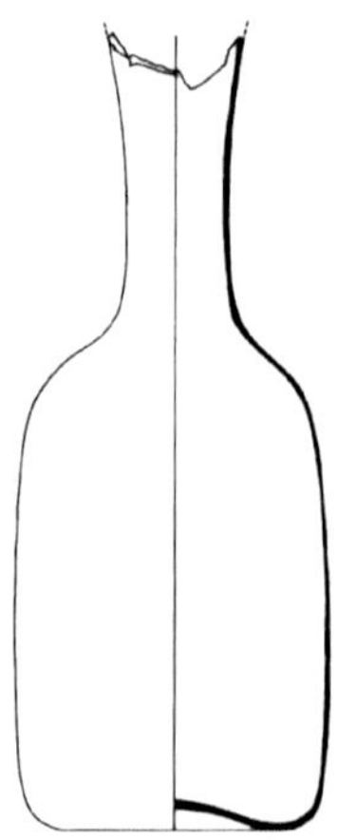

**374**

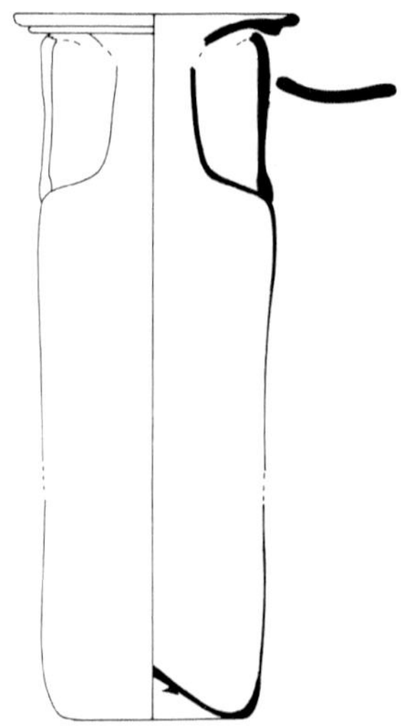

**375**

# Plate 24

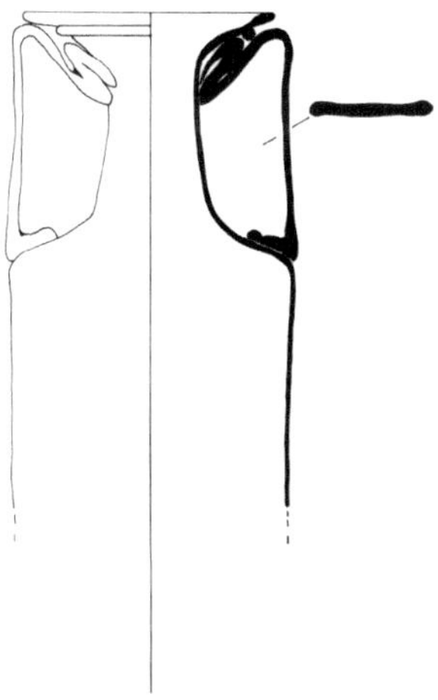
376

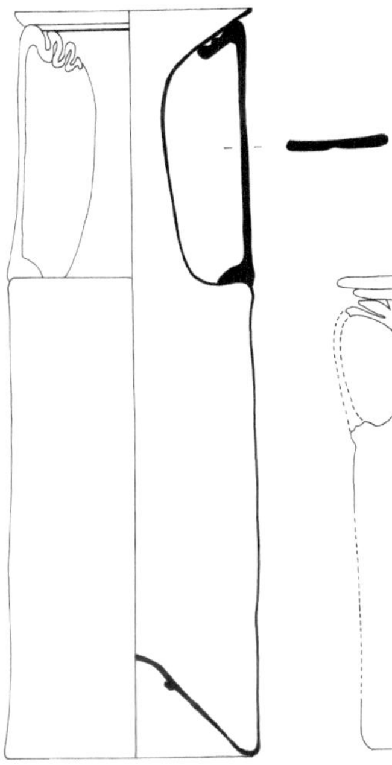
377

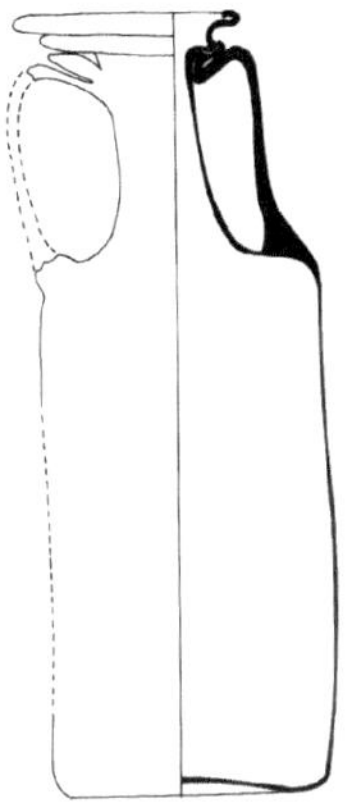
378

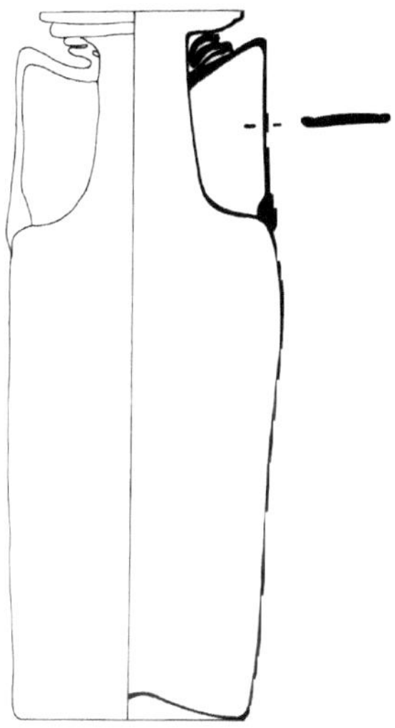
380

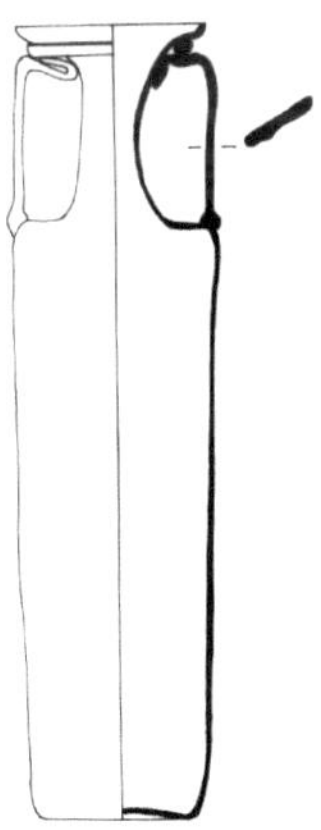
381

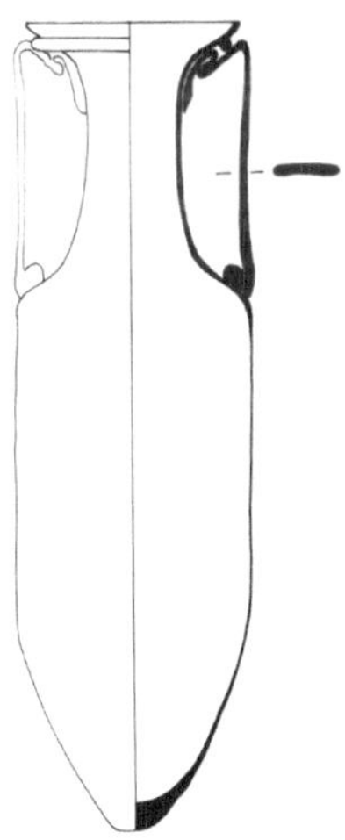
382

383

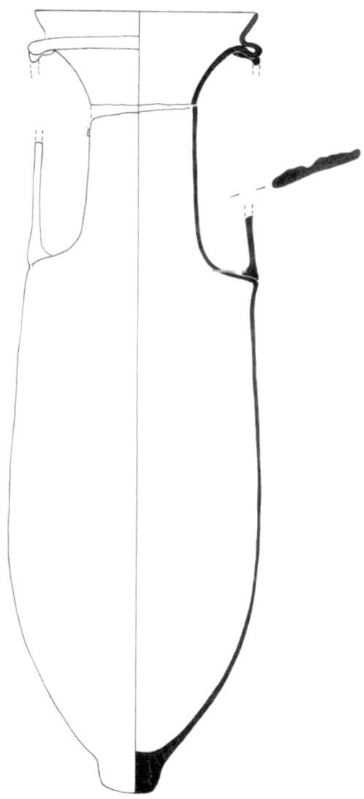
384

# Plate 25

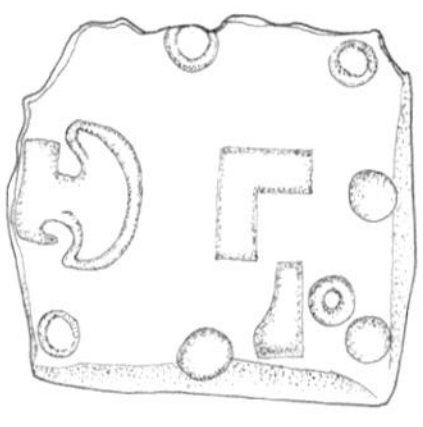
391

392

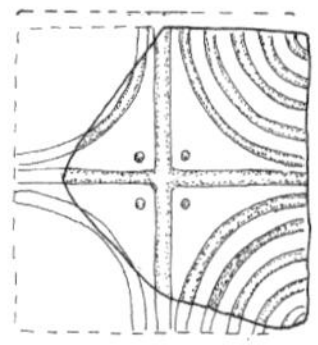
393

394

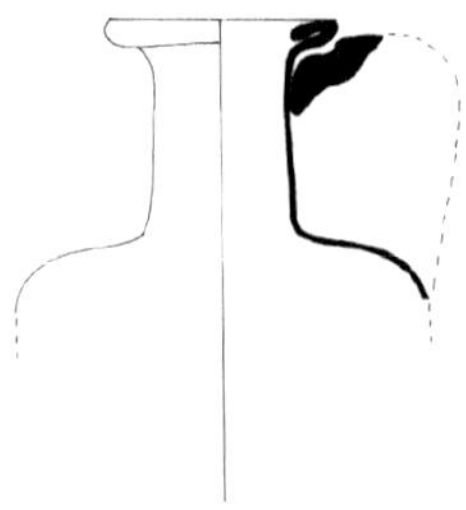
395

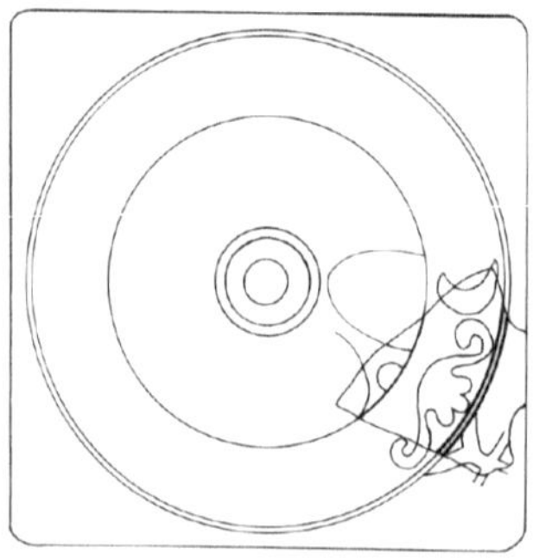
396

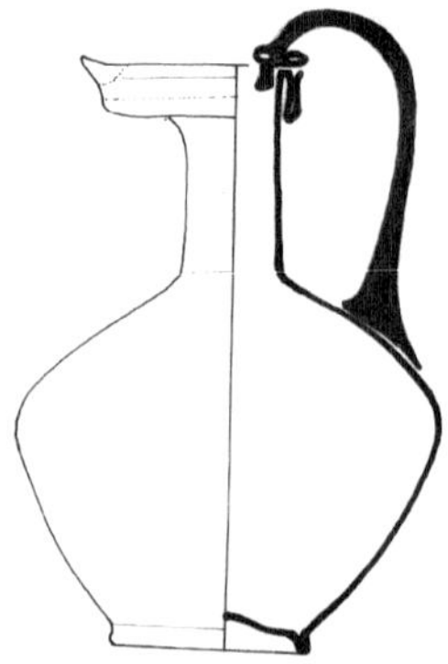
397

403

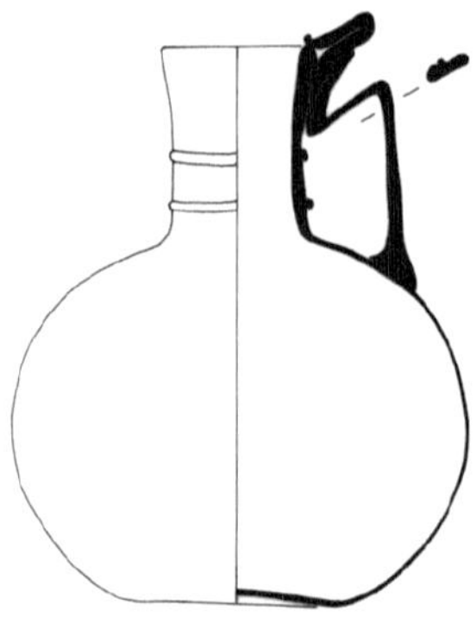
404

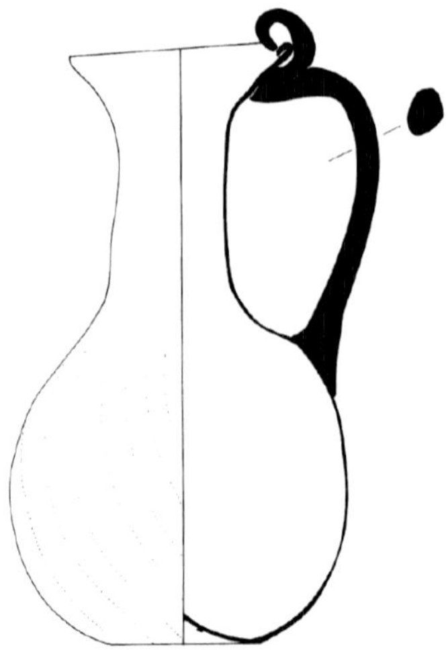
405

# Plate 26

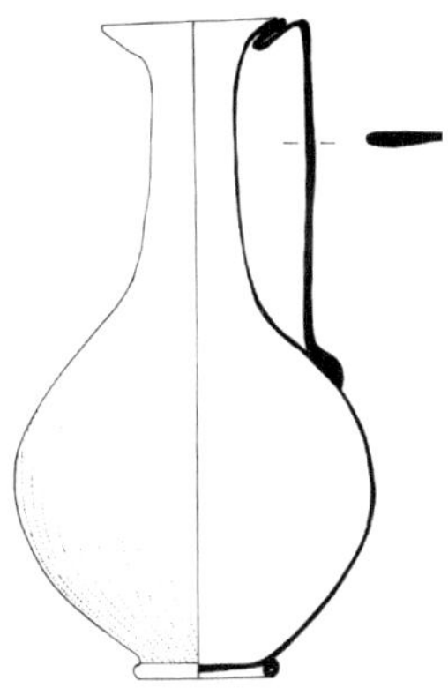

406

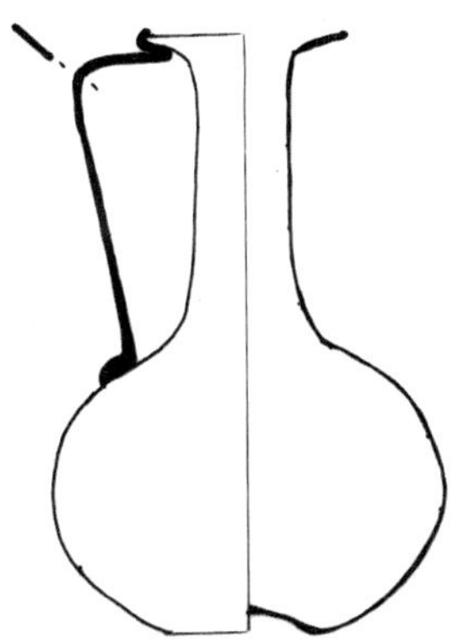

407

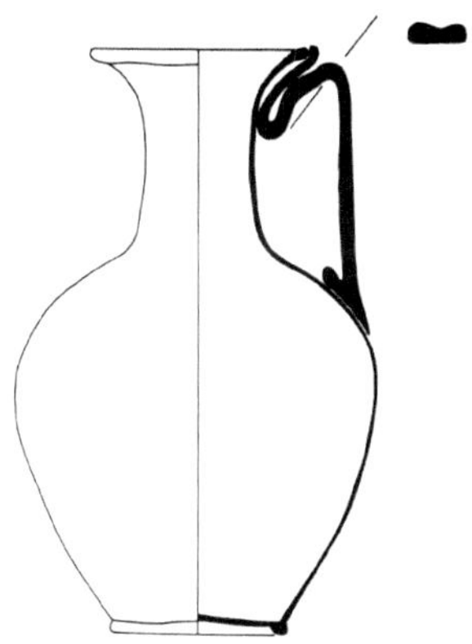

408

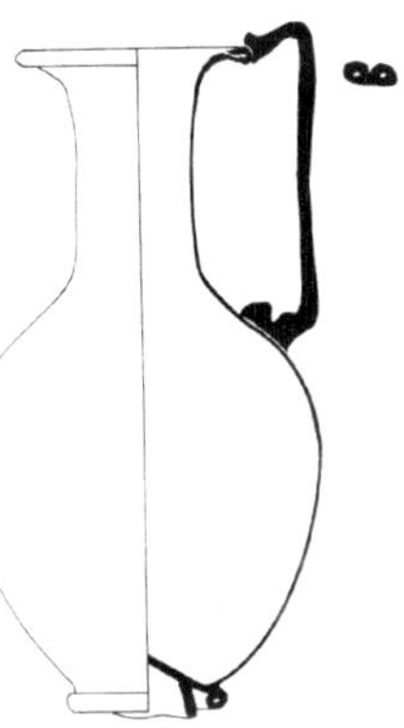

409

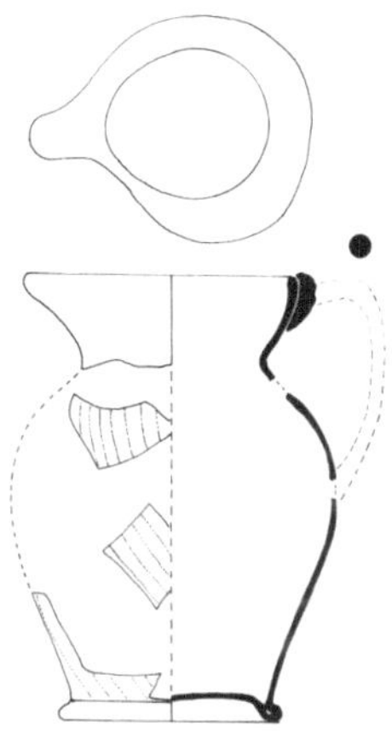

410

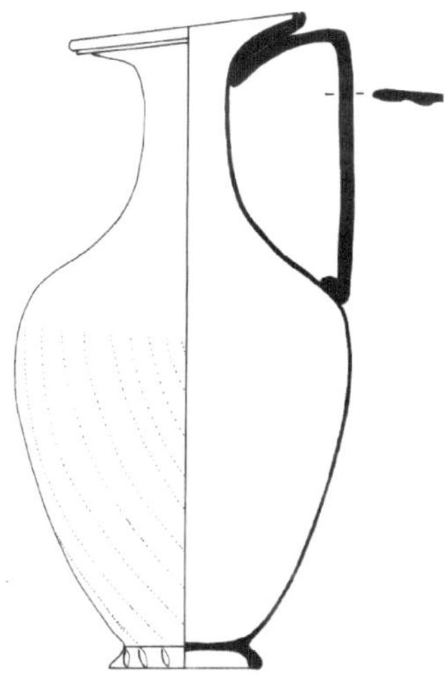

411

# Plate 27

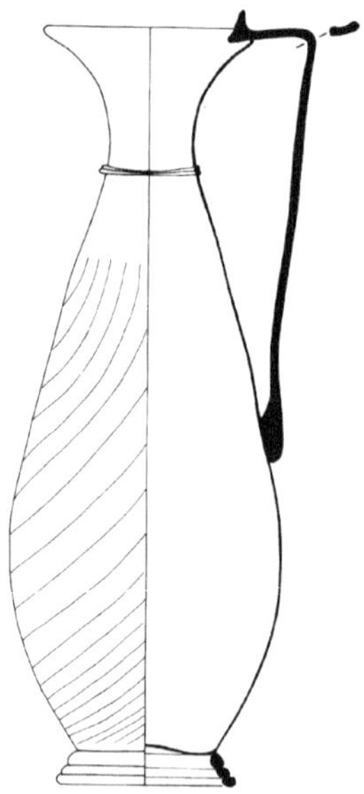

412

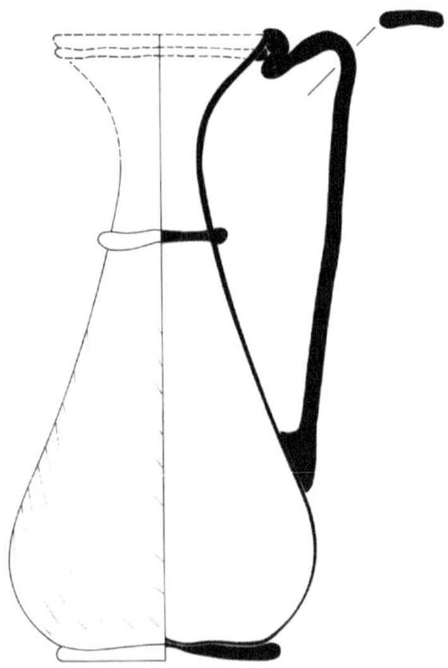

413

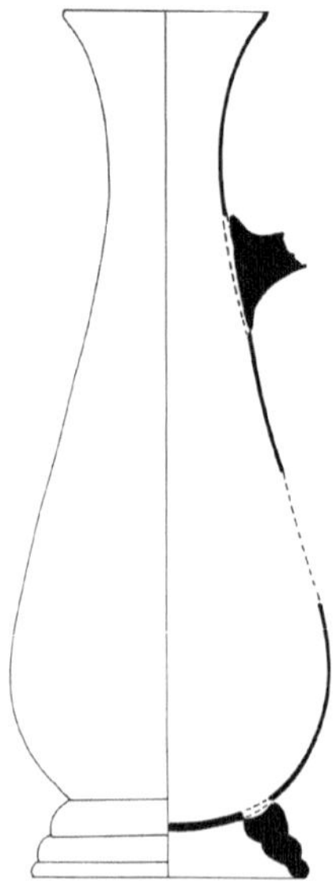

414

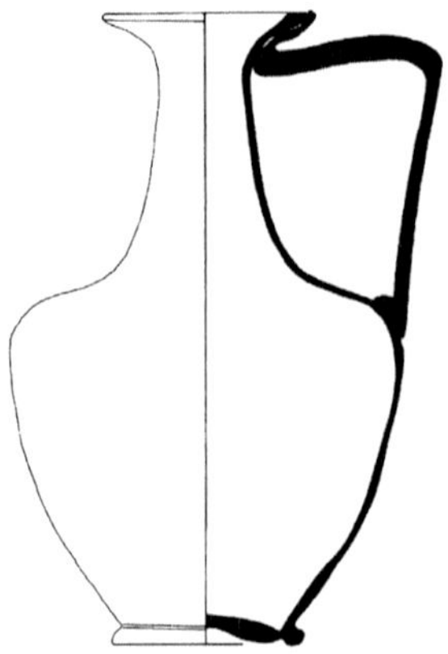

415

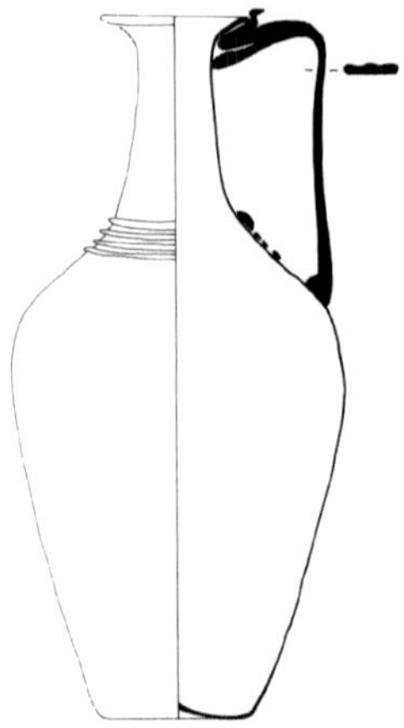

416

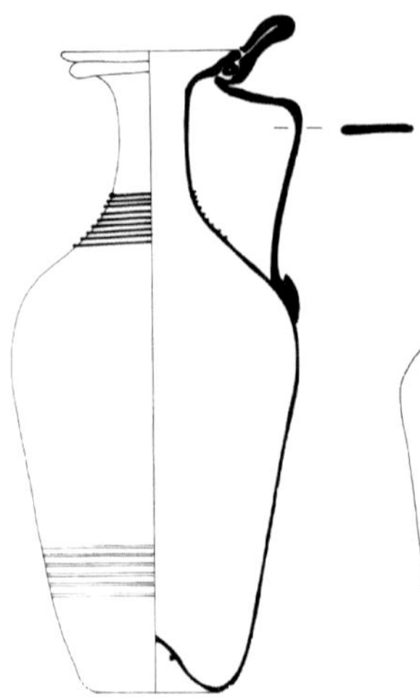

417

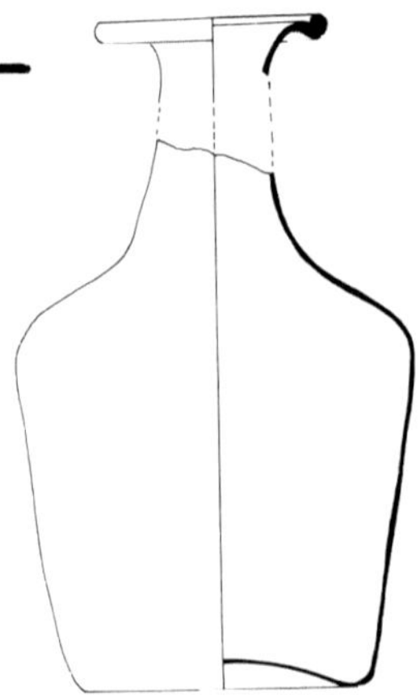

418

# Plate 28

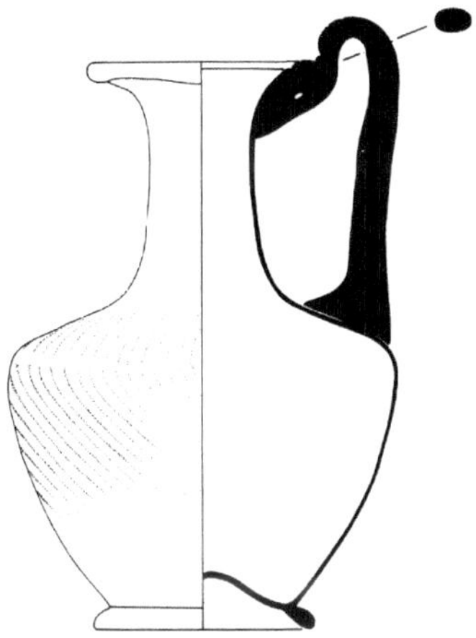

419

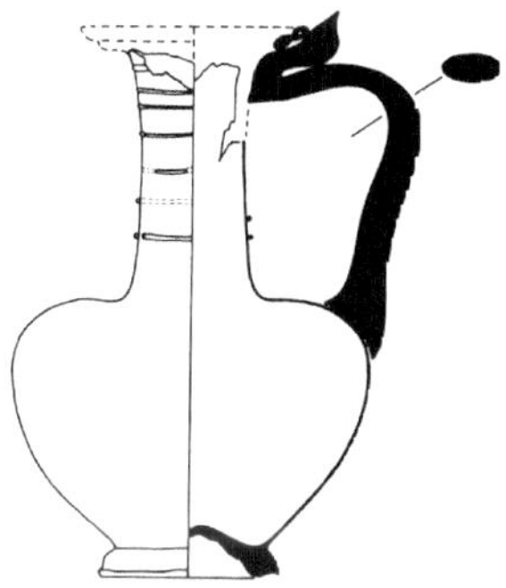

420

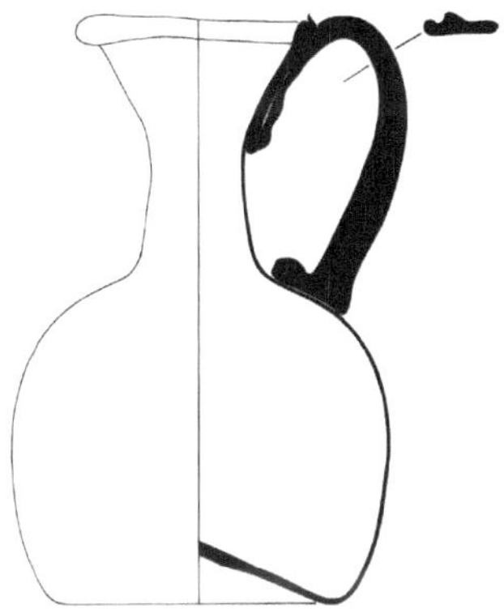

421

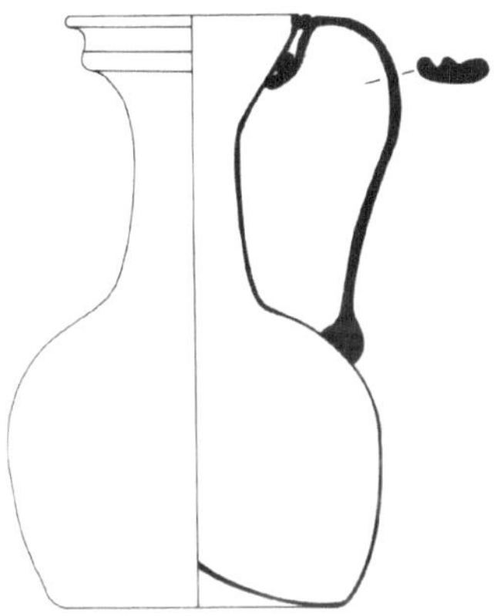

422

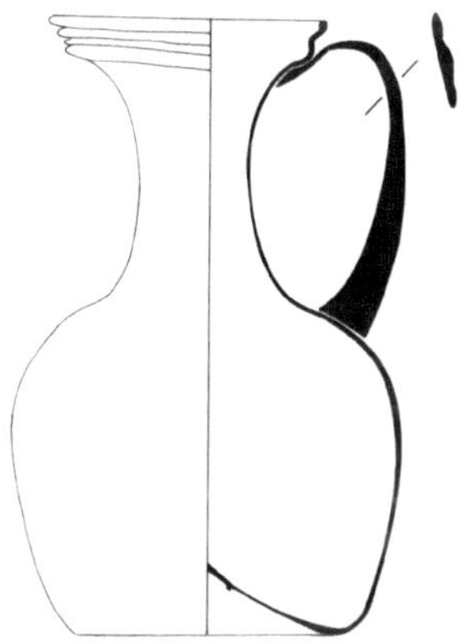

423

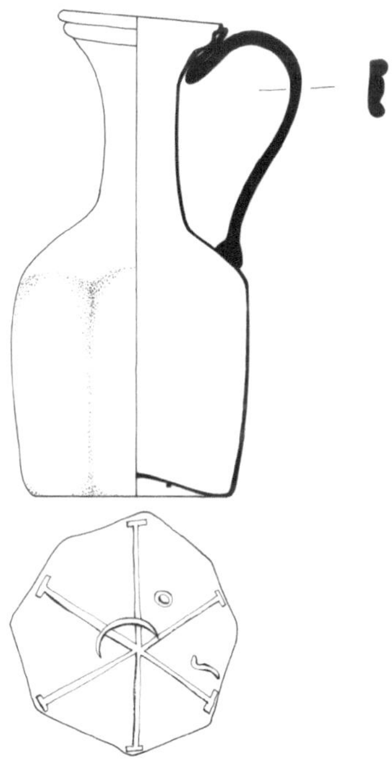

424-425

Plate 29

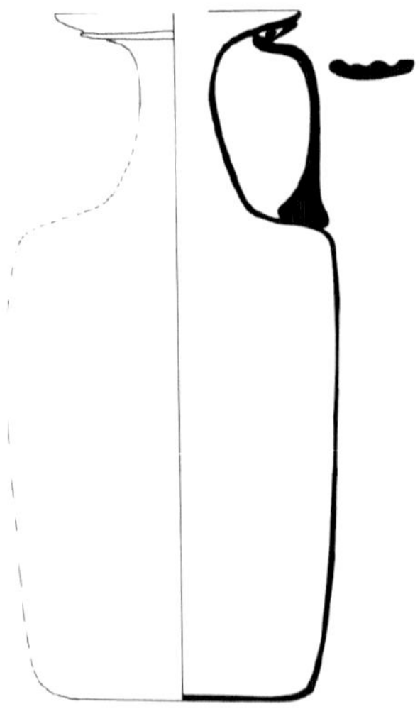

426

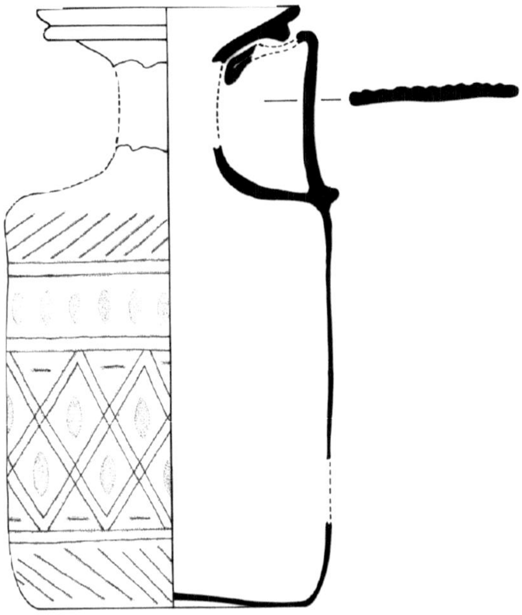

427

428

429

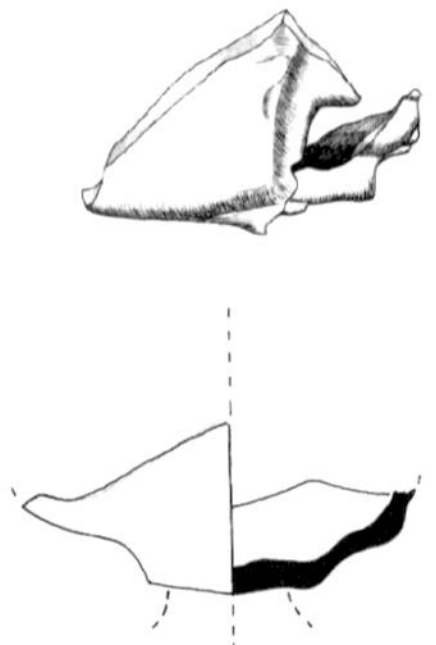

430

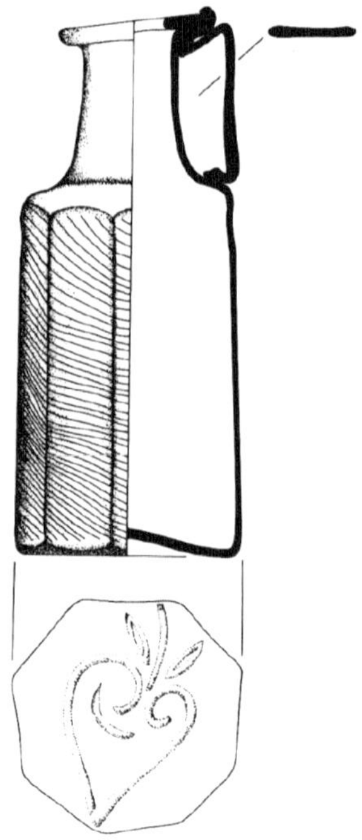

431

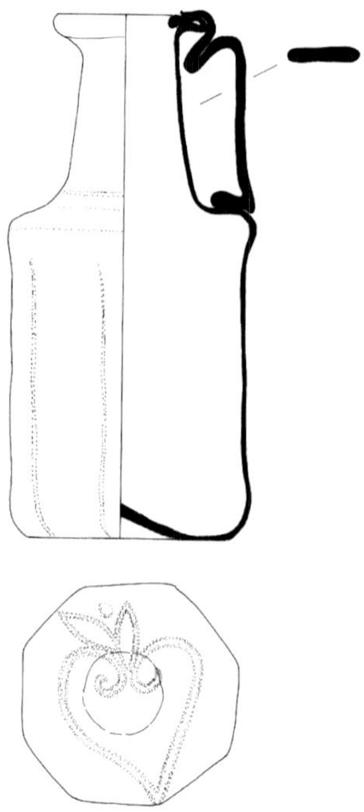

432

# Plate 30

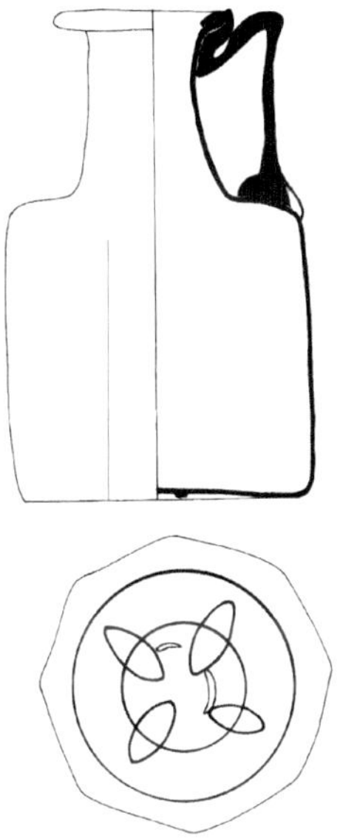

433

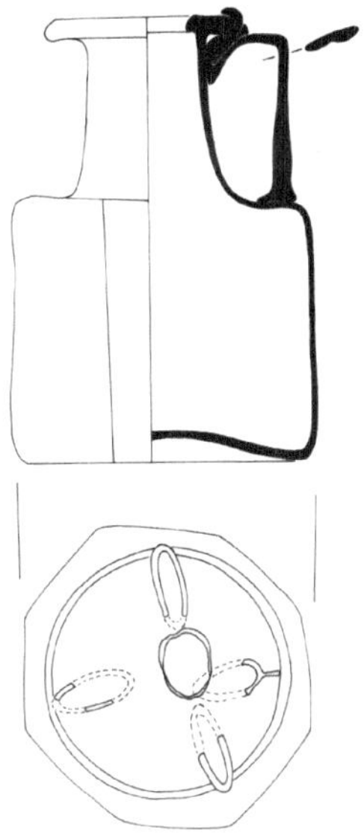

434

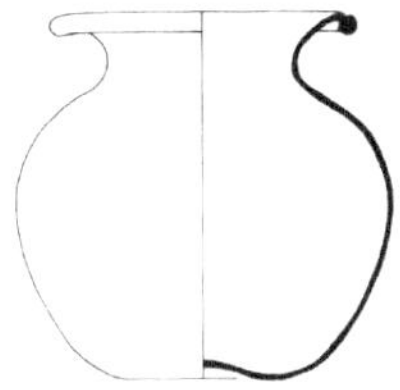

435

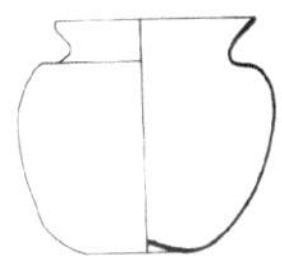

436

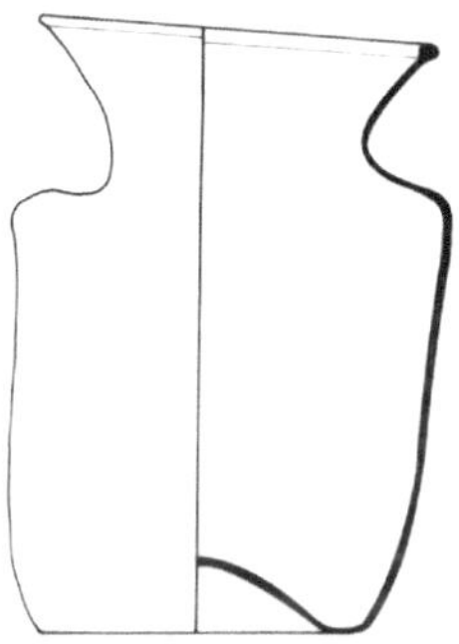

437-438

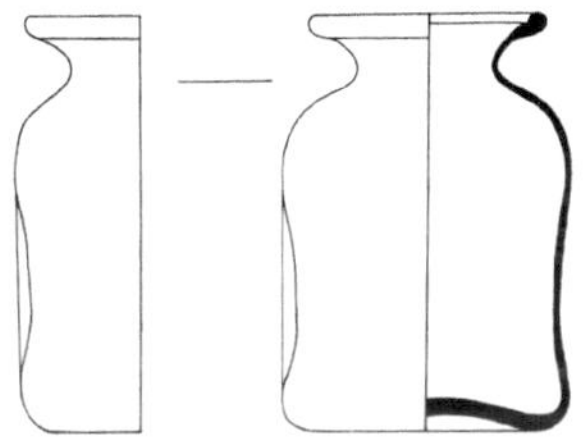

439

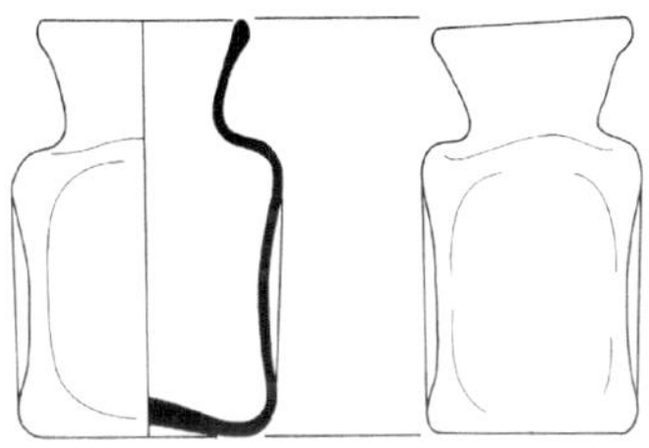

440

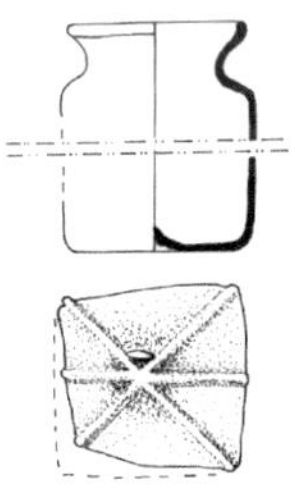

441

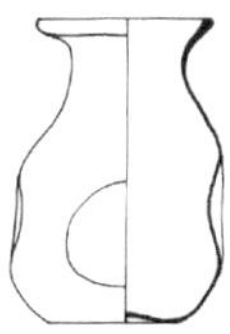

442

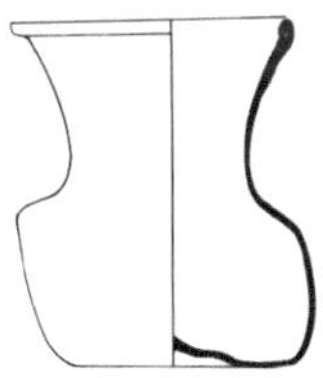

443

445

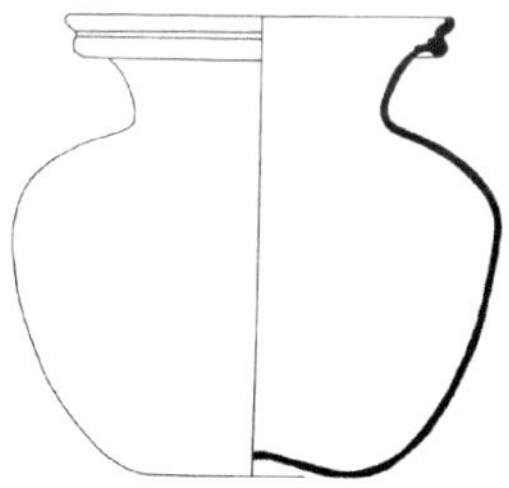

446

# Plate 31

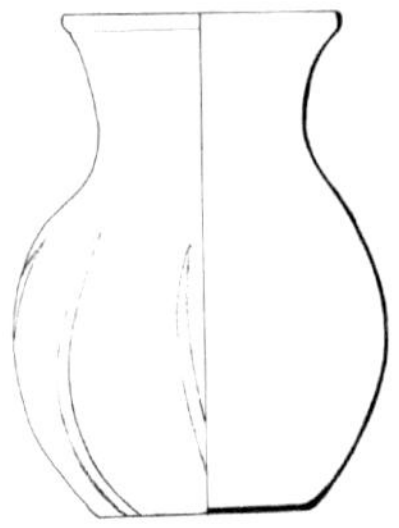
447

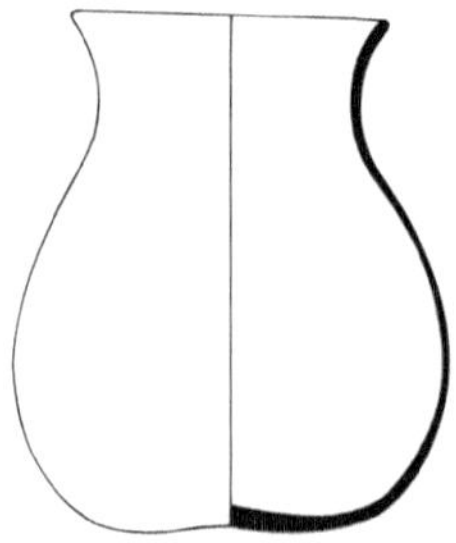
448

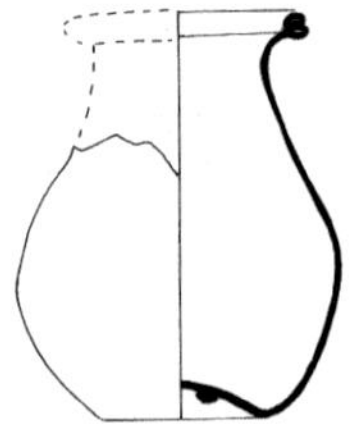
449

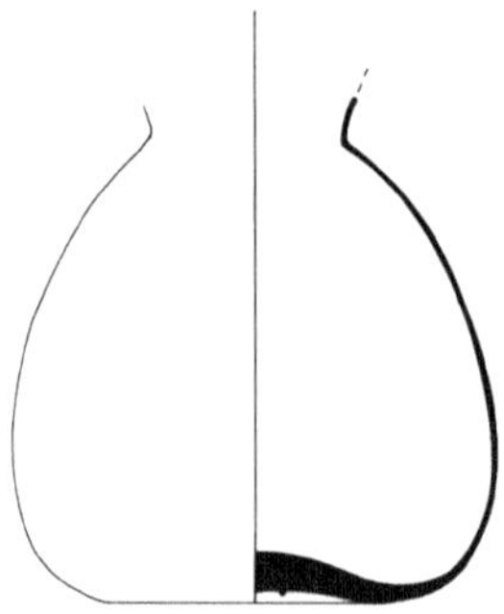
450

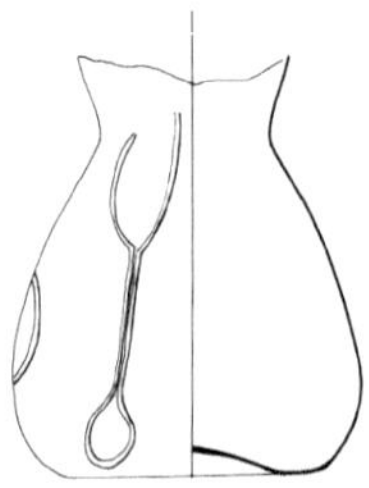
451

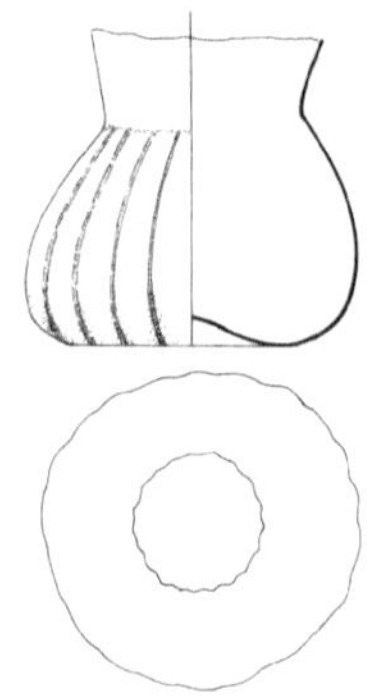
452

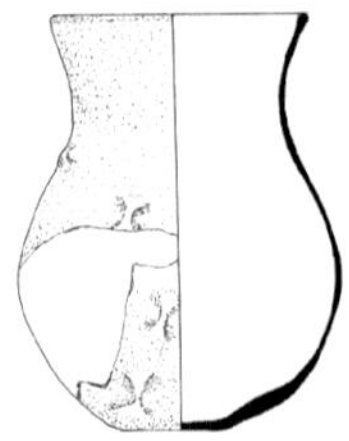
453

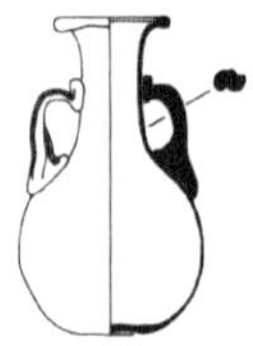
459

460

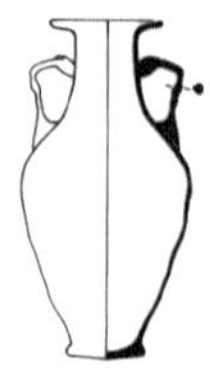
461

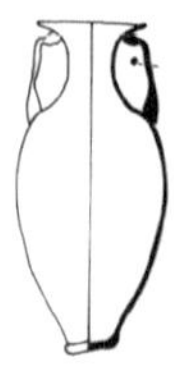
462

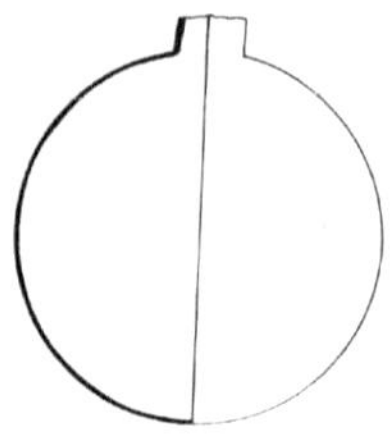
465

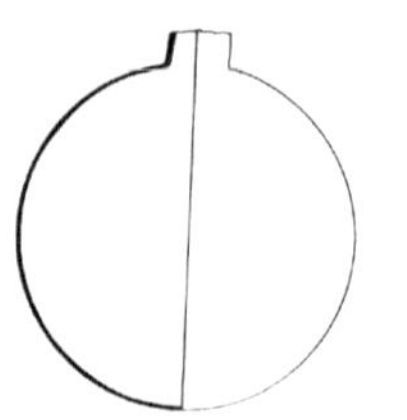
466

# Plate 32

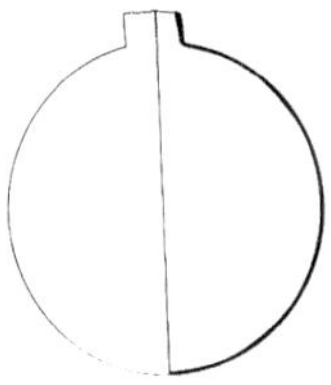

467

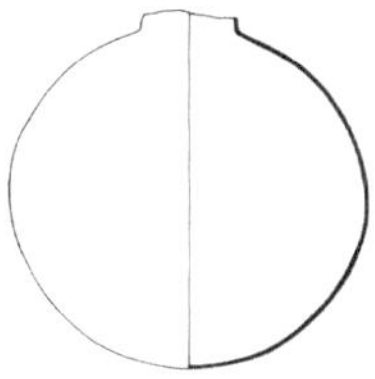

468

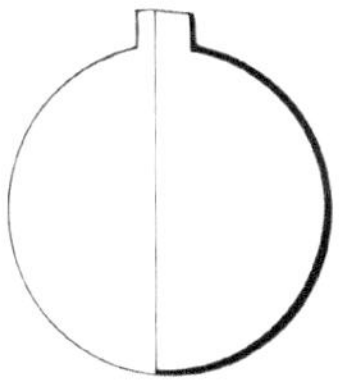

469

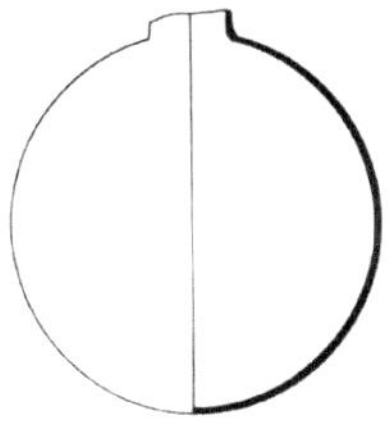

470

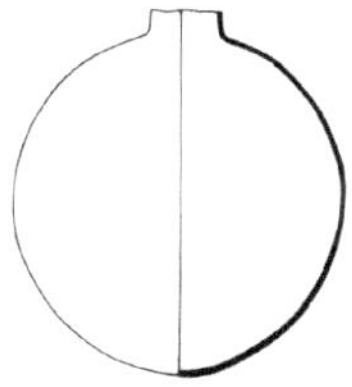

471

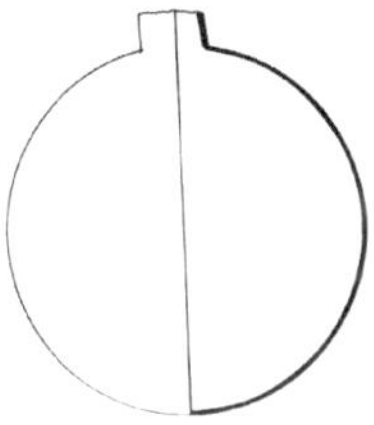

473

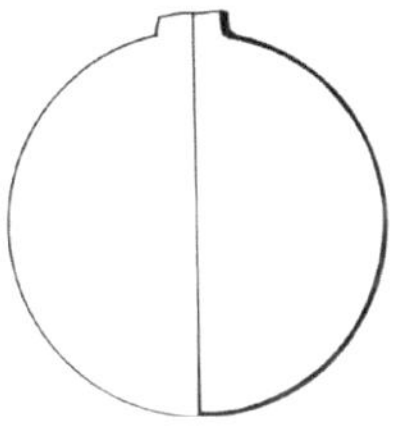

474

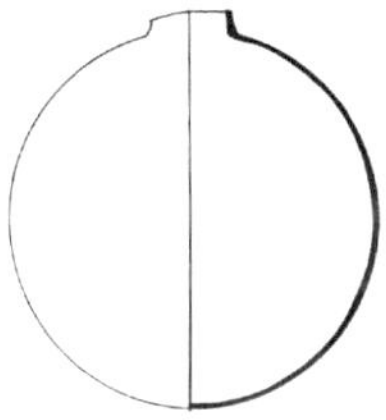

475

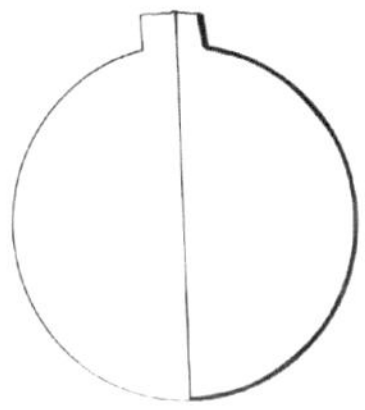

477

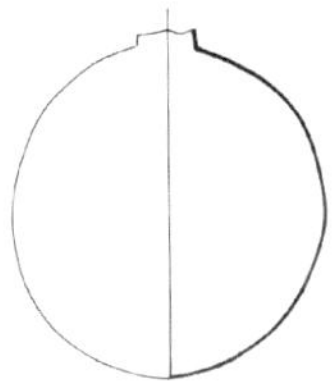

478

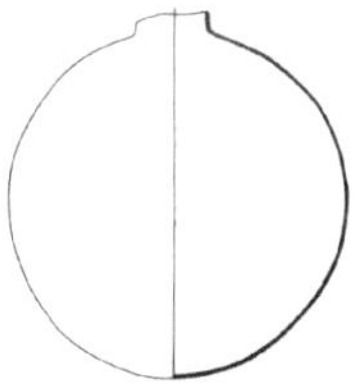

479

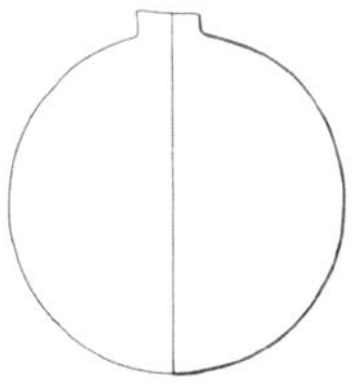

480

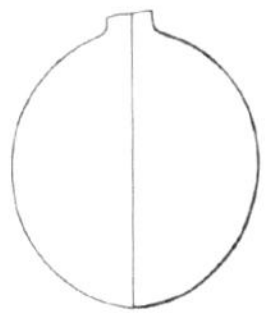

481

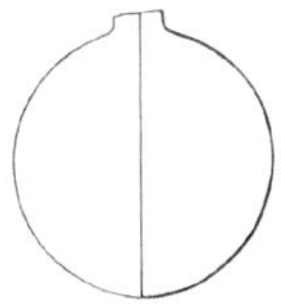

483

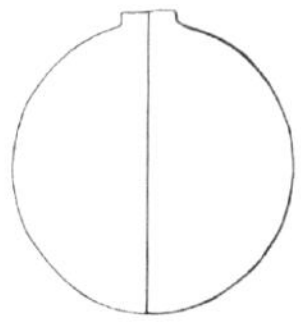

484

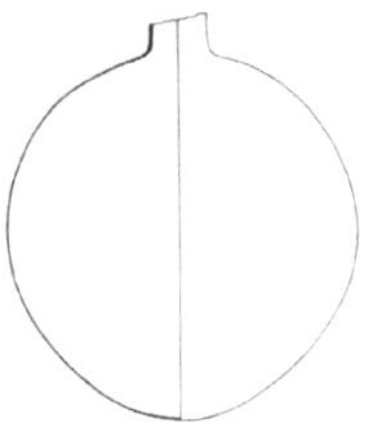

485

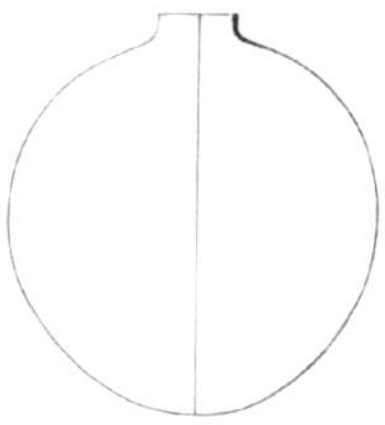

486

497

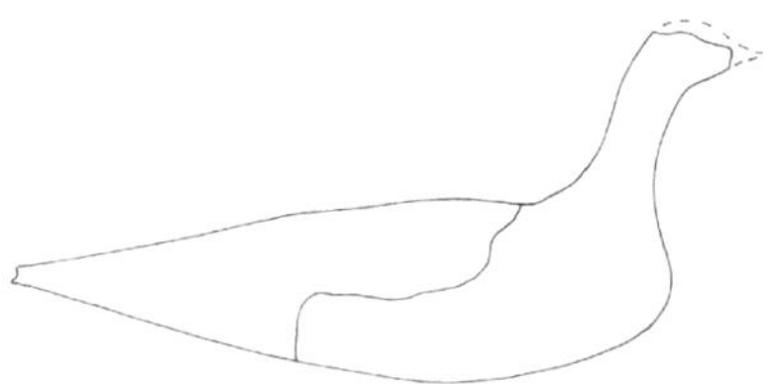

499

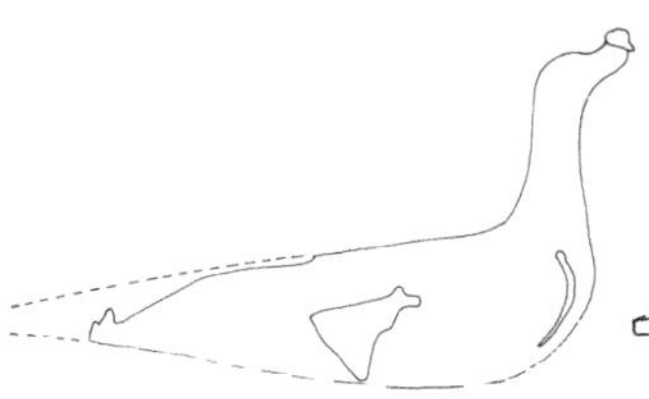

502

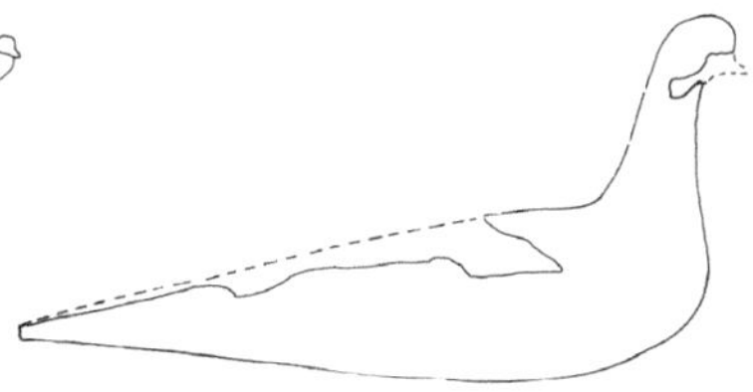

503

504

# Plate 33

505

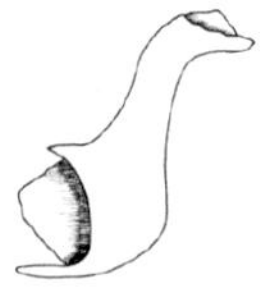
506

508

509

510

511

512

513

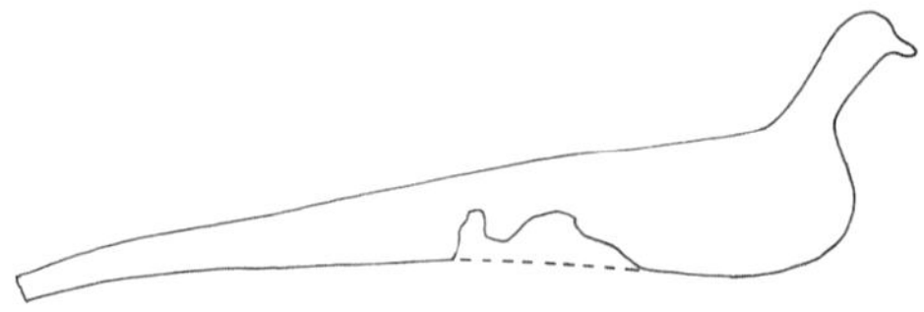
514

516

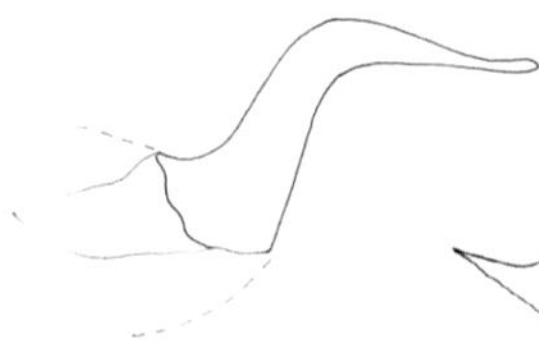
517

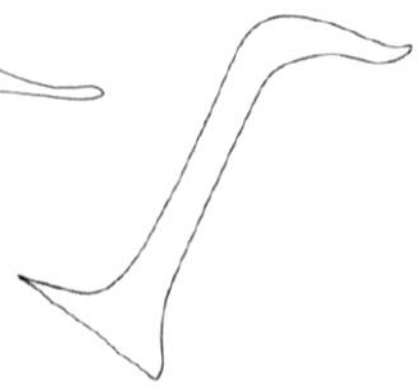
518

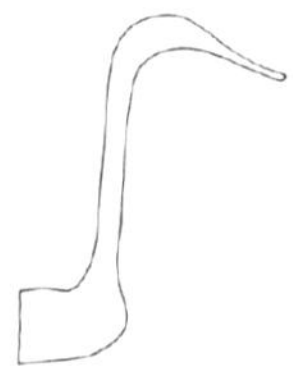
519

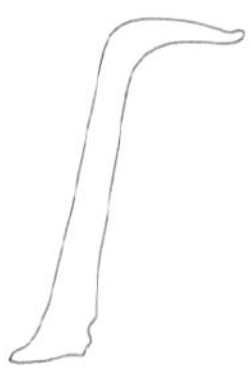
520

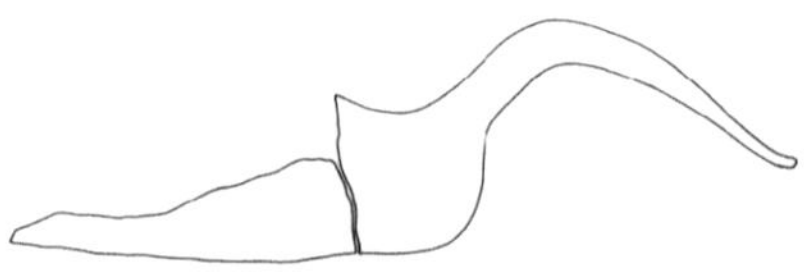
521

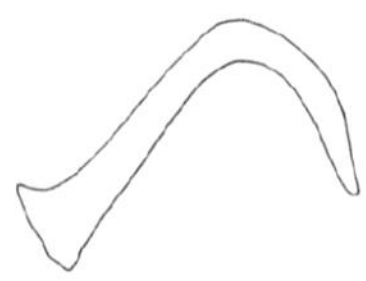
522

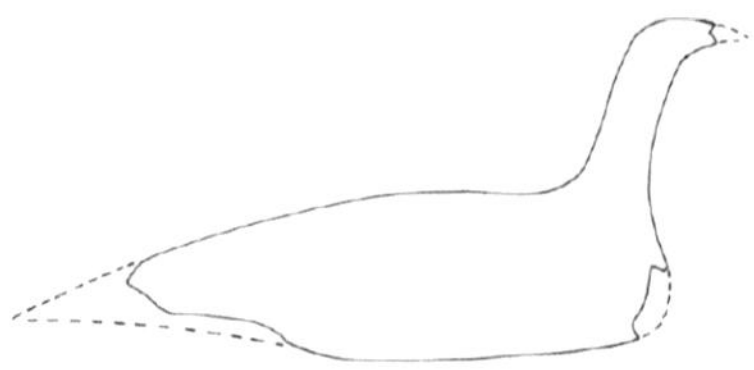
523

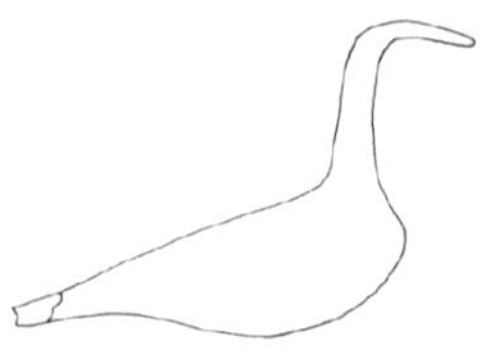
524

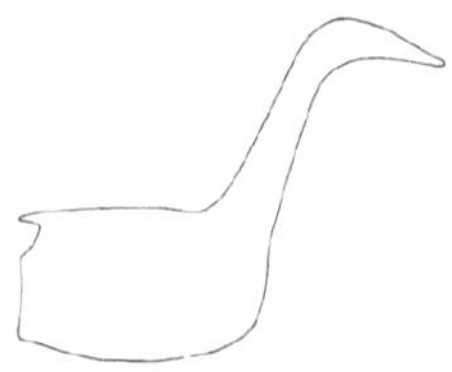
525

526

527

529

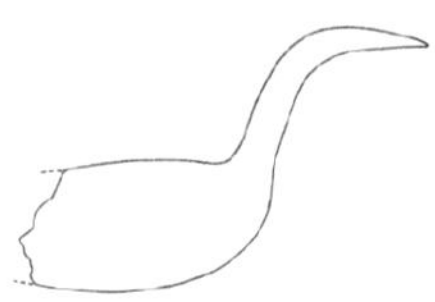
530

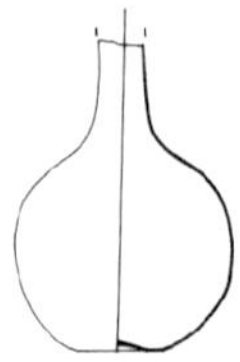
538

# Plate 34

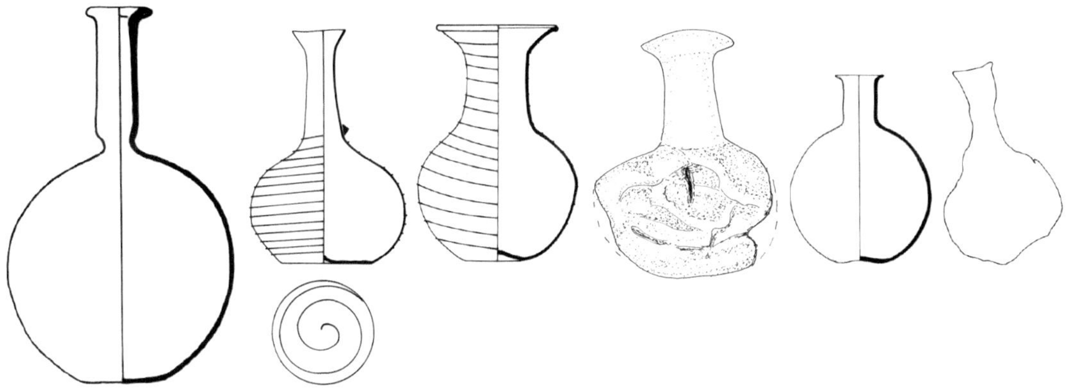

539 540 541 542 543 544

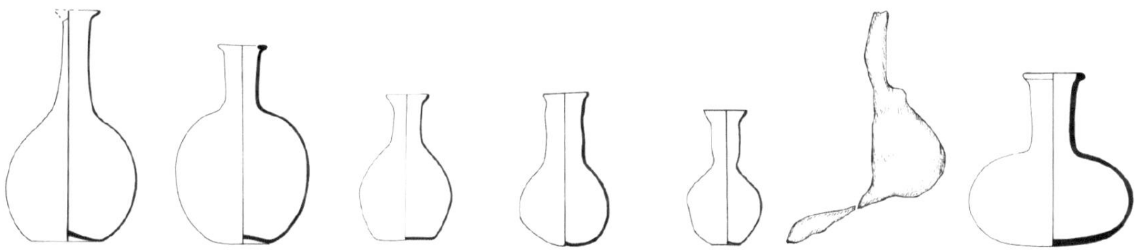

545 546 547 548 549 550 551

552 553 554 555 556 557 558 559

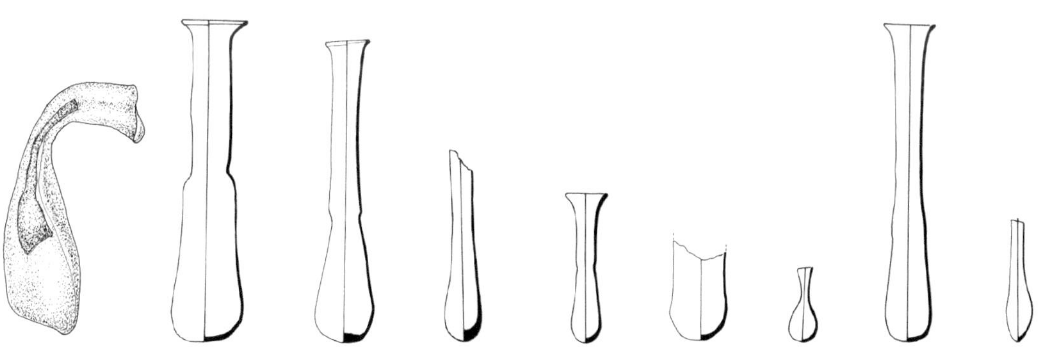

560 561 562 563 565 567 568 569 571

Plate 35

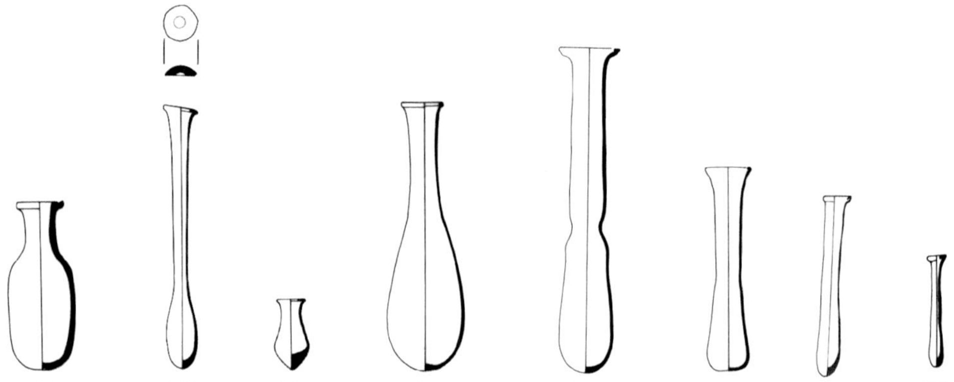

572 573 574 575 577 578 579 580

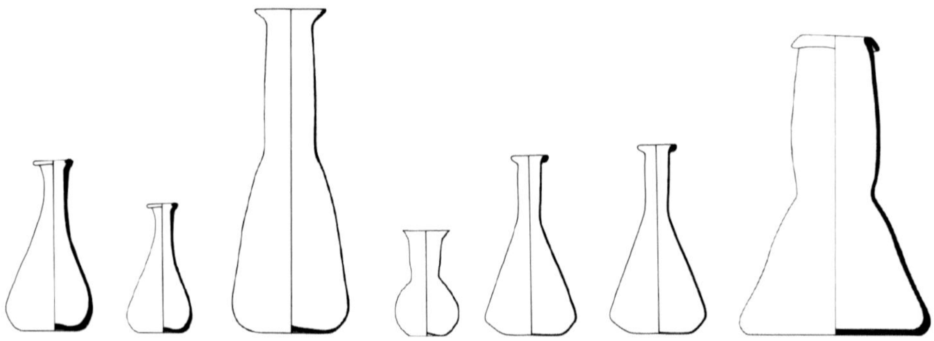

581 583 584 585 586 587 588

589 590 592 593 594

# Plate 36

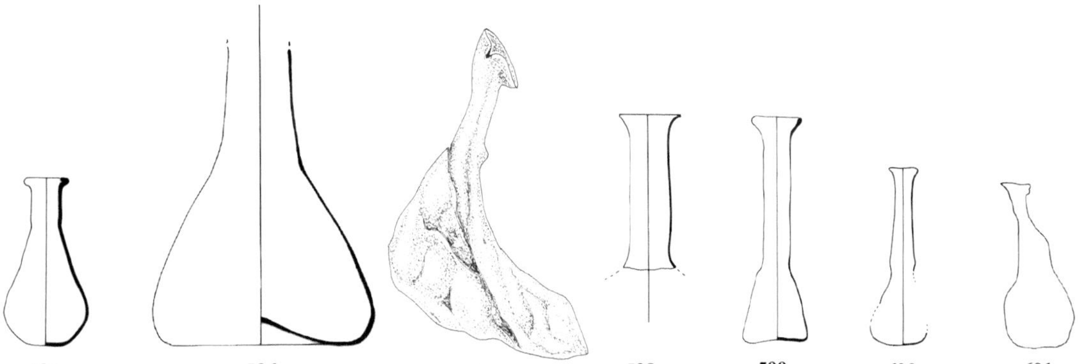

595 596 597 598 599 600 601

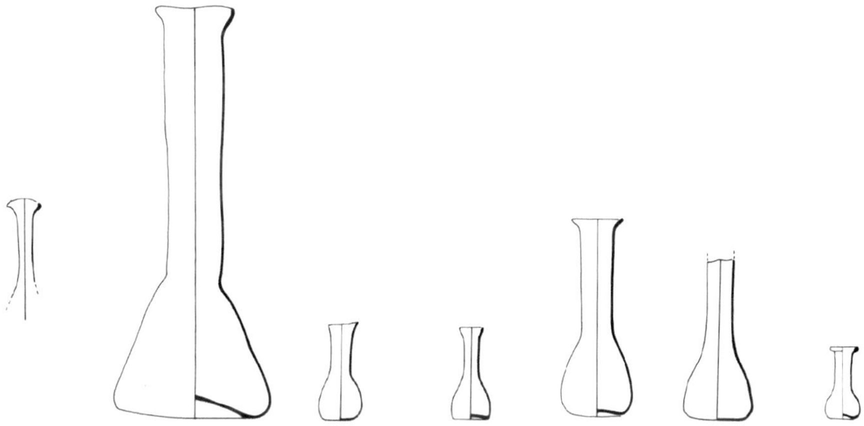

602 603 604 605 606 607 608

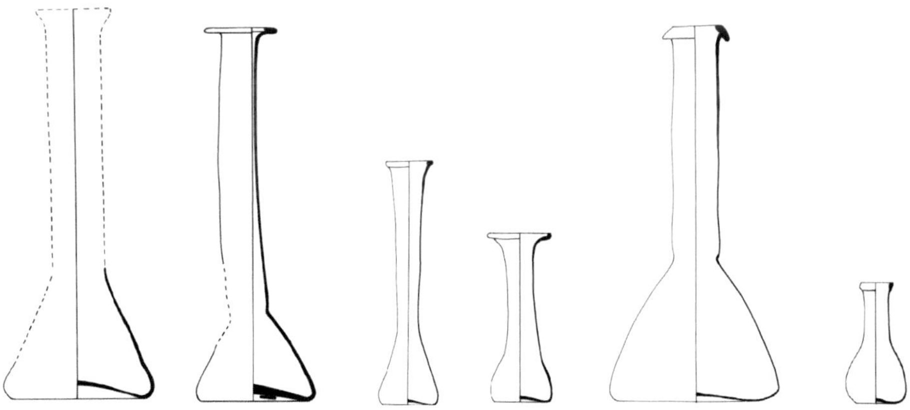

609 610 611 612 613 614

Plate 37

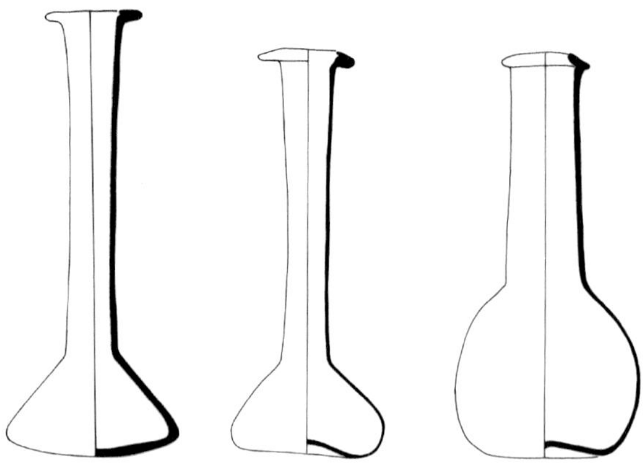

615 616 617 618 619

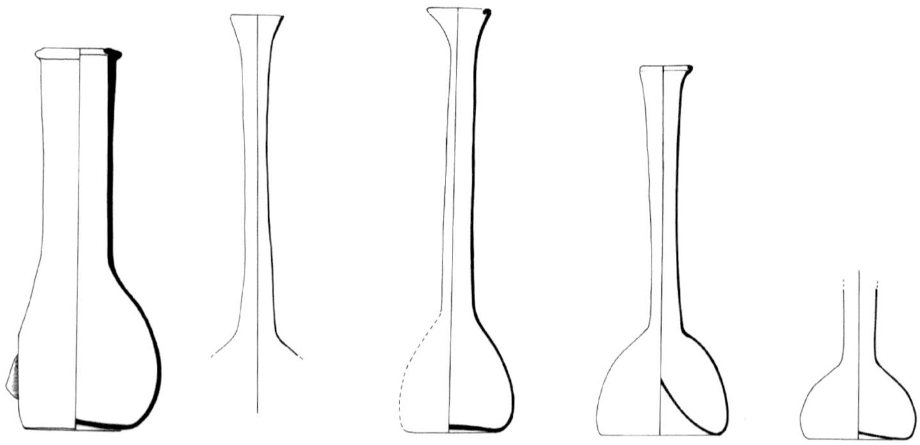

620 621 622 623 624

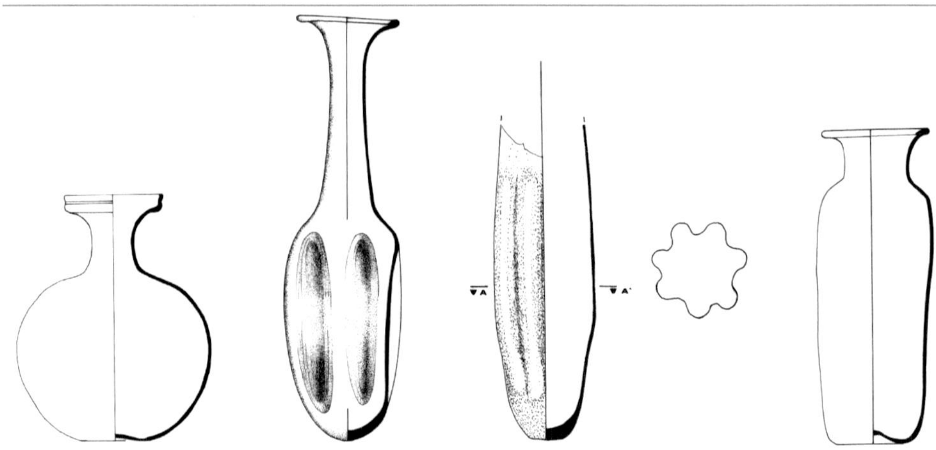

625 626 627 628

# Plate 38

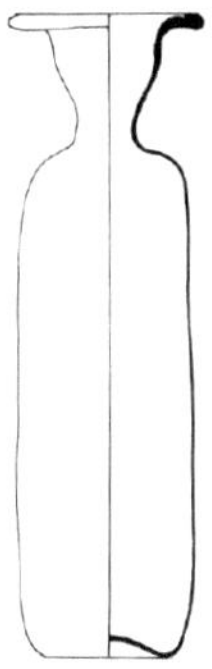
629

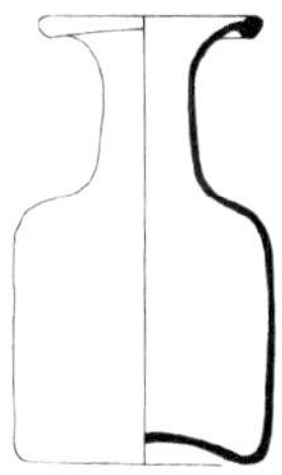
630

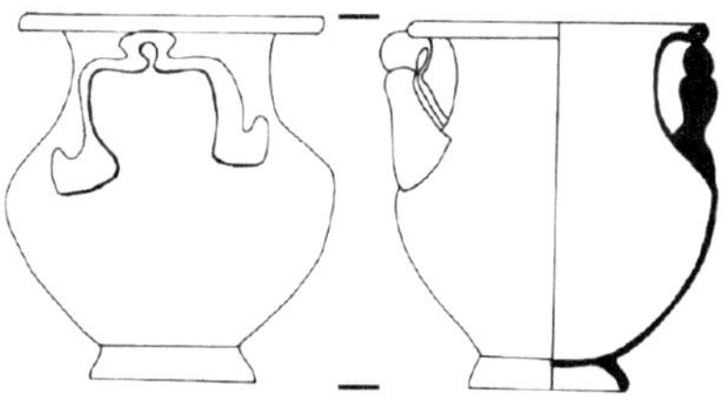
631

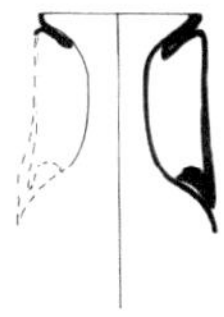
632

633

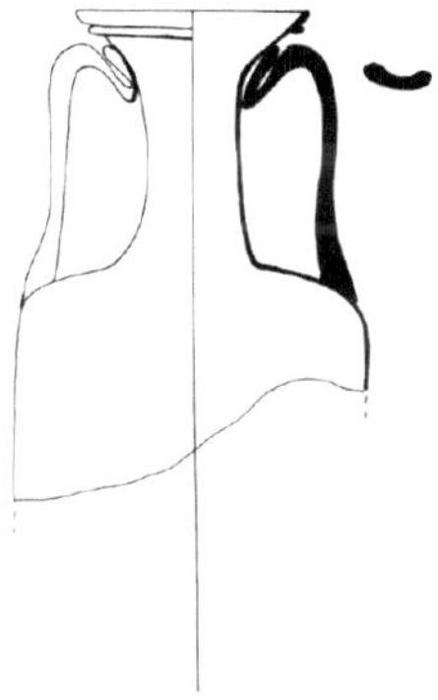
634

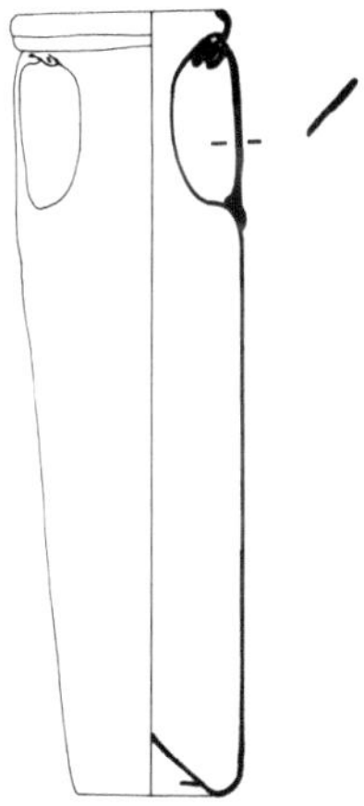
635

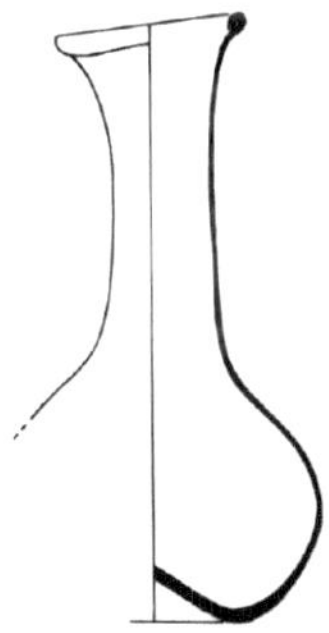
636

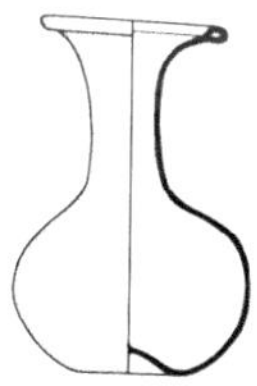
637

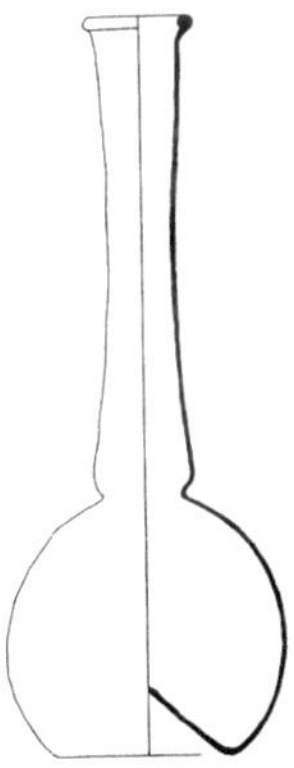
638

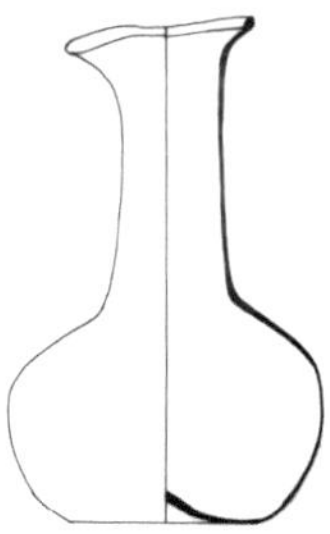
640

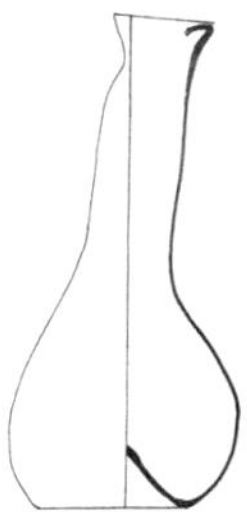
641

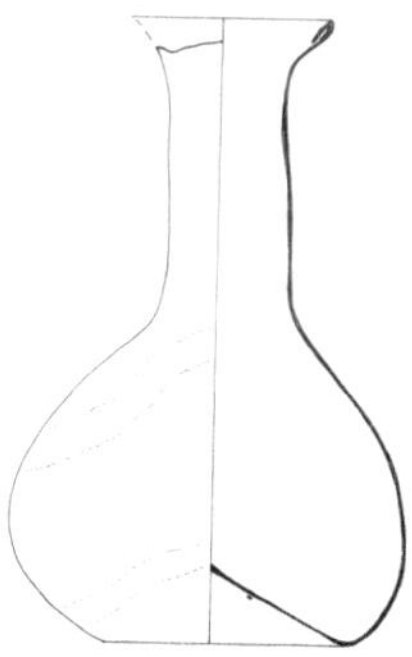
642

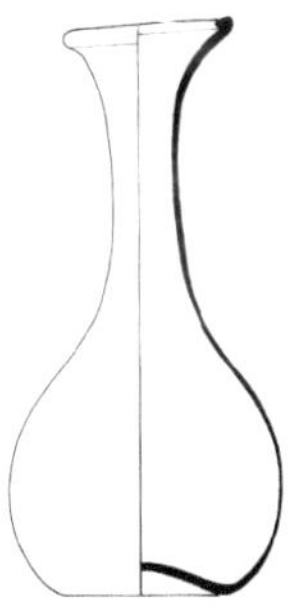
643

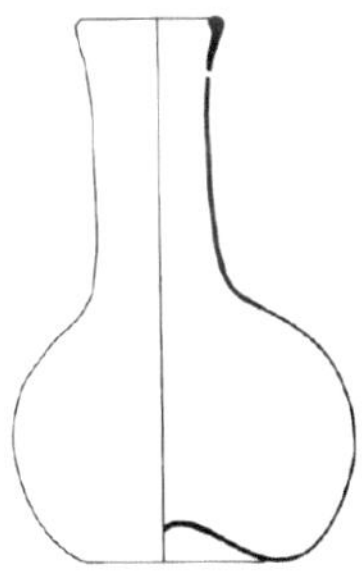
644

Plate 39

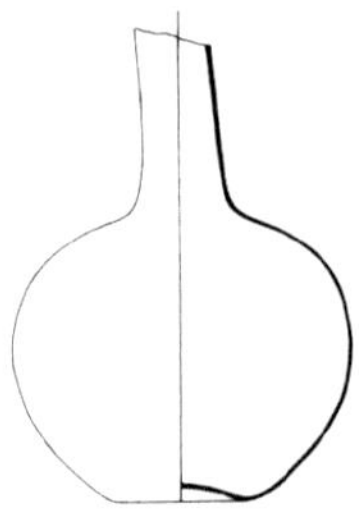
645

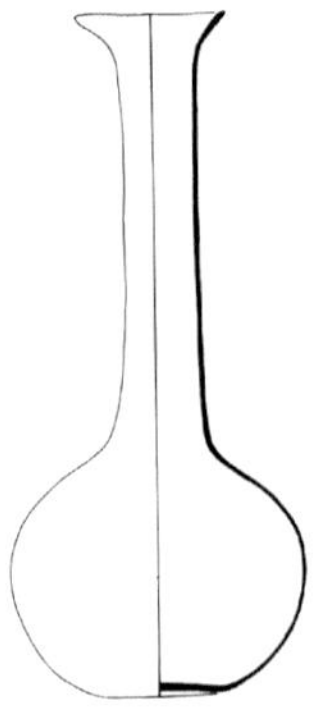
646

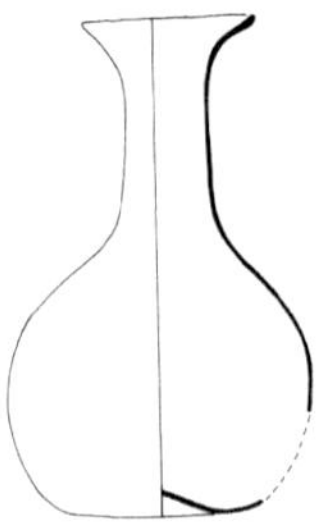
647

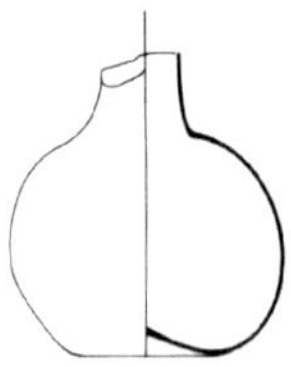
648

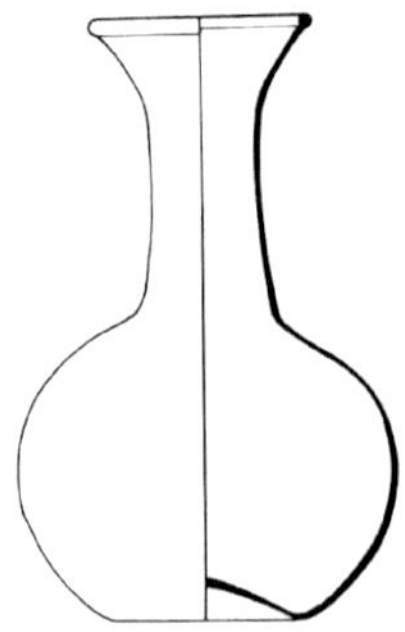
649

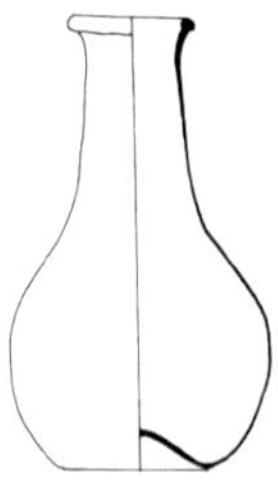
650

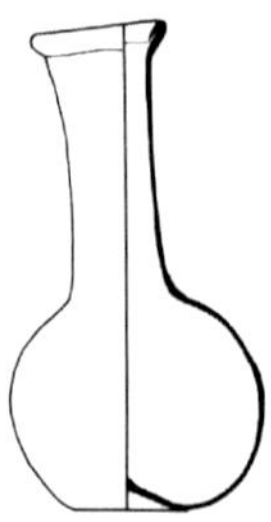
651

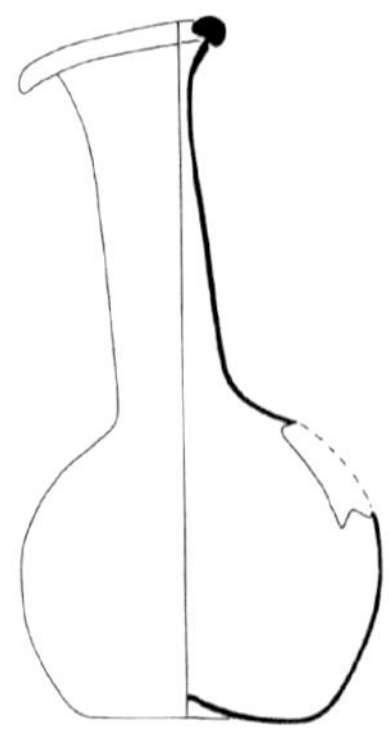
652

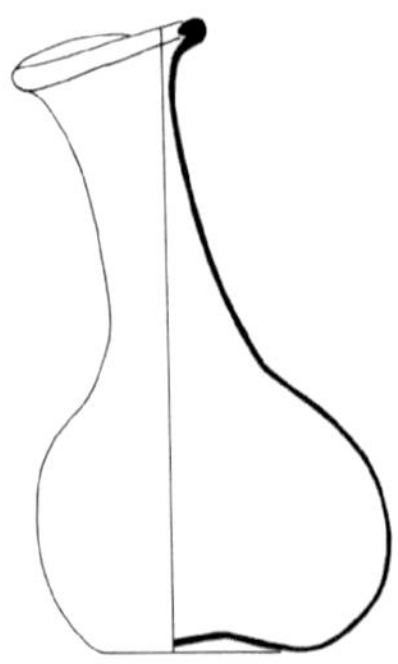
653

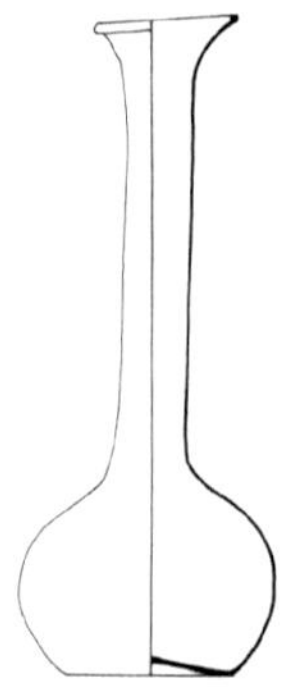
654

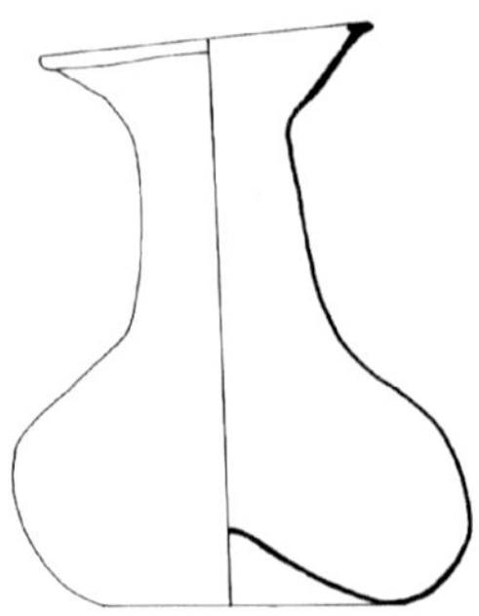
655

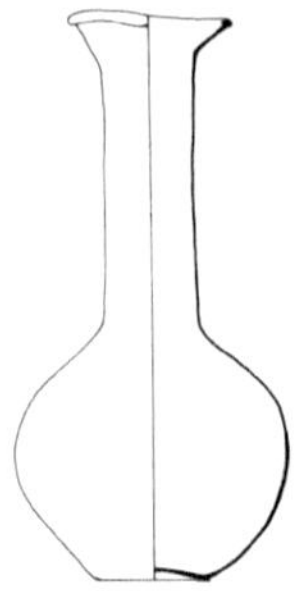
656

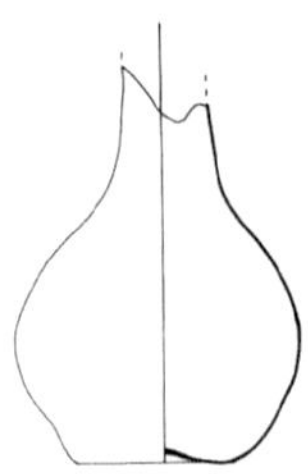
657

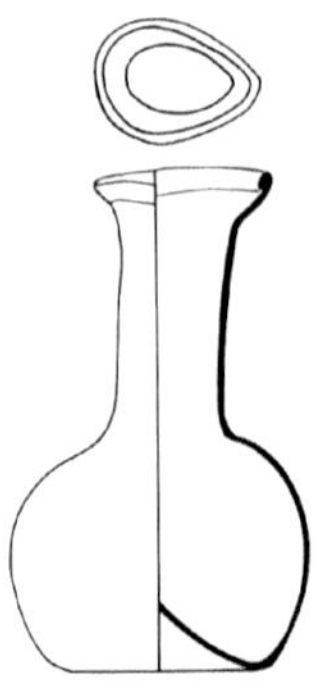
658

# Plate 40

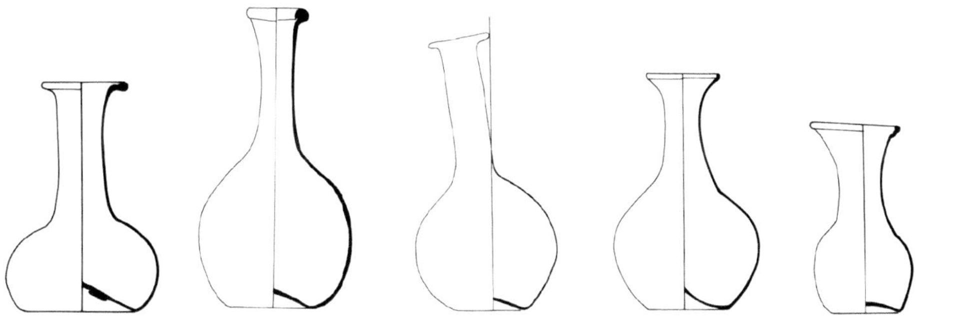

659 660 661 662 663

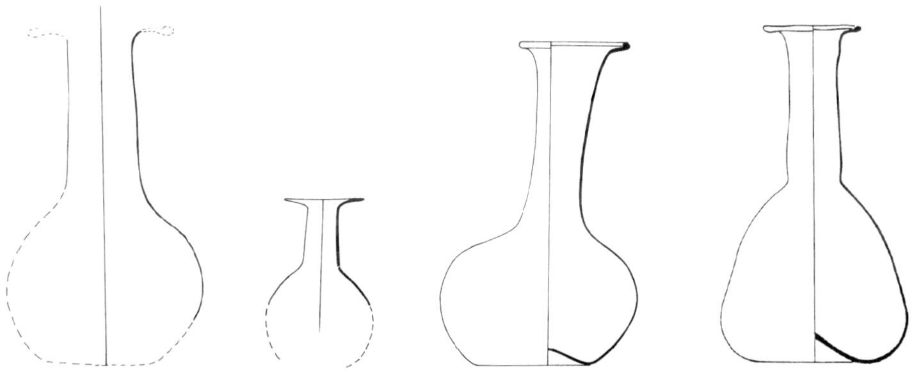

664 665 666 667

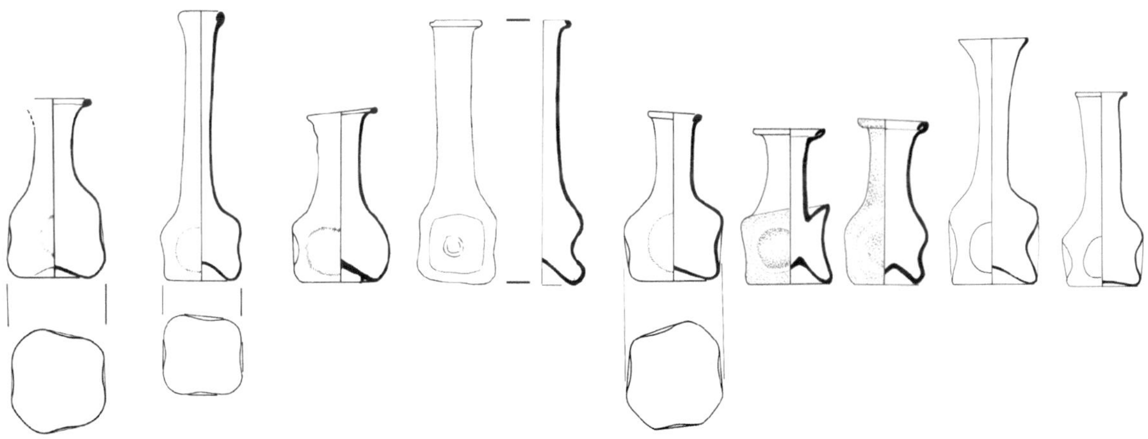

668 669 670 671 672 673 674 675 676

Plate 41

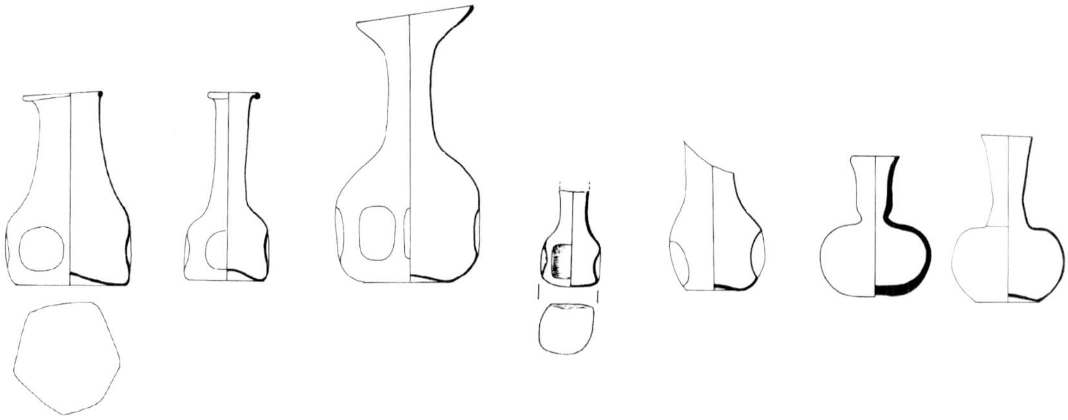

677 678 679 680 681 683 684

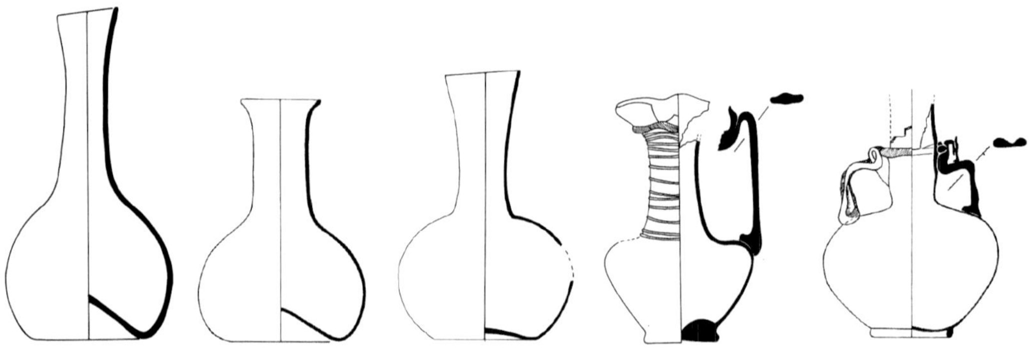

685 686 687 688 689

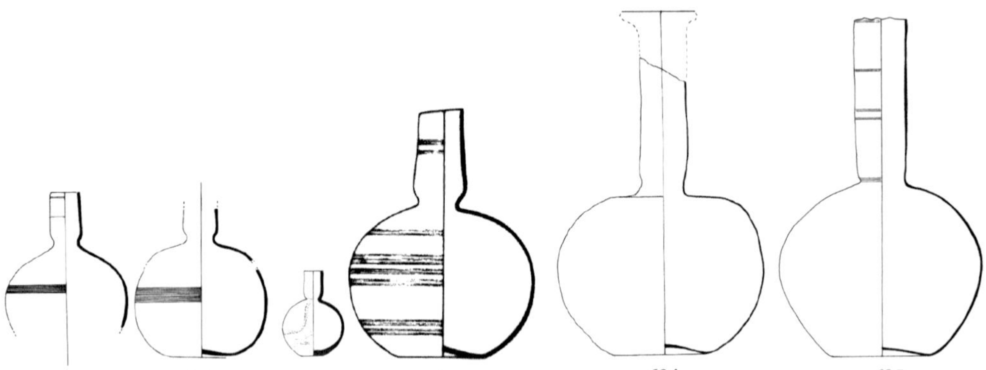

690 691 692 693 694 695

# Plate 42

695

696

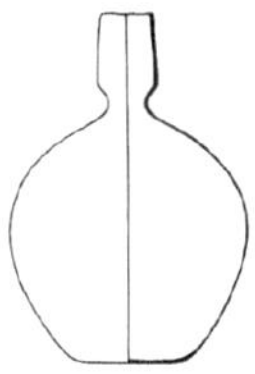
697

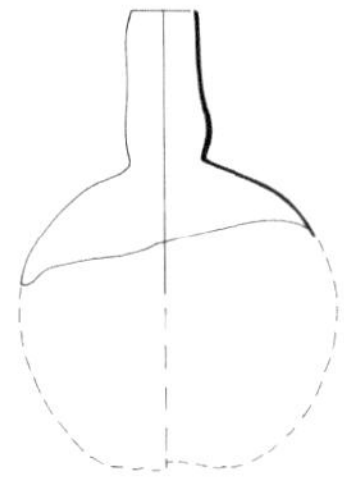
698

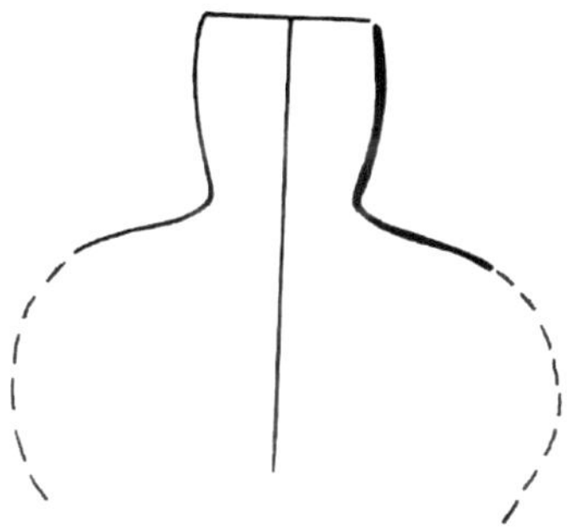
699

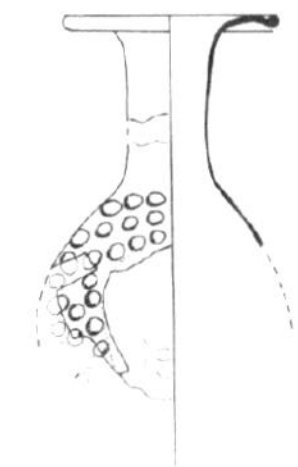
700

701

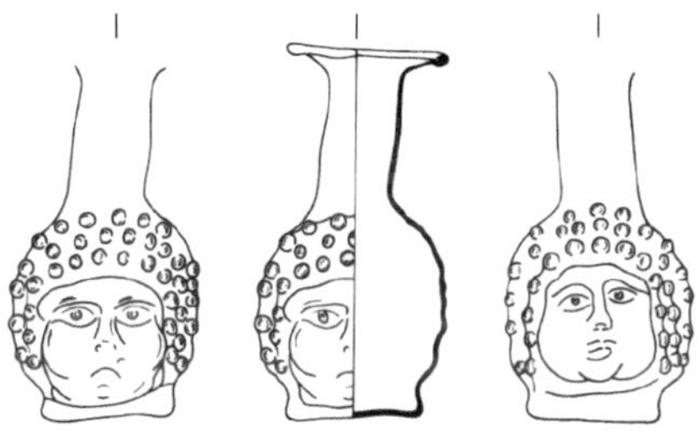
702

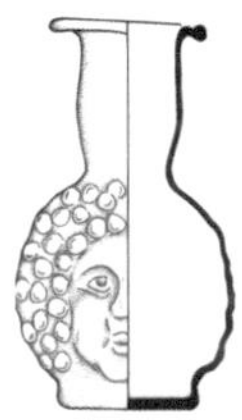
703

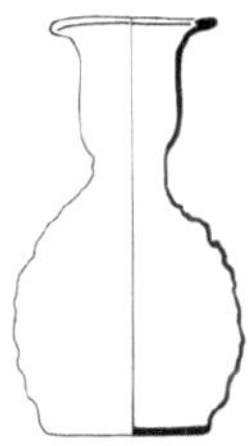
704

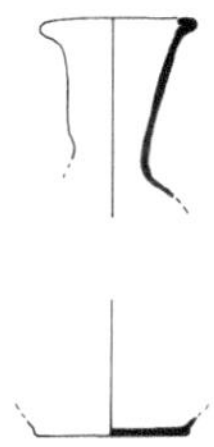
705

712

Plate 43

Scale 1:4

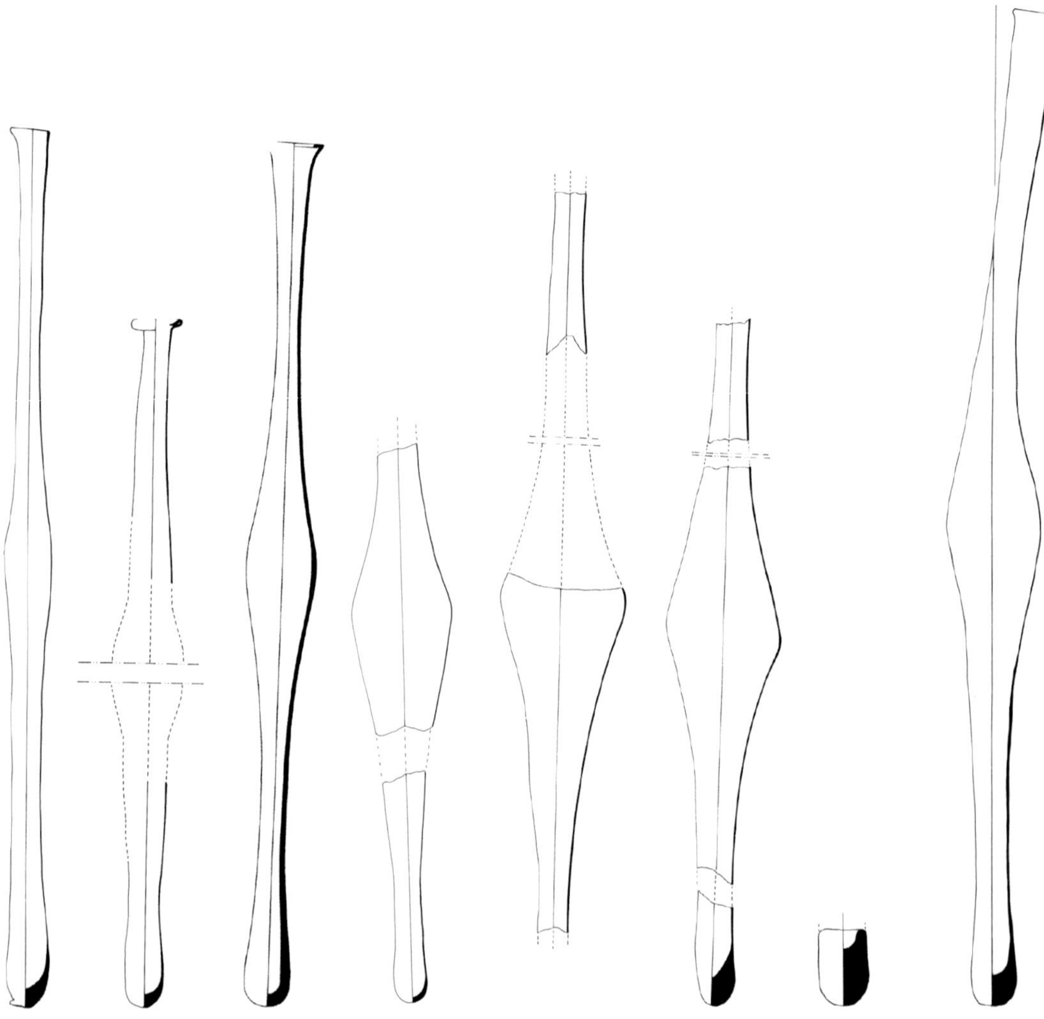

713 714 715 716 718 719 720 722

# Plate 44

Scale 1:4

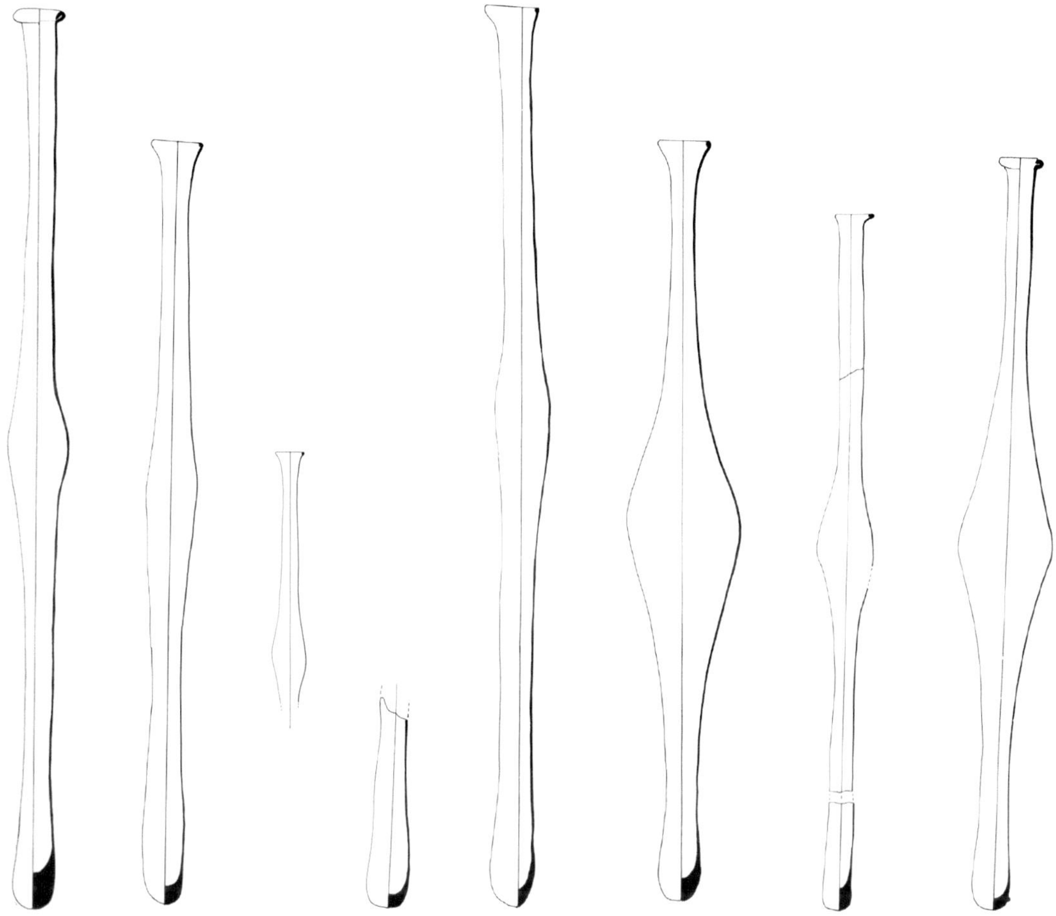

723 724 725 726 727 729 730 731

## Plate 45

Scale: 732–442 1:4; 743 1:3; 735-754 1:2

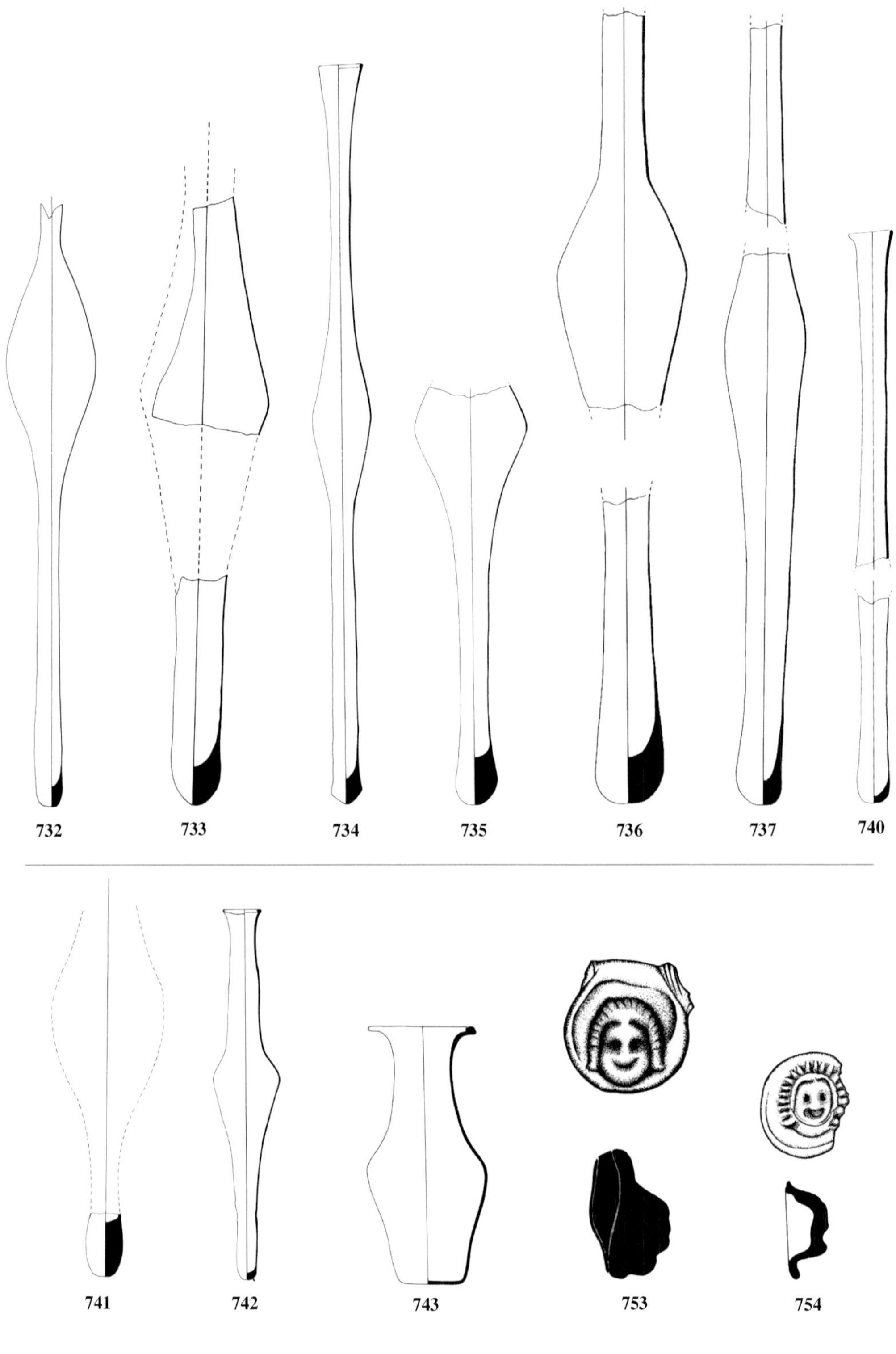

# Plate 1

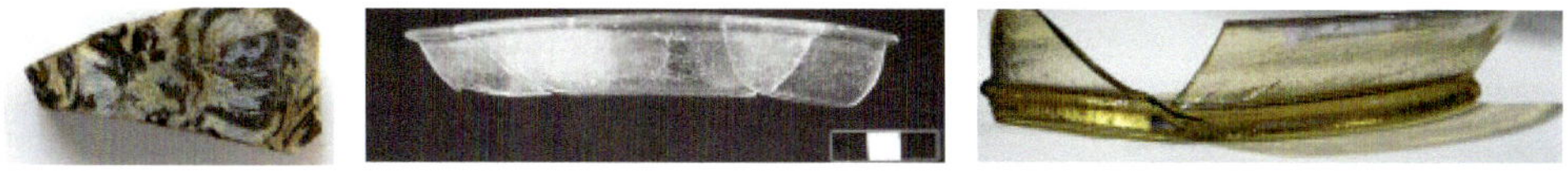

1 2 4

7 8 9-11

12-14 15-17 18

19 20 21 22 23 24

30

25-27 28 29 31

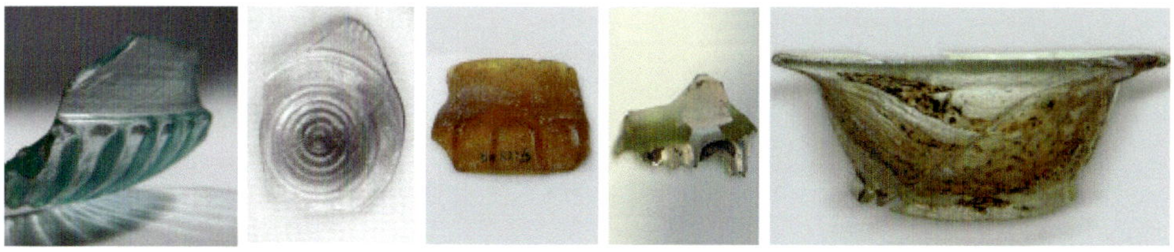

32 33 37 38 39

## Plate 2

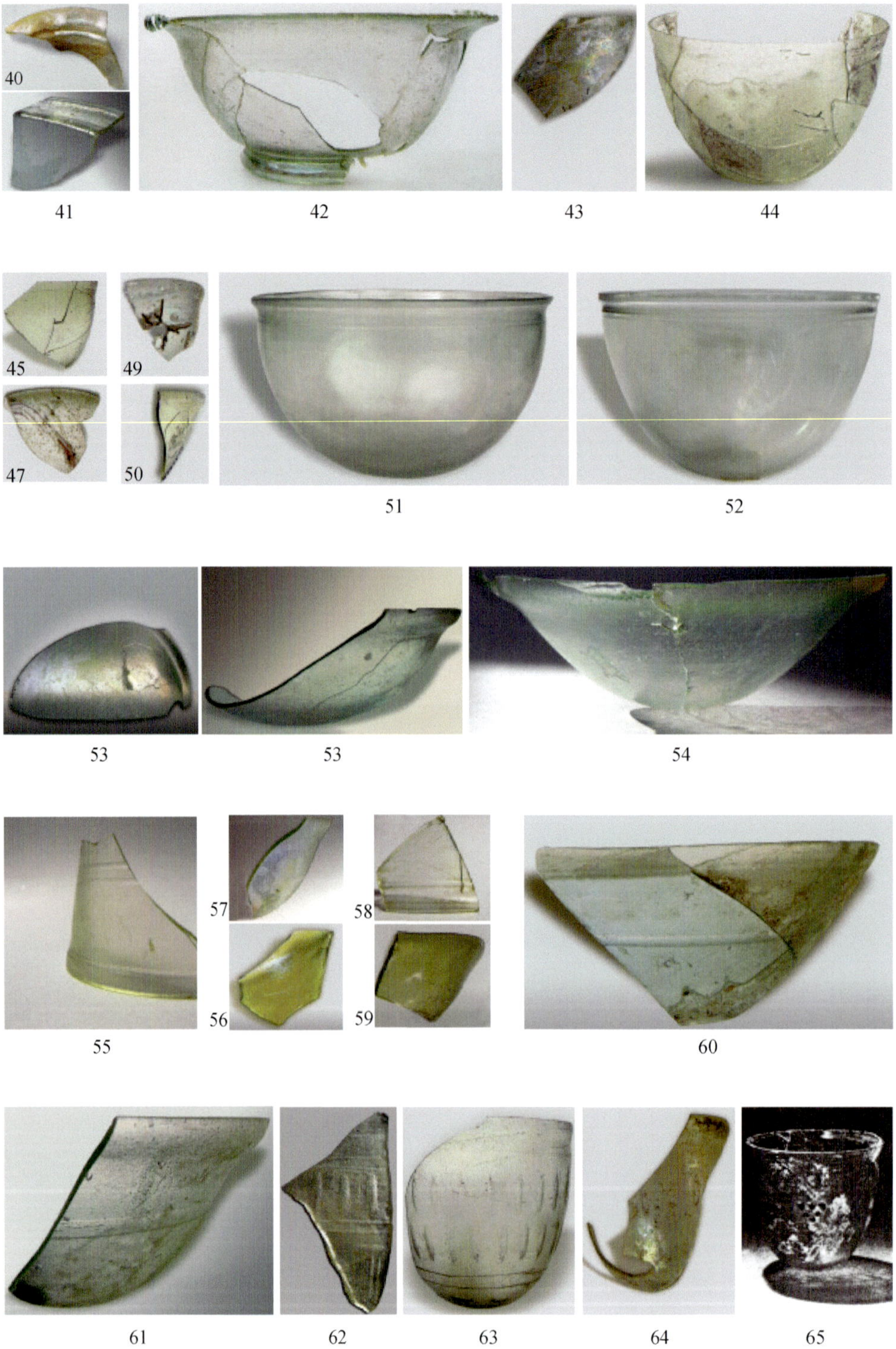
40
41
42
43
44
45
49
47
50
51
52
53
53
54
55
57
58
56
59
60
61
62
63
64
65

# Plate 3

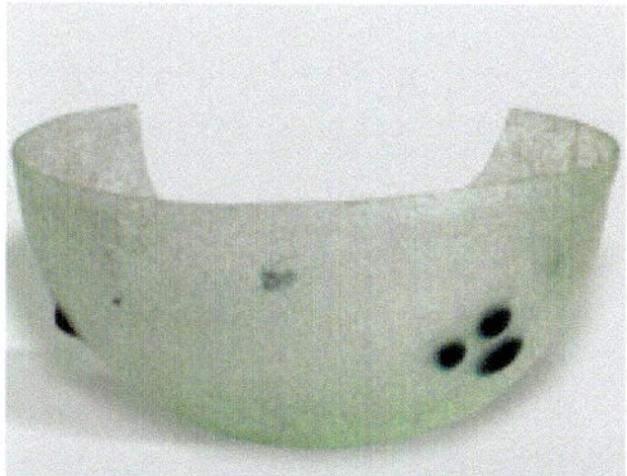
66

67

68

69

70

071

72

73

74

74

75

75

76

77

78

79

80

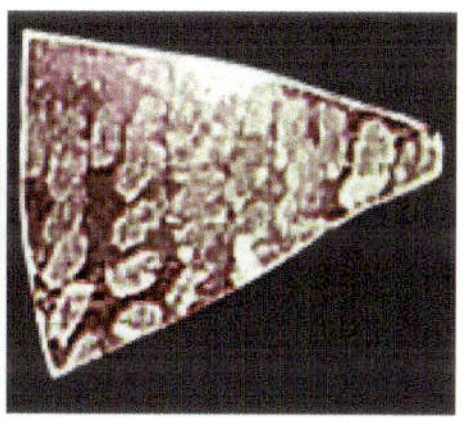
81

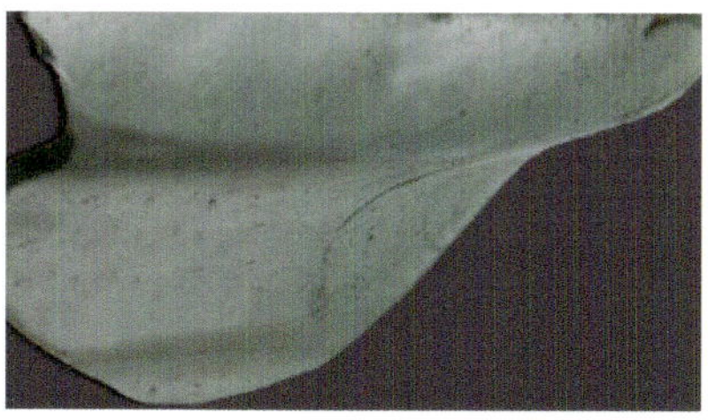
82

83

82-83

## Plate 4

84

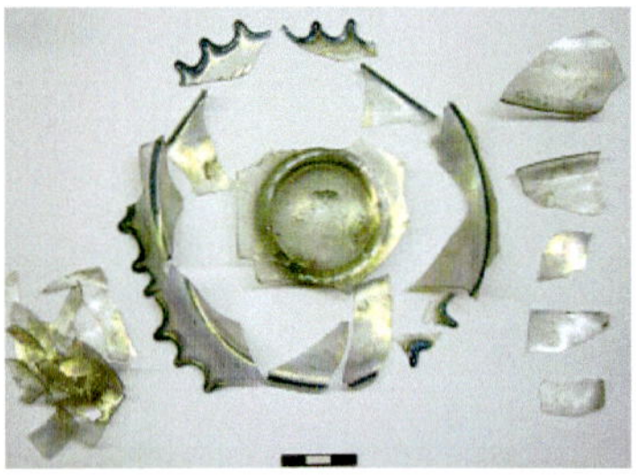

85

86

87

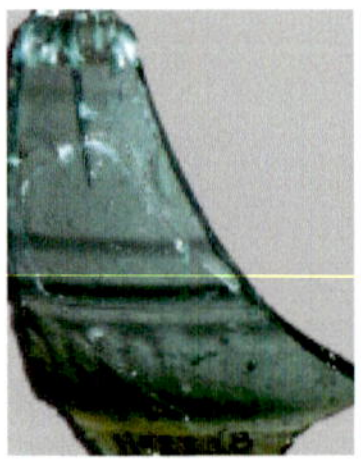

88

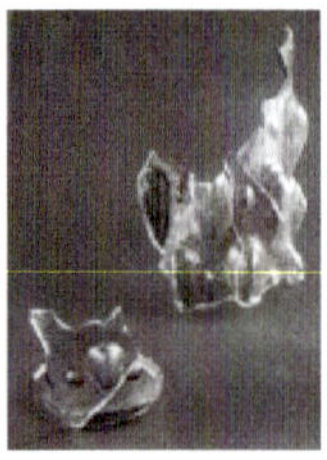

91-92

93

94

95

95

96

98

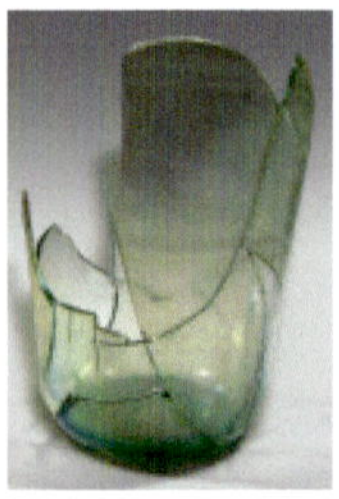

99

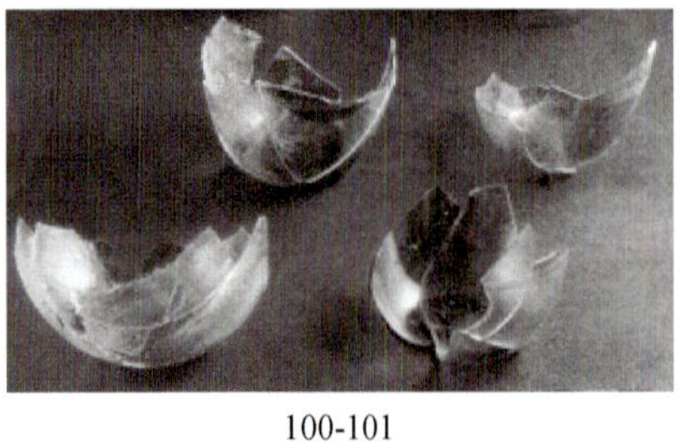

100-101

102

103

104

105

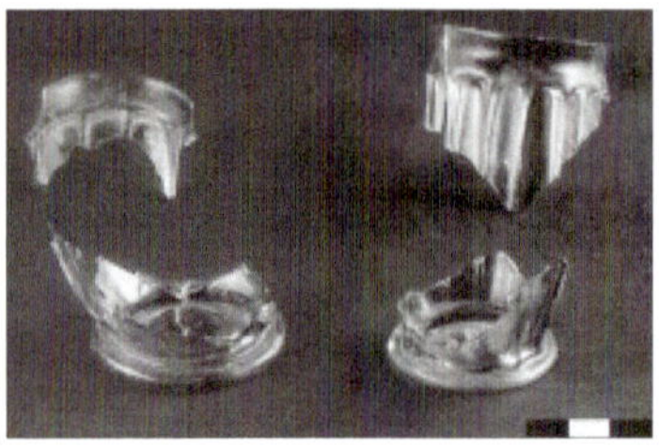

106-107

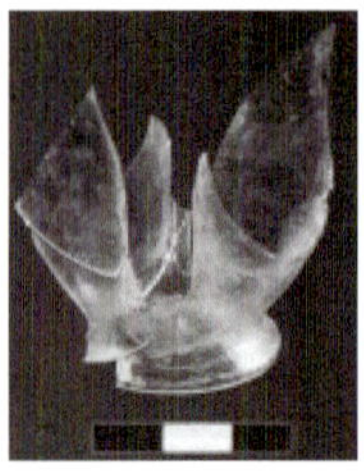

108

# Plate 5

109

110

111

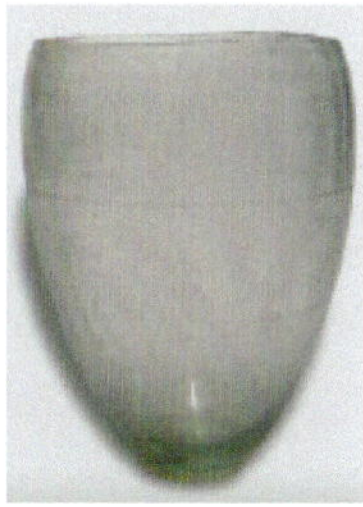
112

113

114

115

117

118

119

120

121

122

123

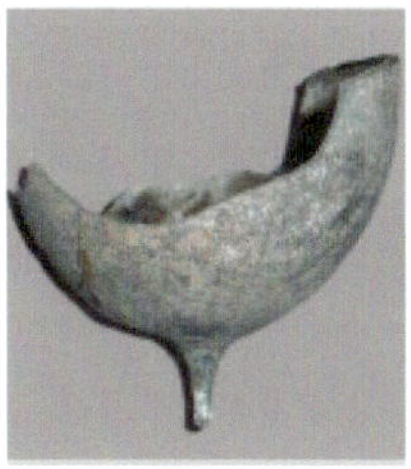
124

125

126

127

128

129

130

131

132

# Plate 6

133-135

136-141

142

144

145

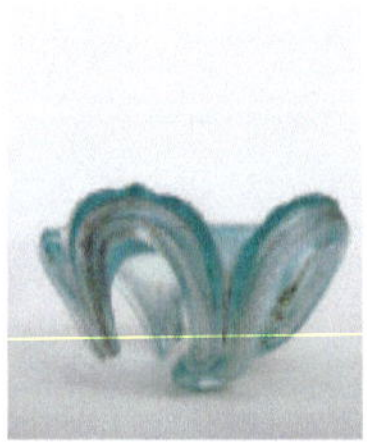

146

147-150 154-161

151

152

153

162-164

164a

164b

164c

164d

165

166

167

168

169

170-172

173

# Plate 7

174

175

175a

175b

175c

176

178

179-184

185

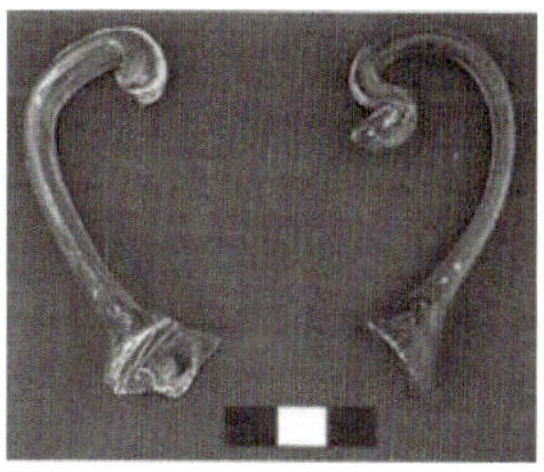
186-187

188

189

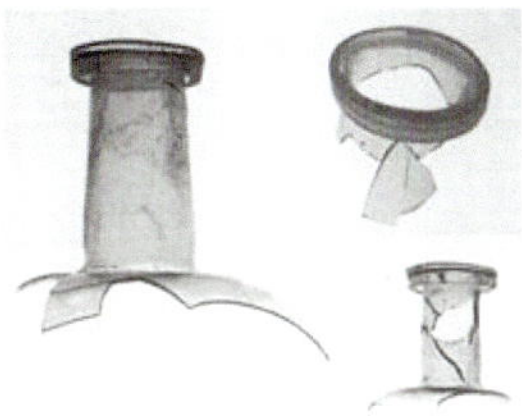
190-192

193

194

195

196

197

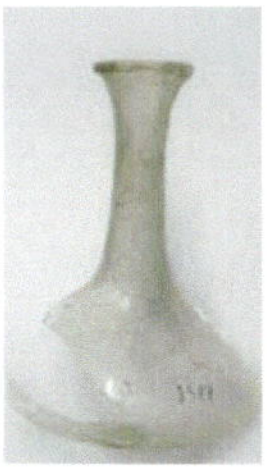
198

199

200

201

202

203

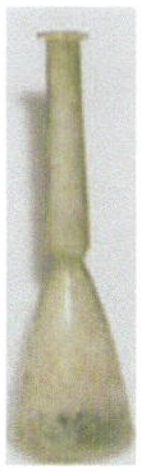
204

205

206

207

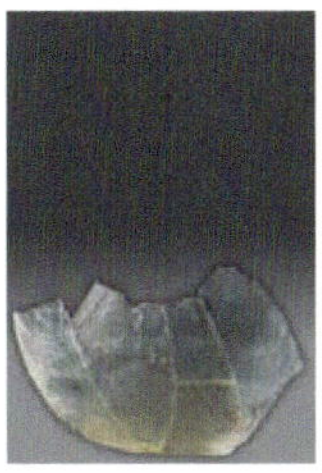
208

209

210-211

212

213

# Plate 8

214

215-216

217

218

219

220

221

222

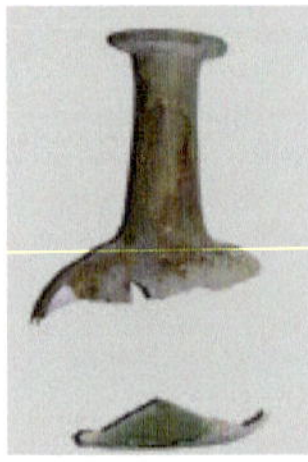
223

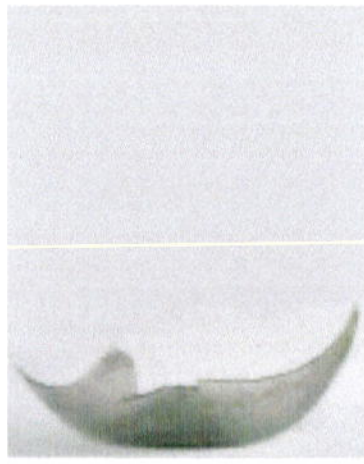
224

225

226

227

229

230

231

232

233

234

235

236

239

240

241

242

243

244

245a

245b

# Plate 9

246

247

248

249

250

251

252

253

254

255

256

257

258

259

260

261

263

264

265

266

267

268

269

270

272

271

273

274

275

276

# Plate 10

277

278

279

280

281

282

283

284

285

286

287

288 & 291

289

290

292

293

294

295

296

297

298

299

300

301

303

# Plate 11

304 305 306 307 308

309 309 310 311 312

313 314 315 316 317 318

319 320 321 322 323 324

325 326 327 328 329 330

# Plate 12

331

332

333

334

335

336

337

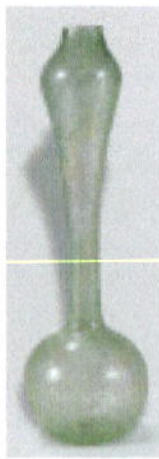
338

339

340

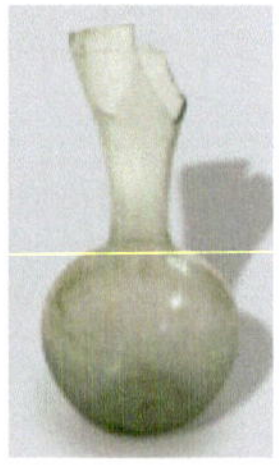
341

342

343

344

345

346

347

348

349

350

351

352

353

354

355

356

357

358

# Plate 13

359 360 361 362 363 364 365

366 367 368 369 370 371 372

373 374 375 376 377 378 380

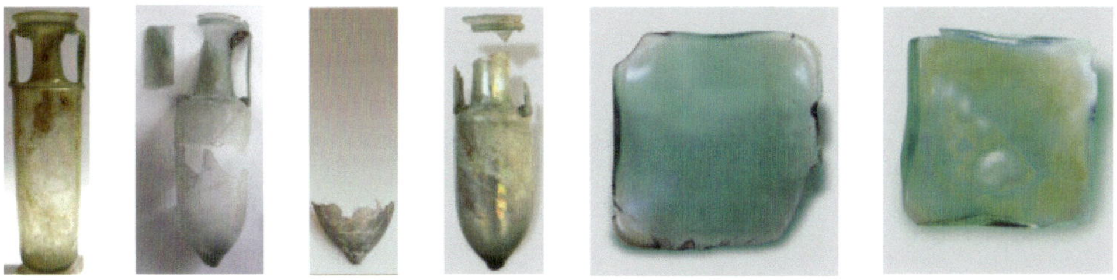

381 382 383 384 385 386

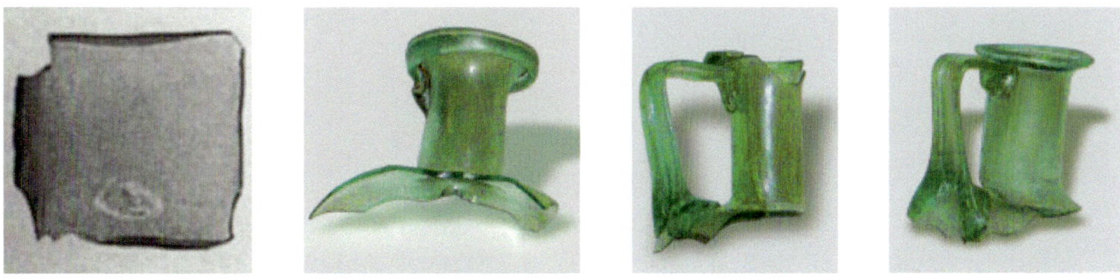

387 388 389 390

## Plate 14

391

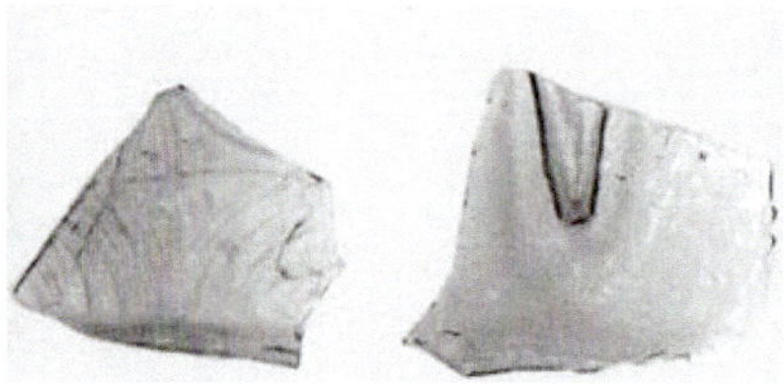
392-393

395

396

397

398

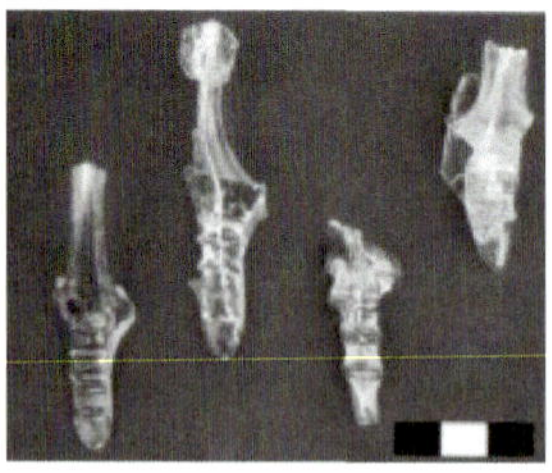
399-402

403

404

405

406

407

408

409

410

411

412

412

413

414

415

416

417

418

419

420

421

# Plate 15

422

423

424

426

427

428

429
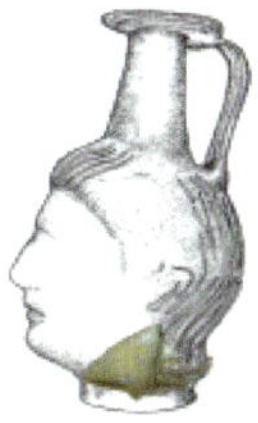
430

430

431

431
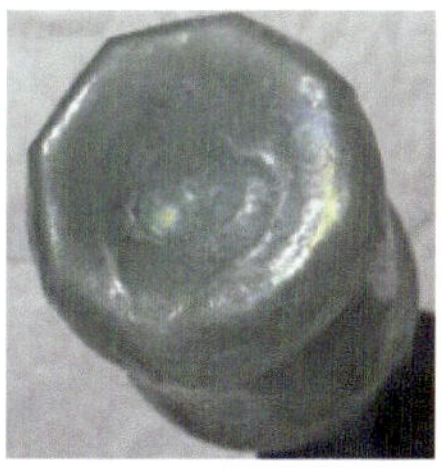
432

432

433

433

434

434

435

436

437-438

439

440

441

442

443

# Plate 16

444 445 446 447 448

449 450 451 452

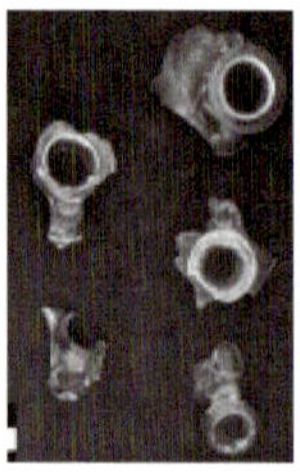

453 454-458 459 460 461 462

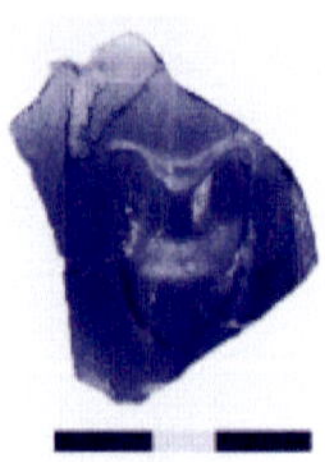

471

463 464 473 472

487-491 493

476 478 479 485 492

# Plate 17

494

495

496

497

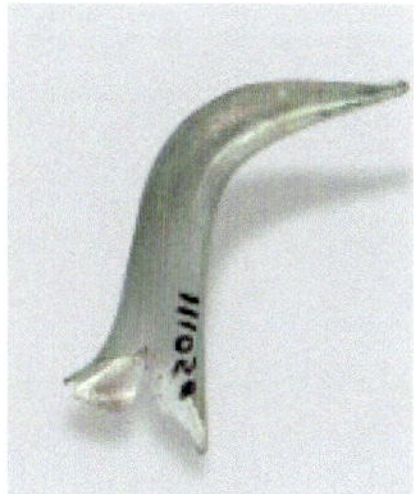
498

499

500

501

502

503

504

505

506

508

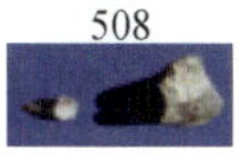
507

511

512

513

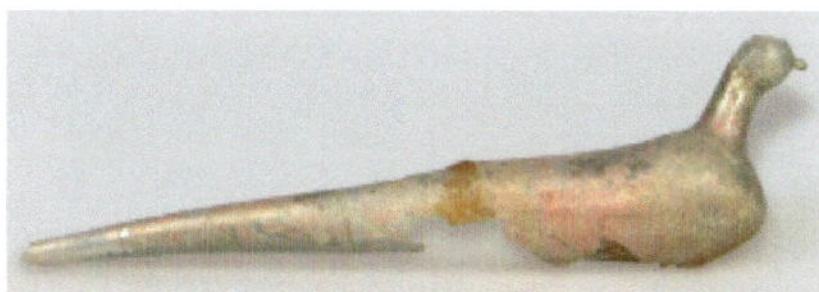
514

515

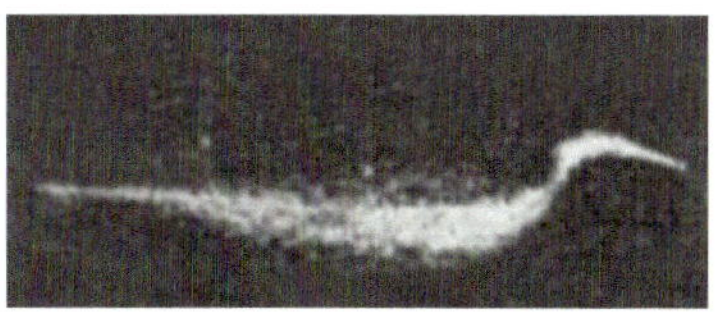
516

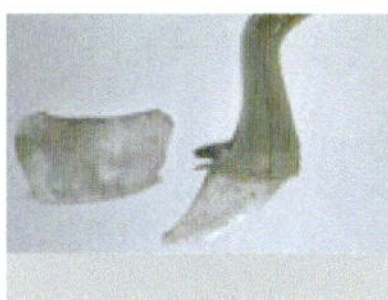
517

518

# Plate 18

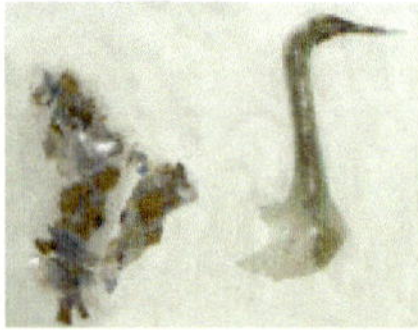
519

520

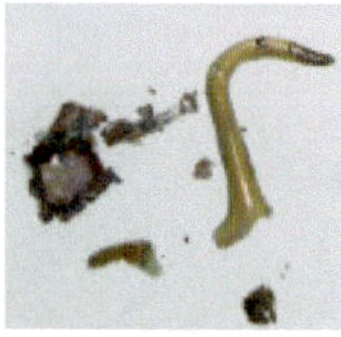
522

523

524

525

526

527

528

530

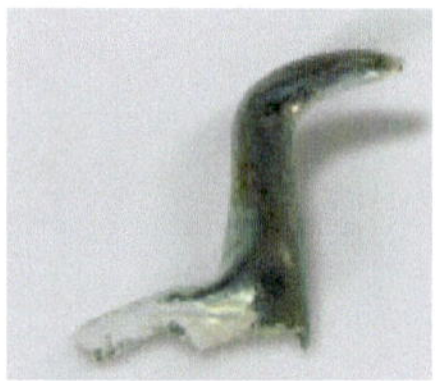
536

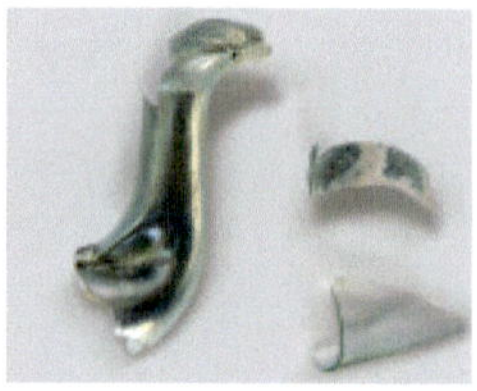
537

538

539

540

541

542

543

544

545

546

548

# Plate 19

549 550 551 552 553

555 556 557 558 559 560 561

562 564 565 566 568 569 571 572

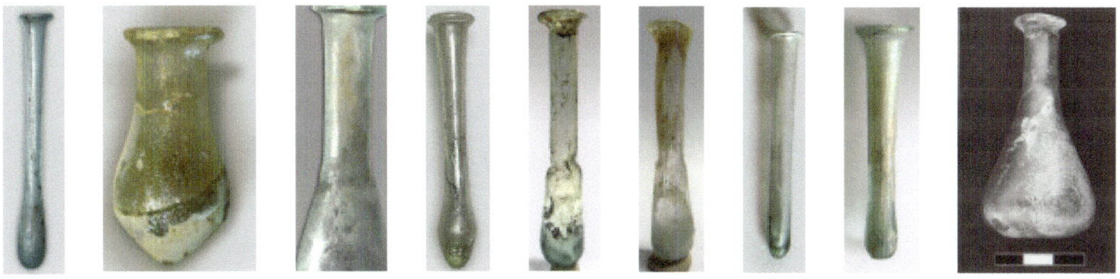

573 574 575 576 577 578 579 580 581

582 583 584 585 586-587 588

# Plate 20

589 590 591 592 593 594 595

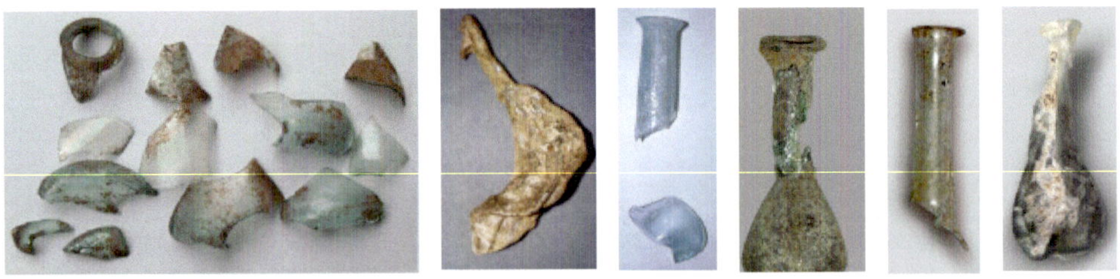

596 597 598 599 600 601

602 603 604 605 606 607 608

609 610 611 612 613 614 615 616 617

618 619 620 621 622 623 625 626 627

# Plate 21

628

629

630

631

632
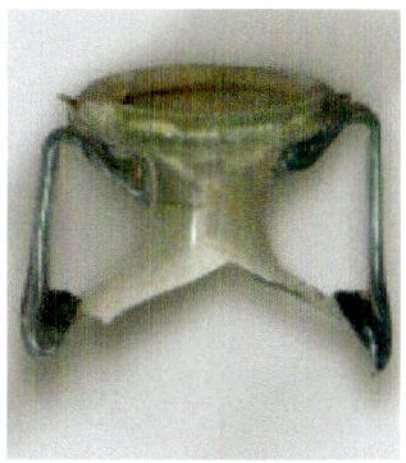
633

634

635

636

637

638

639

640

641
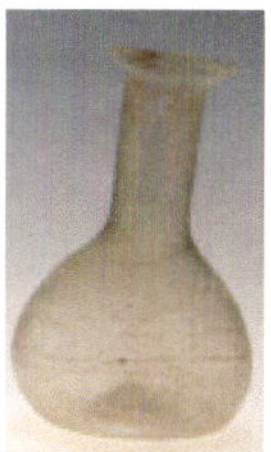
642

643

644

645

646

647

648

649

650

651

652

653

654

655

656

657

658

659

660

661

662

663

# Plate 22

666

667

668

669

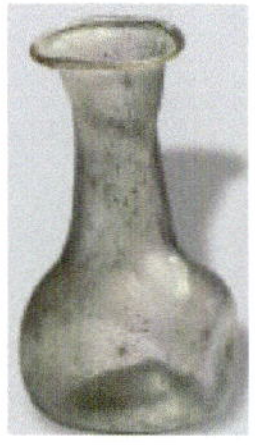
670

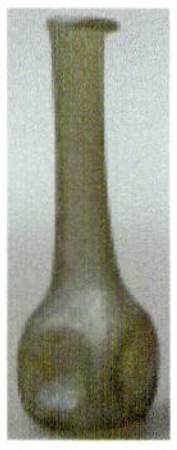
671

672

673

674

675

676

677

678

679

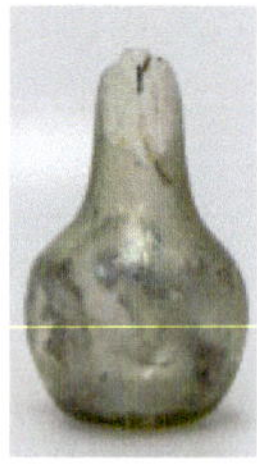
681

682

683

684

685

686

687

688

689

690

691

692

693

694

695

696

698

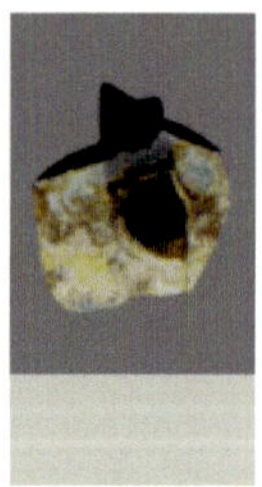
699

# Plate 23

700

701

702

703

704

705

702

703

704

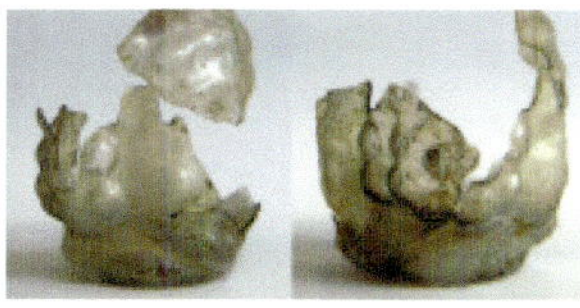

706

707

707

708

711

709

710

# Plate 24

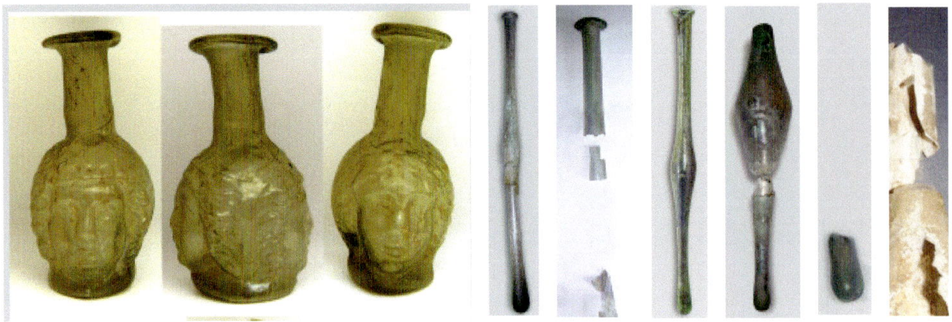

711 713 714 715 716 717 718

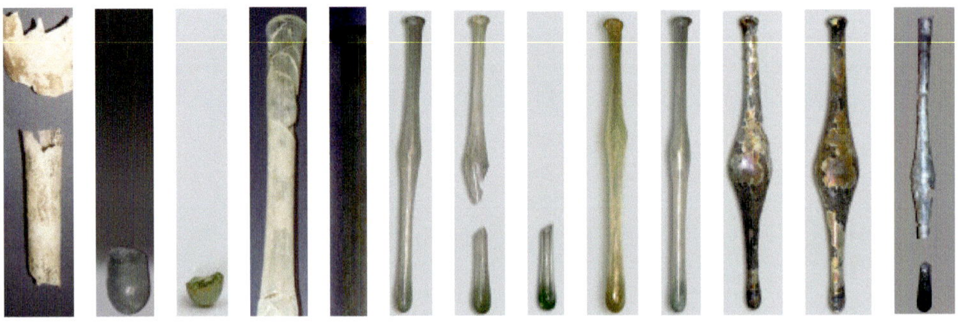

719 720 721 722 723 724 725 726 727 728 729 729 730

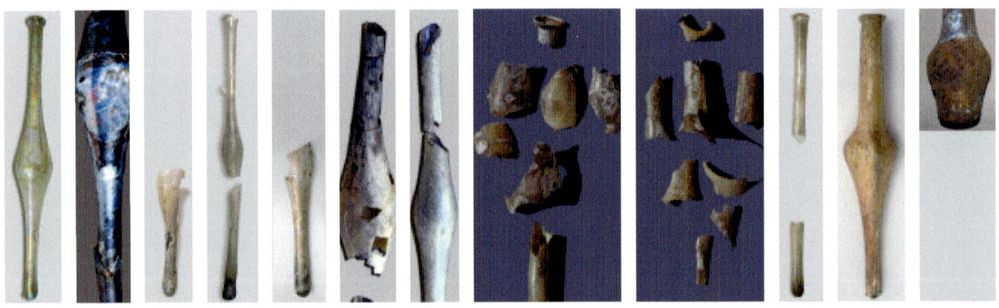

731 732 733 734 735 736 737 738 739 740 742 743

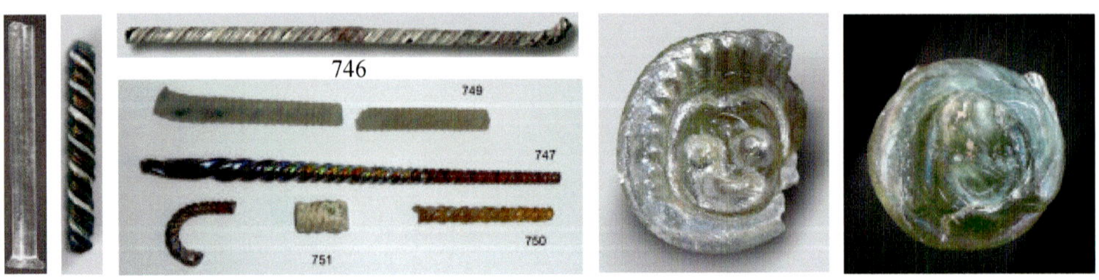


746

744 745 747-752 753 754

# Bibliography and Abbreviations


| | |
|---|---|
| AM | *Archeologia Medievale.* Cultura materiale, insediamenti, territorio, Firenze: Edizioni Clusf. |
| AnnAIHV | Vols. 1-9 Liège: vols. 1-3: Annales du 1er/2e/3e congrès des Journées internationales du Verre; vol. 4: Annales du 4e Congrès International d'Ètude Historique du Verre, vol. 5 ff: Annales du ... Congrès de l'Association Internationale pour l'Histoire du Verre; vols. 10-12 in Amsterdam, vols. 13-14 in Lochem, Netherlands, vol. 15 in Nottingham, UK, vol. 16 in London, vol. 17 in Antwerp, vol. 18 in Thessaloniki, vol. 19 in Piran. |
| *ArchDelt* | *Αρχαιολογικό Δελτίο*, Athens: Ypourgeio Politismou / Tameion Archaiologikon Poron. |
| Archaeologia Bulgarica | *Archaeologia Bulgarica*, Sofia: NOUS Publishers. |
| Archeologia (Warszawa) | *Archeologia,* Rocznik Instytutu Archeologii i Etnologii Polskiej Akademii Nauk, Warszawie: Polskia Akademia Nauk. |
| *ArchEph* | *Αρχαιολογική Εφημερίς*, Athens: He en Athinais Archaiologikē Hetaireia. |
| ArchErgMakThrak | *Αρχαιολογικό Έργο στη Μακεδονία και Θράκη.* Thessaloniki: Aristoteleio Panepistimio Thessalonikis. |
| Archaia Makedonia 4 | *Αρχαία Μακεδονία* IV, Proceedings of the 4th International Conference, Thessaloniki, 1st-25th September 1983, Thessaloniki: Idryma Meleton Chersonisou tou Aimou, 1986. |
| Arheologija (Sofia) | *Arheologija*, Organna Arheologicheskija Instituti Muzej pri B'lgarskata Akademija na Naukite, Sofia: B'lgarskata Akademija na Naukite. |
| ARV2 | J.D. Beazley, *Attic Red-Figure Vase-Painters*, 2nd ed., Oxford1956. |
| Atiqot | *Atiqot.* Journal of the Israel Department of Antiquities, Jerusalem: Israel Antiquities Authority. |
| AV | *ArheološkiVestnik*, Acta Archaeologica, Ljubljana: Slovenska Akademia Znanosti in Umetnosti. |
| Boreas | *Boreas, Münstersche Beiträge zur Archäologie*, Münster: K. Wilmers. |
| BR | *Budapest Régiségei*, A Budapesti Történeti Múzeum Évkönyve, Budapest: A Fovárosi Múzeum Kiadása. |
| BSA | *The Annual of the British School at Athens*, London: MacMillan. |
| BullAIHV | *Bulletin de l'Association internationale pour l'histoire du verre*, vols. 1-9, 1962-1983, Liège: Association internationale pour l'histoire du verre. |
| DeltChrArchEt | *Δελτίο Χριστιανικής Αρχαιολογικής Εταιρείας*, Athens: Christianiki Archaiologiki Etaireia. |
| *EEBΣ* | *Επετηρίς Εταιρείας Βυζαντινών Σπουδών*, Athens: Etaireia Vyzantinon Spoudon. |
| Felix Ravenna | *Felix Ravenna. Rivista di antichità ravennati cristiane e bizantine.* Istituto di Antichità ravennati e bizantine dell'Università degli Studi di Bologna, Faenza: Fratelli Lega. |
| FolArch | *Folia Archaeologica*, Magyar Történeti Múzeum, Budapest: Magyar Nemzeti Múzeum. |
| GAMPlovdiv | *Godišnik na Arheologičeski Muzej*, Plovdiv: Arheologičeski Muzej. |
| GGB | *Godišnjak Grada Beograda*, Beograd: Muzej Grada Beograda. |
| GOTR | *Greek Orthodox Theological Review*, Boston: Holy Trinity. |
| GPSKV | *Gradja za Proučavanje Spomenika Kulture Vojvodine*, Novi Sad: Zavod za zaštitu spomenika kulture. |
| GSAD | *Glasnik Srpskog Arheološkog Društva*, Beograd: Srpsko Arheološko Društvo. |
| GZM | *Glasnik Zemaljskog Muzeja*, Sarajevo: Zemaljski Muzej. |
| GZM(A)ns | *Glasnik Zemaljskog Muzeja Bosne i Hercegovine u Sarajevu*, Arheologija, Nova Serija, Sarajevo: Zemaljski Muzej. |
| IAK | *Izvestiia Arkheologicheskoi komissii*, St. Petersburg. |
| IG | *Inscriptiones Graecae*/consilio et auctoritate Academiae Literarum Regiae Borussicae editae, Berolini: G. Reimer. |
| Iliria | *Iliria*: studime dhe materiale arheologjke, Tirana: Instituti i historise dhe i Gjuhesise. |
| ILS | *Inscriptiones latinae selectae*, ed. H. Dessau, I-III (Berlin, 18921916). |
| INM Varna | *Izvestija na Narodnija Muzej*, Varna: Narodnij Muzej. |
| Instrumentum | *Instrumentum*, Bulletin du Groupe de travail européen sur l'artisanat et les productions manufacturées dans l'Antiquité, Montagnac. |
| JAScien | *Journal of Archaeological Science*, London-New York: Academic Press. |
| JBL | *Journal of Biblical Literature*, Atlanta, GA: The Society of Biblical Literature. |
| LA | *Liber Annuus/Studii Biblici Franciscani*, Jerusalem: Franciscan Printing Press. |

| | |
|---|---|
| LIMC | *Lexicon Iconographicum Mythologiae Classicae* (Zurich and Munich 1974- ). |
| Materialy | *Materialy i Issledovaniya po Arkheologii SSSR*, Moskva: izdatelstvo Akademii Nauk SSSR. |
| MAA | *Macedoniae Acta Archaeologica*, Skoplje: Makedonsko Arheološko Naučno Društvo. |
| NZ | *Niški zbornik*: Publikacija za društveno-istorijska i kulturna pitanja Niša i regiona, Niš: Gradina. |
| Oxford Latin Dictionary | P.G.W. Glare, *Oxford Latin dictionary*, Oxford: Clarendon Press, 1982. |
| OpArch | *Opuscula archaeologica: Radovi Arheološkog Zavoda*, Zagreb: Arheološki Zavod. |
| PAE | *Πρακτικά Αρχαιολογικής Εταιρείας*, Athens: He en Athinais Archaiologikē Hetaireia. |
| Poikila Meletimata | *Ποικίλα Μελετήματα*, Athens: Institouto Ellinikis kai Romaikis Archaiotitos, KEPA, Ethniko Idryma Erevnon. |
| Pontica | *Pontica*, Constanţa: Muzeul de Istorie Naţională şi Arheologie. |
| Rad Vojvodjanskih Muzeja | *Rad Vojvodjanskih Muzeja*, Novi Sad: Vojvodjanski Muzej. |
| Razkopki i Prouchvania | *Razkopki i Prouchvania*, Sofia: B'lgarska Akademiia na Naukite, Arheologicheski Institut i Muzei. |
| Régészeti Füzetek | *Régészeti Füzetek*. Ser. 2, Budapest: Magyar Nemzeti Múzeum Történeti Múzeum. |
| *REG* | *Bulletin épigraphique* de la Revue des Études *grecques, Paris France.* |
| Singidunum | *Singidunum*, Beograd: Arheološki Institut. |
| Sirmium | *Sirmium: Archaeological Investigations in Syrmian Pannonia*, Beograd: Arheološki Institut/ New York: Denison University; The City University of New York. |
| StIr | *Studia Iranica*, Paris: Geuthner. |
| Thessaloniki | *Η Θεσσαλονίκη*, Thessaloniki: Kentro Istorias Thessalonikis, Dimos Thessalonikis. |
| THIE | *Θρησκευτική και Ηθική Εγκυκλοπαίδεια*, Athens: A. Martinos, 1962-1968. |
| Makedonika: *Μακεδονικά*, | σύγγραμμα περιοδικόν της Εταιρείας Μακεδονικών Σπουδών,Thessaloniki: Etaireia Makedonikon Spoudon. |
| Trudy Ermitaza | *Trudy Gosudarstvennogo Ermitaza*, Sankt-Peterburg: Izd-vo Gosudarstvennogo Ermitaza. |
| VAHD | *Vjesnik za Arheologiju i Historiju Dalmatinsku*, Split: Arheološki Muzej. |
| VAPD | *Vjesnik za Arheologiju i Povijest Dalmatinsku*, Split: Arheološki Muzej. |
| Viminacium | *Viminacium*, Beograd: Arheološki Institut. |
| ŽA | *Živa Antika*, Skoplje: Društvo za antički studii na SRM, Seminar na klasična filologija, Filosofski fakultet. |
| ZRNMB | *Zbornik Radova Narodnog Muzeja*, Beograd: Narodni Muzej. |

### Primary Sources

***Acta Sanctorum Iuli***: *Synaxarium Ecclesiae Constantinopoleos, Synaxarium mensis Julii*. Ed. H. Delehaye. *Acta Sanctorum* 62. Brussels, 1902 (repr. Wetteren, Belgium, 1985).

***Anthologia Graeca***: *Anthologia graeca: Epigrammatum Palatina cum Planudea*. Vols. I-III. Ed. Hugo Stadtmüller. Leipzig, 1894, 1899, 1906.

**Bede: *Homilies 17, in Natale Sancti Benedicti Episcopi***: Beda Venerabilis. *Homiliae 17, in Natale Sancti Benedicti Episcopi. Patrologia Latina* 94, cols. 224-28.

***CIL***: *Corpus inscriptionum latinarum*. 18 vols. Berlin, 1862-1989.

**Clement of Alexandria: *Paedagogus***: Clemens Alexandrinus. *Paedagogus. Clément d'Alexandrie. Le pédagogue*. Ed. M. Harl, H.-I. Marrou, C. Matray, and C. Mondésert. 3 vols. *Sources chrétiennes* 70 (1960); 108 (1965); 158 (1970). Paris.

***Codex Justinianus***: *Corpus Juris civilis, Editio Stereotypa Altera: Institutiones, Digesta, Codex Justinianus, Novellae & Opus Schoelli Morte Interceptum*. Ed. Theodorus Mommsen, Rudolf Schoell, Wilhelm Kroll, and Paulus Krueger. 3 vols. Berlin: Weidmann, 1895.

***Codex Theodosianus***: *Theodosiani libri XVI cum constitutionibus Sirmondianis*. 2 vols. in 3 parts. Ed. T. Mommsen and M. Meyer. Berlin, 1905.

***Corpus Glossariorum Latinarum***: *Corpus glossariorum Latinarum*/a Gustavo Loewe incohatum; auspiciis Academiae Litterarum Saxonicae. Composuit, recensuit, edidit Georgius Goetz. Leipzig, 1888-1923.

**Didymus the Blind: *Commentary on Psalms 40–44***: Didymus Caecus. *Commentarii in Psalmos* 40-44.4. *Didymos der Blinde. Psalmenkommentar*, pt. 5, 2-244. Ed. M. Gronewald. Papyrologische Texte und Abhandlungen 12. Bonn, 1970.

**Dioscorides: *De materia medica***: Dioscorides Pedanius. *De materia medica. Pedanii Dioscuridis Anazarbei de materia medica libri quinque*. Ed. M. Wellmann. 3 vols. Berlin, 1907 (vol. 1); 1906 (vol. 2); 1914 (vol. 3) (repr. 1958).

**Evagrius, *Ecclesiastical History***: *The Ecclesiastical History of Evagrius with the Scholia*. Ed. J. Bidez and L. Parmentier. London, 1898 (repr. 1979).

**Georgius Cedrenus, *Synopsis of History***: *Georgius Cedrenus, Compendium historiarum: Georgius Cedrenus Ioannis Scylitzae ope*. Ed. I. Bekker. 2 vols. Corpus scriptorum historiae Byzantinae. Bonn, 1838-1839.

**Hesychius of Alexandria, *Lexicon*:** *Hesychii Alexandrini lexicon*. Ed. K. Latte. 2 vols. Copenhagen, 1953 and 1966.

***IG*:** *Inscriptiones Graecae*. Consilio et auctoritate Academiae Literarum Regiae Borussicae editae. Berlin, 1873-.

***ILS*:** *Inscriptiones latinae selectae*. Vols. I-III. Ed. H. Dessau. Berlin, 1892-1916.

**John Moschus, *Spiritual Meadow*:** Joannes Moschus. *Pratum spiritual. Patrologiae cursus completus (series Graeca) (MPG)* 87.3, 2852-3112. Ed. J.-Migne. Paris, 1857-1866.

**John of Alexandria: *Commentary on Hippocrates' On the Nature of the Child*:** Joannes Alexandrinus. *Commentarii in Hippocratis librum de natura pueri. Scholia in Hippocratem et Galenum*. Ed. F.R. Dietz. Vol. 2, 205-235. Königsberg, 1834 (repr. Amsterdam, 1966).

**Lamprid. *Alex. Severus*:** Aelius Lampridius. *Historia Augusta, Alexander Severus. The Life of Severus Alexander*. Translated by David Magie for the Loeb Classical Library. *Historia Augusta*, vol. 2, 179-215. Cambridge, Mass., 1924.

**Leontius Episcopus Neapolitanus, *Vita S. Symeonis Sali, confessoris*:** Leontius Episcopus Neapolitanus. *Vita S. Symeonis Sali, confessoris. Patrologiae cursus completus (series Graeca) (MPG)* 93, 1669-1746. Ed. J.-Migne. Paris, 1857-1866.

**Martial, *Epigr.*:** M. Val. Martialis. *Epigrammata selecta*. Ed. W.M. Lindsay. Oxford, 1903.

**Michael Glycas, *Chronicle of Events*:** Michael Glycas. *Annales. Michaelis Glycae annales*. Ed. I. Bekker. Corpus scriptorum historiae Byzantinae, 3-457, 460-625. Bonn, 1836.

**Nicephorus Blemmydes, *Making Gold*:** Nicephorus Blemmydes. *Peri tēs ōochrysopoiias. Collection des anciens alchimistes grecs*. Ed. M. Berthelot and C.É. Ruelle. Vol. 2, 452-457. Paris, 1888 (repr. 1963).

**Olympiodorus, *Commentary on Aristotle's* Meteorology:** Olympiodorus. *In Aristotelis meteora commentaria. Olympiodori in Aristotelis meteora commentaria (Commentaria in Aristotelem Graeca* 12.2). Ed. G. Stuve. Berlin, 1900.

**Oribasius, *Medical Compilations*:** Oribasius. *Collectiones medicae. Oribasii collectionum medicarum reliquiae*. Ed. J. Raeder. Vols. 1-4. [*Corpus medicorum Graecorum* 6.1.1–6.2.2.] Leipzig, 1928-1933.

**Paulus Aegineta, *Medical Compendium*:** Paulus. *Epitomae medicae libri septem*. J.L. Heiberg, ed. *Paulus Aegineta*. 2 vols. [*Corpus medicorum Graecorum* 9.1 and 9.2.] Leipzig, 1921-1924.

***PG*:** *Patrologia Graeca*: J.-Migne, ed. *Patrologiae cursus completus (series Graeca)*. 161 vols. in 166 parts. Paris, 1857-1866.

***PL*:** *Patrologia Latina*: J.-Migne, ed. *Patrologiæ Cursus Completus seu Bibliotheca Universalis, Integra, Uniformis, Commoda, Oeconomica, Omnium SS. Patrum, Doctorum Scriptorumque Ecclesiasticorum. Series Latina*. 221 vols. in 222 parts. Paris, 1844-1880.

**Pliny, *Nat. Hist.*:** *C. Plini Secundi naturalis historiae libri XXXVII* recognovit atque indicibus instruxit Ludovicus Janus. Vol. 5, Libb. XXXIII-XXXVII. Leipzig, 1878.

***POxy*:** Bernard Grenfell and Arthur S. Hunt. *The Oxyrhynchus Papyri*. London, 1898.

**Pseudo-zonaras, Lexicon:** *Iohannis Zonarae, lexicon ex tribus codicibus manuscriptis* (ed. J.A.H. Tittman, Leipzig: Crusius1808.

***Roman Inscriptions of Britain*:** R.G. Collingwood, R.Wright (eds), *The Roman Inscriptions of Britain*. Oxford, 1965- .

**Seneca, *Epistulae morales*:** Lucius Annaeus Seneca. *Epistulae morales ad Lucilium*. L.D. Reynolds, ed. *L. Annaei Senecae: Ad Lucilium Epistulae Morales*. 2 vols. Oxford, 1965.

**Statius, *Silvae*: Papini Stati.** *Silvae. Krohni copiis usus iterum edidit Alfredus Klotz*, Leipzig, 1911.

**Strabo, *Geog.*:** Strabo. *Geographica*. A. Meineke, ed. *Strabonis geographica*. 3 vols. Leipzig, 1877 (repr. 1969).

**Suida:** Suidae *Lexicon*, Ed. I. Bekker, Berlin 1854.

**Paulus Silentiarius, *Descriptio Sanctae sophiae*:** Paulus Silentiarius, *Descriptio Sanctae Sophiae et Ambonis*. Ed. I. Bekker Bonn, (*CSHB* 38) 1837.

**Vopiscus, *Historia Augusta*:** *Scriptores historiae Augustae*. Ed. Ernestus Hohl. Leipzig, 1927.

**Zosimos of Panopolis, *Le premier livre*:** *Zōsimos Thēvaios, To prōton vivlion tēs teleutaias apohes Zōsimou Thēvaiou (e cod. Paris. B.N. gr. 2327, fol. 251v)*. M. Berthelot and C.É. Ruelle (eds) *Collection des anciens alchimistes grecs*. Vol. 2, 239-246. Paris, 1888 (repr. 1963).

### Books and Articles

**5i Ellinistiki keramiki:** 5η Επιστημονική Συνάντηση για την Ελληνιστική Κεραμική: Χρονολογικά προβλήματα - Κλειστά σύνολα - Εργαστήρια. Κείμενα και πίνακες (Chania 1997), Athens: Tameion Archaiologikon Poron, 2000.

**6000 years 1997:** *6000 Years of Tradition. Greek Jewelry*. Exhibition Catalogue, Thessaloniki Villa Bianca 21-12-1997/21-2-1998, Athens: Tameion Archaiologikon Poron, 1997.

**9th EBA 2001:** Σωστικές ανασκαφές *1999-2000*, Λουτρά Παράδεισος (Bey Hamam), Thessaloniki: Hellenic Ministry of Culture, 9th Ephorate of Byzantine and Post-Byzantine Antiquities, 2001.

**9th EBA 2002:** Σωστικές ανασκαφές *2001*, Λουτρά Παράδεισος (Bey Hamam), Thessaloniki: Hellenic Ministry of Culture, 9th Ephorate of Byzantine and Post-Byzantine Antiquities, 2002.

**9th EBA 2003:** Σωστικές ανασκαφές *2002*, Λουτρά Παράδεισος (Bey Hamam), Thessaloniki: Hellenic

Minstry of Culture, 9th Ephorate of Byzantine and Post-Byzantine Antiquities, 2003.

**A travers le verre 1989:** AA.VV., *A travers le verre du Moyen Age à la Renaissance*, Rouen, Musées et monuments départementaux de la Seine-Maritime, 1989.

**Abadie, Spyropoulos 1985:** C. Abadie, Th. Spyropoulos, Fouilles à Helléniko (Eva de Thyréatide), *BCH* 109, 1985, 385-454.

**Abadie-Reynal, Sodini 1992:** C. Abadie-Reynal, J.P. Sodini, *La céramique paléochrétienne de Thasos (Aliki, Delkos, fouilles anciennes)*, École Française d'Athènes, Études Thasiennes XIII, Athènes: École Française d'Athènes, 1992.

**Abdul Hak 1965:** S. Abdul Hak, Contribution d'une découverte archéologique récente à l'étude de la verrerie syrienne à l'époque romaine, *JGS* 7, 1965, 26-34.

**Abramić 1959:** M. Abramić, Eine römische Lampe mit Darstellung des Glasblasens, *BJb* 159, 1959, 149-151.

**Adam-Veleni 1997:** P. Adam-Veleni, Βαλανείο προγενέστερο της Αγοράς Θεσσαλονίκης 1989-1999, *ArchErgMakThrak* 11, 1997, 351-364.

**Adam-Veleni 2001:** P. Adam-Veleni, Η δεκάχρονη πορεία εργασιών στην Αρχαία Αγορά Θεσσαλονίκης, in *Adam-Veleni (ed.) 2001*, 15-38.

**Adam-Veleni 2003:** P. Adam-Veleni, Θεσσαλονίκη: Ιστορία και πολεοδομία, in *Grammenos 2003*, 121-162.

**Adam-Veleni *et al.* 1996:** P. Adam-Veleni, Georgaki, I. Zografou, V. Kalavria, K. Boli, G. Skiadaresis, Αρχαία Αγορά Θεσσαλονίκης: Η στρωματογραφία και τα κινητά ευρήματα, *ArchErgMakThrak* 10B, 1996, 501-531.

**Adam-Veleni *et al.* 1998:** P. Adam-Veleni, I. Zografou, V. Kalavria, Aik. Mavromichali, Aik. Boli, Αρχαία Αγορά Θεσσαλονίκης: Η οικοσκευή του βαλανείου, *ArchErgMakThrak* 12, 1998, 85-102.

**Adam-Veleni *et al.* 2000:** P. Adam-Veleni, Georgaki, V. Kalavria, Aik. Boli, Κλειστά σύνολα ελληνιστικών χρόνων από την αγορά της Θεσσαλονίκης, *5i Ellinistiki keramiki*, 2000, 275-297.

**Adam-Veleni (ed.) 2001:** P. Adam-Veleni (ed.), *Αρχαία Αγορά Θεσσαλονίκης 1, πρακτικά διημερίδας για τις εργασίες των ετών 1989-1999*, Thessaloniki: University Studio Press, 2001.

**Adam-Veleni, Konstantoulas 1991:** P. Adam-Veleni, K. Konstantoulas, Ρωμαϊκοί τάφοι σε τμήμα του δυτικού νεκροταφείο Θεσσαλονίκης στην οδό Λαγκαδά, *ArchErgMakThrak* 5, 1991, 221-228.

**Adam-Veleni, Poulaki, Tzanavari 2003:** P. Adam-Veleni, E. Poulaki, K. Tzanavari, *Αρχαίες αγροικίες σε σύγχρονους δρόμους. Κεντρική Μακεδονία/Ancient Country Houses on Modern Roads, Central Macedonia*, Athens: Tameion Archaiologikon Poron, 2003.

**Agallopoulou 1973/74:** P. Agallopoulou, Πάτρα, οδός Κωνσταντινουπόλεως 82, Στρατοκλέους και Καρόλου, *ArchDelt* 29, 1973/74, Chronika B2, 369, pl. 235β.

**AIECMM, Bakirtzis 2003:** Association internationale pour l'étude des céramiques médiévales en Méditerranée, Ch. Bakirtzis, *VIIe Congrès International sur la Céramique médiévale en Méditerranée, Thessaloniki, 11-16 Octobre 1999: actes*, Athens: Tameion Archaiologikon Poron, 2003.

**Akat, Firatli, Kocabaş 1984:** Y. Akat, N. Firatli, H. Kocabaş, *Catalogue of Glass in the Hüseyin Kocabaş Collection*, Istanbul: Arkeoloji ve Sanat Yayınları, 1984.

**Akrivopoulou 2014:** S. Akrivopoulou, Η ανασκαφή στην οδό Βασιλέως Ηρακλείου 45 στη Θεσσαλονίκη, [The excavation at 45 Vaslileos Irakleiou Street, Thessaloniki] *AEMT* 24, 255-262.

**Aktseli, Manakidou 1997:** D. Aktseli, E. Manakidou, Αρωματικά φυτά, αρώματα και μυροδοχεία: χρήσεις τους στην καθημερινή και θρησκευτική ζωή κατά τους αρχαϊκούς και κλασικούς χρόνους, in *Farmakeftika kai aromatika fyta. Paradosiakes chriseis kai dynatotites axiopoiisis tous. 7th triimero ergasias*, Cyprus, Paralimni, 21-25 March 1997, Athens: Politistiko Technologico Idryma of ETBA, 83-109.

**Al 'Ush 1964:** A.F. Al 'Ush, Les verres conservés au département de l'art arabe musulman du Musée National de Damas, *BullAIHV* 3, 1964, 55-64.

**Alexandrescu 1966:** P. Alexandrescu, Necropola tumulară. Săpături 1955-1961, in *Histria. Monografia arheologica*, vol. II, Bucuresti: Academia R.S.R., 1966, 133-295.

**Alexandri 1963:** O. Alexandri, Άργος, οικόπεδο Νικ. Π. Κούρου, *ArchDelt* 18, 1963, Chronika B1, 61-62, pl. 78α-β.

**Alexandri 1967a:** O. Alexandri, Αθήνα, οδός Κωνσταντινουπόλεως και Ελευσινίων, *ArchDelt* 22, 1967, Chronika B1, 97.

**Alexandri 1967b:** O. Alexandri, Αθήνα, οδός Κωνσταντινουπόλεως και Λένορμαν, *ArchDelt* 22, 1967, Chronika B1, 98.

**Alexandri 1968:** O. Alexandri, Athina, οδός Μ. Αλεξάνδρου και Σαλαμίνος 47, *ArchDelt* 23, 1968, Chronika B1, 75.

**Alexandri 1969:** O. Alexandri, Αθήνα, οδός Πυλάρου, *ArchDelt* 24, 1969, Chronika B1, 70.

**Alexandri 1970:** O. Alexandri, Αθήνα, οδός Αχιλλέως και Σαλαμίνος, *ArchDelt* 25, 1970, Chronika B1, 43.

**Alexandri 1972:** O. Alexandri, Αθήνα, οδός Κερατσινίου 54 και Πλάτωνος (Οικόπεδο Ζορμπά), *ArchDelt* 27, 1972, Chronika B1, 115-118, pl. 76.

**Alexandri 1973/74a:** O. Alexandri, Θεσσαλονίκη, Πανεπιστημιούπολις, Χώρος Κτηρίου Διοικήσεως, *ArchDelt* 29, 1973-1974, Chronika B3, 669-670, pls. 482-483.

**Alexandri 1973/74b:** O. Alexandri, Θεσσαλονίκη, οδός Φιλικής Εταιρείας 39 και Δημ. Μαργαρίτη, ArchDelt 29, 1973-1974, Chronika B3, 673-674, pl. 486 α-β.

**Alexandri 1973/74c:** O. Alexandri, Αθήνα, οδός Διαμαντοπούλου, Καλλιρρόης και Περραιβού, *ArchDelt* 29, 1973-1974, Chronika B1, 128-131.

**Alexiou 1966:** St. Alexiou, Ηράκλειον, Κρήτη, *ArchDelt* 21, 1966, Chronika B2, 408, pl. 437στ'.

**Alexiou 1975:** St. Alexiou, Ηράκλειο, Χερσόνησσος, *ArchDelt* 30, 1975, Chronika B2, 340-341, pl. 246α΄β.

**Allamani-Souri 1991a:** V. Allamani-Souri, Βέροια, οδός Σταδίου, *ArchDelt* 46, 1991, Chronika B2, 301-302.

**Allamani-Souri 1991b:** V. Allamani-Souri, Βέροια, οδός Ακροπόλεως, *ArchDelt* 46, 1991, Chronika B2, 302-303.

**Allamani-Souri 1991c:** V. Allamani-Souri, Τιμητικές επιγραφές από την αρχαία Βέροια, *ArchErgMakThrak* 5, 1991, 32-48.

**Amitos 1987:** *Αμητός: Τιμητικός τόμος για τον καθηγητή Μανόλη Ανδρόνικο*, Thessaloniki: Aristoteleio Panepistimio Thessalonikis, 1987.

**Amrein, Nenna 2006:** H. Amrein, M.-D. Nenna, Inventaire des moules destinés à la fabrication des contenants en verre, in *Foy, Nenna (eds) 2006b*, 491-502.

**Andreiomenou 1973/74:** A. Andreiomenou, *Ερέτρια*, *ArchDelt* 29, 1973/74, Chronika, 465, pl. 301ζ.

**Angelkou 2002:** E. Angelkou, Χωροθέτηση παραγωγικών και εργαστηριακών δραστηριοτήτων στο επισκοπικό συγκρότημα των Λουλουδιών Πιερίας, in *Αρχαιολογικά τεκμήρια βιοτεχνικών εγκαταστάσεων κατά τη βυζαντινή εποχή, 5ος-15ος αι.* Ειδικό θέμα *22ου Συμποσίου Βυζαντινής και Μεταβυζαντινής Αρχαιολογίας και Τέχνης*, Αθήνα 17-19 Μαΐου, 2002, Athens 2002, 61-72.

**Antonaras 2001:** A. Antonaras, Early Christian silk fabrics from Thessaloniki, *Museum of Byzantine Culture* 8, 2001, 38-44.

**Antonaras 2002a:** A. Antonaras, Late Antique gold embroidered silks. New finds from Thessaloniki, in *8th Annual Meeting of European Association of Archaeologists*, 24-29 September 2002, Thessaloniki, Program of Abstracts, Thessaloniki, 18-19.

**Antonaras 2002b:** A. Antonaras, Υαλοποιία και υαλουργία στο ρωμαϊκό και παλαιοχριστιανικό κόσμο. Υαλουργική δραστηριότητα στη Θεσσαλονίκη, *ArchDelt* 57, 2002 (2009), Meletes, 237-260.

**Antonaras 2003a:** A. Antonaras, The use of Glass in Byzantine jewelry- the evidence from Northern Greece (4th-16th centuries), *16th AnnAIHV*, London, 2003, 331-334.

**Antonaras 2003b:** A. Antonaras, Πώς ένας λαγός από τη Συρία βρέθηκε στο Δίον, Θεσσαλονικέων Πόλις, period 2nd, vol. 10, April 2003, 77-80.

**Antonaras 2005:** A. Antonaras, Γυάλινα πτηνόμορφα αγγεία, *AAA* 35-38 (2002-2005), 2005, 241-248.

**Antonaras 2006a:** A. Antonaras, New Glass Finds with Printed Bases from Thessaloniki, in *Foy, Nenna (eds) 2006b*, 413-420.

**Antonaras 2006b:** A. Antonaras, Γυάλινα μεσοβυζαντινά βραχιόλια. Συμβολή σε θέματα διάδοσης, παραγωγής, τυπολογίας και χρήσης, *DeltChrArchEt*, period 4th, vol. 27, 2006, 423-434.

**Antonaras 2006c:** A. Antonaras, Ενδείξεις δραστηριότητας εργαστηρίων υαλουργίας στη Θεσσαλονίκη, 2nd International Conference, Ancient Greek Technology, Athens: TEE, 17-21 October 2005, 247-253.

**Antonaras 2006d:** A. Antonaras, A Hare-in-a-Flask from Dion, Macedonia, *17th International Congress of the International Association for the History of Glass*, Antwerp, 2006, 70-73.

**Antonaras 2007:** A. Antonaras, Early Christian Glass Finds from the Museum's Basilica, Philippi, *JGS 49, 2007*, 47-56.

**Antonaras 2008a:** A. Antonaras, Glass and Obsidian Plaques from the Apostle Paul's Basilica at Kephalari, Argos, *JGS*, 50, 2008, 298-302.

**Antonaras 2008b:** A. Antonaras, Glass Lamps of the Roman and Early Christian Periods. Evidence from the Thessaloniki Area, in: C.-A. Roman, N. Gudea (eds), Lychnological Acts 2. Acts of the 2nd International Congress on Ancient and Middle Age Lighting Devices (Zalău – Cluj -Napoca, 13th-18th May 2006). Trade and Local Production of Lamps from Prehistory until the Middle Ages, Cluj-Napoca 2008, 23-30.

**Antonaras 2009:** A. Antonaras, Glass Vessels from Roman and Early Christian Thessaloniki and its surroundings (1st cent. BC - 6th cent. AD), in 17th *AnnAIHV*, 74-79.

**Antonaras 2010a:** A. Antonaras, Glass Doves and Globes from Thessaloniki. North Italian Imports or Local Products? in *Quaderni Friulani Di Archeologia* 19, 2010 (*Atti del Convegno Intorno all'Adriatico. La diffusione e la produzione di vetro sulle sponde del mare Adriatico nell'antichità*), Udine: Società Friulana di Archeologia, 27-33.

**Antonaras 2010b:** A. Antonaras, Glass-working activities in Late Roman and Early Christian Thessaloniki. Local Workshops and Vessels, in *Drauschk,e Keller (eds) 2010*, 93-105.

**Antonaras 2010c:** A. Antonaras, Glassware in Late Antique Thessalonikē, in *Nasrallah, Bakirtzis, Friesen (eds) 2010*, 301-334.

**Antonaras 2010d:** A. Antonaras, Early Christian and Byzantine Glass Vessels: Forms and Uses, in *Daim, Drauschke (eds) 2010*, 383-430.

**Antonaras 2010e:** A. Antonaras, Γυάλινα και κεραμικά αγγεία 3ου-7ου αι. μ.Χ. από τη Θεσσαλονίκη. Δύο όψεις του ίδιου νομίσματος; (*Glass and ceramic vases of the 3rd-7th c. AD. Two sides of the same coin?*) in *Papanikola-Bakirtzi, Kousoulakou (eds) 2010*, 309-319.

**Antonaras 2010f:** A. Antonaras, Window Glass, in *Glass Kosmos 2010*, 112–13.

**Antonaras 2010g:** A. Antonaras, Roman and Early Christian Mold-blown Vessels from Thessaloniki and its Region. 1st-5th c. AD, in *Fontaine (ed.) 2010*, 241-252.

**Antonaras 2011a:** A. Antonaras, Production of glass lamps in Early Christian Thessaloniki, in *Motsianos Bintsi (eds) 2011*, 126-128.

**Antonaras 2011b:** A. Antonaras, Stemmed beakers with stamped bases from an Early Christian glass workshop at the centre of Thessaloniki, in *Foy, Nenna (eds) 2011*, 239-242.

**Antonaras 2012:** A. Antonaras, *Fire and Sand. Ancient Glass in the Princeton University Art Museum*, Yale University Press, 2012.

**Antonaras 2014a:** A. C. Antonaras, An Early Christian glass workshop at 45, Vasileos Irakleiou Street in the centre of Thessaloniki, in Keller, Price, Jackson (eds) 2014, 95-113.

**Antonaras 2014b:** A. Antonaras, The Production and Uses of Glass in Byzantine Thessaloniki, in *Entwistle, James (eds) 2014*, 189-198.

**Antonaras 2014c:** A. Antonaras, Ορειχάλκινα ευρήματα από το ιερό της βασιλικής κάτω από την Αγία Σοφία Θεσσαλονίκης, (Bronze finds from the apse of the basilica under Haghia Sophia in Thessaloniki), in *34th Symposium of Byzantine and Post Byzantine Archaeology and Art held by Christian Archaeological Society, Athens, May 9-11, 2014, Program of abstracts*, Athens 2014, 21-22.

**Antonaras 2014d:** A. Ch. Antonaras, Βασιλέως Ηρακλείου 45, Θεσσαλονίκη. Το παλαιοχριστιανικό εργαστήριο υαλουργίας, (45 Vasileos Irakleiou street, Thessaloniki: The Early Christian Glass Workshop), *ArchErgMakThrak* 24, 2010, 263-272.

**Antonaras 2014e:** A. Ch. Antonaras, Posude iz provincijske balkanske bazilike, Bazilika Solinos na Halkidikiju, Grčka/The Glass Finds From a Provincial Early Christian Balkan Basilica. The Solinos Basilica in Chalkidiki, Greece, *Nova antička Duklja* V, 2014, 97-110.

**Antonaras 2016a:** A. Ch. Antonaras, *Arts, Crafts and Trades in Ancient and Byzantine Thessaloniki*, Mainz: RGZM.

**Antonaras 2016b:** A. Ch. Antonaras, Ορειχάλκινα ευρήματα από το ιερό της βασιλικής κάτω από την Αγία Σοφία Θεσσαλονίκης, in M. Giannopoulou, Ch. Kallini, (eds), *Τιμητικός τόμος για τη Στέλλα Δρούγου, ήχάδιν*, τ. I Athens: Tameion Archaiologikon Poron, 2016, 117-132.

**Antonaras in press 1:** A. Ch. Antonaras, Ανασκαφή επί της οδού Βασιλίσσης Σοφίας 44. Τα ανασκαφικά συμφραζόμενα καλυκωτής γυάλινης λυχνίας, *ArchErgMakThrak* 26, 2013, in print.

**Antonaras in press 2:** A. Ch. Antonaras, A three-handled, calyx-shaped glass lamp from Thessaloniki and its archaeological context, in *Lighting in Byzantium*. 4th International ILA Round Table. Museum of Byzantine Culture, Thessaloniki, Greece, 11-14 October 2011. International Lychnological Association and Archaeological Institute of Macedonian and Thracian Studies, in preparation.

**Antonaras, Anagnostopoulou-Chatzipolychroni 2002:** A. Antonaras, I. Anagnostopoulou-Chatzipolychroni, Γυάλινα ευρήματα από το Αρχαίο Ίωρο, in *Kordas (ed.) 2002*, 113-122.

**Antonaras, Chrysostomou 2012:** A. Ch. Antonaras, A. Chrysostomou, A Secondary Glass Workshop in Ancient Edessa, *19th AnnAIHV*, 293-303.

**Antonaras, Keramaris 2012:** A. Ch. Antonaras, A. Keramaris, Evidence on the sealing of glass globes (Isings Form 10). A short note, *19th AnnAIHV*, 94-97.

**Antonaras, Kordas (eds) 2002:** A. Antonaras, G. Kordas, *Ιστορία και Τεχνολογία του Αρχαίου Γυαλιού*, Athens: *GLASNET*, 2002.

**Antonaras, Lamprothanasi in press 1:** A. Antonaras, E. Lamprothanasi, Ανασκαφή δυτικής νεκρόπολης Θεσσαλονίκης. Οικόπεδο οδού Μοναστηρίου 53-55. Τα γυάλινα ευρήματα (Excavation of the Western Necropolis of Thessaloniki. 53-55 Monastiriou Street. The Glass Finds), *AAA*, in press.

**Archontidou-Argyri 1988:** A. Archontidou-Argyri, Μυτιλήνη, οδός Πέτρας, *ArchDelt* 43, 1988, Chronika B2, 458-459.

**Arveiller-Dulong, Arveiller 1985:** V. Arveiller-Dulong, J. Arveiller, *Le Verre d'époque romaine au Musée archéologique de Strasbourg*, Paris: Ministère de la Culture. Éditions de la Réunion des Musées Nationaux, 1985.

**Arveiller-Dulong, Nenna 2000:** V. Arveiller-Dulong, M.-D. Nenna, *Les verres antiques I, contenants à parfums en verre moulé sur noyau et vaisselle moulée. VIIe siècle avant J.-C.-Ier siècle après J.-C.*, Paris: Réunion des Musées Nationaux, 2000.

**Arveiller-Dulong, Nenna 2011:** V. Arveiller-Dulong, M.-D. Nenna, *Les Verres antiques du Musée du Louvre. 3, Parures, instruments et éléments d'incrustation*, Paris: Éditions Somogy.

**Ascough 2004:** R. Ascough, A Question of Death: Paul's Community-building language in 1 Thessalonians 4: 13-18, *JBL* 123/3 (2004), 509-530.

**Asimakopoulou-Atzaka 1998:** P. Asimakopoulou-Atzaka, *Σύνταγμα των παλαιοχριστιανικών ψηφιδωτών δαπέδων της Ελλάδος III. Μακεδονία-Θράκη 1. Τα ψηφιδωτά δάπεδα της Θεσσαλονίκης* (Byzantine Monuments 9), Thessaloniki: ΟΡΡΕΡΤΗ'97-ΑΡΤΗ/ΚΒΕ, 1998.

**Aupert 1980:** P. Aupert, *Objets de la vie quotidienne à Argos en 585 ap. J.-C.*, in *Études Argiennes*, Bulletin de Correspondance Hellénique, supplément VI, Athènes: EFA, 1980, 395-457.

**Auth 1973:** S.H. Auth, The Eugene Shaefer Collection of Ancient Glass at the Newark Museum: a New Look at an Old Collection, *AnnAIHV* 6, 1973, 25-30.

**Auth 1975:** S.H. Auth, Roman Glass, in *Clairmont 1975*, 145-175.

**Auth 1976:** S.H. Auth, *Ancient Glass at the Newark Museum from the Eugene Schaefer Collection of Antiquities*, New Jersey: Newark Museum, with assistance from The Ford Foundation and The National Endowment for the Arts, 1976.

**Auth 1996:** S.H. Auth, DRINK MAY YOU LIVE! Roman motto glasses in the context of Roman life and death, *AnAIHV 13*, 103-112.

**Avezou, Picard 1912:** Ch. Avezou, Ch. Picard, La nécropole de Thessalonique, *Mélanges d'archéologie et d'histoire* XXXII, 1912, 337-361.

**Avigad (ed.) 1976:** N. Avigad, *Beth She'Arim, Report on the excavations during 1953-1958*, vol. III: catacombs 12-23, New Brunswick, N.J., Rutgers Univ. Press on behalf of the Israel Exploration Society and the Institute of Archaeology, Hebrew University, 1976.

**Avshalom-Gorni, Tacher 2002:** D. Avshalom-Gorni, A. Tacher, *Excavations at Khirbet el-Shubeika 1991, 1993, Eretz Zafon, Studies in Galilean Archaeology*. Jerusalem: Israel Antiquities Authority, 2002.

**Ayalon 1994:** E. Ayalon, A Roman-Byzantine Mausoleum at Kh. Sabiya, Kefar Sava, *Atiqot* XXV, 1994, 27-39, 189-190.

**Bagati, Milik 1958:** B. Bagatti, J.T. Milik, *Gli Scavi del Dominus Flevit, I*, Jerusalem: Pubblicazioni dello Studium Biblicum Franciscanum, 1958.

**Bagatti 1967:** B. Bagatti, I vetri del Museo Francescano di Nazaret, *LA* XVII (1967), 220-240.

**Bailey 1992:** D. M Bailey, A Grave Group from Cyzicus, *JGS* 34, 1992, 27-34.

**Băjenaru, Bâltâc 2000-2001:** C. Băjenaru, A. Bâltâc, Depozitul de candele din sticlă descoperit la bazilica episcopalâ de la Histria, *Pontica* 33-34, 2000-2001, 469-513.

**Bakirtzis 1983:** Ch. Bakirtzis, Ρωμαϊκός λουτρών και η Αχειροποίητος της Θεσσαλονίκης, *Αφιέρωμα στη μνήμη του Στυλιανού Πελεκανίδη*, Thessaloniki: Etaireia Makedonikon Spoudon, 1983, 310-329.

**Baldoni 1987:** D. Baldoni, Una lucerna romana con raffigurazione di officina vetraria: alcune considerazioni sulla lavorazione del vetro soffiato nell'antichità, *JGS* 29, 1987, 22-29.

**Baldoni, Berti 1996:** D. Baldoni, F. Berti, Il vetro di Iassos (Caria) nel quadro delle produzioni nel Mediterrraneo Orientale, in *Il Vetro dall' Antichità all' Età Contemporanea: Aspetti Technologici, Funzionali e Commerciali*, Atti 2e Giornate Nazionali di Studio AIHV-Comitato Nazionale Italiano, Milano 14-15 Dicembre 1996, [s.l., s.n.], 73-86.

**Ballance *et al.* 1989:** M. Ballance, J. Boardman, S. Corbet, S. Hood, *Excavations in Chios 1952-55: Byzantine Emporio*, [Athens]: British School of Archaeology at Athens; [London]: Thames & Hudson, 1989.

**Baluta 1978:** C.L. Baluta, Considérations sur la production et la diffusion du verre dans la Dacie Supérieure, *AnnAIHV 7*, 1978, 97-109.

**Baluta 1981:** C.L. Baluta, Fond de moule romain pour la fabrication de bouteilles carrées trouvé à Apulum (Dacie), *AnnAIHV* 8, 1981, 111-114.

**Barag 1963:** D. Barag, The Glassware, in *Yadin 1963*, 101-110.

**Barag 1967a:** D. Barag, The Glass Vessels, in *Dothan, Fredman 1967*, 36-37.

**Barag 1967b:** Dan Barag, review on C.W. Clairmont, *The excavations at Dura-Europos, Final Report IV, Part V: The Glass Vessels*, New Haven, Dura Europos Publications, 1963, *Israel Exploration Journal* 17, 1967, 196-198.

**Barag 1967c:** D. Barag, The Glass, in *Prausnitz* 1967, 65-70.

**Barag 1970a:** D. Barag, Syropalestinian Flasks within Flasks, *Atiqot* series in Hebrew 6, 1970, 8, 74-75.

**Barag 1970b:** D. Barag, Glass Pilgrim Vessels from Jerusalem, Part I, *JGS* 12, 1970, 35-63.

**Barag 1970c:** D. Barag, *Glass Vessels of the Roman and Byzantine Periods in Palestine*, unpublished PhD thesis (in Hebrew), Hebrew University, Jerusalem, 1970.

**Barag 1971:** D. Barag, Glass Pilgrim Vessels from Jerusalem, Part II-III, *JGS* 13, 1971, 45-63.

**Barag 1976:** D. Barag, Glass Vessels, in *Avigad (ed.) 1976*, 198-213.

**Barag 1978:** D. Barag, *Hanita Tomb XV. A Tomb of the Third and Early Fourth Century CE, Atiqot XIII*, 1978.

**Barag 1985a:** D. Barag, *Catalogue of Western Asiatic Glass in the British Museum*, vol. 1, London: British Museum Publications, 1985.

**Barag 1985b:** D. Barag, Recent Important Epigraphic Discoveries Related to the History of Glassmaking in the Roman World, *AnnAIHV* 10, 1985, 109-116.

**Barag 1991:** D. Barag, The Contribution of Masada to the History of Early Roman Glass, in *Newby, Painter (eds) 1991*, 137-140.

**Baramki 1937:** D.C. Baramki, An Early Byzantine Synagogue near Tell es Sultan, Jericho, *QDAP* 6, 1937, 73-77.

**Baratte (ed.) 1988:** F. Baratte (ed.), *Argenterie romaine et byzantine*, Actes de la table ronde, Paris 11-13 octobre 1983, organisée sous le patronage du Centre National de la Recherche Scientifique et de l'Université de Paris-Sorbonne, avec une contribution du Ministère des Affaires Étrangères, Paris: CNRS, 1988.

**Baratte 1989:** F. Baratte, La verrerie dans l'Afrique romaine: état des questions, *KölnJb* 22, 1989, 141-149.

**Barkóczi 1966-7:** L. Barkóczi, Die datierten Glasfunde aus dem II. Jahrhundert von Brigetio, *FolArch* XVIII, 1966-7, 67-89.

**Barkóczi 1968:** L. Barkóczi, Die datierten Glasfunde aus dem 3-4 Jahrhundert von Brigetio, *FolArch* XIX, 1968, 59-86.

**Barkóczi 1971:** L. Barkóczi, Plastisch verzierte Spätromische Glasfunde aus Pannonien, *FolArch* XXII, 1971, 71-83.

**Barkóczi 1986:** L. Barkóczi, Geschliffene Gläser aus der ersten Hälfte des 3. Jhs. in Pannonien, *AErt* 113, 1986, 166-189.

**Barkóczi 1988:** L. Barkóczi, *Pannonische Glasfunde in Ungarn*, Budapest: Akadémiai Kiadó, 1988.

**Barkóczi 1996:** L. Barkóczi, *Antike Gläser*, Roma: 'L'Erma' di Bretschneider, 1996.

**Bass 1986:** G.F. Bass, A Bronze Age Shipwreck at Ulu Burun (Kaş): 1984 Campaign, *AJA 90.3*, (Jul., 1986), 269-296.

**Baur 1938:** P.V.C. Baur, Glassware, in *Kraeling (ed.) 1938*, 505-546.

**Bavant, Kondić, Spieser (eds) 1990:** B. Bavant, V. Kondić, J.-M. Spieser (eds), *Caričin Grad II. Le quartile sud-ouest de la ville haute*, Belgrade-Rome: Institut Archéologique de Belgrade, École Française de Rome, 1990.

**Baziopoulou-Valavani:** E. Baziopoulou-Valavani, Αντίκυρα, οικόπεδο Ι. Φαράντου, *ArchDelt* 35, 1980, Chronika B1, 262-263, pl. 114.

**BCH, Reduction of:** La Rédaction, Chroniques des fouilles et découvertes archéologiques dans l'Orient hellénique, *BCH* 45, 1921, 487-568.

**Belivanova 1999:** A. Belivanova, Early Roman Glass from Bulgaria (1st-the First half of the 2nd century), *Archaeologia Bulgarica* III, 1999, 3-49.

**Benkő 1962:** P. Benkő, Űvegcorpus, *Régészeit Fűzetek*, ser. II, 11 szám, Budapest: Nemzeti Múzeum, Történeti Múzeum, 1962.

**Beretta (ed.) 2004:** M. Beretta (ed.), *When Glass Matters, Studies in the History of Science and Art from Graeco-Roman Antiquity to Early Modern Era*, Florence: Leo S. Olschki Editore, 2004.

**Berger 1960:** L. Berger, *Römische Gläser aus Vindonissa*, (Veröffentlichungen der Gesellschaft Pro Vindonissa IV), Basel: Verlag Birkhäuser, 1960.

**Bermont-Montanari (ed.) 1983:** G. Bermont-Montanari (ed.), *Ravenna e il porto di Classe. Venti anni di ricerche archeologiche tra Ravenna e Classe*, Bologna: University Press, 1983.

**Bernhard 1979:** H. Bernhard, *Römische Gläser in Worms*, Worms: Museum der Stadt Worms, 1979.

**Bezborodov, Abdurazakov 1964:** M.A. Bezborodov, A.A. Abdurazakov, Newly Excavated Glassworks in the USSR, 3rd-14th centuries AD, *JGS* 6, 1964, 64-69.

**Bianchi *et al.* 2002:** R.S. Bianchi, B. Schlick-Nolte, G.M. Bernheimer, and D. Barag (eds). *Reflections on Ancient Glass from the Borowski Collection: Bible Lands Museum, Jerusalem*, Mainz: von Zabern, 2002.

**Biaggio-Simona 1991:** S. Biaggio-Simona, *I vetri romani provenienti dalle terre dell' attuale Cantone Ticino*, 2 vv., Locarno: Armando Dadò Editore, 1991.

**Biaggio-Simona 2002:** S. Biaggio-Simona, Bicchieri in vetro della necropolis romana di Moghegno (Canton Ticino), *Quaderni Ticinesi di Numismatica e Antichità Classiche* 31, 2002, 287-314.

**Biernacki *et al.* 1976:** A. Biernacki *et al.*, Novae-Sector Zachodni, 1975, *Archeologia* (Warszawa) 27, 1976, 137-170.

**Biernacki *et al.* 1991:** A. Biernacki *et al.*, Novae-Sector Zachodni, 1988, *Archeologia* (Warszawa) 41, 1991, 106-111.

**Bikić 1997:** V. Bikić, Rezultati zaštitnih arheoloških iskopavanja u Knez Mihailovoj ulici br: 46-48, *Singidunum* 1, 1997, 157-168.

**Bikić, Nikolić-Djordjević, Simić 2001:** V. Bikić, S. Nikolić-Djordjević, Z. Simić Istraživanja na lokalitetu Studenski Trg br. 9, *Singidunum* 3, 2001, 183-220.

**Billewicz 1971:** E. Billewicz, Novae-Sektor Zachodni 1970, *Archeologia* (Warszawa) 21, 1971, 136-137.

**Bimson, Freestone 1987:** M. Bimson, I. Freestone, The Discovery of an Islamic Glassmaking Site in Middle Egypt, *AnnAIHV* 10, 1987, 237-243.

**Bitrakova 1966:** V. Bitrakova, Rimski grob od Zajas, *Zbornik na Arheološkiot Muzej-Skopje* 1961-1966, IV-V, 95-97.

**Boli, Skiadaresis 2001:** A. Boli, G. Skiadaresis, H στρωματογραφία στη Νότια Πτέρυγα, in *Adam-Veleni (ed.) 2001*, 87-104.

**Bommelaer, Grandjean 1972:** J.F. Bommelaer, Y. Grandjean, Quartier sud d'Argos, *BCH* 96, 1972, 155-228.

**Bonnet-Borel 1997:** F. Bonnet-Borel, *Le verre d'époque romaine à Avenches-Aventicum, Typologie générale*, Documents du Musée romain d'Avenches 3, Montreux: Imprimerie Corbaz S.A., 1997.

**Boosen 1984:** M. Boosen, *Antike Gläser*, Kassel: Staatliche Kunstsammlungen, 1984.

**Boura, Parani 2008:** L. Boura, M.G. Parani, *Lighting in Early Byzantium*, Washington, DC: Dumbarton Oaks Research Library and Collection, 2008.

**Boyd 1988:** S. Boyd, A Bishop's Gift: Openwork Lamps from the Sion Treasure, in *Baratte (ed.) 1988*, 191-209.

**Boyd 1992:** S. Boyd, A "Metropolitan" Treasure from a Church in the Provinces: An Introduction to the Study of the Sion Treasure, in *Boyd, Mundell-Mango (eds) 1992*, 5-38.

**Boyd, Mundell-Mango (eds) 1992:** S. Boyd, M. Mundell-Mango, *Ecclesiastical Silver Plate in Sixth-Century Byzantium*. Papers of the Symposium held May 16-18, 1986 at The Walters Art Gallery, Baltimore and Dumbarton Oaks, Washington, DC. Organized by S.A. Boyd, M. Mundell-Mango and G. Vikan/ edited by Susan A. Boyd and Marlia Mundell Mango, Washington, DC: Dumbarton Oaks Research Library and Collection Washington, DC, 1992.

**Braat 1963:** W.C. Braat, The Glass Collection of the Rijksmuseum van Oudheden at Leiden, *OudhMeded* 44, 1963, 103-112.

**Braun, Haevernick 1981:** K. Braun, T.E. Haevernick, *Das Kabirenheiligtum bei Theben Bd. IV, Bemalte Keramik* *und Glas aus dem Kabirenheiligtum bei Theben*, Berlin: Walter de Gruyter & Co, 1981.

**Brill 1967:** R.H. Brill, A Great Slab from Ancient Galilee, *Archeology* 20, 1967, 88-95.

**Brill 1999:** R.H. Brill, *Chemical Analyses of Early Glasses*, 1. Catalogue, 2. The Tables, Corning, NY: Corning Museum of Glass Press, 1999.

**Brosh 2003:** N. Brosh, Glass in the Islamic Period, in *Israeli 2003*, 319-383.

**Brouskari 2002:** E. Brouskari, Παλαιοχριστιανικό εργαστήριο υαλουργίας στην Κω: Μία πρώτη παρουσίαση, in *Themelis (ed.) 2002*, 135-141.

**Brun 1999:** J.-P. Brun, Laudatissimum fuit antiquitus in Delo insula. La maison IB du Quartier du stade et la production des parfums à Délos, *BCH* 123.1 (1999), 87-155.

**Brun 2000:** J.-P. Brun, The Production of Perfumes in Antiquity: The Cases of Delos and Paestum, *AJA* 104 (2000), 277-308.

**Bruneau 1970:** P. Bruneau, Tombes d'Argos, *BCH* 94, 1970, 437-531.

**Buck, Mattingly (eds), 1985:** J. Buck, D.J. Mattingly (eds), *Town and Country in Roman Tripolitania*. Papers in Honour of Olwen Hackett, BAR International Series 274. Society for Libyan Studies Occasional Papers II, Oxford: Society for Lybian Studies, 1985.

**Buckton (ed.) 1984:** D. Buckton (ed.), *Byzantium: Treasures of Byzantine Art and Culture from British Collections*, London: British Museum Press, 1994.

**Bucovală 1968:** M. Bucovală, *Vase antice de sticlă la Tomis*, Constanţa: Muzeul de arheologie, 1968.

**Bucovala 1979:** M. Bucovala, Pièces romaines d'importation sur le littoral Ouest-Pontique, *AnnAIHV* 8, 1979, 89-95.

**Bucovală 1984:** M. Bucovală, Roman Glass Vessels Discovered in Dobrudja, *JGS* 26, 1984, 59-63.

**Bucovală, Paşca 1992:** M. Bucovală, C. Paşca, Cercetări în necropola Romană devesta Tomisului (1992), *Pontica* XXV, 1992, 241-272.

**Buechner 1960:** T.S. Buechner, The Glass from Tarrha, *Hesperia* 29, 1960, 109-117.

**Bujukliev 1984:** H. Bujukliev, Trakiiska mogilna grobnica ot s. M'drec, starozagorski okr'g, *Razkopki i Prouchvania* 10, 1984, 7-33.

**Bujukliev 1986:** H. Bujukliev, Trakiiskiat mogilen nekropol pri Chatalka, starozagorski okr'g, *Razkopki i Prouchvania* 16, 1986, 5-151.

**Bulat 1976:** M. Bulat, Antičko staklo u muzeju Slavonije, *AV* 25, 1974, (1976), 89-101.

**Buljević 2002a:** Z. Buljević, Stakleni Štapići, in *Marin (ed.) 2002*, 297-312.

**Buljević 2002b:** Z. Buljević, Stakleni Balzamariji, in *Marin (ed.) 2002*, 383-455.

**Buljević 2009:** Z. Buljević, Traces of Glassmmakers in the Roman Province of Dalmatia, *Quaderni Friulani di Archeologia* 19 (2009)(*Atti del Convegno Intorno all'Adriatico. La diffusione e la produzione di vetro sulle sponde del mare Adriatico nell'antichità. Triest-Piran, 30-31 maggio 2009*), 35-50. Trieste: Editreg, 2009.

**Burger 1966:** A.Sz. Burger, The Late Roman Cemetery at Ságvár, *ActaArchHung* XVIII, 1966, 99-235.

**Burić, Čače, Fadić 2001:** T. Burić, S. Čače, I. Fadić, *Sv. Juraj od Putalja*, Split: Muzej Hrvatskih Arheoloških Spomenika Split, 2001.

**Cabrol, Leclerq:** F. Cabrol, H. Leclerq, *Dictionnaire d'archéologie chrétienne et de liturgie*, Paris: Letouzey et Ané, 1907-1953.

**Calvi 1968:** M.C. Calvi, *I vetri romani del Museo di Aquileia*, Aquileia: Associazione nazionale per Aquileia, 1968.

**Cambi 1976:** N. Cambi, Neki kasnoantički predmeti od stakla s figurlanim prikazima u Arheološkom Muzeju u Splitu, *AV* 25, 1974, (1976), 139-154.

**Campbell 1997:** Sh. Campbell (ed.), *The Malcove Collection. A Catalogue of the Objects in the Lillian Malcove Collection of the University of Toronto*, Toronto-Buffalo-London: University of Toronto Press 1997.

**Carazzetti, Biaggio-Simona 1988:** R. Carazzetti, S. Biaggio-Simona, *Vetri Romani del Cantone Ticino*, exhibition catalogue in Museo Civico e Archeologico Castello Visconteo 2-9/13-11-1988, Locarno: Museo civico e Archeologico, 1988.

**Carboni 2001:** S. Carboni, *Glass from Islamic Lands*, New York, in association with the al-Sabah Collection, Dar al-Athar al-Islamiyyah, Kuwait National Museum, 2001.

**Carboni, Whitehouse 2001:** S. Carboni, D. Whitehouse with contributions by R.H. Brill and W. Gudenrath, *Glass of the Sultans*, The Metropolitan Museum of Art, New York - The Corning Museum of Glass, Corning - Benaki Museum, Athens, New Haven and London: Yale University Press/New York: The Metropolitan Museum of Art, 2001.

**Carducci, 1968:** C. Carducci, *Arte romana in Piemonte*, Turin: Istituto Bancario San Paolo di Torino, 1968.

**Carington-Smith 1982:** J. Carington-Smith, A Roman Chamber Tomb on the South-East Slopes of Monasteriaki Kephala, Knossos, *BSA* 77, 1982, 255-293.

**Caron 1997:** B. Caron, Roman Figure-Engraved Glass in the Metropolitan Museum of Art, *Metropolitan Museum Journal* 32, 1997, 19-50.

**Cermanović-Kuzmanović 1971:** A. Cermanović-Kuzmanović, Oblici i hronologija rimskog staklenog materijala iz nekropole Municipia S..., *ŽA* XXI/1, 1971, 287-301.

**Cermanović-Kuzmanović 1976:** A. Cermanović-Kuzmanović, Pregled i razvitak rimskog stakla u Crnoj Gori, *AV* 25, 1974, (1976), 175-190.

**Cermanović-Kuzmanović 1981a:** A. Cermanović-Kuzmanović, Antičko staklo iz Sirmijuma, *GPSKV* X, 1981, 7-29.

**Cermanović-Kuzmanović 1981b:** A. Cermanović-Kuzmanović, *Rimska Keramika*, Beograd: Privredno finansijski vodič, 1981.

**Cermanović-Kuzmanović 1987:** A. Cermanović-Kuzmanović, *Rimsko staklo*, Beograd: Naučna Knjiga, 1987.

**Cermanović-Kuzmanović 1992:** A. Cermanović-Kuzmanović, Einige Typen der Glasgefasse aus der römischen Nekropole in Kolovrat bei Prijepolje, *ZRNMB* 16/1, 1992, 387-390.

**Cermanović-Kuzmanović *et al.* 1975:** A. Cermanović-Kuzmanović, O. Velimirović-Žižić, D. Srejović, *Antička Duklja Nekropole*, Cetinje: Obod, 1975.

**Cesnola 1903:** L. Palma di Cesnola, *A Descriptive Atlas of the Cesnola Collection of Cypriote Antiquities in the Metropolitan Museum of Art, New York*, v. 3, New York: the Metropolitan Museum of Art, 1903.

**Charleston 1964:** R.J. Charleston, Wheel-Engraving and -Cutting: Some Early Equipment. I. Engraving, *JGS* 6, 1964, 83-100.

**Charleston 1965:** R.J. Charleston, Wheel-Engraving and -Cutting: Some Early Equipment. II Water-Power and Cutting, *JGS* 7, 1965, 41-54.

**Charlesworth 1966:** D. Charlesworth, Roman Square Bottles, *JGS* 8, 1966, 26-40.

**Chasiotis (ed.) 1997:** I. Chasiotis (ed.), *Tois agathois vasilevousa*. Thessaloniki Istoria kai politismos, Thessaloniki: OPPETH'97/Paratiritis, 1997.

**Chatzi 1972:** D. Chatzi, Μυτιλήνη, οδός Π. Καβέτσου και Ξενοφώντος (οικόπεδο Ι. Κλειδαρά), *ArchDelt* 27, 1972, Chronika B2, 593.

**Chatzi 1973:** D. Chatzi, Μυτιλήνη, Στρατόπεδο Κουρτζή, *ArchDelt* 28, 1973, Chronika B2, 509, pl. 476.

**Chavane 1975:** M.-J. Chavane, *Salamine de Chypre, VI, Les petits objets*, Paris: Université Lyon II-Maison de l'orient Méditerranéen ancien, 1975.

**Chéhab 1986:** M.H. Chéhab, *Fouille de Tyr, La nécropole IV, Description des fouilles, Bulletin du Musée de Beyrouth* 36, 1986.

**Cherneva-Tilkian 1997:** S. Cherneva-Tilkian, Tipologia i hronologia na stklenite balsamarii ot Filipopol, *GAMPlovdiv* VIII, 1997, 42-56.

**Chevalier 1997-98:** P. Chevalier, Les luminairies paléochrétiens de Dalmatie romaine, *VAHD* 90-91, 1997-98, 161-199.

**Choremis 1973:** Ang. Choremis, Χαλκίς, οικόπεδο Εω. Γερωνίτη, *ArchDelt* 28, 1973, Chronika B1, 299-301.

**Choremis 1974:** Ang. Choremis, Ειδήσεις εξ Ευβοίας. Ι. Αρχαίον νεκροταφείον εις Άγιον Στέφανον Χαλκίδος, *AAA* 7/1, 1974, 27-34.

**Crowfoot *et al.* 1957:** J.W. Crowfoot, G.M. Crowfoot, K.M. Kenyon, *Samaria-Sebaste: Reports of the Work of the joint Expedition in 1931-1933 and of the British Expedition in 1935, v. 3. The Objects from Samaria*, London: Palestine Exploration Fund, 1957.

**Chrysostomou 1984:** P. Chrysostomou, Ανασκαφή στο ΝΔ νεκροταφείο της Νικόπολης, *Ipeirotika Chronika* 26, 1984, 9-37.

**Chrysostomou 1999-2001:** P. Chrysostomou, Το ταφικό ιερό μυστών του Διονύσου στη Μενιήδα Βοττιαίας, *AAA* 23-24, 1999-2001, 195-220.

**Chrysostomou 2000:** P. Chrysostomou, Το ταφικό ιερό μυστών του Διονύσου στη Μενιήδα Βοττιαίας: η ανασκαφή του 2000, *ArchErgMakThrak* 14, 2000, 455-471.

**Chrysostomou A. 2000:** A. Chrysostomou, *Αρχαία Έδεσσα: Τα νεκροταφεία*, unpublished PhD thesis submitted to the Aristotle University of Thessaloniki, 2000.

**Chrzanovski 2003:** L. Chrzanovski, *Lumière ! L'éclairage dans l'antiquité*. Exhibition catalogue, Musée romain de Nyon, Suisse, 15 mai 2003-30 avril 2004, Milano: Musée romain, Service de la Culture, 2003.

**Clairmont 1963:** C.W. Clairmont, *The Excavations at Dura-Europos, Final Report* 4, pt. 5, *The Glass Vessels*, New Haven, Connecticut: Dura-Europos publications, 1963.

**Clairmont 1975:** C.W. Clairmont, *Excavations at Salona, Yugoslavia (1969-1972)*, Park Ridge, NJ: Noyes Press, 1975.

**Clairmont 1977:** C.W. Clairmont, *Catalogue of Ancient and Islamic Glass*, Athens: Benaki Museum, 1977.

**Cool 1995:** H.E.M. Cool, Glass Vessels of the Fourth and Early Fifth Century in Roman Britain, in *Foy (ed.) 1995*, 11-24.

**Cool, Jackson, Monaghan 1999:** H.E.M. Cool, C.M. Jackson, J. Monaghan, Glass-Making and the sixth Legion at York, *Britannia* 30, 1999, 147-162.

**Cool, Price 1995:** H.E.M. Cool and J. Price, *Colchester Archaeological Report 8: Roman vessel glass from excavations in Colchester, 1971-85*, Colchester: Colchester Archaeological Trust Ltd and English Heritage, 1995.

**Coscarella 1994:** A. Coscarella, I vetri dell'Antiquarium di Bosra, *XLI Corso di Cultura Sull'Arte Ravennate e Bizantina (12-16 settembre 1994)*, Ravenna 1994, 387-407.

**Čremošnik 1965:** I. Čremošnik, Rimska vila u Višićima, *GZM (A)* ns 20, 1965, 147-260.

**Crowfoot 1957:** G.M. Crowfoot, *Glass*, in *J.W. Crowfoot et al. 1957*, 403-422.

**Crowfoot, Harden 1931:** G.M. Crowfoot, D.B. Harden, Early Byzantine and Later Glass Lamps, *JEA* 17, 1931, 196-208.

**Curina 1983:** R. Curina, Vetri, in *Bermont-Montanari (ed.) 1983*, 166-170.

**Czurda-Ruth 1979:** B. Czurda-Ruth, *Die Römischen Gläser von Magdalensberg*, Kärntner Museumsschriften, no. 65 and Archäologische Forschungen zu den Grabungen auf dem Magdalensberg, no. 6, Klagenfurt: Verlag des Landesmuseums für Karnten, 1979.

**Daffa-Nikonanou 1987:** A. Daffa-Nikonanou, Δύο επείσακτα αγγεία από τον τάφο της Νεάπολης Θεσσαλονίκης, *Amitos* 1, 1987, 263-276.

**Dahari *et al.* 2000:** U. Dahari, with contributions by R. Calderon, W.D. Cooke, Y. Gorin-Rosen and O. Shamir, *Monastic Settlements in South Sinai in the Byzantine Period. The Archaeological Remains*, Jerusalem: Israel Antiquities Authority, 2000.

**Daim, Drauschke (eds) 2010:** F. Daim, J. Drauschke, *Byzanz - Das Römerreich im Mittelalter*, vol. 1, *Welt der Ideen, Welt der Dinge*, Monographien des Römisch-Germanischen Zentralmuseums 84. Mainz: Römisch-Germanischen Zentralmuseums, 2010.

**Dakoronia 1987:** F. Dakoronia, Άγιος Ιωάννης Θεολόγος, οικόπεδο Μ. Τρανούλη, *ArchDelt* 42, 1987, Chronika B1, 228-231, pl. 134.

**Damaskos (ed.) 2004:** D. Damaskos (ed.), *Χάρις Χαίρε. Μελέτες στη μνήμη της Χάρις Κάντζια*, vol.A'-B', Athens: ΥΠΠΟ, Archaiologiko Instituto Aigiakon Spoudon, 2004.

**Damevski 1976:** V. Damevski, Pregled tipova staklenog posudja iz italskih, galskih, mediteranskih i porajnskih radionica na području Hrvatske u doba rimskog carstva, *AV* 25, 1974, (1976), 62-87.

**Dauphin, Edelstein 1984:** C. Dauphin, G. Edelstein, *L' Église byzantine de Nahariya (Israël). Étude archéologique*, Thessaloniki: KBE/ΑΠΘ, 1984.

**Dauterman-Maguire, Maguire, Duncan-Flowers (eds) 1989:** E. Dauterman-Maguire, H. Maguire, M.J. Duncan-Flowers (eds), *Art and Holy Powers in the Early Christian House* (exhibition catalogue), Kranert Art Museum of the University of Illinois at Urbana-Champaign, 25 August - 1 October 1989, The Kelsey Museum of Archeology of Michigan in Ann Arbor, 27 October 1989 - 29 April 1990, Urbana: Krannert Art Museum, University of Illinois at Urbana-Champaign: University of Illinois Press, 1989.

**Daux 1958:** G. Daux, Chroniques de fuilles, *BCH* 82, 1958, 644-830.

**Davaras 1985 :** K. Davaras, Ρωμαϊκό νεκροταφείο Αγίου Νικολάου, *ArchEph* 1985, 130-211.

**Davidson 1952:** G.R. Davidson, *Corinth: Results of Excavations Conducted by The American School of Classical Studies at Athens*, v. 12, *The Minor Objects*, Princeton, New Jersey: ASCSA, 1952.

**Davidson-Weinberg 1972:** G. Davidson-Weinberg, Mold-Blown Beakers with Mythological Scenes, *JGS* 14, 1972, 26-47.

**De Franciscis 1963:** A. De Franciscis, Vetri Antichi Scoperti ad Ercolano, *JGS* 5, 1963, 137-139.

**De Maine 1979:** M.R. De Maine, *Roman Glass from Diocletian's Palace at Split, Yugoslavia*, unpublished PhD thesis submitted in the University of Minnesota, 1979.

**De Maine 1985:** M.R. De Maine, Ancient Glass Distribution in Illiricum: The Emona Ladles, *AnnAIHV* 10, 1985, 135-160.

**De Maine 1988:** M.R. De Maine, The Northern Necropolis at Emona, *AnnAIHV* 11, 1988, 129-144.

**De Matteis, De Tommaso 2001:** L. De Matteis, G.D. De Tommaso, Vetri, in *Di Vita 2001*, V.3*. t. I, 190-250, V.3** tav. XXXV-XXXVIII.

**De Ridder 1909:** A. de Ridder, *Collection de Clercq, Catalogue. Tome VI: Les terres cuites et les verres*, Paris: Ernest Leroux, 1909.

**De Tommaso 1990:** G.D. De Tommaso, *Ampullae vitreae. Contenori in vetro di unguenti e sostanze aromatiche dell'Italia Romana (I sec. A.C. - III sec. D.C.)*, Roma: G. Bretschneider, 1990.

**De Tommaso 2001:** G.D. De Tommaso, Il vetro e i profumi, in *Massabò 2001*, 141-172.

**Dekoulakou 1975a:** If. Dekoulakou, Πάτρα, οδός Έλληνος Στρατιώτου 73, *ArchDelt* 30, 1975, Chronika B1, 113-114.

**Dekoulakou 1975b:** If. Dekoulakou, Πάτρα, οδός Καραϊσκάκη116, *ArchDelt* 30, 1975, Chronika B1, 115.

**Dekoulakou 1975c:** If. Dekoulakou, Πάτρα, Μεσολογγίου 32, *ArchDelt* 30, 1975, Chronika B1, 115-117.

**Dekoulakou 1975d:** If. Dekoulakou, Μαμουσιά Αιγιαλίας, *ArchDelt* 30, 1975, Chronika B1, 120.

**Dekoulakou 1976:** If. Dekoulakou, Πάτρα, οδός Ασημάκη Φωτήλα, *ArchDelt* 31, 1976, Chronika B1, 103-104.

**Dell'Acqua and Silva (eds) 2001:** F. Dell'Acqua, R. Silva (eds). *Il Colore nel medioevo: arte, simbolo, tecnica. La Vetrata in Occidente dal IV all'XI secolo. Atti delle giornate di studi, Lucca, 23 - 25 settembre 1999.* Lucca, 2001.

**Delougaz, Haines 1960:** P. Delougaz, R.C. Haines, *A Byzantine Church at Khirbat al-Karak*, University of Chicago Oriental Institute Publications, v. 85, Chicago: University of Chicago Press, 1960.

**Deroy 1985:** L. Deroy, Sur deux poèmes grecs d'époque romaine décrivant un verrier au travail, *AnnAIHV 10*, 1985, 183-184.

**Deroy 1989:** L. Deroy, Sur deux poèmes grecs d'époque romaine décrivant un verrier au travail, *L'Antiquité Classique* 58, 1989, 178-184.

**Despoini 1976:** Aik. Despoini, Μακρύγιαλος, αγρός 493 (Χαρ. Αθανασιάδη), *ArchDelt* 31, 1976, Chronika B2, 249-250, pl. 198.

**Despoini 1982:** Aik. Despoini, Κεραμικοί κλίβανοι Σίνδου, *ArchEph* 1982, 61-84.

**Di Vita (ed.), 2001:** A. Di Vita (ed.), *Gortina v. 3. Lo scavo del Pretorio (1989-1995). I materiali-Testo*, Monografie della Scuola Archeologica Italiana di Atene e delle Misioni Italiane in Oriente, Padova: Bottega d'Erasmo, Aldo Ausilio Editore, 2001.

**Di Vita, Martin (eds) 1997:** A. Di Vita, A. Martin (eds) *Gortina II. Pretorio. Il materiale degli scavi Colini, 1970-1977*, Monografie della Scuola Archeologica Italiana di Atene e delle Missioni Italiane in Oriente, v. 7. Padova: Bottega d'Erasmo, Aldo Ausilio Editore, 1997.

**Dilly, Mahéo 1997:** G. Dilly, N. Mahéo, *Verreries antiques du Musée de Picardie*, Amiens: Musée de Picardie/ Paris: Éditions d'art Somogy, 1997.

**Djingov 1964:** G. Djingov, Le verre dans l'Antiquité et au Moyen Age en Bulgarie, *AnnAIHV* 3, 1964, 104-115.

**Doppelfeld 1959:** O. Doppelfeld, Der Muschelpokal von Köln, *BJb* 159, 1959, 152-166.

**Doppelfeld 1966:** O. Doppelfeld, *Römischer und fränkisches Glass in Köln*, Köln: Greven, 1966.

**Dörig 1975:** J. Dörig, *Art Antique, Collections privées de Suisse Romande*, Genève: Éditions Archéologiques de l'Université de Genève, 1975.

**Dothan, Fredman 1967:** M. Dothan, D.N. Fredman (eds), *Ashdod I, The First Season of Excavations 1962*, Atiqot VII, Jerusalem: Israel Antiquities Authority, 1967.

**Dragendorff 1903:** H. Dragendorff, *Thera II. Theräische Gräber*, Berlin: von Georg Reimer, 1903.

**Drauschke, Keller (eds) 2010:** J. Drauschke and D. Keller (eds), *Glass in Byzantium - Production, Usage, Analyses.* RGZM - Tagungen 8, Mainz: Römisch-Germanisches Zentralmuseum. 2010.

**Drča 2000:** S. Drča, Rimsko staklo u Niškom Muzeju, *GSAD* 15-16 (1999-2000), 209-229.

**Drosogianni 1963:** F. Drosogianni, Θεσσαλονίκη, Αγία Σοφία, *ArchDelt* 18, 1963, Chronika B2, 235-242.

**Du Plat Taylor, Megaw 1981:** J. Du Plat Taylor, A.H.S. Megaw, Excavations at Ayios Philon. The Ancient Carpasia. Part II. The Early Christian buildings, *RDAC*, 1981, 209-250.

**Duncan-Jones 1994:** J. Duncan-Jones, Glass from the Roman Circus at Carthage: Preliminary Report and Comments on a Fragmentary Roman Head-shaped Mold-blown Vessel, *Archeological News* 19, 1994, 11-14.

**Dusenbery 1967:** E. Dusenbery, Ancient Glass from the Cemeteries of Samothrace, *JGS* 9, 1967, 34-49.

**Dusenbery 1971:** E. Dusenbery, Ancient Glass in the Collections of Wheaton College, *JGS* 13, 1971, 9-33.

**Dusenbery 1998:** E. Dusenbery, *Samothrace, The Nekropoleis, Catalogues of Objects by Categories*, Bollingen Series LX.11, Princeton, NJ: Princeton University Press, 1998.

**Dussart 1998:** O. Dussart, *Le verre en Jordanie et en Syrie du sud*, Beyrouth: Institut Français d'Archéologie du Proche-Orient, Beyrouth-Damas-Aman. Bibliothèque Archéologique et historique. T. CLII, 1998.

**Dyczek 1998:** P. Dyczek, A Glass Atelier from sector IV Novae, in *Von Bülow, Milcheva (eds) 1998*, 99-104.

**Edgar 1905:** C.C. Edgar, *Catalogue général des antiquités égyptiennes du Musée du Caire. Nos 32401-32800. Graeco-Egyptian Glass*, Cairo: Institut Français d'Archéologie Orientale, 1905.(photostatic reprint: Otto Zeller Verlag. Osnabrück, 1974.)

**Eisen 1927:** G. A. Eisen (assisted by F. Kouchakji), *Glass: Its Origins, History, Chronology, Technique and Classification to the Sixteenth Century*, New York: William Edwin Rudge, 1927.

**Eiwanger 1981:** J. Eiwanger, *Demetrias IV. Keramik und Kleinfunde aus der Damokratia - Basilika in Demetrias*, Bonn: R. Habelt, 1981.

**Eleftheratou (ed.) 2006:** S. Eleftheratou (ed.), *Το Μουσείο και η ανασκαφή. Ευρήματα από τον χώρο ανέγερσης του Νέου Μουσείου Ακρόπολης* (exhibition catalogue), 3-7-2006/-, Athens: Organismos Anegersis Neou Mouseiou Akropolis, 2006.

**Eleftheriadou 1987:** K. Eleftheriadou, Θεσσαλονίκη, Δημοτικό Νοσοκομείο "Άγιος Δημήτριος", *ArchDelt* 42, 1987, Chronika B2, 407.

**Eleftheriadou 1988:** K. Eleftheriadou, Θεσσαλονίκη, Δημοτικό Νοσοκομείο "Άγιος Δημήτριος", *ArchDelt* 43, 1988, Chronika B2, 393-394.

**Eleftheriadou 1989a:** K. Eleftheriadou, Ανασκαφή χριστιανικού κοιμητηρίου στην περιοχή του νοσοκομείου "Άγιος Δημήτριος", στη Θεσσαλονίκη, *ArchErgMakThrak* 3, 1989, 271-282.

**Eleftheriadou 1989b:** K. Eleftheriadou, Θεσσαλονίκη, Κοιμητήριο Δημοτικού Νοσοκομείου (Ανατολικό νεκροταφείο), *ArchDelt* 44, 1989, Chronika B2, 336-337.

**Eleftheriadou 1990:** K. Eleftheriadou, Θεσσαλονίκη, οδός Λαπηθών 7, *ArchDelt* 45, 1990, Chronika B2, 332-334.

**Engemann 1969:** J. Engemann, Bemerkungen zu spätrömischen Gläsern mit Goldfoliendekor, *JbAC* 11/12, 1968/1969, 7-25, pls. 1-10.

**Entwistle, James (eds) 2014:** C. Entwistle, L. James (eds), *New Light on Old Glass: Recent Research on byzantine Mosaics and Glass*, London: British Museum 2014.

**Erath, Lehner, Schwarz (eds) 1997:** G. Erath, M. Lehner and G. Schwarz (eds), *Komos Festschrift für Thuri Lorenz zum 65. Geburtstag*, Vienna 1997.

**Erdmann 1977:** E. Erdmann, Die Glasfunde von Mezad Tamar (Kasr Gehainije) in Israel, *Saalburg Jahrbuch* 34, 98-146.

**Fadić 1986:** I. Fadić, Četverokutni i poligonalni stakleni antički recipijenti iz Arheološkog Muzeja Istre, *Istraživanja u Istri i Hrvatskom primorju*, Izdanja Hrvatskog Arheološkog Društva, 11/2, 1986, 99-118.

**Fadić 1988:** I. Fadić, Antičko staklo Asserije iz Arheološkog Muzeja u Splitu, in *Benkovački kraj kroz vjekove, Benkovac, Zbornik* 2, 1988, 29-70.

**Fadić 1989:** I. Fadić, *Rimsko staklo Argyruntuma* (exhibition catalogue), Arheološki Muzej u Zadru, Zadar: Arheološki Muzej, 1989.

**Fadić 1994a:** I. Fadić, Kasnoantičko staklo u Gatima, in *Jeličić-Radonić et al. 1994*, 213-225.

**Fadić 1994b:** I. Fadić, I vetri tardoantichi. Il corredo ecclesiastico e la liturgia, in Acta XIII Congressus Internationalis Archaeologiae Christianae (Split-Poreč 25-9/1-10-1994), v. III, 241-250.

**Fadić 1997:** I. Fadić, Invenzione, produzione e tecniche antiche di lavorazione del vetro, in *Trasparenze imperiali*, 75-92.

**Fadić 1999-2000:** I. Fadić, Žigovi kriptograma i križa na dnu kvadratičnih staklenih vrčeva, *OpArch*, 23-24, 1999-2000, 185-194.

**Fadić 2001:** I. Fadić, Putalj u kasnoj antici, in *Burić, Čače, Fadić 2001*, 115-151.

**Fadić 2004:** I. Fadić, Antičke staklarske radionice u Hrvatskoj, in *Lazar (ed.) 2004b*, 95-106.

**Feissel 1983:** D. Feissel, *Recueil des inscriptions chrétiennes de Macédoine du IIIe au VIe siècle*, Bulletin de Correspondance Hellénique, supplément VIII, Athènes: EFA, 1983.

**Felten 1975:** *Die spätrömische Akropolismauer: Architektur und Spolien*/von Wolfgang W. Wurster; *Inschriften*/ von Florens Felten; *Die christliche Siedlung*/von Florens Felten, Alt-Ägina I. 2., Mainz a.R.: von Zabern, 1975.

**Filarska 1952:** B. Filarska, *Szkła starożytne Katalog Naczyn*, Warszawa: Museum Narodowew Warszawie, 1952.

**Fitzgerald 1931:** G.M. Fitzgerald, *Beth Shean Excavations 1921-1923; the Arab and Byzantine Levels*, Philadelphia: the University Press for the University of Pennsylvania Museum, 1931.

**Fleming 1999:** S.J. Fleming, *Roman Glass. Reflections on Cultural Change*, Philadelphia: University of Pennsylvania Museum of Archaeology and Anthropology, 1999.

**Follmann-Schulz 1988:** A.B. Follmann-Schulz, *Die Römischen Gläser aus Bonn*, Köln: Rheinland/Bonn: R. Habelt, 1988.

**Follmann-Schulz 1989a:** A.B. Follmann-Schulz, Ein römischer Grabfund des 4. Jahrhunderts n. Chr. aus Zürlich-Enzen, Ehreinland, *KölnJb* 22, 1989, 49-68.

**Follmann-Schulz 1989b:** A.B. Follmann-Schulz, Fours de verriers romains dans la province de Germanie Inférieure, in *Foy, Sennequier (eds) 1989*, 35-40.

**Follmann-Schulz 1992:** A.B. Follmann-Schulz, *Die Römischen Gläser im Rheinischen Landesmuseum Bonn*, (Kunst und Altertum am Rhein, Band 138), Köln: Rheinland-Verlag; Bonn: Habelt, 1992.

**Follmann-Schulz 2001:** A.B. Follmann-Schulz, Die Produkte der Glashütten in Hambacher Forst, *AnnAIHV* 14, 2001, 61-67.

**Follmann-Schulz 2010:** A.B. Follmann-Schulz, Formen für Vierkantgefässe aus dem Römischen Legionslager in Bonn, I. Jahrhundert n. Chr., in *Fontaine (ed.) 2010*, 235-240.

**Fontaine (ed.) 2010:** Fontaine-Hodiamont, Chantal (ed.), with Catherine Bourguignon and Simon Laevers. *D'Ennion au Val Saint-Lambert: Le Verre soufflé-moulé. Actes des 23ème Rencontres de l'Association française pour l'archéologie du verre. Colloque internationale, Bruxelles - Namur, 17-19 octobre 2008*. Scientia Artis 5. Brussels: Institut royal du patrimoine artistique, 2010.

**Fontaine, Foy 2007:** S.D. Fontaine, D. Foy, L'épave Ouest-Embiez 1, Var. Le commerce maritime du verre brut et manufacture en Méditerranée occidentale dans l'Antiquité, *Revue Archéologique de Narbonnaise* 40, 2007, 235-268.

**Fontaine, Margos 2010:** P. Fontaine, R. margos, Un gobelet «épicurien», inédit au Musée du Verre de Charleroi, Milieu du 1er siècle apr. J.-C. Étude de l'inscription, in *Fontaine (ed.) 2010*, 79-83.

**Foy (ed.) 1995:** D. Foy (ed.), *Le verre de l'Antiquité tardive et du haut Moyen Age. Typologie-Chronologie-Diffusion*, Guiry-en-Vexin: Association Française pour l'Archéologie du Verre et Musée Archéologique Départemental du Val d' Oise, 1995.

**Foy (ed.) 2003:** D. Foy (ed.), *Cœur de verre. Production et diffusion du verre antique* (exhibition catalogue), Musée gallo-romain de Lyon-Fouvriére, 19-12-2003/25-4-2004, Gollion: Infolio/Rhône le Département, 2003.

**Foy 1995:** D. Foy, Le verre de la fin du IVe au VIIe siècle en France méditerranéenne, premier essai de typo-chonologie, in *Foy (ed.) 1995*, 187-242.

**Foy 2000:** D. Foy, Un atelier de verrier à Beyrouth au début de la conquête islamique, *Syria* 77, 2000, 239-290.

**Foy 2001:** D. Foy, Secteur nord de Tebytnis (Fayoum). Le verre byzantine et islamique, *Annales Islamoloqigues* 35, 2001, 465-489.

**Foy 2003:** D. Foy, L'incessant Mouvement du commerce: la circulation des matières, des hommes, des ateliers et des modèles, in *Foy (ed.) 2003*, 138-147.

**Foy *et al.* 1989:** Foy *et al.*, Ateliers de verriers de l'Antiquité et du haut Moyen Age en France, in *Foy, Sennequier (ed.) 1989*, 55-69.

**Foy, Bonifay 1984:** D. Foy, M. Bonifay, Éléments d'évolution des verreries de l'Antiquité tardive à Marseille d'après les fouilles de la Bourse, 1980, *Revue Archéologique de Narbonnaise* 17, 1984, 289-308.

**Foy, Hochuli-Gysel 1995:** D. Foy, A. Hochuli-Gysel, Le verre en Aquitaine du IVe au IXe siècle, in *Foy (ed.) 1995*, 151-176.

**Foy, Nenna (eds) 2006a:** D. Foy et M.-D. Nenna (eds), *Corpus des signatures et marques sur verre, vol. 1: La France*, Aix-en-Provence, Lyon: Association Française pour l'Archéologie du Verre, 2006.

**Foy, Nenna (eds) 2006b:** D. Foy et M.-D. Nenna (eds), *Corpus des signatures et marques sur verre, vol. 2; Belgique, Luxembourg, Allemagne, Autriche, Suisse, Slovénie, Hongrie, Croate, Espagne, Portugal, Maghreb, Grèce, Chypre, Turquie, mer Noire, Proche-Orient, Égypte, Soudan, Cyrénaique, France (addenda)*, Aix-en-Provence, Lyon: Association Française pour l'Archéologie du Verre, 2006.

**Foy, Nenna (eds) 2011:** D. Foy and M.-D. Nenna (eds), *Corpus des signatures et marques sur verres antiques 3*, Aix-en-Provence/Lyon: Association Française pour l'Archéologie du Verre, 2011.

**Foy, Nenna 2001:** D. Foy, M.-D. Nenna, *Tout feu tout sable. Mille ans de verre antique dans le Midi de la France.*

Musée d'Histoire de Marseille, Aix-En-Provence: Musées de Marseille/Éditions Édisud, 2001.

**Foy, Sennequier (ed.) 1989:** D. Foy, G. Sennequier (ed.), *Ateliers de verriers de l'Antiquité à la période préindustrielle*, Actes des 4èmes Rencontres de l' Association Française pour l'Archéologie du Verre, 24-25 novembre 1989, Rouen: Association Française pour l'Archéologie du Verre et les auteurs, 1991.

**Fremersdorf 1933:** F. Fremersdorf, *Der Römische Gutshof Köln-Münsgersdorf* (Römisch-Germanische Forschungen 6), Berlin-Leipzig: Walter de Gruyter & Co., 1933.

**Fremersdorf 1951:** F. Fremersdorf, *Figürlich Geschliffene Gläser. Eine Kölner Werkstatt des 3. Jahrhunderts*, Köln: Verlag von W. De Gruyter & Co., 1951.

**Fremersdorf 1958a:** F. Fremersdorf, *Das naturfarbene sogenannte blaugrüne Glas in Köln* (Die Denkmäler des römischen Köln, IV), Köln: Verlag der Löwe, 1958.

**Fremersdorf 1958b:** F. Fremersdorf, *Römisches Buntglas in Köln* (Die Denkmäler des römischen Köln, III), Köln: Verlag der Löwe, 1958.

**Fremersdorf 1959:** F. Fremersdorf, *Römische Gläser mit Fadenauflage in Köln*, Schlangenfadengläser und Verwandtes (Die Denkmäler des römischen Köln V), Köln: Verlag der Löwe, 1959.

**Fremersdorf 1961:** F. Fremersdorf, *Römisches geformtes Glas in Köln* (Die Denkmäler des römischen Köln, VI), Köln: Verlag der Löwe, 1961.

**Fremersdorf 1962:** F. Fremersdorf, *Die römischen Gläser mit aufgelegten Nuppen in Köln* (Die Denkmäler des römischen Köln, VI), Köln: Verlag der Löwe, 1962.

**Fremersdorf 1967:** F. Fremersdorf, *Die Römischen Gläser mit Schliff, Bemalung und Goldauflagen aus Köln* (Die Denkmäler des römischen Köln, VIII), Köln: Verlag der Löwe, 1967.

**Fremersdorf 1975:** F. Fremersdorf, *Antikes, islamisches und mittelalterliches Glas, sowie Kleinere Arbeiten aus Stein, Gagat und verwandten Stoffen in den Vatikanischen Sammlungen Roms* (Museo Sacro, Museo Profano, Museo Egizio, Antiquarium Romanum), Città del Vaticano: Biblioteca Apostolica Vaticana, 1975.

**Fremersdorf, Polónyi-Fremersdorf 1984:** F. Fremersdorf, E. Polónyi-Fremersdorf, *Die farblosen Gläser der Frühzeit in Köln, 2. und 3. Jahrhundert* (Die Denkmäler des römischen Köln, IX), Bonn: Dr. Rudolf Hambelt GmbH, 1984.

**Frend, Johnston 1962:** W.H.C. Frend, D.E. Johnston, The Byzantine Basilica Church at Knossos, *The Annual of the British School at Athens* 57, 1962, 186-238.

**Füllep 1969:** F. Füllep, Későrómai temető Pécs-Geisler Eta u. 8. sz. Alatt, *AErt* 96, 1969, 3-42.

**Fünfschilling 1986:** S. Fünfschilling, Römische Gläser aus Baden-Aquae Helveticae (aus den Grabungen 1892-1911), *Jahresbericht 1985 der Gesellschaft Pro Vindonissa*, 1986, 81-160.

**Gaitzsch 1999:** W. Gaitzsch, Spätrömische Glashütten im Hambacher Forst. Die Werkstatt des ECVA-Produzenten, in *Polfe (ed.) 1999*, 125-149.

**Gaitzsch *et al.* 2000:** W. Gaitzsch, A.-B. Follmann-Schulz, K.H. Wederpohl, G Hartmann, U. Tegtmeier, Spätrömische Glashütten im Hambacher Forst - Produktionsort der ECVA-Fasskrüge, *BJb* 200, 2000, 83-241.

**Galavaris 1970:** G. Galavaris, *Bread and Liturgy. The Symbolism of Early Christian and Byzantine Bread Stamps*, Madison-Milwaukee-London: The University of Wisconsin Press, 1970.

**Gasparetto 1979:** A. Gasparetto, Matrici e aspetti della vetreria veneziana e veneta medievale, *JGS* 21, 1979, 76-97.

**Gavriilaki 1989:** E. Gavriilaki, Σφακάκι Παγκαλοχωρίου, οικόπεδο Πολιουδάκη, *ArchDelt* 44, 1989, Chronika B2, 457-460.

**Gehrig 1967:** U. Gehrig, *Hildesheim Silberfund Antikenabteilung, Staatliche Museen Berlin*, Berlin: Stiftung Preussischer Kulturbesitz, 1967.

**Geraskli 2004:** E. Geraskli, Παλαιοχριστιανικό νεκροταφείο στον Βροντή Καρπάθου: πρώτη παρουσίαση, in *Damaskos (ed.) 2004*, 389-402.

**Gerousi 2002:** E. Gerousi, Glass Vessels from a Roman and Early Christian Cemetery, in *Kordas (ed.) 2002*, 133-140.

**Giacchero 1974:** M. Giacchero, *Edictum Diocletiani et collegarum de pretiis rerum venalium*, Genoa: Pubblicazioni dell'Istituto di Storia Antica & Scienze Ausiliarie dell'Università degli Studi di Genova, 8, 1974.

**Gialouris 1966:** N. Gialouris, Φραγκονήσι Ηλείας, *ArchDelt* 22, 1966, Chronika B1, 171.

**Giardina (ed.) 1993:** A. Giardina (ed.), *The Romans.* Chicago: The University of Chicago Press, 1993.

**Glass Cosmos 2010:** *Glass Cosmos*, P. Adam-Veleni (gen. ed.), D. Ignatiadou (scientific co-ordination), Archaeological Museum, Thessaloniki (temporary exhibition), Thessaloniki: Archaeological Museum, 2010.

**Glass from the Ancient World 1957:** *Glass from the Ancient World: The Ray Winfield Smith Collection*, Corning, NY: The Corning Museum of Glass, 1957.

**Glassammlungen der DDR 1973-1976:** AA.VV., Glassammlungen der DDR, *BullAIHV* 7, 1973-1976, 25-138.

**Glušćević 1991:** S. Glušćević, Kvadratične staklene boce s pečatom iz rimske luke u Zatonu, *Diadora 13*, 1991, 147-168.

**Glušćević 1998:** S. Glušćević, Roman glass in Zadar: Certain characteristic forms, *AnnAIHV 14*, 1998, 182-189.

**Goethert-Polaschek 1977:** K. Goethert-Polaschek, *Katalog der Römischen Gläser des Rheinischen Landesmuseums Trier*, Trieren Grabungen und

Forschungen no. IX, Mainz a. R.: Verlag Philipp von Zabern, 1977.

**Goitein 1967:** S.D. Goitein, *A Mediterranean Society 1: Economic Foundations*, Berkeley: University of California Press, 1967.

**Gojković 2015:** V. Gojković, Glass Finds from Poetovio Grave at Ljudski Vrt, *AnnAIHV 19*, Piran 2015, 264-270.

**Goldman (ed.) 1950:** H. Goldman (ed.), *Excavations at Gözlu Kule, Tarsus I: The Hellenistic and Roman Periods*, Princeton: Princeton University Press, 1950.

**Goldstein 2001:** S.M. Goldstein (ed.), *Ancient Glass*, exhibition catalogue in Miho Museum, 15 March - 19 August 2001, Japan: Miho Museum, 2001.

**Goldstein 2004:** S.M. Goldstein, with contributions by J.M. Rogers, M. Gibson, and J. Kröger, *Glass: from Sasanian Antecedents to European Imitations. The Nasser D. Khalili Collection of Islamic Art*, vol. 15. London and New York: The Nour Foundation, in association with Azimuth Editions, 2004.

**Golofast 2001:** L.A. Golofast, Steklo Rannevizantijskogo Hersonesa, *Materiay i po Arheologii, Istoriii Etnografii Tavrii*, VIII, Simferopol 2001, 97-260.

**Golofast 2009:** L.A. Golofast, Early Byzantine Glass from the Tauric Chersonesos (Crimea), in *Laflı (ed.) 2009*, 301-335.

**Gomolka 1978:** G. Gomolka, Spätantiken und frühmittelalterlicher Gläser aus dem Limeskastell Jatrus bei Krivina (bez. Ruse) Nordbulgarien, *AnnAIHV 7*, 1978, 25-30.

**Gomolka 1979:** G. Gomolka, Die Spätantiken und frühmittelalterlichen Gläser aus Iatrus, in *Wendel, Böttger, Bartosiewicz (eds) 1979*, 145-166.

**Gorbunova (ed.) 1973:** K.S. Gorbunova (ed.), *Pamiatniki antichnogo prikladnogo iskusstva Sbornik statej*, Leningrad: Avrora, 1973.

**Gorin-Rosen 1995:** Y. Gorin-Rosen, Hadera, Bet Eli 'ezer, *Excavations and Surveys in Israel* 13, 1995, 42-43.

**Gorin-Rosen 1998:** Y. Gorin Rosen, *Ancient Glass from the Holy Land*, Fine Arts Museum of San Francisco and Israel Antiquities Authority, San Francisco: Fine Arts Museum of San Francisco, 1998.

**Gorin-Rosen 2000:** Y. Gorin-Rosen, The Ancient Glass Industry in Israel, Summary of the Finds and New discoveries, in *Nenna (ed.) 2000*, 49-63.

**Gorin-Rosen 2000a:** Y. Gorin-Rosen, Glass from Monasteries and Chapels in South Sinai, *in Dahari et al. 2000*, 233-245.

**Gorin-Rosen 2002a:** Y. Gorin-Rosen, A Group of Glass Vessels from Nir Gallim, *Atiqot* 43, 2002, 120-126.

**Gorin-Rosen 2002b:** Y. Gorin-Rosen, The Glass Vessels, in *Avshalom-Gorni, Tacher 2002*, 186-187, 288-322.

**Gorin-Rosen 2004:** Y. Gorin-Rosen, The Glass Vessels from the Cemetery at Horbat Rimmon, *Atiqot* 46, 2004, 113-124.

**Gorin-Rosen 2006:** Y. Gorin-Rosen, The Glass Vessels from Khirbat Ka'Kul, *Atiqot* 54, 2006, 107-112.

**Gounaris 1995-2000:** G. Gounaris, Πανεπιστημιακή ανασκαφή Φιλίππων 1997-99, *Egnatia* 5, 1995-2000, 323-356.

**Gounaris 2002:** G. Gounaris, Πανεπιστημιακή ανασκαφή Φιλίππων. Εγκαταστάσεις υαλουργείου, in πρόγραμμα και περιλήψεις εισηγήσεων και ανακοινώσεων *22ου Συμποσίου Βυζαντινής και Μεταβυζαντινής Αρχαιολογίας και Τέχνης*, Athens 17, 18 and 19 May 2002 (Athens 2002), 28.

**Gounaris, Velenis 1996:** G. Gounaris, G. Velenis, Πανεπιστημιακή ανασκαφή Φιλίππων 1988-1996, *ArchErgMakThrak* 10B, 1996, 719-731.

**Gourevitch *et al.* (ed.) 2003:** D. Gourevitch, A. Moirin, N. Rouquet (eds), *Maternité et petite enfance dans l'Antiquité romaine* (exhibition catalogue), Bourges, Muséum d'Histoire Naturelle, 6 novembre 2003-28 mars 2004, Bourges: Éditions de la ville de Bourges, Service d'Archéologie Municipal, 2003.

**Grammenos (ed.) 2003:** D. Grammenos (ed.), *Ρωμαϊκή Θεσσαλονίκη* (exhibition catalogue), Thessaloniki Tellogleio Idryma Technon, Thessaloniki: Archaeological Museum of Thessaloniki and Tameion Archaiologikon Poron, 2003.

**Greifenhagen 1962:** A. Greifenhagen, Ancient Glass in the Berlin-West Museum, *JGS* 4, 1962, 61-66.

**Grose 1973-76:** D.F. Grose, Découvertes archéologiques récentes de verre: Italie, *BullAIHV* 7, 1973-76, 175-182.

**Grose 1977:** D.F. Grose, Early Blown Glass: The Western Evidence, *JGS* 19, 1977, 9-29.

**Grose 1982:** D.F. Grose, The Hellenistic and Early Roman Glass from Morgantina (Serra Orlando), Sicily, *JGS* 24, 1982, 20-29.

**Grose 1983:** D.F. Grose, The Formation of the Roman Glass Industry, *Archaeology* 36.4, July-August 1983, 38-45.

**Grose 1984:** D.F. Grose, Glass Forming Methods in Classical Antiquity: Some Considerations, *JGS* 26, 1984, 25-34.

**Grose 1989:** D.F. Grose, *Early Ancient Glass. Core-Formed, Rod-Formed, and Cast Vessels and Objects from the Late Bronze Age to the Early Roman Empire, 1600 B.C. to A.D. 50*, NY-Toledo: Hudson Hills Press/Toledo Museum of Art, 1989.

**Grose 1991:** D.F. Grose, Early Imperial Roman Cast Glass: The Translucent Coloured and Colourless Fine Wares, in *Newby, Painter (ed.) 1991*, 1-18.

**Guštin (ed.) 1991:** M. Guštin (ed.), *Capodistria tra Roma e Venezia, Gli scavi nel conveto dei Cappuccini* (exhibition catalogue), Museo Regionale di Capodistria 8-31 maggio 1989, Ljubljana: Joža Moškrič, 1991.

**Haberey 1966:** W. Haberey, Zur Herstellung der römischen Wabenbecher, *BJb* 166, 1966, 208-212.

**Hadad 1998:** S. Hadad, Glass Lamps from the Byzantine through Mamluk Periods, *JGS* 40, 1998, 63-76.

**Haevernick 1967:** T.E. Haevernick, Die Verbreitung der Zarten Rippenschalen, *JbRGZM* 14, 1967, 153-166.

**Haevernick 1978:** T.E. Haevernick, Modioli, *Glastechnische Berichte* 51 (1978), 328-330.

**Haevernick 1981:** T.E. Haevernick, Kleinfunde aus Glas und Stein, in *Braun, Haevernick 1981*, 97-139.

**Haevernick, Pfeffer 1958:** T.E. Haevernick, W. von Pfeffer, Zarte Rippenschalen, *Saalbourg Jahrbuch* 17, 1958, 76-91.

**Haevernick, Pfeffer 1981:** T.E. Haevernick, W. von Pfeffer, Zarte Rippenschalen, in T.E. Haeverninck, Beitrage zur Glasforschung 1938 bis 1981, Mainz a.R.: Philipp von Zabern, 1981, XI-XXVIII.

**Hakim 2001:** B.S. Hakim, Julian of Ascalon's Treatise of Construction and Design Rules from Sixth-Century Palestine, *Journal of Society of Architectural Historians*, 60/1, March 2001, 4-25.

**Hamilton 1939:** R.W. Hamilton, The Domed Tomb at Sebastya, *QDAP* 6, 1939, 64-71.

**Hampe and Winter 1965:** R. Hampe, A. Winter, *Bei Töpfern und Zieglern in Süditalien, Sizilien und Griechenland*, Mainz, Verlag des Römisch-Germanischen Zentralmuseums Mainz; in Kommission bei R. Habelt, Bonn, 1965.

**Han (ed.) 1975:** V. Han (ed.), *Srednjovekovno staklo na Balkanu (V-XV. vek)*, Zbornik radova sa medjunarodnog savetovanja održanog od 22. do 24. Aprila 1974. U Beogradu, Beograd: Srpska Akademija Nauka i Umetnosti, 1975.

**Han 1981:** V. Han, *Tri veka dubrovačkog staklarstva* (XIV-XVI. vek), Beograd: Srpska Akademija Nauka i Umetnosti, 1981.

**Han 1986:** V. Han, Staklo (Rtkovo-Glamija I), *Djerdapske Sveske* III, Beograd: Arheološki Institut, Beograd; Narodni Muzej, Beograd; Odeljenje za Arheologiju Filozofskog Fakulteta u Beogradu, 1986, 92-94.

**Harden 1935:** D.B. Harden, Romano-Syrian Glasses with Mould-Blown Inscriptions, *JRS* 25, 1935, 163-186.

**Harden 1936:** D.B. Harden, *Roman Glass from Karanis Found by the University of Michigan Archaeological Expedition in Egypt. 1924-29*. University of Michigan Studies, Humanistic Series, v. 41, Ann Arbor, Michigan: University of Michigan Studies, 1936.

**Harden 1944-1945:** D.B. Harden, Two Tomb Groups of the First Century A.D. from Yamur, Syria, and a Supplement to the List of Romano-Syrian Glasses with Mold-Blown Inscriptions, *Syria* 24, 1944-45, 81-95, 291-292.

**Harden 1947:** D.B. Harden, *The Glass*, in *Hawkes, Hull 1947*, 287-307.

**Harden 1959:** D.B. Harden, The Highdown Hill Glass Goblet with Greek Inscription, *Sussex Archaeological Collections Relating to the History and Antiquities of the County*, XCVII, 1959, 3-20.

**Harden 1960:** D.B. Harden, The Wint Hill Hunting Bowl and Related Glasses, *JGS 2*, 1960, 44-81.

**Harden 1966:** D.B. Harden, Some Tomb Groups of Late Roman Date in the Amman Museum, *AnnAIHV 3*, 1966, 48-55.

**Harden 1969:** D.B. Harden, Ancient Glass, II: Roman, *AJ* 126 (1969), 44–77.

**Harden 1975:** D.B. Harden, Some Lombard glasses of the 6th and 7th centuries, in *Han (ed.) 1975*, 11-29.

**Harden *et al.* 1968:** D.B. Harden, K. Painter, R.H. Pinter-Wilson, H. Tait, *Masterpieces of Glass*, London: Trustees of the British Museum, 1968.

**Harden *et al.* 1987:** D.B. Harden, H. Hellenkemper, K. Painter, and D. Whitehouse, *Glass of Caesars*, Milan: Olivetti, 1987.

**Harter 1999:** G. Harter, *Römische Gläser des Landesmuseums Mainz,* Weisbaden: Reichter Verlag, 1999.

**Hawkes, Hull 1947:** C.F.C Hawkes, M.R. Hull, *Camulodunum: First Report on the Excavations at Colchester, 1930-1939*, Reports of the Research Committee of the Society of Antiquaries of London, no. 14, London: Committee of the Society of Antiquaries of London, 1947.

**Hayes 1971:** J.W. Hayes, Four Early Roman Groups from Knossos, *BSA* 66, 1971, 249-275.

**Hayes 1972:** J.W. Hayes, *Late Roman Pottery*, London: The British School at Rome, 1972.

**Hayes 1975:** J.W. Hayes, *Roman and Pre-Roman Glass in the Royal Ontario Museum*, Toronto: The Royal Ontario Museum, 1975.

**Hayes 1976:** J.W. Hayes, *Roman Pottery in the Royal Ontario Museum, A Catalogue*, Ontario: Royal Ontario Museum, 1976.

**Hayes 1978:** J.W. Hayes, Glass finds from the 1975 season, in *Humphrey (ed.) 1978*, 187-193.

**Hayes 1990:** J.W. Hayes, The Glass Finds (1990), *in Stevens et al. 1993*, 289-294.

**Hayes 1993:** J.W. Hayes, *Excavations at Saraçhane in Istanbul, vol. 2: The Pottery*, Princeton: Princeton University Press and Dumbarton Oaks Research Library and Collection, 1993.

**Henein 1974:** N.H. Henein, *Le verre soufflé en Égypte*, Cairo: Institut Français d'Archéologie Orientale, 1974.

**Hilgers 1969:** W. Hilgers, *Lateinische Gefassnamen. Bezeichnungen, Funktion und Form römischer Gefäße nach den antiken Schriftquellen*, Düsseldorf: Rheinland-Verlag, 1969.

**Hourmouziadis, Asimakopoulou-Atzaka, Makris 1982:** G. Hourmouziadis, P. Asimakopoulou-Atzaka, K.A. Makris, *The Story of a Civilization, Magnesia*, Athens: R. and M. Capon, 1982.

**Humphrey (ed.) 1978:** J.H. Humphrey (ed.), *Excavations at Carthage 1975 conducted by the University of Michigan*, vol II, Ann Arbor: Kelsey Museum/The University of Michigan, 1978.

**Hurst, Roskams 1984:** H.R. Hurst, S.P. Roskams, *Excavations at Carthage: The British Mission*, v. 1, pt.

1, Sheffield: University of Sheffield for the British Academy, 1984.

**Ignatiadou 2002:** D. Ignatiadou, Έρευνα για το *χαλαστραίο νίτρο*, *ArchErgMakThrak* 16, 2002, 241-248.

**Ignatiadou 2013:** D. Ignatiadou, *Διαφανής ύαλος για την αριστοκρατία της αρχαίας Μακεδονίας/Colourless glass for the élite in ancient Macedonia*, Thessaloniki: Archaeological Institute of Macedonian and Thracian Studies, 2013.

**Ignatiadou *et al.* 2003:** D. Ignatiadou, E. Dotsika, A. Kouras, Y. Maniatis, Nitrum Chalestricum. The Natron of Macedonia, *AnnAIHV* 16, 2003, 64-68.

**Ignatiadou, Antonaras 2007:** D. Ignatiadou, A.C. Antonaras, *Glassworking, Ancient and Medieval, Terminology, Technology, and Typology. A Greek-English, English-Greek Dictionary*, Centre of Greek Language, Thessaloniki, 2007.

**Iliffe 1933:** J.H. Iliffe, A Tomb at El Bassa c. A.D. 396, *QDAP* III.2, 1933, 81-91.

**Intzesiloglou 1980:** Ch. Intzesiloglou, Μητρόπολη Καρδίτσας, οικόπεδο Αθ. Παιδά, *ArchDelt* 35, 1980, Chronika B1, 269, pl. 117γ.

**Intzesiloglou 1987:** A. Intzesiloglou, Κάτω Κτημένη, *ArchDelt* 42, 1987, Chronika B1, 272, pl. 160.

**Isings 1957:** C. Isings, *Roman Glass from Dated Finds*, Groningen/Djakarta: J.B. Wolters, 1957.

**Isings 1964:** C. Isings, A Fourth Century Glass Jar with Applied Masks, *JGS* 6, 1964, 59-63.

**Isings 1971:** C. Isings, *Roman Glass in Limburg* (Archaeologica Trajectina IX), Groningen: Wolters Noordhoff Publishing, 1971.

**Isings 1976:** C. Isings, Excavated for Sulphur, in *Zadoks-Josephus, Boersma (eds) 1976*, 353-356.

**Isler 1969:** H.P. Isler, Heraion von Samos: Eine frühbyzantinische Zisterne, *Mitteilungen des Deutschen Archäologischen Instituts. Athenische Abteilung* 84, 1969, 202-230.

**Israeli 1991:** Y. Israeli, The Invention of Blowing, in *Newby, Painter (ed.) 1991*, 46-55.

**Israeli 2003a:** Y. Israeli, What did Jerusalem's first-century BCE glass workshop produce?, *AnnAIHV 16*, 2003, 54-58.

**Israeli 2003b:** Y. Israeli, *Ancient Glass in the Israel Museum. The Eliahu Dobkin Collection and Other Gifts*, Jerusalem: The Israel Museum, 2003.

**Ivachtchenko 1995:** Y. Ivachtchenko. Le verre proto-byzantin. Recherches en Russie (1980-1990), in *Foy (ed.) 1995*, 319-330.

**Ivanišević 2009:** V. Ivanišević, *Nekropole iz epohe Seobe naroda u Singidunumu= Necropolises from the Migration Period in Singidunum*, Belgrade: JP Beogradska tvrdjava, 2009.

**Ivanišević, Kazanski 2009:** V. Ivanišević, M. Kazanski, Nouvelle nécropole de Grandes Migrations de Singidunum, *Starinar*, n.s. 57, (2007), 2009, 117-139.

**Ivanišević, Nikolić-Djordjević 1997:** V. Ivanišević, S. Nikolić-Djordjević, Novi tragovi antičkih fortifikacija u Singidunumu-Lokalitet Knez Mihailova 30, *Singidunum* 1, 1997, 65-150.

**Ivanova, Goloubev 1961:** M.I. Ivanova, P.M. Goloubev, Nachodki v Loo, *SovA* 1961/3, 284-290.

**Jackson, Cool, Wager 1998:** C.M. Jackson, H.E.M. Cool, E.C.W. Wager, The Manufacture of Glass in Roman York, *JGS* 40, 1998, 55-62.

**Jackson, Nicholson, and Gneisinger 1998:** C.M. Jackson, P.T. Nicholson, and W. Gneisinger, Glassmaking at Tell el-Amarna: An Integrated Approach, *JGS* 40 (1998), 11-23.

**Jacopi 1939:** G. Jacopi, *Clara Rhodos, vol. VI-VII-p.I. Esplorazione archeologica di Camiro-II*. Rhodes: Istituto Storico-Archeologico-Rodi, 1932-1933-1939.

**Jeličić-Radonić *et al.* 1994:** J. Jeličić-Radonić, Crnović, Čače, Katić, Fadić, Bonacić-Mandinić, Kovačić, *Gata. Crkva Justinijanova doba*, Split: XIII Medjunarodni Kongres za Starokršćansku Arheologiju, 1994.

**Jennings 2004:** S. Jennings, Vessel Glass from Beirut BEY 006, 007 and 045, *Berytus* 48-49, 2004-05.

**Josifofska 1956:** B. Josifofska, Dve poznorimske grobnice iz Dračeva, *ŽA* 1956, 6/2, 277-292.

**K. Kallinikos 1969:** Protopresvyterou Konstantinou Kallinikou, Ο χριστιανικός ναός και τα τελούμενα εν αυτώ, Athens: Grigoris, 1969.

**Kaba 1958:** M. Kaba, Azaquincumi üvegek, *BR* XVIII, 1958, 425-448.

**Kalamartzi-Katsarou, Tampaki (eds) 2001:** Ai. Kalamartzi-Katsarou, S. Tampaki (eds), *Αφιέρωμα στη μνήμη του Σωτήρη Κίσσα*, Thessaloniki: Elliniki Etaireia Slavikon Meleton/University Studio Press, 2001.

**Kaltsas 1983:** N. Kaltsas, Από τα ελληνιστικά νεκροταφεία της Πύλου, *ArchDelt* 38, 1983, Meletes, 1-75.

**Kanaki 2002:** E. Kanaki, Γυάλινα αντικείμενα από τα Μάταλα Πυργιωτίσης, in *Themelis (ed.) 2002*, 103-108.

**Kanonidis 1996b:** I. Kanonidis, 9η Εφορεία Βυζαντινών Αρχαιοτήτων. Ανασκαφικές εργασίες. Ν. Θεσσαλονίκης. Αγάπης 3, *ArchDelt* 51, 1996, B2, 490-493.

**Kanonidis 1996b:** I.O. Kanonidis, Η περιοχή του Διοικητηρίου στα παλαιοχριστιανικά και βυζαντινά χρόνια, *ArchErgMakThrak* 10β, (1996), 558-570.

**Kanonidis 2002:** I.O. Kanonidis, Ανασκαφή εργαστηρίου υαλουργίας στην Πλατεία Διοικητηρίου της Θεσσαλονίκης, in *Themelis (ed.) 2002*, 143-153.

**Kanonidis 2003:** I. Kanonidis, Μεσοβυζαντινή εφυαλωμένη κεραμική με λευκό πηλό από ανασκαφές οικοπέδων στη Θεσσαλονίκη, in *AIECMM, Bakirtzis 2003*, 71-76.

**Karagianni 2001:** F. Karagianni, Κινητά ευρήματα της πρώτης περιόδου: δείγμα των ετών 1970-73, in *Adam-Veleni (ed.) 2001*, 147-160.

**Karagiorga 1971:** Th. G. Karagiorga, Αρχαία Ιλίς, *ArchDelt* 26, 1971, Chronika B1, 142.

**Karamitrou-Mentesidi 1982:** G. Karamitrou-Mentesidi, Ποντοκόμη Κοζάνης, *ArchDelt* 37, 1982, Chronika B2, 298.

**Karamitrou-Mentesidi 2000:** G. Karamitrou-Mentesidou, Νομός Κοζάνης 2000. Ανασκαφές εν *οδοίς και παρόδοις*, *ArchErgMakThrak* 14, 2000, 607-640.

**Karamitrou-Mentesidou 1985:** G. Karamitrou-Mentesidou, Ο μακεδονικός τάφος στην Σπηλιά Εορδαίας, *ArchDelt* 40, 1985, Meletes, 242-280.

**Karamitrou-Mentesidou 1993:** G. Karamitrou-Mentesidou, Κοζάνη, πόλη Ελιμιώτιδος. *Αρχαιολογικός οδηγός*, Thessaloniki: published by the author and Municipality of Kozani, 1993.

**Karamperi, Christodoulidou, Kaïafa 1996:** M. Karamperi, E. Christodoulidou, A. Kaïafa, Το ανασκαφικό έργο στο Γαλεριανό συγκρότημα, *ArchErgMakThrak* 10a, 1996, 533-544.

**Karović 1999:** G. Karović, Rimsko staklo Marguma, *Viminacium* 10, 1999, 75-103.

**Karydas 1994:** N. Karydas, Θεσσαλονίκη, Ευξείνου και Παναγοπούλου, Αμπελόκηποι, *ArchDelt* 49, 1994, Chronika B2, 501.

**Katsnelson, Jackson-Tal 2004:** N. Katsnelson, R. Jackson-Tal, The Glass Vessels from Ashqelon, Semadar Hotel, *Atiqot* 48, 2004, 99-109.

**Keller 2006:** D. Keller, Die Gläser aus Petra, in *Keller, Grawehr 2006*, 1-256.

**Keller 2008:** D. Keller, Preliminary Report on the Late Roman, Byzantine and Early Islamic Glass Finds from Elephantine, in *Von Günter et al. 2008*, 137-148.

**Keller, Grawehr 2006:** D. Keller, M. Grawehr, *Petra ez Zantur III*. Terra Archaeologica V, Mainz: Verlag Philipp von Zabern, 2006.

**Keller, Price, Jackson (eds) 2014:** D. Keller, J. Price, and C. Jackson (eds), *Neighbours and Successors of Rome: traditions of glass production and use in Europe and the Middle East in the later first millennium AD*, **Oxford: Oxbow, 2014.**

**Keramaris, Papagianni 2000:** A. Keramaris, E. Papagianni, Υστερορωμαϊκό νεκροταφείο στη βιομηχανική περιοχή της Σίνδου, *ArchErgMakThrak* 14, 2000, 145-152.

**Keramopoulos 1917:** A.D. Keramopoulos, Θηβαϊκά. Α' μέρος- Ανασκαφαί εν Θήβαις, νεώτεροι τάφοι, *ArchDelt* 3, 1917, 210-252.

**Kisa 1908:** A. Kisa, *Das Glas im Altertume*, Leipzig: Verlag von Karl W. Hiersemann, 1908 (Photostatic reprint Roma: L'Erma di Bretschneider, 1968).

**Kissas 1988:** S. Kissas, Η ανασκαφή στην Κολχίδα Κιλκίς, *ArchErgMakThrak* 2, 1988, 207-212.

**Kitzinger 1960:** E. Kitzinger, A Marble Relief of the Theodosian period, *DOP* 14 (1960), 17-42.

**Klein 1995:** M.J. Klein, Römische Luxusgläser aus Höhen-Sülzen (Rheinland-Pfalz, Deutschland), *AnnAIHV* 13, 1995, 151-162.

**Kojcheva 1990:** K. Kojcheva, Ranovizantiiska st'klarska rabotilnica, *Arheologija (Sofia)* 32/4, 1990, 36-46.

**Kolonas 2002:** L. Kolonas, Τα γυάλινα αγγεία της Πάτρας, in *Themelis (ed.) 2002*, 109-134.

**Koltes 1982:** J. Koltes, *Catalogue des collections archéologiques de Besançon*, v. 7, *La verrerie gallo-romaine*, Annales Littéraires de l'Université de Besançon, v. 270, Paris: Les Belles Lettres, 1982.

**Kondić 1992:** J. Kondić, Kasnoantičko staklo iz nekropole kod crkve Sv. Nikole (Kuršumlija), *ZRNMB* XIV/1, 1992, 411-418.

**Konstantinopoulos 1965:** Gr. Konstantinopoulos, Ρόδος, οδός Λίνδου έναντι αρ. 41, (Οικόπεδο Παυλίδη), *ArchDelt* 20, 1965, Chronika B3, 598-599, pl. 762.

**Konstantinopoulos 1969:** Gr. Konstantinopoulos, Ρόδος, οικ. Τρ. Νικολίτση (Νότια νεκρόπολη Ρόδου), *ArchDelt* 24, 1969, Chronika B2, 472, pl. 472δ.

**Kordas (ed.) 2002:** G. Kordas (ed.), *1st International Conference. Hyalos Vitrum Glass. History Technology and Conservation of Glass and Vitreous Materials in the Hellenic World*, Athens: Glasnet, Publication, 2002.

**Korošec 2004:** P. Korošec, Sledovi steklarske delavnice v Rebalčji vasi na Ptuju, in *Lazar (ed.) 2004b*, 67-70.

**Kotzias 1953-54:** N. Kotzias, Ανασκαφαί Βεροίας και τα εξ αυτών ευρήματα, *ArchEph* 1953-54, (1961), vol. 3, 167-175.

**Koukoules 1940:** F. Koukoules, Βυζαντινών νεκρικά έθιμα, *EEVS*, 16, 1940, 3-80.

**Koukoules 1948-52:** F. Koukoules, Βυζαντινών Βίος και Πολιτισμός, vol. I-V, Athens: Institut Français d'Athènes, 1948-52.

**Koukouli-Chrysanthaki 1975:** Ch Koukouli-Chrysanthaki, Φίλιπποι, *ArchDelt* 30, 1975, Chronika B1, 285, pl. 193δ.

**Kourkoutidou 1970:** E. Kourkoutidou-Nikolaidou, Ανασκαφαί Θεσσαλίας. Νέα Αγχίαλος. 3. Ρωμαϊκόν νεκροταφείον, *ArchDelt 24*, 1969, Chronika B2 (1970), 229-235.

**Kourkoutidou-Nikolaidou 1990:** E. Kourkoutidou-Nikolaidou, 9η Εφορεία Βυζαντινών Αρχαιοτήτων, *ArchDelt* 45, 1990, Chronika B2, 323.

**Kourkoutidou-Nikolaidou, Mavropoulou-Tsioumi, 1978:** E. Kourkoutidou-Nikolaidou, Ch. Mavropoulou-Tsioumi, Θεσσαλονίκη, Φρούριο Βαρδαρίου, *ArchDelt* 33, 1978, Chronika B2, 239-240.

**Kourkoutidou-Nikolaidou, Mavropoulou-Tsioumi 1983:** E. Kourkoutidou-Nikolaidou, Ch. Mavropoulou-Tsioumi, Θεσσαλονίκη, οδός Κωνσταντίνου Παλαιολόγου 18, *ArchDelt* 38, 1983, Chronika B2, 285.

**Kourkoutidou-Nikolaidou, Mavropoulou-Tsioumi, Pazaras 1976:** E. Kourkoutidou-Nikolaidou, Ch. Mavropoulou-Tsioumi, Th. Pazaras, Θεσσαλονίκη, οδός Απολλωνιάδος 18, *ArchDelt* 31, 1976, Chronika B2, 268, pl. 210α.

**Kousoulakou 1993:** D. Kousoulakou, Ανασκαφή ανατολικού νεκροταφείου Θεσσαλονίκης, *ArchErgMakThrak* 7, 1993, 311-314, 317-319.

**Kraeling (ed.) 1939:** C.H. Kraeling (ed.), *Gerasa, City of Decapolis*, New Haven, Connecticut: American School of Oriental Research, 1938.

**Kröger 1995:** J. Kröger, *Nishapur. Glass of the Early Islamic Period*, New York: The Metropolitan Museum of Art, 1995.

**Kunina 1973:** N.Z. Kunina, Sirijskie vidutye v forme steklianye sosudy iz nekropolia Pantikapeia, in *Gorbunova (ed.) 1973*, 101-154.

**Kunina 1984:** N.Z. Kunina, K voprosu o zapadnom importe stekla na Bospor (po materialam nekropoli Pantikapei na sobranii otdela Antichnogo mira Ermitazha), *Trudy Ermitaza* 24, 1984, 147-164.

**Kunina 1997:** N. Kunina, *The Art Treasures of Russia, Ancient Glass in the Hermitage Collection*, St. Petersburg: The State Hermitage and ARS Publishers Ltd, 1997.

**Kunina, Sorokina 1972:** N. Kunina, N.Sorokina, Stekliannye balzamarii Bospora, *Trudy Ermitaza* 13, 1972, 146-177.

**Künzl 1997:** S. Künzl, Römisches Tafelsilber-Formen und Verwendung, in *von Prottwitz, Gaffron, Mielsch 1997*, 9-30.

**Künzl 1997a:** S. Künzl, Sagalassos Red Slip Ware im Römisch-Germanischen Zentralmuseum Mainz, in *Waelkens, Poblome (eds)1997*, 469-489.

**Kyparissis 1926:** N. Kyparissis, Αθήναι, Ιεράς Οδού και Τριπτολέμου (οικοπεδο Ν. Διαμαντοπούλου), *ArchDelt* 10, 1926, 65-73.

**Kyriakakis 1974:** J. Kyriakakis, Byzantine Burial Customs. Care for the Deceased from Death to the Prothesis, *GOTR* XIX/1, 1974, 37-72.

**La Baume 1961:** P. La Baume, Die Sarkophage von der König-Luise-Schule in Köln, *KölnJb* 5, 1961, 85-88.

**La Baume 1964:** P. La Baume, *Römisches Kunstgewerbe zwischen Christi Geburt und 400*, Braunschweig: Klinkhardt und Biermann, 1964.

**La Baume 1973:** P. La Baume, *Glas der antiken Welt I*. Wissenschaftliche Kataloge des Römisch-Germanischen Museums Köln, Bd. I, Köln: Römisch-Germanischen Museums, 1973.

**La Baume, Salomonson 1976:** P. La Baume and J.W. Salomonson, *Römischen Kleinkunst. Sammlung Karl Löffler*. Wissenschaftliche Kataloge des Römisch-Germanischen Museums Köln Bd. III, Köln: Römisch-Germanischen Museums, 1976.

**Laflı (ed.) 2009:** E. Laflı (ed.) *Late Antique/Early Byzantine Glass in the Eastern Mediterranean*, Izmir: Hürriyet Matbaası. 2009.

**Lamm 1929-1930:** C.J. Lamm, *Mittelalterliche Gläser und Steinschnittarbeiten aus dem Nahen Osten*, I-II. Forschungen zur islamischen Kunst 5. Berlin: D. Reimer, 1929-1930.

**Lancel 1967:** S. Lancel, *Verrerie antique de Tipasa*, Paris: E. de Boccard, 1967.

**Lang, Price 1975:** J. Lang, J. Price, Iron Tubes from a Late Roman Glassmaking Site at Mèrida (Badajoz), in Spain, *JAScien* 2.4, Dec. 1975, 289-296.

**Laskaris 2000:** N.G. Laskaris, *Monuments funèraires palèochétiens (et byzantins) de Grèce*, Athens: Les Èditions Historiques, Stèfanos D. Vasilopoulos, 2000.

**Lazar (ed.) 2004a:** I. Lazar (ed.), *Rimljani: steklo, glina, kamen* (katalog razstav), Celje: Pokrajinski muzej; Maribor: Pokrajinski muzej; Ptuj: Pokrajinski muzej, 2004.

**Lazar (ed.) 2004b:** I. Lazar (ed.), *Drobci antičnega stekla. Fragments of Ancient Glass*, Koper: Univerza na Primorskem, Znanstveno-raziskovalno središče, Institut za dediščino Sredozemlja, 2004.

**Lazar 2003:** I. Lazar, *Rimsko steklo Slovenje*, Ljubljana: Z ZRC SAZU, 2003.

**Lazar 2004:** I. Lazar, Najdbe rimskega stekla iz Logatca (Longaticum), in *Lazar (ed.) 2004b*, 57-66.

**Lazar 2004a:** I. Lazar, Odsevi davnine, anticno steklo v Sloveniji, in *Lazar (ed.) 2004a*, 13-80.

**Lazar 2005:** I. Lazar, An oil Lamp depicting a Roman glass furnace - a new find from Slovenia, *Instrumentum* 22, Dec. 2005, 17-18.

**Lazar 2006a:** I. Lazar, An oil lamp from Slovenia depicting a Roman glass Furnace, *VAPD* 99, 2006, 227-234.

**Lazar 2006b:** I. Lazar, Glass finds in Slovenia and neighbouring areas, *JRA* 19, 2006, 329-342.

**Lazaridis 1965:** D. Lazaridis, Ἄβδυρα, *ArchDelt* 20, 1965, Chronika B3, 453-461.

**Lazaridis 1973/74:** P. Lazaridis, Θεσσαλονίκη, Παλαιοχριστιανικοί τάφοι εις Πύργον Τριγωνίου, *ArchDelt* 29, 1973-74, Chronika B3, 729, pl. 524γ.

**Lazić (ed.) 2001:** M. Lazić (ed.), *Vestigatio Vetustatis Aleksandrini Cermanović-Kuzmanović od prijatelja, saradnika i učenika* (Univerzitet u Beogradu Filozofski Fakultet, Centar za Arheološka Istraživanja knjiga 20), Beograd: Centar za Arheološka Istraživanja, 2001.

**Lehmann, Lehmann (eds), 1969:** K. Lehmann, P.W. Lehmann (eds), *Samothrace* 3, Text II, *The Hieron*, Excavations Conducted by the Institute of Fine Arts, New York University Bollingen Series LX.3, Princeton: Princeton University Press, 1969.

**Lehrer 1979:** G. Lehrer, *Ennion: A First Century Glassmaker* (exhibition catalogue), Tel Aviv, Haaretz Museum, Ramat Aviv, 1979.

**Lehrer-Jacobson 1986:** G. Lehrer-Jacobson, A Signed Roman Glass Juglet, *MUSE*, Annual of the Museum of Art and Archaeology, 20, 1986, 46-49.

**Lehrer-Jacobson 1992:** G. Lehrer Jacobson, Greek Names on Prismatic Jugs, *JGS* 34, 1992, 35-43.

**Lieberman 1971:** S. Lieberman, The book of Julian of Ascalon; the law and the habits in the region of Palestine (in Hebrew), *Tarbits* 40, 1971, 409-417.

**Lierke 1991:** R. Lierke, Glass Bowls made on the Potter's Wheel, *Glastechnische Berichte* 64, 1991, 310-317.

**Lierke 1999:** R. Lierke, *Antike Glastöpferei, Ein vergessenes Kapitel der Glasgeschichte*, Mainz: von Zabern, 1999.

**Lierke 2001:** R. Lierke, With "Trial and Error", through Ancient Glass Technology, in *Kordas (ed.) 2001*, 181-186.

**Lierke, Lindig 1997:** R. Lierke, M.R. Lindig, Recent investigations of Early Roman cameo glass, *Glastechnische Berichte* 70/6, 1997, 189-197.

**Lightfoot 1989:** C.S. Lightfoot, *A Catalogue of Glass Vessels in Afyon Museum*, BAR International Series 530, Oxford: B.A.R., 1989.

**Lightfoot 1991:** C. Lightfoot, Recent Finds of Roman Glass from the Tigris in South-East Turkey, *AnnAIHV* 12, 1991, 89-98.

**Lightfoot 1993:** C.S. Lightfoot, A Catalogue of the Glass Finds. Sagallassos 1990, in *Waelkens* (ed.) *1993*, 173-193.

**Lightfoot, Arslan 1992:** C.S. Lightfoot, M. Arslan, *Ancient Glass of Asia Minor: The Yüksel Erimtan Collection*, Ankara: Ünal Offset, 1992.

**Liddell and Scott:** H.G. Liddell, R. Scott, *A Greek-English lexicon*. Oxford: Clarendon Press.

**Ljubinković 1983:** M. Ljubinković, Aspects de la verrerie médiévale d'influence byzantine en Serbie, *AnnAIHV* 9, 1983, 181-193.

**Loeschcke, Willers 1911:** S. Loeschcke, H. Willers, *Beschreibung römischer Altertümer gesammelt von Carl-Anton Niessen britischem Konsul in Köln a. Rh.*, Band I-II, Köln: von Greven und Bechtold, 1911.

**Loukatos 1940:** D. Loukatos, Λαογραφικαί περί τελευτής ενδείξεις παρά Ιωάννη τῷ Χρυσοστόμῳ, *Επετηρίς του Λαογραφικού Αρχείου Ακαδημίας Αθηνών* 2, 1940, 30-117.

**Ložar 1935:** R. Ložar, Ein Glasbecher mit Götterfiguren aus Črnelo, *Glasnik* 16, 1935, 97-105.

**Maccabruni 1991:** C. Maccabruni, Vetri romani nel teritorio di Ticinum(Pavia), *Helvetia Archaeologica* 22, 87/88, 1991, 111-124.

**Mackworth-Young 1949:** G. Mackworth-Young, Excavations in Siphnos. The Roman Graves of the First Century AD, *BSA* 44, 1949, 80-92.

**MacMahon, Price (eds) 2005:** A. MacMahon, J. Price (eds), *Roman Working Lives and Urban Living*, Oxford: Oxbow Books, 2005

**Magnusson (ed.) 1997:** B. Magnusson (ed.), *Ultra terminum vagari. Scritti in onore di Carl Nylander*, Rome: Edizioni Quasar, 1997.

**Mandruzzato, Marcante 2005:** L. Mandruzzato, A. Marcante, *Vetri Antichi del Museo Archeologico Nazionale di Aquileia. Il vasellame da mensa*, Corpus delle Collezioni del Vetro nel Friuli Venezia Giulia, 2, Pasian di Prato (UD), Venezia: Comitato nazionale italiano dell'AIHV, 2005.

**Makaronas 1941/52:** Ch. Makaronas, Χρονικά Αρχαιολογικά. Ανασκαφαί, έρευναι και τυχαία ευρήματα εν Μακεδονίᾳ και Θράκη κατά τα έτη 1940-50, *Makedonika* 2, 1941/52, 590-678.

**Makaronas 1951:** Ch. Makaronas, Ανασκαφή νεκροπόλεως εν Θεσσαλονίκη, *PAE*, 1951, 164-172.

**Makropoulou 1983:** D. Makropoulou, Ο παλαιοχριστιανικός ναός έξω από τα ανατολικά τείχη της Θεσσαλονίκης, *Makedonika* 23, 1983, 25-46.

**Makropoulou 1987:** D. Makropoulou, Θεσσαλονίκη, Ανατολικό παλαιοχριστιανικό νεκροταφείο (Νοσοκομείο 'Άγιος Δημήτριος'), *ArchDelt* 42, 1987, Chronika B2, 396-397.

**Makropoulou 1988:** D. Makropoulou, Θεσσαλονίκη. Οδός Μοναστηρίου 93, *ArchDelt* 43, 1988, Chronika B2, 379-380.

**Makropoulou 1990a:** D. Makropoulou, Ανασκαφή παλαιοχριστιανικών ταφών Θεσσαλονίκης, *Makedonika* 27, 1990, 190-208.

**Makropoulou 1990b:** D. Makropoulou, Θεσσαλονίκη, οικόπεδο συγκροτημάτων κατοικιών "Αξιός", *ArchDelt* 45, 1990, Chronika B2, 335-337.

**Makropoulou 1991:** D. Makropoulou, Δυτικό νεκροταφείο Θεσσαλονίκης 1991: Αρχαιολογικές έρευνες στην οδό Λαγκαδά, *ArchErgMakThrak* 5, 1991, 257-270.

**Makropoulou 1992a:** D. Makropoulou, Θεσσαλονίκη, Νότιο οικόπεδο του συγκροτήματος "Αξιός", στην οδό Λαγκαδά (Δυτικό νεκροταφείο), *ArchDelt* 47, 1992, Chronika B2, 425-426.

**Makropoulou 1992b:** D. Makropoulou, Θεσσαλονίκη, Συμβολή οδών Αριστάτου-Παναγοπούλου-Πατρόκλου στους Αμπελοκήπους (Δυτικό νεκροταφείο), *ArchDelt* 47, 1992, Chronika B2, 426-428.

**Makropoulou 1994:** D. Makropoulou, Θεσσαλονίκη. Οδός Δοξάτου 8, Αμπελόκηποι, *ArchDelt* 49, 1994, Chronika B2, 507-510.

**Makropoulou 1995:** D. Makropoulou, Θεσσαλονίκη, Μαργαροπούλου 90, *ArchDelt* 50, 1995, Chronika B2, 527-530.

**Makropoulou 1996:** D. Makropoulou, Θεσσαλονίκη, Πελειάδου 6, *ArchDelt* 51, 1996, Chronika B2, 504.

**Makropoulou 1997:** D. Makropoulou, Θεσσαλονίκη, Συνταγματάρχου Φρυζή 31, *ArchDelt* 52, 1997, Chronika B2, 701-702.

**Makropoulou 2001:** D. Makropoulou, Ταφικά ευρήματα, νομίσματα και νομισματικοί θησαυροί στα παλαιοχριστιανικά κοιμητήρια της Θεσσαλονίκης, in *Kalamartzi-Katsarou, Tampaki (eds) 2001*, 263-272.

**Makropoulou 2007:** D. Makropoulou, *Τάφοι και ταφές από το δυτικό νεκροταφείο της Θεσσαλονίκης (β' μισό 3ου αιώνα - 6ος αιώνας μ.Χ.)-Δομικά χαρακτηριστικά-Ταφικές πρακτικές-Κινητά ευρήματα*, unpublished PhD thesis submitted to the National Capodistrian University of Athens 2007.

**Malama 2000:** P. Malama, Νεώτερα στοιχεία από το ανατολικό νεκροταφείο της Αμφίπολης στα πλαίσια του έργου Διαπλάτυνση του δρόμου Αμφίπολης-Μεσολακκιάς, *ArchErgMakThrak* 14, 2000, 55-70.

**Malama, Christopoulou, Ntarakis 2000:** P. Malama, M. Christopoulou, K. Ntarakis, Νεότερα στοιχεία από τις έρευνες στην περιοχή του Ν. Κερδυλλίων κατά τις εργασίες κατασκευής της Εγνατίας οδού, *ArchErgMakThrak* 14, 2000, 71-85.

**Malama, Ntarakis 2008:** P. Malama, K. Ntarakis, Νεκροταφείο ρωμαϊκών χρόνων στα Νέα Κερδύλλια *Σερρών*, Thessaloniki: IH EPKA/AIMTHS, 2008.

**Mano 1975:** A. Mano, La nécropole d'Apollonie, *Iliria* III, 1975, 163-245.

**Mansel 1940:** A.M. Mansel, Les fouilles de Thrace, *Belleten* 4, 1940, 115-139.

**Marijanović 1990:** I. Marijanović, Prilog problemu datiranja nekih starokršćanskih crkava Bosne i Hercegovine, *GZM (A)* n.s. 45, 1990, 109-136.

**Marijanović-Vujović 1973:** G. Marijanović-Vujović, Dva ranahrišćanska polijeleja iz Narondong Muzeja, *ZRNMB* 7, 1973, 13-24.

**Marin (ed.) 2002:** Marin, E. (ed.), *Longae Salonae*, Split: Arheološki Muzej, 2002.

**Marki 1981:** E. Marki, Το σταυρικό μαρτύριο και οι χριστιανικοί τάφοι της οδού Γ' Σεπτεμβρίου στη Θεσσαλονίκη, *ArchEph* 1981, 53-69.

**Marki 1985:** E. Marki, Ο ανώνυμος σταυρικός ναός της οδού Αγίου Δημητρίου στη Θεσσαλονίκη, *Η Θεσσαλονίκη* 1, 1985, 159-188.

**Marki 1996:** E. Marki, Συμπεράσματα απο τις ανασκαφές της 9ης Εφορείας Βυζαντινών Αρχαιοτήτων στη Βόρεια Πύδνα, *ArchErgMakThrak* 10A (1996), 239-258.

**Marki 2000:** E. Marki, Θεσσαλονίκη, οδός Μπόρου 4, *ArchDelt* 55, 2000, Chronika B2, 745-747.

**Marki 2002:** E. Marki, Χωροθέτηση παραγωγικών και εργαστηριακών δραστηριοτήτων στο επισκοπικό συγκρότημα των Λουλουδιών Πιερίας, in *πρόγραμμα και περιλήψεις εισηγήσεων και ανακοινώσεων 2ου Συμποσίου Βυζαντινής και Μεταβυζαντινής Αρχαιολογίας και Τέχνης*, Athens 17-19 May 2002 (Athens 2002), 65-66.

**Marki 2006:** E. Marki, *Η νεκρόπολη της Θεσσαλονίκης στους υστερορωμαϊκούς και παλαιοχριστιανικούς χρόνους (Από τα μέσα του 3ου έως τον 7ο αι. μ.Χ.)*, Athens: Tameion Archaiologikon Poron, 2006.

**Marki 2013:** E. Marki, Το αρχαιολογικό εύρημα ως ιστορικό τεκμήριο. Η εγκατάσταση των Γότθων στην πεδιάδα της Κεντρικής Μακεδονίας και η ίδρυση του επισκοπικού συγκροτήματος στις Λουλουδιές (The Archaeological Find as Historical Evidence. The Settlement of Goths in the Plain of Central Macedonia and the Establishment of the Episcopal Complex at Louloudies), *DeltChrArchEt* 34, 2013, 1-10.

**Marki, Akrivopoulou 2003:** E. Marki, S. Akrivopoulou, Ανασκαφή αγρέπαυλης στο Παλιόκαστρο Ωραιοκάστρου, *ArchErgMakThrak* 17, 2003, 283-298.

**Marki, Karydas 1994:** E. Marki, N. Karydas, Βόρεια επέκταση κτηρίου Κεντρικής Βιβλιοθήκης του ΑΠΘ (ανατολικό νεκροταφείο), *ArchDelt* 49, 1994, Chronika B2, 523.

**Martin 1995:** C. Martin, Le verre de l'Antiquité Tardive en Valais. Notes préliminaires, in *Foy (ed.) 1995*, 93-107.

**Massabò 2001:** B. Massabò (ed.), *Magiche Trasparenze. I vetri dell' antica Albingaunum* (exhibition catalogue), Aquileia, Museo Civico del Patriarcato, 2 maggio - 2 settembre 2001, Milano: Anthelios, 2001.

**Matheson 1980:** S.B. Matheson, *Ancient Glass in The Yale University Art Gallery*, New Haven: Yale University Press, 1980.

**Mavromichali 2001:** Ai. Mavromichali, Τα γυάλινα ευρήματα του βαλανείου, in *Adam-Veleni (ed.) 2001*, 131-143.

**Mavropoulou-Tsioumi, Kourkoutidou-Nikolaidou 1981:** Ch. Mavropoulou-Tsioumi, E. Kourkoutidou-Nikolaidou, Θεσσαλονίκη, Τάφοι Θεολογικής Σχολής, *ArchDelt* 36, 1981, Chronika B2, 311.

**Mavropoulou-Tsioumi, Kourkoutidou-Nikolaidou 1983:** Ch. Mavropoulou-Tsioumi, E. Kourkoutidou-Nikolaidou, Θεσσαλονίκη, Λεωφόρος Καλλιθέας 5, *ArchDelt* 38, 1983, Chronika B2, 285-287.

**Mavropoulou-Tsioumi, Kourkoutidou-Nikolaidou, Pazaras 1977a:** Ch. Mavropoulou-Tsioumi, E. Kourkoutidou-Nikolaidou, Th. Pazaras, Θεσσαλονίκη. Αγίου Νέστορος 8, *ArchDelt* 32, 1977, Chronika B2, 233.

**Mavropoulou-Tsioumi, Kourkoutidou-Nikolaidou, Pazaras 1977b:** Ch. Mavropoulou-Tsioumi, E. Kourkoutidou-Nikolaidou, Th. Pazaras, Θεσσαλονίκη, Τάφοι μεταξύ Φιλοσοφικής και Θεολογικής Σχολής, *ArchDelt* 32, 1977, Chronika B2, 233.

**Maxwell 1979:** *The Constable-Maxwell Collection of Ancient Glass* (auction catalogue), Sotheby Parke Bernet, London 4-5 June 1979.

**McClellan 1983:** M. McClellan, Recent Finds from Greece of First-Century A.D. Mold-blown Glass, *JGS* 25, 1983, 71-78.

**McCray, Kingery (eds) 1998:** McCray, W.D. Kingery (eds), *Prehistory and History of Glassmaking Technology* (Ceramics and Civilization 8), Westerville, Ohio: The American Ceramic Society, 1998.

**McFadden 1946:** G.H. McFadden, A Tomb of the Necropolis of Ayios Ermoyenis at Kourion, *AJA* 50/4, 1946, 449-489.

**McNicoll *et al.* 1982:** A.W. McNicoll, R.H. Smith, J.B. Hennessy, *Pella in Jordan 1, An Interim Report of the Joint University of Sidney and College of Wooster Excavations at Pella 1979-1981*, Canberra: Australian National Gallery, 1982.

**Meconcelli-Notarianni 1979:** G. Meconcelli-Notarianni, *Vetri antichi nelle collezioni del Museo Civico Archeologico di Bologna*, Bologna: Istituto per la Storia di Bologna, 1979.

**Megaw 1971:** A.H.S. Megaw, Excavations at "Saranda Kolones", Paphos. Preliminary Report on the 1966-67 and 1970-71 Seasons, *RDAC*, 1971, 117-146.

**Megaw 2007:** A.H.S. Megaw, *Kourion: Excavations in the Episcopical Precinct* (Dumbarton Oaks Studies XXXVIII), Washington, DC: Dumbarton Oaks Research Library and Collection, 2007.

**Meimaris 1990:** Y. Meimaris, The monastery of Saint Euthymios the Great at Khan El-Ahmar, in the wilderness of Judea: Rescue excavations and basic protection measures, 1976-1979, *Poikila Meletimata* 10, 1990, 397-430.

**Mentzos 1981:** A. Mentzos, Symvoli stin erevna tou archaioterou naou tis Agias Sophias Θεσσαλονίκης, *Makedonika* 21, 1981, 201-220.

**Merker 1988:** G. Merker, The Coins, in *Weinberg* 1988, 103-115.

**Meyer 1987:** C. Meyer, Glass from the North Theater Byzantine Church, and Soundings at Jerash, Jordan, 1982-1983, *Bulletin of the American Schools of Oriental Research, Supplement, no. 25*, 1987, 175-222.

**Miatev 1928:** K.I. Miatev, Balgarski mogilni pogrebeniya ot XIV v. pri s. Radyuvene (Vrachansko), *Izvestija na Bulgarskija Arheologicheski Institut* 5, 1928.

**Michailidis 1967:** M. Michailidis, Βορείως Πανεπιστημιουπόλεως, *ArchDelt* 22, 1967, Chronika B2, 437-438, pl. 324.

**Michailidis 1969:** M. Michailidis, Θεσσαλονίκη. Αποκάλυψις παλαιοχριστιανικών ταφών, *ArchDelt* 24, 1969, Chronika B2, 361.

**Michaud 1971:** J.J. Michaud, Chroniques et rapport: Thessalie. Macédoine (Central. Occidental. Oriental). Thrace, *BCH* 95, 1971, 932-991.

**Mikulčić 1976:** I. Mikulčić, Antičko staklo iz Scupia I ostali makedonski gradovi, *AV* 25, 1974, (1976), 191-210.

**Milchev, Petkov 1965:** A. Milchev, D. Petkov, Novootkriti nahodki ot Mihailovgrad (Montana), *Arheologia-Sofia* VII.2, 1965, 43-52.

**Milošević 1973:** P. Milošević, Grobnice IV veka iz Čalme kod Sremske Mitrovice, *Sirmium III*, 1973, 85-96.

**Milošević 1979:** P. Milošević, Tri ranohrišćanska grobna objekta u Sremu, *Rad Vojvodjanskih Muzeja* 25, 1979, 115-120.

**Minchev 1988:** A. Minchev, Antichko steklo ot zapadnoto Chernomorie (I-VIv.), II: Chashi, *INMVarna* 24 (39) 1988, 35-59.

**Minchev 1990:** A. Minchev, Antichko steklo ot zapadnoto Chernomorie (I-VI v.), IV: Butilki, *INMVarna* 26 (41) 1990, 56-78.

**Minchev 1992:** A. Minchev, Spätantike Gläser von der Westlichen Schwarzmeerküste, in *Pilinger, Pülz (ed.) 1992*, 72-80.

**Minchev, Georgiev 1981:** Al. Minchev, Georgiev, Rannohristiianska grobnica s' stenopisi kraj s. Osenovo, Varnensko, *INMVarna* 17 (32), 1981, 9-16.

**Mirchev 1953:** M. Mirchev, Razkopki na trakijskoto selishte kraj s. Galata, *Izvestiia na arheologicheskoto druzhestvo v gr. Stalin*, IX, 1953, 1-28.

**Morel 1993:** J.-Morel, The Craftsman, in *Giardina (ed.) 1993*, 214-44.

**Morel *et al.* 1992:** J. Morel, H. Amrein et al., Un atelier de verrier du milieu du 1er siécle apr. J.C. á Avenches, *ASchw* 15, 1992, 9-17.

**Morin-Jean 1922-1923:** P. Morin-Jean, *La verrerie en Gaule sous l'empire romain*, Paris: Société de Propagation des Livres d'Art, 1922-23.

**Morou 1981:** E. Morou, Ἄργος, οδός Ηρακλέους και πάροδός της, *ArchDelt* 36, 1981, Chronika B1, 105-109.

**Moschonisiotou 1988:** S. Moschonisiotou, Θεσσαλονίκη, Χώρος ΔΕΘ, *ArchDelt* 43, 1988, Chronika B2, 354, pl. 204[α].

**Mossakowska-Gaubert 2000:** M. Mossakowska-Gaubert, Question d'éclairage: ermitage no 44 à Naqlun (Fayoum), *Annales Islamologiques* 34, 2000, 335-357.

**Motsianos 2011:** I. Motsianos, *Φως Ιλαρόν: Ο τεχνητός φωτισμός στο Βυζάντιο,* Unpublished PhD thesis submitted to the University of Volos in 2011.

**Motsianos, Bintsi (eds) 2011:** I. Motsianos, E. Bintsi (eds), *Light on Light: An Illuminating Story* (exhibition catalogue), Thessaloniki, Folklore and Ethnological Museum of Macedonia and Thrace, 31 October 2011 - 11 June 2012, Thessaloniki: Folklore and Ethnological Museum of Macedonia and Thrace, 2011.

**Moutsopoulos 1996:** N. Moutsopoulos, Ο Βυζαντινός οικισμός της Ρεντίνας, *ArchErgMakThrak* 10β, 1996, 605-36.

**Moutsopoulos 2000:** N. Moutsopoulos, *Ρεντίνα IV, Οι εκκλησίες του βυζαντινού οικισμού*, Thessaloniki: Etairia Makedonikon Spoudon, 2000.

**Müller 1911:** S. Müller, Juellinge-Fundet og den romerske Periode, *Nordiske fortidsminder* 2.1, 1911, 1-46.

**Mundell-Mango 1992:** M. Mundell-Mango, Monetary Value of Silver Revetments and Objects Belonging to Churches A.D. 300-700, in *Boyd, Mundell-Mango (eds) 1992*, 123-136.

**Mylonas 1931-32:** G. Mylonas, Ελευσινιακά πάρεργα (Ελευσίνα, κτήμα Ἀγγελου Κορίζη), *AD* 14, 1931-1932, 41-48.

**Mylonas 1975:** G. Mylonas, *Το δυτικόν νεκροταφείον της Ελευσίνος*, vol. Α'-Γ', Athens: Archaiologiki Etaireia, 1975.

**Mylonas, Raymond (eds) 1953:** G. Mylonas, D. Raymond (eds), *Studies presented to David Moore Robinson on his Seventieth Birthday*, vol. I-II, Saint Luis, MI: Washington University, 1953.

**Nalpantis 1987:** D. Nalpantis, Θεσσαλονίκη, οδός Κολωνίαρη, Γαλανάκη και Παπαθανασίου, *ArchDelt* 42, 1987, Chronika B2, 403-406.

**Nalpantis 1988:** D. Nalpantis, Θεσσαλονίκη, οικόπεδο Βυζαντινού Μουσείου, *ArchDelt* 43, 1988, Chronika B2, 381.

**Nalpantis 1989:** D. Nalpantis, Θεσσαλονίκη, οικόπεδο Βυζαντινού Μουσείου (Ανατολικό νεκροταφείο), *ArchDelt* 44, 1989, Chronika B2, 346.

**Nalpantis 1992:** D. Nalpantis, Νεώτερα ευρήματα από σωστικές ανασκαφές στη Θεσσαλονίκη, *ArchErgMakThrak* 6, 1992, 311-326.

**Nalpantis 2003:** D. Nalpantis, *Ανασκαφή στο οικόπεδο του Μουσείου Βυζαντινού Πολιτισμού στη Θεσσαλονίκη. Ταφές και ευρήματα*, Athens: Tameion Archaiologikon Poron, 2003.

**Nasrallah, Bakirtzis, Friesen (eds) 2010:** L. Nasrallah, Ch. Bakirtzis and S.J. Friesen (eds), *From Roman to Early Christian Thessalonikē. Studies in Religion and Archaeology*. Harvard Theological Studies 64, Cambridge, MA: Harvard University Press, 2010.

**Nenna (ed.) 2000:** M.-D. Nenna (ed.), *La route du verre. Ateliers primaires et secondaires du second millénaire av. J.C. au Moyen Âge*, Lyon: Maison de l'Orient Méditerranéen-Jean Pouilloux, 2000.

**Nenna 1998:** M.-D. Nenna, Exploration Archéologique de Délos, Fascicule XXXVII, *Les Verres*, Paris: École Française d'Athènes, Dépositaire de Boccard éditions-Diffusion, 1998.

**Nenna 2003:** M.-D. Nenna, Les ateliers traditionnels d'aujourd'hui: des modèles pour l'archéologie?, in *Foy (ed.) 2003*, 52-59.

**Nenna 2006:** M.-D. Nenna, Marques sur verre découvertes au Proche et au Moyen-Orient (Irak, Israël, Jordanie, Liban et Syrie), in *Foy, Nenna (Eds) 2006a*, 431-446.

**Nenna 2007:** M.-D. Nenna, *Glassmakers of Papanaidupet*. Film, CNRS, Harpocrates Publishing, Alexandia, Egypt, 2007.

**Nenna 2008a:** M.-D. Nenna, Nouveaux acquis sur la production et le commerce du verre antique entre Orient et Occident, *Zeitschrift für Schweizerische Archäologie und Kunstgeschichte* 65.1, 2008, 61-66.

**Nenna 2008b:** M.-D. Nenna, Production et commerce du verre à l'époque impériale: nouvelles découvertes et problématiques, *Facta: A Journal of Roman Material Culture Studies* 1 (2007 [2008]), 125-147.

**Nenna, Picon, Vichi 2000:** M.-D. Nenna, M. Picon et M. Vichy, Ateliers primaires et secondaires en Égypte à l'époque gréco-romaine, in *Nenna (ed.) 2000*, 97-112.

**Newby, 2008:** M. Newby, *Byzantine Mould-Blown Glass from the Holy Land with Jewish and Christian Symbols*, London: Shlomo Moussaieff, 2008.

**Newby, Painter (eds) 1991:** M. Newby, K. Painter (eds), *Roman Glass. Two Centuries of Art and Invention*, London: The Society of Antiquaries of London, 1991.

**Newman, Lie 1992:** R. Newman, H. Lie, Technical Examination and Conservation of Objects, in *Boyd, Mundell-Mango (eds) 1992*, 77-88.

**Nicholson 2007:** P.T. Nicholson, *Brilliant Things for Akhenaten: The Production of Glass, Vitreous Materials and Pottery at Amarna Site 0.45.1 (Excavation Memoirs)*. London: Egypt Exploration Society, 2007.

**Nicholson, Jackson, Trott 1997:** P.T. Nicholson, C.M. Jackson, K.M. Trott, The Ulu Burun Glass Ingots, Cylindrical Vessels and Egyptian Glass, *JEA* 83, 1997, 143-153.

**Nigdelis 1994:** P.M. Nigdelis, Synagoge(n) und Gemeinde der Juden in Thessaloniki: Fragen aufgrund einer neuen jüdischen Grabinschrift der Kaiserzeit, *ZPE* 102, 1994, 297-306.

**Nikolić-Đorđević 1990:** S. Nikolić-Đorđević, Rimske staklene posude sa beogradske tvrdjave, *GGB* XXXVII, 1990, 39-54.

**Nikolić-Đorđević 1999:** S. Nikolić-Đorđević, Staklene posude, in *Popović 1999*, 108-109.

**Nikolov, Bujukliev 1967:** D. Nikolov, H. Bujukliev, Novi trakijskii mogil'ni pogrebeniia ot Chatalka, Starozagorska, *Arheologiia* (Sofiia) 9.3, 1967, 10-25.

**Nikolovski 2006:** Z. Nikolovski, Nekropolata Drezga s. Lopate-Kumanovsko, *MAA* 17, 1999-2001, 2006, 231-260.

**Olczak 1978:** J. Olczak, *Novae-Sector zachodni 1974* Czesc 1, Piec szklarski, *Archeologia (Warszawia)* 28, 1978, 127-137.

**Olczak 1998:** J. Olczak, *Produkcja szkła w rzymskim i wczesnobizantyjskim Novae*, Torun: Uniwersytet Mikolaja Kopernika, 1998.

**Oliver 1980:** A. Oliver Jr., *Ancient Glass in The Carnegie Museum of Natural History, Pittsburgh*, Pittsbourgh: Carnegie Museum of Natural History, 1980.

**Oliver 1983:** A. Oliver Jr., Tomb 12 at Episkopi, *RDAC* 1983, 245-256, pls. 39-41.

**Oliver 1984:** A. Oliver Jr., Early Roman Faceted Glass, *JGS* 26, 1984, 35-58.

**Opait 1985:** C. Opait, Unele considerații privind vasele de sticlă din burgul roman tirziu de la Topraichioi, *Pontica* 18, 1985, 249-257.

**Oppenheim *et al.* 1970:** A.L. Oppenheim, R.H. Brill, D. Barag and A. von Sladern, *Glass and Glass-making in Ancient Mesopotamia*, Corning NY: The Corning Museum of Glass, 1970.

**Özet 2000:** A. Özet, *Sparkles from the DeeGlass Vessels of the Bodrum Museum of Underwater Archaeology*, Istanbul: Bericap, 2000.

**Pallas 1977:** D. Pallas, *Les Monumens paléochrétiens de Grèce découverts de 1959 à 1973*, Città del Vaticano: Pontifico Istituto di Archeologia Cristiana, 1977.

**Pantermalis 2001:** D. Pantermalis, Δίον 2001. Διέων υαλά σκεύη, *ArchErgMakThrak* 15, 2001, 347-354.

**Panti 2010:** A. Panti, Κεραμική της Ύστερης Αρχαιότητας από το ανατολικό νεκροταφείο Θεσσαλονίκης. Βόρεια πύλη της ΔΕΘ 1998, in *Papanikola-Bakirtzi, Kousoulakou (eds) 2010*, 466-485.

**Paolucci 1997:** F. Paolucci, *I vetri incisi dall'Italia settentrionale e dalla Rezia nel periodo medio e tardo imperiale*, Firenze: All'Insegna del Giglio, 1997.

**Paolucci 2002:** F. Paolucci, *L'arte del vetro inciso a Roma nel IV secolo D.C.*, Firenze: All'Insegna del Giglio, 2002.

**Papachristodoulou 1968:** I. Papachristodoulou, Άργος, Κτήμα Κομνηνού, *ArchDelt* 23, 1968, Chronika B1, 130.

**Papadopoulou 1963:** F. Papadopoulou, Ανασκαφή Πλατείας Δικαστηρίων Θεσσαλονίκης, *ArchDelt* 18, 1963, Chronika B2, 196-199.

**Papadoupoulou 1964a:** F. Papadopoulou, Θεσσαλονίκη, Πλατεία Δικαστηρίων, *ArchDelt* 19, 1964, 329-331.

**Papadopoulou 1964b:** F. Papadopoulou, Θεσσαλονίκη, Ανασκαφή περιοχής Ραμόνας παρά την αρχήν της οδού Λαγκαδά, *ArchDelt* 19, 1964, Chronika B3, 334, drawing 6, pl. 381α-β.

**Papangeli 1987:** P. Papangeli, Ελευσίνα, Γήπεδο Πανελευσινιακού (δυτική νεκροταφείο), *ArchDelt* 42, 1987, Chronika B1, 32.

**Papangelos 2005:** I. Papangelos, Η βασιλική Σωλήνος στην Κασσάνδρα, *ArchErgMakThrak* 19, 2005 (Oral presentation given in the conference but not included in the printed volume of the acts.)

**Papanikola-Bakirtzi 2002 (ed.):** D. Papanikola-Bakirtzi (ed.), *Everyday Life in Byzantium* (exhibition catalogue), Thessaloniki, White Tower, October 2001 - January 2002, Athens: Hellenic Ministry of Culture, 2002.

**Papanikola-Bakirtzi, Kousoulakou (eds) 2010:** D. Papanikola-Bakirtzi, D. Kousoulakou (eds), Επιστημονική συνάντηση Η κεραμική της ύστερης αρχαιότητας στον ελλαδικό χώρο *(3ος - 7ος αι.)*, 12-16 Νοεμβρίου 2006, Thessaloniki: Institute of Macedonian and Thracian Studies, 2010.

**Papapostolou 1983:** I.A. Papapostolou, Κτερίσματα ταφής σε ρωμαϊκό μαυσωλείο στην Πάτρα, *ArchEph* 1983, 1-34.

**Paškvalin 1976:** V. Paškvalin, Antičko staklo s područija Bosne i Herzegovine, *AV* 25, 1974, (1976), 109-138.

**Pastorino 1989:** A.M Pastorino, *Vetri antichi nelle collezioni del Museo Civico di Archeologia Ligure di Genova-Pegli, Bolletino Musei Civici Genovesi* Anno IX-N. 31-Gennaio/Aprile 1989, Genova: Museo Civico di Archeologia Ligure di Genova-Pegli, 1989.

**Pastorino 1998:** A.M. Pastorino, Vetri romani dagli scavi urbani di Albenga, *AnnAIHV* 14, 1988, 108-112.

**Pazaras 1981:** Th. Pazaras, Δύο παλαιοχριστιανικοί τάφοι από το δυτικό νεκροταφείο της Θεσσαλονίκης, *Makedonika* 21, 1981, 373-389.

**Pejić 1982:** P. Pejić, Prilog topografiji Remesiane, *NZ* 11, 1982, 79-83.

**Pelekanidis 1960:** S. Pelekanidis, Μεσαιωνικά Μακεδονίας. Θεσσαλονίκη, *ArchDelt* 16, 1960, Chronika, 222-225, pl. 194β, 195α, 196β.

**Pelekanidis 1961/62a:** S. Pelekanidis, Μεσαιωνικά Μακεδονίας. Α. Θεσσαλονίκη, 1. Βασιλική Αχειροποιήτου, *ArchDelt* 17, 1961/62, Chronika B, 252-253.

**Pelekanidis 1961/62b:** S. Pelekanidis, Μεσαιωνικά Μακεδονίας. Α. Θεσσαλονίκη, 2. Αγία Σοφία, *ArchDelt* 17, 1961-62, Chronika B', 253-256.

**Pelekanidis 1961/62c:** S. Pelekanidis, Μεσαιωνικά Μακεδονίας. Θεσσαλονίκη, 5. Παλαιοχριστιανικοί τάφοι, *ArchDelt* 17, 1961/2, Chronika B, 257, pl. 314α-β.

**Pelekanidou 1992:** E. Pelekanidou, Θεσσαλονίκη, Ανασκαφή βορείως Κοιμητηρίου Ευαγγελιστρίας (Ανατολικό νεκροταφείο), *ArchDelt* 47, 1992, Chronika B2, 439-440.

**Pelekanidou 1993:** E. Pelekanidou, Νέα ευρήματα στο ανατολικό νεκροταφείο της Θεσσαλονίκης, *ArchErgMakThrak* 7, 1993, 373-387.

**Pelekanidou 1994:** E. Pelekanidou, Θεσσαλονίκη. Ανασκαφή βόρεια του κοιμητηρίου της Ευαγγελίστριας (ανατολικό νεκροταφείο), *ArchDelt* 49, 1994, Chronika B2, 526-529, pls. 158, 159.

**Pelekanidou 1995:** E. Pelekanidou, Θεσσαλονίκη, Κοιμητήριο Ευαγγελίστριας (ανατολικό νεκροταφείο), *ArchDelt* 50, 1995, Chronika B2, 534-535.

**Perdrizet 1908:** P. Perdrizet, *Fouilles de Delphes t. V. Monuments figurés, petits bronzes, terres-cuites, antiquités diverses*, Paris: Al. Fontemoing, 1908.

**Pérez-Arantegui, Paz-Peralta and Ortiz-Palomar 1996:** J. Pérez-Arantegui, J.Á. Paz-Peralta and E. Ortiz-Palomar, Analysis of the Products Contained in Two Roman Glass *unguentaria* from the Colony of *Celsa* (Spain), *Journal of Archaeological Science* 23, 1996, 649-655.

**Perlzweig 1961:** J. Perlzweig, *The Athenian Agora, vol. VII, Lamps of the Roman Period. First to Seventh Centuries After Christ*, Princeton, NJ: ASCSA, 1961.

**Pernice 1893:** E. Pernice, Σίφων, *JdI* 8 (1893), 180-184.

**Petkos 2000:** A. Petkos, Ανασκαφή νοτιοδυτικού τμήματος τειχών Διοκλητιανουπόλεως στο Άργος Ορεστικό Καστοριάς, *ArchErgMakThrak* 14, 2000, 581-590.

**Petković *et al.* 2005:** S. Petković, M. Ružić, S. Jovanović, M. Vuksan, Zs. K. Zoffmann, *Roman and Medieval Necropolis in Ravna near Knjaževac*, Belgrade: Arheološki Institut, 2005.

**Petković, Ružić 2005:** S. Petković, M. Ružić, Roman Necropolis, in *Petković et al. 2005*, 23-155.

**Petridis 2002:** P. Petridis, Βιοτεχνικές εγκαταστάσεις της πρώιμης βυζαντινής περιόδου στους Δελφούς, in πρόγραμμα και περιλήψεις εισηγήσεων και ανακοινώσεων 22ου Συμποσίου Βυζαντινής και Μεταβυζαντινής Αρχαιολογίας και Τέχνης, Αθήνα 17-19 Μαΐου 2002, (Athens 2002), 90-91.

**Petrie 1925:** W.M.F. Petrie, *Tombs of the Courtiers at Oxyrhynkhos*, British School of Archaeology in Egypt and Egyptian Research Account, v. 37, London: British School of Archaeology in Egypt, 1925.

**Petru 1972:** S. Petru, *Emonske Nekropole*, Ljubljana: Narodni Muzej Slovenie, 1972.

**Petru 1975:** S. Petru, Das spätantike Glas in Slowenien, in *Han (ed.)1975*, 5-10.

**Petru 1976:** S. Petru, Rimsko steklo Slovenije, *AV* 25, 1974 (1976), 13-34.

**Petsas 1961/62:** F. Petsas, Κοζάνη, Λιγνάδες, *ArchDelt* 17, 1961/62, Chronika, 215, pl. 253δ.

**Petsas 1964:** F. Petsas, Βέροια, *ArchDelt* 19, 1964, Chronika B2, 353, pl. 414δ.

**Petsas 1966:** F. Petsas, Θεσσαλονίκη, Ανασκαφή Πανεπιστημίου. Θεολογική Σχολή, *ArchDelt* 21, 1966, Chronika B2, 334-339, pls. 345-359.

**Petsas 1966/67:** F. Petsas, Χρονικά Αρχαιολογικά, *Makedonika* 7, 1966-67, 277-368.

**Petsas 1967a:** F. Petsas, Θεσσαλονίκη, Ανασκαφή Πανεπιστημίου, παρά την Νομική Σχολή, *ArchDelt* 22, 1967, Chronika B2, 391-92, 298, pl. 296ε, drawing 12ε.

**Petsas 1967b:** F. Petsas, Θεσσαλονίκη, οικόπεδο Σ. Γεωργιάδου, *ArchDelt* 22, 1967, Chronika B2, 393-396, pls. 299-300.

**Petsas 1967c:** F. Petsas, Θεσσαλονίκη, Πλατεία Δικαστηρίων, *ArchDelt* 22, 1967, Chronika B2, 383.

**Petsas 1968a:** F. Petsas, Η αγορά της Θεσσαλονίκης, *AAA* 1, 1968, 156-161.

**Petsas 1968b:** F. Petsas, Θεσσαλονίκη, Ιπποδρόμιο, Οικόπεδο Νάστου, *ArchDelt* 23, 1968, Chronika B2, 332-334, pls. 273β,ε, 279α-γ.

**Petsas 1969a:** F. Petsas, Θεσσαλονίκη, οικόπεδο 'Axios', *ArchDelt* 24 1969, Chronika B2, 300, pl. 309.

**Petsas 1969b:** F. Petsas, Θεσσαλονίκη. 'Αλλα ευρήματα εντός οικοπέδων (Θεσσαλονίκη, οδός Σωκράτους 58-62 (οικοπεδο Βελίκα)), *ArchDelt* 24, 1969, Chronika B2, 303-304, pl. 313β.

**Petsas 1969c:** F. Petsas, Χρονικά Αρχαιολογικά, *Makedonika* 9, 1969, 101-224.

**Petsas 1970a:** F. Petsas, Θεσσαλονίκη. Ανατολικό νεκροταφείο. Στρατόπεδο Γ' Σώματος Στρατού, *ArchDelt* 25, 1970, Chronika B2, 347-351.

**Petsas 1970b:** F. Petsas, Θεσσαλονίκη. Δυτικόν νεκροταφείον, *ArchDelt* 25, 1970, Chronika B2, 351-352, pl. 303 ε, στ.

**Petsas 1971:** F. Petsas, 'Αγιο, δίπλα στο Γυμνάσιο Αρρένων, *ArchDelt* 26, 1971, Chronika B1, 177.

**Petsas 1974:** F. Petsas, Χρονικά Αρχαιολογικά 1968-1970, *Makedonika* 14, 1974, 212-381.

**Philippe 1964:** J. Philippe, *Initiation à l'histoire du verre*, Liège: Musée du verre, 1964.

**Philippe 1970:** J. Philippe, *Le Monde byzantin dans l'histoire de la verrerie (Ve-XVIIe siècle)*. Bologna, 1970.

**Picozzi, Carinci (ed.) 1996:** M.G. Picozzi, F. Carinci (ed.), Studi Miscellanei 30. *Studi in memoria di Lucia Guerrini*, Roma: L'Erma, 1996.

**Pilinger, Pülz (eds) 1992:** R. Pilinger, A. Pülz, H. Vetters (eds), *Die Schwarzmeerküste in der Spätantike und im Frühen Mittelalter* (Schriften der Balkan-Kommission der ÖAW, Antiquarische Abteilung), Wien: Österreichische Akademie der Wissenschaften, 1992.

**Pinder-Wilson, Scanlon 1973:** R.H. Pinder-Wilson, G.T. Scanlon, Glass Finds from Fustat: 1964-1971, *JGS* 15, 1973, 12-30.

**Pirling 1959:** R. Pirling, Gräber des frühen 5. Jahrhunderts aus Krefeld-Gellep, *BJb* 159, 1959, 215-242.

**Pirling 1966:** R. Pirling, *Das römisch-fränkische Gräberfeld von Krefeld-Gellep, 1960-1963* (Germanische Denkmäler der Vörkerwanderungszeit. Serie B, Band 2), Berlin: Gebr. Mann, 1966.

**Pirling 1977:** R. Pirling, Die Römischen und Byzantinischen Glasfunde von Apamea in Syrien, *AnnAIHV* 7, 1977, 137-147.

**Pitsakis 1971:** Konstantinos Armenopoulos, Πρόχειρον Νόμων ή *εξάβιβλος*, edit: K. G. Pitsakis, Athens: Dodoni - E.K. Lazos, 1971.

**Pitsakis 1992:** K. G. Pitsakis, Η σημασία του νομικού έργου του Κωνσταντίνου Αρμενόπουλου, in 2° συμπόσιο Η Μακεδονία κατά την εποχή των Παλαιολόγων (Th. Zisis, P. Asimakopoulou-Atzaka, V. Katsaros eds), Thessaloniki 14-20 December 1992, Thessaloniki 2002, 239-258.

**Platz-Horster 1976:** G. Platz-Horster, *Antike Gläser* (exhibition catalogue), November 1976 - February 1977, Berlin: Antikenmuseum Berlin Staatliche Museen Preussischer Kulturbesitz, 1976.

**Plesničar-Gec 1972:** L. Plesničar-Gec, *Severnoemonsko grobišče*, Ljubljana: Narodni Muzej Slovenie, 1972.

**Plesničar-Gec 1976:** L. Plesničar-Gec, Steklene zajemalke iz severnega emonskega grobišča, *AV 25*, 1974, (1976), 35-37.

**Plesničar-Gec 1983:** L. Plesničar-Gec, *Starokrščanski Center v Emoni*, Ljubljana: Narodni Muzej Slovenije, 1983.

**Polfe (ed.) 1999:** M. Polfe (ed.), *Artisanat et productions artisanales en milieu rural dans les provinces du Nord-ouest de l'empire romain*. Actes du colloque organisé à Erpeldange (Luxembourg) les 4 et 5 mars 1999 par le Séminaire d'Études Anciennes du Centre Universitaire de Luxembourg et Instrumentum, Montagnac: Editions Monique Mergoil, 1999.

**Pologiorgi 1985:** M. Pologiorgi, Κίσσαμος: Τοπογραφία μίας αρχαίας πόλης, *AAA* 1985, 65-80.

**Pologiorgi 1998:** M. Pologiorgi, *Μνημεία του δυτικού νεκροταφείου του Ορωπού*, Athens: Tameion Archaiologikon Poron, 1998.

**Pop-Lazić 2002:** S. Pop-Lazić, Nekropole rimskog Singidunuma, *Singidunum* 3, 2002, 7-100.

**Popović 1999:** M. Popović, *Tvrdjava Ras*, Beograd: Arheološki Institut, 1999.

**Prausnitz 1967:** M.W. Prausnitz, *Excavations at Shavei Zion: the Early Christian Church*, Rome: Centro per la antichità e la storia del vicino oriente, 1967.

**Preda 1980:** C. Preda, *CALLATIS Necropola romano-bizantină*, Bucureşti: Editura Academiei Republicii Socialiste Romania, 1980.

**Preka-Alexandri 1992:** K. Preka-Alexandri, Κατάσχεση στην Κέρκυρα, *ArchDelt* 47, 1992, Chronika B1, 333, pl. 100,

**Price 1981:** J. Price, *Roman Glass in Spain*, unpublished PhD thesis, University of Wales, Cadriff, 1981.

**Price 1985:** J. Price, Early Roman Vessel Glass from Burials in Tripolitania: A Study of Finds from Forte Della Vite and other sites now in the collections of the National Museum of Antiquities in Tripoli, in *Buck, Mattingly (eds) 1985*, 67-106.

**Price 1988a:** J. Price, An Egyptian Terracotta Group Showing Eros beside a Glass Furnace, *The Antiquaries Journal* 68, 1988, 317-319.

**Price 1988b:** J. Price, A Survey of the Hellenistic and Early Roman Vessel Glass Found on the Unexplored Mansion Site at Knossos in Crete, *AnnAIHV* 11, 1988, 27-36.

**Price 1988c:** J. Price, Glass from the Argentière and Aiguières sites, Fréjus, *2e Journées d'étude de l'AFAV* (Rouen 1987), Rouen 1988, 24-39.

**Price 1974:** J. Price, A Roman Mould-blown Negro-head Glass Beaker from London, *Antiquaries Journal* 54, 1974, 291–292.

**Price 1991:** J. Price, Decorated Mould-Blown Glass Tableware in the First century AD, in *Newby, Painter (eds) 1991*, 56-75.

**Price 1992:** J. Price, Hellenistic and Roman Glass, in *Sackett et al. (ed.) 1992*, 415-490, pls. 336-353.

**Price 1998:** J. Price, The social context of glass production in Roman Britain, in *McCray, Kingery (eds), 1998*, 331-347.

**Price 2002:** J. Price, Glass in Benghazi, Knossos and Mytilene: comparison of finds of the mid-1st century AD, in *Kordas (ed.) 2002*, 123-126.

**Price 2005:** J. Price, Glass Working and Glassworkers in Cities and Towns, in *MacMahon, Price (eds) 2005*, 167-190.

**Price, Cool 1989:** J. Price, H.E.M. Cool, The evidence for the production of glass in Roman Britain, in *Foy, Sennequier (eds) 1989*, 23-30.

**Price, Cottam 1998a:** J. Price, S. Cottam, *Romano-British Glass Vessels: A Handbook*, York: Council for British Archaeology, 1998.

**Price, Cottam 1998b:** J. Price, S. Cottam, Glass tableware in use at Mytilene in Lesbos in the early-mid 1st century AD, *AnnAIHV* 14, 1998, 58-62.

**Radić and Jurišic 1993:** I. Radić, M. Jurišić, Das Antike Schliffswrack von Mljet, Kroatien, *Germania* 71, 1993, 113-138.

**Radnóti 1957:** A. Radnóti, Glasgefässe und Glasgegenstände, *AH* XXXVI, *Intercisa* II, 1957, 141-163.

**Raknić 1968:** Ž. Raknić, Nekoliko novih spaljenih grobova iz Zadra, *Diadora* 4, 1968, 211-213.

**Ravagnan 1994:** G. L Ravagnan, *Vetri antichi del Museo Vetrario di Murano. Collezioni dello Stato*, Corpus delle collezioni archeologiche del vetro nel Veneto, v. 1, Venezia: Comitato Nazionale Italiano dell'Association International pour l'Histoire du Verre, 1994.

**Rehren, Pusch 1997:** T. Rehren, and E.B. Pusch. New Kingdom Glass-melting Crucibles from Qantir-Piramesses, *JEA* 83 (1997), 127-141.

**Reimenschneider 1989:** U. Reimenschneider, Antike Gläser im Museum Altenessen, *Boreas* 12, 1989, 159-198.

**Reut 1973:** M. Reut, Le verre soufflé de Hérat, *StIr* 2.1, 1973, 96-112.

**Rizakis 1983:** A. Rizakis, Η κοινότητα των συμπραγματευομένων Ρωμαίων, της Θεσσαλονίκης και η ρωμαϊκή οικονομική διείσδυση στη Μακεδονία, in *Archaia Makedonia* 4, 1983, 511-521.

**Robert 1937:** L. Robert, *Études Anatoliennes*, Paris: E. de Boccard, 1937.

**Robert 1974:** L. Robert, Les inscriptions de Thessalonique, *RPh* 48, 1974, 242-243.

**Robinson 1959:** H.S. Robinson, The Athenian Agora: Results of Excavations Conducted by The American School of Classical Studies, v. 5, *Pottery of the Roman Period. Chronology*, Princeton, NJ: ASCSA, 1959.

**Roffia 1993:** E. Roffia, *I vetri antichi delle civiche raccolte archeologiche di Milano*. Comune di Milano, Settore Cultura e Spettacolo Civiche Raccolte Archeologiche, Milano: Comune di Milano, 1993.

**Romiopoulou 1977a:** Aik. Romiopoulou, Θεσσαλονίκη, Τάφοι. Αγίου Νέστορος 8, *ArchDelt* 32, 1977, Chronika B2, 197, pl. 126 α.

**Romiopoulou 1977b:** Aik. Romiopoulou, Θεσσαλονίκη, Αμπελόκηποι, οδός Ελευθερίας 13, *ArchDelt* 32, 1977, Chronika B2, 197, pl. 126β.

**Romiopoulou 1978:** Aik. Romiopoulou, Θεσσαλονίκη, Νομική Σχολή Πανεπιστημίου Θεσσαλονίκης, *ArchDelt* 33, 1978, Chronika B2, 234.

**Romiopoulou 1979:** Aik. Romiopoulou, Θεσσαλονίκη, Περιοχή Πανεπιστημίου - Διεθνούς Εκθέσεως, *ArchDelt* 34, 1979, Chronika B2, 275.

**Romiopoulou 1981:** Aik. Romiopoulou, Σίνδος, *ArchDelt* 36, 1981, Chronika B2, 302.

**Romiopoulou 2002:** K. Romiopoulou, Glass Finds from Graves at Amphipolis, in *Kordas (ed.) 2002*, 71-77.

**Ropkić, 2006:** A. Ropkić, Stakleni instrumenti-mešalice iz Viminaciuma, *GSAD* 22, 2006, 375-380.

**Rouquet 2003:** N. Rouquet, Les biberons, les tire-lait ou les tribulations d'une tubulure peu commune, in *Gourevitch et al. (ed.) 2003*, 164-170.

**Rütti 1991:** B. Rütti, *Die römischen Gläser aus Augst und Kaiseraugst*. Forschungen in Augst 13/1-2, Augst: Römermuseum, 1991.

**Ružić 1994:** M. Ružić, *Rimsko staklo u Srbiji*, Beograd: Filozofski Fakultet Centar za Arheološka Istraživanja, 1994.

**Ružić 1996:** M. Ružić, Prilog proučavanju zvonolikih boca od stakla, *GSAD* 12, 1996, 191-196.

**Ružić 2001:** M. Ružić, Stakleni pehar sa medaljonima iz Municipiuma S..., in *Lazić 2001*, 413-429.

**Sackett *et al.* (ed.) 1992:** I.H. Sackett *et al.* (ed.), *Knossos from Greek City to Roman Colony, Excavations at the Unexplored Mansion II*, London: B.S.A., 1992.

**Sagadin 2004:** M. Sagadin, Poznoantička steklarska delavnica (?) v Kranju, in *Lazar (ed.) 2004*, 107-114.

**Saguì 1996:** L. Saguì, Un piatto di vetro inciso da Roma: contributo ad un inquadramento delle officine vetrarie tardoantiche, in *Picozzi, Carinci (eds) 1996*, 337-358.

**Saguì 1998:** L. Saguì, Produzioni vetrarie a Roma tra V e VII secolo. Nuovi dati archeologici, *AnnAIHV* 14, 1998, 203-207.

**Sakellarakis 1965:** G. Sakellarakis, Ηράκλειον, Πλατεία Δασκαλόγιαννη (οικόπεδο Α. Μανασάκη), *ArchDelt* 20, 1965, Chronika B3, 562, pl. 709.

**Saldern 1968:** A. von Saldern, *Ancient Glass in the Museum of Fine Arts, Boston*, Boston: The Museum of Fine Arts, 1968.

**Saldern 1974:** A. von Saldern, *Glas, Glassammlung Hentrich, Antike und Islam* I,3, Dusseldorf: Kunstmuseum, 1974.

**Saldern 1980a:** A. von Saldern, *Ancient and Byzantine Glass from Sardis*, Archaeological Exploration of Sardis, Monograph 6, Cambridge Massachusetts, and London, England: Harvard University Press, 1980.

**Saldern 1980b:** A. von Saldern, *Glas von der Antike bis zum Jugendstil, Sammlung Hans Cohn Los Angeles/Cal.*, Mainz am Rhein: Verlag Philipp von Zabern, 1980.

**Saldern *et al.* 1974:** A. von Saldern, B. Nolte, P. La Baume, T.E. Haevernick, *Gläser der Antike, Sammlung Erwin Oppenländer*, Mainz am Rhein: P. von Zabern, 1974.

**Salibi 1979:** N. Salibi, Verres provenant de la côte phénicienne, *AnnAIHV* 8, 1979, 133-144.

**Šaranović-Svetek 1986:** V. Šaranović-Svetek, *Antičko staklo u Jugoslovenskom Delu Provincije Donje Panonije*, Novi Sad: Vojvodjanski Muzej, 1986.

**Sazanov 1995:** A. Sazanov, Verres à décor de pastilles bleues provenant de fouilles de la Mer Noire. Typologie et chronologie, in *Foy (ed.) 1995*, 331-342.

**Scanlon, Pinder-Wilson 2001:** G.T. Scanlon, R. Pinder-Wilson, *Fustat Glass of the Early Islamic Period. Finds excavated by the American Research Center in Egypt 1964-1980*, London: Altajir World of Islam Trust, 2001.

**Scatozza-Höricht 1986:** L.A. Scatozza-Höricht, *I vetri romani di Ercolano*, Roma: L'Erma di Bretschneider, 1986.

**Scatozza-Höricht 1988:** L.A. Scatozza-Höricht, Die Verwendung der Römischen Gläser im 1. jh. n. Chr. und die Fundkomplexe der Vesuvstäde, *AnnAIHV* 11, 1988, 43-48.

**Scatozza-Höricht 1991:** L.A. Scatozza-Höricht, Syrian Elements among the Glass from Pompeii and Herculaneum, in *Newby, Painter (eds) 1991*, 76-85.

**Schäfer 1936:** E. Schäfer, Die Heilige mit dem Kreuz in altchristliche Kunst, *Römische Quartalschrift für christliche Altertumskunde und Kirchengeschichte, Freiburg*, 44, 1936, 67-104.

**Schauer 2001-2002:** C. Schauer, Γυάλινα αγγεία από την Ολυμπία, in 6° Διεθνές Συνέδριο Πελοποννησιακών Σπουδών, Τρίπολις 24-29 Σεπτεμβρίου 2000, vol.Α'-Β', Athens: Etaireia Peloponnisiakon Spoudon, 33-48.

**Scheibler 1983:** I. Scheibler, *Griechische Töpferkunst: Herstellung, Handel und Gebrauch der antiken Tongefässe*, Munich 1983.

**Schlick-Nolte, Lierke 2002:** B. Schlick-Nolte, and R. Lierke. From Silica to Glass. On The Track of the Ancient Glass Artisans, in *Bianchi et al. 2002*, 9-40.

**Sennequier 1985:** G. Sennequier, *Verrerie d'époque romaine.* Collections des musées départementaux de Seine Maritime, Rouen: Musée des Antiquités de Rouen, 1985.

**Sennequier 1986:** G. Sennequier, Un certain Amarantus (ou Amaranthus?), verrier installé en Bourgogne au Ier siècle de notre ère, *JGS* 28, 11-18.

**Sennequier 1994:** G. Sennequier, Roman Glass Found in Upper Normandy, *JGS* 36, 1994, 56-67.

**Sève 1980:** M. Sève, Un puits argien du Haut-Empire, *Études Argiennes, BCH supl* VI (1980), 317-318.

**Silvano 1999:** F. Silvano, Università di Pisa-Musei di Ateneo, Catalogo delle collezioni Egittologiche, III, *Vetri Bizantini dall'Egitto. Medinet Madi 1988*, Pisa: Edizioni ETS, 1999.

**Sironen 1990-91:** E. Sironen, Another Glass-worker in an Early Christian Epitaph from the Athenian Acropolis, *Horos* 8-9, 1990-91, 83-85.

**Skarlatidou 1981:** E. Skarlatidou, Αλεξανδρούπολη, συμβολή Βασιλέως Γεωργίου, Μιαούλη και Ειρήνης, *ArchDelt* 36, 1981, Chronika B2, 351-353, pl. 242γ.

**Skordara, Gounaris, Maniatis 2002:** M. Skordara, G. Gounaris, Y. Maniatis, Analysis of Glass of the Early Christian Period from the Workshop and Settlement of Philippi, Northern Greece: Preliminary Results, in *Kordas (ed.) 2002*, 321-326.

**Skordara 2014:** M. Skordara, M.A. Skordara, *Palaiochristianiko ergastirio yalourgias stous Filippous: Technologiki, typologiki kai morfologiki meleti tis topikis paragogis* (Early Christian Glass Workshop at Philippi: Technological, typological and morphological study of the local production). PhD study submitted to the University of Athens in 2014.

**Slane 1994:** K.W. Slane, Tetrarchic Recovery in Corinth: Pottery, Lamps, and Other Finds from the Peribolos of Apollo, *Hesperia* 63, 1994, 127-168.

**Smith 1957:** R. Winfield Smith, *Glass from the Ancient World: the Ray Winfield Smith collection*, Corning: Corning Museum of Glass, 1957.

**Snow 1992:** C. Snow, From Ingot to Object: Fabrication Techniques Used in the Manufacture of the Hama Silver, in *Boyd, Mundell-Mango (eds) 1992*, 197-202.

**Sokolovska 1986:** V. Sokolovska, *Isar-Marvinci i povardarjeto vo antičko vreme*, Skopje: Muzej na Makedonija, 1986.

**Sorokina 1962:** N.P. Sorokina, Steklo iz raskopok Pantikapeija 1945-59 gg., *Materialy* 103, 1962, 210-236.

**Sorokina 1965:** N.P. Sorokina, Steklanie sosudi iz Tanaisa, *Materialy* 127, 1965, 202-248.

**Sorokina 1967:** N.P. Sorokina, Das antike Glas der Nordschwarzmeerküste, *AnnAIHV* 4, 1967, 67-79.

**Sorokina 1970:** N.P. Sorokina, Die Nuppengläser von der nordküste des Schwarzen Meeres, *AnnAIHV* 5, 1970, 71-79.

**Sorokina 1971:** N.P. Sorokina, O steklijanih sosudah s kaplami sinego stekla iz Pričernomorija, *Sovietskaia Arkheologia*, Moskva 4.1971, 85-101.

**Sorokina 1977a:** N.P. Sorokina, Antichnye stekljannye sosudy iz paskopole nekropolja bosporskovo goroda Kepy na Tamanskom Polyostrove, *Antichny Mir i Archeologia*, no. A72, 1977, 115-144.

**Sorokina 1977b:** N.P. Sorokina, Facettenschliffgläser des 2-3 Jhd u. Z. aus dem Schwarzmeergebiet, *AnnAIHV* 7, 1977, 111-122.

**Sorokina 1987:** N.P. Sorokina, Glass Aryballoi (First-Third centuries A.D.) from the Northern Black Sea Region, *JGS* 29, 1987, 40-46.

**Sorokina 1988:** N.P. Sorokina, Die Kleinasiatische Gruppe von Gläser der 1.-2. Jh. n. Chr., *AnnAIHV* 11, 1988, 55-60.

**Sotiriou 1929:** G.A. Sotiriou, Αἱ Χριστιανικαί Θῆβαι τῆς Θεσσαλίας, *ArchEph* 1929, 1-150.

**Sotiriou 1939:** G.A. Sotiriou, Ἀνασκαφαί ἐν Νέα Ἀγχιάλῳ, *PAE* 1939, 53-72.

**Spaer 1988:** M. Spaer, The Pre-Islamic Glass Bracelets of Palestine, *JGS* 30, 1988, 51-62.

**Spaer 2001:** M. Spaer, *Ancient Glass in the Israel Museum, Beads and Other Small Objects*, Jerusalem: The Israel Museum, 2001.

**Spartz 1967:** E. Spartz, *Antike Gläser*, Kataloge den Staatlichen Kunstsammlungen Kassel, no. 1, Kassel: Staatliche Kunstsammlungen, 1967.

**Spasić-Djurić 2006:** D. Spasić-Djurić, *Lectus Genialis* iz Viminacijuma, *GSAD* 22, 2006, 295-310.

**Spyropoulos 1980:** Th. Spyropoulos, Κυνουρίας Ελληνικό / Τείχιο, *ArchDelt* 35, 1980, Chronika B1, 147-152.

**Srejović 1975:** D. Srejović, Grobnice Aurelije Maksimine i Pekonije Montane u Kominima (Municipium S...), *ZRNMB* VIII, 1975, 177-190.

**Stamenković, Ivanišević, Pešić 2016:** S. Stamenković, V. Ivanišević, J. Pešić, A Roman Cemetery at Mala Kopašnica, in S. Perić, A. Bulatović (eds), *Archaeological Investigations along the Highway Route E75 (2011-2014)*, Belgrade: Institute of Archaeology, 2016, 17-45.

**Stavropoulos 1964:** F. Stavropoulos, Αθήνα, οδός Πλαταιών και Παραμυθίας, *ArchDelt* 19, 1964, Chronika B1, 60-61.

**Steckner 1988:** C. Steckner, *Phanokantharoi* und *Kylikeia*, Dionysische Lichtgefässe in architektonischem Kontext, *AnnAIHV* 11, 1988, 257-270.

**Stern 1977:** E.M. Stern, *Ancient Glass at the Fondation Custodia (Collection Frits Lugt), Paris*, Archaeologia Traiectina, v. 12, Groningen: Wolters-Noordhoff, 1977.

**Stern 1984:** E.M. Stern, Antikes Glas in der Südtürkei, *Glastechnische Berichte* 57/5, 1984, 132-139.

**Stern 1985:** E.M. Stern, Ancient and Medieval Glass from the Necropolis Church at Anemurium, *AnnAIHV* 9, 1985, 35-60.

**Stern 1989:** E.M. Stern, The Production of Glass Vessels in Roman Cilicia, *KölnJb* 22, 121-128.

**Stern 1995:** E.M. Stern, *The Toledo Museum of Art. Roman Mold-blown Glass. The First through Sixth Centuries*, Rome: 'L'Erma' Di Bretschneider/Toledo, OH: Toledo Museum of Art, 1995.

**Stern 1997:** E.M. Stern, Neikais - A Woman Glassblower in the First Century AD?, in *Erath, Lehner, Schwarz (ed.) 1997*, 129-132.

**Stern 1998:** E.M. Stern, Interaction between Glassworkers and Ceramists, in *McCray, Kingery (eds) 1998*, 183-204.

**Stern 1999:** E.M. Stern, Roman Glassblowing in a Cultural Context, *AJA* 103/3, 1999, 441-484.

**Stern 2001:** E.M. Stern, *Roman, Byzantine and Early Medieval Glass 10 BCE-700 CE. Ernesto Wolf Collection*, Ostfildern-Ruit: Hatje Cantz Publishers, 2001.

**Stern 2002:** E.M. Stern, The Ancient Glassblower's Tools, in *Kordas (ed.) 2002*, 159-167.

**Stern 2004:** E.M. Stern, The Glass Banausoi of Sidon and Rome, in *Beretta (ed.) 2004*, 77-120.

**Stern 2007:** E.M. Stern, Ancient Glass in a Philological Context, *Mnemosyne* 60, 2007, 341-406.

**Stern 2010:** E.M. Stern, *Audacis plebeia toreumata vitri*. Souffler le verre danas des moules, in *Fontaine (ed.) 2010*, 25-37.

**Stern E. 1997:** Edna J. Stern, Burial Caves at Kisra, *Atiqot* XXXIII, 1997, 17, 103-135.

**Stern, Schlick-Nolte 1994:** E.M. Stern, B. Schlick-Nolte, *Early Glass of the Ancient World, 1600 B.C.- A.D. 50, Ernesto Wolf Collection*, Ostfildern: Gerd Hatje, 1994.

**Sternini 1988:** M. Sternini, Verrerie byzantine de Gortyne, *IIIèmes Journées d'Ètude, Association Française pour l'archéologie du verre*, Lattes, Octobre 1988, 18-25.

**Sternini 1989:** M. Sternini, *Una manifatura vetraria di V sec. a Roma*, Firenze: All' Insegna del Giglio, 1989.

**Sternini 1993:** M. Sternini, Verres tardifs de Roma, *AnnAIHV* 12, 1993, 121-128.

**Sternini 1995a:** M.Sternini, Il vetro in Italia tra Ve IX secoli, in *Foy (ed.) 1995*, 243-289.

**Sternini 1995b:** M. Sternini, *La fenice di sabbia. Storia e tecnologia del vetro antico*, Bari: Edipuglia, 1995.

**Sternini 1997:** M. Sternini, *Vetri*, in *Di Vita, Martin (eds) 1997*, 185-263.

**Sternini 1998:** M. Sternini, Les verres romains d'après les fouilles italiennes à Carthage (1973-1977), *AnnAIHV* 14, 1998, 137-141.

**Sternini 1999:** M. Sternini, I vetri provenienti dagli scavi della missione italiana a Cartagine (1973-1977), *JGS* 41, 1999, 83-104.

**Sternini 2001:** M. Sternini, Reperti in vetro da un deposito tardoanitco sul colle Palatino, *JGS* 43, 2001, 21-76.

**Stevens *et al.* 1993:** S. T. Stevens *et al.*, *Bir el Knissia at Carthage: a rediscovered cemetery Church. Report no. 1* (JRA Suppl. no.7), Ann Arbor, MI, Kelsey Museum, 1993.

**Stiaffini, Borghetti 1994:** D. Stiaffini, B. Borghetti, *I vetri romani del museo archeologico nazionale di Cagliari*, Mediterraneo tardoantico e medievale, scavi e richerche, v. 9, Orisano: Editrice S' Alvure, 1994.

**Stroszeck 2000:** J. Stroszeck, Kerameikosgrabung 1999, *Archäologischer Anzeiger* 2000, 455-493.

**Šubic 1976:** Z. Šubic, Tipološki in kronološki pregled rimskega stekla v Poetovioni, *AV* 25, 1974 (1976), 33-51.

**Sussman 1976:** V. Sussman, A Burial Cave at Kefar' Ara, *Atiqot* 11 (English Series), 1976, 92-101.

**Sverkos 2001:** I. K. Sverkos, Η Θεσσαλονίκη υπό την κυριαρχία των Ρωμαίων, *Thessalonikeon Polis 4*, 2001, 91-102.

**Syon 1998:** D. Syon, A Wine press at Akhziv, *Atiqot* 34, 7, 85-100.

**Tait (ed.) 1991:** H. Tait (ed.), *Five Thousand Years of Glass*, London: British Museum Press, 1991.

**Tasia *et al.* 1996:** A. Tasia, Z. Liola, A. Bachlas, A. Stagkos, Το ανασκαφικό έργο της ΙΣΤ Εφορείας στον Διοικητήριο, *ArchErgMakThrak* 10β, 1996, 545-557.

**Tatton-Brown 1984:** V.A. Tatton-Brown, The Glass, in *Hurst, Roskams 1984*, 194-212.

**Tatton-Brown 1991:** V.A. Tatton-Brown, The Roman Empire, in *H. Tait (ed.)1991*, 62-97.

**Tek 2001:** A.T. Tek, Prismatic Glass Bottles with Greek Inscriptions from Arycanda in Lycia, *AnnAIHV* 15, 2001, 82-87.

**Terzopoulou 1998:** D. Terzopoulou, Ταφικός τύμβος στη Λάδη του Διδυμοτείχου, *ArchErgMakThrak* 12, 1998, 19-24.

**Terzopoulou 2013:** D. Terzopoulou, *Ο ταφικός τύμβος της Μικρής Δοξιπάρας-Ζώνης Νομού Έβρου. Τα ανασκαφικά δεδομένα και τα κτερίσματα από τις καύσεις των νεκρών*, unpublished phd thesis submitted to Aristotle University of Thessaloniki 2013.

**Themelis (ed.) 2002:** P. Themelis (ed.), Το γυαλί από την αρχαιότητα έως σήμερα, 2° συνέδριο Μαργαριτών Μυλοποτάμου Ρεθύμνης Κρήτης, Μαργαρίτες Μυλοποτάμου, 26-28 Σεπτεμβρίου 1997, Athens: Etaireia Messiniakon Archaiologikon Spoudon, 2002.

**Themelis 1963:** P.G. P. Themelis, Η σαρκοφάγος του Τιβερίου Κλαυδίου Λύκου βουλευτού, *Makedonika* 5, 1963, 438-447.

**Themelis 1969:** P. Themelis, Εύβεια, Κάτω Βάθεια, *ArchDelt* 24, 1969, Chronika B1, 204, pl. 213.

**Themelis 1997:** P. Themelis, Εργαστήρια τορευτικής, in *Themelis, Touratsoglou 1997*, 170-185.

**Themelis, Touratsoglou 1997:** P. Themelis, I. Touratsoglou, *Οι τάφοι του Δερβενίου*, Athens: Tameion Archaiologikon Poron, 1997.

**Theocharidou 1985-86:** K. Theocharidou, Συμβολή στη μελέτη της παραγωγής οικοδομικών κεραμικών προϊόντων στα βυζαντινά και μεταβυζαντινά χρόνια, *DeltChrArchEt*, 4th period, vol. 13, 1985/86, 97-111.

**Theocharidou 1994:** K. Theocharidou, Η αρχιτεκτονική του ναού της Αγίας Σοφίας στη Θεσσαλονίκη, Athens: Tameion Archaiologikon Poron, 1994.

**Thessaloniki apo ta proïstorika mechri ta christianika chronia, 1986:** *Θεσσαλονίκη από τα προϊστορικά μέχρι τα χριστιανικά χρόνια* (exhibition catalogue), Thessaloniki, Archaiologico Mouseio, Athens: Tameion Archaiologikon Poron, 1986.

**Thessaloniki Istoria kai Techni 1986:** *Θεσσαλονίκη, Ιστορια και Τέχνη* (exhibition catalogue), Thessaloniki, White Tower, Athens: Tameion Archaiologikon Poron, 1986.

**Thomas 1962:** E.B. Thomas, Die Gläser des Espelmayrfeldes, *Forschungen in Lauriacum 8*, Linz: Institut für Landeskunde von Oberösterreich, 1962, 99-111.

**Thompson 1949:** H.L. Thompson, Excavations in the Athenian Agora: 1948, *Hesperia*, 18/3, Jul.-Sept. 1949, 211-229.

**Threpsiadis 1972:** I. Threpsiadis, Ανασκαφή Γαλαξιδίου, *ArchEph* 1972, 184-207.

**Threpsiadis, Platonos 1963:** I. Threpsiadis, N. Platonos, Αθήνα, Αιόλου 33-37, *ArchDelt* 18, 1963, Chronika B1, 35.

***TMA, Art in Glass 1969*:** The Toledo Museum of Art, *Art in Glass: A Guide to the Glass Collections*, Toledo: The Toledo Museum of Art, 1969.

**Tölle-Kastenbein 1974:** R. Tölle-Kastenbein, *Samos* XIV, *Das Kastro Tigani. Die Bauten und Funde griechischer, römischer und byzantinischer Zeit*, Bonn: R. Habelt, 1974.

**Tomić 1996:** B. Tomić, Bronzani pojasni jezicci u obliku stilizovanih amfora, *GSAD* 1996, 197-206.

**Touchais, Huber, Philippa-Touchais 2001:** G. Touchais, S. Huber, A. Philippa-Touchais, Chroniques des fouilles en Grèce en 2001, *BCH* 125, 2001, 779-1046.

**Touratsoglou, Loverdou-Tsigarida 1976:** I. Touratsoglou, Aik. Loverdou-Tsigarida, Ρωμαϊκοί και βυζαντινοι θησαυροί, από το βορειοελλαδικό χώρο στην Εφορεία Βυζαντινών Αρχαιοτήτων της Θεσσαλονίκης, *ArchDelt* 31, 1976, Meletai, 23-48.

**Trakosopoulou 2002:** E. Trakosopoulou, Glass Grave Goods from Acanthus, in *Kordas (ed.) 2002*, 79-90.

**Trakosopoulou-Salakidou 2000:** E. Trakosopoulou-Salakidou, Γομάτι Χαλκιδικής: Μία νέα αρχαιολογική θέση, *ArchErgMakThrak* 14, 2000, 359-366.

***Trasparenze imperiali 1997:*** AA.VV., *Trasparenze Imperiali. Vetri romani dalla Croazia* (exhibition catalogue), Milan: Skira, 1997.

**Triantafyllidis 1997:** P.Z. Triantafyllidis, Ενδείξεις και εργαστήρια υαλοτεχνικής παραγωγής στη Ρόδο από τους προϊστορικούς χρόνους, in *Themelis (ed.) 2002*, 41-55.

**Triantafyllidis 1998:** P.Z. Triantafyllidis, *Τα γυάλινα αντικείμενα από τη Μινώα Αμοργού. Συμβολή στην μελέτη της υαλουργίας στις Κυκλάδες κατά την ελληνιστική και ρωμαϊκή περίοδο*, unpublished PhD thesis submitted to the University of Ioannina, 1998.

**Triantafyllidis 2000:** P.Z. Triantafyllidis, *Ροδιακή υαλουργία I. Τα εν θερμώ διαμορφωμένα διαφανή αγγεία πολυτελείας. Οι κλασικοί και οι πρώιμοι ελληνιστικοί χρόνοι*, Athens: Ypourgio Aigaiou, 22i Ephoria Proistorikon kai Klasikon Arheotiton, 2000.

**Triantafyllidis 2006:** P.Z. Triantafyllidis, Base Marks from Greece, Cyprus and Turkey, in *Foy, Nenna (eds) 2006a*, 369-402.

**Triantafyllos, Karadima 1991:** D. Triantafyllos, Ch. Karadima, Δίκαια, τύμβος, *ArchDelt* 46, 1991, Chronika B2, 336, pl. 131β.

**Tronchetti 1983:** *Ceramica attica a figure nere: grandi vasi: anfore, pelikai, crateri*, Roma 1983.

**Trowbridge 1930:** M.L. Trowbridge, *Philological Studies in Ancient Glass*, University of Illinois Studies in Language and Literature, v. 13, nos 3-4, Urbana, IL: The University of Illinois, 1930.

**Tsakos 1977:** K. Tsakos, Σάμος, Δυτική νεκρόπολις, *ArchDelt* 32, 1977, Meletes, 344-420.

**Tsakos 1980:** K. Tsakos, Σάμος, Πυθαγόρειο, αγρός Ν. Καρμίρη, *ArchDelt* 35, 1980, Chronika B2, 463, pl. 272.

**Tsigarida, Vasileiou, Naoum 2009:** E. Tsigarida, S. Vasileiou, E. Naoum, Νέα στοιχεία για την οργάνωση και την οικονομία της Κασσάνδρας κατά την ελληνιστική και ρωμαϊκή περίοδο, *ArchErgMakThrak* 23, 2009, 377-397.

**Tsigaridas 1972:** E.N. Tsigaridas, Θεσσαλονίκη, Κλειστόν κολυμβητήριον Πανεπιστημιουπόλεως, *ArchDelt* 27, 1972, Chronika, 559-564, pls. 509α, 510γ.

**Tsigaridas 1973a:** E. Tsigaridas, Θεσσαλονίκη, Βυζαντινά τείχη. Ανατολικό Τείχος παρά το Τριγώνιον, *ArchDelt* 28, 1973, Chronika B2, 478-480, 496.

**Tsigaridas 1973b:** E. Tsigaridas, Θεσσαλονίκη, Βασιλική Πανοράματος, *ArchDelt* 28, 1973, Chronika B2, 498-501.

**Tsigaridas, Loverdou-Tsigarida 1979:** E. Tsigaridas, K. Loverdou-Tsigarida, *Κατάλογος χριστιανικών επιγραφών στα μουσεία της Θεσσαλονίκης*, Thessaloniki: Etaireia Makedonikon Spoudon, 1979.

**Tsimpidou-Auloniti, Lioutas 1994:** M. Tsimpidou-Auloniti, A. Lioutas, Θεσσαλονίκη, οδός Μελενίκου και Χατζηανδρέου, *ArchDelt* 49, 1994, Chronika B2, 433-436.

**Tsimpidou-Avloniti 2003:** M. Tsimpidou-Avloniti, Το ανατολικό νεκροταφείο της Θεσσαλονίκης και πάλι, *ArchErgMakThrak* 17, 2003, 233-244.

**Tsimpidou-Avloniti, Galiniki, Anagnostopoulou 2001:** M. Tsimpidou-Avloniti, St. Galiniki, K. Anagnostopoulou, ΔΕΘ και αρχαιότητες: μία σχέση με το παρελθόν και μέλλον. Η ανασκαφική έρευνα στο χώρο επέκτασης του ΜΜΣΤ, *ArchErgMakThrak* 15, 2001, 215-228.

**Tsimpidou-Avloniti, Kaïafa 2004:** M. Tsimpidou-Avloniti, A. Kaïafa, Θεσσαλονίκη, οδός Κ. Μελενίκου. Η συνέχεια ..., *ArchErgMakThrak* 18, 2004, 269-278.

**Tsouris 1996-1998:** K. Tsouris, Σωστική ανασκαφή στην Αγία Ειρήνη, Λιμένος Θάσου, *AAA* 31, 1996-98, 57-65.

**Turno 1989:** A. Turno, Römische und Frühbyzantinische Gläser aus Novae in Bulgarien, *KölnJb* 22, 1989, 163-170.

**Tzachili 1997:** I. Tzachili, *Υφαντική και υφαντές στο προϊστορικό Αιγαίο 2000-1000 π.Χ.*, Herakleion: Panepistimiakes Ekdoseis Kritis, 1997.

**Tzanavari 2003:** K. Tzanavari, Λητή, μία πόλη της αρχαίας Μυγδονίας, in *Adam-Veleni, Poulaki, Tzanavari* 2003, 71-90.

**Tzanavari, Filis 2000:** K. Tzanavari, K. Filis, Αγροτικές εγκαταστάσεις στη χώρα της αρχαίας Λητής, *ArchErgMakThrak* 14, 2000, 153-169.

**Tzedakis 1978:** G. Tzedakis, Καστέλλι Κισάμου, οικόπεδο Ι και Μ. Φαντάκη, *ArchDelt* 33, 1978, Chronika B1, 370-373.

**Tziafalias 1976:** A. Tziafalias, Μητρόπολη Καρδίτσας, αγρός Βούζα, *ArchDelt* 31, 1976, Chronika B1, 183, pl. 130 .

**Tziafalias 1988:** A. Tziafalias, Λάρισα, οδός Ηρώων Πολυτεχνείου 167, *ArchDelt* 43, 1988, Chronika B1, 260-263, pl. 138β.

**Uboldi 1995:** M. Uboldi, Diffusione delle lampade vitree in etá tardoantica e altomedievale e spunti per una typologia, *AM* 22, 1995, 93-145.

**Uboldi 1996:** M. Uboldi, Vetri di etá tardoantica e altomedievale dagli scavi del monastero di S. Giulia a Brescia, in *Il Vetro dall'Antichità all'Età Contemporanea: Aspetti Technologici, Funzionali e Commerciali*, Atti 2e Giornate Nazionali di Studio AIHV-Comitato Nazionale Italiano, 14-15 Dicembre 1996, Milano: AIHV-Comitato Nazionale Italiano, 181-188.

**Uboldi 1998:** M. Uboldi, Recipienti in vetro de età altomedievale da recenti scavi archeologici in Lombardia, *AnnAIHV* 14, 1998, 208-212.

**Uboldi 2009:** M. Uboldi, Lo sviluppo del calice altomedievale con elementi a colonnine, tra gambo e coppa, *Quaderni Friulani di Archeologia* XIX/2009, 221-228.

**Valavanidou 2001:** A. Valavanidou, Εργαστηριακές χρήσεις στο χώρο της Αρχαίας Αγοράς Θεσσαλονίκης, in *Adam-Veleni (ed.) 2001*, 119-130.

**Vandiver 1983:** P.B. Vandiver, Glass Technology at the Mid-Second-Millennium B.C. Hurrian Site of Nuzi, *JGS* 25 (1983), 239-247.

**Van Geesbergen 1999:** D. Van Geesbergen, Les ateliers de verriers dans le Nord de la Gaule et en Rhénanie (1er-4e siècle après J.-C.), in *Polfe (ed.) 1999*, 105-124.

**Van Lith 1977:** S.M.E. van Lith, Römisches Glas aus Velsen, *OudhMeded* 58, 1977, 1-62.

**Van Lith 1978/79:** S.M.E. van Lith, Römisches Glas aus Volkenburg Z.H., *OudhMeded* 59-60, 1978-79, 1-150.

**Van Lith 1987:** S. van Lith, *Glas aus Asciburgium* (Funde aus Asciburgium, v. 10), Duisburg, Niederrheinsiches Museum der Stadt Duisburg, 1987.

**Vavritsas 1970:** A. Vavritsas, Θεσσαλονίκη. Οδός Αγίου Δημητρίου, *ArchDelt* 25, 1970, Chronika B2, 363-364.

**Vavritsas 1971:** A. N. Vavritsas, Θεσσαλονίκη, Πλατεία Σιντριβανίου, *ArchDelt* 26, 1971, Chronika B2, 373-382, pls. 367-379.

**Vavritsas 1972:** A. Vavritsas, Θεσσαλονίκη, οδός Μαβίλη17, *ArchDelt* 27, 1972, Chronika, 505.

**Velenis 1996a:** G. Velenis, Συμπραγματευόμενοι Ρωμαίοι σε μία νέα επιγραφή της Θεσσαλονίκης, *Tekmiria* 2, 1996, 8-15.

**Velenis 1996b:** G. Velenis, Νομισματοκοπείο στην Αρχαία Αγορά της Θεσσαλονίκης, *CHARAKTIR, afieroma stin M. Oikonomidou*, 1996, 49-60.

**Veličković 1976:** M. Veličković, Tipologija i hronologija rimskog stakla iz Budve u zbirci Narodnog Muzeja u Beogradu, *AV* 25, 1974, (1976), 165-174.

**Verbanck and Massar 2013:** A. Verbanck, N. Massar, Follow the scent... Marketing perfume vases in the Greek world, in A. Tsingarida, D. Viviers, and Z. H. Archibald (eds), *Pottery markets in the Ancient Greek World: (8th - 1st centuries B.C.): proceedings of the international symposium held at the Université libre de Bruxelles, 19-21 June 2008*, Brussels: CReA-Patrimoine, 2013, 273-300.

**Vessberg 1952:** O. Vessberg, Roman Glass in Cyprus, *OpArch* 7, 1952, 109-165.

**Vessberg 1953:** O. Vessberg, Notes on the Chronology of Roman Glass in Cyprus, in *Mylonas, Raymond (ed.) 1953*, 163-167.

**Vessberg 1956:** O. Vessberg, Glass: Typology-Chronology, in *Vessberg, Westholm (eds) 1956*, 128-74, 193-219.

**Vessberg, Westholm (eds) 1956:** O. Vessberg and A. Westholm (eds), *The Swedish Cyprus Expedition, vol. 4, pt. 3, The Hellenistic and Roman Periods in Cyprus*, Stockholm: The Swedish Cyprus Expedition, 1956.

**Vickers 1973:** M. Vickers, Fifth century Brickstamps from Thessaloniki, *BSA* 68, 1973, 285-294.

**Vickers 1992:** M. Vickers, The metrology of gold and silver plate in classical Greece. Economics of cult in the Ancient Greek World, *Boreas* 21, Uppsala 1992, 53-72.

**Vickers 1996:** M. Vickers, Rock crystal: The key to cut Glass and *diatreta* in Persia and Rome, *JRA* 9, 1996, 48-65

**Vitti 1996:** M. Vitti, *Η πολεοδομική εξέλιξη της Θεσσαλονίκης. Από την ίδρυσή της έως τον Γαλέριο*, Athens: En Athinais Archaiologiki Etaireia, 1996.

**Vlizos 2003:** S. Vlizos, Μία φρυγική στήλη στην Αθήνα, *ΜΟΥΣΕΙΟ ΜΠΕΝΑΚΗ* 2, 2002 (2003), 25-35.

**Vokotopoulou 1971:** I. Vokotopoulou, Νικόπολις, οδός Υδατοπύργου Μύτικα, *ArchDelt* 26, 1971, Chronika B2, 335, pl. 314.

**Vokotopoulou 1980:** I. Vokotopoulou, Ξηροκρήνη, Οδός Σταματίου Ρεγκούκου 14, *ArchDelt* 35, 1980, Chronika B2, 365-367, pl. 215γ-δ.

**Von Boeselager 1989:** D. von Boeselager, Zur Datierung der Gläser aus zwei Graben an der Luxemburger Strasse in Köln, *KölnJb* 22, 1989, 25-35.

**Von Bülow, Milcheva (eds) 1998:** G. Von Bülow, A. Milcheva (eds), *Der Limes an der Unteren Donau von Diokletian bis Heraklios*, Vorträge der Internationalen Konferenz Svishtov, Bulgarien (1.-5. September 1998), Sofia-Berlin: RGK DAI Frankfurt, 1999.

**Von Günter *et al.* 2008:** Von Günter Dreyer *et al.*, Stadt und Tempel von Elephantine. 33./34./35. Grabungsbericht, *Mitteilungen des Deutschen Archäologischen Institutes Abteilung Kairo* 64, 2008, 63-146.

**Von Prottwitz, Gaffron, Mielsch 1997:** H.-H. von Prottwitz, Gaffron, H. Mielsch, *Das Haus lacht vor Silber. Die Prunkplatte von Bizerta und das römische Tafelgeschirr*, Köln: Rheinland-Verlag GmbH/Bonn: Dr. Rudolf Habelt GmbH, 1997.

**Voutyras 1997:** E. Voutyras, Η Θεσσαλονίκη στην εποχή της Ρωμαιοκρατίας, in *Chasiotis 1997*, 78-87.

**Waelkens (ed.) 1993:** M. Waelkens (ed.), *Sagalassos I*, First General Report on the Survey (1986-1989) and Excavations (1990-1991), Leuven: Leuven University Press, 1993.

**Waelkens, Poblome (eds) 1997:** M. Waelkens, J. Poblome (eds), *Sagalassos IV, Report on the Survey and Excavation Campaigns of 1994 and 1995*, Leuven: Leuven University Press, 1997.

**Walter 1940:** O. Walter, Archäologische Fünde in Griechenland von Frühen 1940 bis Erbst 1941, *AA* 1940, cols. 172-175.

**Wamser (ed.) 2004:** L. Wamser (ed.), *Die Welt von Byzanz-Europas Östliches Erbe. Glanz, Krisen und Fortleben einer tausendjährigen Kultur*, Exhibition of the Archäologischen Staatssammlung- Museum für Vor- und Frühgeschichte, München, in collaboration with Staatlichen Museen zu Berlin-Preussischer Kulturbesitz 22.10.2004-3.4.2005, Stuttgart: Theiss, 2004.

**Wardle 1972:** K.A. Wardle, Two Notes from Knossos, *BSA* 67, 1972, 271-278.

**Warner Slane 1986:** K. Warner Slane, Two Deposits from the Early Roman Cellar Building, Corinth, *Hesperia* 55/3, 1986, 271-318.

**Wealth of the Roman World 1977:** AA.VV., *Wealth of the Roman World. Gold and Silver AD 300-700*, exhibition catalogue in British Museum, London: British Museum Publication Ltd, 1977.

**Wedepohl, Gaitzsch, Follmann-Schulz 2001:** K.H. Wedepohl, W. Gaitzsch, A.B. Follmann-Schulz, Glassmaking and Glassworking in Six Roman Factories in the Hambach Forest, Germany, *AnnAIHV* 15, 2001 56-61.

**Weinberg 1962:** G.D. Weinberg, Glass Manufacture in Ancient Thessaly, *AJA* 66, 1962, 129-133.

**Weinberg 1963:** G.D. Weinberg, A Parallel to the Highdown Hill Glass, *JGS* 5, 1963, 25-28.

**Weinberg 1988:** G.D. Weinberg, *Excavations at Jalame: Site of a Glass Factory in Late Roman Palestine*, Columbia, Missouri: University of Missouri Press, 1988.

**Weinberg, McClellan 1992:** G.D. Weinberg, M.C. McClellan, *Glass Vessels in Ancient Greece, their History Illustrated from the Collection of the National Archaeological Museum, Athens*, Athens: Tameion Archaiologikon Poron, 1992.

**Weinberg, Stern 2009:** G.D. Weinberg, E.M. Stern, *Athenian Agora XXXIV: Vessel Glass*, Princeton, NJ: The American School of Classical Studies at Athens, 2009.

**Welker 1974:** E. Welker, *Die römischen Gläser von Nida-Heddernheim*, Frankfurt a.M.: Waldemar Kramer, 1974.

**Welker 1985:** E. Welker, *Die römischen Gläser von Nida-Heddernheim II*, Bonn: Rudolf Habelt, 1985.

**Wendel, Böttger, Bartosiewicz (eds) 1979:** M. Wendel, B. Böttger, L. Bartosiewicz, *Iatrus-Krivina: spätantike Befestigung und frühmittelalterliche Siedlung an der unteren Donau 1 1.* Berlin: Akad.-Verl., 1979.

**Whitehouse 1997:** D. Whitehouse, *Roman Glass in the Corning Museum of Glass, Volume One*, Corning, New York: Corning Museum of Glass, 1997.

**Whitehouse 1997b:** D. Whitehouse, A Distinctive Group of Late Roman Glass Vessels, in *Magnusson (ed.) 1997*, 367-375.

**Whitehouse 2001:** D. Whitehouse, *Roman Glass in the Corning Museum of Glass, Volume Two*, Corning, New York: Corning Museum of Glass, 2001.

**Whitehouse 2001a:** D. Whitehouse, Window Glass between the First and the Eighth Century, in *Dell'Acqua and Silva (eds) 2001*, 31-43.

**Whitehouse 2003a:** D. Whitehouse, *Roman Glass in the Corning Museum of Glass, Volume Three*, Corning, New York: Corning Museum of Glass, 2003.

**Whitehouse 2003b:** D. Whitehouse, A Mold-Blown Bottle from the Workshop of Titianus Hyacinthus, *JGS* 45, 2003, p179-180.

**Whitehouse, Carboni 2001:** S. Carboni, D. Whitehouse, with contributions by R.H. Brill and W. Gudenrath, *Glass of the Sultans* (exhibition catalogue), New York: Metropolitan Museum of Art; Corning: Corning Museum of Glass; Athens: Benaki Museum; New Haven: Yale University Press, c. 2001.

**Wight 1994:** K. Wight, Mythological Beakers: A Re-examination, *JGS* 36, 1994, 24-55.

**Wilkinson 1971:** L. Wilkinson, *Egeria's Travels*, London: S.P.C.K., 1971.

**Wiseman 1969:** J. Wiseman, Excavations in Corinth, The Gymnasium Area, 1967-68, *Hesperia* 38/1, 1969, 64-106.

**Xanthopoulou 2010: M.** Xanthopoulou, *Les lampes en bronze à l'époque paléochrétienne* (Bibliothèque De l'antiquité tardive 16), Turnhout: Brepols, 2010.

**Xyngopoulos 1941/52:** A. Xyngopoulos, Περί την Αχειροποίητον Θεσσαλονίκης, *Makedonika* 2, 1941/52, 472-487.

**Yadin 1963:** Y. Yadin, *The Finds from the Bar-Kokhba Period in the Cave of Letters*, Jerusalem: The Israel Exploration Society, 1963.

**Young 1993:** S. Young, A Preview of Seventh-Century Glass from the Kourion Basilica, Cyprus, *JGS* 35, 1993, 39-47.

**Young 2007:** S.H. Young, Glass, in *Megaw 2007*, 485-526.

**YPPO-ERME 2003:** ΥΠΠΟ-ΕΡΜΕ, Αρχαιολογικές έρευνες και μεγάλα δημόσια έργα. Λεύκωμα πινακίδων. Αρχαιολογική συνάντηση εργασίας, Επταπύργιον, Thessaloniki, 18-20 September 2003.

**Zachariadou 2003:** O. Zachariadou, Γ' Εφορεία Προϊστορικών και Κλασικών Αρχαιοτήτων, Πλατεία Συντάγματος, in *YPPO-ERME 2003*, 16.

**Zadoks-Josephus, Boersma (eds) 1976:** J.A. Zadoks-Josephus, J. Boersma, *Festoen: opgedragen aan A.N. Zadoks-Josephus Jitta bij haar zeventigste verjaardag*, (Scripta Archaeologica Groningana 6), Groningen: Tjeenk Willink, 1976.

**Zafeiropoulou 1972:** F. Zafeiropoulou, Μήλος, Τρυπητή, *ArchDelt* 27, 1972, Chronika B2, 605.

**Zafeiropoulou 1975:** F. Zafeiropoulou, Υάλινα αγγεία και πήλινοι λύχνοι εκ της αρχαιολογικής συλλογής Γαλαξιδίου, *ArchEph* 1975, 54-58.

**Zafeiropoulou 1977:** F. Zafeiropoulou, Νάξος, οικόπεδο Αξαοπούλου, *ArchDelt* 32, 1977, Chronika B2, 308-311, pl. 180γ.

**Zafeiropoulou 1982:** F. Zafeiropoulou, Το Κάλλιο στην Ύστερη Αρχαιότητα, *ArchEph* 1982, 1-13.

**Zafeiropoulou 1999:** F. Zafeiropoulou, Γυάλινα φυσητά αγγεία από τα νεκροταφεία της Νάξου και της Πάρου, *ArchDelt* 54, 1999, Meletes, 229-233.

**Zampieri 1998:** G. Zampieri, *Vetri antichi del Museo Civico Archeologico di Padova*, Corpus delle collezioni archeologiche del vetro nel Veneto, v. 4, Padova: Comitato Nazionale Italiano dell'AIHV, 1998.

**Zecchin 1987:** L. Zecchin, *Vetro e Vetrai di Murano. Studi sulla Storia del Vetro I-III*, Venice: Arsenale, 1987-1990.

**Zervoudaki 1971:** E. Zervoudaki, Κως, Στενή Αντιμάχεια, *ArchDelt* 26, 1971, Chronika B2, 545-549.

**Zervoudaki 1972:** E. Zervoudaki, Κως, Αντιμάχεια, *ArchDelt* 26, 1972, Chronika B2, 683-685, pl. 638.

**Živić 2003:** M. Živić, *Felix Romuliana- 50 godina odgonetanja= Felix Romuliana- 50 years of Solving*, Zajecar: Narodni Muzej, 2003.

**Zoridis 1981:** P. Zoridis, Μέγαρα, οδός Θεοκρίτου 5, *ArchDelt* 36, 1981, 39.

**Zoridis 1983:** P. Zoridis, Μέγαρα, οδός Μινώας 76 και Πλάτωνος, *ArchDelt* 38, 1983, Chronika B1, 33-38.

**Zotović, Jordović 1990:** Lj. Zotović, C. Jordović, *Viminacium: nekropola Više Grobalja*, Beograd: Arheološki Institut, 1990.

# Index of Places

Sifnos, 55, 93, 137–138, 148, 150, 152–153, 167